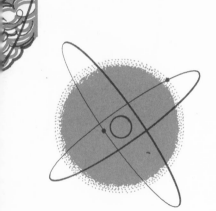

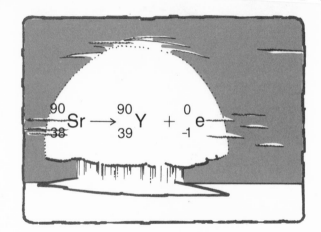

$$^{90}_{38}Sr \rightarrow\ ^{90}_{39}Y + ^{0}_{-1}e$$

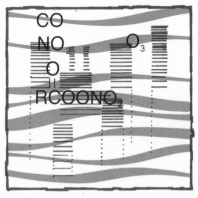

CO
NO$_2$
O
RCOONO$_2$
O$_3$

CO
C$_6$H$_6$
NO$_x$

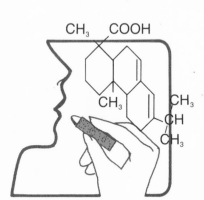

CH$_3$ COOH
CH$_3$
CH$_3$
CH
CH$_3$

SALE
Consumer Products

C
CaCO$_3$
Fe$_2$O$_3$
Fe

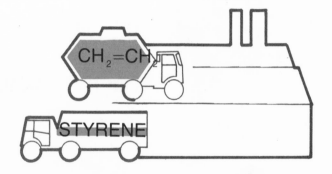

CH$_2$=CH$_2$
STYRENE

MARK M. JONES

Professor of Chemistry
Vanderbilt University, Nashville, Tennessee

JOHN T. NETTERVILLE

Chairman, Department of Chemistry
David Lipscomb College, Nashville, Tennessee

DAVID O. JOHNSTON

Professor of Chemistry
David Lipscomb College, Nashville, Tennessee

JAMES L. WOOD

Associate Professor of Chemistry
David Lipscomb College, Nashville, Tennessee

JOHN R. BLACKBURN

Georgetown College
Georgetown, Kentucky

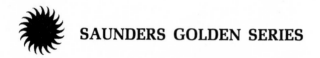

SAUNDERS GOLDEN SERIES

CHEMISTRY MAN AND SOCIETY

W. B. SAUNDERS COMPANY

PHILADELPHIA · LONDON · TORONTO · 1972

W. B. Saunders Company: West Washington Square
Philadelphia, Pa. 19105

12 Dyott Street
London, WC1A 1DB

833 Oxford Street
Toronto 18, Ontario

Chemistry, Man and Society   ISBN 0-7216-5219-0

Print No: 9 8 7 6 5 4 3

PREFACE

This text is designed for either a one-semester or a two-semester course in college chemistry. The material provided covers a much greater variety of topics than is customary in texts for such courses. We believe that this will allow the instructor a greater opportunity to select topics for coverage which reflect his own opinion of the proper topics in a course for students who will take only one course in chemistry. As a one-semester text, this book is highly flexible. By an appropriate choice of chapters, the one-term course can be a current survey of the chemistry-society interface based on a cursory survey of atoms and molecules, a fairly rigorous development of the train of human thought leading to modern applications of chemistry, or a number of different combinations between these approaches.

The organization and goals of this text differ from those of most other books of this type by virtue of the greater emphasis placed upon the use of chemistry to gain understanding both of personal needs (the personal pleasure of understanding natural phenomena and the ability to appreciate consumer chemistry) and of the broader problems facing humanity as a whole (air and water pollution, toxic substances in the environment, and a rational approach to the use of chemical and energy resources).

Although an effort has been made to tailor the text to suit the student with a minimum of scientific training, the approach is sufficiently different to challenge and interest the student with a high school chemistry background. Chemical theory is shown as a continuing development through emphasis of the historical and philosophical aspects of the growth of chemistry into a modern science. Most of the topics covered in the last half of the text are not presented in high school texts.

The first 11 chapters in the book deal primarily with the theoretical concepts needed to understand typical chemical phenomena in terms of bonding and submicroscopic structures. Emphasis is placed on the innate curiosity of man and his use of experimentation in the evolution of the modern theories of chemistry. The reader is asked to examine critically the basic assumption that there is a submicroscopic structure of matter which can be employed to explain and predict events. Chapter 12, an essay, seeks to bring the study of chemistry into focus in the student's life. In the last 13 chapters, which can be built on a very thoughtful analysis of chemical theory or a simple notion of atoms, molecules, and a few molecular geometries, and interactions, the student is shown some

V

of the many ways in which chemical knowledge has been applied to practical problems.

The material in Chapters 13 through 18 covers various aspects of descriptive and theoretical chemistry and includes material on industrial chemistry, organic chemistry, biochemistry, and solution chemistry. It is interspersed with material that can be related directly to the student's interests. Chapters 19 through 25 represent an innovative approach to the problem of teaching students something about the chemical processes which touch upon their daily lives. In these chapters are discussions of the *chemical aspects* of (1) *toxic substances,* (2) *water pollution,* (3) *air pollution,* (4) *foods,* (5) *medicines,* (6) *beauty aids,* (7) *cleaning agents,* (8) *surface coatings,* (9) *photography,* and (10) *nuclear processes.* We believe that it is possible to bring the student's appreciation of chemistry to a new level by illustrating a few of the many ways in which it enters his daily life. The examples we have selected *are not those typically found in freshman chemistry* texts; they are much closer to the daily life of the average person.

The teacher who wishes to emphasize chemical applications can skip Chapters 3 through 8, replacing these with one or two lectures on the consequences of atomic structure in chemical bonding. Chapter 9—Chemical Bonding—and Chapter 10—Molecular Shapes—can be presented in summary wherein only the salient ideas are presented. Since many examples of these ideas will be encountered in later chapters, the detailed examples in these two chapters can be omitted unless there is serious student interest.

Numerous exercises have been provided at the end of each chapter. Although some of these exercises test the reader's recall of facts and understanding of concepts, a few open-ended exercises have been added to stimulate the bright student and to involve him in some of the inevitable controversies of chemical science. References for further reading are included to encourage the student to develop a taste for studying in depth a particular point of interest.

We wish to acknowledge the assistance of Mr. Grant Lashbrook and Mr. Alexis Kelner in developing the illustrative material and of Mrs. Jolly Waynick who helped secure the photographs. The manuscript has been read in whole or in part by a large number of experts whose help we wish to acknowledge. These include Professors David J. Wilson and Wayland J. Hayes, Jr., of Vanderbilt University and Professor James W. Thomas of David Lipscomb College. Others who have furnished valuable assistance in this manner are Professors Eugene G. Rochow, Harvard University; William L. Masterton, University of Connecticut; Gordon Evans, Tufts University; and Norman Juster, Pasadena City College. We also wish to acknowledge with thanks the work of Misses Debbie Washam, Peggy Patton, Pat Palmer, and Brenda Knowles in the preparation of the final manuscript. It is to our wives that we dedicate this effort and gratefully acknowledge support and critically needed assistance during the preparation of this manuscript.

MARK M. JONES

JOHN T. NETTERVILLE

DAVID O. JOHNSTON

JAMES L. WOOD

CONTENTS

CONTENTS

THE CHEMICAL VIEW OF MATTER

A careful examination of the things we use in our daily life emphasizes the fact that most of them are very different from the materials which are an obvious part of our natural surroundings. Practically everything we use has been transformed from a natural state of little or no utility to one of very different appearance and much greater utility. The processes by which the materials of nature can be transformed and a detailed description of such changes are highly intriguing. Wherever we look we see matter undergoing change. This is what chemistry is all about: *matter*, and the *changes* it undergoes.

An understanding of our environment and the changes which are possible in it is an indispensable goal of education. Such an understanding is necessary if we are to know the consequences of our acts on our environment and even more critical if we are to undertake the successful repair of previous damage. Since life itself involves very intricate sequences of changes, each of us has a more personal reason for valuing knowledge of such processes—a desire to gain some insight into the manner in which chemicals in our food, air, and water can influence our individual health, happiness, and behavior.

While all changes in matter are of concern, our attention here will be focused on a particular kind of change called *chemical change*. In any chemical change the starting material is changed into a different *kind* of matter. Matter may also undergo another kind of change which does not produce a new kind of matter, but simply results in a new form of the same material. This kind of change is a *physical change*.

Since the feature used to characterize chemical changes is the production of a different kind of matter, it is necessary for a thorough study of chemical change to be able to classify matter into different types. It must be noted that the types of matter are usually mixed together in a natural source and that the separation of such mixtures had to precede their systematic classification. After an examination of the methods of separating such mixtures into their components, a more detailed discussion of chemical change is possible. As the methods are developed, it becomes possible to appreciate some of the problems involved in an accurate definition of the terms "kinds of matter" and "chemical change."

1

In spite of the difficulties experienced in establishing the exact bounds of chemistry, it will be helpful to think of *chemistry* as *the study of the kinds of matter and the changes that transform one kind of matter into another.*

PURE SUBSTANCES

Most samples of matter are complex mixtures. Sometimes it is easy to see the various ingredients in a mixture, such as the bits of sand and clay in a sample of soil, or the glittering crystals (mica and quartz) and dark areas (feldspar and magnetite) in a piece of granite. More often, however, it is not apparent to the casual observer that the mixture is made up of distinctly separate ingredients. It is only by an experiment that we can establish that the air we breathe is a mixture.

When a mixture is separated into its components, the components of the mixture are said to be purified. However, most efforts at separations are incomplete in a single operation or step, and repetition of the purification process results in a better separation. Ultimately in such a process, the experimenter arrives at pure substances, samples of matter that cannot be further purified. For example, if sulfur and iron powder are ground together to form a mixture, the iron can be separated from the sulfur by repeated stirrings of the mixture with a magnet. When the mixture is stirred the first time and the magnet removed, much of the iron is removed with it, leaving the sulfur in a higher state of purity. However, after just one stirring the sulfur may still have a dirty appearance due to a small amount of iron that remains. Repeated stirring with the magnet, or perhaps the use of a very strong magnet, will finally leave a bright yellow sample of sulfur that apparently cannot be purified further by this technique. In this purification process a property of the mixture, its color, is a measure of the extent of purification. After the bright yellow color is obtained, it could be assumed that the sulfur has been purified. Drawing a conclusion based on one property of the mixture is dangerous, however, because other methods of purification might change some other properties of the sample. It is only safe to call the sulfur a pure substance when all possible methods of purification fail to change its properties. This assumes that all pure substances have a set of properties by which they can be recognized just as a person can be recognized by a set of characteristics. A *pure substance, then, is a kind of matter with properties that cannot be changed by further purification.*

There are some naturally occurring pure substances. Rain is very nearly pure water except for small amounts of dust and air. Gold, diamonds, and sulfur are also found in very pure form. These substances are special cases. Man, a complex assemblage of mixtures, lives in a world of mixtures—eating them, wearing them, living in houses made of them, and making most of his tools out of them.

Although naturally occurring pure substances are not common, it has been possible to produce many pure substances from natural mixtures. A number of ingenious methods have been designed to separate mixtures into their component pure substances. Furthermore, processes have been developed to change such pure substances into a multitude of other pure substances, many of which do not occur in nature.

Relatively pure substances are now very common as a consequence of the development of modern purification techniques. Common examples are table salt (sodium chloride), copper, sodium bicarbonate, nitrogen, dextrose, ammonia, uranium, and carbon dioxide—to mention just a few. In all, about two million pure substances have been identified.

DEFINITIONS: OPERATIONAL AND THEORETICAL

The preceding definition given for a pure substance is an operational definition, or a definition in terms of specific experiments or operations. That is, if any further purification effort is unsuccessful in changing the properties of a substance, it is said to be a pure substance. It is evident then that *operational* definitions result from performing operations or tests on matter and summarizing the results in a statement. For example, pure sulfur is a yellow substance that boils at 833°F and has a density of 129 pounds per cubic foot. When all the properties of pure sulfur have been listed, we find that the pure substance has been characterized in a way that distinguishes it from any other pure substance.

A pure substance also can be defined in theoretical terms; that is, in terms of the molecules, atoms, and subatomic particles that compose it. Both types of definitions are important in the study of chemistry and both will be used in the presentation of this text. The theoretical definition will follow the development of the theory on which it is based.

SEPARATION OF MIXTURES INTO PURE SUBSTANCES

The separation of mixtures is usually more difficult than the magnetic separation of iron and sulfur described previously. Most beginning chemistry students would find the separation of a piece of granite into the pure substances that compose it a bewildering task. Indeed, a chemist would find this a difficult assignment. Since each of the pure substances in the granite has a set of properties unlike those of any other pure substance, it should be possible, in theory at least, to use these properties to separate the pure substances, just as the attraction of iron to a magnet is used to separate it from sulfur.

The numerous methods that have been devised for the *physical* separation of mixtures exploit a difference in a given property (such as boiling point or solubility) of the pure substances involved. Of equal importance are the *chemical* methods which have been developed for separations. They are based upon changing one substance into another one with different properties. Chemical methods of separation are discussed later in this text. At this point, an examination of some of the physical separation methods will be used to convey an understanding of their usefulness and their limitations.

FILTRATION

Often two solids which are intimately mixed together can be separated by making use of the fact that one of them will dissolve in a certain liquid, while

3

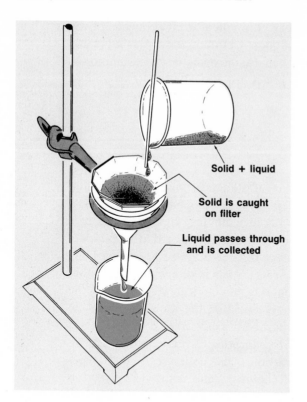

Solid + liquid

Solid is caught
on filter

Liquid passes through
and is collected

FIGURE 1-1   The separation of a solid from
its mixture with a liquid by filtration.

the other will not. The solid that will not dissolve can then be caught on a filter
(Figure 1–1) while the one that dissolved will pass through the filter with the
liquid. Salt mixed with sand can be separated by stirring the mixture in water.
The salt will dissolve, leaving the sand as the only remaining solid. The sand
then can be removed by filtration. Another mixture which can be separated by
filtration is the mixture of iron and sulfur mentioned earlier. In this case the
sulfur can be dissolved in a liquid, carbon disulfide, in which iron is completely
insoluble. Filtration can then be used to separate the iron powder from the sulfur
solution. Most of the solid material in river water that is used for drinking is
removed in water purification plants by passing the mixture through a filter bed
composed in part of fine sand.

DISTILLATION

The boiling point of a liquid is the lowest temperature at which bubbles
of its vapor first form throughout it. Usually the components in a mixture will
have different boiling points. If this is the case, a separation can be based on
this difference. In a simple example, common table salt can be separated from
its solution in water because the boiling point of water is very much lower than
that of salt. The water can simply be boiled away, leaving the dry salt (Figure
1–2). To reclaim the water as well as the salt, the gaseous water can be passed
through a cool tube where it will liquefy. The process of boiling followed by
the condensation of the vapor into a separate vessel is called *distillation*. In some

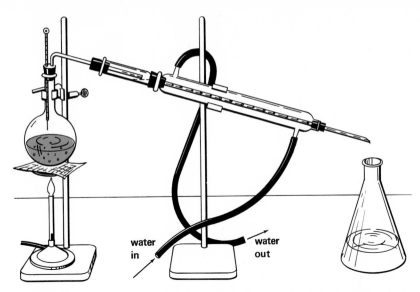

FIGURE 1-2   Distillation. Sodium chloride dissolves in water to form a clear solution. When heated above the boiling point (indicated by thermometer), water will pass into vapor and into the condenser. Cool water injected into the glass jacket of the condenser circulates over the inner tube causing the steam to liquefy and collect in the flask. In this simple example pure water collects in the receiving flask while the salt remains in the boiling flask.

mixtures the boiling points are relatively close together and a single distillation will only partially separate them. In such cases repeated distillations, called *fractional distillations*, yield relatively pure substances. The components of crude petroleum are partially (in some special cases completely) purified in this manner.

Air is a mixture of nitrogen and oxygen along with small amounts of several other gases. If air is liquefied by cooling, the nitrogen, oxygen, and other gases

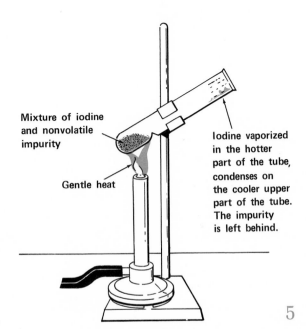

FIGURE 1-3   Iodine, because of its volatility, can be separated from nonvolatile impurities by sublimation.

5

can be separated from each other by taking advantage of their different boiling points. As the liquid air is warmed, the nitrogen, which has a lower boiling point than oxygen, will tend to boil off, leaving the oxygen in a higher state of purity. This process can be repeated until the nitrogen and oxygen are obtained in a very pure form.

A similar type of separation is based on a rather unusual property of some solids, i.e., they sublime (pass directly into the gaseous state without melting). A solid that will sublime mixed with one that will not can be removed from a mixture of the two by simply heating the mixture. Dry ice, which is solid carbon dioxide, and iodine are familiar examples of solids that will sublime. The process of purification by sublimation is illustrated in Figure 1–3.

EXTRACTION

Some pairs of liquids, such as oil and water, will separate into two layers after being shaken together and allowed to stand. Such a pair of liquids is said to be *immiscible*. This is in contrast to the pair gasoline and kerosene, which will mix in all proportions, forming a single layer; gasoline and kerosene are said to be *miscible*. Solid iodine will dissolve in both carbon tetrachloride and water, two solvents which are immiscible with each other. If water, carbon tetrachloride, and a small amount of iodine are shaken together, one finds that the iodine is much more soluble in the carbon tetrachloride (85 times as much)

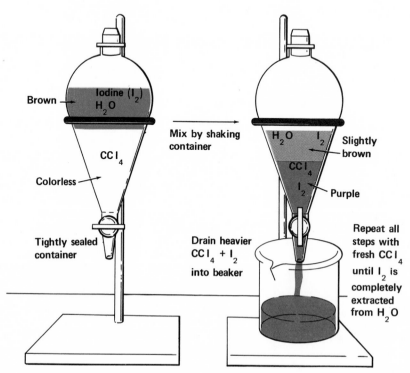

FIGURE 1-4   The extraction of iodine from water by carbon tetrachloride, CCl_4.

than in the water. Iodine dissolved in water can be effectively removed by shaking the mixture with carbon tetrachloride. If this process is repeated a few times, essentially all of the iodine can be removed from the water.

This technique of removing a substance from one liquid by making use of its greater tendency to dissolve in another liquid is called *extraction,* and it can be used to separate mixtures. For example, if one has a water solution containing both sodium chloride (table salt) and iodine, it would be easy to extract the iodine with carbon tetrachloride, leaving the salt in the water, since the salt is not soluble in carbon tetrachloride. The technique used in extraction is shown in Figure 1–4.

RECRYSTALLIZATION

Solid substances often are more soluble in a hot liquid than in the same liquid when cold. In these cases a hot solution which contains the maximum amount of substance that will dissolve at that temperature will form crystals as the solution is cooled, since the cooler liquid cannot hold as much of the pure substance in solution.

It is often possible to take advantage of this behavior to separate the components of a solid mixture. Consider the case in which two solid substances are present in a mixture; both of the pure substances are very soluble in a hot liquid, and only one of the two is very soluble in the cold liquid. It is evident, then, that the two can be separated by dissolving the mixture in a minimum amount of the hot liquid, allowing the solution to cool (crystals of the less soluble pure substance form), and filtering the newly formed crystals of the less soluble material from the liquid. This process is called *recrystallization* (Figure 1–5). By its use, impure crystals often can be dissolved and reformed in a higher state of purity. The purification process is assisted by the fact that crystal formation often selectively removes only one kind of substance from solution and rejects the other.

Recrystallization can be demonstrated in the purification of sodium bicarbonate. In an important commercial process, the Solvay process, crystals of sodium bicarbonate are formed in a solution containing large amounts of another substance, ammonium chloride. As a result, the sodium bicarbonate is contaminated with ammonium chloride. Recrystallization is an ideal method for the purification of sodium bicarbonate in this case, since it is sparingly soluble in cold water but more soluble in hot water, whereas the ammonium chloride is quite soluble even in cold water.

Zone melting is an interesting form of recrystallization that has recently been developed and is capable of producing substances in an extremely high state of purity. Pure substances with less than one part per billion of an impurity can be obtained using this technique. This degree of purity is required of materials used in the components of solid-state electronic devices such as the transistor. In zone melting, a heater ring around a tube is slowly moved along the tube containing the material to be purified (Figure 1–6). The material melts in the region of the heater, only to solidify again after the heater has moved to an adjacent region. If the impurity is more soluble in the molten (liquid) region,

7

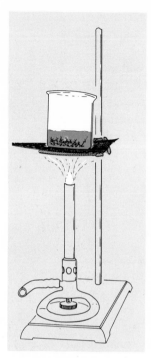

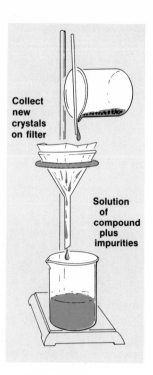

Collect
new
crystals
on filter

Solution
of
compound
plus
impurities

A. Dissolve solid
   in minimum quantity
   of hot solvent.

B. Cool solution
   (generally in ice + water),
   New crystals form.

C. Collect new,
   purified crystals on filter.

D. Repeat process
   if necessary.

FIGURE 1-5   Recrystallization can be used to separate some solid mixtures.

which is most often the case, the impurity will be partially carried along with the moving molten region. Hence, the material left behind will be in a higher state of purity. Many passes of the heater over the same sample result in very high purity of the solidified material; the impurities are concentrated in the moving molten zone. At the end of the process, the impure zone is finally allowed to solidify and is then cut off the bar.

CHROMATOGRAPHY

Absorbent papers, such as paper towels, attract water to the extent that water will flow against the pull of gravity to wet the paper. If a spot of ordinary washable ink is placed in the path of the moving water the ink will be extracted by and will move along with the water. However, some substances in the ink will move faster than others since they are more easily extracted by the water. Bands of color will develop on the absorbent paper as the ink mixture separates. This technique of separation is called *paper chromatography* (Figure 1–7). The term chromatography is taken from the Greek word "chromos," which means color. The original experiments involved the separation of colored substances from plants. However, this method of separation has been widely extended to

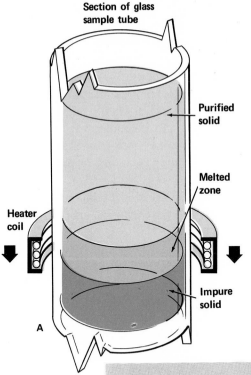

Section of glass
sample tube

Purified
solid

Melted
zone

Heater
coil

Impure
solid

A

FIGURE 1-6   *A.* Zone melting purification. A heater coil is moved slowly along the solid sample to be purified, resulting in a moving liquid zone. Impurities, which are more soluble in the liquid, are thereby swept out of the sample. *B.* A 1-inch diameter crystalline sample of the element silicon (Si) prepared by the zone melting method. This is one of the purest samples of a solid element that has been prepared. (Courtesy of K. E. Benson, Bell Telephone Laboratories, Allentown, Pennsylvania.)

B

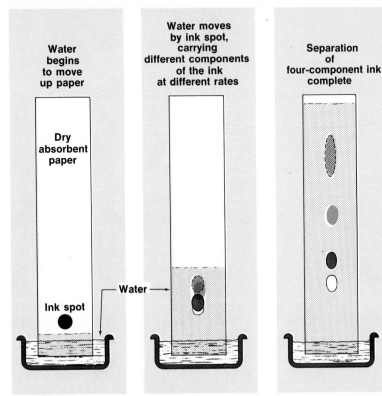

Water
begins
to move
up paper

Water moves
by ink spot,
carrying
different components
of the ink
at different rates

Separation
of
four-component ink
complete

Dry
absorbent
paper

Water →

Ink spot

FIGURE 1-7   Paper chromatography of ink. Owing to the absorbent character of paper, water moves against gravity, *a*, and carries the ink dyes along its path, *b*. If the ink dyes move at different rates, they will be separated in the developed chromatogram, *c*.

a   b   c include colorless mixtures, such as the amino acids from egg white protein, so that now the idea of color is not directly connected with the concept.

The scope of chromatography, developed as a major separation method only since 1950, has been extended in a general way to any two phases of matter that are moving relative to each other. The term "phase" is used to refer to a homogeneous mass of matter with distinct boundaries which separate it from other phases. Thus water exists in a solid phase (ice), a liquid phase (water), and a gaseous phase (steam). A mixture of gases can be separated by passing it in a stream of another gas over a solid, if the solid tends to adsorb one gas more than the other. The solid, in the form of small particles, is packed in a tube and the mixture of gases is passed through the tube. Under ideal conditions the faster moving gas will completely pass out of the end of the tube before the slower gas begins to come out. Similar separations can be made by passing mixtures of gases over liquid surfaces.

ELEMENTS AND COMPOUNDS

Experimentally, pure substances can be classified into two categories: those that can be broken down by chemical change into simpler pure substances and those that cannot. Table sugar (sucrose), a pure substance, will decompose when heated in the oven, leaving carbon, another pure substance. No *chemical* operation has ever been devised that will decompose carbon into simpler pure substances. Obviously sucrose and carbon belong to two different categories of pure substances. Only 90 substances that cannot be reduced chemically to simpler substances are found in nature; 15 others are available artificially. These 105 substances are called *elements*. Pure substances that can be decomposed into two or more different pure substances are referred to as *compounds*. Even though there are presently only 105 elements, there appears to be no practical limit to the number of compounds that can be made.

Elements, the basic building blocks of the world in which we live, include such substances as oxygen, hydrogen, iron, copper, and sodium. Table 1–1 lists some common elements; a complete list of the elements is found inside the back cover of this text. A number of elements are found as the elementary substance in nature; examples include gold, silver, oxygen, nitrogen, carbon (graphite and diamond), copper, platinum, sulfur, and the noble gases (helium, neon, argon, krypton, xenon, and radon). Many more elements, however, are found chemically combined with other elements in the form of compounds.

Elements in compounds no longer show their characteristic physical properties, such as color, hardness, and melting point. Consider sucrose, $C_{12}H_{22}O_{11}$, as an example. It is made up of three elements: carbon (which is usually a black powder), hydrogen (the lightest gas known), and oxygen (a gas necessary for respiration). The compound, sucrose, is completely unlike any of the three elements; it is a white crystalline powder which, unlike solid carbon, is readily soluble in water.

A careful distinction should be made between a compound of two or more elements and a mixture of the same elements. The two gases, hydrogen and oxygen, can be mixed in all proportions, forming various mixtures of these

TABLE 1-1   SOME COMMON ELEMENTS°

Metals

(A metal is a good conductor of electricity, can have a shiny or lustrous surface, and in the solid form can be deformed without breaking.)

NAME	SYMBOL	PROPERTIES OF PURE ELEMENT
Iron Latin, ferrum	Fe	strong, malleable, corrodes
Copper Latin, cuprum	Cu	soft, reddish-colored, ductile
Sodium Latin, natrium	Na	light silvery metal, very reactive, soft, low melting point
Silver Latin, argentum	Ag	shiny white metal, relatively unreactive, best conductor of electricity
Gold Latin, aurum	Au	heavy yellow metal, not very reactive, ductile
Chromium	Cr	resistant to corrosion, hard grayish-white

Nonmetals

(A nonmetal is often a poor conductor of electricity, normally lacks a shiny surface, and is brittle in crystal-solid form.)

Hydrogen	H	colorless, odorless (H_2), occurs as a very light gas, burns in air
Oxygen	O	colorless, odorless gas (O_2), reactive, constituent of air
Sulfur	S	yellow solid (S_8), low m.p., burns in air
Nitrogen	N	colorless, odorless gas (N_2), rather unreactive
Chlorine	Cl	greenish yellow gas (Cl_2), very sharp choking odor, poisonous
Iodine	I	dark purple solid (I_2), sublimes easily

° Chemists usually use the symbol rather than the name of the element. In addition to denoting the element, the chemical symbol has a very specialized meaning which is described later in this chapter. A complete list of the elements with the symbols can be found inside the back cover of this book.

References explaining the historical development of chemical symbols are given in the suggestions for further reading at the end of Chapter 1.

elements. However, these two elements can and do react chemically to form the compound *water*. Not only does water exhibit properties peculiar to itself and different from hydrogen and oxygen, but it also has a definite percentage composition by weight (88.8 per cent oxygen and 11.2 per cent hydrogen). This is a second distinct difference between compounds and mixtures. Compounds have a definite percentage composition by weight of the combining elements.

CHEMICAL CHANGES AND PHYSICAL CHANGES

Chemical and physical changes are the two categories into which changes of matter are placed. A chemical change occurs when substances are used up

11

in the production of different ones. For example, the burning of carbon (as charcoal) to form carbon dioxide is a chemical change. A physical change produces no new substances; it only alters the original state of the elements or compounds. If one boils water, it is transformed from liquid water to gaseous water; this is in many ways a typical physical change. Some additional examples of chemical and physical changes are the following:

Chemical Changes	*Physical Changes*
Rusting of iron	Evaporation of a liquid
Burning gasoline in an automobile engine	Melting of iron
Preparation of caramel by heating sugar	Freezing of water
Preparation of iron from its ores	Grinding or pulverizing of a solid
Solution of copper in nitric acid	

It is important to catalogue the properties of a new substance, since they serve to identify it. This type of information for all known compounds can be found on the reference shelves of chemical libraries. It is useful to classify the properties of a substance into physical properties and chemical properties.

A chemical property of an element or compound describes its *ability* to undergo chemical change. It follows, then, that to determine a chemical property we must carry out a chemical change. Physical properties, on the other hand, are characteristics of a substance which can be determined without chemically changing the element or compound into something else. Physical properties of a pure substance include such things as melting temperature, boiling temperature, density (weight of the substance in a given volume), color, physical state (whether gas, liquid, or solid, at a specified temperature), crystalline form, and magnetic properties.

It should be emphasized that not all of the properties of a sample need to change in order to establish that a chemical change has occurred. For example, if table salt is treated with sulfuric acid and heated strongly, the solid product of this reaction is much like table salt in its appearance, but it has a different melting point and different chemical properties. A change in only one significant property usually indicates that a newly formed substance has resulted from a chemical change.

While most changes can easily be classified on the basis of whether a new substance is produced or not, some changes are difficult to put into such neat categories. Examples are those which accompany the solution of a solid in a liquid. We often apply a simple test to classify these. We test to see if the solid can be recovered unchanged by the evaporation of the liquid. Thus, table salt or sugar dissolves in water, but it can be recovered by careful evaporation, so that such solutions are said to be formed by *physical changes* even though the interaction between constituents is considerable. An example of a chemical change resulting in solution is observed when hydrogen chloride dissolves in water. In this case, the properties of the solution are very different from those of the constituents, and evaporation of the solution is not a feasible way to recover the starting materials completely. In general, however, solution processes are of such complexity that their detailed discussion is inappropriate at this point.

The classification of solution changes illustrates some difficulties about exact definitions as used in chemistry. Like the stamp collector arranging stamps and the housewife arranging foods in her kitchen, we will find our work easier if we group similar things and phenomena into categories. This makes it easier to organize knowledge. Of course, we try to define the limits of our categories so that there will be no question about what fits into each classification, but nature sometimes presents situations that are very hard to classify. These are borderline cases, and all efforts to be specific and limiting will usually meet with an item or idea that defies exact classification. There are some changes in matter that are difficult to classify as either physical or chemical, and all of our definitions are subject to this kind of limitation. However, a few exceptions do not stop us from originating simple categories for the things we observe.

WHY AN INTEREST IN PURE SUBSTANCES AND THEIR CHEMICAL CHANGES?

Perhaps by now you are wondering why we should be interested in elements and compounds and their chemical properties. There are two reasons. The first is the belief that the knowledge of chemical substances and chemical changes will allow us to bring about desired changes in the nature of everyday life. Two hundred years ago most of the materials surrounding a normal person were only physically changed from the way the material occurred in nature. Only a few of his useful materials, such as iron, were the product of man's control over chemical change. By contrast, synthetic fibers, plastics, drugs, latex paints, detergents, new and better fuels, and photographic films are but a few of the materials produced by controlled chemical change. (We will return to examine the chemistry of many of these later.) You will discover that it is difficult to find more than a few objects in your home that have not been altered by a desirable chemical change. The second reason for chemical studies is even more important—the curiosity of man. Chemicals and chemical change are a part of nature that is open to investigation, and, like the mountain climber, we will find this task both interesting and challenging. If we hope to understand matter, it is obviously necessary to discover the simplest forms of matter and to study their interactions.

STRUCTURE OF MATTER AND RESULTING PROPERTIES

For reasons which are partly theoretical and partly practical, we are deeply interested in the structure of matter, that is, the minute parts of matter and how these parts are fitted together to make larger units. We want to be able to explain chemical change on the basis of changes in structure. A very large

13

portion of the research in chemistry being done today is aimed at sorting out and elucidating the structure of matter. Indeed, the basic theme of this text is the relationship between the structure of matter and its properties. This theme of structure and related properties is of great interest because if we know exactly how and with what strength the minute parts of matter are put together, we can discover exact relationships between structure and properties. Armed with this understanding, we can make changes that result in new substances and predict the properties of these substances. Such knowledge can save many hours of trial and error which otherwise may be required to prepare a product with the desired qualities. While this day of predicting chemical changes based on structural characteristics has not completely arrived, such significant advances have been made that the practice of modern chemistry would not be possible without such knowledge.

Samples of matter large enough for ordinary laboratory experiments are called *macroscopic* samples in contrast to *microscopic* samples, which are so small that they have to be viewed with the aid of a microscope. The structure of matter that really interests us, however, is at the *submicroscopic* level. Our senses have no direct access into this small world of structure, and any conclusions about it will have to be based on circumstantial evidence gathered in the macroscopic world. You may, at this point, raise the objection that it is only an assumption that there really is a relationship between submicroscopic structure and the macroscopic properties of matter. The authors would only ask that you keep an open mind on this point as the circumstantial evidence is developed in the following chapters.

We can now extend our concept of the science of chemistry. It is that science which deals with the separation of natural mixtures and the determination of the physical and chemical properties of the resulting pure substances. It is also deeply concerned with structure, both *macrostructure and submicrostructure,* in an effort to give plausible reasons for chemical change.

FACTS, LAWS, AND THEORIES

A scientific fact is an observation about nature that can usually be reproduced at will. For example, carbon in some forms will burn in the presence of air. If you have any doubt about this fact it is easy enough to set up an experiment that will readily demonstrate the fact anew. You would only need some carbon, air, and a source of heat. The repeatability of a scientific fact distinguishes it from a historical fact, which obviously cannot be reproduced. To be certain, some scientific facts that are also historical facts—such as the movement of heavenly bodies—are not repeatable at will.

Often a large number of scientific facts can be summarized into broad, sweeping statements called natural laws. The law of gravity is a classic example of a natural law from the science of physics. When you state this law, that all bodies in the universe have an attraction for all other bodies which is directly proportional to the product of their masses and inversely related to the square of their separation distance, you are summarizing in one sweeping statement an unlimited number of facts. You imply that any heavy object lifted a short distance from the surface of the earth will fall back if released. Such a natural law can only be established in the mind of man by inductive reasoning; that is, you conclude that the law applies to all possible cases, since it applied in all of the cases studied. A well-established law allows us to predict future events. When convinced of the generality of a law, a scientist reasons deductively, based on his belief that if the law holds for all situations, it will surely hold for the events in question.

The same procedure is used in the establishment of chemical laws, as can be seen from the following example. Suppose an experimenter carried out hundreds of different chemical changes in closed, leakproof containers, and that he weighed the containers and their contents before and after each of the chemical changes. Also, suppose that in every case he found that the container and its contents weighed exactly the same before and after the chemical change occurred. Finally, suppose that he repeated the same experiments over and over again, obtaining the same results each time, until he was absolutely sure that he was dealing in reproducible facts. It can be understood then that the experimenter would reasonably conclude: *"All chemical changes occur without any detectable loss or gain in weight."* This is indeed a basic chemical law and serves as one of the foundations of modern scientific theory.

After man has firmly established a natural law in his mind, his curiosity causes him to wonder why the law is true. Chemists are not satisfied until they have explained chemical laws logically in terms of the submicroscopic structure of matter. This is indeed a difficult process, and the progress, until recent times, has been painfully slow because of man's lack of direct access into the submicroscopic structure of matter with his physical senses. All he can do is to collect information in the macroscopic world in which he lives, and try, by circumstantial reasoning, to visualize what the submicroscopic world must be like in order to explain his macroscopic world. Such a visualization of the submicroscopic world is called a *theoretical* model. If the theoretical model is successful in explaining a number of chemical laws, a major scientific theory is built around it. The atomic theory and the electron theory of chemical bonding are two such major theories and both will be discussed in relation to chemical laws in later chapters.

Consider again the chemical law concerning the conservation of weight in chemical changes. What is a possible theoretical model that could explain this law? If we assume that matter is made up of atoms and that these atoms are arranged in a particular way in a given pure substance, we can reason that a chemical change is the rearrangement of these atoms into new groupings, consequently, into new substances. If the same atoms are still there, they should have the same weight. Therefore, as there is no detectable weight change in the reaction, the law can be explained in terms of an atomic model.

For a scientific theory to have much value, it must not only explain the pertinent facts and laws at hand, but it must also be able to explain new facts and laws that are obviously related. If the theory cannot consistently perform in this manner, it must be revised until it is consistent, or, if this is not possible, it must be completely discarded. You must not allow yourself to think that this process of trying to understand nature's secrets is nearing completion. The process is a continuing one.

15

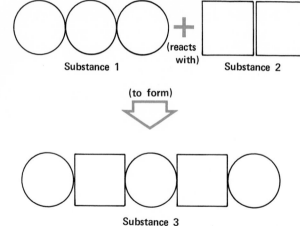

Substance 1   (reacts with)   Substance 2

(to form)

Substance 3

FIGURE 1-8   Atomic explanation for weight conservation in chemical changes. If substances are composed of atoms, if chemical changes are atomic rearrangements, and if atoms have fixed weights, the products in a reaction must weigh the same as the starting materials; hence, the law of conservation of weight is explained in terms of a theoretical model.

The word *theory* is often used in a different sense than the one discussed above. If a student is absent from the chemistry class, his neighbor may say, "I do not know why he is absent, but my theory is that he is sick and unable to come to class." This use of the word *theory* is similar to the concept of the general scientific theory in that there is an element of speculation in both uses. But the speculative guess of the student about his absent friend is vastly different from the broad theoretical picture that is able to explain a number of laws. The reader should be alert for the considerable amount of confusion that has resulted from the different meanings associated with this word. In this book the word *hypothesis* is used when speaking of a speculation about a particular set of data, reserving the word *theory* for the broad imaginative concepts that have gained wide acceptance.

THE LANGUAGE OF CHEMISTRY

Symbols, formulas, and equations are used in chemistry to convey ideas quickly and concisely. These shorthand notations are merely a convenience, and contain no mysterious concepts that cannot be expressed in words. Certain characters are used often, and a general familiarity with them is desirable.

It will be necessary to speak in theoretical terms of atoms and molecules to give a full meaning to chemical symbolism. This is done with the understanding that you are already acquainted in a general way with these terms. The evidence that has convinced the scientific world that these submicroscopic, theoretical particles must indeed exist will be presented later.

A chemical symbol for an element is a one- or two-letter term, the first letter a capital and the second a lower case letter. The symbol is a sign for three concepts. First, the symbol stands for the element in general. H, O, N, Cl, Fe, Pt are shorthand notations for the elements hydrogen, oxygen, nitrogen, chlorine, iron, and platinum, respectively, and it is customary to substitute these symbols for the words themselves in describing chemical changes. Some symbols originate from Latin words, such as Fe, from ferrum, the Latin word for iron. Second, the chemical symbol stands for a single atom of the element. The *atom* is the smallest particle of the element that can enter into chemical combinations. Third, the elemental symbol stands for a mole of the atoms of the element. The *mole* is a term that has evolved in chemical usage (it is derived from the Latin for "a pile of" or "a quantity of") and has come to mean a particular number, just as a dozen means a number; just as a dozen apples would be 12 apples, a mole of atoms would be 602,000,000,000,000,000,000,000 atoms or 6.02×10^{23} atoms. It turns out that a mole of atoms is usually a convenient amount for laboratory work. Thus, the symbol Ca can stand for the element calcium, a single calcium atom, or a mole of calcium atoms. It will be evident from the context which of these meanings is implied.

Atoms can unite (bond together) to form molecules. A molecule is the smallest particle of an element or a compound that can have a stable existence in the close presence of like molecules. One or more of the same kind of atom can make a molecule of an element. For example, two atoms of hydrogen will form a molecule of ordinary hydrogen, and eight sulfur atoms will form a single molecule. Subscripts in a chemical formula show the number of atoms involved: H_2 means a hydrogen molecule is composed of two atoms, and S_8 means a sulfur molecule is composed of eight atoms. The noble gases, such as helium, He, have monatomic molecules.

When unlike atoms combine, as in the case of water (H_2O) or sulfuric acid (H_2SO_4) the formulas tell what atoms and how many of each are present. For example, H_2SO_4 molecules are composed of two hydrogen atoms, one sulfur atom, and four oxygen atoms. A formula not only can stand for the molecule itself, it can also stand for a mole of such molecules, or for the substance in general, depending on the context.

When elements or compounds undergo a chemical change, the formulas, arranged in the form of a chemical equation, can present the information in a very concise fashion. For example, carbon can react with oxygen to form carbon monoxide. Like most solid elements, carbon is written as though it had one atom per molecule; oxygen exists as diatomic (two-atom) molecules, and carbon monoxide molecules contain two atoms, one each of carbon and oxygen. Furthermore, one oxygen molecule will combine with two carbon atoms to form two carbon monoxide molecules. All of this information is contained in the equation

$$2C + O_2 \longrightarrow 2\,CO$$

The arrow should be read "yields"; the equation then states the following information:

(a) carbon plus oxygen yields carbon monoxide
(b) two monatomic molecules of carbon plus one diatomic molecule of oxygen yield two molecules of carbon monoxide
(c) two moles of carbon atoms plus one mole of diatomic oxygen molecules yield two moles of carbon monoxide

The number written before a formula, the coefficient, gives the amount of the substance involved while the subscript is a part of the definition of the pure substance itself. Changing the coefficient only changes the amount of the element or compound involved, whereas changing the subscript would necessarily involve changing from one substance to another. For example, 2 CO means either two molecules of carbon monoxide or two moles of these molecules.

MEASUREMENT

The heart and soul of reliable scientific facts, and, consequently, the basis for laws and theories, is the accurate measurement of weights, volumes, times, lengths, and other quantities. Perhaps more than any other person, Antoine Lavoisier, who lived in the late 1700's, put chemistry on a quantitative basis by very accurately measuring the weights of materials used in chemical reactions. Because of the significance of his work, Lavoisier, a French chemist, has been called the Father of Modern Chemistry. The reproducible data which he obtained led directly to the formulation of several laws concerning chemical change.

The system of measurement used worldwide by scientists is the *metric system*. This system was developed in France in 1790 and is now the official system of weights and measures in most countries. Proposals for the conversion of the United States economy from the English system to the metric system are presently under study by Congress. By 1975 the United States may be the only major country in the world which does not use the metric system and its products will be harder to sell elsewhere for that reason.

The metric system has one great advantage over older systems: it is simple. Its simplicity lies in its organized structure and its decimal counting which require relatively few terms to be remembered. For example, units such as those of length are defined in such a way that a larger unit is ten (or some power of ten) times larger than a smaller unit. A meter, a unit of length equal to 39.4 inches, is 100 *centi*meters and a centimeter is 10 *milli*meters. To make a conversion from one unit of length to another in the metric system merely involves shifting the decimal point. Compare this to the complexity of the conversion of miles to inches wherein one would probably employ the factors 12 and 5280. Instead of having a number of different root words for length as in the English system (league, mile, rod, yard, foot, inch, mil), the metric system has just one, the meter. With suitable prefixes, the meter can be expressive of a unit useful to the watchmaker (millimeter = 0.001 meter) or to the distance runner (kilometer = 1000 meters).

While the convenience of the metric system can scarcely be overstated, it should be emphasized that the science of chemistry need not rest on this system of measurement. Measurements made in the English system can be and are just as accurate. However, in the metric system only four basic units will be required: a unit of length, the meter; a unit of volume, the liter; a unit of mass (weight), the gram; and a temperature unit, the Celsius degree. English equivalents of the first three are:

$$1 \text{ meter} = 39.4 \text{ inches}$$
$$1 \text{ liter} = 1.06 \text{ quarts}$$
$$1 \text{ gram} = 0.0352 \text{ ounce}$$

Temperature scales are defined in terms of the behavior of samples of matter. For example, the familiar Fahrenheit scale defines the temperature at which water freezes to be 32°F and that at which it boils to be 212°F. Thus there are 180 (212 − 32) Fahrenheit degrees between the freezing and boiling points of water. Most scientists prefer the Celsius scale which defines the freezing point of water to be 0°C and the boiling point, 100°C. There are 100 Celsius degrees in this same temperature range.

For the interested reader and those who need to understand additional aspects of the metric system to support laboratory work, more information is presented in Appendices A, B, and C.

CONCLUSION AND PREVIEW

You have seen briefly how a chemical view of the world can be developed to deal with certain kinds of changes in the world around us. This is derived from the assumption that all the matter in the world is composed of a limited number of elements. The pieces of our world which we actually see consist of very large numbers of atoms which generally are combined with each other in definite ratios to form compounds that are intermingled with each other. It is these elements and compounds which are rearranged and combined to produce the chemical changes of nature. In later chapters we will center our attention on chemical changes. Not only will we examine the methods used to study and characterize natural chemical changes, but we will also see how the concepts developed can be used to produce new types of chemical changes.

We will find that behind all effects associated with chemical changes are causes related to the basic patterns of the minute bits of matter we call atoms; that is, to the submicroscopic structure of matter. We begin with a detailed discussion of the structure of matter and the steps that lie behind the development of this knowledge. We will conclude by applying our knowledge of structure to explain chemical phenomena that are intimately related to everyday life.

QUESTIONS

1. Name as many materials as you can that you have used during the past day that were not in any way the result of a man-made chemical change.

2. Identify the following as physical or chemical changes. Explain in terms of the operational definitions for these types of changes.

   (a) formation of snow flakes
   (b) dissolving oxygen gas in water
   (c) purple cabbage turning red in vinegar
   (d) fashioning a table leg from a piece of wood
   (e) fermenting grapes

3. Chemical changes can be both useful and destructive to man's purposes. Cite a few examples of each kind with which you have had personal experience. Also give observed evidence that each is indeed a chemical change and not a physical change.

19

4. Classify each of the following as a physical property or a chemical property. Explain in terms of operational definitions.

(a) density
(b) melting temperature
(c) substance decomposes into two elements upon heating
(d) electrical conductivity of a solid
(e) the substance does not react with sulfur
(f) ignition temperature of a piece of paper

5. Classify the following as element, compound, or mixture. Justify each answer.

(a) mercury   (e) ink
(b) milk   (f) iced tea
(c) pure water   (g) pure ice
(d) a tree   (h) carbon

6. Which of the materials listed in Problem 5 are pure substances?

7. Explain how the operational definition of a pure substance allows for the possibility that it is not actually pure.

8. Suggest a method for purifying water slightly contaminated with a dissolved solid.

9. Distinguish between theory and law as used in chemistry.

(a) Which has a better chance of being true? Why?
(b) Which summarizes? Which explains?

10. Given the sentence below, express a chemical reaction in terms of symbols that the chemist uses to convey the same information. "One nitrogen molecule containing two nitrogen atoms per molecule reacts with three hydrogen molecules containing two hydrogen atoms per molecule, to produce two ammonia molecules containing one nitrogen and three hydrogen atoms per molecule."

11. Aspirin is a pure substance, a compound of carbon, hydrogen, and oxygen. If two manufacturers produce equally pure aspirin samples, what can be said of the relative worth of the two products?

12. Explain the statement: "An operational definition rests on decisions concerning classifications whereas a theoretical definition rests on intuitive interpretations." Which type of definition has the greater chance of being in error?

13. How is the salt content of the sea related to the purity of rain water? What method of purification does nature employ in the purification of rain water?

14. Suggest a hypothesis in terms of the purification process of recrystallization whereby one might explain the occurrence in nature of relatively pure salt deposits.

15. Two boys begin in front of a marching column of boy scouts. At the end of the march the playful fellow brings up the rear, having taken time to observe many of his surroundings, while the dedicated marcher remains in front. How is this analogous to chromatographic separations?

16. The chemist believes that there are submicroscopic structures that will explain macroscopic properties in a simple, straightforward way. Do you think the fact that there are only 90 naturally occurring elements argues for the "simplicity idea" concerning matter? Why?

17. How many times do you think a given experiment should give a result before a scientific fact is established? How many failures would you require before rejecting the "fact?"

18. Suppose a wife and children discover the family car missing on returning from an afternoon ball game, even though the father rode to work with a neighbor that morning. The mother

says to the children, "Don't worry, Dad must have had an unexpected need for the car and got it after lunch." Would you call the statement made by the mother a theory or a hypothesis? Why?

19. From the molecular formulas given below, tell what kind of atoms and how many of each kind are present in a molecule. (You may look up the names of any elements whose symbols you do not know in the list printed inside the cover.)

$$SO_3, \ HCl, \ NH_3, \ H_2S$$

20. Describe in words, the chemical process which is summarized in the following equation:

$$2Na + Cl_2 \longrightarrow 2NaCl$$

21. How many *atoms* are present in each of the following:

 (a) one mole of He
 (b) one mole of Cl_2
 (c) one mole of O_3

NOTE: The following problems should be attempted after a more thorough study of the metric system as presented in Appendices A, B, and C.

22. Determine:

 (a) the mass in grams of a half-pint of water weighing 8 ounces.
 (b) the mass in kilograms of a 1 pound loaf of bread.
 (c) the length in centimeters of a pencil 9 inches long.
 (d) the volume in gallons of 10 liters of cider.
 (e) the speed in kilometers per hour of a car going 45 miles per hour.

23. If you prefer 70°F for your living quarters, what would be your preference on the Celsius scale?

24. While traveling in Europe, you have antifreeze put into your car radiator. The attendant tells you that your car will now be safe at temperatures down to −10°C. When you return to the United States, you hear on the radio that a low temperature of 20°F is expected for the coming evening. Do you need to worry about your car? Why?

25. A doctor in the United States expects the body temperature of a normal person to be 98.6°F.

 (a) If the doctor uses a European clinical thermometer with a Celsius scale, what temperature is normal?
 (b) Does a person with a body temperature of 40°C have a fever?

26. Is the numerical value of the temperature in Fahrenheit degrees always larger than the numerical value in Celsius degrees?

27. At what temperature is the numerical value on the Fahrenheit scale exactly twice the numerical value on the Celsius scale?

SUGGESTIONS FOR FURTHER READING

Chemical Viewpoint

Ihde, A. J., "The Development of Modern Chemistry," Harper and Row, New York, 1964, pp. 3–31.

Kesselman, B., "The Skeptical Chemist," *Chemistry*, Vol. 40, No. 1, p. 9 (1967).

Hildebrand, J. H., "It Ain't Necessarily So," *Chemistry*, Vol. 40, No. 10, p. 19 (1967).

Garrett, A. B., "The Discovery Process and the Creative Mind," *Journal of Chemical Education*, Vol. 41, p. 479 (1964).

Benjamin, A. C., "Science, Technology, and Human Values," University of Missouri Press, Columbia, Mo., 1965.

Nampel, G. C., "Aspects of Scientific Explanations," Free Press, New York, 1965.

Braithwaite, R. B., "Scientific Explanations," Cambridge University Press, Cambridge, England, 1953.

Chemical Symbols

Szokefalvi-Nagy, Z., "How and Why of Chemical Symbols," *Chemistry*, Vol. 40, No. 2, p. 21 (1967).

Dinga, G. P., "The Elements and the Derivation of Their Names and Symbols," *Chemistry*, Vol. 41, No. 2, p. 20 (1968).

Relevance of Chemical Knowledge

Keller, E., "The DDT Story," *Chemistry*, Vol. 43, No. 2, p. 8 (1970).

SOME PRINCIPLES
OF CHEMICAL REACTIVITY

There are over 100 different chemical elements, each with its own characteristics. In addition to the broad classifications of metals and nonmetals mentioned in Chapter 1, smaller groups or families of elements can be recognized on the basis of similar chemical behavior. In this chapter we shall observe some general characteristics of chemical reactions and note some groupings of reactions that help to classify the elements; for, after all, if we are to study chemistry as it relates to us, we must seek to simplify the task wherever possible.

CHARACTERISTICS OF CHEMICAL REACTIONS

As it is in all in-depth studies, a thorough consideration of chemical change involves more than is immediately evident in a cursory survey. In addition to the obvious production of new chemical substances, other phenomena are also present which should be studied and understood. *Energy* changes are as definite and reproducible as substance changes in a chemical reaction. The rate at which a chemical change occurs is affected by several factors which can be controlled. A chemical reaction is a two-way street; many chemical reactions are *reversible* in practice, and all reactions are reversible in theory. To an observer, a reversible reaction appears to proceed to a point at which both starting materials and product materials are present. Finally, most chemical reactions are not completely unique. For example, if substance A is chemically similar to substance B, it will be found that they will enter into similar chemical reactions. We shall now consider in detail some of these characteristics of reactions.

Reactants Become Products

In all chemical reactions some pure substances disappear and others appear. A familiar example is the change of shiny steel to a red-brown iron rust (Figure 2–1). This is a very important reaction since it has been estimated that the dollar loss from corrosion of iron and steel in the United States is slightly over $30 per person per year.

23

FIGURE 2-1   Chemical changes produce new substances, often with properties very different from the starting material. The strong steel with which this automobile fender was made has been converted by oxidation to powdered rust.

Some other chemical changes are given in the following equations:

Reactants		Products	Heat effect*
CaO + H$_2$O	$\longrightarrow$	Ca(OH)$_2$	+ Heat [15.6 kcal per mole of Ca(OH)$_2$]
Calcium oxide (quicklime) Water		Calcium hydroxide (slaked lime)	
2Na + Cl$_2$(gas)	$\longrightarrow$	2NaCl	+ Heat [98.2 kcal per mole of NaCl]
Sodium Chlorine		Sodium chloride (table salt)	
H$_2$(gas) + I$_2$(gas)	$\longrightarrow$	2HI(gas)	− Heat [6.20 kcal per mole of HI]
Hydrogen Iodine		Hydrogen iodide	

From these reactions we can note our first important point concerning chemical reactions:

In chemical reactions, reactants become products; some substances are consumed and new ones appear.

Quantitative Energy Changes

In the first reaction, calcium oxide (quicklime) reacts with water to give calcium hydroxide (slaked lime) with the evolution of heat. In the second reac-

---

*Heat energy can be measured in calories (cal). A kilocalorie (kcal) is 1000 calories. A *calorie* is the amount of heat required to heat one gram of water one degree centigrade.

tion, metallic sodium reacts with the greenish yellow gas, chlorine, to give sodium chloride or table salt. If a piece of hot sodium is put into a flask containing chlorine, the sodium burns, liberating a great deal of heat and light, to produce white crystals of sodium chloride. In the last reaction, gaseous hydrogen reacts with gaseous iodine to produce gaseous hydrogen iodide, with the absorption of heat. These facts, then, lead to a second point:

A given amount of a particular chemical change corresponds to an energy change of a fixed amount.

To illustrate, the preparation of one mole of $Ca(OH)_2$ from CaO and H_2O releases 15.6 kcal; for two moles of $Ca(OH)_2$, 2×15.6 or 31.2 kcal of heat is released.

Figure 2–2 shows photographs of three chemical reactions, two of which release heat energy, while the third absorbs energy. The reaction of sodium with chlorine, part B, is a spectacular reaction releasing energy in the forms of heat, light, and sound. In part C, the electrolysis of water is displayed. In this reaction, in which water is decomposed, electrical energy must be supplied in order to make the reaction proceed.

Sometimes energy changes in reactions are difficult to observe because of the very slow rate of reaction. An example is the rusting of iron. Here the reaction involved is complicated, but we can represent it by the simplified equation:

$$4Fe + 3O_2 + 6H_2O \longrightarrow 4Fe(OH)_3 + 788 \text{ kcal}$$

Iron   *Moist air*   *Iron Hydroxide + heat*
(rust)

The rusting of iron occurs so slowly that the liberation of heat is perceptible only with the aid of special instruments. The total amount of heat evolved in rusting is considerable, but it typically takes place over a long period of time.

Temperature Effect on Reaction Rate

We noted earlier that sodium reacts more rapidly with chlorine than iron does with oxygen of the atmosphere. The whole notion of how fast or how slow chemical reactions proceed is one which can be put on a quantitative basis by the concept of the *reaction rate*. The rate of a reaction is always defined in terms of the changes in the amounts of chemical substances present per unit of time. Thus, if we consider the burning of sulfur to produce sulfur dioxide,

$$S + O_2 \longrightarrow SO_2$$

we can discuss the rate of the reaction in terms of the amount of SO_2 formed per minute or of the amount of S or O_2 consumed per minute.

It is possible to alter the rate of a chemical reaction by changing the temperature. If the temperature is raised, the rates of chemical reactions are increased; if the temperature is reduced, they are decreased. We make use of this principle in cooking foods (a roast will cook at a faster rate at a higher temperature) and in preserving foods (foods spoil less quickly if refrigerated).

FIGURE 2-2   *A*, Sodium metal burning in air (20 per cent oxygen), forming sodium oxide. *B*, Sodium metal burning in pure chlorine, forming sodium chloride (salt). *C*, Electrical energy is required to decompose water into hydrogen (right tube) and oxygen (left tube). Note that there are about two volumes of hydrogen produced for each volume of oxygen.

FIGURE 2-3   The biochemical processes of decomposition occur more rapidly at higher temperatures. Half the peach shown in this photograph was refrigerated, while the other half was kept warm. The refrigerated half on the right shows little discoloration, while the other one shows the typical sign of decay.

Figure 2–3 illustrates the effect of temperature on the biochemical reactions of decomposition that take place in fruit.

We conclude that another general characteristic of chemical reactions is:

The rate of a typical chemical reaction is increased if the temperature is increased.

Effect of Concentration on Reaction Rate

It is also possible to alter the rate of a reaction by changing the concentration of the reactants. For example, in the reaction of sulfur and O_2 given earlier, if air replaces the oxygen, the reaction will proceed at a slower rate, since air is a mixture of oxygen and nitrogen. The rusting of iron can be decreased by painting or galvanizing the surface of the metal to cut down on the concentration of the oxygen and moisture at the surface. Figure 2–4 contrasts the oxidation of iron when oxygen is in relatively small or in relatively large concentrations. These facts illustrate still another characteristic of chemical reactions:

The rate of a chemical reaction depends on the concentrations of the reactants (Figure 2–4).

FIGURE 2-4   Effect of concentration on reaction rate. Steel wool held in the flame of a gas burner is oxidized. It is in contact with air, which is 20 per cent oxygen. When the red hot metal is placed in pure oxygen in the flask, it oxidizes much more rapidly.

A molecular explanation can be given for these observed or empirical results. In the reaction of O_2 and sulfur, the reaction proceeds more slowly in air because there are fewer oxygen molecules per volume of space to react with the sulfur. This example illustrates the useful technique of reducing the rate of a reaction by reducing the concentration of the reactants. Figure 2–5 illustrates the theoretical (molecular) explanation for both concentration effect and the temperature effect on the rate of chemical reactions. The explanation, a bit sketchy here, will be more fully developed as the kinetic molecular theory is presented in Chapter 5.

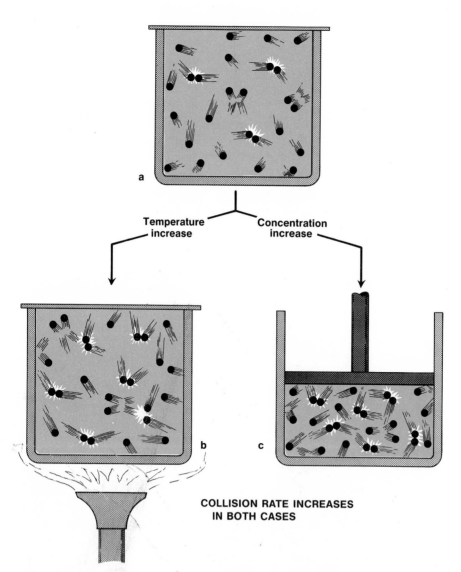

**COLLISION RATE INCREASES
IN BOTH CASES**

FIGURE 2-5   Effects of temperature and concentration on rates of chemical reactions. At the higher temperature, more collisions occur between molecules, and a greater percentage of the collisions produce a chemical reaction. At the higher concentration (no temperature change), more collisions occur, but the percentage of effective collisions remains the same.

Reversibility of Chemical Reactions

Most chemical processes are capable of being reversed under suitable conditions. When a chemical reaction is reversed, some of the products are converted back into reactants. For example, heating calcium hydroxide will drive off some of the water. This process is the reverse of adding water to quicklime.

$$Ca(OH)_2 + Heat \longrightarrow CaO + H_2O$$

Other methods can be used to reverse chemical reactions. Thus, if we put calcium hydroxide in a vacuum, there will soon be water vapor in the space around the solid. It is sometimes possible to reverse chemical reactions when they are associated with small heat changes. For the sodium-chlorine reaction, it is posssible to observe the process of gaseous NaCl molecules breaking apart into atoms,

$$NaCl(gas) + Heat \longrightarrow Na + Cl$$

but only at the very elevated temperatures at which the compound NaCl is vaporized. Since sodium chloride is very stable compared to sodium and chlorine, the reaction forming sodium chloride is reversed only under conditions similar to those in which water is decomposed into hydrogen and oxygen using electrical energy (Figure 2–2). The fact that many reactions can be approached from either direction leads to the conclusion:

Chemical reactions are generally reversible.

Figure 2–6 shows the reverse reaction to the electrolytic decomposition of water. Hydrogen when burned in air produces water, which can readily be condensed (Figure 2–6).

A bit of theoretical explanation on this empirical rule may also be of interest. If atoms in one arrangement—as in the case of water—can be decomposed into another arrangement, i.e., hydrogen and oxygen, there would be every reason to believe that the atoms could be rearranged into the original structure. After all, a set of building blocks can be made into a building, then into a bridge and then back into the building again. This concept will be discussed in detail later.

There are many reversible chemical reactions which are important to human life. One of these is involved in the transport of atmospheric oxygen from the lungs to the various parts of the body. This task is carried out by hemoglobin, a complex molecule which is found in the blood. This molecule takes up oxygen while in the lungs.

$$Hemoglobin + O_2 \rightleftharpoons Oxyhemoglobin$$

The oxyhemoglobin is then carried by the bloodstream to the various parts of the body where it releases the oxygen for use in metabolic processes.

While these have been examples of reversible chemical reactions, it is appropriate to add here that many physical changes are also reversible. The physical change, ice to water, is reversible and can be written as:

$$ice \longrightarrow liquid - energy$$

29

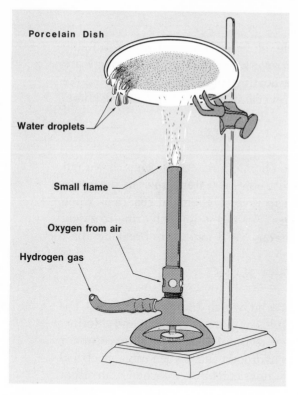

FIGURE 2-6   Hydrogen and oxygen burning to produce water in the gaseous state. The water is condensed on the cooler porcelain dish.

or

$$\text{liquid} \longrightarrow \text{ice} + \text{energy}$$

This process, and many other physical processes, can be reversed an indefinite number of times by the repeated addition or removal of energy.

Reactions by Chemical Groups

When a survey is made of the known types of chemical reactions, a diversity is found which is illustrated by the following examples:

$$C \quad + \quad 2F_2 \quad \longrightarrow \quad CF_4$$

Carbon   Fluorine   Carbon tetrafluoride

$$PCl_3 \quad + \quad 3H_2O \quad \longrightarrow \quad H_3PO_3 \quad + \quad 3HCl$$

Phosphorus   Water   Phosphorous   Hydrogen
trichloride   acid   chloride

$$SO_3 \quad + \quad H_2O \quad \longrightarrow \quad H_2SO_4$$

Sulfur   Water   Sulfuric acid
trioxide

$$C \quad + \quad 2S \quad \xrightarrow{\text{Heat}} \quad CS_2$$

Carbon   Sulfur   Carbon disulfide
              vapor

$$Fe_2O_3 \quad + \quad 3CO \quad \xrightarrow{\text{Heat}} \quad 2Fe \quad + \quad 3CO_2$$

Iron(III)   Carbon   Iron   Carbon dioxide
oxide   monoxide

30

These reactions are not presented for you to learn at this point but only to show the wide variety of reactions and substances encountered in chemistry. Indeed, the number of known compounds runs into the millions and the number of reactions by which they are produced into multimillions. However, becoming familiar with chemical reactions is not as hopeless as it may seem at first. There is much order to be found in the apparent chaos. For example, even though there is little order that is obvious to the beginner in the reactions given above, consider the following reactions:

$$2\text{Na} + \text{F}_2 \longrightarrow 2\text{NaF}$$
Sodium Fluorine Sodium fluoride

$$2\text{Na} + \text{Cl}_2 \longrightarrow 2\text{NaCl}$$
Chlorine Sodium chloride

$$2\text{Na} + \text{Br}_2 \longrightarrow 2\text{NaBr}$$
Bromine Sodium bromide

$$2\text{Na} + \text{I}_2 \longrightarrow 2\text{NaI}$$
Iodine Sodium iodide

Note that the group of elements, fluorine, chlorine, bromine, and iodine, all react with sodium in much the same way; that is, products with similar formulas are produced. Once we have learned one reaction in such a group of elements, we have some idea about what is likely to happen to other members in the group. Even the names of the compounds involved become easier to remember, because a system of names is obviously involved.

There are elements that undergo similar chemical reactions. Elements showing such similarities are identified as a chemical group or family. The elements fluorine, chlorine, bromine and iodine form one such family.

Another such family includes the group of elements oxygen, sulfur, and selenium, all of which react with calcium in much the same way.

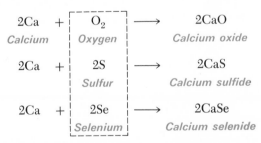

$$2\text{Ca} + \text{O}_2 \longrightarrow 2\text{CaO}$$
Calcium Oxygen Calcium oxide

$$2\text{Ca} + 2\text{S} \longrightarrow 2\text{CaS}$$
Sulfur Calcium sulfide

$$2\text{Ca} + 2\text{Se} \longrightarrow 2\text{CaSe}$$
Selenium Calcium selenide

Not only do elements react as members of groups but groups of atoms may act as a single unit. Consider the following:

$$\text{Mg} + \text{H}_2\text{SO}_4 \longrightarrow \text{MgSO}_4 + \text{H}_2$$
Magnesium Sulfuric acid Magnesium sulfate Hydrogen

$$\text{Ca} + \text{H}_2\text{SO}_4 \longrightarrow \text{CaSO}_4 + \text{H}_2$$
Calcium Calcium sulfate

31

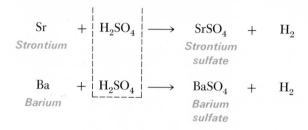

The sulfate group, SO_4, in this series of reactions acts very much like the oxygen, sulfur, and selenium in the group of reactions with calcium given previously.

Two major topics growing out of this discussion of the general characteristics of the chemical reactions are the periodic classification of the elements and chemical equilibrium. Usually these topics are developed much later in a chemistry course because of their complexity. However, these ideas in their simpler forms go back to the early history of chemistry, and we feel they should be carried along and expanded as an overview of the science is obtained.

The Periodic Chart

Many of the chemical and physical properties of the elements can be correlated and presented in a Periodic Chart of the elements (Figure 2–7). This particular arrangement of the elemental symbols is dictated by nature, but it is also the result of man's ability to integrate factual information and theoretical models in an organized way. The interpretation of the periodic chart can be carried out at various levels. You can expect this chart to be meaningful from both a practical and a theoretical point of view, as well as being a memory aid in obtaining an overview of chemistry.

Elements with similar chemical properties are placed in vertical groups (also known as *families*). For example, Group I, the Alkali Metal Group, is composed of lithium, sodium, potassium, rubidium, cesium, and francium. All of these elements are metals with general characteristics that distinguish them from all the other elements. Each of them reacts with chlorine to form chlorides with the general formula:

$$MCl$$

where M can be either Li, Na, K, Rb, Cs, or Fr.

It is obvious that the elements within a group are not the same in all respects or they would be the same element. To illustrate, sodium reacts more vigorously with chlorine than does lithium. Note that sodium is found under lithium in this table. Also, potassium reacts more vigorously than sodium. In fact, there is actually a trend of reactivity from the top of the group to the bottom. For the alkali metals, the trend is to react more vigorously with chlorine as one moves down the group. Such trends of reactivity exist in all of the chemical groups.

In the periodic chart, you will note that each element is assigned a number. This is the *order number* of the element and is based, from a historical point of view, on all the known chemical and physical information about the element. Later we shall find that the order number of an element is identical to its *atomic number*, which is a fundamental concept derived from atomic structure. The

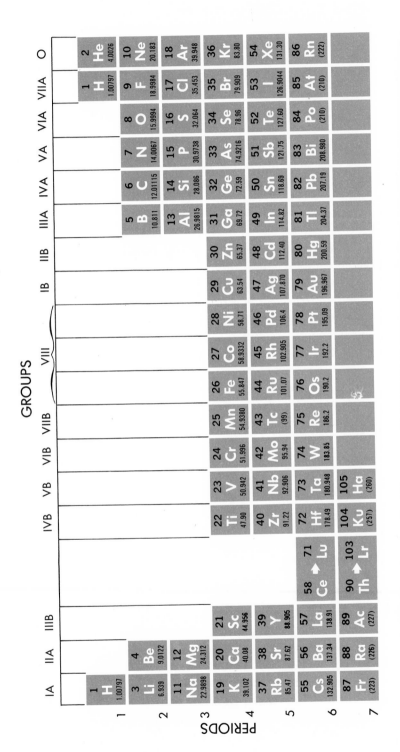

FIGURE 2–7   The periodic chart of the elements. Each block contains the symbol for the element, its atomic number, and its atomic weight.

chart is "periodic" because when the elements are arranged in the order of their atomic numbers, periodically we encounter an alkali metal and periodically we encounter members of each of the other elemental groups. Observe that the first period has only two elements, hydrogen and helium. (Hydrogen is unique in that it can fit into two chemical groups. The second and third periods have eight elements:

Li Be B C N O F Ne
Na Mg Al Si P S Cl Ar

The fourth and fifth periods have 18 elements each, and the situation becomes more complex as we proceed. But the point to be made is that there are periods of such length that similar elements do fall beneath one another in the chart.

Chemical Equilibrium

When a chemical reaction proceeds to a point at which the amounts of the reactants and products do not change any further, the reaction has reached a state of *chemical equilibrium*. At the beginning of a reaction, we usually are dealing only with reactants. As the reaction proceeds, the amounts of reactants decrease while the amounts of products increase until the reaction arrives at equilibrium. At equilibrium there are still reactants remaining in the presence of products. The amounts of products may be greater than, equal to, or less than the amounts of reactants.

As an example of chemical equilibrium, consider a laboratory experiment in which N_2O_4 (dinitrogen tetroxide) is the reactant and NO_2 (nitrogen dioxide) is the product, according to the equation:

$$N_2O_4 \rightleftharpoons 2NO_2 - 13.9 \text{ kcal}$$

Let us examine what happens if the reaction is carried out in a suitable container in which one mole of N_2O_4 is present initially, as is shown in Figure 2–8. We can follow the progress of the reaction by measuring the color of the system, since NO_2 is brown and N_2O_4 is colorless. The intensity of the color of the system is thus related to the amount of NO_2 present. As time passes, the brown color develops and then reaches a steady intensity in the reaction vessel. Since the intensity does not change beyond this point, as long as the container remains leakproof and is held at a constant temperature, one might be inclined to believe that the reaction had gone to completion. However, if one tests the system, he finds that the vessel contains a mixture of NO_2 and unreacted N_2O_4. Since some N_2O_4 remains, and since the amounts of N_2O_4 and NO_2 do not change with time, we say that the reaction has reached a state of equilibrium.

If the previous experiment is changed by starting with two moles of NO_2, rather than one mole of N_2O_4, the same shade of brown will eventually appear, provided, of course, the conditions of container size and temperature are the same (Figure 2–8). This emphasizes the fact that the same position of chemical equilibrium can be attained by approaching from either direction.

Further investigations indicate that chemical equilibrium is *dynamic* in the

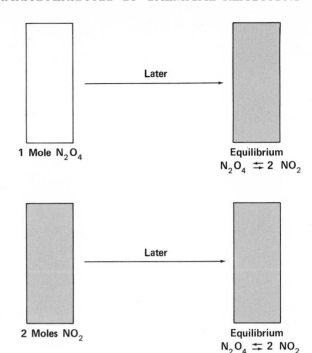

1 Mole N_2O_4

Later

Equilibrium
$N_2O_4 \rightleftharpoons 2\ NO_2$

FIGURE 2-8   The equilibrium between N_2O_4 and NO_2 can be approached, from either direction because the reactions are easily reversible. The intensity of the color indicates the amount of NO_2 present in the system.

2 Moles NO_2

Later

Equilibrium
$N_2O_4 \rightleftharpoons 2\ NO_2$

sense that molecular transformations occur continuously when a system is at equilibrium. Although the relative amounts of reactants and products do not change with time, experimental studies indicate that product molecules are reverting to reactants and that reactant molecules are becoming products. *However, both processes occur at the same rate, so the overall amounts of products and reactants remain the same.* To illustrate, consider the reaction, $N_2 + 3H_2 \rightleftharpoons 2NH_3$. If we allow nitrogen, hydrogen, and ammonia to come to equilibrium, then remove some of the hydrogen (H_2) and replace it with exactly the same number of molecules of deuterium (D_2), we can show the dynamic nature of equilibrium. (Deuterium is a heavier form of hydrogen with similar properties, distinguishable from ordinary hydrogen by its weight.) After a few moments, we analyze the reaction mixture, using a mass spectrograph (Chapter 6). Perhaps to our surprise, we find the species HD, NH_2D, NHD_2, and ND_3, as well as the H_2, N_2, and NH_3 present in the original equilibrium mixture. The only way to explain this assortment of molecular species is to assume that hydrogen (D_2 and H_2) is reacting with nitrogen to form ammonia simultaneously with its decomposition into nitrogen and hydrogen. This can be generalized into a fundamental characteristic of chemical equilibrium.

At equilibrium, the rate of formation of products from reactants is equal to the rate at which products revert back to reactants.

It is the dynamic aspect of chemical equilibrium that we can associate with the usual meaning of equilibrium—balance. In chemical equilibrium, there is a balance (or equality) between the forward and reverse reaction rates. This balance in rates gives rise to constant relative amounts of reactants and products.

CONCLUSION AND PREVIEW

At every point in the accumulation of empirical information one is tempted to ask, "Why is it this way?" Although unraveling the why of nature is a never-ending process, it is extremely satisfying to discover explanations for some of the accumulated observations. Thus far we have examined a few aspects of chemistry that are reproducible and observable with our five senses. We conclude that there is factual chemical information, that it is readily classified in an organized way, and that there must be explanations for the chemistry we observe. In the succeeding chapters, we shall build on the most fundamental assumption in chemistry: that there is structure in the submicroscopic world in the form of atoms and molecules, and what we observe by our senses can be explained in terms of these structures. We shall now trace the development of modern atomic theory. Later, we shall consider the consequences of the science upon our society.

QUESTIONS

1. Based on information presented in this chapter on likenesses of elements in groups, predict the formulas of the products of the following reactions.

   $Rb + Cl_2 \longrightarrow$ $\qquad$ $K + Cl_2 \longrightarrow$

   $Ba + O_2 \longrightarrow$ $\qquad$ $Mg + Se \longrightarrow$

   $Na + Br_2 \longrightarrow$ $\qquad$ $Mg + S \longrightarrow$

   $Sr + S \longrightarrow$ $\qquad$ $Be + S \longrightarrow$

2. In the periodic chart the elements are arranged in the order of their _____.

3. If you were arranging the elements in a row, A, B, C, D, . . . , and came upon an element, G, which had properties similar to A, where would you place it?

4. When a chemical reaction reaches equilibrium, what does this mean with respect to:

   (a) the relative amounts of reactants and products?
   (b) the cessation of chemical reaction?

5. Name some ways energy plays a role in chemical reactions.

6. Give an example of a chemical reaction whose rate is fast and one whose rate is slow.

7. List three characteristics of all chemical reactions and give an example of each.

8. Using the periodic chart, select elements which can be expected to have chemical properties similar to:

   (a) Ca (calcium), atomic number, 20
   (b) Fe (iron), atomic number, 26
   (c) Sn (tin), atomic number, 50
   (d) S (sulfur), atomic number, 16

9. In 1968 an Apollo spacecraft cabin fire killed three astronauts. The fact that pure oxygen was used as the cabin atmosphere contributed to the fire. How?

10. Give an example of a chemical reaction that would be very difficult to reverse. Why would it be difficult to reverse?

11. The element helium is very unreactive chemically. What type of behavior would you expect for argon? Study their relationship on the periodic chart.

12. The recycling of many by-products in our society, such as paper and glass, involves the principles of reversibility. Outline how paper and glass may be recycled for further use.

13. Consider what would be the relative rates of rusting of iron in a dry climate as opposed to a damp climate. What principle(s) of reaction rate is(are) involved?

14. Copper forms a chloride salt, CuCl. Is this consistent with the group number into which copper is placed in the periodic chart?

15. Fires have been started by water seeping into bags in which quicklime was stored. Why would this produce a fire?

SUGGESTIONS FOR FURTHER READING

"An Oscillating Chemical Reaction," *Chemistry*, Vol. 41, No. 5, p. 26 (1968).
Schaff, J. F., and Westmeyer, P., "Dynamic Nature of Chemical Equilibrium," *Chemistry*, Vol. 41, No. 7, p. 48 (1968).
Videen, Tom, "Burning Rates of Candles," *Chemistry*, Vol. 39, No. 8, p. 26 (1966).

THE DALTONIAN ATOM

CHAPTER 3

MATTER: CONTINUOUS OR DISCONTINUOUS?

A basic question that has concerned man from antiquity is: Can a sample of matter be subdivided indefinitely without losing its identity? For example, would one expect the division of gold into smaller and smaller particles to leave always a sample of gold, or should he expect that he would ultimately arrive at a tiny particle of gold which, if further subdivided, would yield something other than gold? Although most students would probably agree with the particle nature of matter, a thoughtful consideration of the problem would force them to admit that a direct observation made with one or more of their physical senses is not sufficient to demonstrate the discontinuous nature of matter beyond all reasonable doubt.

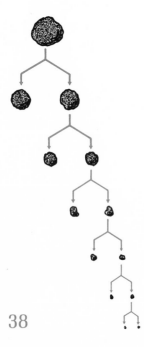

FIGURE 3-1   Is matter continuous or discontinuous?

THEORETICAL ATOMS VISUALIZED: THE GREEK INFLUENCE

The earliest known advocate of the atomic (discontinuous) nature of matter was a Greek philosopher, Leucippus, who lived in the fifth century B.C. However, it was his student, Democritus (460–370 B.C.), who apparently aroused the interest of the natural philosophers of his own and of later times in this concept of matter. Democritus used the word *atom*, meaning "that which cannot be further divided," to describe the ultimate particles of matter. He believed that there was no limit to the kinds of atoms and that all of them were made of the same basic material. He also believed that the various kinds of atoms differed only in their shapes and sizes. Democritus even sought to explain the properties of substances like lead and iron in terms of the way in which the different kinds of atoms associated in forming the macrostructure. He reasoned that the shape of the lead atom enables the close packing that agrees with the formation of a very dense but soft material, whereas iron atoms fit together in a rigid network pattern accounting for iron's lower density and greater strength.

It is unjust to report that the atomic concepts of the Greeks were based purely on intuitive feeling rather than on scientific observation. The student is reminded that a physical theory of matter has value if it can explain some, even though not all, of the macroscopic facts at hand, in terms of submicroscopic models. Although the atomic theory of Democritus was limited in its scope and application, it did indeed explain in simple terms some well-known phenomena, such as evaporation (the drying of clothes), condensation (the moisture that appears on the outside of a glass of ice water), diffusion (the movement of an odor through a room), and the growth of new material (crystal growth from a solution).

Based solely on philosophical thought, Aristotle (384–322 B.C.), Plato (427–327 B.C.), and consequently those in the mainstream of enlightened thought rejected the atomistic theories of Democritus. More than two thousand years were to pass before John Dalton (1766–1844), an English schoolteacher, forcefully reintroduced the idea of the atom. Later Dalton's ideas captured the imagination

John Dalton (1766–1844), a self-taught English school teacher, moved to Manchester in 1793 and devoted the rest of his life to scientific investigations. His presentation of atomic theory in the early part of the 19th Century served as the basis from which modern chemical theories have grown.

39

of the scientific community and provided a solid basis for modern atomic theory. Prior to Dalton, however, the intuitive feeling that matter was somehow composed of particles had showed up again and again in scientific thought. Two centuries before Dalton, Galileo reasoned that the appearance of a new substance resulting from chemical change could be explained in terms of a rearrangement of parts too small to be seen. Robert Boyle (1627–1691) and Isaac Newton (1642–1727) used atomic concepts in their speculations concerning chemical and physical laws. Even before this, Sir Francis Bacon (1561–1626) had speculated that heat might be a form of motion in submicroscopic matter.

SELECTED PHYSICAL AND CHEMICAL INFORMATION ACCUMULATED BY 1810

What were the reasons that prompted Dalton to advance his new ideas of the atomic nature of matter? We shall look at these in some detail for two reasons. First, an awareness of the chemical and physical laws explained by Dalton is important to modern science; indeed, if the laws that he examined are true statements, they cannot be nullified, but only extended perhaps to more general considerations. Second, this offers an opportunity to gain insight into the most fundamental process in man's quest for knowledge and understanding—the observation of reproducible facts and the interpretation of these facts in terms of concepts new to the mind of man. Both chemical and physical changes and properties were used by Dalton as bases for his concepts of the atom.

While Dalton's atomic theory primarily sought to explain the chemical laws that will be discussed in this chapter, there were certain physical properties of matter that undoubtedly affected his thinking. It is worth noting that a successful theory in the natural sciences must be able not only to explain laws in one science, but it must also, at the same time, be consistent with laws and well-established theories of interrelated sciences.

STATES OF MATTER

Matter occurs in three well-defined forms called *states:* solid, liquid, and gas (Figure 3–2). A solid has a definite shape and volume, a liquid has a definite volume but no definite shape, and a gas has neither definite shape nor definite volume. Under ordinary conditions substances generally exist in just one of the three states. However, by proper manipulation of pressure (force per unit area) and temperature, substances can be made to change state. For example, oxygen at room temperature and atmospheric pressure is a gas, but at $-200°C$, oxygen is a liquid. At temperature and pressure conditions that are easily attained, water can exist as solid ice, liquid water, or gaseous steam.

Long before Dalton's time, scientists knew of the states of matter and understood, at least partly, how to bring about the physical changes from one state to another. Not understood, however, was exactly what was happening

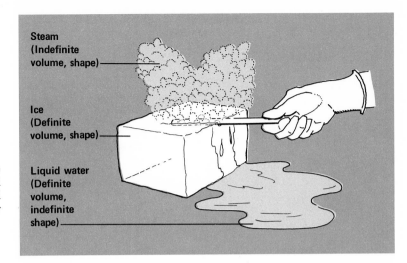

FIGURE 3–2 The three states of matter illustrated by solid water (ice), liquid water, and gaseous water (steam).

Steam (Indefinite volume, shape)

Ice (Definite volume, shape)

Liquid water (Definite volume, indefinite shape)

within the sample of matter as it changed from one state to another. This was particularly true in the case of gaseous matter, and it was the study of gases that, in a large measure, led Dalton to his understanding of the structure of matter. Just how do the physical properties of gases and the physical laws that describe these properties suggest an atomic nature of matter? Let us look at some of the properties and laws and see.

BOYLE'S LAW

Robert Boyle, an English physicist of the 17th century, studied the relationship between the pressure of a gas and its volume at constant temperature. A simple way to measure the change in volume accompanying a change in pressure is to use an apparatus like that shown in Figure 3–3. The sample of gas trapped in the cylinder is subjected to various pressures by changing the weights. For each pressure, the piston will adjust until the pressure exerted by the weights is equal to the pressure exerted by the entrapped gas. The volume is then measured for each pressure. Some typical data emerging from this kind of apparatus could be as follows:

Pressure (Pounds/square inch)	Volume (Cubic inches)	Pressure × Volume
20	100	2000
25	80	2000
30	67	2010
40	50	2000

Is there any regular change in the volume as the pressure is changed? How can this relationship be expressed? It appears from these data that the product of the pressure times the volume is approximately constant (within the accuracy of the measurements) and that doubling the pressure halves the volume. In essence, these are statements of *Boyle's law*, which can be stated in several alternate equivalent ways.

41

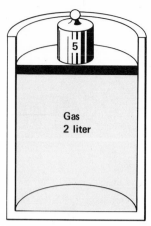

Condition:
Temperature constant
no gas gained or
lost.

5

Gas
2 liter

5   5

Gas
1 liter

FIGURE 3-3   Illustration of Boyle's law. If the pressure is doubled, the
volume is halved at constant temperature and amount of gas.

For a series of pressure and volume measurements on the same sample
of gas at constant temperature, the volume of a gas varies inversely
with the pressure on that sample of gas, or

$$\text{pressure} \times \text{volume} = \text{a constant}$$
$$PV = \text{constant}$$

For any two sets of measurements, P_1V_1 and P_2V_2, on the same sample
of gas,

$$P_1V_1 = P_2V_2$$

Boyle's observations and subsequent studies have shown that this relationship
is a good *approximation* for describing the behavior of most gases. From the
fact that gases can be squeezed into a smaller space, Boyle theorized that gases
are made of particles much like submicroscopic threads of wool or small coiled
springs, and that the pressure is due to these particles being whirled about so
vigorously by the ether (a hypothetical medium proposed to support the particles
of a gas) that each particle endeavors to beat off all others from coming within
its own little sphere. The interpretation of the properties of matter in terms
of small particles becomes more and more attractive as the knowledge of matter
becomes more detailed.

DALTON'S LAW OF PARTIAL PRESSURES

Many of Dalton's experiments involved measurements on air. Scientists of
his time believed that air was a compound composed of nitrogen, oxygen, and
water. Dalton's own analyses of air samples gathered at different locations
showed that it is always composed of 21 parts oxygen to 79 parts nitrogen,

but the percentage of water varies, increasing with temperature. In addition to showing that air is a mixture, his experiments showed that the pressure exerted by a sample of dry air increases when water vapor is added to it, and the increase is by an amount equal to the pressure exerted by the water vapor alone at that temperature.

This type of behavior is observed for other gaseous mixtures as well. As a typical example, 32 grams of oxygen at 0°C exerts 14.7 pounds per square inch (1 atmosphere) pressure when confined in 22.4 liters, and 28 grams of nitrogen exerts the same pressure (1 atmosphere) in the same volume at the same temperature. When the 32 grams of oxygen and the 28 grams of nitrogen are placed together in a 22.4 liter container, the pressure becomes 2 atmospheres (Figure 3–4). This is typical of many gaseous mixtures and is known as *Dalton's law of partial pressures*. How can this example be concisely stated and generalized into a statement of the law?

$$P_{\text{Oxygen}} + P_{\text{Nitrogen}} = P_{\text{Gaseous Mixture}}$$

Or, in general,

The pressure exerted by a gaseous mixture is the sum of the pressures exerted by the individual gases when each is confined separately in the same volume.

Studies Dalton performed in 1801 led him to believe that gases must be somehow composed of particles, capable of diffusing through a volume already occupied by another gas, with little regard for the gas already there. This is a reasonable idea if the particles of the gas are extremely small relative to the great distance between them.

HENRY'S LAW

In addition to studying the pressures of mixed gases, Dalton also investigated solutions of gases in liquids. The amounts of nitrogen that will dissolve in 100

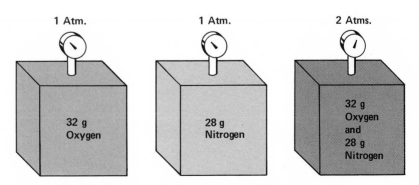

FIGURE 3–4   Illustration of Dalton's law of partial pressures. The pressure exerted by a gaseous mixture is the sum of the pressures of the individual gases.

liters of water at 20°C at various pressures are as follows:

> 1.9 g of nitrogen at 1 atmosphere of pressure
> 3.8 g   2 atmospheres
> 5.7 g   3 atmospheres
> 7.6 g   4 atmospheres

In 1803, William Henry, a close friend of Dalton, summarized observations like these on several gases and liquids in the following way: the amount of gas that will dissolve in a given quantity of liquid is dependent upon the pressure on the gas when exposed to the surface of the liquid. A more precise statement of this relationship, which has come to be known as *Henry's law* is:

> The weight of a gas that dissolves in a definite volume of a liquid is directly proportional to the pressure on the gas above the liquid at a constant temperature.

Carbonated beverages illustrate Henry's law. When the container is opened, gas is evolved because as the pressure on the gas is reduced, the liquid can dissolve less gas.

Dalton's interpretation of this behavior was again based upon particles, yet he had no clear-cut description of the particle. As we will see later in this chapter, it was chemical evidence that finally led Dalton to his concept of the atom. There is, however, another physical law of gases that should be examined because it plays a prominent part in the development of the structure of matter. This law expresses the effect of temperature on the volume of a gas.

CHARLES' LAW

The volume of a sample of gas at various temperatures and at constant pressure can be measured in an apparatus such as that shown in Figure 3–5. Typical data obtained with that apparatus are:

Temperature (°C)	Volume (ml)
27	600
54	654
127	800
227	1000
327	1200

Note that as the temperature is increased, the volume is increased (Figure 3–5). This generalization is true for all gases. However, a simple relationship between volume and Celsius temperature is not evident from these data. The data show that doubling the Celsius temperature from 27 to 54 degrees (100 per cent

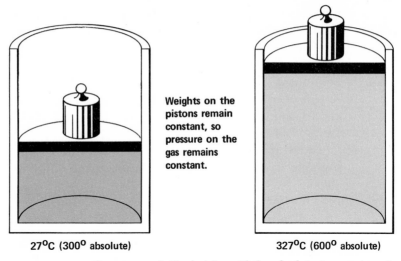

Weights on the pistons remain constant, so pressure on the gas remains constant.

27°C (300° absolute)   327°C (600° absolute)

FIGURE 3-5   Illustration of Charles' law. If the absolute temperature is doubled, the volume is doubled at constant pressure and amount of gas.

increase) results in only an 8.8 per cent $\left(\dfrac{654 - 600}{600} \times 100\%\right)$ increase in volume.

Often, in the study of natural science, a simple relationship exists but is not immediately obvious. Further observations concerning the temperature-volume relationship for a gas lead to an interestingly simple result. However, before this result can be appreciated fully, it will be necessary to introduce a new temperature scale.

It is observed that a sample of gas at 0°C decreases by $\frac{1}{273}$ of its original volume when cooled 1°C. If the gas sample is cooled 10 Celsius degrees, it decreases by $\frac{10}{273}$ of its starting volume. It appears then that a gas sample would lose all $\left(\frac{273}{273}\right)$ of its volume if it were cooled from 0°C to $-273°$C (Figure 3–6). In reality, this zero volume is never observed because all real gases first

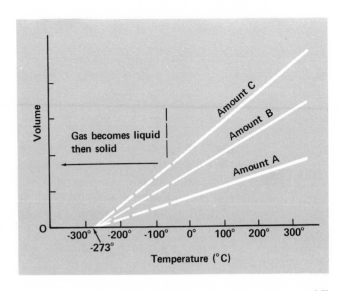

FIGURE 3-6   The variance of the volume of different amounts of a gas with temperature. The temperature corresponding to a hypothetical zero volume is $-273°$C, absolute zero.

liquefy and most solidify before reaching $-273°C$. However, on the basis of this behavior of gases, and for other reasons as well, $-273°C$ is termed *absolute zero*, and a new temperature scale measured in degrees absolute (°A) has been established. Temperatures measured on this scale are often called Kelvin temperatures (°K), named after the originator of the scale, Lord Kelvin. On the Kelvin or absolute temperature scale, the freezing point of water is $+273°A$ and its boiling point is $+373°A$. The relationship between °C and °A then is $°A = °C + 273$.

Now, if the same set of data as used before is expressed in degrees absolute, a simple temperature-volume relationship at constant pressure is immediately obvious.

Temperature		Volume
(°C)	(°A)	(ml)
27	300	600
54	327	654
127	400	800
227	500	1000
327	600	1200

It is evident now that the volume of the sample is directly proportional to the temperature when the temperature is expressed in absolute degrees, because when the temperature is increased from $300°A$ to $600°A$ (doubled), the volume is doubled from 600 ml to 1200 ml.

In 1787, Jacques Charles, a French scientist, studied the temperature-volume relationship for gases, and in 1801, another French scientist, Joseph Gay-Lussac, published his findings, which were similar to Charles' but more refined. The result, named after the first discoverer, is known as *Charles' law:*

At constant pressure, the volume of a gas is directly proportional to the absolute temperature of the gas. That is,

$$V = \text{constant } (°A)$$

For any two sets of measurements, V_1, T_1, and V_2, T_2, on the same sample of gas, held at constant pressure,

$$\frac{V_1}{T_1} = \frac{V_2}{T_2}$$

More frequently, we heat gas samples in essentially fixed volumes; under these conditions increased temperatures bring about higher pressures. A common example of this is the increased pressure in an automobile tire after a long, fast trip. The air pressure in the hot tire will be greater than the pressure at a cooler temperature.

While the physical laws of gases provided some basis for the logical assumption of the particle nature of matter, chemical laws of nature were the primary evidence for Dalton's concept of the atom. It is to these chemical laws, known during Dalton's time, that we now turn.

THE LAW OF CONSERVATION OF MATTER

While many had suspected the existence of atoms that merely became rearranged in a chemical change but were never destroyed (see Figure 1–8, page 16), it remained for Lavoisier, a French nobleman, to establish firmly, in 1785, that matter is neither created nor destroyed in chemical transformations.

Lavoisier found that he could cause mercury (Hg) to combine with a portion of the air, oxygen, to form mercuric oxide. This reaction is expressed by the equation:

$$\text{mercury} + \text{oxygen} \longrightarrow \text{mercuric oxide}$$

$$2Hg + O_2 \longrightarrow 2HgO$$

On heating mercuric oxide, an orange powder, Lavoisier found that he could recover exactly, within the limits of his ability to measure weights, the same amount of mercury used to prepare the mercuric oxide originally.

$$\text{mercuric oxide} \xrightarrow{\text{Heat}} \text{mercury} + \text{oxygen}$$

$$2HgO \xrightarrow{\text{Heat}} 2Hg + O_2$$

Furthermore, he found that the amount of gas consumed in the production of a sample of mercuric oxide was exactly equal to the amount produced in the destruction of the sample.

An examination of all chemical reactions that lend themselves to the kind of quantitative study carried out by Lavoisier leads to the generalization known as the *law of conservation of matter:*

In a chemical reaction, the sum of all of the weights of the products is exactly equal to the sum of all of the weights of the substances that enter into the reaction.

It has been observed that substances can be created or destroyed in a chemical process but matter cannot. As a further example, consider Figure 3–7.

THE LAW OF CONSTANT COMPOSITION

Sodium chloride (common table salt is a slightly impure form of this compound) has been found always to contain 39.4 per cent by weight sodium and 60.6 per cent by weight chlorine. Sucrose or table sugar, another common compound in man's daily environment, is 42.1 per cent by weight carbon, 6.4 per cent by weight hydrogen, and 51.5 per cent by weight oxygen. A compound of these three elements with any other composition by weight could not possibly be sucrose.

Information of this type concerning the compositions of various compounds

47

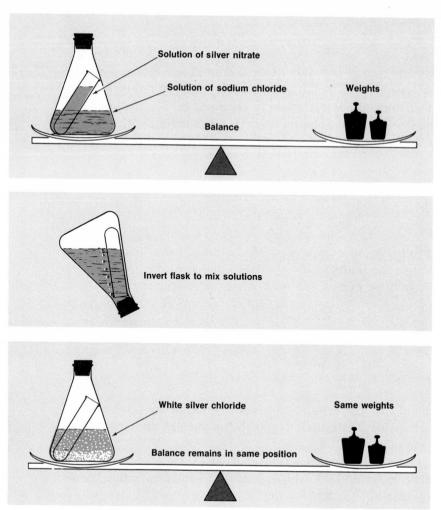

FIGURE 3-7   Mixing a solution of sodium chloride with a solution of silver nitrate produces a new substance, insoluble silver chloride, but the total weight of the matter remains the same.

led to the law of constant composition:

> When two or more elements combine to form a given compound, the ratio of the weights of the elements involved is always the same.°

If anyone wishes to prepare pure water, a compound that is 88.8 per cent by weight oxygen and 11.2 per cent by weight hydrogen, he will find that, if he is fortunate enough to react 11.2 grams of hydrogen and 88.8 grams of oxygen, he will obtain 100 grams of water with no appreciable amount of hydrogen or oxygen left over. If he uses 15.2 grams of hydrogen he will not obtain more water because the 88.8 grams of oxygen will make only 100 grams of water—he will

---

° Recently, chemists have encountered complex substances with varying compositions, substances that they choose to call compounds. Ordinary rust is an example. Such substances resemble compounds more than mixtures; as with many definitions of categories in nature, terms become difficult to apply in borderline cases.

simply have 4 grams of hydrogen left over. This further illustrates the fact that a pure compound always has the same chemical composition.

Since elements combine in fixed proportions to form compounds, it necessarily follows that compounds have characteristic compositions. One can readily see that the per cent composition of a compound can be useful in identifying that compound. It must be remembered, however, that per cent composition alone is not sufficient. Physical properties such as melting point, boiling point, solubility, and so forth, are also needed for positive identification.

Let us look at some simple examples of the usefulness of the law of constant composition.

Example Problem 1. Carbon dioxide, the gaseous product of respiration, is 27 per cent carbon and 73 per cent oxygen. This compound can be produced by burning pure carbon in air. What weight of oxygen will be needed to combine with 54 g of carbon to produce carbon dioxide?

Answer. 73 grams of oxygen react with 27 grams of carbon to produce 100 grams of carbon dioxide. Therefore, since 54 grams of carbon is 2×27 grams, 146 grams of oxygen (that is, 2×73 grams) will be needed to combine with the 54 grams of carbon.

Example Problem 2. Sodium chloride, 61 per cent chlorine, is a commercial source of chlorine. What weight of sodium chloride would be required to produce 183 pounds of chlorine?

Answer. 100 pounds of table salt produces 61 pounds of chlorine. Since 183 pounds of chlorine is desired (that is, 3×61 pounds of chlorine), it will take 3×100 pounds or 300 pounds of table salt.

Example Problem 3. Charcoal is a form of carbon made from wood. Why is it not possible to calculate the weight of charcoal obtainable from 100 g of wood?

Answer. Since wood and charcoal are mixtures, different samples of wood may contain different fractions of carbon.

SOURCES OF PERCENTAGES BY WEIGHT

The curious student at this point may ask, "How did early chemists establish that water is 88.8 per cent oxygen, carbon dioxide is 27 per cent carbon, and sodium chloride is 61 per cent chlorine?" The answers are found through quantitative analyses. Two possible methods of arriving at per cent composition—synthesis and decomposition—will now be discussed to answer this question.

Per Cent Composition by Synthesis

When exactly the correct weighed amounts of two elements, A and B, are combined to form a compound, the weights of the two parts (wt. A and wt. B) are known, and their sum (wt. A + wt. B) would be the weight of the compound produced. Since per cent is the $\dfrac{\text{part}}{\text{whole}} \times 100$, the per cent of each element in the compound (the per cent composition by weight) can be calculated in a straightforward manner.

$$\left(\frac{\text{wt. A}}{\text{wt. A} + \text{wt. B}}\right) \times 100 = \left(\frac{\text{wt. A}}{\text{wt. compound}}\right) \times 100 = \%A \text{ (by weight)}$$

$$\left(\frac{\text{wt. B}}{\text{wt. A} + \text{wt. B}}\right) \times 100 = \left(\frac{\text{wt. B}}{\text{wt. compound}}\right) \times 100 = \%B \text{ (by weight)}$$

With a slight modification, the analysis of carbon dioxide can be achieved in a similar way. A very pure sample of the element carbon, which is a black solid in its usual form, can be obtained easily and then weighed accurately on the analytical balance. Though conceivably possible, it would be extremely difficult to select just the right amount of oxygen for a fixed amount of carbon and then weigh this sample of gas. It would be much easier to burn the weighed sample of carbon in excess oxygen, weigh the gaseous product, carbon dioxide, and, from the difference, obtain the weight of oxygen.

Pure oxygen is passed into the series of glass tubes, as is shown in Figure 3–8. A weighed sample of carbon is heated (usually indicated in such diagrams by the Greek letter *delta*, Δ) until it is completely reacted. Because there is a tendency for some of the carbon to be converted only to carbon monoxide, CO, instead of carbon dioxide, CO_2, the solid copper oxide is heated to aid the complete conversion of carbon monoxide to carbon dioxide. Next in line is a tube filled with a drying agent, such as calcium chloride, which selectively absorbs water vapor but allows excess oxygen and the newly formed carbon dioxide to pass through. This step is necessary, since it is difficult to keep the chemicals and apparatus perfectly dry. The water must be removed first or it will be absorbed along with the carbon dioxide. Potassium hydroxide in the next

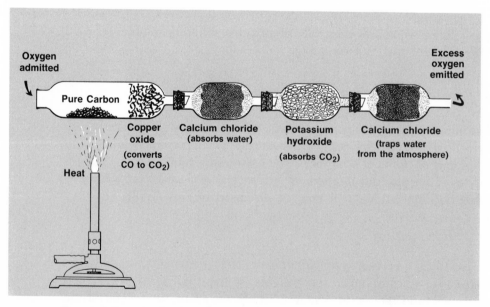

FIGURE 3–8   A schematic diagram of an apparatus for burning carbon in oxygen and trapping the carbon dioxide produced.

tube absorbs the carbon dioxide and allows the excess oxygen to pass. The potassium hydroxide is weighed before and after the experiment; the gain in weight is equal to the weight of carbon dioxide produced from the original amount of carbon. The last tube is filled with a drying agent to prevent water in the atmosphere from entering the tube containing the potassium hydroxide. Connections are placed between the tubes so they can be separated and closed easily. A set of data for this experiment might appear as follows:

Wt. of carbon: 0.105 g

Wt. of carbon dioxide produced: 0.385 g

Wt. of oxygen (by difference, 0.385 g − 0.105 g): 0.280 g

$$\% \text{ carbon} = \frac{\text{wt. carbon}}{\text{wt. compound, } CO_2} \times 100\% = \frac{0.105 \text{ g}}{0.385 \text{ g}} \times 100\% = 27.3\%$$

$$\% \text{ oxygen} = \frac{\text{wt. oxygen}}{\text{wt. compound, } CO_2} \times 100\% = \frac{0.280 \text{ g}}{0.385 \text{ g}} \times 100\% = 72.7\%$$

Per Cent Composition by Decomposition

The analysis of mercuric oxide is a classic experiment in chemistry and offers a very simple illustration of determining per cent composition by decomposing a chemical compound.

If a weighed sample of mercuric oxide, an orange powder, is strongly heated, it decomposes, liberating mercury and gaseous oxygen ($2HgO \xrightarrow{\Delta} 2Hg + O_2$). The resulting mercury can be collected and weighed, yielding the weight of oxygen as the difference between the weight of the original mercuric oxide and the remaining mercury. Typical data might be:

Wt. of mercuric oxide: 100.0 g

Wt. of mercury obtained: 92.6 g

Wt. of oxygen by difference: 7.4 g

$$\% \text{ oxygen} = \frac{7.4 \text{ g}}{100.0 \text{ g}} \times 100\% = 7.4\%$$

$$\% \text{ mercury} = \frac{92.6 \text{ g}}{100.0 \text{ g}} \times 100\% = 92.6\%$$

These calculations are presented to show that by simple experiments and the ability to recognize a pure substance, the early chemists could obtain the percentage composition for compounds even before they had definite ideas about atoms, molecules, or molecular formulas.

THE LAW OF MULTIPLE PROPORTIONS

It is often observed that two elements can form more than one compound. For example, carbon and oxygen can combine to form carbon monoxide, a

51

poisonous gas, and carbon dioxide, a gas produced in respiration. Hydrogen and oxygen form both water and hydrogen peroxide, and nitrogen and oxygen combine to form several different compounds. In an extreme case, hydrogen and carbon form compounds, called hydrocarbons, which number into the hundreds of thousands. In every case in which the compounds are composed of the same elements, each compound has a characteristic set of properties which distinguishes it from other compounds of these elements.

You might quickly guess that the weight of oxygen needed to convert a fixed weight of carbon, say 10 g, into carbon monoxide would be different from the amount needed to convert the 10 g of carbon to carbon dioxide. You would reason that since carbon monoxide and carbon dioxide are different compounds, each with its own fixed composition, the weight of oxygen for a fixed weight of carbon in the two compounds would necessarily have to be different.

A set of experimental data for carbon monoxide and carbon dioxide is the following:

0.75 g carbon and 1.00 g oxygen form 1.75 g carbon monoxide;
0.75 g carbon and 2.00 g oxygen form 2.75 g carbon dioxide.

Note that a fixed weight of carbon (0.75 g) is used in the formation of each compound, but that different amounts of oxygen are required. In fact, the ratio of the weights of oxygen in the two compounds combined with the fixed weight of 0.75 g of carbon is a ratio of 2:1, a ratio of small whole numbers.

In 1804 Dalton recognized this as a general phenomenon, and a statement summarizing these facts is known as the *law of multiple proportions:*

The weights of one element that combine with a fixed weight of a second element when forming two or more compounds are in a ratio of small whole numbers (integers) such as 2:1, 3:1, 3:2, or 4:3.

Figure 3–9 is an illustration of this law as it applies to the two compounds of carbon with oxygen. The law is also illustrated by the compounds sulfur dioxide and sulfur trioxide. Twenty-five grams of sulfur trioxide can be prepared by combining 10 g of sulfur with 15 g of oxygen. Now, let us ask ourselves, in preparing sulfur dioxide, which of the following amounts of oxygen would most likely combine with 10 g of sulfur? (That is, which weight of oxygen would give a small whole number ratio with the 15 g of oxygen used for the sulfur trioxide?) (a) 8 g, (b) 10 g, (c) 13 g. The answer is (b), since the amounts of oxygen combining with the fixed weight (10 g) of sulfur in the two compounds are in the ratio of 15 to 10, or 3:2, a ratio of small whole numbers. No such simple ratio is possible using answers (a) or (c).

In fact, ratios such as 3.4:1 or 5.7:3 are never observed for ordinary chemical compounds. If Dalton was to advance a satisfactory theory of matter in terms of atoms, he had to be able to explain this ratio.

LAW OF COMBINING VOLUMES

Gay-Lussac carried his study of the volume of gas samples a step further than did Charles, in that he studied the volumes of gases involved in chemical

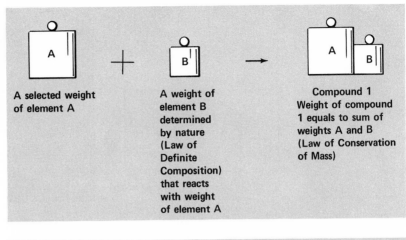

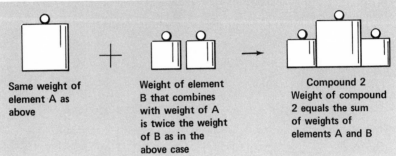

FIGURE 3-9   An illustration of the law of multiple proportions. The laws of definite composition and conservation of matter are also illustrated.

reactions. When gases were involved either as reactants or products in a chemical reaction, Gay-Lussac wanted to know the relative amounts of the gaseous substances. At this point you should be careful to remember that because of the nature of the gaseous state, one must know not only the volume of the sample but also the temperature and pressure, in order to know the amount of matter involved. Gay-Lussac measured the amounts of his gaseous samples in terms of volumes, being careful to measure these volumes at constant temperature and pressure, or to make suitable calculations according to Boyle's and Charles' laws for variation in these factors.

Some examples will illustrate the type of data obtained by Gay-Lussac.

Two liters of hydrogen react with one liter of oxygen to form two liters of steam (Figure 3–10).

One liter of hydrogen will combine with one liter of chlorine gas to form two liters of hydrogen chloride gas (Figure 3–11).

One liter of nitrogen will combine with three liters of hydrogen to form two liters of ammonia gas (Figure 3–12).

Observations of this type led Gay-Lussac to state the rather simple relationship known as the *law of combining volumes of gases:*

When measured at constant temperature and pressure, the volumes of gases involved in a reaction are always in a ratio of small whole numbers.

53

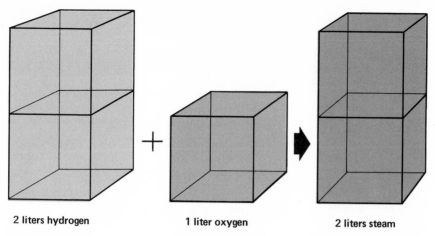

2 liters hydrogen   **1 liter oxygen**   **2 liters steam**

FIGURE 3-10   Volume ratios in the reaction of hydrogen and oxygen to form water (steam).

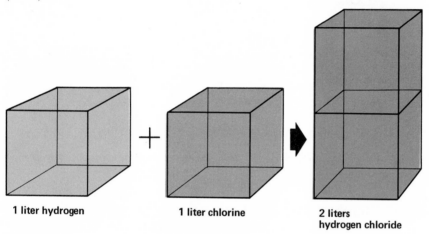

1 liter hydrogen   **1 liter chlorine**   **2 liters hydrogen chloride**

FIGURE 3-11   Volume ratios in the reaction of hydrogen and chlorine to form hydrogen chloride.

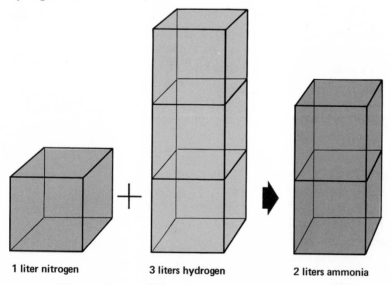

1 liter nitrogen   **3 liters hydrogen**   **2 liters ammonia**

FIGURE 3-12   Volume ratios in the reaction of nitrogen and hydrogen to form ammonia.

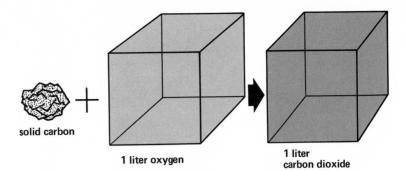

FIGURE 3-13 Volume ratio of oxygen to carbon dioxide in the reaction of carbon and oxygen to form carbon dioxide.

solid carbon

1 liter oxygen

1 liter carbon dioxide

In the previous examples, these ratios are $2:1:2$ for the formation of steam, $1:1:2$ for the formation of hydrogen chloride, and $1:3:2$ for the formation of ammonia.

All of the reactants or products do not have to be gases. A certain weight of carbon, a solid, will combine with one liter of oxygen gas to form one liter of carbon dioxide gas (Figure 3–13).
For the two substances that are gases, the volume ratio is $1:1$, in agreement with the law of combining volumes.

DALTON'S ATOMIC THEORY

Dalton was aware of the physical and chemical laws described previously, some of which he had observed in his own experimental work. Desiring to offer an explanation for these chemical laws, he postulated the following points concerning a submicroscopic world of atoms and molecules (Figure 3–14). His conclusions about elements and atoms were:

1. A sample of an element is composed of an array of identical particles called atoms.

2. The atom of each element is different in weight from the atoms of all elements.

3. An elemental sample can be subdivided only to the point of yielding individual atoms. Further division of the element is impossible.

4. It is impossible to create or destroy an atom of an element.

About compounds and molecules:

1. The ultimate particle of a compound is a molecule° which is made up of one or more atoms of at least two elements.

2. When atoms combine to form molecules of a given compound, they form identical molecules, each having the same ratio of the combining atoms.

3. Two or more kinds of atoms may combine in different ways to form more than one kind of molecule.

4. Since atoms are permanent unchanging bodies, atoms combine in simple numerical ratios, such as one to one, one to two, two to three, in the formation of molecules.

° For want of a better term Dalton called these molecules "atoms" but indicated these "atoms" to be compounded of elemental atoms. You should understand that in this presentation the thoughts of early workers are presented in modern terms.

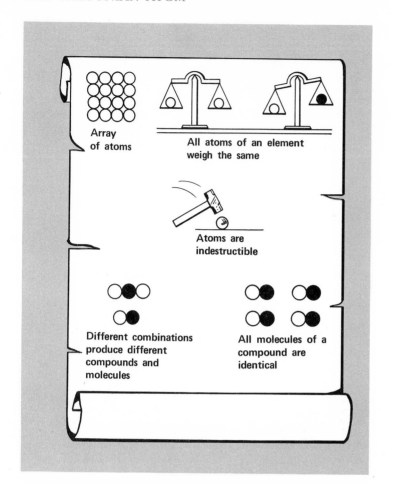

Array
of atoms

All atoms of an element
weigh the same

Atoms are
indestructible

Different combinations
produce different
compounds and
molecules

All molecules of a
compound are
identical

FIGURE 3-14   Features of
Dalton's atomic theory.

5. The most stable and abundant compound of two elements consists of molecules made up of one atom of each element.

The significantly new concept of the atom as proposed by Dalton is the definite distinguishing weight attached to atoms of a given element. However, in the light of present chemical knowledge and theory, most of the original statements of Dalton require some modification. The most obvious example is the statement made by Dalton that atoms cannot be created or destroyed. It is now well known that atoms of various elements can be fused together to create new elements (fusion), and that certain atoms can be split apart (fission) to produce other elements. However, all of Dalton's statements except the last one concerning compounds and molecules approached more closely the realities of nature than did previous concepts. This is precisely the way an understanding of nature grows.

New experiments are suggested, new facts are gathered and recorded, and theoretical concepts are modified in keeping with the new findings.

Indeed, it would be truly surprising if the English schoolteacher had been able to define the theoretical atom in such a way that his statements would be in complete agreement with all later discoveries. The genius of Dalton is displayed in that all points except the last one contain sufficient truth that each of them can be accepted today if slightly modified.

The last point, that molecules that are most stable are made up of atoms in a 1:1 ratio, serves as an excellent example of the frailties of theoretical concepts. Not only are theories limited by insufficient data, but errors in human judgment also are often a significant limitation. Dalton firmly believed that water, for example, was composed of molecules with the formula HO. His reasoning dismissed the possibility that H_2O might be the formula.

Dalton did not propose his atomic theory to explain the physical laws of matter, such as Boyle's law, Charles' law, Henry's law, and so forth. However, these physical laws were indispensible to Dalton in experiments and reasoning that led to his atomic theory. In Chapter 5 we shall see that by proper extensions of Dalton's atomic theory, these physical laws concerning gases can also be explained. Let us now turn our attention to Dalton's atomic theory as it explained the chemical laws we have previously discussed.

DALTON'S THEORY EXPLAINS THREE OF FOUR CHEMICAL LAWS

The conservation of matter in chemical transformations and the composition of a compound were well established by Lavoisier and others prior to the major contributions of Dalton. Soon thereafter, the law of multiple proportions was clearly stated, based on laboratory experiments. All three of these laws were beautifully explained if one accepted the postulates of Dalton's theory. However, the fourth law discussed previously, Gay-Lussac's law of combining volumes of gases, was not at first considered by Dalton, nor was it explained. Indeed, Gay-Lussac's law was in conflict with Dalton's idea that the most stable molecules of compounds containing only two elements were made up of just two atoms, one from each element.

LAW OF CONSERVATION OF MATTER

If atoms are indestructible, as Dalton suggested, it follows that the appearance of new substances in chemical changes is simply the result of a new arrangement of the same atoms already present. The total weight of the products would have to be exactly the same as the weight of the reactants, since the very same atoms are involved. If a child uses all of his blocks in building first a fort and then a bridge, the two displays would necessarily weigh the same (Figure 3–15).

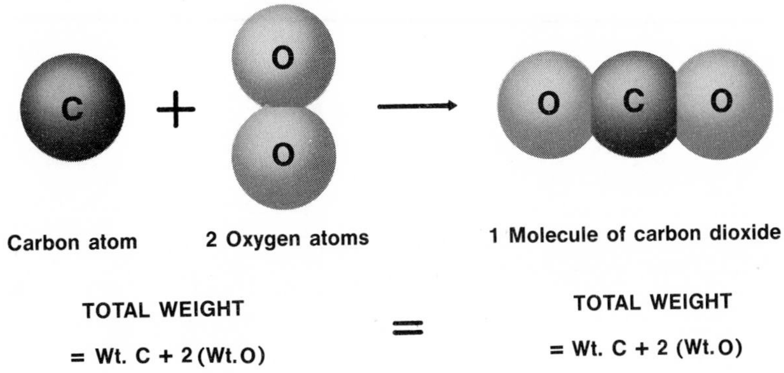

FIGURE 3–15 According to Dalton's theory, atoms are rearranged in a chemical reaction. Matter is conserved.

Carbon atom   **2 Oxygen atoms**   **1 Molecule of carbon dioxide**

TOTAL WEIGHT

= Wt. C + 2 (Wt. O)

=

TOTAL WEIGHT

= Wt. C + 2 (Wt. O)

LAW OF CONSTANT COMPOSITION

Carbon monoxide is a chemical compound composed of carbon and oxygen. Dalton explained the fact that carbon monoxide always gives the same analysis, 42.5 per cent carbon and 57.5 per cent oxygen by weight, by stating that a sample of carbon monoxide is simply an array of carbon monoxide molecules each of which is 42.5 per cent carbon and 57.5 per cent oxygen. Since there is only one way that carbon and oxygen atoms can combine to form a carbon monoxide molecule, all carbon monoxide molecules would have to be the same. Therefore, any group of carbon monoxide molecules would have to yield the same analysis as would the single molecule. If a party is made up only of couples, each of which is composed of a boy and a girl, the ratio of boys to girls at the party would have to be $1:1$ regardless of the size of the party.

LAW OF MULTIPLE PROPORTIONS

In the earlier discussion of this law we stated that 0.75 g of carbon combines with 1.00 g of oxygen to form 1.75 g of carbon monoxide and that 0.75 g of carbon combines with 2.00 g of oxygen to form 2.75 g of carbon dioxide. A Daltonian explanation would say this is a simple matter if one considers that a carbon atom, weighing 0.75 unit of weight, combines with an oxygen atom, 1.00 unit of weight, to form a carbon monoxide molecule containing one carbon atom and one oxygen atom, together having a combined weight of 1.75 units. According to Dalton, all oxygen atoms have the same weight, all carbon atoms have the same weight, and whole atoms are combined to form molecules. Therefore, consistent with the units of weight assumed for the carbon monoxide molecule, the 2.75 units for a carbon dioxide molecule would necessarily require two atoms of oxygen, each weighing the same 1.00 unit, and one atom of carbon, weighing 0.75 unit of weight. In order to satisfy the weight requirements of 2.75 units for carbon dioxide and 1.75 units for carbon monoxide, it is obvious that the weight of oxygen in a carbon dioxide molecule, CO_2, is twice $(2:1)$ the weight of oxygen in a carbon monoxide molecule, CO, when there is the fixed weight of carbon in the two compounds $(1:1)$. Since this small whole number ratio holds at the molecular level, one would expect the same ratio to exist between macroscopic samples of these compounds, which are simply collections of large numbers of the two kinds of molecules (Figure 3–16).

AVOGADRO'S HYPOTHESIS RESOLVES A DIFFICULTY

An empirical law as simple as Gay-Lussac's law of combining volumes called for a simple theoretical explanation, but Dalton's atomic theory was inadequate primarily because it formulated water as HO and ammonia as NH. This discrepancy was so disturbing to Dalton that he even suggested that Gay-Lussac's data were incorrect. "The truth is," Dalton maintained, "that gases do not unite in equal or exact measures in any one instance; when they appear to do so, it is

Carbon dioxide   **Carbon monoxide**

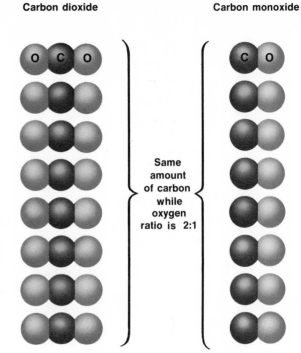

Same
amount
of carbon
while
oxygen
ratio is 2:1

FIGURE 3-16   Illustration of the law of multiple proportions. Eight molecules of carbon dioxide contain the same number of carbon atoms as do eight molecules of carbon monoxide whereas the number of oxygen atoms is twice as great (2:1) in the carbon dioxide. Thus, we have an atomic explanation for the 2:1 weight relationships that holds on the macroscopic level.

owing to the inaccuracy of our experiments." While other contemporary experimenters proved that Gay-Lussac's law of combining volumes was correct, and Jöns J. Berzelius, a most eminent and influential chemist of the time, wrote Dalton that he needed to alter his thinking on Gay-Lussac's data, Dalton was not convinced and the result was utter confusion.

In 1811, Amedeo Avogadro, an Italian physicist, proposed hypotheses that were capable of resolving the dilemma by adequately explaining Gay-Lussac's law of combining volumes while retaining Dalton's concept of unbreakable atoms.

Equal volumes of all gases under the same conditions of temperature and pressure contain the same number of molecules.
Molecules of some elements are diatomic, that is, the molecule is composed of two atoms.

The first hypothesis means that regardless of the size of the molecules of gases, equal volumes contain equal numbers of molecules. Large molecules have less free space between them than smaller molecules have (Figure 3–17). The second hypothesis means that certain gases, such as hydrogen, chlorine, nitrogen, oxygen, and fluorine, are packaged two atoms per molecule.

What evidence did Avogadro have for these bold assumptions? In essence, he had none. He could not count molecules; neither could he weigh them. It remained for Jean Baptiste Perrin, Robert Andrews Millikan, and other, later experimenters to devise methods to count the number of molecules in a sample of gas, and for Irving Langmuir, over a century later, to show that the hydrogen molecule is composed of two atoms. Although Avogadro did not have experi-

59

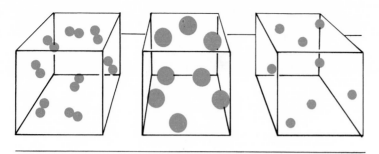

FIGURE 3-17   Avogadro's hypothesis. Some elements have diatomic molecules. Equal volumes of gases have equal numbers of molecules.

Same volume temperature, and pressure

mental evidence to support his hypotheses, he had a very consistent and workable explanation for Gay-Lussac's law of combining volumes, and for the fact that water must be H_2O, not HO. The following examples will illustrate.

Example 1.   What are the formulas for hydrogen and chlorine molecules based on the experimental results given in Figure 3–18? These results indicate, since the volume ratios are believed to be the same as the molecular ratios, that one molecule of hydrogen reacts with one molecule of chlorine to form two molecules of hydrogen chloride. Now each of the two molecules of hydrogen chloride must contain at least one atom of hydrogen. This dictates that the molecule of hydrogen must contain at least two atoms, in which case the formula is H_2. Even though later evidence discounts the possibility, one has to admit that the above argument allows for the possibility that hydrogen molecules could be even larger than H_2, perhaps H_4; or H_6, and so forth. Believing in the inherent simplicity of nature, Avogadro chose the simplest and, in this case, correct possibility. The same argument can be made to show that chlorine molecules are diatomic. A molecular representation of the reaction might appear as given in Figure 3–19.

Example 2.   Assuming the formula H_2 for hydrogen molecules, what is the formula for water if 2 volumes of hydrogen react with one volume of oxygen to produce 2 volumes of steam? (See Figure 3–20.)

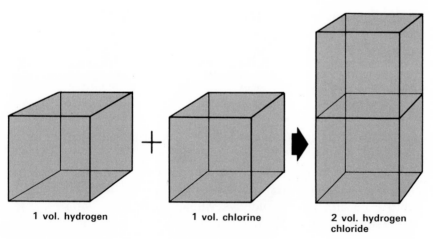

1 vol. hydrogen   1 vol. chlorine   2 vol. hydrogen
                                                           chloride

FIGURE 3-18   Volume ratios in reaction of hydrogen and chlorine to form hydrogen chloride.

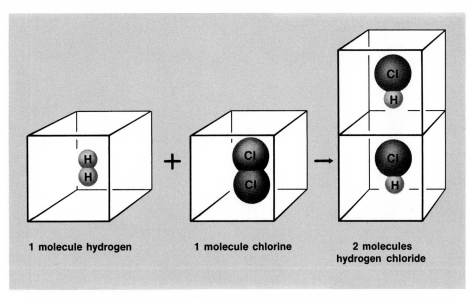

FIGURE 3-19   Molecular ratios are the same as volume ratios in the reaction of hydrogen and chlorine to form hydrogen chloride.

Two hydrogen molecules (4 atoms) plus one oxygen molecule (2 atoms) yield two water molecules, each of which would have to contain at least one oxygen atom. The simplest way to explain the volume ratios is to assume that the formula of water is H_2O.

Unfortunately, Avogadro's explanation was not widely accepted until nearly 50 years later. Stanislao Cannizzaro, at the Karlsruhe Conference in 1860, which was called in an effort to clarify conflicting information about atomic weights,

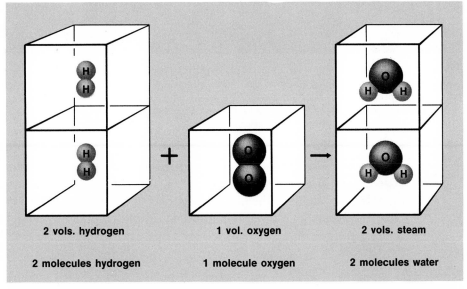

FIGURE 3-20   In the reaction of hydrogen and oxygen to form steam, the ratio of two volumes of hydrogen to one volume of oxygen to two volumes of steam suggests that two molecules of hydrogen react with one molecule of oxygen to form two molecules of water (steam).

FIGURE 3-21   Chains of thorium atoms separated by an organic molecule. These chains were placed on a thin carbon film a tenth of a millionth of an inch thick. In each chain the smallest white dots represent single thorium atoms. The larger white dots are probably aggregates of a few thorium atoms very close together. (Courtesy of Professor Albert V. Crewe, Department of Physics, Enrico Fermi Institute, University of Chicago.)

championed Avogadro's hypothesis so successfully that the matter was finally settled. As will be shown in Chapter 4, the determination of atomic weights (the relative weights of the individual atoms of different elements) could proceed when the formulas for water and other simple compounds were firmly established.

MODERN EVIDENCE

Modern evidence in support of the atomic theory is massive and varied. Much of it is too complex for presentation at this point. Where it seems appropriate, brief presentations, such as the one concerning mass spectroscopy in Chapter 6, will add additional evidence for our belief in atoms. However, a thorough and complete elucidation of all supporting evidence is beyond the purposes and scope of this text.

For emphasis and to show that the labor of gathering evidence goes on, the work of Albert V. Crewe is cited. Dr. Crewe, of the University of Chicago, announced in 1970 that finally he had achieved the long sought for "photograph" of the atom (Figure 3–21). Not to be confused with ordinary photographs produced by light, such remarkable photographs were achieved with a scanning electron microscope. Atoms of the element thorium, which are relatively large and heavy, were adsorbed on a very thin carbon film. When a beam of electrons (see Chapter 6) was passed through this film, a pattern was projected onto a screen, which is believed to show the actual position of the thorium atoms. Figure 3–21 is an ordinary photograph of the image on the screen.

CONCLUSION AND PREVIEW

In this chapter, we have seen how some of the physical and chemical laws of matter served in the development of Dalton's atomic theory. The importance of the theory lies in its workable explanations for three of the four major chemical laws and its concept of a definite, distinguishing weight for the atoms of a given element. The implications of the latter contribution will be treated in Chapter 4. With suitable modifications Dalton's atomic theory can explain the physical laws of gases as well. It will be a purpose of Chapter 5 to look at explanations for those physical laws.

Dalton's atomic theory stands as a monumental starting point for the modern concept of the atom. However, as we study modern atomic theory, we will see that the atom is much more complex than envisioned by Dalton.

QUESTIONS

1. Contrast the Greeks' method and Dalton's method of justifying belief in atoms.

2. Which gas laws most closely relate to the following effects?

   (a) A soft drink fizzes when the top is removed.
   (b) A pressure cooker blows up if it is heated too hot and the pressure valve stops up.
   (c) If you hold your finger over the end of a bicycle pump, a rapid push on the plunger raises your finger off the outlet.
   (d) As a balloon rises into the air, the volume becomes greater.
   (e) If a jar is inverted over a burning candle arranged on a wooden block floating on water, the water rises into the bottle.

3. How does Dalton's atomic theory explain:

   (a) the law of conservation of matter?
   (b) the law of constant composition?
   (c) the law of multiple proportions?

4. The pressure on 6 liters of gas is 2 atmospheres. What will be the volume (in liters) when the pressure is doubled without changing the temperature?

5. How much does the volume of a gas change if the temperature is changed from 200°A to 600°A at a constant pressure of 1 atmosphere.

6. Change to degrees absolute (°A):

    (a) 50°C
    (b) −30°C
    (c) 77°F

7. For the reaction of hydrogen with fluorine, one volume of hydrogen gas combines with one volume of fluorine gas, yielding two volumes of hydrogen fluoride gas.

    (a) Use Avogadro's hypotheses to explain these observations.
    (b) Why would Dalton's ideas be incapable of explaining these observations?

8. The following data were obtained when sodium reacted with oxygen under different conditions. How can the two sets of data be explained?

Wt. of sodium used	Wt. of compound formed
10.0 g	17.0 g
5.00 g	6.75 g

9. List one specific contribution of each man to atomic theory: Democritus, Boyle, Dalton, Avogadro.

10. What are two ways in which the percentage composition of compounds may be determined?

11. In the chemical reaction, $2Ag_2O \longrightarrow 4Ag + O_2$, 10.7 g of silver oxide (Ag_2O) will decompose into 10.0 g of silver metal (Ag). According to the law of conservation of matter, how much oxygen must be formed?

12. In the preparation of sulfur trioxide by the oxidation of sulfur, the following data were obtained in a series of experiments:

	Wt. of sulfur used	Wt. of oxygen used	Wt. of sulfur trioxide obtained
Experiment 1	1.0 g	1.5 g	2.5 g
Experiment 2	2.0	3.0	5.0
Experiment 3	3.0	4.5	7.5

    Determine if these data are in accord with the law of definite composition.

13. An experiment in which volumes of gases (all under standard conditions) were allowed to react with each other showed that 3 volumes of gaseous H_2 reacted with 1 volume of a gas X to produce 3 volumes of gaseous H_2X. How many atoms are present in a molecule of gaseous X?

14. Twelve grams of carbon are burned in a closed container which originally had 80 grams of gaseous oxygen present. Forty-four grams of carbon dioxide are formed. How much oxygen remains unreacted in the container?

15. Assuming that nitrogen and hydrogen gases are diatomic, determine the simplest formula for ammonia, if 1 volume of nitrogen reacts with 3 volumes of hydrogen to form 2 volumes of ammonia.

16. The laws of gases and the laws of chemical change presented in this chapter are often referred to as empirical laws. What does "empirical" mean? How does empirical differ from theoretical?

17. Although there may not be a very reliable way to check the conservation of matter in a large explosion of dynamite, what leads us to believe that the law of conservation of matter is obeyed?

18. The pressure on 2.00 liters of gas is 0.91 atmosphere. If the pressure is raised to 1.45 atmospheres on this sample without changing the temperature, what will the new volume have to be?

19. What will be the new volume of a 5.00 liter sample of a gas if it is cooled from $80.0°C$ to $0.0°C$ at constant pressure?

20. What is the basic assumption made when it is argued that one molecule of nitrogen reacts with three molecules of hydrogen to form two molecules of ammonia because, in the laboratory, one volume of nitrogen reacts with three volumes of hydrogen to form two volumes of ammonia? The volumes are measured at constant temperature and pressure.

21. A hydrogen atom has a weight of 1.0 unit and an oxygen atom has a weight of 16.0 units. Water is composed of molecules having two atoms of hydrogen and one atom of oxygen per molecule. Hydrogen peroxide, another compound of hydrogen and oxygen, is composed of molecules having two atoms of hydrogen and two atoms of oxygen in each molecule. If these statements can be assumed to be true, how do they illustrate and explain the Law of Multiple Proportions?

22. The density (weight per unit volume) of carbon dioxide, CO_2, is 1.96 g per liter and the density of ammonia, NH_3, is 0.76 g per liter at $0°C$ and 1 atmosphere pressure. Hence, the density of carbon dioxide is 1.96/0.76 or about 2.6 times that of ammonia. How many times heavier is a carbon dioxide molecule than an ammonia molecule? Explain.

SUGGESTIONS FOR FURTHER READING

Ihde, A. J., "The Development of Modern Chemistry," Harper and Row, New York, 1964.
Jaffe, B., "Crucibles: The Story of Chemistry," Fawcett World Library, New York, 1957.
Kesselman, B., "The Skeptical Chemist (Robert Boyle)," *Chemistry*, Vol. 39, No. 1, p. 9 (1966).
Lucretius, "The Nature of the Universe," Penguin Books, Inc., New York, 1959.
Spritzer, M. S., "Testing Boyle's Law," *Chemistry*, Vol. 43, No. 9, p. 29 (1970).
Spritzer, M. S., and Markham, J., "Charles' Law: Estimating Absolute Zero," *Chemistry*, Vol. 42, No. 8, p. 24 (1969).
Szabadvary, F., "Great Moments in Chemistry. Part I: A Visit with Antoine Lavoisier," *Chemistry*, Vol. 42, No. 4, p. 14 (1969).
Szabadvary, F., "Great Moments in Chemistry. Part II: From Thales to Bohr," *Chemistry*, Vol. 42, No. 11, p. 6 (1969).

ATOMIC WEIGHTS
AND STOICHIOMETRY

CHAPTER 4 ━━━━━━━━━━━━━━━━━━━

A major contribution of Dalton's atomic theory was the assumption that atoms of a given element have a unique weight, which is different from the weights of atoms of other elements. This assumption set into motion an enthusiastic search for the weights of the atoms. However, the ensuing 40 years following the proposal of Dalton's theory brought more confusion over atomic weights than agreement about them. The primary hindrances to accepting a consistent scale of atomic weights were the lack of publicity and acceptance of Avogadro's hypothesis (it was practically unnoticed even by the time of his death in 1856), and the confusion between the terms "atom" and "molecule." Dalton used these terms interchangeably and believed that one atom (molecule) of oxygen and one atom (molecule) of hydrogen would combine to form one molecule of water.

$$H + O \longrightarrow HO$$

In contrast, Avogadro, as a result of his understanding of Gay-Lussac's law of combining volumes, believed that oxygen and hydrogen molecules are composed of two atoms each, O_2 and H_2. Furthermore, he believed that one molecule of oxygen reacts with two molecules of hydrogen to form two molecules of water, the molecular ratio being exactly the same as the ratio of combining volumes of gases.

$$2H_2 + O_2 \longrightarrow 2H_2O$$

As will be shown in the following section, Dalton's hypothesis led to the conclusion that an oxygen atom is eight times as heavy as a hydrogen atom, whereas Avogadro's hypothesis required an oxygen atom to be 16 times as heavy. Obviously, such a major point of confusion had to be resolved before an acceptable scale of atomic weights could be generally accepted.

RELATIVE WEIGHTS OF HYDROGEN
AND OXYGEN ATOMS—WHY THE DILEMMA?

By 1860 experimental methods had been developed by which the weights of each element in a sample of a compound could be calculated. Two examples were given in Chapter 3. However, to state that water is 88.8 per cent oxygen and 11.2 per cent hydrogen tells nothing concerning the relative weights of the oxygen and hydrogen atoms. However, if one added to this the belief, as did Dalton, that the molecule of water has the formula HO, it would follow that 88.8 per cent of the weight of each molecule is due to the oxygen atom and 11.2 per cent due to the hydrogen atom. The ratio of weight of the oxygen atom to the hydrogen atom would then be 88.8:11.2. In other words, the oxygen atom would be $\frac{88.8}{11.2}$ or 7.9 times as heavy as the hydrogen atom. Now as Avogadro believed the formula for water is not HO, but H_2O, the thought process involved is a bit more complex. If two hydrogen atoms are 11.2 per cent of the weight of the molecule, one hydrogen atom would weigh one-half of 11.2 or 5.6 per cent of the total weight of a molecule. Since the one oxygen atom in the molecule is 88.8 per cent of the weight, the ratio of the weight of the oxygen atom to the weight of the hydrogen atom would be $\frac{88.8}{5.6}$ or 15.9, the approximate value which is accepted today.

In 1860 it was apparent that if an atomic weight scale was to be firmly established, it would be necessary to know the molecular formulas for compounds as well as to know the per cent composition of the compounds. The latter was readily available but the former remained in doubt in spite of Avogadro's efforts.

DILEMMA RESOLVED— THE KARLSRUHE CONFERENCE

In September, 1860, many of the most brilliant minds in chemistry met in Karlsruhe, Germany, to settle the atomic weight crisis. After several of the 140 chemists present had spoken on and debated the distinction between molecule and atom, Stanislao Cannizzaro raised his voice to combat their ideas and to clearly revive the ideas of the dead and forgotten Avogadro. The young orator described how Avogadro's molecules consistently explained Gay-Lussac's law of combining volumes and how the acceptance of molecules of hydrogen composed of two atoms would clarify the atomic weight dilemma. Although it took several years for the effect to be seen, the chemical community finally recognized the great contribution made by Avogadro, the atomic weight scales were made consistent, and the surge forward in atomic theory began again.

It now could be said with general agreement that the oxygen atom weighs approximately 16 times as much as the hydrogen atom. Evidence developed later in this text leaves little room to doubt this assertion. Man's ability to make this statement is truly one of his most notable accomplishments, for he is describing the relative weights of particles so small that it is impossible for him to visualize their size and weight in terms of the world that he experiences with his physical senses.

THE ATOMIC WEIGHT SCALE—AT LAST

Using an approach similar to that described previously for hydrogen and oxygen, we can learn that a carbon atom weighs 12 times as much as a hydrogen atom, nitrogen 14 times as much, chlorine 35.5 times as much, and sulfur 32 times as much. Even though these numbers tell us the relative weights of the atoms, it is important to realize that they are not the actual weights of atoms in any previously known units. If a specific number is assigned as the weight of any particular atom, this immediately fixes relative numbers to the weights of all other atoms. Since no atom has been found to be lighter than the hydrogen atom, a satisfactory system of numbers could be obtained based on the assignment of the number 1 as the weight of the hydrogen atom. This would result in all other atomic weights being greater than 1. On this scale the atomic weight of oxygen would be slightly less than 16 since the ratio of the weight of oxygen to hydrogen is not 16 to 1 but 15.873 to 1. A more popular scale in chemistry for many years set oxygen at exactly 16.000, which resulted in hydrogen having the value of 1.008.

In 1961, a new atomic weight scale was adopted for use in both chemistry and physics based upon the carbon-12 isotope (isotopes are atoms of the same element having different weights; see Chapter 6) having a value of exactly 12.000 awu. Arbitrarily, a label has been attached to this weight, that of *atomic weight unit* (awu). The carbon-12 isotope is said to have a weight of 12.000 awu. This unit of atomic weight is just as much a unit of weight as is the gram. Its extremely small relative size is convenient for the description of the weight of atoms, just as, in another instance, inches are more convenient than miles for measuring household articles. For example, the actual weight of a hydrogen atom is 1.67×10^{-24} g or 1.008 awu. It is obvious that the awu is the more convenient unit to manipulate mathematically.

With extensions of the reasoning that led to the working out of the relative weights of oxygen and hydrogen, relative atomic weights of other elements have been determined. In recent years new experimental techniques have been developed that measure the actual atomic weights of the elements very accurately (see discussion of mass spectroscopy, Chapter 6). A complete listing of the atomic weights based on carbon-12 is shown in Table 4–1.

The consequences of atomic weights are far reaching in the field of chemistry. Indeed, modern chemistry would be impossible without this cornerstone, and the characterization of every new substance depends on the atomic weights of the atoms involved.

HOW ARE FORMULAS DERIVED FROM ATOMIC WEIGHTS?

Once the atomic weight scale was firmly established, it was then possible to determine the simplest formula for *any* compound, provided the elemental per cent analysis had been determined. Obviously, this is of major importance in the study of new compounds. To illustrate, let us determine the formula of silicon dioxide, the principal compound in ordinary sand. Quantitative analyses of silicon dioxide show it to have an average of 46.7 per cent silicon and 53.3 per cent oxygen by weight. The per cent analysis tells us the ratio of the weight of the two elements in the compound; for each 100 grams of silicon dioxide, we expect to find 46.7 grams of silicon and 53.3 grams of oxygen. Other units

TABLE 4-1   TABLE OF ATOMIC WEIGHTS (Based on Carbon-12 = 12.0000 awu)°

° Values in parentheses are estimates and denote, in most cases, isotopes of longest half-life.

	SYMBOL	ATOMIC No.	ATOMIC WEIGHT		SYMBOL	ATOMIC No.	ATOMIC WEIGHT
Actinium	Ac	89	(227)	Mendelevium	Md	101	(256)
Aluminum	Al	13	26.9815	Mercury	Hg	80	200.59
Americium	Am	95	(243)	Molybdenum	Mo	42	95.94
Antimony	Sb	51	121.75	Neodymium	Nd	60	144.24
Argon	Ar	18	39.948	Neon	Ne	10	20.179
Arsenic	As	33	74.9216	Neptunium	Np	93	(237.0482)
Astatine	At	85	(210)	Nickel	Ni	28	58.71
Barium	Ba	56	137.34	Niobium	Nb	41	92.9064
Berkelium	Bk	97	(249)	Nitrogen	N	7	14.0067
Beryllium	Be	4	9.01218	Nobelium	No	102	(254)
Bismuth	Bi	83	208.9806	Osmium	Os	76	190.2
Boron	B	5	10.81	Oxygen	O	8	15.9994
Bromine	Br	35	79.904	Palladium	Pd	46	106.4
Cadmium	Cd	48	112.40	Phosphorus	P	15	30.9738
Calcium	Ca	20	40.08	Platinum	Pt	78	195.09
Californium	Cf	98	(249)	Plutonium	Pu	94	(242)
Carbon	C	6	12.01115	Polonium	Po	84	(210)
Cerium	Ce	58	140.12	Potassium	K	19	39.102
Cesium	Cs	55	132.9055	Praseodymium	Pr	59	140.9077
Chlorine	Cl	17	35.453	Promethium	Pm	61	(145)
Chromium	Cr	24	51.996	Protactinium	Pa	91	(231.0359)
Cobalt	Co	27	58.9332	Radium	Ra	88	(226.0254)
Copper	Cu	29	63.546	Radon	Rn	86	(222)
Curium	Cm	96	(245)	Rhenium	Re	75	186.2
Dysprosium	Dy	66	162.50	Rhodium	Rh	45	102.9055
Einsteinium	Es	99	(253)	Rubidium	Rb	37	85.467
Erbium	Er	68	167.26	Ruthenium	Ru	44	101.07
Europium	Eu	63	151.96	Samarium	Sm	62	150.4
Fermium	Fm	100	(254)	Scandium	Sc	21	44.9559
Fluorine	F	9	18.9984	Selenium	Se	34	78.96
Francium	Fr	87	(223)	Silicon	Si	14	28.086
Gadolinium	Gd	64	157.25	Silver	Ag	47	107.868
Gallium	Ga	31	69.72	Sodium	Na	11	22.9898
Germanium	Ge	32	72.59	Strontium	Sr	38	87.62
Gold	Au	79	196.9665	Sulfur	S	16	32.064
Hafnium	Hf	72	178.49	Tantalum	Ta	73	180.947
Hahnium	Ha	105	(260)	Technetium	Tc	43	(98.9062)
Helium	He	2	4.00260	Tellurium	Te	52	127.60
Holmium	Ho	67	164.9303	Terbium	Tb	65	158.9254)
Hydrogen	H	1	1.00	Thallium	Tl	81	204.37
Indium	In	49	114.82	Thorium	Th	90	232.0381
Iodine	I	53	126.9045	Thulium	Tm	69	168.9342
Iridium	Ir	77	192.22	Tin	Sn	50	118.69
Iron	Fe	26	55.847	Titanium	Ti	22	47.90
Kurchatovium	Ku	104	(239)	Tungsten	W	74	183.85
Krypton	Kr	36	83.80	Uranium	U	92	238.029
Lanthanum	La	57	138.91	Vanadium	V	23	50.9414
Lawrencium	Lw	103	(257)	Xenon	Xe	54	131.30
Lead	Pb	82	207.19	Ytterbium	Yb	70	173.04
Lithium	Li	3	6.939	Yttrium	Y	39	88.9059
Lutetium	Lu	71	174.97	Zinc	Zn	30	65.37
Magnesium	Mg	12	24.312	Zirconium	Zr	40	91.22
Manganese	Mn	25	54.9380				

of weight would do as well; 46.7 awu of silicon would combine with 53.3 awu of oxygen. Since the oxygen and silicon atoms do not have the same weight, we will have to consider this fact when determining the atomic ratio for the formula from the weight ratio expressed by the per cent composition. This can be done as follows.

1. Divide the weight of each element per 100 awu of compound by the corresponding relative atomic weight to obtain a ratio of atoms in the molecule.

$$\text{Si:} \quad 46.7 \text{ awu Si} \times \frac{1 \text{ atom Si}}{28.09 \text{ awu Si}} = 1.66 \text{ atoms Si}$$

$$\text{O:} \quad 53.3 \text{ awu O} \times \frac{1 \text{ atom O}}{16.0 \text{ awu O}} = 3.34 \text{ atoms O}$$

Note that the ratio of 46.7 to 53.3 is a weight ratio of silicon to oxygen in silicon dioxide, whereas the ratio of 1.66 to 3.34 refers to that of the number of silicon atoms to oxygen atoms. Since we must consider whole atoms, the ratio of 1.66 to 3.34 must be changed to its equivalent form in whole numbers.

2. Convert to a whole number ratio by
   a. dividing each number in the ratio by the smaller number, and
   b. (if necessary) convert the resulting decimal fractions to whole numbers.

$$\frac{1.66 \text{ atoms Si}}{1.66} = 1$$

$$\frac{3.34 \text{ atoms O}}{1.66} = 2$$

3. Write the formula.

$$\text{SiO}_2$$

In summary, divide the percentage for each element by the respective atomic weight. The resulting numbers are proportional to the number of atoms involved and a simple whole number ratio can be obtained by multiplication and/or division by a constant factor.

If the ratio of atoms is reduced to the smallest whole number ratio possible, the simplest formula for the compound is thus obtained. It should be carefully understood at this point that the simplest formula is not necessarily the molecular formula. While it is often the case that the simplest formula is the molecular formula (H_2O, HCl, CO, CO_2, NO), there are numerous examples in which this is not the case. Hydrogen peroxide has the simplest formula HO, but there is ample evidence (see discussion on mass spectroscopy, Chapter 6) to show that its molecules are composed of two atoms of hydrogen and two atoms of oxygen. Its molecular formula is H_2O_2. Likewise, benzene is C_6H_6 (not CH), and ethane is C_2H_6 (not CH_3).

MOLECULAR FORMULAS— WHY C_2H_6 INSTEAD OF CH_3?

We observed in Chapter 3 how Avogadro's hypothesis could be used to deduce the molecular formulas indicated in Table 4–2. It should be admitted

TABLE 4-2   FORMULAS AND MOLECULAR WEIGHTS FOR SOME
ELEMENTS AND COMPOUNDS

SUBSTANCE	MOLECULAR FORMULA	RELATIVE MOLECULAR WEIGHT	GRAM MOLECULAR WEIGHT
Hydrogen	H$_2$	2 awu	2 g
Chlorine	Cl$_2$	71 awu	71 g
Hydrogen chloride	HCl	36.5 awu	36.5 g
Nitrogen	N$_2$	28 awu	28 g
Ammonia	NH$_3$	17 awu	17 g
Oxygen	O$_2$	32 awu	32 g
Nitric oxide	NO	30 awu	30 g
Water	H$_2$O	18 awu	18 g

at this point that all the formulas in Table 4–2 could be doubled without doing any violence to Gay-Lussac's law or Avogadro's hypothesis. However, experiments in mass spectroscopy (see Chapter 6) and other chemical evidence developed later leave little doubt that these are indeed the correct molecular formulas for these elements and compounds.

The relative molecular weights for the species in Table 4–2 can be obtained simply by adding up the relative weights of the atoms in the molecule. For example, the relative weight of the ammonia molecule, NH$_3$, is the sum of the weights of one nitrogen atom and three hydrogen atoms ($14 + 3 \times 1 = 17$).

One molecule of ammonia is a conceptual, submicroscopic particle which has a weight of 17 awu. However, 17 grams of ammonia is very real to man's senses and is a convenient amount of ammonia for small-scale laboratory experiments.

The gram molecular weight of an element or a compound is obtained by taking the numerical value of the relative molecular weight and attaching the unit of grams to it.

A very interesting relationship develops when a study of the volumes of gram molecular weights of different gases is made. Two grams of hydrogen occupy 22.4 liters when measured at 0°C and 1 atm pressure (standard conditions or STP), and 71 grams of chlorine occupy 22.4 liters at these same conditions. Furthermore, the gram molecular weights of all of the gases listed in Table 4–2 occupy this same volume, 22.4 liters, at standard conditions. This volume is the *gram molecular volume* (Figure 4–1).

The concept of the gram molecular volume gives us an easy method for determining the molecular weight of a gaseous substance. We simply find out how much 22.4 liters of the gas weighs at standard conditions. For example, natural gas contains several gaseous compounds, one of which is ethane. Per cent analysis and relative atomic weights indicate the simplest formula for ethane is CH$_3$. But the formula could be CH$_3$ or C$_2$H$_6$ or C$_3$H$_9$, etc., and we would still have the same ratio of carbon to hydrogen. However since 22.4 liters of ethane has a weight of 30 grams, the molecular weight must be 30, and the molecular formula must by C$_2$H$_6$ [$(2 \times 12) + (6 \times 1) = 30$].

A more timely use of the gram molecular volume is to estimate the maxi-

71

FIGURE 4-1   Gram molecular volume. STP = standard temperature and pressure (0°C and 1 atmosphere pressure).

mum STP volume of carbon monoxide put into the air by burning 1 gallon of gasoline. Since the density of gasoline is about 0.70 g/ml and 1 gallon is 4 quarts or 3788 ml, the weight of 1 gallon of gasoline is about 2650 grams.

$$\text{Weight of gasoline} = 1 \text{ gal} \times \frac{4 \text{ qts}}{\text{gal}} \times \frac{1000 \text{ ml}}{1.06 \text{ qt}} \times \frac{0.70 \text{ g gasoline}}{\text{ml}}$$
$$= 2650 \text{ grams}$$

The maximum weight of carbon monoxide (CO) formed from this amount of gasoline is about 5210 grams (see example 3 on page 75). Since the gram molecular weight of CO (28 grams) occupies 22.4 liters at STP, 5210 grams CO will occupy about 4170 liters.

$$\text{Volume of CO} = 5210 \text{ g CO} \times \frac{22.4 \text{ liters CO}}{28 \text{ g CO}}$$
$$= 4170 \text{ liters CO at STP}$$

This is the space enclosed by a cubical box that is about your height (5 feet 4 inches) on each side.

AVOGADRO'S NUMBER

Since the weight of a hydrogen molecule is 2 awu and the gram molecular weight of hydrogen is 2 grams, you might wonder about the relative size of the awu and the gram. In order to decide this, we first need to know how many molecules of hydrogen constitute 2 grams. A number of indirect methods have been developed since Avogadro's time that actually determine the number of molecules in one gram molecular weight of any element or compound. All the methods give the same value, within the range of experimental error. One method, explained in Chapter 6, involves measuring the volume of helium produced in a radioactive decay reaction while counting the actual number of atoms of helium producing this volume.

The number of molecules in one gram molecular weight of any substance (called *Avogadro's number*), is 6.02×10^{23}, that is, 602,000,000,000,000,-000,000,000. Recall that this number was also defined in Chapter 1 as a *mole*. Therefore, Avogadro's number, or a mole, is the number of molecules required to weigh the same as the gram molecular weight of a substance. Two grams of hydrogen (H_2), 18 grams of water (H_2O), 28 grams of nitrogen (N_2), or the gram molecular weight of any element or compound all contain the same number of molecules, 6.02×10^{23}. This is a tremendously large number. Compared with the number of molecules in 44 grams of CO_2 (one gram molecular weight of CO_2), the earth's population, about three billion (3×10^9) people, is exceedingly small. Indeed, it would require 200 trillion planets with the earth's population to have a total human population of 6×10^{23}. Since the mole is a number, it is perfectly consistent to speak of moles of atoms, moles of groups of atoms, and moles of subatomic particles as well as moles of molecules.

Now a comparison between awu and gram can be made. The weight of a single molecule can be calculated using Avogadro's number. Since Avogadro's number of water molecules weighs 18 grams, one water molecule weighs

$$\frac{18 \text{ grams}}{6.02 \times 10^{23} \text{ molecules}} = 2.99 \times 10^{-23} \text{ gram/molecule, or}$$

0.0000000000000000000000299 gram per water molecule. This weight is so small that we would need about 10^{16} or 10 million billion molecules of water in one sample before we could weigh them on even the most sensitive analytical balance. Since one molecule of water also weighs 18 awu, the number of awu per gram is

$$\frac{18 \text{ awu/H}_2\text{O molecule}}{2.99 \times 10^{-23} \text{ grams/H}_2\text{O molecule}} = 6.02 \times 10^{23} \text{ awu/gram.}$$

That is, there is one mole of awu's per gram of any substance.

STOICHIOMETRY

The word *stoichiometry* comes from the Greek stoicheion (element) and literally means the measurement of amounts of elements. In its chemical usage, stoichiometry refers to the weights of elements and compounds that are consumed or produced in a chemical reaction. For example, an important stoichiometric question is: How much aluminum can be produced from one ton of aluminum oxide? Once the atomic weight scale was established, such important calculations could readily be accomplished. Stoichiometric problems are presented in the following paragraphs; these problems involve simple number relationships and illustrate the considerations involved in such calculations. More challenging problems are presented in Appendix D.

Since the law of conservation of matter states that matter is neither lost nor gained in a chemical reaction, the weight of the reactants in a chemical reaction must be the same as the weight of the products. For example, according to the law of conservation of matter, when hydrogen reacts with chlorine to form hydrogen chloride, the weight of the reactants, hydrogen and chlorine, used in the reaction must equal the weight of the product, hydrogen chloride. The reaction can be written more concisely by using the symbolism of a chemical equation.

$$H_2 + Cl_2 \longrightarrow 2HCl$$

Note that a coefficient, 2, has been placed in front of the hydrogen chloride so there will be 2 atoms of hydrogen and 2 atoms of chlorine represented in both the reactants and the products; that is, none will be gained or lost. What is the meaning of the symbolism of the equation? Here are two alternate but equally meaningful ways to express its significance:

  1 molecule of hydrogen reacts with 1 molecule of chlorine to form 2 molecules of hydrogen chloride;

or,

  1 mole of hydrogen molecules reacts with 1 mole of chlorine molecules to form 2 moles of hydrogen chloride molecules.

Once the equation is balanced, the relative number of moles for each substance involved is simply given by the respective coefficients. Furthermore, since 1 mole weighs 1 gram molecular weight, a set of weights for all substances involved in the reaction can easily be obtained by simply adding up the atomic weights in each formula.

These facts allow several types of calculations to be made for chemical reactions. The following examples will illustrate some of these.

Example 1. How many moles of nitrogen (N_2) are required to react with 6 moles of hydrogen (H_2) in the formation of ammonia (NH_3)?
  a. Write and balance the equation:

$$N_2 + 3H_2 \longrightarrow 2NH_3$$

  b. Since 1 mole of N_2 reacts with 3 moles of H_2, how many moles of N_2 will react with 6 moles of H_2? The answer is 2 moles of N_2. If the number of moles of H_2 is doubled, the number of moles of N_2 must be doubled to keep the same ratio of nitrogen and hydrogen that react with each other.

Example 2. How many moles of nitrogen dioxide (NO_2) will be produced by 4 moles of oxygen reacting with nitric oxide (NO)?

$$NO + O_2 \longrightarrow NO_2$$

a. Balance the equation.

$$2NO + O_2 \longrightarrow 2NO_2$$

b. From the balanced equation we see that the number of moles of NO_2 produced is twice that of the oxygen reacting. Therefore 4 moles of oxygen would produce 8 moles of NO_2.

Example 3. How many grams of carbon monoxide (CO) can be produced by burning 2650 grams of gasoline (C_8H_{18}), assuming the gasoline burns according to the following equation?

$$C_8H_{18} + O_2 \longrightarrow CO + H_2O$$

a. Balance the equation.

$$2C_8H_{18} + 17O_2 \longrightarrow 16CO + 18H_2O$$

b. The balanced equation states that 2 moles of gasoline produces 16 moles of CO. Since the molecular weight of C_8H_{18} is 114, i.e., $(8 \times 12) + (18 \times 1)$, 2 moles would weigh 228 grams. With a molecular weight of 28 for CO, 16 moles would weigh $16 \times 28 = 448$ grams. Thus, 228 grams of gasoline would produce a maximum of 448 grams of CO, or the weight of CO produced is about twice the weight of gasoline burned. This means our 2650 grams of gasoline should produce about 5300 grams of CO. To be more exact:

$$\text{Weight of CO} = 2650 \text{ g } C_8H_{18} \times \frac{448 \text{ g CO}}{228 \text{ g } C_8H_{18}}$$

$$= 5210 \text{ g CO}$$

Example 4. What volume of hydrogen gas would be required to react with 10 liters of oxygen gas at the same temperature and pressure in the formation of water?

$$H_2 + O_2 \longrightarrow H_2O$$

a. Balance the equation.

$$2H_2 + O_2 \longrightarrow 2H_2O$$

b. Avogadro's hypothesis tells us that the volume ratios are the same as the molecular-number ratios if pressure and temperature are held constant.

75

Since 2 molecules of hydrogen react with one molecule of oxygen, it follows that 2 liters of hydrogen react with one liter of oxygen and 20 liters of hydrogen are required to react with 10 liters of oxygen.

After working through these examples, it should be obvious that a consistent set of atomic weights, backed up by the law of conservation of matter, makes possible the calculation of amounts used and produced in a chemical reaction. Obviously, such calculations are much easier and less expensive than actually doing the experiments to gain such information.

CONCLUSION AND PREVIEW

John Dalton excited the chemical world by suggesting that atoms had a particular weight for each different element. Once the confusion between molecule and atom had been resolved, a set of consistent atomic weights emerged. Relative atomic weights provide a sound basis for chemical arithmetic. They allow the calculation of formulas when percentage composition by weight is known. Along with Avogadro's number, atomic weights can be used to calculate the actual weight of a molecule. Atomic weights can be summed to obtain molecular weights. Atomic weights are extremely valuable in the calculation of the amounts of substances either used or produced in chemical reactions. Without relative atomic weights, the accounting of chemistry would be drastically different.

By adding the concept of ceaseless motion to the concepts of Dalton's atoms and Avogadro's molecules, the physical laws of gases can be explained. Chapter 5 will deal with this contribution to atomic theory and with related effects of energy on matter.

QUESTIONS

1. If a debate had been held between Dalton and Avogadro on the subject of the molecular formula of water, would it have been possible for Avogadro to prove Dalton wrong?

2. What is meant by "relative atomic weight?"

3. Suppose two kinds of corn (A and B) have seeds that are essentially the same size and that they pack equally well in a bushel container. If a bushel of seeds of corn A weighs only $\frac{3}{4}$ as much as a bushel of seeds of corn B, what can be said about the relative weights of a typical seed of each kind of corn? How is this analogous to the determination of the molecular weights of gases?

4. Suppose you wish to establish a new atomic weight scale and you define the atomic weight of oxygen to be one. What would be the atomic weights of hydrogen, helium, sulfur and copper on your new scale? Round your numbers off to not more than two figures.

5. If a change is made in the reference for the atomic weight scale, would this change the number of molecules that weigh a gram molecular weight of a substance? Why?

6. What is the difference between the molecular weight of water and its gram molecular weight?

7. What is the molecular weight (formula weight) of CH_4 (methane), $C_{12}H_{22}O_{11}$ (sucrose), $Fe_2(SO_4)_3$ (ferric sulfate), $(NH_4)_3PO_4$ (ammonium phosphate)?

8. Determine the correct simplest formulas for compounds with the following percentage by weight:

   (a) 64.86% C, 13.51% H, and 21.63% O.
   (b) 27.48% Mg, 23.66% P, and 48.85% O.
   (c) 38.79% Cl and 61.21% O.
   (d) 22.9% Na, 21.5% B, and 55.7% O.

9. How can the molecular weight of a gas be determined?

10. From the percent composition by weight and the relative weights of hydrogen and carbon, it can be calculated that there are twice as many hydrogen atoms as carbon atoms in propylene (a compound of just these two elements). Thus, the molecular formula must be CH_2 or C_2H_4, C_3H_6. . . . Since propylene is a gas, how could you experimentally determine its molecular weight and thereby recognize the correct formula?

11. (a) What is a mole?
    (b) What is the weight of one mole of hydrogen molecules? Of hydrogen atoms?

12. If one mole of oxygen molecules (O_2) weighs 32 grams, by what number would 32 have to be divided to obtain the weight of a single oxygen molecule?

13. Why is it necessary to balance a chemical equation before it can serve as the basis of a stoichiometric calculation?

14. An important source of hydrogen, used as a rocket fuel, is from the decomposition of water by electrical energy. The equation for this reaction is:

$$H_2O \xrightarrow{\text{Electrical energy}} H_2 + O_2$$

   (a) Balance the equation.
   (b) What weight of water is necessary to produce 2 grams of hydrogen?
   (c) How many grams of oxygen would be produced as a by-product?
   (d) How much water would be necessary to produce 2 tons of hydrogen?

15. The carbon dioxide (CO_2) produced while making bread causes the bread to rise. Suppose you add 16.8 grams of baking soda ($NaHCO_3$) to a batch of bread dough. What volume of CO_2 at STP could you expect from this amount of soda?

$$2NaHCO_3 \longrightarrow Na_2CO_3 + H_2O + CO_2$$

   Compare the volume of CO_2 produced to the volume of an ordinary bread pan ($4'' \times 4'' \times 10''$) and to the volume of an ordinary kitchen oven ($24'' \times 24'' \times 24''$).

16. Cite other pivotal moments in United States and world history similar to the Karlsruhe Conference in which one man came forth and solved a dilemma that had baffled mankind and stifled progress for many years.

17. The Pacific Ocean covers one-third of the 196,950,000 square miles of the earth's surface. The average depth of the ocean is about 12,500 feet. Is the volume of the water in the Pacific Ocean more or less than the volume of 1 mole of sugar cubes that are 1 centimeter on each edge?

18. If we now know that the weight of a carbon atom is 2.0×10^{-23} grams, why retain the concept of the relative weight of 12 for a carbon atom?

19. A major source of sulfur dioxide (SO_2) in the air is burning coal and oil that contain sulfur. If a shipment of coal contains 3% sulfur by weight, what weight of SO_2 is available to pollute the air if one ton of this coal is burned? The sulfur burns according to the following equation.

$$S + O_2 \longrightarrow SO_2$$

20. How is it possible for 1 mole of the larger propane molecules (C_3H_8, molecular weight 44) to occupy the same space as 1 mole of the much smaller hydrogen molecules (H_2, molecular weight 2) at STP?

21. Complete the following table for the reaction indicated. Be consistent with the one value given.

	NaH	+ H$_2$O	$\longrightarrow$ NaOH +	H$_2$
Moles:	4 moles	+ _____	$\longrightarrow$ _____ +	_____
Grams:	24 grams	+ _____	$\longrightarrow$ _____ +	_____
Atoms of hydrogen:	1.02×10^{23} atoms	+ _____	$\longrightarrow$ _____ +	_____

22. Which of the following are conserved during a chemical reaction? Which are not? Color, mass, volume, atoms, moles of substances, solubility.

23. The formula for vitamin C is $C_6H_8O_6$. What is the percentage of carbon in this compound?

24. Naturally occurring carbon, as in soot, is 98.89 per cent carbon atoms with atomic weight of 12.00000 awu and 1.11 per cent carbon atoms with atomic weight of 13.00335 awu. Calculate the average atomic weight of naturally occurring carbon. Compare your value with the one given in Table 4–1.

SUGGESTIONS FOR FURTHER READING

MacNevin, W. M., "Berzelius—Pioneer Atomic Weight Chemist," *Journal of Chemical Education,* Vol. 31, p. 207 (1954).

Wichers, E., "Report of The International Commission on Atomic Weights—1961," *Journal of the American Chemical Society,* Vol. 84, p. 4175 (1962).

Coward, H. F., "John Dalton (b. 1766, d. 1844)," *Journal of Chemical Education,* Vol. 4, p. 22 (1927).

Douville, J. A., "Determining the Molar Volume: A Safe Method," *Chemistry,* Vol. 43, No. 1, p. 25 (1970).

"Hydrogen Trioxide," *Chemistry,* Vol. 43, No. 6, p. 20 (1970).

Labbauf, A., "The Carbon-12 Scale of Atomic Masses," *Journal of Chemical Education,* Vol. 39, p. 282 (1962).

Sunier, A. A., "Some Methods of Determining Avogadro's Number," *Journal of Chemical Education,* Vol 6, p. 299 (1929).

Guggenheim, E. A., "The Mole and Related Quantities," *Journal of Chemical Education,* Vol. 38, p. 86 (1961).

Kieffer, W. F., "The Mole Concept in Chemistry," Reinhold Publishing Corp., New York, 1964.

Zink, G. E., "Aspects of Nonstoichiometry," *Chemistry,* Vol. 41, No. 11, p. 13 (1968).

ENERGY IN MATTER

For many years it has been stated that matter "occupies space and has weight" and that energy is the "ability to do work." Since 1945 it has been widely recognized that matter and energy are actually two forms of the same reality that are interchangeable according to Einstein's famous equation, $E = mc^2$. No one needs to be told that energy transfers occur when matter is changed, because we all have felt the warmth of the burning fire, the cold feeling when water evaporates from our skin, and many other indications that a storage or a depletion of energy is associated with changes in matter. The charged battery, for example, has a measure of energy that the same battery does not have when "dead," and, similarly, the water above the electrical generator of a dam has a measure of energy that the same water does not have when it is below the dam.

In understanding the properties of substances and the changes that transform one substance into another kind of substance (or into another form of the same substance), account must be made of the many associated energy effects. The modern chemist seeks to explain macroscopic changes in matter and associated energy changes in terms of structural and energy changes in submicroscopic matter. Since submicroscopic particles have not been directly observed, nor, certainly, has their time-dependent motion, the most obvious way to describe the interactions of these particles is to treat them analogously to macroscopic matter which we can observe and measure. A typical analogy of this kind is the application of the laws describing the behavior of macroscopic pieces of matter to the attractions holding submicroscopic particles together.

An interlude will be taken to develop an understanding of two important kinds of energy that objects can have (kinetic and potential), which are fundamental to the kinetic molecular theory. These energy concepts will be needed to understand the theories developed later to explain the behavior of atoms and molecules.

KINETIC ENERGY

Our everyday experiences offer ample evidence that there is energy possessed by a moving piece of matter that the matter does not have when at rest.

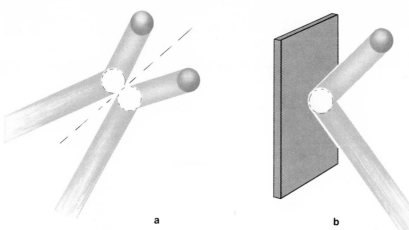

a   b

FIGURE 5-1   A diagrammatic representation of perfectly elastic collisions
(A) between two similar bodies and (B) between a small body and a stationary
object. Kinetic energy is conserved in both instances.

Kinetic energy results from matter being in motion. While the motion is necessary, the mass of the matter is an important part of the kinetic energy of the
object. A ping-pong ball and a golf ball traveling at the same speed could have
decidedly different effects on a picture window.

The batted ball illustrates, as do many other ordinary experiences, that the
kinetic energy of one object can be transferred to another object through a
collision. A careful study of such collisions indicates that in the absence of friction
and other energy losses, the total kinetic energy of the colliding bodies is the
same after the collision as it was before. This describes a perfectly elastic collision
(Figure 5–1). As early as the 17th century, it was recognized that when two
bodies collide in an elastic collision, the sum total of the mass times the velocity
squared values (mv^2) for the bodies is the same after the collision as it was before,
although each body may be moving with a new velocity. Later the term *kinetic
energy* was introduced for the term, $\dfrac{mv^2}{2}$. Thus, the kinetic energy for any
particle in motion is given by the equation

$$\text{kinetic energy} = \frac{mv^2}{2} \tag{1}$$

Actually there are no perfectly elastic collisions in the macroscopic world
even though this ideal can be closely approached. The concept is introduced
at this point for two reasons: first, the loss of some kinetic energy in inelastic
collisions and the resulting change in the matter, such as a change in temperature
or a change in structure, indicate that matter can contain energy in ways other
than by motion; and second, the idea of the perfectly elastic collision, with
absolutely no loss in kinetic energy, is assumed at the molecular level in the
theoretical discussion of the structure of matter.

POTENTIAL ENERGY

An object may contain energy by virtue of its position or composition. Such
energy is called potential energy. A rock balanced on a mountain ledge has the

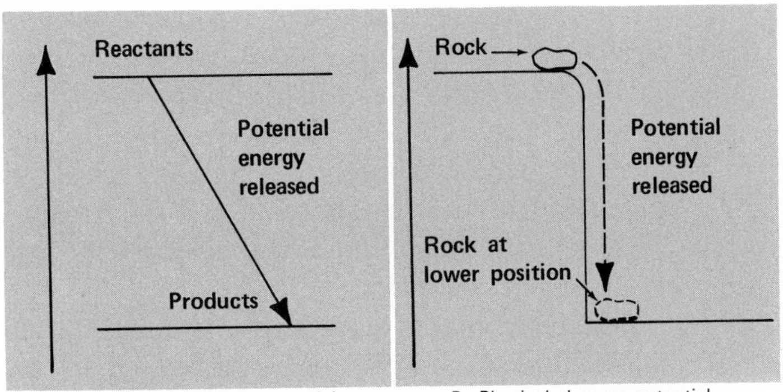

FIGURE 5-2 The similarity between chemical and physical changes and the energy involved is illustrated in two examples. What happens, energetically, when these processes are reversed?

A. Chemical change: potential energy is released as heat, light, electricity, etc.

B. Physical change: potential energy is released as kinetic energy as the rock descends

potential energy to kill a mountain climber, although the rock would pose no threat if lying on the valley floor below. A piece of wood and the oxygen around it may release energy in the form of heat and light when they combine in a process called *combustion*. New substances—ashes and gases of combustion that weigh the same as the destroyed wood and oxygen but have less energy of composition—are left. It seems reasonable to believe that the rock's change in position and the chemical change are similar (Figure 5–2). Just as the rock has less energy after it has fallen, so the molecular structure of the wood and oxygen rearrange, through combustion, into products of less energy. Both of these processes result in a loss of *potential* (or stored) *energy*, which is *the energy contained in a sample of matter due to its position or composition that can be released by means of a physical or chemical change.*

CONSERVATION OF ENERGY

Kinetic energy can be converted into potential energy. A boy may give kinetic energy to his favorite ball only to find this energy converted into potential energy as the ball resides high in a tree. Likewise potential energy can be converted into kinetic energy as a squirrel nudges the ball from its lodgment in the tree. Many energy transformations are possible. However, there is no loss or gain in the amount of energy in any of these transformations. If energy appears in one form it has to come from an equal amount of energy in another form or the destruction of an equivalent amount of matter. These changes of energy into alternate forms without loss are generalized into the law of conservation of energy:

The sum total of the energy in the universe is constant.

ELECTRICAL CHARGE

In order to understand how energy can be stored in the structure of matter, it is necessary to understand something about the ability of matter to interact

81

with other matter at a distance. Of special interest in chemistry are the forces that are exerted through space between electrical charges. Indeed, the forces that appear to hold matter together can be explained in terms of these forces between electrical charges. Macroscopic amounts of electrical charge are not at all uncommon in ordinary experience. The bolt of lightning, the spark between comb and hair in dry weather, the shock on touching the door handle when certain clothes are brushed across the seat covers in a car, are all the result of electrical charge.

Experiments indicate that charges either attract or repel each other. The fact that there are two and only two effects suggests that there are only two kinds of electrical charge, positive and negative. In Figure 5–3 a very sensitive apparatus is illustrated, which can be used not only to show the two types of charges, but also to measure the forces of interaction. A wire under tension supports a small mirror and an arm connected both to a metal ball through an electrical insulator and a damping device, which is a paddle in a container of water used to reduce quickly any motion about the wires. A bright line of light, produced by a slit in front of a parallel light beam, shines on the mirror and is reflected to a screen marked with a linear scale. A very slight twist in the wire produces a much larger change in the position of the light on the screen. An electrical charge can be produced on a hard rubber rod by rubbing it vigorously for a few moments with animal fur (Figure 5–3). Part of the charge on the rubber rod can be transferred to metal ball A simply by touching the two together. Similarly, the same charge can be placed on ball B. Now, if B is placed near A, the balls will move away from each other. The resulting twist of the wire is clearly evident by the movement of the light on the screen. Exactly the same visible results can be obtained if a glass rod and a silk cloth are substituted for the rubber rod and fur. The charge transferred from the glass rod to the metal balls causes them to repel each other. But, if the charge from the rubber rod is transferred to one of the metal balls and the charge from the

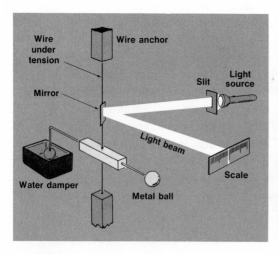

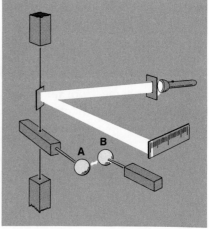

a   b

FIGURE 5–3   Apparatus for the demonstration of electrical charges and their inter-

action.

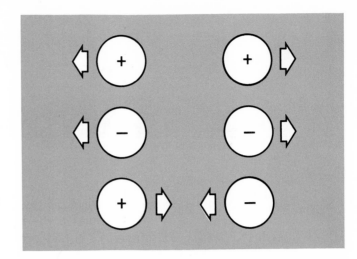

FIGURE 5-4   Like charges repel
and unlike charges attract.

glass rod to the other, the two balls attract and move closer together. Further-more, any source of electrical charge will do one of two things when brought close to the charge from the rubber rod: (a) It will repel the rubber-rod charge just as this charge repels itself; or (b) it will attract the rubber-rod charge just as the glass-rod charge attracts it. Since all charges behave exactly the same as the rubber-rod charge or the glass-rod charge, we conclude that there are but two kinds of electrical charge and name them negative (rubber-rod) and positive (glass-rod). Also, we can conclude from these experiments that a positive charge attracts a negative charge, and that if both charges are positive or both negative, they will repel each other (Figure 5–4). *Like charges repel. Unlike charges attract.*

The attraction between unlike charges or the repulsion between like charges is directly proportional to the magnitude of the charges. When both balls of Figure 5–3 are charged with the same charge, one-half of the charge on one ball can be removed by touching it with another identical ball mounted on a suitable insulator. When the third ball is removed, the light beam shifts the proper amount to indicate the repulsive force is one-half as great. In another experiment, we find that doubling the charges on both the original balls makes the force four times as great. The effect of distance between the charges can be observed by placing ball B at various fixed distances from A without changing the charges. It is found that doubling the distance reduces the force to one-quarter and tripling the distance reduces the force to one-ninth of its initial value. All of this information is summarized by the equation:

$$F = \frac{q_1 \times q_2}{d^2} \tag{2}$$

where F is force, q_1 and q_2 are the sizes of the two charges and d is the distance between the charges.

ENERGY OF CHARGED SYSTEMS

Just as it requires an effort to separate small magnets, a similar effort is required to separate a positive and a negative charge. The energy that is used

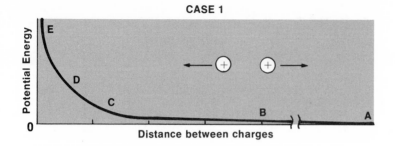

CASE 1

Potential Energy

Distance between charges

Figure 5-5   Potential energy of a system composed of two positive charges at different distances. Similar results are observed for two negative charges.

to separate the unlike charges is then stored in the system as potential energy. This energy, stored by virtue of the structure of the system, is potential energy because it can and will be released if the two charges are allowed to come together. The bolt of lightning is a display of the energy released when opposite charges come together.

Potential energy diagrams are very useful in understanding the energy stored in atomic and molecular systems; Figure 5–5 is one kind of potential energy diagram. Three cases will be considered.

Case I. Imagine two units of positive charge that are isolated in space from any other influence. Since the positive charge will repel another like charge, this force will cause them to move apart. In space there will be no resistance to this separation so the distance will increase as long as there is any repulsive force. Equation (2) tells us the force becomes very small as the distance of separation increases. However, there remains a real force even if it is extremely small, as it will be at very large distances. The result is that the two charges will separate until there is an infinite distance between them (to the point where they have absolutely no influence on each other). Of course, such an idealized experiment is not possible in man's environment, but experiments give every indication that these would be the results of the idealized case. With the like charges infinitely separated, any effort to close the gap between them would require energy. For the time being, we will simply take the point at which the charges are infinitely separated as our zero point reference on a potential scale. This infinite distance of separation is indicated by Point A in Figure 5–5. Energy is required from outside the system to bring the two charges closer together, as indicated by Points B, C, and D. What happens if the separating distance is reduced to zero? Equation (3)* tells us that we reach an impossible situation; the fraction representing the potential energy becomes infinite or undefined when the denominator becomes zero.

$$\text{potential energy} = \frac{q_1 \times q_2}{d} \tag{3}$$

---

*Equation (3) can be derived from equation (2) since:
a. work = force × distance
b. energy is the ability to do work
c. therefore, energy = force × distance, or

$$\text{energy} = \frac{q_1 \times q_2}{d^2} \times d = \frac{q_1 \times q_2}{d}$$

84

The energy necessary to move against the very large forces at very small values of d is also very large and indicates the charges will have much energy stored at point E, where d is very small.

Case II. For this case, consider an isolated unit of positive charge and a unit of negative charge (Figure 5-6). At an infinite distance apart, point A, the two charges would have their maximum stored energy because it *requires* energy to separate the attracting charges. It is customary to assign zero potential energy to the system at infinite separation, as was done with the two charges that repel. It is not strange that both sets of charges have the same potential energy at infinite separation because, at this distance, one charge has no effect on the other regardless of whether we are dealing with like or unlike charge. If zero potential energy is assigned to the unlike charges when they are at infinite distances, then all other potential values possible for the system are negative; that is, they are less than the maximum value of zero. As the attracting charges approach each other, they release energy and contain progressively less stored energy at Points B, C, and D (the larger the negative value is, the lower the potential energy of the system). Point E, where the separating distance is very small, represents a system with relatively little potential energy, although the value of the stored energy would be a large negative value.

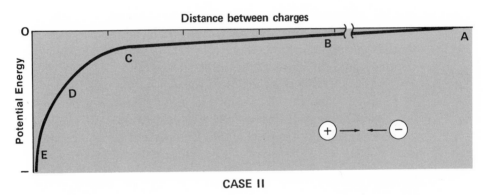

FIGURE 5-6   Potential energy of a system composed of a positive charge and a negative charge at different distances.

Case III. An extreme of Case II, in which the *distance* of separation between opposite charges is reduced to *zero*, is impossible in any real situation. According to Equation (2), the force of attraction between the charges is infinite for a zero value of distance. It would take infinite energy to separate the charges when together, and an infinite amount of energy would be released if the charges came together. Since this would involve the release of more energy than the universe contains, it becomes necessary to postulate a repulsive force between unlike charges at very small distances. The attractive force is still present, but at a given distance it is counterbalanced by this new, though somewhat mysterious, repulsive force. A macroscopic, physical system that is somewhat analogous will help to clarify this point. A steel cannon ball elevated 10 feet above the floor has considerable energy due to the arrangement of the system. However, all of this energy cannot be released from the system if the floor is covered with

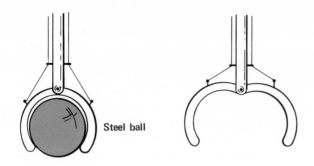

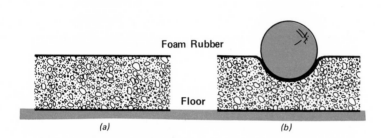

FIGURE 5-7   Resting on the floor would represent a lower potential energy for the steel ball, but the presence of the foam rubber pad prevents this from happening. A new position (b) is now a minimum potential energy position.

a 1-foot-thick pad of foam rubber. The steel ball, when dropped, will bounce a while and come to rest at a distance somewhat less than 1 foot above the floor (Figure 5–7). An amount of potential energy has been released, but the short-range upward force exerted by the compressed rubber prevents a closer approach and the release of even more energy.

An idealized energy-distance diagram for Case III is shown in Figure 5–8. At Point A, when the charges are far apart, there is a relatively large amount of potential energy in the system. The system contains less energy at B as the charges are closer together. Point C represents a separation distance for the system when it is at its lowest possible energy level (sometimes called the bottom of the energy well). At Point C any effort to increase the distance between the charges will require energy to overcome the electrostatic force of attraction, and any effort to reduce the distance will require energy to overcome the short-range force of repulsion. Unless energy is supplied from some outside

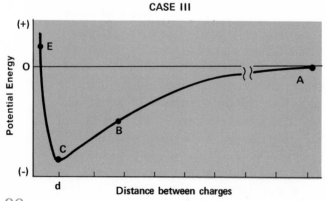

FIGURE 5–8   Potential energy of a system composed of a positive charge and a negative charge at different distances when short-range repulsive forces are present.

source, the two unlike charges will lose all energy possible and will exist as a pair separated by the distance, d.

The analogy of macroscopic particles is used in the next chapter to explain the structure of the simplest of all atoms, the hydrogen atom. The Bohr theory postulates, among other things, that atoms are made up of charged particles, and that the hydrogen atom is made up of one particle of unit positive charge and one particle of unit negative charge. If these two particles are placed in an isolated system, they will attract each other as a result of their opposite charges. The force of attraction will cause the two particles to move closer together, yielding previously stored potential energy. Finally, they will approach so closely that the short-range repulsive force exactly counterbalances the electrostatic attractive force, and a stable partnership results. Energy is released in the formation of this stable arrangement—the hydrogen atom—and energy will be required to tear it apart.

EFFECT OF THE ADDITION OF ENERGY TO MATTER

Energy acting upon matter can be used to alter its structure by changing the distance between particles. This sometimes brings about a chemical change and at other times causes a physical change. If the particles are at their lowest energy (distance d in Figure 5–8), the additional energy can either push the particles closer together or separate them more. Most commonly, the energy is used to distort or break the bond between the charged particles. A logical extension of the idea of this bond between charged particles can explain how atoms are held together in molecules. Provided that suitable interatomic forces can be found to explain effects in the macroscopic world, we should be able to account for an everyday phenomenon as the action of a material like glue, which causes the bonding of one substance to another.

To illustrate how added energy can produce either physical or chemical changes by separating particles which attract, consider the addition of energy to a pure substance composed of molecules at zero degrees absolute (0°A). This is a point of very low energy at which the molecules are as close together as possible. The addition of energy spreads the molecules slightly apart. The continued addition of energy eventually melts the solid, heats the liquid, and vaporizes it into a gas. Each process results in a further separation of the particles which attract each other. All of these changes are physical changes, since the same pure substance survived through all of the additions of energy. However, if energy continues to be added, the gas is heated to the point where the molecules break apart into atoms. At this point, a chemical change occurs, since the original substance composed of molecules is changed into another substance composed of atoms. Continued additions of energy can cause atoms to be separated into their component parts, producing additional chemical changes. Throughout the period of change from molecules at absolute zero to subatomic particles at higher temperatures, the addition of energy causes a separation of particles which attract each other. This absorbed energy is stored in the system

87

by means of its expanded structure. At any point, the total energy contained in a sample of matter is called its *internal energy*.

Having seen a consequence of the attraction between submicroscopic particles, we will now examine some physical changes produced when energy affects a pure substance. We will speculate on the attendant, underlying interactions of the molecules. It is through this type of thoughtful approach that matter has been better understood and has been put to better use.

ENERGY CHANGES IN A PURE SUBSTANCE

The internal energy of a sample of a pure substance can be increased by heating the sample or by doing work on it. Heating can be accomplished by placing the sample in contact with other matter which is at a higher temperature. Work can be done on the sample by beating on it, stretching it, compressing it, or in any way forcing the parts of the sample to move against forces that tend to hold these parts in place. Since internal energy can be increased by either heat or work, we might suspect that heat and work are related in some way. Indeed, two of the signal advancements in all of physical science were the demonstration by Count Rumford (Benjamin Thompson), in about 1800, that heat and work are actually two expressions of energy, and the verification of James Prescott Joule, about 40 years later, that a given quantity of work is equivalent to and interchangeable with a given amount of heat.

What are some of the results of increasing the internal energy of a pure substance? *Either the temperature will increase or there will be a change in state.*

In order to deal with specific changes in a familiar substance in terms of given energy changes, it will be necessary to become acquainted with a unit used to measure heat or work. There are a number of such defined units: erg, joule, foot-pound, British thermal unit, kilowatt-hour, and calorie. Any of these units can be used to measure quantities of energy just as units of length (foot, inch, yard, meter) are used to measure distance. In this text we will use the calorie unit exclusively. The calorie is defined as the amount of heat necessary to raise the temperature of one gram of water one Celsius degree at a starting temperature of about 15°C. This calorie (cal) is one-thousandth of the large calorie (Cal), which is commonly used to measure the potential energy available in foods.

It is also important to understand the difference between heat and temperature. Without being concerned with definitive statements, we can say simply that *heat* is energy and is measured in any of the units given in the last paragraph, whereas *temperature* is a measure of the amount of heat in a given substance and is measured in degrees. (More accurately, temperature is a measure of the driving force behind heat transfer.) The heat from a burning match will raise the temperature of an iceberg only a very small part of one degree. Although the burning match has a much higher temperature, it furnishes very little total heat compared to the large iceberg's capacity to store heat. Heat will flow from a body of higher temperature into a body of lower temperature (Figure 5–9) regardless of the amount of heat already present in each body. You should also

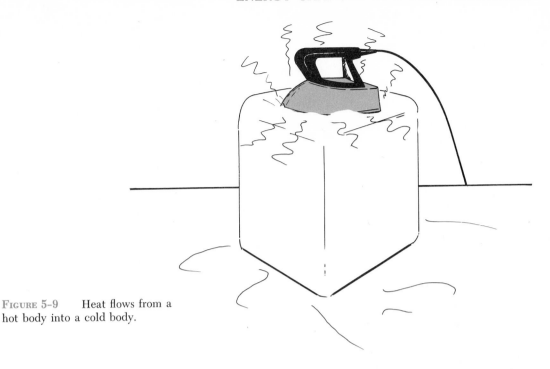

FIGURE 5-9   Heat flows from a
hot body into a cold body.

keep in mind the fact that substances differ from one another in their capacity
to store heat. While it takes 1 calorie to heat 1 gram of water 1 degree, it only
takes one-half of a calorie to heat 1 gram of ice 1 degree. Iron is even more
responsive in its temperature changes when energy is applied. It takes only
0.12 calorie to raise the temperature of 1 gram of iron 1 degree. These numbers
are called the *specific heats* for these materials; the specific heat of iron, 0.12
$\frac{\text{cal}}{\text{g deg}}$ is read 0.12 calorie per gram per degree.

When a pure substance undergoes a change of state, there is a relatively
large amount of energy involved. It requires about 80 calories of heat to melt
one gram of ice at 0°C. Similarly, about 80 calories of heat have to be removed
from one gram of water at 0°C to convert it into 1 gram of ice at 0°C. The
80 cal/g is called the *heat of fusion* for ice, and each pure substance can be
expected to have its own characteristic heat of fusion. The *heat of vaporization*
of a liquid is the amount of heat necessary to convert one gram of the liquid
to one gram of the gas at the same temperature. The heat of vaporization of
water is about 540 cal/g.

In order to clarify the points just made, consider Figure 5–10. Here, the
interaction of energy with 1 gram of ice at 0°C is followed in such a way that
one can easily see the results of fusion (melting of the ice), further heating
(0°C ⟶ 100°C), and vaporization (boiling the water). An application of the
law of conservation of energy leads to the conclusion that 1 gram of water at
0°C has more energy than 1 gram of ice at 0°C, 1 gram of water at 100°C
has more energy than 1 gram of water at 0°C, and so on. This energy difference
is illustrated for 1 gram of water (as ice, liquid, and steam) in Figure 5–11.

All pure substances have characteristic heats of fusion and vaporization and
specific heats, which are dependent on their submicroscopic properties. Each

89

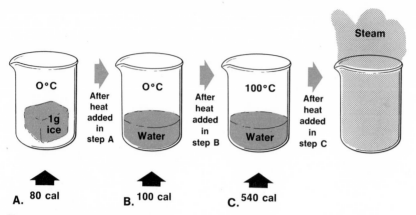

FIGURE 5-10   The interaction of energy with matter. The processes in steps (A), (B), and (C) are *idealized*, since, in practice, some vaporization losses would always take place.

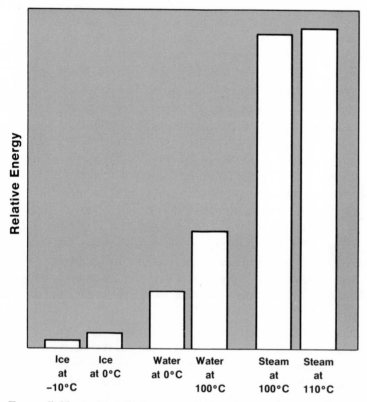

FIGURE 5-11   A graphic representation of the relative internal energies of equal weight samples of the compound H_2O at various temperatures.

can be expected to exhibit behavior similar to that of water at characteristic temperatures.

THE KINETIC-MOLECULAR THEORY EXPLAINS ENERGY CHANGES IN A PURE SUBSTANCE

The kinetic-molecular theory developed rapidly after 1860 as a sophisticated model of the structure of gases. Later, when some of the theoretical predictions (such as molecular sizes) were found to agree with independent experimental results, it was strong evidence that scientists were on the right track in sorting out the submicroscopic picture of matter.

Although we will deal with the theory as a picture model, it can be expressed in mathematical terms. Not only does it offer a qualitative explanation for the gas laws as presented in Chapter 3 in mathematical form, but the theory can also be used to derive the gas laws from basic theoretical concepts. This helps to confirm our belief in molecules. The mathematical part of the theory allows one to calculate the velocities of molecules, the frequency of molecular collisions, and the distances traveled by molecules between collisions. Although it is beyond the scope of this presentation to develop the mathematics involved, perhaps you will begin to see why the physical scientist is impressed with the power of the kinetic theory. It provides a molecular explanation of existing gas laws, as well as an avenue for obtaining information about molecules and their behavior in gases that was not available from direct laboratory experiments.

The kinetic-molecular theory visualized a gas in the following way (only those points that are important to this discussion are included):

(1) A gas is made up of molecules (monatomic molecules in the case of gases like helium).
(2) The molecules are very small relative to the normally great distance between them.
(3) The molecules have no appreciable attraction for each other at these great distances.
(4) The molecules are moving with very high velocities, causing many intermolecular collisions and collisions with the walls of the container at each second.
(5) Collisions at the molecular level are perfectly elastic. There is, therefore, no total loss of energy resulting from these collisions, and no tendency for the molecules to slow down and stop, as do billiard balls on a table.
(6) Molecules move faster as the temperature is increased.

These assumptions are quite effective in explaining the gas laws that were discussed in Chapter 3. The following are examples:

Boyle's Law. Boyle said that the pressure exerted by a gas goes down as the volume goes up at constant temperature. This is readily understood if one conceives of the pressure on the walls of a balloon resulting from the molecules beating against the walls. If the balloon were suddenly made *larger* the same number of molecules moving at the same speed would hit the walls less often, resulting in a *reduced* pressure.

Dalton's Law of Partial Pressures. The additivity of gas pressures is easily understood when one notes that gas molecules have little attraction for one another and that the space between molecules is rather large. Hence, there is always "room for one more" molecule. Since the molecules do not interact they act independently upon the walls of the container and the pressure increases accordingly.

Henry's Law. The increased solubility of a gas in a liquid can be explained in terms of Boyle's law and the collisions that take place between the surface of the liquid and the gas molecules. As the pressure of a gas increases, the number of molecules in a given volume increases. This results in an increased number of collisions of gas molecules with the liquid surface, some of which lead to gas molecules going into the liquid (dissolving). The greater the pressure, the more collisions, and the greater the solubility.

Charles' Law. Charles' law states that the volume of a gas is increased as the temperature is increased, if the pressure is held constant. If increased temperature causes molecules to move faster, as is asserted by the kinetic-molecular theory, increasing the temperature of a gas would cause the molecules to hit the walls of the container more often. This would result in an increased pressure. If the pressure is to be held constant, it follows that the container will have to be made larger so that the faster moving molecules will hit the walls with the same energy as the slower moving ones did in the small volume.

The kinetic-molecular theory is not as successful in mathematical form with liquids. While notable efforts have been made in trying to calculate from theory the properties observed in liquids, the results fall far short of those obtained in the study of gases. The reason for this difficulty lies in the fact that the liquid molecules are closely packed (Figure 5–12), resulting in relatively larger inter-molecular forces which cannot be discounted in the mathematical description

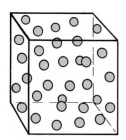

Some particles in gaseous state (not drawn to scale); particles far apart, completely fill container.

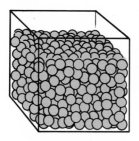

Some particles in a liquid state; not in fixed positions (liquid flows), definite volume, usually somewhat larger than when in solid state.

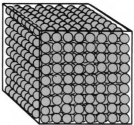

Particles in a solid; fixed positions, definite volume.

FIGURE 5–12   An illustration of the three states of matter. Molecules move more rapidly with increased temperature. This usually results in a liquid being less dense than the solid it was obtained from as it has more space not occupied by molecules.

of this model. Accurate mathematical expressions for these forces are difficult to devise and manipulate. The picture is complicated further by the shifting structure within the liquid. The result is that the theory is more qualitative (pictorial) and less quantitative (mathematical) than in the study of gases. The essential points needed for liquids are:

(1) Molecules are close-packed, essentially in contact with each other.
(2) There are many solid-like crystal structures within the body of the liquid. These are constantly changing in size and in the particular molecules involved.
(3) Molecules move more rapidly with increased temperature. This usually results in a liquid being less dense than the solid it was obtained from as it has more space not occupied by molecules.

In solids, as well as liquids, the kinetic-molecular theory thus far is incapable of providing the basis for exact equations in terms of molecular events that can be used to calculate the observed properties of solids. Again, the difficulty lies in the complexity of relating intermolecular forces (tending to hold the solid together) and the kinetic energy of the individual molecules (tending to cause the solid to fly apart). However, the study of structure in solids is much easier than in liquids because the molecules of the solid hold their position rather well relative to the molecules around them. This is true in spite of the fact that the molecules are vibrating about fixed positions (see Figure 5–13), and some even have enough kinetic energy to escape the binding intermolecular forces, moving from place to place within the solid or even out of the solid.

Using the assumptions of the kinetic-molecular theory, the behavior of a pure substance as energy is added can be explained in terms of the kinetic and potential energy of the individual molecules. As a pure substance is heated below the point at which it changes state, the increased internal energy can be stored in one of two ways. Primarily, the increased energy is stored as kinetic energy in the faster-moving molecules. However, some expansion usually occurs when a substance is heated. This expansion results in a movement of the molecules

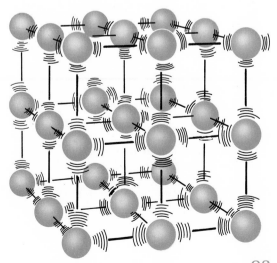

FIGURE 5-13   Vibrations of particles in their fixed positions in a solid lattice.

away from the restraining intermolecular forces, resulting in increased potential energy. In the change of state of a pure substance there is usually a dramatic change in volume. In going from a liquid to a gas the increase in volume is most pronounced. There is usually a relatively small increase in volume in going from the solid to the liquid phase. (Water is an exception to this rule.) Since the temperature does not change during a change of phase, it is reasonable to conclude that heats of fusion and heats of vaporization are stored as potential energy in the molecular system. It then becomes easy to understand why heats of vaporization are larger than the corresponding heats of fusion. It takes much more to move the molecules through the relatively larger distances involved in the gas structure, while changes in distances that occur in going from a solid to a liquid are relatively small.

CONCLUSION AND PREVIEW

We have now surveyed the concepts of energy, kinetic energy, and potential energy, and examined the role of these in producing both physical and chemical changes in pure substances. We are now ready to use these same concepts in the further development of atomic theory. Dalton's concepts and the kinetic-molecular theory are based on the particulate nature of matter. These were sufficient models to explain *some* macroscopic laws of chemical change and *some* physical properties of matter, but a number of astounding discoveries were made near the beginning of this century that show these models were *far too simple*. Chapter 6 launches into this exciting development.

QUESTIONS

1. Which of the following are exhibiting primarily potential energy and which kinetic energy?

    (a) wound clock spring
    (b) a moving automobile
    (c) water stored in a water tower
    (d) a rock just beginning its tumble down a mountainside

2. Most steel bridges have expansion joints in them to allow for variations in the weather. Why?

3. Distinguish between gas, liquid, and solid with respect to the differences between the kinetic and potential energies of the particles in the three states.

4. In what important aspect are electrical and magnetic interactions different from gravitational interactions?

5. How many calories are required to melt 10 grams of ice at 0°C and raise the temperature of the liquid water to 40°C?

6. Explain why water in a boiler (at sea level) will heat up and then boil at 100°C until all of it is vaporized, as heat is added to it.

7. Describe the potential energy and kinetic energy relationships as a rock tumbles off a cliff.

8. Explain in detail how the law of conservation of energy applies to the following processes:
    (a) Water flows from the storage pool behind a dam through conduits to a turbine which

turns a generator; the energy produced turns a motor, furnishes a light, rings an electrical bell, and heats a house, in addition to sustaining frictional losses.

(b) A piece of wood is burned.

9. Draw a diagram showing how the potential energy of a positive particle and a negative particle changes as the two particles approach each other from a great distance apart, touch, and because of outside force, are caused to dent each other.

10. How would the kinetic molecular theory explain (in terms of molecules):

(a) the absorption of the heat of fusion in melting a solid without changing the temperature.
(b) the absorption of the heat of vaporization in vaporizing a liquid without changing the temperature.
(c) why a liquid expands as the temperature is increased.

11. Could two containers of water have the same temperature but contain different quantities of heat? Explain.

12. Two slowly moving automobiles crash into each other head-on. They recoil only slightly. What happened to the rest of the kinetic energy which they possessed prior to their crash?

13. Mercury has a melting point of $-38.87\,°C$ and a boiling point of $356.58\,°C$. What changes would you expect to find in the internal energy of a sample of mercury as you heated it from $-50\,°C$ to $400\,°C$?

14. Which has greater kinetic energy: a freight train that weighs 2000 tons, which is moving at 40 miles per hour, or an airplane that weighs 20 tons and is moving at 1000 miles per hour?

15. Name some materials which are commonly used to provide energy because of the potential energy stored in their molecular structure.

16. Design a system that you could use to obtain potential energy from the following sources of kinetic energy.

(a) the tidal motion of the ocean
(b) wind

SUGGESTIONS FOR FURTHER READING

Ihde, A. J., "The Development of Modern Chemistry," Harper and Row, New York, 1964, p. 395.

Farber, E., "The Evolution of Chemistry," Ronald Press Co., New York, 1952, p. 204.

Sabine, D. B., "Count Rumford," *Chemistry*, Vol. 39, No. 8, p. 18 (1966).

Peisen, H. S., and Torgesen, J. L., "Crystal Growth as Chemical Research," *Chemistry*, Vol. 38, No. 9, p. 15 (1965).

Lonsdale, K., "Disorder in Solids," *Chemistry*, Vol. 38, No. 12, p. 14 (1965).

"The Geometry of Liquids," *Chemistry*, Vol. 38, No. 12, p. 20 (1965).

Angrist, S. W., and Hepler, L. G., "Order and Chaos; laws of energy and entropy," Basic Books, N.Y., 1967.

"A Sun Powered Furnace," *Chemistry*, Vol. 43, No. 10, p. 23 (1970).

ATOMIC THEORY—
DALTON TO RUTHERFORD

CHAPTER 6 ━━━━━━━━━━━━━━━━━━━━━━━━━

Up to the beginning of the 20th century, the Daltonian concept of the atom was used largely in its original form. The model was successful because it explained most of the chemical and many of the physical laws of interest during this period. About the end of the 19th century, however, certain experimental facts, which were in direct conflict with the atomic model presented by Dalton, became generally known. These facts necessitated a rather severe modification in atomic theory, but many of Dalton's ideas were retained and incorporated into the more elegant theory. The resulting atomic theory, as refined in 1913 by Niels Bohr, was not only able to explain the laws that occupied the attention of Dalton, but was also able to explain many other experimental facts as well. It is the purpose of this chapter to look at some of these experimental facts in detail and see how they led scientists to a more realistic, even though still imperfect, model of the atom.

ELECTRICAL DISCOVERIES

The fact that matter can be charged electrically has been observed since ancient times, but it was not until 1600 that William Gilbert coined the word electric (from the Greek *elektron*, for amber, fossilized pine gum). It was Benjamin Franklin, 1747, who set our present definitions—that glass rubbed with silk becomes *positive* and that amber or rubber rubbed with fur becomes *negative*. In 1791, Luigi Galvani, a physician, noted the convulsions in a partially dissected frog leg when a static electric charge was allowed to flow through the leg nerves. Subsequent experiments revealed that the static charge was unnecessary and that the mere contact of nerves with pieces of metals produced the same effect. Although it was not realized at first, Galvani was successful in passing a current through an electrical circuit (the network of nerves) by making use of the energy derived from a chemical change. Galvani concluded that he was observing "animal electricity" and that it was characteristic only of living tissues. A few years later, in 1800, Alessandro Volta made the first battery, a discovery that captured the imagination of many scientists and sparked widespread investigations into electrical phenomena. Using an electric current furnished by a battery, Nicholson and Carlisle were successful in that same year in decomposing water into hydrogen and oxygen. Sir Humphrey Davy and also Michael Faraday soon followed suit with studies of the electrical decomposition (electrolysis) of numerous compounds. These experiments and other, similar ones, unfolding during the time of John Dalton, strongly suggested that a fundamental relationship exists between electricity and the structure of matter. However, Dalton's atomic theory did not encompass this relationship, and as a result, atomic and electrical theory developed concurrently, with little interaction, until about 1900.

CATHODE RAYS AND THE ELECTRON

As early as 1748, scientists were involved in the study of electrical discharges in partially evacuated glass tubes. Later, Eugene Goldstein, Sir William Crookes, and others applied high voltage to metal discs (electrodes) connected through the ends of the tubes. In this type of simple discharge tube (Figure 6–1), when the voltage is high enough, streaks of light can be observed streaming between the electrodes. Interesting and changing patterns of light are observed as the amount of gas in the tube is varied. Changing from air to other gases results in light of different colors. If the gas is almost completely removed, the colored light practically disappears. However, if a sample of matter is placed between the electrodes in such an evacuated tube, there is clear evidence that an invisible ray is passing between the electrodes. For example, a thin piece of metal foil

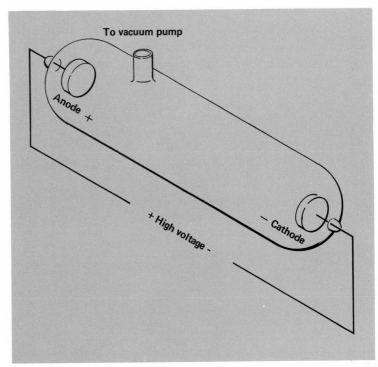

FIGURE 6–1 Simple discharge tube.

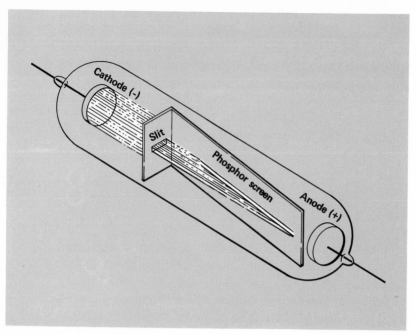

FIGURE 6-2   Discharge tube with slit and phosphor screen to show path of cathode ray.

will become red hot when subjected to the ray, and a phosphor, such as zinc sulfide, will glow with light, as do the gas samples, when exposed to the ray. These results clearly indicate that considerable energy is transmitted in the ray.

These rays are called cathode rays. They always pass from the negative electrode (cathode) to the positive electrode (anode). (In discharge tubes, the negative electrode is the cathode, and the positive electrode is the anode.) As illustrated in Figure 6–2, the path of the ray can be blocked partially with a

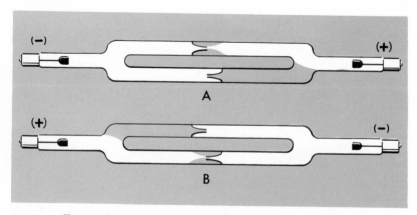

FIGURE 6-3   Discharge tube containing small amount of gas to show directional effect of cathode ray. In (A) the glowing gas shows the cathode ray to be "funneled" through the upper branch of the tube; (B) shows the reverse effect of the gas when the position of the negative electrode is reversed.

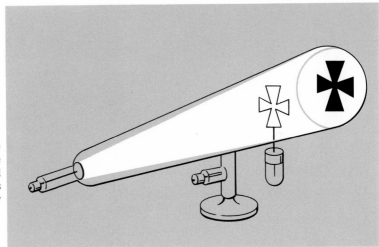

FIGURE 6-4   Discharge tube to demonstrate that cathode rays travel in straight lines and cast sharp shadows. The cross in the path of the cathode ray casts a sharp shadow.

metal plate containing a narrow slit. A metallic screen covered with phosphor is placed at an oblique angle beyond the slit. The ray that passes through the slit then strikes the screen in a straight line, and this results in a straight line of bright light on the phosphor coating. If the other electrode is made negative, this line does not appear. In this case, the entire screen glows, indicating that the ray always comes from the negative electrode. This experiment not only shows that the rays are negative in origin, but also shows that the rays travel in straight lines and cast sharp shadows. These points are also demonstrated in specially designed tubes as shown in Figure 6–4.

If a positively charged plate is placed above a cathode ray and a negatively charged plate below it, the ray will bend upward toward the positive charge. This means that the cathode ray is not light, since light is not deflected from its path in such a weak electric field. Hence, since charged particles of matter are deflected by electric fields (see Chapter 5), it must be concluded that the cathode ray is negatively charged. This same conclusion can be drawn from the behavior of the cathode ray in a magnetic field. The ray curves downward when the north pole of a magnet is brought up to the side of the tube and upward when the south pole is brought near. This is in agreement with the established behavior of a negatively charged particle moving through a magnetic field. The magnetic deflection of the cathode ray is illustrated in Figure 6–5.

A careful study of the light from a phosphor screen in the path of a cathode ray shows that the light is emitted in minute flashes. This suggests that the cathode ray is made up of discrete particles, each of which (due to its kinetic energy) is able to cause a phosphor particle to glow. Numerous experiments have shown that the cathode ray is the same, regardless of the material of which the cathode is made. This leads one to the conclusion that the cathode ray consists of particles basic to all matter. All the properties of the cathode ray can be explained if one assumes that the ray is composed of negatively charged particles with very high kinetic energy streaming out from the negative electrode. These particles are called *electrons*.

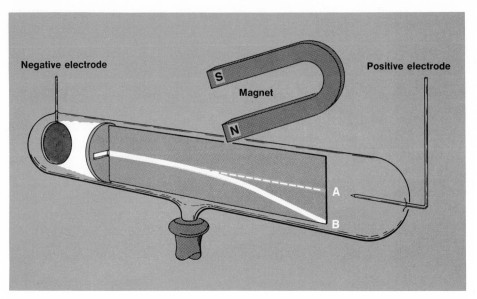

FIGURE 6-5   Magnetic deflection of cathode ray from position A to position B when magnetic field is applied.

WHAT IS THE ELECTRON REALLY LIKE?

The direct observation of the cathode-ray tube leaves two important questions unanswered: What is the mass° of the electron? What is the size of the negative charge that it carries?

In 1897, Sir Joseph John Thomson made a significant step toward answering these questions with an ingenious experiment. He made use of the basic fact that the path of a charged particle will curve when passing through a magnetic or electric field. Thomson passed a cathode ray through a magnetic field and, as expected, the stream of charged particles curved away from a straight-line path, as shown in Figure 6–6. One can readily show that the path followed, described by the radius of curvature, depends on the charge, velocity, and mass of the electron, as well as on the strength of the magnetic field. Both a larger amount of charge on the electron and a stronger magnetic field tend to increase the curvature of the path, whereas an increase in the mass of the particle or a greater velocity tends to reduce the curvature. In addition to the magnetic field, Thomson applied an electric field of such strength that the electric effect exactly opposed the magnetic effect (Figure 6–7). Hence, the electrons traveled in a straight path as indicated on the zinc sulfide detecting screen. In this balanced state a rather simple relationship (equation) could be derived from basic laws in magnetism and electricity. The equation is

$$\frac{e}{m} = \frac{E}{H^2 r} \qquad (1)$$

° Mass is a measure of the amount of matter in a body, whereas weight is a measure of the attraction of the body for the earth. The weight of the body varies with its distance from the earth, whereas the mass remains constant. Unless otherwise stated, the term "weight" refers to the earth-surface weight.

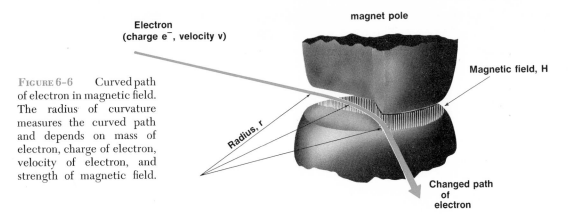

Electron
(charge e⁻, velocity v)

magnet pole

Magnetic field, H

Radius, r

FIGURE 6-6   Curved path of electron in magnetic field. The radius of curvature measures the curved path and depends on mass of electron, charge of electron, velocity of electron, and strength of magnetic field.

Changed path of electron

and relates five variables under the balanced conditions. They are: e, charge on the electron; m, mass of the electron; E, strength of electric field; H, strength of magnetic field; and r, the radius of curvature when only the magnetic field is applied. Thomson was able to measure E, H, and r, and as a result was able to calculate the charge-to-mass ratio (e/m) for the electron. Since he still had two unknowns (e and m) and only one equation, he was unable, from this experiment, to determine either. However, he did know the quotient of e divided by m (and if either value could be obtained elsewhere, he could calculate the other). The value of the ratio in this case is 5.273×10^{17} esu per gram. The electrostatic unit (esu) is a unit of charge and the gram is a unit of mass.

Thomson and a co-worker, J. Townsend, made a notable effort to measure the charge of a single electron by studying clouds composed of negatively charged

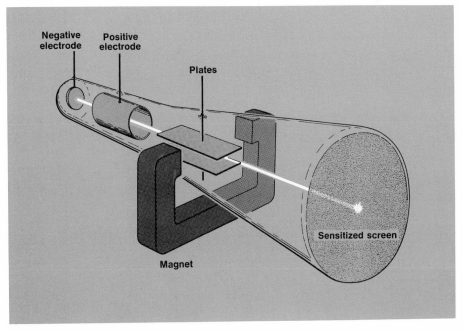

Negative electrode   **Positive electrode**

Plates

Sensitized screen

Magnet

FIGURE 6-7   J. J. Thomson experiment. Electric field, applied by plates, and magnetic field, applied by magnet, cancel each other to allow cathode ray (electron beam) to travel in straight line.

101

droplets. The weight of each droplet was estimated by the rate at which it fell through the air. The total weight of the cloud could be obtained by weighing the condensed water from the cloud. As a result, the number of drops in the cloud could be calculated:

$$\text{Total weight of water} \div \frac{\text{weight of water}}{\text{drop}} = \text{number of drops} \qquad (2)$$

Thomson *assumed* that *each drop* of water contained a negative charge composed of *a single electron*. Therefore, if he knew the number of drops, he also knew the number of electrons. Now the total charge on all the drops could be obtained by allowing the cloud to condense on an electrometer (a charge-measuring meter). Knowing the total charge and the number of electrons, he could divide to obtain:

$$\frac{\text{Total charge}}{\text{Number of electrons}} = \text{charge per electron} \qquad (3)$$

The value obtained was 6.5×10^{-10} esu.

The assumption that Thomson made of one electron per drop prevented complete acceptance of the work and it remained for Robert Millikan, with his famous oil-drop experiment in 1909, to firmly establish the charge on a single electron. Millikan's experiment consisted of measuring the number of electric charges carried by tiny drops of oil when suspended in an electric field (Figure 6–8). By means of an aspirator, Millikan distributed the oil droplets in a test chamber. Gravitational forces caused the droplets to settle slowly through the air. High energy x-rays were passed through the chamber to charge the droplets negatively (the x-ray caused the air molecules to give up electrons to the oil). By means of a beam of light and a small telescope, Millikan could study the motion of single droplets. When the electric field opposing the effect of gravity was increased enough, the droplets could be made to move upwards, and a careful adjustment of the electric field allowed Millikan to observe a droplet in a motionless state. At this point, he was able to equate the gravitational force

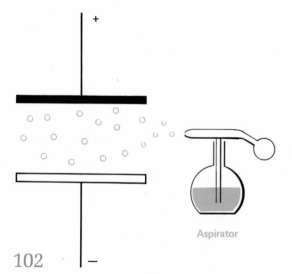

Aspirator

FIGURE 6–8   Charged oil drops, suspended as a result of opposing gravitational and electrostatic forces, provided Millikan with the means of calculating the charge on the electron.

with the electrostatic force and, hence, was able to calculate the charge carried by the droplet in order to achieve this balance. Millikan found different sizes of negative charge on different drops, but the charge measured each time was always a whole-number multiple of a very small basic unit of charge. The largest common divisor of all charges measured is 4.8080×10^{-10} esu. Millikan assumed this to be the charge on the electron.

Millikan's determination of the charge of the electron is analogous to the assignment given to a student who was asked to determine the weight of a marble when given eight sacks, each containing an unknown number of identical marbles. He was not allowed to open the sacks to count the marbles inside. Assuming the sacks to weigh a negligible amount, he weighed each sack of marbles and found them to weigh 8, 14, 18, 20, 24, 28, 36, and 40 grams. Does each marble weigh 1 gram? This is not likely, since chances are that one of the sacks would contain an odd number of marbles and weigh an odd number of grams. Weights of 3, 4, 5, 6, 7, and 8 grams are not possible since the weights of the different sacks are not multiples of any of these numbers. However, if each marble weighs 2 grams, the weights of the sacks could be explained in terms of the sacks containing 4, 7, 9, 10, 12, 14, 18, and 20 marbles respectively. The number 2 is the largest factor in *all* of the weights measured.

Now, with a good estimate of the charge of the electron and the ratio of charge-to-mass as determined by Thomson, the mass of the electron can be calculated.

$$\frac{e}{m} = 5.273 \times 10^{17} \frac{esu}{gram} \tag{4}$$

$$m\left(5.273 \times 10^{17} \frac{esu}{gram}\right) = e \tag{5}$$

$$m = \frac{e}{5.273 \times 10^{17} \frac{esu}{gram}} \tag{6}$$

$$e = 4.808 \times 10^{-10} \ esu \tag{7}$$

$$m = \frac{4.808 \times 10^{-10} \ esu}{5.273 \times 10^{17} \frac{esu}{gram}} \tag{8}$$

$$m = 9.109 \times 10^{-28} \ gram \tag{9}$$

The hydrogen atom, the lightest atom, has a mass 1837 times as large as the electron. Since no charge has ever been found smaller than the charge on the electron, since no particle of ordinary matter has ever been found which has less mass, and since electrons can be obtained from any sample of matter, it must be assumed that the electron is a universally distributed, subatomic particle, which must be considered basic in any modern atomic theory.

POSITIVE RAYS AND THE PROTON

As stated above, many of the experiments performed with cathode-ray tubes were made with tubes that contained small amounts of some gas, such as hydrogen, oxygen, or carbon dioxide. Such tubes emit colored light characteristic of the particular gas. While experimenting with such tubes in 1886, Eugene Goldstein made an interesting discovery. While using a tube with a perforated negative electrode (Figure 6–9 shows a single hole for simplicity), he noticed a glow outside the region between the two electrodes. Goldstein suspected that these rays might be made up of positively charged particles, since such particles

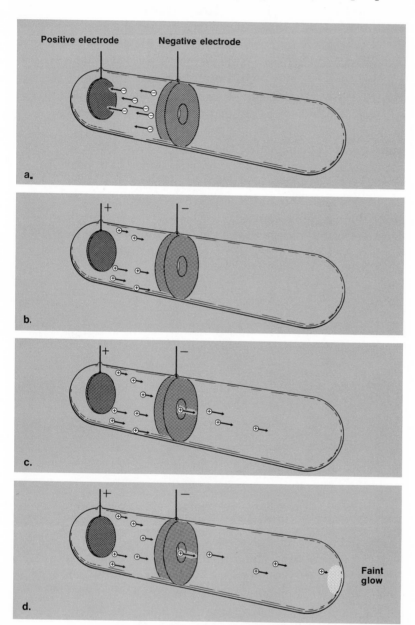

FIGURE 6-9   a, Electrons rush from negative electrode to the positive electrode due to high voltage. b, electrons collide with gas molecules to produce positive ions which are accelerated toward the negative electrode. c, Some of the positive ions escape capture by the electrode and rush through opening due to their kinetic energy. d, Some of the positive ions in the positive ray collide with the gas molecules to produce a characteristic glow.

could be produced from the gas. On rushing toward the negatively charged electrode, some of them would coast through the opening. The positive character of the ray could easily be demonstrated by its deflection toward a negatively charged plate.

By properly constructing the experimental tube, W. Wein was able to obtain a well-defined positive ray (Figure 6–9). As a result, he determined e/m values (charge-to-mass ratios) for the positive ray, just as Thomson had done for the cathode ray. He found the e/m values to be dependent on the gas used in the tube, in contrast to the e/m value for the cathode ray, which remained the same regardless of the materials of construction or the gas that happened to be in the tube. Wein also found that the e/m values for different positive rays were all thousands of times smaller than the e/m value for the cathode ray or the electron. The largest e/m value for a positive ray, however, was found when hydrogen, the lightest atom known, was used as the filler gas.

How could the observations made by Wein be explained? If the high energy electron, moving from negative electrode to positive electrode, has enough energy to break apart (ionize) the molecules and atoms composing the filler gas, fragments of the atoms will result. In the case of helium gas, for example, a moving electron may knock another electron out of a helium atom, leaving a positively charged ion (an atom or a group of atoms carrying a charge is called an *ion*). This reaction could be expressed:

$$\text{He} + e^- \xrightarrow{\text{High voltage}} \text{He}^+ + 2e^- \tag{10}$$

The resulting helium ions, some of which would move through the opening in the negative electrode, constitute the positive ray in this example.

If, as indicated in the equation above, a helium atom can be ionized (broken) into a He^+ ion and an electron, it is reasonable to conclude that the size of the positive charge on the helium ion is the same as the size of the negative charge on the electron. This is a reasonable conclusion since the two particles, when combined, form a neutral particle. Now, if the charges on the He^+ ion and the electron are the same, the e/m value for the He^+ ion would have to be smaller, since its m value is larger for the same value of e. By analogy, consider the fraction, $\frac{1}{2}$. If the denominator is made larger without changing the numerator, the size of the fraction is smaller: $\frac{1}{3}$ is smaller than $\frac{1}{2}$. If He^+ ions are thousands of times heavier than electrons, then the relative e/m values are satisfactorily explained.

The fact that hydrogen gives rise to a positive ray with the largest e/m value obtainable in such experiments suggests that the positive ion from the hydrogen atom, H^+, is a fundamental subatomic particle. This particle is called the *proton* and has essentially the same weight (obtained from e/m ratio) as the hydrogen atom. Relative weights based on the atomic weight scale are: hydrogen atom 1.007826 awu; hydrogen ion (proton), 1.007277 awu; and the electron, 0.000549 awu.

In contrast to the Daltonian, indestructible atoms, the study of positive rays offers ample evidence that atoms can be broken down and provides a basis for the characterization of another fundamental, subatomic particle, the proton. Later experiments in nuclear chemistry show that protons can be obtained from the more massive atoms, establishing the proton as a particle that must be accommodated in a modern atomic theory.

THOMSON'S ATOMIC MODEL, 1900

By 1900, the scientific community was interested in revising Dalton's model of the atom so that static electricity, cathode rays, and positive rays could be explained, at the same time retaining those ideas of Dalton which accounted for the fundamental laws of chemical combination. A rather large number of atomic models, some of which were quite reasonable, were suggested and, in turn, rejected for various reasons.

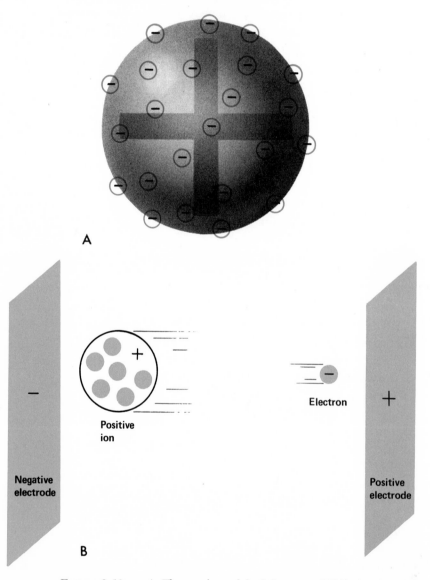

A

B

FIGURE 6-10   A, Thomson's model of the atom (1900). A sphere of a positive charge with electrons floating on the surface. B, Under electrical pressure (high voltage) Thomson's atom ionizes to form positive ion and electron.

J. J. Thomson offered an interesting model; he suggested that the atom was composed of a positively charged sphere, in which the bulk of the atom resided and upon the surface of which floated the electrons (Figure 6–10A, B). Static electricity could be explained in terms of the separation of the electrons from the bulk of the atom. Cathode rays would involve the ejection and movement of the electrons under an electrical potential, and the positive rays were streams of charged atoms (minus one or more electrons per atom). This model of the atom did not offer a satisfying explanation for the source of alpha particles which were characterized a few years later (1909).

Thomson's model is introduced, not because it is believed to be a major milestone in atomic theory on par with the Daltonian or Bohr theories, but because it is hoped that the reader will see the constant drive by the scientist to explain laboratory observations in terms of submicroscopic structure. The Thomson atom did, however, set the stage for an experiment by Rutherford that resulted in the concept of the nuclear atom.

NATURAL RADIOACTIVITY

In 1896, a French chemist named Henri Becquerel (1852–1908) was experimenting with some compounds of the element uranium. He was interested in the visible fluoresence (light emission) of these materials. In his experiments Becquerel measured the amount of light emitted from a sample by exposing the sample to a photographic film, developing the film, and observing the amount of darkening produced. On one occasion, apparently by accident, he left a sample of uranium ore in contact with an undeveloped film which was still enclosed in its light-tight envelope of black paper. When the film was developed there was evident exposure to the uranium ore. Evidently, the ore sample was emitting, in addition to the fluorescent light, a radiation that was quite capable of passing through the black paper. Further experimentation proved that it was the element *uranium* that was the source of the mysterious radiation. Later, the element *thorium* was found to exhibit the same property, and Pierre and Marie Curie followed with the famed discovery of the element *radium*. These elements, which were the source of a radiation even more penetrating than visible light, were termed *radioactive elements*.

As would be expected, there was considerable interest in the determination of the exact nature of natural radiation after the discovery of radioactive elements. It was soon shown that the radiation from radium was actually composed of three separate rays, which were called α (alpha), β (beta), and γ (gamma) rays. These rays can be demonstrated in the experiment outlined in Figure 6–11. A radioactive sample, such as uranium, is placed deep in a block of lead; the lead is necessary since it absorbs stray radiation and hence helps to focus the rays. Slits are used to select a very narrow beam of radiation. The radiation then passes between charged plates and separates, as indicated by a photographic exposure or a phosphor screen. The beta ray is evidently a stream of negative particles, since it is deflected toward a positive plate. Its behavior in a magnetic field leads to the same conclusion. A study of this ray showed that the negative particles were actually electrons, and, except for the different kinetic energies of the particles, the beta ray was the same as the cathode ray.

107

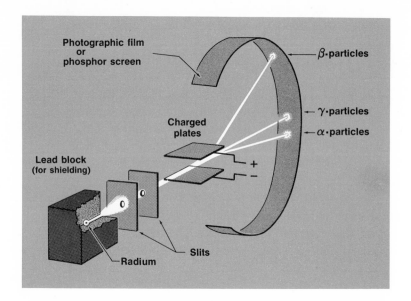

FIGURE 6-11   Separation of alpha, beta, and gamma rays by electrostatic field.

The charge-to-mass ratio was determined for the alpha ray particles. The value determined indicated these particles to be helium ions (helium atoms that have lost two electrons). This idea was supported by W. Ramsey and F. Soddy, who discovered that the element helium is always found to be present with naturally occurring radium. In 1909, Ernest Rutherford, who had first demonstrated alpha and beta rays 10 years earlier, carried out an experiment that proved beyond all doubt that alpha particles were actually helium ions. Rutherford placed the radioactive element, radon, which was known to be an alpha emitter, in a closed container and trapped the gas that was produced. The gas had the properties of helium, so Rutherford concluded that the alpha particles were helium ions and that these ions picked up electrons from the environment to form helium atoms.

$$He^{++} + 2e^- \longrightarrow He \tag{11}$$

Lord Rutherford (Ernest Rutherford, 1871–1937) was Professor of Physics at Manchester when he and his students discovered the scattering of α particles by matter. Such scattering led to the postulation of the nuclear atom. For this work he received the Nobel prize in 1913. In 1919, Rutherford discovered and characterized nuclear transformations.

TABLE 6–1   CHARACTERISTICS OF ALPHA, BETA, AND GAMMA RAYS

RAY	PARTICLE	WEIGHT	CHARGE	ENERGY
α	Helium ion	4 awu	$+2$	Low
β	Electron	0.000549 awu	-1	Medium
γ	Radiant energy	none	none	High

The gamma ray, first noted in 1900 by P. Villard, in France, has no detectable mass. It is unaffected by electric or magnetic fields and is extremely penetrating. Whereas alpha radiation can be stopped by aluminum foil and beta rays by sheet aluminum, heavy lead sheets are required to stop gamma rays. Table 6–1 summarizes some of the information about alpha, beta, and gamma rays.

The discovery of natural radioactivity left little doubt that atoms were divisible: some naturally occurring atoms spontaneously decompose, producing fragments which are either atoms of lighter elements or subatomic particles.

Natural radioactivity offers a straightforward approach to the determination of Avogadro's number, that is, the number of particles in a mole (6.02×10^{23} molecules of helium per 4 grams). If a radioactive element is an alpha emitter, it is possible to set up an experiment wherein the charges of the alpha particles produced are neutralized by electrons to produce helium gas.

$$He^{++} + 2e^- \longrightarrow He$$

By counting the number of alpha particles emitted and measuring the volume of helium gas produced, Avogadro's number can be calculated.

$$\frac{x \text{ atoms}}{\text{mole}} = \frac{\text{He atoms counted}}{\text{measured volume of He}} \times \frac{22.4 \text{ liters}}{\text{mole}}$$

ISOTOPES AND THE NEUTRON

An extension of the earlier studies of cathode and positive rays led to the development of the mass spectrometer (Figure 6–12). In this instrument, positive ions are produced from molecules or atoms by subjecting them to an electric discharge or some other source of high energy. The positive ions are accelerated by means of an electric field and then passed through a slit into a magnetic field. The slit serves to select a beam of ions. As expected, the charged particles follow a curved path in the magnetic field which is determined by the charge-to-mass ratio of the ion. The heavier ions are more difficult to deflect. If the chamber is highly evacuated, there will be no air to interfere with the ion paths. When two ions with the same charge travel through the tube, the one with the greater mass will tend to follow the wider circle. By increasing the strength of the magnetic field, it becomes possible to bring first the less massive ions, and then the more massive ions, through the second slit, producing a signal at the collector plate.

In 1912, Thomson was investigating the positive rays associated with the elemental gas neon. According to the Daltonian concept that all atoms of an element weigh the same, it would have been reasonable for Thomson to have

109

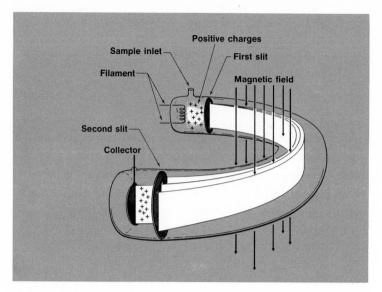

FIGURE 6-12   Mass spectrometer. Sample to be studied is injected near filament. Electrodes (not shown) subject sample to electron beam which ionizes a part of the sample. Electrodes are arranged to accelerate positive ions toward first slit. The positive ions that pass the first slit are immediately put into a magnetic field perpendicular to their path and follow a curved path determined by charge-to-mass ratio of the ion. A collector plate, behind the second slit, allows the experimenter to determine electronically when charged particles are passing through the second slit. The relative sizes of the electronic signals are a measure of the numbers of the different kinds of positive ions.

expected separate beams for Ne^+, Ne^{++}, and Ne^{+++} ions. This is true because the charge-to-mass ratio would be twice as big for the Ne^{++} ion. These expected results were obtained, but a surprising, additional observation was also made. After the Ne^+ ions were focused onto the collector, Thomson found that by increasing the magnetic field strength slightly, he picked up another signal with an e/m value that corresponded to mass 22 on the atomic weight scale. Neon has an atomic weight of 20. The amount of signal for the mass-22 particle was small and, at first, Thomson thought it was an impurity in the sample of neon. Later, Thomson and his student, F. W. Aston, considered the possibility that there were actually two kinds of neon atoms. In an experiment designed to answer this question, Aston allowed neon to diffuse through a porous clay pipe. If some neon atoms were more massive than others, they would tend to move more slowly through the pipe, and, as a result, concentrate in the gas *remaining in the pipe.* Surely enough, the gas remaining in the tube gave a much increased signal at mass 22. Aston concluded that not all neon atoms weigh the same: *most* of them have a mass of 20 and a *few* of them have a mass of 22 on the atomic weight scale; and he concluded that the average value for all of them is an intermediate value of 20.18. The flood of experimental data that followed proved beyond all doubt that the mass 22 signal was not an impurity and, furthermore, that other elements like neon have more than one kind of atom. *Atoms of the same element that differ in mass are called* isotopes.

110   As a result of some of these experiments, Rutherford concerned himself with

the problem of how it was possible to have two like-charged positive ions of neon (mass 20 and mass 22) with different masses. If the charges were the same, there would be an equal number of positive particles, protons. The appearance of different masses apparently meant the presence of a different number of some heavy, uncharged particles. In order to explain these mass differences Rutherford utilized a neutral particle called the *neutron*, first suggested by Sir J. J. Thomson. Thus, two isotopes of the same mass would contain the same number of protons but a different number of neutrons. The neutron turned out to be a rather elusive particle.

It was not until 1932 that a student of Rutherford, James Chadwick, actually detected the neutron. Chadwick found that when he subjected beryllium (and some other elements as well) to a high-energy ray of alpha particles, massive, uncharged particles were emitted. The particles discovered have essentially the same mass as a proton or a hydrogen atom. These new particles are called *neutrons* and have a mass of 1.0087 awu on the atomic weight scale.

It was evident that the solid, indestructible Daltonian atom must give way to structured atoms composed of subatomic particles. The theory must allow for heavy atoms losing or gaining charge to form ions, for two atoms of the same element having different masses, and for the source of subatomic particles. The positively charged proton with a relative mass of 1, the negatively charged electron with essentially a mass of zero, and the neutral neutron with a mass of 1 awu provide the subatomic particles for a revised atomic theory. The question remains: In what numbers and arrangements are these particles arrayed to form atoms of the elements?

RUTHERFORD'S GOLD FOIL EXPERIMENT—WHY BELIEVE IN THE NUCLEUS?

After the characterization of the alpha particle by Rutherford and his students in 1909, considerable interest was expressed in the study of interactions between alpha rays and various kinds of matter. It was found that an alpha ray could pass through a very thin piece of gold foil (4×10^{-5} cm thick) almost as if the gold foil were not present. The alpha particles were detected with a phosphor (scintillation) screen (Figure 6–13). The passage of the alpha particles through the gold foil was at first explained in terms of the Thomson atom. The alpha particles, believed to be small compared to the gold atoms, were assumed to pass through the atoms of gold in much the same way as a bullet passes through a wire fence.

However, a close examination of the intensity of the light from the scintillation screen led Rutherford to the conclusion that some of the alpha particles were being scattered or deflected from their original path (Figure 6–14). When the gold foil was removed from the path of the beam, the light image was relatively sharp and clear in contrast to the slightly larger, and fuzzier, image obtained when the foil was in place. It seemed reasonable to believe that some of the alpha particles were being deflected because they were passing through certain regions of the atoms that caused this effect, whereas other regions of the atom were different in structure, and, consequently, did not produce the

111

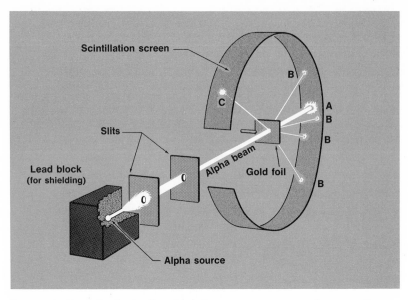

FIGURE 6-13   Rutherford's gold foil experiment. A circular, scintillation screen is shown for simplicity; actually, a movable screen was employed. Most of the alpha particles pass straight through the foil to strike the screen at point A. Some alpha particles are deflected to points B, and some are even "bounced" backwards to points such as C.

effect. The alpha ray beam then became a tool, or a probe, with which, hopefully, something could be learned concerning the structure of atoms.

Rutherford asked two of his students, Hans Geiger and Ernest Marsden, to make a systematic study of the angles of deflection of the alpha particles. To their surprise they found that some of the alpha particles were deflected at rather large angles; some of the particles were even "bounced" back toward their source (Figure 6–14). Rutherford expressed his surprise by stating he would have been no more surprised if someone had fired a 15-inch artillery shell into tissue paper and then found it in flight back toward the cannon. At first, one might consider the possibility that the positively charged alpha particle might be deflected out of its path by its attraction for a negatively charged electron on the surface of the atom. However, since the alpha particle is about 8000 times more massive than the electron, this would be somewhat similar to suggesting that a bird could knock an automobile off the highway.

Rutherford reasoned that a massive particle with a large positive charge must be somewhere within the atom if alpha particle scattering was to be explained. As a result the concept of the nuclear atom was born. If the atom had an extremely small nucleus which contained all the positive charge and most of the atomic mass, and if the negative charges of the electrons were distributed in the relatively vast space around the nucleus, most alpha particles would tend to pass through the atoms without ever coming close to a gold nucleus (Figure 6–14). If the alpha particle passed close to a gold nucleus, which is about 50 times more massive and carries a much higher positive charge than the alpha

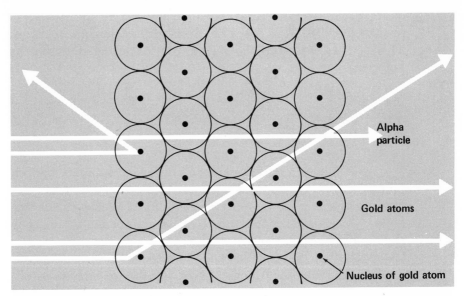

Figure 6-14   Rutherford's interpretation of how alpha particles interact with atoms in a thin gold foil. Actually, the gold foil was about 1000 atoms thick. For illustration purposes, points are used to represent the gold nuclei and the alpha particles are drawn larger than scale.

particle, the repulsion forces between the positive particles would cause the lighter alpha particle to be deflected.

How could an alpha particle be returned toward its source? The particle with a very large amount of kinetic energy will slow down if approaching a positively charged nucleus almost directly, and the kinetic energy will be stored as potential energy between the two positive particles. When all of the kinetic energy is gone, the two particles would be very close together, repelling each other with tremendous force, and with nothing restraining them from flying apart. Hence, the much lighter alpha particle will rush backwards and the stored potential energy will be converted again to kinetic energy.

Knowing the mass of the alpha particle, a good estimate of its kinetic energy, and the magnitude of the charge on both the alpha particles and the gold nucleus, Rutherford was able to calculate just how close the two positive particles would have to approach each other in order to store all of the kinetic energy as potential energy. The distance was 10^{-12} cm (0.000000000001 cm). This distance was compared to the size of the entire atom. It was a simple matter to measure the volume of one mole (197 g) of gold, and, knowing the number of atoms in a mole, find the volume of a single atom. From the volume of the space occupied by a single atom, the radius was calculated to be approximately 10^{-8} cm. This means that the radius of the atom is 10,000 times $\left(\dfrac{10^{-8}}{10^{-12}} = 10^4 = 10{,}000\right)$ greater

than the radius of the nucleus. In terms of objects the sizes of which are easily understood, if an average atomic nucleus were the size of a basketball, the electron would be about two miles from the nucleus. Rutherford concluded that the volume occupied by an atom is mostly empty space; all the positive charge and practically all the mass of the atom are contained in a nucleus at the center

113

of the atom. The electrons, equal in number to the positive charges in the nucleus, are distributed in the relatively vast reaches of space outside the nucleus. This picture is not completely in accord with the data that were uncovered later, but it was the best model at that time.

THE PERIODIC LAW AND THE ATOMIC NUMBER

After Rutherford's introduction of the nuclear atom, the problem of determining the size of the nuclear charge arose. Rutherford had an approximation of the amount of charge on the gold and silver nuclei, for example, because he was able to make an estimate of the size of the force necessary to deflect the rapidly moving, alpha particles.

At this point, as can be seen in retrospect, there occurred one of those great intellectual syntheses which brought together a great mass of previously unrelated information. Further development of atomic theory produced a model of the atom that began to explain both the physical behavior of atoms and their chemical properties from a unified viewpoint. Here we must note that Rutherford and his co-workers were fully aware of the relationships among the elements which are summarized in the periodic table or chart of the elements. This periodic table was developed over a period of fifty years (1820–1870), by a number of chemists, and ultimately was produced in a near final form by the brilliant Russian chemist, Dmitri Mendeleev. The basic notions of this table are that the chemical properties of the elements are periodic or recurring and that when the elements are arranged properly, the chemical behavior of any element can be predicted with a fair degree of accuracy. Mendeleev began with the lightest element and arrayed the elements in the order of increasing atomic weights, in recurring periods of appropriate length. When this is done properly, all the elements in a given vertical column (group) of the table have very similar chemical properties. Almost simultaneously with Mendeleev, the German physi-

Dmitri Mendeleev (1834–1907). Born in Siberia, Mendeleev rose to Professor of Chemistry at St. Petersburg (now Leningrad) and then to director of the Russian bureau of Weights and Measures. Although a prolific writer, a versatile chemist and inventor, and a popular teacher, the fame of this brilliant scientist rests on his discovery of the periodic law.

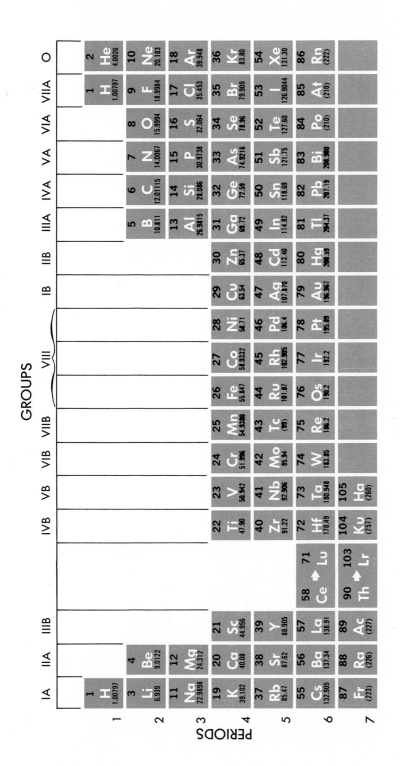

FIGURE 6-15   Periodic chart of the elements.

cist, Lothar Meyer, showed that many of the physical properties of the elements exhibit the same kind of periodicity and can be correlated with the same periodic table. A modernized version of this table is presented in Figure 6–15.

Mendeleev realized that the order of increasing atomic weights was not perfect in that there were some pairs of adjacent elements in which the element of higher atomic weight had properties expected for the preceding element in the sequence. In such cases, he always relied on the *chemical* properties in making the final decision on the exact position of such elements in the periodic table. Subsequent work has completely justified his decisions. Mendeleev saw that some elements had not yet been discovered. For a number of these he predicted detailed chemical properties which greatly facilitated their discovery.

Some of these features can be appreciated from a detailed consideration of the way the elements are arranged in Figure 6–15.

If the elements are arranged in the order of their atomic weights beginning with hydrogen, the element with the smallest atomic weight, a remarkable repetition of chemical properties occurs. Following *consecutively* the series of Li, Be, B, C, N, O, F, and Ne in order is the series of Na, Mg, Al, Si, P, S, Cl, and Ar. What is remarkable about these two series is the coincidence of chemical properties between similarly positioned members in each series of elements. For example, the first two members of one family, Li and Na, have very similar chemical properties (for example, they both form chlorides of formula MCl); Be and Mg are similar ($BeCl_2$ and $MgCl_2$), and likewise, B and Al are similar (BCl_3 and $AlCl_3$, B_2O_3 and Al_2O_3). Therefore, when the second series of elements is placed beneath the first series, each element beneath has chemical properties similar to the element directly above.

Placing subsequent series beneath the first two series, as in Figure 6-15, results in three notable exceptions to the vertical line-up of elements with similar chemical properties if atomic weight order is taken. Note that K comes after Ar, even though the atomic weight of K is greater than that of Ar. A similar reversal holds for Ni–Co and I–Te. These exceptions suggest there is a more fundamental ordering property among the elements than atomic weight. This property of the elements is the *atomic number*. It was shown by Henry Moseley in 1914 that the atomic number is truly an ordering property among the elements. This means that a listing of elements in order of atomic number is in better agreement with the systematic arrangement of the elements according to their chemical properties than is a listing in order of atomic weight. For example, the atomic number order tells us to put potassium (K) with the other active metals rather than in a family of elements all of which are gases.

Rutherford suggested that the *atomic number* might be the same as the positive charge on the nucleus, as well as the number of protons in the nucleus.

CONCLUSION AND PREVIEW

We have seen how the continual probing of the atom gradually revealed more and more information about its composition and structure. The insight of Rutherford provided a preliminary synthesis of a large number of previously unconnected pieces of information to produce a model of the atom. This model had almost all of the atomic mass concentrated in a relatively minute nucleus. The atom is also shown to be made up of a small number of more fundamental particles. The electron is present in all atoms and most of the volume occupied by an atom results from its electrons. The nucleus of the atom is composed of particles much heavier than the electron; these are the proton (a hydrogen atom nucleus) and the neutron.

The number of protons in a nucleus will be seen to determine its atomic number and the position of that element in the periodic table. Moseley's work (see Chapter 7), involving a study of the x-rays emitted from a variety of metallic electrodes, produces striking support for this. X-rays, like gamma rays and visible light, are a part of the electromagnetic spectrum. Before discussing Moseley's work, then, we will digress for some background information concerning electromagnetic energy.

QUESTIONS

1. What experimental evidence indicates that

   (a) cathode rays have considerable energy?
   (b) cathode rays have mass?
   (c) cathode rays have charge?
   (d) cathode rays are a fundamental part of all matter?
   (e) two isotopes of neon exist.
   (f) atoms are destructible.

2. Describe in detail Rutherford's gold foil experiment under the following headings:

   (a) experimental set-up
   (b) observations.
   (c) interpretations.

3. Why was Thomson's charge-to-mass ratio determination for electrons very significant although he did not determine either the charge or the mass of the electron?

4. Suppose Millikan had determined the following charges on his oil drops:

$$4 \times 10^{-10} \text{ esu}$$
$$8 \times 10^{-10} \text{ esu}$$
$$10 \times 10^{-10} \text{ esu}$$
$$14 \times 10^{-10} \text{ esu}$$
$$24 \times 10^{-10} \text{ esu}$$

   What would be the fundamental charge on an electron, assuming these are representative data?
   Suppose, some time later, someone had determined a charge of 13×10^{-10} esu; what effect would this have on the fundamental charge?

5. What part do ions play in explaining positive rays?

6. How do the following discoveries indicate that Daltonian atoms are inadequate?

   (a) cathode rays.   (c) nucleus.   (e) isotopes.
   (b) positive rays.   (d) natural radioactivity.

7. Characterize the three types of emissions from naturally radioactive substances as to charge, relative mass, and the relative penetrating power.

8. When a charged particle is sent through a magnetic field, which member of the pair would have a more curved path?

(a) an ion of charge $+1$ or an ion of charge $+2$ (both have same mass)?
(b) an ion of charge $+1$ or an uncharged particle?
(c) an ion of 2 awu or an ion of 1 awu (both have same charge)?

9. Suppose an isotope of aluminum has an atomic weight of 27.0 awu. How many protons, neutrons, and electrons are in an atom of this isotope? What is the charge on the nucleus?

10. Use the information on the periodic chart to answer the following:

(a) the nuclear charge on cadmium, Cd.
(b) the atomic number of arsenic, As.
(c) the atomic weight of an isotope of bromine, Br, having 46 neutrons.
(d) the number of electrons in an atom of barium, Ba.
(e) the number protons in an isotope of zinc, Zn.
(f) the number of protons and neutrons in an isotope of strontium, Sr, atomic weight of 88 awu.
(g) the number of electrons in a magnesium (Mg) ion of charge $+2$.
(h) the number of electrons in an oxygen (O) ion of charge -2.
(i) an element forming similar compounds to gallium, Ga.

11. Explain what the following terms mean:

(a) isotopes of an element.
(b) atomic number.
(c) an alpha emitter.
(d) chemical group.

12. If the nucleus of a gold atom is 10^{-12} cm in diameter and the diameter of the atom itself is 3×10^{-8} cm, what percentage of the volume of the atom is occupied by the nucleus? The volume of a sphere is $\frac{4}{3}\pi r^3$.

13. What causes lightning?

14. If electrons are a part of all matter, why are we not shocked continually by the abundance of electrons about and in us?

15. Sodium metal reacts violently with water and forms hydrogen gas in the process. Magnesium metal will react with water only when the water is very hot. Copper metal does not react with water. Suppose you find a bottle containing a lump of metal in a liquid and a label, "Cesium (Cs) Metal." Based on your knowledge of the periodic table, what danger is there, if any, of disposing of the metal by throwing it into a barrel of water?

16. Protons are to different atoms what _____ are to different isotopes.

17. There are more than 1000 atoms with different weights, yet there are only 105 elements. How does one explain this in terms of fundamental particles?

18. What is a practical application of cathode ray tubes?

SUGGESTIONS FOR FURTHER READING

Birks, J. B. (ed.), "Rutherford at Manchester," W. A. Benjamin, Inc., New York, 1963. (A memorial account of Rutherford's work with articles written by Bohr, Marsden, and others. Also contains reprints of Rutherford's original papers.)
Greenaway, F., "John Dalton and the Atom," Cornell University Press, Ithaca, New York, 1966.
Rich, R., "Periodic Correlations," W. A. Benjamin, Inc., New York, 1965.
Duggan, J. L., and Yeggl, J. F., "A Rutherford Elastic Scattering Experiment," *Journal of Chemical Education*, Vol. 45, p. 85 (1968).
Seaborg, G. T., "Some Recollections of Early Nuclear Age Chemistry," *Journal of Chemical Education*, Vol. 45, p. 278 (1968).

Clark, H. M., "The Origin of Nuclear Science," *Chemistry*, Vol. 40, No. 7, p. 8 (1968).

Young, J. A., and Malik, J. G., "Chemical Queries," *Journal of Chemical Education*, Vol. 45, p. 254 (1968). (The derivations of all the equations for the Millikan oil drop experiments are given.)

"Electron Microscope," *Chemistry*, Vol. 43, No. 2, p. 27 (1970).

Keller, O. L., "Predicted Properties Of Elements 113 And 114," *Chemistry*, Vol. 43, No. 10, p. 8 (1970).

Allen, W. M., "The Diagonal Periodic Relationship," *Chemistry*, Vol. 43, No. 4, p. 22 (1970).

"Spiral Form Of The Periodic Table," *Chemistry*, Vol. 43 , No. 1, p. 27 (1970).

"A Naked Electron," *Chemistry*, Vol. 42, No. 6, p. 25 (1969).

Morris, D. L., "Music Of New Spheres," *Chemistry*, Vol. 42, No. 11, p. 10 (1969).

"The Periodic Table, 1869–1969," *Chemistry*, Vol. 42, No. 5, p. 26 (1969). Zimmerman, J., "Mendeleev—His Own Man," *Chemistry*, No. 42, No. 11, p. 32 (1969).

Wallace, H. G., "The Atomic Theory—A Conceptual Model," *Chemistry*, Vol. 40, No. 10, p. 8 (1967).

Milliken, Robert, "Electrons—What They Are And What They Do," *Chemistry*, Vol. 40, No. 4, p. 13 (1967).

DEVELOPMENT OF THE QUANTUM ATOM

CHAPTER 7

Although the Rutherford model of the atom provides an explanation of the way atoms scatter alpha particles, it is not capable of explaining the observations on the ways in which atoms can pick up or give off energy. The Danish physicist, Niels Bohr, modified the Rutherford model *and accounted for the energy changes* which atoms can undergo. This was accomplished shortly after a reliable method for the determination of atomic numbers was provided by Moseley. Since both of these developments required a more detailed treatment of the electromagnetic spectrum, we must now turn to this topic and see how scientists characterize light.

A source of light is a source of energy, since light is radiant energy in motion. Light requires no medium through which to move; it can penetrate space occupied by no matter at all (a vacuum), at the very high speed of 186,000 miles per second, or 3×10^{10} centimeters per second. When light travels through forms of matter, it does so at lesser speeds, depending on the kind of matter, and some of it is absorbed. This point, coupled with the fact that light is only produced in and by matter, suggests that an understanding of how the absorption and emission are accomplished will reveal a great deal of information on the structure of matter.

The exact nature of light remains somewhat mysterious; no single theory of light has been developed to explain all of its observable properties. It appears

Niels Bohr (1885–1962) received his doctor's degree in the same year that Rutherford discovered the atomic nucleus, 1911. After studying with Thomson and Rutherford in England, Bohr formulated his model of the atom. Bohr returned to the University of Copenhagen and, as Professor of Theoretical Physics, directed a program that produced a number of brilliant theoretical physicists.

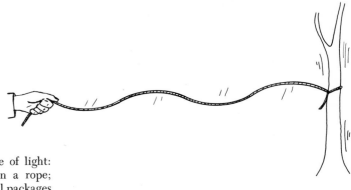

FIGURE 7-1   The dualistic nature of light: wave nature is similar to waves in a rope; particle nature is similar to individual packages of energy from a machine gun.

to have a dual character. Some of the properties of light, such as color, can be explained if light is considered to be a series of waves, like waves on the surface of water. This wave theory considers the valleys and peaks of the waves to be determined by vibrations in electrical and magnetic fields arranged at right angles to each other; hence the name, electromagnetic energy. Other properties of light can be explained only when light is considered to be a stream of particles. The discrete particles or units of energy are called *photons* or *quanta* (Figure 7-1). More will be said concerning this theory later in this chapter.

THE QUANTIZATION OF ENERGY AND THE BOHR ATOM

White light, passed through a prism, is dispersed into the various colors of the rainbow (Figure 7-2). The resulting panorama of color is called the continuous visible spectrum (see Plate II). The triangular prism, which can be

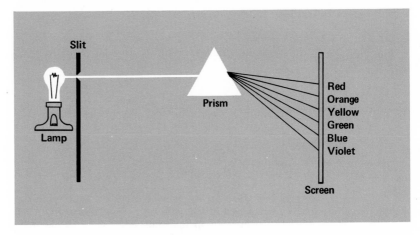

FIGURE 7-2   Spectrum from white light.

made of any transparent material, bends the path of the violet light the most and that of red light the least.

According to the wave theory of light, the different colors have different wavelengths. In traveling a given distance, one of the wavelengths in the blue region, having a shorter wavelength than red light, will trace out more waves than red light (Figure 7–3). Since all colors of light travel at the same speed, the number of waves per second, called frequency, will also be greater for blue light. Since the frequency, ν (the Greek letter nu), refers to the number of waves per second, the product of the frequency times the length of one wave, λ (lambda), is the distance traveled per second or the speed of light (c). It is important for the reader to see in this relationship that the frequency and the wavelength for a particular color of light are inversely proportional: as one increases, the other must necessarily decrease in order that their product remain equal to the speed of light (c). This relationship is given by the equation

$$c = \nu\lambda \tag{1}$$

It is important to remember that as one looks toward the blue end of the spectrum, the frequency, ν, increases and the wavelength, λ, decreases. The value of ν or λ varies continuously, so that there is not just one wavelength of red light, but many. One color then blends smoothly into the next as the spectrum is viewed (Figure 7–2).

The visible portion of the spectrum is only a very small part of the electromagnetic spectrum of radiant energy. Ultraviolet light has a higher frequency than visible light. X-rays and γ rays (mentioned earlier as an emission in radioac-

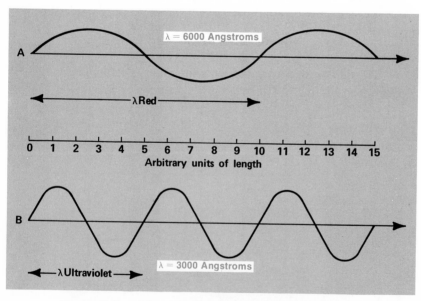

FIGURE 7–3   The wave theory of light considers light to be waves vibrating at right angles to their path of motion. Red light, A, completes one vibration or wave in the same distance and time that involve two complete waves of a given wavelength of ultraviolet light. A = Angstrom = 10^{-8} centimeter.

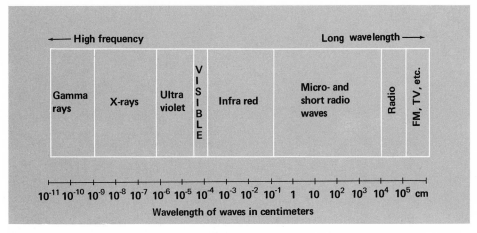

High frequency   Long wavelength

| Gamma rays | X-rays | Ultra violet | V I S I B L E | Infra red | Micro- and short radio waves | Radio | FM, TV, etc. |

Wavelength of waves in centimeters

FIGURE 7-4   Regions of the electromagnetic spectrum.

tive decay of unstable atoms) have even higher frequencies. Infrared light, microwaves, and radio waves have progressively smaller frequencies than visible light. The wavelength of an x-ray is extremely short (about the same length as the distance between atoms in molecules), whereas a radio wavelength may be as long as a room. Figure 7–4 summarizes such information about the electromagnetic spectrum.

MOSELEY'S DETERMINATION OF ATOMIC NUMBERS

X-rays can be produced by placing a metal target in the path of a high energy stream of electrons, as in Figure 7–5 (a typical x-ray tube). In addition to the broad background of radiation produced, each element produces certain wavelengths of x-rays in considerable abundance (called lines in the x-ray spectrum), and these wavelengths are characteristic of the element used as the target in the x-ray tube. Moseley measured these characteristic frequencies in the x-ray spectra of the various elements used as targets and studied the results to see if a relationship existed between the characteristic wavelengths emitted by the element and some known property of the element. Moseley found that if he

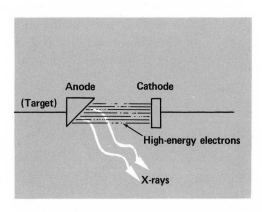

FIGURE 7-5   X-ray production.

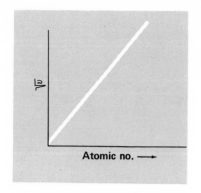

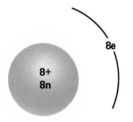

FIGURE 7-6   A plot of the square root of the frequency of certain x-rays characteristic of the anode material versus the atomic number of the element composing the anode.

plotted the square root of the frequency of the x-ray, $\sqrt{\nu}$, against the atomic number of the element, a straight line resulted, as shown in Figure 7–6. It is evident, then, that Moseley found a "probe" (x-rays) that revealed a characteristic in the structure of atoms that truly distinguishes one element from another. This characteristic is called the atomic number of the element and is identified with the positive charge on the nucleus.

With the determination of atomic numbers, it becomes possible to describe the content of the atoms in terms of three subatomic particles: the electron, the proton, and the neutron. For example, the atomic number of 8, for oxygen, indicates that there are eight protons in the nucleus, each of which has a weight of 1 awu. From the atomic weight of oxygen, 16, it can be concluded that oxygen has eight neutrons, each with a weight of 1 awu (16 minus 8) in the nucleus. In order for the atom to be electrically neutral there must be eight electrons, each with a very small weight, somewhere in the volume of the atom outside of the nucleus. This information is sometimes symbolized as shown in Figure 7–7.

FIGURE 7-7   Representation of an oxygen atom.

Consider the case of chlorine, atomic number 17 and atomic weight 35.45. This element consists of two isotopes of masses 35 and 37 (Figure 7–8). Atoms of Cl-35 are more abundant in nature, which explains why the average atomic weight is closer to 35 than to 37.

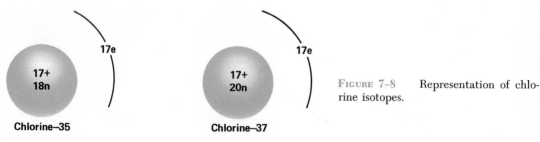

FIGURE 7-8   Representation of chlorine isotopes.

Isotopes are sometimes described by a subscript that gives the charge on the nucleus (this is redundant because the symbol of the element also gives this information) and a superscript to give the atomic weight such as:

$$^{35}_{17}Cl \text{ and } ^{37}_{17}Cl$$

These are also abbreviated as Cl-35 and Cl-37 on occasion.

PROBLEMS WITH THE RUTHERFORD ATOM

With the Rutherford model in mind (a small heavy nucleus containing the positive charge and extranuclear electrons at relatively great distances), a question immediately arises concerning the stability of the atom. Why do the negative electrons not collapse into the positive nucleus, since oppositely charged particles attract each other? To avoid this difficulty, Rutherford pictured the atom as a very small solar system; that is, the electrons orbit about the nucleus in much the same way the planets orbit about the sun. However, this is not a completely satisfying explanation because the orbiting charge presents another dilemma. In 1865, an English physicist, James Clerk Maxwell, showed by some very elegant mathematical reasoning that a moving charged body, when changing its direction in space, will radiate electromagnetic energy. A German physicist, H. Hertz, showed in 1879 that Maxwell's theory was correct. He caused charges to move back and forth (oscillate) in a wire and thereby produced the first radio waves controlled by man. A radio station operating at 600 kilocycles will have electric charges oscillating 600,000 cycles each second in its sending antenna. As a result, electromagnetic waves are emitted by the antenna. According to this established theory, the orbiting (oscillating) electron within the atom should radiate energy and run down, spiraling into the nucleus. Obviously, this is not the case.

One of the ways out of the dilemma is to postulate that Maxwell's theory and Hertz's experiments apply only to macroscopic units of charge and not to atoms. We are then forced to conclude that the macroscopic laws of the world in which man lives are not completely adequate to explain the theoretical world of atoms and molecules.

EMISSION SPECTRUM OF HYDROGEN

The beautiful colors of fireworks are produced by mixing various elements (usually in the form of one of their compounds) into the gunpowder charge. Copper in a flame produces green light; sodium, yellow; calcium, brick red; strontium, scarlet red; and so forth. If this is the case, the implication is strong that if the atoms of various elements are capable of giving off only certain wavelengths of light, this fact must somehow be related to the structures of the atoms of these elements.

In order to study the characteristic light emitted by an element, a prism can be used to disperse the light so that the actual wavelengths of the light

125

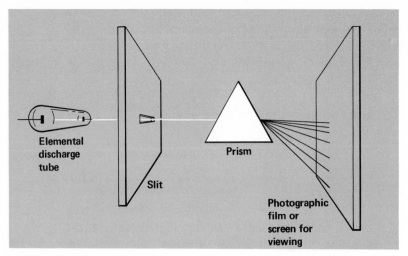

FIGURE 7-9   Basic components of spectrograph (or spectroscope) for viewing atomic spectra. The gaseous element is energized by electrical discharge. A beam of the emitted light passes through a slit. Each wavelength of light follows a different path through the prism, resulting in an image of the slit on the screen for each wavelength of light. These images are called spectral lines. From the optics of the prism, the wavelength of each line of light can be calculated.

emitted can be measured. An electrical discharge tube can be used to provide energy to the atoms so that they in turn can emit light. An instrument combining these two devices (Figure 7–9) is called a *spectroscope* if the different wavelengths (colors) of light are observed with the eye, or a *spectrograph,* if the spectrum is photographed on film.

The hydrogen spectrum is the simplest of all of the elemental emission spectra and, since hydrogen is the simplest of all the atoms, this is another clue that there is a fundamental relationship between atomic structure and atomic spectra. Notice that in the hydrogen spectrum (see Color Plate I), the lines are grouped closer together toward the blue end of the spectrum. If the photograph is extended beyond the visible region into the ultraviolet (photographic paper can "see" where the eye fails), this series of lines is observed to approach a limiting value (see line A in Figure 7–10). This series of lines is called the *Balmer series* in honor of Johann Balmer, who, in 1885, discovered a simple mathematical relationship involving the wavelengths of this series of lines in the hydrogen emission spectrum and certain integers. The Balmer equation relating these is

$$\lambda_n = b\frac{n^2}{n^2 - 4} \tag{2}$$

where λ_n = the wavelength of a spectral line

   b = a constant term = 3645.6×10^{-8}

   n = an integral number with a value of 3, 4, 5, 6, or greater. Each value of n gives a different spectral line

This relationship can also be given as

$$\frac{1}{\lambda_n} = R\left(\frac{1}{4} - \frac{1}{n^2}\right) \qquad \text{where n} = 3,4,5, \ldots \tag{3}$$

and R is a constant (109,678 cm^{-1}). If the value n = 3 is substituted into either

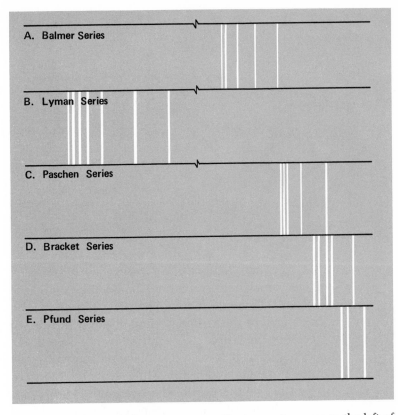

FIGURE 7-10   The hydrogen spectrum. Frequency increases to the left of the figure. The symbol ⌇ indicates that the Lyman series, in the ultraviolet region, is considerably removed from the Balmer series in the visible region.

equation and the wavelength calculated, it is found to be 6564.7×10^{-8} cm, and this is found to be the spectral line of longest wavelength in this series of spectral lines. As the value of n is increased, wavelengths corresponding to other observed lines are obtained. This is truly a remarkable achievement, especially for those who believe there is an inherently simple relationship between the structure of atoms and their properties. There is some property of the hydrogen atom which allows it to emit only certain wavelengths of light, and these wavelengths are related by a simple algebraic equation. The fact that n can have only integral values suggests that there must be a simple orderly change occurring in the atom.

Other spectroscopists—Lyman, Paschen, Brackett, and Pfund—extended the study of the hydrogen spectrum into the infrared and ultraviolet regions of the spectrum. They discovered four more series of lines similar to the Balmer series. In each series a similar, simple mathematical relationship holds; in each series integers are involved just as with the n in the Balmer equation.

PLANCK'S QUANTUM HYPOTHESIS

The observation that hot objects give off visible light is as old as man's knowledge of fire. Hot masses are also capable of giving off infrared and ultravi-

127

Max Planck (1858–1947), a German physicist, explained how energy radiates from a hot object. In 1900 he announced his quantum hypothesis which was later used to explain other natural phenomena by Einstein (photoelectric effect) and Bohr (hydrogen spectral lines). He received the Nobel Prize in 1918.

olet electromagnetic radiation. In addition, it has been observed that black objects are the most efficient absorbers and emitters of electromagnetic energy in the region of the visible portion of the spectrum. Another observation is that the color of the visible light emitted by hot objects varies with the temperature. For example, as it is heated, a piece of iron will glow first dull red, then bright red, then orange, and finally white as its temperature is increased.

Just before the turn of the twentieth century, scientists became interested in the study of just what frequencies of light and how much of each frequency is emitted by a "black body radiator" at different temperatures. A black body radiator is essentially an enclosed space containing a small opening through which the radiation in the space can be examined. Plots were made of the energy of light at each wavelength for a black body radiator as shown for two temperatures in Figure 7–11. Note two things about these curves: first, as the temperature

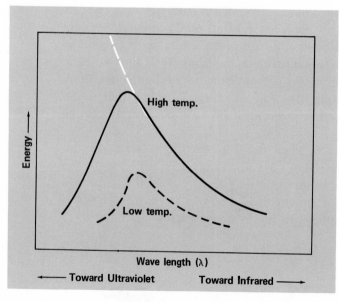

FIGURE 7–11   Energy of black body radiation versus wavelength at two temperatures. White dashed line is predicted by the wave theory.

increases, the energy at every wavelength is increased; second, shorter wavelengths (toward the ultraviolet) appear in appreciable amounts only at higher temperatures. The fact that an object has to be heated to white heat in order for the light to be richly emitted at the blue end of the spectrum is related to the fact that more energy is associated with the blue wavelengths than with the red. These plots and the color changes which occur as an object is heated can be explained if it is assumed that light is made up of discrete units, called *photons*, and that a photon of blue light contains considerably more energy than a photon of red light. In 1900, Max Planck made these assumptions in proposing a mathematical model which predicts the shape of the curves in Figure 7–11.

In 1900, it was generally believed that solid matter contained charged parts at the submicroscopic level that could oscillate. It was assumed that each of the individual oscillators could oscillate at only one frequency, and that there were enough of them to represent all the frequencies of light. This leads to the conclusion that energy could be added or lost in any amount by the oscillator. When these ideas were combined into a mathematical model, the result was the prediction that the energy of the light emitted should get greater continuously toward the ultraviolet region. Obviously, this is not the case, as the energy falls after reaching a maximum, as is shown in Figure 7–11.

Planck added one point to the black body radiator theory that solves the problem. According to Planck, the submicroscopic oscillator can receive only discrete quantities called *quanta* of energy and cannot accept intermediate values. Furthermore, the quanta of energy that can be accepted or given up bears a very simple relationship to the frequency of oscillation. The relationship is

$$\Delta E = h\nu \tag{4}$$

where ΔE (delta E or change in E) is the amount of energy in an energy change, ν is the frequency of the oscillator, and h is a constant, Planck's constant. A given oscillator can have energy of oscillation, $h\nu$, $2h\nu$, $3h\nu$, and so forth, but it can never have intermediate values. This is Planck's quantum hypothesis, wherein the *quantum* of energy (increment of energy) is the discrete package of energy $(h\nu_1)$ that can be accepted by the particular oscillator with frequency, ν_1. Mathematically, the energy of a submicroscopic oscillator is given by

$$E = nh\nu \tag{5}$$

where E is the energy of the oscillator, h is Planck's constant, ν is the frequency of the oscillator, and n is an integer (1, 2, 3, and so on).

With the quantum hypothesis substituted for the wave theory of light in black body radiation theory, Planck was able to show, theoretically, that the energy of the light should decrease at shorter wavelengths in complete agreement with experimental results (Figure 7–11). Perhaps you can see this qualitatively. Red light has a lower frequency than blue light, which means, in theory, that the oscillator which can absorb and emit red light oscillates more slowly than the blue-light oscillator. A quantum of energy, $h\nu_{red}$, then, is a smaller jump than $h\nu_{blue}$. Consequently, at the temperature of a red hot iron, the quanta available, for the most part, are just not large enough to activate the blue-light oscillator. Hence, the blue light oscillator does not receive energy and in turn does not give off light. As the temperature is increased, larger quanta are available which will now activate the blue-light oscillators. Still there will not be quanta large enough to activate ultraviolet oscillators ($h\nu_{ultraviolet}$), ν is even larger since the frequency is larger and, again, even at the higher temperatures, the energy of the light absorbed will fall sharply as one moves toward the ultraviolet (higher frequencies and shorter wavelengths).

Planck's quantum hypothesis is a remarkable idea, and there are few things analogous to it in the macroscopic world. Here is a fair comparison: Consider a town with evenly square blocks. At each intersection there is a traffic light, and all the lights in town change color (red to green) simultaneously, staying lit 1 minute on red and 1 minute on green. If a car travels at the rate of one block in 2 minutes, it could proceed through the town without stopping. However, this is not the only speed that will accomplish the nonstop trip. A speed of one block in 4 minutes will also work, but any speed between these will result in having to stop at a red light. In this example, the permissible speeds for avoiding stop lights are "quantized."

THE PHOTOELECTRIC EFFECT

As described earlier, a sufficiently high voltage will cause electrons to move through space between electrodes in a vacuum tube. If the voltage is increased almost, but not quite, to the value to cause a discharge, a photoelectric effect can be observed. If light energy is directed on the negative electrode in such a charged system, current will flow (electrons stream from the negative electrode

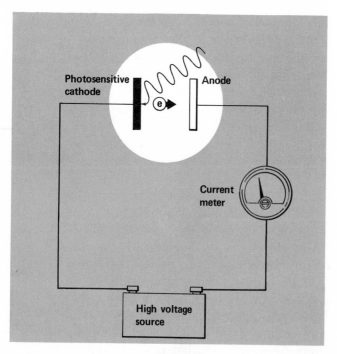

FIGURE 7-12   Photoelectric effect. Light, with sufficient energy per quantum, strikes a photosensitive cathode and causes electrons to be emitted. A high voltage source provides voltage just under the value for discharge due to voltage alone, and a current meter is used to measure current (electron flow).

130

to the positive electrode only in the presence of light) (Figure 7–12). For a particular charged photocell (just under the voltage required for a discharge caused by voltage alone), the wavelength of the light required for the discharge is a critical factor. For example, a light beam of a given intensity (brightness) in the infrared region may not cause a particular tube (photocell) to discharge, whereas another beam of equal intensity having a shorter wavelength (higher frequency, perhaps in the visible region) will cause the tube to discharge.

The photoelectric effect can readily be explained if one assumes the infrared light to be composed of small-energy photons ($h\nu_{infrared}$) and the visible light to be composed of larger-energy photons ($h\nu_{visible}$). The removal of an electron from the positive electrode requires a certain amount of energy, and the quantum of energy in the infrared photon of light is just not large enough. Some higher frequency of light would then be expected to have the correct quantum of energy, and this is the case. It would also be expected that different metals would lose electrons with differing degrees of difficulty for the same condition of electrical charge, and thus respond to different frequencies of light. This also is experimentally verified.

THE BOHR THEORY

In 1913, a Danish physicist, Niels Bohr, made a dramatic assumption that seemed, almost overnight, to answer the questions that had bothered the spectroscopists (scientists who study spectra). Why do excited atoms of an element give up only certain wavelengths of light and why are the lines in the visible, hydrogen spectrum so neatly ordered by the Balmer equation? Bohr assumed that an atom can exist in specified energy states only. The energy states possible for a given atom are separated by increments or quanta of energy. If a photon of light having the correct amount of energy ($E = h\nu$) is taken up by the atom, the atom is raised to the next highest energy level. It cannot have a stable existence between the levels, and if this quantum of light is emitted, the atom returns to the lower energy level. The wavelength of the light radiated depends on the amount of energy involved in the jump from the higher energy level to the lower one according to the equation:

$$\Delta E = h\nu \tag{6}$$

The explanation advanced by Bohr turned out to be brilliantly accurate in relating the spectrum of hydrogen and quantum theory.

Making use of Rutherford's ideas, Bohr assumed that in a stable atom an electron can travel in a given orbit about the nucleus without any loss or gain in energy. For an atom to move to the next higher energy level means that the electron must jump to another orbit further from the nucleus. In doing so, a quantum of energy would be absorbed. The size of the quantum and, thereby, the frequency of light involved, would be determined by the properties of the atom. A return of the electron to the lower orbit would result in the emission of a photon of the same energy and same wavelength. The various possible energy levels are called quantum levels. Now, if the hydrogen atom has a number of possible quantum levels, and if there is a tendency to emit energy and return

131

to lower levels, a number of energy jumps and hence, light frequencies, are possible.

If the hydrogen atom is limited to only five excited quantum levels, we would have a hypothetical atom with 15 possible downward quantum jumps. If we had multiplied millions of these excited atoms in an electrical discharge, we would expect all possible jumps to occur, and this would result in a spectrum containing 15 lines. Furthermore, we would expect the lines to be arranged in series (5 of them, with one "series" containing only one line) based on whether the downward transitions were to the lowest, second, third, and so on, energy level, as in Figure 7–13. A Bohr atom contains more energy if its electrons are more distant from the nucleus. Our hypothetical atom with six possible energy levels can then be illustrated as in Figure 7–14 with six possible radii. If this hypothetical atom is extended to just a few more quantum levels and if the quantum levels are properly spaced, by using Bohr's theory, we can account for the hydrogen spectrum in detail.

Using the Bohr structure of the hydrogen atom and basic equations of physics, it is possible to calculate the energy for the hydrogen electron in the various possible energy levels. The differences between these energy levels, ΔE in Equation 6, can be converted into light frequencies by rearranging Equation 6 into the form:

$$\nu = \frac{\Delta E}{h}$$

It is evident that we can obtain the frequencies by dividing the energy changes

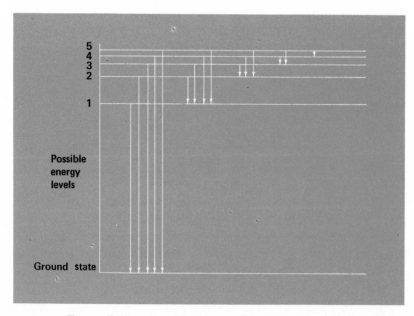

FIGURE 7–13   Possible downward energy jumps (15) for a hypothetical atom with one electron and five excited quantum levels possible in addition to the ground (unexcited) state.

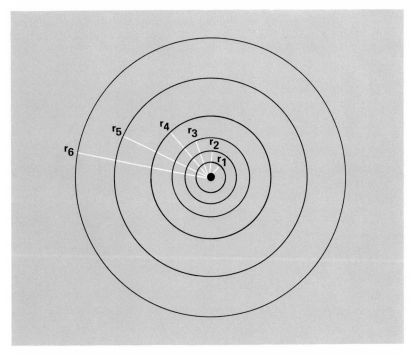

FIGURE 7-14   Six possible radii for hypothetical atom, containing one proton and one electron, and having five possible excited states.

by Planck's constant. It is important at this point to realize that the frequencies thus calculated are based on theoretical considerations. Obviously, if laboratory determinations of frequencies agree with those from theory, the theory seems accurate.

Actually, the agreement is remarkable. Table 7–1 compares the theoretical predictions and the experimental results for seven of the lines in the hydrogen

TABLE 7-1   AGREEMENT BETWEEN BOHR'S THEORY AND THE HYDROGEN SPECTRUM

QUANTUM JUMPS (IN QUANTUM NUMBERS)	SERIES	FREQUENCY PREDICTED BY BOHR'S THEORY $\left(\dfrac{\text{CYCLES}}{\text{SEC}}\right)$	FREQUENCY DETERMINED FROM LABORATORY MEASUREMENT $\left(\dfrac{\text{CYCLES}}{\text{SEC}}\right)$	SPECTRAL REGION
$2 \rightarrow 1$	Lyman	2.467×10^{15}	2.4660×10^{15}	Ultraviolet
$3 \rightarrow 1$	Lyman	2.925×10^{15}	2.9228×10^{15}	Ultraviolet
$4 \rightarrow 1$	Lyman	3.084×10^{15}	3.0827×10^{15}	Ultraviolet
$3 \rightarrow 2$	Balmer	0.4569×10^{15}	0.45681×10^{15}	Visible-red
$4 \rightarrow 2$	Balmer	0.6167×10^{15}	0.61669×10^{15}	Visible-blue-green
$5 \rightarrow 2$	Balmer	0.6908×10^{15}	0.69069×10^{15}	Visible-blue
$4 \rightarrow 3$	Paschen	0.1599×10^{15}	0.15988×10^{15}	Infrared

NOTE: These lines are typical; other lines could be cited as well with equally good agreement between theory and experiment.

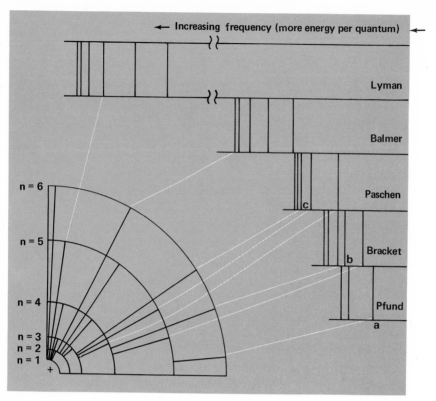

FIGURE 7-15   Theoretical transitions in Bohr hydrogen atom related to experimentally determined spectral lines. Only six quantum levels are shown for the
atom, and only some of the spectral lines are shown in each series. For simplicity,
only some of the transitions are connected with white lines to the spectral lines.

spectrum. This excellent agreement between theory and experimental results
is truly one of the significant milestones in the development of atomic theory.

The kinetic molecular theory allows the calculation of the diameter of the
H_2 molecule. This determination is completely independent of subatomic theory
and results in a value of 2.2×10^{-8} cm. for the diameter of the hydrogen molecule. If one calculates the radius from the Bohr theory, for $n = 1$, and then
doubles it to obtain the diameter of the atom, the value obtained is 1.06×10^{-8} cm. With two atoms to the hydrogen molecule, the agreement, this time
between kinetic molecular theory and the estimate from the Bohr theory, is again
reassuring.

The five series of lines in the hydrogen spectrum are readily explained in terms
of electron energy jumps to each of the first five quantum energy levels. Figure
7-15 illustrates these transitions and relates them to spectral lines.

CONCLUSION AND PREVIEW

Seemingly unrelated experiments which were prompted by the innate
curiosity of man were integrated by great scientific minds to form the basis for
modern atomic theory. Experiments in static and current electricity, electrical

discharge tubes, natural and induced radioactivity, and the photoelectric effect left little question concerning the existence of subatomic particles common to all atoms. An ordered subatomic structure through which the characteristic properties of the various atoms can be explained then becomes an immediate goal. Alpha particle scattering can be explained in terms of a nuclear atom, and x-ray emission by the elements places them in their atomic order (according to increasing nuclear charge). Matter produces and absorbs electromagnetic energy only in discrete units; the energy content of these units is related to the frequency by Planck's constant. The measured frequencies of the hydrogen spectrum are differences in atomic energy levels. These energy levels can be explained in terms of electron orbits at different distances from the hydrogen nucleus.

It will be seen in Chapter 8 that the Bohr theory is unable to explain all the properties of atoms containing many electrons. Even though this theory fails to predict exactly the spectral lines of more massive atoms, its basic pattern of electron energy levels will be preserved to give electron configurations for all the atoms. These electron structures will explain the properties of the atoms, why and how they combine, and the shapes of resulting groups of atoms. The extension of these structures in turn provides an understanding of the macro-structures (the structure in a salt crystal or a bakelite telephone case) in the macroscopic world.

QUESTIONS

1. What is a quantum? What is a photon?

2. What experimental evidence indicates that electromagnetic energy is composed of quanta?

3. How do black body radiation and the photoelectric effect indicate that electromagnetic energy is composed of quanta?

4. Distinguish between atomic number and atomic weight.

5. Which has more energy per photon, red light or blue light? How is it possible for a beam of red light to contain more energy than a beam of blue light?

6. What part do the following play in formulating the Bohr hydrogen atom?

   (a) electrons
   (b) Rutherford's nuclear concept
   (c) Planck's quantum theory
   (d) the hydrogen spectrum
   (e) Maxwell's theory

7. Distinguish between a continuous spectrum and a bright line spectrum under the two headings:

   (a) general appearance
   (b) source

8. How does the Bohr theory explain the emission of light by matter?

9. In the hydrogen atom which electron is easier to remove from the atom, an electron in the ground state or in energy level number 5? Why?

10. What is unusual about Planck's method of arriving at quanta of energy and Bohr's method of assuming energy levels in the atom?

11. Would it be reasonable to argue that macroscopic energy changes are quantized, but the quantum jumps are too small to be observed?

12. Three possible stable energies for the hydrogen atom are $-313,700$, $-78,400$, and $-34,900$ cal/g atom. In which energy level does the atom have the highest energy?

SUGGESTIONS FOR FURTHER READING

Moore, R., "Niels Bohr: The Man, His Science and the World They Changed," Alfred A. Knopf, Inc., New York, 1966.

Ihde, A. J., "The Development of Modern Chemistry," Harper and Row, New York, 1964, p. 493–507.

Gamow, G., "The Atom and its Nucleus," Prentice-Hall, Inc., Englewood Cliffs, N.J., 1961, p. 50.

Hochstrasser, R. M., "Behavior of Electrons in Atoms," W. A. Benjamin, Inc., New York, 1964.

MODERN ATOMIC THEORY

By the early 1920's, it was apparent that the Bohr theory of the atom had several serious shortcomings. Although it provided a very attractive picture of the hydrogen atom, it was not in accord with the spectra of atoms containing several electrons, and it proved incapable of explaining how atoms could form molecules. Part of its inadequacy is related to the fact that the picture which the Bohr atom develops is simply too detailed. It specifies exact orbits, velocities, and energies of the electron in the atom.

WAVE NATURE OF THE ELECTRON

The Bohr theory of the atom was very clearly a theory which was based upon the laws of physics developed to describe the behavior of particles. Often, however, problems in physics are treated not only in terms of a particle theory but also in terms of a theory based upon the behavior of waves. In many cases it is possible to develop two theories to describe a phenomenon, one based upon the behavior of particles and another based upon the behavior of waves. For example, as considered in Chapter 6, the behavior of light under some conditions is best treated by considering it to consist of a beam of photons or light "particles" (as for the photoelectric effect); and under other conditions, light is best treated by considering it to consist of a wave with a characteristic wavelength (as in diffraction, the bending of light to separate it into its component wavelengths).

If the properties of light, which have been explained for the most part in terms of a wave theory, sometimes require an explanation based on discrete units (photons), it would not be entirely unreasonable to require some wave theory in the explanation of the behavior of a particle of matter such as the electron. The basic duality in the theories describing physical behavior was first extended to particles by the French physicist, Louis de Broglie, who suggested that particles should have a characteristic wavelength determined by their momentum. He proposed the relationship:

$$\lambda = \frac{h}{mv} \qquad (1)$$

137

where m is the mass of the particle, v is its velocity, mv is its momentum, h is Planck's constant, and λ is the wavelength characteristic of the particle.

In 1927, the wave nature of moving electrons was experimentally verified independently by C. J. Davisson and L. H. Germer in America and by G. P. Thomson in Britain. In both cases, moving electrons were shown to give patterns similar to those produced when light is sent through or reflected from a diffraction grating (Chapter 7), or when x-rays are sent through crystals (Figure 8–1). Davisson and Germer reflected a beam of electrons off a nickel crystal; Thomson shot a stream of high-speed electrons through a very thin crystal of gold. From these studies it was evident, that electrons can be diffracted exactly in the same manner as light. Since the diffraction of light or x-rays is very convincing evidence for their wave nature, the diffraction of a stream of electrons indicates

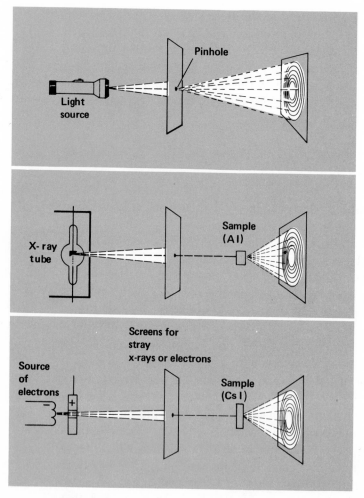

FIGURE 8–1   Similar patterns shown by light, x-rays, and electrons as each is diffracted. Diffraction is the bending and spreading of wave motion around edges. The effect is prominent when the wavelength is large compared to the size of the obstacle and small when the wavelength is short compared to the size of the obstacle. Similar effects from light, x-rays, and electrons indicate a common property to all: each has a wave nature.

at least a partial wave character for electrons. Consequently, electrons have a characteristic wavelength and frequency for a given energy.

HEISENBERG UNCERTAINTY PRINCIPLE

Another important development in the replacement of the Bohr theory was the formulation of the Heisenberg uncertainty principle by Werner Heisenberg in 1926. This very important principle sets the limits of accuracy which may be attained in simultaneous measurements of the position and velocity of submicroscopic particles such as electrons and atoms, or in simultaneous measurements of the energy and the time for such systems. It can best be appreciated from a consideration of the problems which come up when an attempt is made to measure the position and the velocity of an electron with great accuracy. In order to determine the position of the electron, it is necessary to shine light on it or bounce some particle of matter off it. The energy of the light or the collision with an electron will lead to the jarring of the electron, which produces a subsequent movement and a change in its position. The basic problem is that to measure the position of an object, it must be "touched" with some sensing device which will, if the object is small enough, disturb it. There is no sensing device yet known that can sense an electron without disturbing its position (Figure 8–2). *The limitations caused by these disturbances set the limits within which any physical theory must be set.*

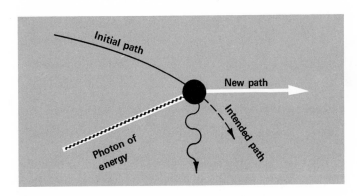

FIGURE 8–2   Illustration of Heisenberg uncertainty principle. As energy is used to sense a particle, the position, the velocity of the particle, or both, will be changed.

One of the basic flaws of the Bohr theory is that it specified the behavior of atomic systems to a far higher degree of accuracy than is physically attainable. A basic requirement for a valid theory of the atom is that it be consistent with the uncertainty principle; this is fulfilled in the development of the wave theory of the atom.

THE SCHRÖDINGER EQUATION AND STANDING WAVES

In 1926, Erwin Schrödinger developed an equation consistent with the wave properties of the electron. His basic equation was similar to an equation that is used to describe standing waves in general. An understanding of standing waves

139

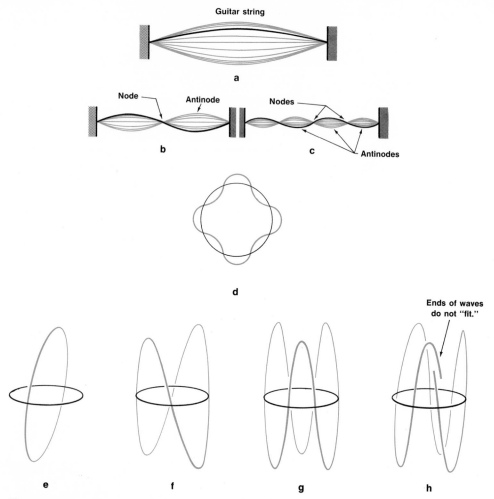

FIGURE 8-3   Standing waves, nodes, and antinodes. *a* and *e* portray fundamentals; *b*, *c*, *d*, *f*, and *g* are examples of overtones. *h* is not a standing wave. Destructive interference would destroy the antinodes.

can be appreciated from the behavior of a specific kind of wave, such as a sound wave. When a string on a guitar is plucked, it doesn't produce a variety of tones, but rather, a single tone determined by the length, thickness, and elasticity of the string. This tone is related to the standing wavelengths which are possible when a node of the wave is at both ends of the string (Figure 8–3). Both ends of the string must be fixed for standing waves to be produced. These waves are called "standing waves" because the string appears to be permanently shaped into nodes and antinodes—that is, the waves appear to be standing still (Figure 8–3). The Schrödinger equation, then, considers the electron in the atom as a standing wave of energy.

QUANTUM NUMBERS AND ENERGY LEVELS

When the concepts described above (the Heisenberg uncertainty principle, the electron as a standing wave, and the de Broglie relationship) are combined, the result is the wave-mechanical theory of the atom, usually expressed in the form of the Schrödinger wave equation.

Solving the Schrödinger equation yields energy levels and wave functions for electrons in particular orbitals in an atom, specifically in the hydrogen atom. Each orbital can contain a maximum of two electrons, and each orbital is a different stable energy level which these electrons can have within the atom. It is important to realize that specifying the orbital in which an electron is located tells us nothing about the path of an electron; if an electron is in a particular orbital, this simply means that the electron has a particular energy and a probable location within the atom. The energies of the orbitals are sub-divisions of the energies of the Bohr orbits.

Probabilities of locating electrons within an atom and quantum numbers are two properties which emerge from the solutions of the Schrödinger equation. Quantum numbers are used to describe the energy and the assignment of an electron to possible orbitals within the atom. Probabilities give the chance of finding an electron at a given location within an atom. Much of the next two chapters is devoted to showing how helpful these aspects of the wave theory are in explaining the chemical properties of matter, the periodic table, the structures of molecules, and other chemical phenomena. In the rest of this chapter, we will try to gain a clear concept of quantum numbers, orbitals, probabilities, and the consistency of the theory with the arrangement of the periodic table.

The actual solution of the Schrödinger equation involves rather sophisticated calculus and, consequently, we shall not solve it. However, several very interesting results come from its solution. In order to solve the equation for the various energies of an electron in a hydrogen atom, three integers must be substituted into the equation. Each set of three integers determines an orbital and represents an energy level of an electron in the atom. The three integers, n, ℓ, and m, are called *quantum numbers*. The value of n specifies a main energy level such as in the Bohr theory (Figure 8–4).

In its most stable state (called the *ground state*), the electron in a hydrogen atom will occupy the n = 1 level and will be strongly attracted by the nucleus. If the atom picks up enough energy the electron is excited to the n = 2 level. If it picks up still larger quanta of energy, it can be excited to a higher energy

TABLE 8-1   SOME CONSISTENT VALUES FOR *n*, ℓ, AND *m* QUANTUM NUMBERS

VALUE OF n	VALUE OF ℓ	VALUE OF m
1	0	0
2	0	0
	1	$+1, 0, -1$
3	0	0
	1	$+1, 0, -1$
	2	$+2, -1, 0, -1, -2$
4	0	0
	1	$+1, 0, -1$
	2	$+2, +1, 0, -1, -2$
	3	$+3, +2, +1, 0, -1, -2, -3$

141

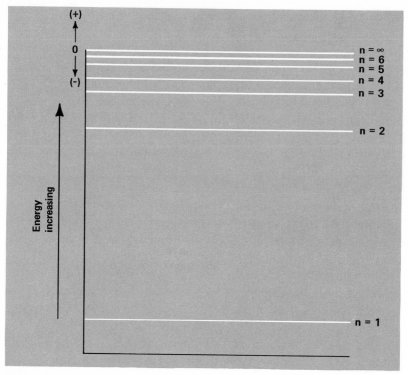

FIGURE 8-4   The energy levels of an electron for various values of the quantum number, n.

level or even removed completely from the atom, thus ionizing the atom and producing an ion plus an electron.

$$\text{H atom} \longrightarrow \text{H}^+ \text{ ion} + \text{e}^- \tag{3}$$

Values of l are limited by the value of n, and the values of m are limited by the value of l. The value of l can be any integer from 0 to $n - 1$, and m can have any integral value from $-l$ to $+l$. The limitations on l and m required for proper solutions of the Schrödinger equation are summarized in Table 8–1.

A consistent set of values for n, l, and m designates a particular orbital of an electron. For example, orbitals could be designated by n, l, and m values of 1,0,0, or 2,0,0, or 2,1,−1, or 2,1,0, or 2,1,1 but not by the combinations 2,2,0 or 2,1,2. The latter two sets would not give meaningful energies upon solution of the Schrödinger equation.

The symbol, $n l_m{}^x$, which combines n, l, m and the number of electrons (x = 1 or 2) in the orbital, is sometimes used to designate an orbital. Oftentimes the m is omitted and the symbol becomes $n l^x$. In both symbols, l is written as a letter instead of a number, as has been used up to this point. In the symbol without m, the number of electrons is the number in all orbitals having the stated values of n and l. There are $2l + 1$ orbitals in a set of orbitals having given n and l values. The letters to be used for l and the maximum number of electrons for each value of l are as follows:

Value of l	Corresponding letter designation for l	Number of orbitals	Maximum number of electrons
0	s	1	2

1	p	3	6
2	d	5	10
3	f	7	14

The letters s, p, d, f come from terms which are descriptive of spectral lines: sharp, principal, diffuse, and fundamental. Orbitals designated by comparable symbolism are shown in the following examples.

n	ℓ	m	$n\ell^x$	
1	0	0	$1s^1$	(for 1 electron)
1	0	0	$1s^2$	(for 2 electrons)
2	1	-1	$2p^2$	
2	1	0	$2p^2$	$2p^6$
2	1	1	$2p^2$	

A fourth quantum number, s, arises when the Schrödinger equation is solved by relativistic mechanics. In this case s is defined as the relative phase angle of an electron wave (that is, $\Delta S = 1$ means a phase angle lag of 180 degrees between waves of the same n, ℓ, and m). Historically, however, the concept of s arose because an electron in an atom gives effects similar to a charged particle which spins on its axis, producing a magnetic field about it. There are only two possible spin quantum numbers, $+\frac{1}{2}$ and $-\frac{1}{2}$. A given orbital can contain only two electrons, and in order to do so, the two electrons are thought to be spinning in opposite directions, that is, one has $s = +\frac{1}{2}$ and the other $s = -\frac{1}{2}$.

Now that we have a way to name the orbitals emerging from solving the Schrödinger equation, it is in order to ask, "What are the relative energies of the various orbitals?" For isolated hydrogen atoms, the energy of an electron in a main energy level, n, increases as the number of the level increases but is independent of the value of ℓ. This is in complete agreement with the Bohr theory, in which electron transitions between main energy levels are sufficient to explain completely the lines in the emission spectrum of hydrogen. However, for atoms containing several electrons, the orbital divisions within a main energy level are not necessarily of the same energy. For isolated gaseous atoms, the approximate order of increasing energy of the orbitals is 1s, 2s, 2p, 3s, 3p, 4s, 3d, 4p, 5s, 4d, 5p, 6s, 4f, 5d, 6p, 7s, 5f, 6d (Figure 8–5). The higher energy levels

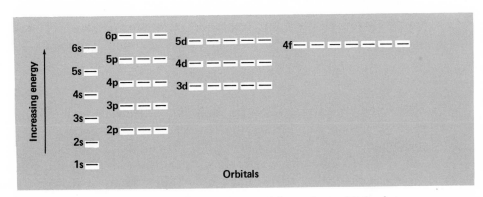

FIGURE 8–5   Diagram of the relative energies of the various orbitals of energy in an atom. Dashes represent the orbitals within a given set.

are very close together, and when occurring in atoms of high atomic number, these energy levels are often in different relative positions from the order listed. Nevertheless, this order is a good indication of relative energies of electron orbitals for the ground state of many atoms. The order of filling the orbitals beginning with the orbital of lowest energy can be written down readily by using the guide shown in Figure 8–6.

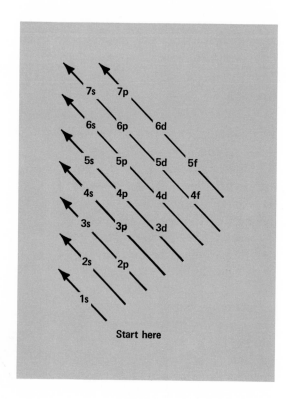

Start here

FIGURE 8–6   Guide for writing the order of filling groups of orbitals.

To deduce the order in which the various orbitals are filled, write the types of orbitals as shown in the figure and draw lines diagonally across, from lower right to upper left. Then, to derive the orbital electron structure of a particular atom, simply fill up each type of orbital in order starting with the orbital of lowest energy, 1s. For example, arsenic, As (atomic number 33), has 33 electrons, and these electrons would be placed into the following orbitals.

$$1s^2 2s^2 2p^6 3s^2 3p^6 4s^2 3d^{10} 4p^3$$

Notice the 4p set of orbitals has only the 3 electrons required to give a total of 33 electrons. This means the last set of orbitals need not be filled to its maximum.

A slightly different way of showing the order of filling the orbitals may be given schematically, as in Figure 8–5. Both systems of writing the electronic configuration of an atom can be used to construct its energy level diagram. For example, a nitrogen atom ($1s^2 2s^2 2p^3$, by the previous notation) has the following

144

diagram, where an arrow represents an electron:

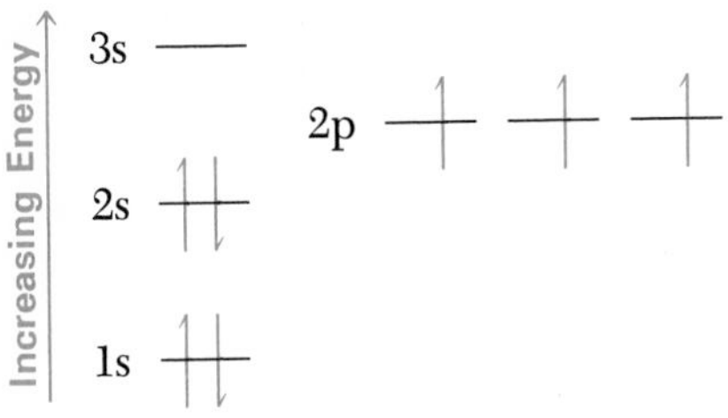

Paired electrons in an orbital are indicated by arrows pointing in opposite directions. Fluorine (atomic number 9), $1s^2 2s^2 2p^5$, is diagrammed as follows:

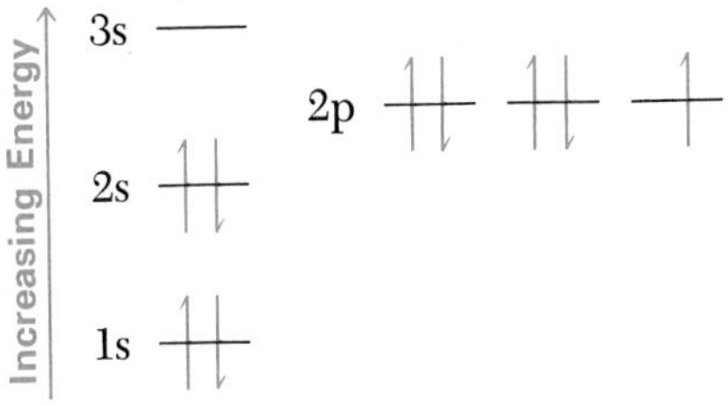

When one considers atoms in subsequent periods of the periodic table, the same schemes are used. For example, silicon (atomic number 14) is $1s^2 2s^2 2p^6 3s^2 3p^2$ and its energy levels are occupied as follows:

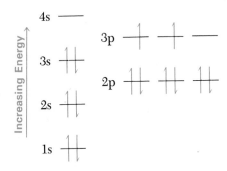

Notice that the 2p electrons of nitrogen and the 3p electrons of silicon are not paired. A state of somewhat lower energy will result if electrons are unpaired in a set of orbitals because energy is required to pair electrons (that is, to give them opposite spins). Thus, electrons will not pair in orbitals at the same energy level (same n and ℓ values) as long as there is an empty orbital in that set.

THE ORDER OF FILLING ORBITALS PREDICTS PERIODS IN THE PERIODIC TABLE

Although the ordering of the energy of orbitals has its limitations, it can predict the divisions of the periodic table remarkably well. In Chapter 5, it was pointed out that the periodic chart was originally arranged on the basis that

145

chemical and physical properties show periodicity provided that the elements are taken in the order of their atomic numbers. If the orbitals of the wave theory are filled in the order stated previously (Figures 8–5 and 8–6), periodic changes in electronic structure coincide with the periodic changes in chemical properties.

Consider the memory guide (Figure 8–5 or 8–6). Main energy level no. 1 contains only one orbital, 1s, and can contain only two electrons; hence, two elements are possible. Period 1 in the chart contains but two elements, hydrogen and helium. Main energy level (n = 2) contains two orbital sets, 2s and 2p, and can contain eight electrons; hence, eight elements are possible. Period 2 in the chart contains eight elements, lithium through neon. Note that both periods 1 and 2 begin with elements formed by adding to an s orbital; period 2 ends with an element, neon, formed by completing the filling of p orbitals.

Period 3 also contains eight elements, sodium through argon. The next orbitals in line to be filled are the 3s and 3p orbitals which can hold the eight electrons for forming these elements. Again, the period begins by adding to an s orbital and ends by filling a set of p orbitals. Period 4 contains 18 elements, potassium through krypton; the orbitals in line to be filled are the 4s (capacity = $2e^-$), 3d (capacity = $10e^-$) and 4p (capacity = $6e^-$) which can contain the 18 electrons for forming these elements. Period 5 contains 18 elements, rubidium through xenon; the orbitals in line to be filled are the 5s, 4d, and 5p, which can contain exactly the 18 electrons necessary to form these elements. Period 6 contains 32 elements; the next groups of orbitals are the 6s, 4f (capacity = $14e^-$), 5d, and 6p, holding exactly the 32 electrons necessary to form these elements. Period 7, thus far, has only 19 elements, which fill up the 7s and 5f orbitals and put 3 electrons in the 6d orbitals.

Each of these periods begins with an element formed by adding an electron to an s orbital. Each period ends with an element formed by completing the filling of a set of p orbitals. Is it any wonder, then, that the periods of elements follow other periods of elements in which similarly positioned members of the various periods have similar chemical properties? The similarly positioned members in the periods have similarly structured atoms. The elements at the first of each period have their highest energy electrons in an s orbital. The elements near the end of each period have their highest energy electrons in a p orbital. Thus, there is a trend of electronic arrangements across the periodic table as there is a trend of chemical properties (Figure 8–7).

We see that the wave theory predicts the arrangements of elements into periods of the same sort that were developed when similarities in chemical properties are considered. It is a striking confirmation of the effectiveness of the theory that it agrees so well with experimental observations.

PROBABLE POSITIONS OF ELECTRONS IN ATOMS

In addition to quantum numbers and energies, the solution of the Schrödinger equation gives information about the probable position of the electron within an atom. Because we must accept the Heisenberg uncertainty principle and the limitations that it places upon the kind of information that

Periodic table (Groups I–VIII, Periods 1–7):

Period	I	II	III	IV	V	VI	VII	VIII (Fe/Co/Ni triads)			I	II	III	IV	V	VI	VII	VIII
1	1 H $1s^1$																	2 He $1s^2$
2	3 Li $2s^1$	4 Be $2s^2$											5 B $2s^2 2p^1$	6 C $2s^2 2p^2$	7 N $2s^2 2p^3$	8 O $2s^2 2p^4$	9 F $2s^2 2p^5$	10 Ne $2s^2 2p^6$
3	11 Na $3s^1$	12 Mg $3s^2$											13 Al $3s^2 3p^1$	14 Si $3s^2 3p^2$	15 P $3s^2 3p^3$	16 S $3s^2 3p^4$	17 Cl $3s^2 3p^5$	18 Ar $3s^2 3p^6$
4	19 K $4s^1$	20 Ca $4s^2$	21 Sc $4s^2 3d^1$	22 Ti $4s^2 3d^2$	23 V $4s^2 3d^3$	24 Cr $4s^1 3d^5$	25 Mn $4s^2 3d^5$	26 Fe $4s^2 3d^6$ / 27 Co $4s^2 3d^7$ / 28 Ni $4s^2 3d^8$			29 Cu $4s^1 3d^{10}$	30 Zn $4s^2 3d^{10}$	31 Ga $4s^2 4p^1$	32 Ge $4s^2 4p^2$	33 As $4s^2 4p^3$	34 Se $4s^2 4p^4$	35 Br $4s^2 4p^5$	36 Kr $4s^2 4p^6$
5	37 Rb $5s^1$	38 Sr $5s^2$	39 Y $5s^2 4d^1$	40 Zr $5s^2 4d^2$	41 Nb $5s^1 4d^4$	42 Mo $5s^1 4d^5$	43 Tc $5s^2 4d^5$	44 Ru $5s^1 4d^7$ / 45 Rh $5s^1 4d^8$ / 46 Pd $4d^{10}$			47 Ag $5s^1 4d^{10}$	48 Cd $5s^2 4d^{10}$	49 In $5s^2 5p^1$	50 Sn $5s^2 5p^2$	51 Sb $5s^2 5p^3$	52 Te $5s^2 5p^4$	53 I $5s^2 5p^5$	54 Xe $5s^2 5p^6$
6	55 Cs $6s^1$	56 Ba $6s^2$	57 La $6s^2 5d^1$ / ⊗ / 72 Hf $6s^2 5d^2$	73 Ta $6s^2 5d^3$	74 W $6s^2 5d^4$	75 Re $6s^2 5d^5$	76 Os $6s^2 5d^6$ / 77 Ir $6s^2 5d^7$ / 78 Pt $6s^1 5d^9$				79 Au $6s^1 5d^{10}$	80 Hg $6s^2 5d^{10}$	81 Tl $6s^2 6p^1$	82 Pb $6s^2 6p^2$	83 Bi $6s^2 6p^3$	84 Po $6s^2 6p^4$	85 At $6s^2 6p^5$	86 Rn $6s^2 6p^6$
7	87 Fr $7s^1$	88 Ra $7s^2$	89 Ac $7s^2 6d^1$ / ⊕ / 104 Ku $7s^2 6d^2$	105 Ha $7s^2 6d^3$														

⊗ - At this point in the periodic table, the 4f orbitals begin to be filled. The series of elements between lanthanum (atomic no. 57) and hafnium (atomic number 72) is called the lanthanides and consists of:

58 Ce $6s^2 4f^2$	59 Pr $6s^2 4f^3$	60 Nd $6s^2 4f^4$	61 Pm $6s^2 4f^5$	62 Sm $6s^2 4f^6$	63 Eu $6s^2 4f^7$	64 Gd $6s^2 4f^8$	65 Tb $6s^2 4f^9$	66 Dy $6s^2 4f^{10}$	67 Ho $6s^2 4f^{11}$	68 Er $6s^2 4f^{12}$	69 Tm $6s^2 4f^{13}$	70 Yb $6s^2 4f^{14}$	71 Lu $6s^2 5d^1 4f^{14}$

⊕ - At about this point in the periodic table the 5f orbitals begin to be filled. The series of elements involved in this process are called the actinides and consists of:

90 Th $7s^2 6d^1 5f^2$	91 Pa $7s^2 6d^1 5f^2$	92 U $7s^2 6d^1 5f^3$	93 Np $7s^2 6d^1 5f^4$	94 Pu $7s^2 6d^1 5f^6$	95 Am $7s^2 5f^7$	96 Cm $7s^2 6d^1 5f^7$	97 Bk $7s^2 5f^9$	98 Cf $7s^2 5f^{10}$	99 Es $7s^2 5f^{11}$	100 Fm $7s^2 5f^{12}$	101 Md $7s^2 5f^{13}$	102 No $7s^2 5f^{14}$	103 Lw $7s^2 6d^1 5f^{14}$

FIGURE 8-7   The periodic table including the atomic numbers and the outermost electrons of each element.

147

we can validly possess, the picture of the atom must reflect this lack of precision. Rather than knowing precisely where the electron is located, only the probability of finding the electron in a given volume of space around the nucleus can be calculated. As a consequence, the representations of the atom portray the different ways in which probabilities of finding the electron can be presented. The mathematics involved allows the calculation of the relative probability of finding the electron at a given distance from the nucleus of the atom, as shown in Figure 8–8.

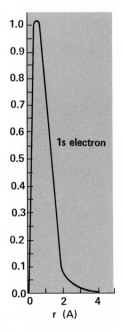

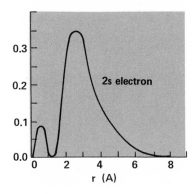

FIGURE 8–8   Relative probability of finding a 1s and a 2s electron at various distances from the nucleus in a hydrogen atom.

In order to obtain a pictorial representation it is usual to plot the surface that will bound the volume in which the electron will be expected to be found most of the time (e.g., 90 per cent of the time). These sketches are important because they tell the preferred locations of the electrons and the way in which these electrons may be expected to influence chemical behavior. It should be emphasized that these shapes are not the paths of the electrons, but are simply the boundaries within which an electron with a particular energy can most likely be found. These shapes are to an electron what a cage might be to a lion. A large percentage of the time, the lion is expected to be found within the bounds of the cage, but giving the bounds does not tell much about the movement of the lion within its bounds. The shapes and spatial orientations of the orbitals are designated by the letters s, p, d, and f and, where necessary, with a subscript which relates to the spatial orientations (Figure 8–9). An important aspect of orbital geometry is that an electron is not as likely to be found in some directions out from the nucleus as in other directions. In the case of the s orbitals, which are spherically symmetrical, there is equal probability of finding the electrons, regardless of the direction taken from the nucleus. However, a 2s orbital is farther from the nucleus than a 1s orbital. The three different p orbitals in a set differ only in their orientations about the nucleus, being mutually perpendicular to

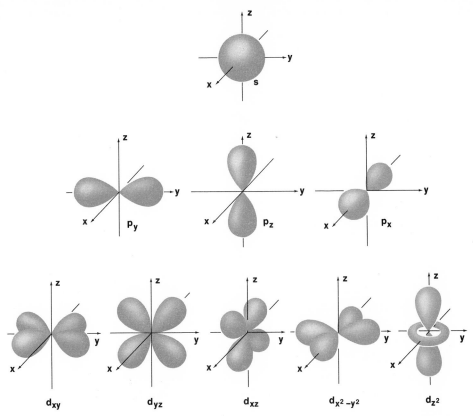

FIGURE 8-9   Spatial orientations of s, p, and d orbitals for hydrogen-like atoms.

each other. The five d orbitals have orientations and shapes as shown in Figure 8-9. In each p and d orbital, it is more probable that one might find an electron in a particular direction away from the nucleus than in other directions. The validity of these theoretical shapes and spatial orientations will be partially established if they can be used to explain how atoms unite to form molecules and, in turn, if they agree with the shapes of these molecules. We shall look for this in Chapters 9 and 10.

SUMMARY OF THE WAVE MECHANICS PICTURE OF THE ATOM

The working model of the atom, according to wave mechanics, involves the following aspects:

(a) An acceptable theoretical treatment of the atom consistent with both the wave nature and the particle nature of the electron is obtained.

(b) Sufficient energy levels for an adequate number of transitions to explain the lines in the spectrum of an element are provided.

(c) It is no longer possible to speak of the exact position of the electrons in an atom, but only of the relative probability of finding an electron in a given volume. In the same manner, there are limits on the accuracy with which the

149

velocity of the electron, its energy, and other properties of the atom can be specified.

(d) According to the Heisenberg uncertainty principle, if one of the two quantities to be measured simultaneously (for example, energy and time, or position and velocity) is specified rather precisely, a large error appears in the other quantity. If a large error in time is allowed, energies can be specified more exactly. This usually allows the specification of the nearly exact energy of the electrons in the various orbitals.

(e) The chief factors of chemical interest about the electronic orbitals are their energies, shapes, and positions relative to each other. The shapes of the orbitals are important in that they tell the directions in which electronic density is concentrated.

(f) The rules which dictate the manner in which orbitals are to be filled are basic in explaining the periodic table as these rules determine the number of elements in each period and the recurrence of similar electronic structures for elements having similar chemical activity.

(g) Since the wave mechanical model is the most consistent theory of the atom yet developed, any modern theory of chemical binding between atoms should be built on this theory. That is, the electronic orbitals which are available should be used as a starting point in getting a picture of any molecule. Any changes which are made on the atomic orbitals to build up the molecular picture must be consistent with the known energy changes occurring in chemical reactions. It will be shown in subsequent chapters that the electrons in the outermost orbitals serve to describe the electronic properties of molecules.

QUESTIONS

1. Give two shortcomings of the Bohr theory.

2. What two new discoveries or concepts made it necessary to originate a new atomic theory to replace the Bohr theory?

3. Give experimental evidence for the wave nature of the electron.

4. Which exhibits standing waves: the ringing of a bell or the breaking of a glass?

5. (a) Draw the general shape of an s orbital and a p orbital, separately, and then draw three p orbitals in proper relationship to each other.
   (b) What do the shapes of these orbitals signify?

6. Using the notations, $1s^2$, and so forth, write the ground state, electronic arrangement of lithium (Li), oxygen (O), sodium (Na), and chlorine (Cl).

7. Distinguish between the meaning of a Bohr orbit and the wave mechanical orbital.

8. (a) Neon, argon, krypton, xenon and radon form a group of elements which are similar in that they form very, very few compounds. From their atomic structures, suggest a reason for this similarity in relative inactivity.
   (b) How is the atomic structure of helium similar to the structures of the other members of this group?

9. In which orbital does the electron have the greater energy?

   (a) 1s or 2s   (c) 5s or 5f
   (b) 2s or 2p   (d) 3s or 3d

10. The Heisenberg uncertainty principle can be written as

$$\Delta p \Delta v = \frac{h}{2\pi m}$$

where Δp is the uncertainty in the position of a particle, and Δv is the uncertainty in the measurement of its velocity, m is the mass of the particle, and h is Planck's constant. What happens to the magnitude of the uncertainties as the body becomes more massive?

11. A student is arrested for speeding. He claims that the Heisenberg uncertainty principle makes it impossible for the arresting officer to determine his speed and his position at the same time. The judge calls you as an expert witness. What would be your expert opinion?

12. How many electrons (maximum number) are there in each of the following?

(a) an s orbital
(b) a d orbital
(c) a p orbital
(d) a set of p orbitals
(e) n = 2 main energy level
(f) n = 4 main energy level

13. What is the significance of this equation?

$$\lambda = \frac{h}{mv}$$

14. What is wrong with the following attempts to write electronic configurations of atoms?

(a) $1s^2 2s^3$
(b) $1s^2 1p^6 2s^2 2p^6$
(c) $1s^2 2s^2 3s^2 3p^6 4s^2 4p^6$
(d) $1s^2 2s^2 2p^8 3s^2 3p^8$

SUGGESTIONS FOR FURTHER READING

Hoffman, B., "The Strange Story of the Quantum, " Dover Publications, Inc., New York, 1959.

Kaufman, E. D., "Advanced Concepts in Physical Chemistry," McGraw-Hill Book Company, New York, 1966, pp. 31–49.

Lambert, F. L., "Atomic Orbitals from Wave Patterns," *Chemistry*, Vol. 41, No. 2, p. 10 (1968).

Sherwin, C. W., "Introduction to Quantum Mechanics," Holt, Rinehart & Winston, Inc., New York, 1959, pp. 1–101.

CHEMICAL BONDING

CHAPTER 9 ━━━━━━━━━━

INTRODUCTION

From our studies in previous chapters we have seen how the wave-mechanical picture of the atom can predict the electronic configurations of multielectron atoms and can explain some of their properties, such as atomic spectra. One very important aspect of atomic theory is its ability to explain the forces that hold atoms together in molecules, molecules together in molecular substances, and ions together in ionic structures. A fixed link between submicroscopic particles, resulting from attractive forces between them, is called a *chemical bond* (Figure 9–1). Our knowledge of chemical bonding has emerged through the interaction of two parallel lines of development: (1) the investigation of the properties of substances, and (2) the outgrowth of theories of chemical bonding based on concepts of atomic structure.

Studies of the properties of substances, such as chemical reactivity, volatility, melting point, electrical conductivity, and color, can often give some indication as to how submicroscopic particles are bonded together. For example, since

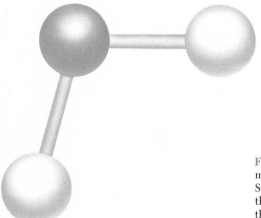

FIGURE 9-1   The two chemical bonds in a water molecule are often represented by a ball-and-stick model. Such models tell which atoms are bonded together and the angles involved, but give no information as to why the atoms are bonded in a particular pattern, the relative sizes of the atoms, or the distances between them.

melting involves a transition to a state in which atoms or molecules are less firmly bound to their neighbors, a high melting point implies that a solid is held together by very stable chemical bonds. As we shall see shortly, salts, metals, and compounds composed of a network of tightly bound ions or atoms tend to have relatively high melting points. The volatility of a substance offers another example. Certain compounds are gases at room temperature, and yet they are composed of molecules made up of several different kinds of atoms bonded together. For example, in the case of carbon dioxide, CO_2, we must assume that the bonding between molecules (intermolecular bonding) is slight, since it takes relatively little energy to break up the solid CO_2, whereas the bonding within the molecules (intramolecular bonding) must be quite great since gaseous CO_2 does not decompose appreciably at temperatures as high as 1000°C.

There has been a continuing effort to develop theories which can explain the nature of chemical bonding. These theories have been advanced in the firm belief that there is a fundamental relationship between the structure of matter and its properties, and that the properties of all macroscopic materials are a direct consequence of the submicroscopic bonds between the atoms, molecules, ions, or some combination of these. Successful theories have been advanced to establish detailed structure-property relationships for known substances and to predict properties for new substances. The theories of bonding that will be discussed in this chapter are based largely on the wave-mechanical theory.

TYPES OF CHEMICAL BONDING

The forces that hold atoms together in various types of substances are of five major types, each having its own characteristics. These types of bonds, along with common materials in which they occur, are given below.

1. Ionic bonding   Salts
2. Covalent bonding   Molecular compounds, such as water, and in polymers, such as polyethylene
3. Hydrogen bonding   Water, ammonia, and large molecules in living organisms
4. Van der Waal's attractions   The forces that hold molecules together in some liquids and solids, such as liquid helium and solid CO_2.
5. Metallic bonding   Metals and alloys

We shall now look at these types of chemical bonds in more detail, with a major emphasis on describing various types of compounds and accounting for some of their properties in terms of the bonding that holds them together.

Ionic Bonding

Ions and Ion Formation

Charged atomic or molecular-sized particles are called *ions*. They may be positively charged, as Na^+ (sodium ion), or negatively charged, as Cl^- (chloride ion); multicharged ions, such as Ca^{2+} (calcium ion) and O^{2-} (oxide ion), are also encountered. Recall the necessity for postulating ions in order to explain the gaseous conductance of electricity and alpha rays in Chapter 6.

Since electrons are in the outermost parts of the atom, it is reasonable to assume that electrons are transferred between particles to form ions. There are then two ways for ions to be formed: electron loss (to form positive ions) and electron gain (to form negative ions). Let us examine the formation of positive ions first.

Metals have electronic configurations in which there are three or less electrons in the outermost principal energy level. These electrons have a high probability of being relatively far from the nucleus, as shown for sodium in Figure 9–2. These outermost electrons are s and p electrons for the IA, IIA, and IIIA Group metals. Since they are far from the influence of the positively charged nucleus, they have the highest energy of any electrons in the atom. Remember that the potential energy of an electron-nucleus system increases as the distance between the nucleus and the electrons is increased. Thus, it would take a minimum amount of additional energy to completely remove these electrons from the atom. Since metal atoms are neutral (uncharged), the loss of a negative charge results in a positive charge on the ion.

$$Na \longrightarrow Na^+ + 1e^-$$

The energy required to completely remove each electron from an atom is called its ionization energy (for the first electron lost, the first ionization energy; for the second electron lost, the second ionization energy; and so on). Metals have low ionization energies; nonmetallic elements generally have higher ionization energies. Figure 9–3 shows how the first ionization energy varies with atomic number.

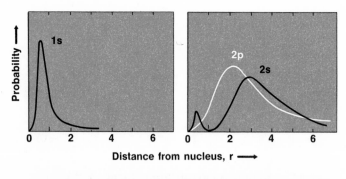

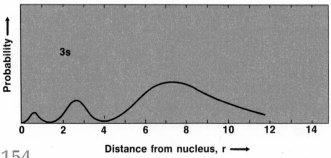

FIGURE 9–2   Probable distances from the nucleus for the 1s, 2s, 2p, and 3s electrons of a sodium atom.

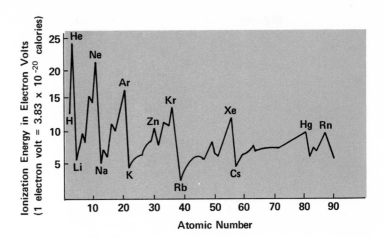

FIGURE 9-3   First ionization
energies of the elements.

This periodic behavior of ionization energy is directly related to the electronic configuration of the elements. For example, the elements known as the noble gases (helium, neon, argon, krypton, xenon, and radon) are at the end of each upward trend. These elements have the highest ionization energies in each period of elements and do not readily enter into chemical change. Indeed, they are called noble gases because of this indifference to chemical combination. It is reasonable to assume that both the high ionization energies and the relative chemical inertness are related to the ns^2np^6 electronic configuration common to this group of elements. (Note that helium has no p electrons; its configuration is $1s^2$.) We must also conclude that the electronic configurations of both metals and nonmetals represent more chemically reactive structures than the ns^2np^6 structure of the noble gas.

In the formation of positive ions from metal atoms, a question might be asked concerning the number of electrons a given metal atom will lose in an ordinary chemical process. The answer to this question should be determined by the electronic configuration of the metal atom. If the metal atom has only one outermost s electron, it could lose that electron, form a singly positive ion, and at the same time achieve a very special electronic configuration, the configuration of the noble gas preceding it in the periodic chart. Consider lithium (Li, atomic number 3), with a configuration of $1s^2 2s^1$. Loss of the 2s electron gives a lithium ion, Li^+, with a $1s^2$ configuration, which is the electronic configuration of the noble gas, helium.

$$Li \longrightarrow Li^+ + 1e^-$$
$$1s^2 2s^1 \qquad 1s^2$$
(Helium configuration)

Likewise, sodium (Na, atomic number 11), by the loss of its $3s^1$ electron, gives a sodium ion, Na^+, with the configuration of its noble gas neighbor, neon.

$$Na \longrightarrow Na^+ + 1e^-$$
$$1s^2 2s^2 2p^6 3s^1 \qquad 1s^2 2s^2 2p^6$$
(Neon configuration)

A study of the electronic configurations of each of the Group I elements, the   155

TABLE 9-1   ELECTRONIC CONFIGURATIONS OF THE ALKALI METALS

ELEMENT	ATOMIC NUMBER	CONFIGURATION
Li	3	$1s^2 2s^1$
Na	11	$1s^2 2s^2 2p^6 3s^1$
K	19	$1s^2 2s^2 2p^6 3s^2 3p^6 4s^1$
Rb	37	$1s^2 2s^2 2p^6 3s^2 3p^6 4s^2 3d^{10} 4p^6 5s^1$
Cs	55	$1s^2 2s^2 2p^6 3s^2 3p^6 4s^2 3d^{10} 4p^6 5s^2 4d^{10} 5p^6 6s^1$

alkali metal family, reveals that each of them has the possibility of losing an ns^1 electron to form a singly charged positive ion (Table 9-1).

It appears from the electronic structures of the alkali metals that ions tend to be formed by attaining a closed-shell configuration (the noble gas configuration) by the loss of electrons. This principle applies to some other metals as well. Calcium (Ca, atomic number 20), for example, does this by the loss of two electrons. The stable Ca^{2+} ion has the configuration of the noble gas argon (Ar, atomic number 18, $1s^2 2s^2 2p^6 3s^2 3p^6$).

$$Ca \longrightarrow Ca^{2+} + 2e^-$$
$$1s^2 2s^2 2p^6 3s^2 3p^6 4s^2 \qquad 1s^2 2s^2 2p^6 3s^2 3p^6$$
(Argon configuration)

All nonmetals have relatively high first ionization energies. The energies available in ordinary chemical change are simply not great enough to form positive ions from nonmetallic atoms. In order for nonmetals to achieve the generally unreactive state of the noble gases, it is reasonable to assume that they would tend to gain electrons. Thus, they could fill their partly empty outer orbitals[*] and, in so doing, attain a stable closed-shell configuration of a noble gas.

Consider an atom of fluorine (F, atomic number 9). Its electronic configuration is $1s^2 2s^2 2p^5$ with one p orbital being half filled ($2p^5 = 2p^2 2p^2 2p^1$). If the fluorine atom gains one electron, a fluoride (F^-) ion would be formed with the closed-shell configuration of neon ($1s^2 2s^2 2p^6$).

$$F + 1e^- \longrightarrow F^-$$
$$1s^2 2s^2 2p^5 \qquad 1s^2 2s^2 2p^6$$
(Neon configuration)

Oxygen (O, atomic number 8) atoms have an electronic configuration $1s^2 2s^2 2p^4$, with two half-filled 2p orbitals ($2p^4 = 2p^2 2p^1 2p^1$). This means that by the gain of two electrons, an oxygen atom could form an oxide (O^{2-}) ion with the stable neon configuration.

$$O + 2e^- \longrightarrow O^{2-}$$
$$1s^2 2s^2 2p^4 \qquad 1s^2 2s^2 2p^6$$
(Neon configuration)

[*] The orbitals in the outermost energy level (highest value of n) are called valence orbitals and the electrons therein are termed valence electrons. The word valence has historically been used to denote chemical bonding.

Thus far we have reasoned, based on theoretical atomic structures, ionization energies, and the inert character of the noble gases, that metals should form positive ions by losing electrons and nonmetals should form negative ions by gaining electrons. More specifically, we reasoned that Na, Ca, F, and O tend to form Na^+, Ca^{2+}, F^-, and O^{2-} ions, respectively. There is ample experimental evidence that this is actually the case. For example, consider the formulas of the compounds that can be formed from these elements and the equations for the reactions involved:

$$NaF: \qquad 2Na + F_2 \longrightarrow 2Na^+ + 2F^-,$$
$$CaF_2: \qquad Ca + F_2 \longrightarrow Ca^{2+} + 2F^-,$$
$$Na_2O: \qquad 4Na + O_2 \longrightarrow 4Na^+ + 2O^{2-},$$
$$CaO: \qquad 2Ca + O_2 \longrightarrow 2Ca^{2+} + 2O^{2-},$$

Note that in each case the formula is compatible with the ideas developed previously, namely:

(a) a sodium atom loses 1 electron to form Na^+,
(b) a calcium atom loses 2 electrons to form Ca^{2+},
(c) a fluorine atom gains 1 electron to form F^-, and
(d) an oxygen atom gains 2 electrons to form O^{2-}.

X-ray studies, as well as chemical properties, indicate that these four compounds are ionic solids. When molten, all four compounds conduct electricity, which is characteristic of mobile ions (see Chapter 11). Mass spectrographic studies leave little room for doubt that these ions have a real and stable existence and that they actually have the charges indicated.

Further considerations of ionic structures along these lines lead to four important conclusions:

(1) There is a tendency for atoms with ns^1, ns^2, or ns^2np^1 configurations (less than four electrons in outer energy shell) to lose electrons and form positive ions with an electronic configuration of a noble gas;
(2) There is a tendency for atoms whose p orbitals have 3, 4, or 5 electrons (5, 6, or 7 electrons in outer energy shell) to gain electrons and form negative ions with an electronic configuration of a noble gas;
(3) When metal atoms collide with nonmetal atoms (except for noble gases) there is a tendency for the metal atoms to lose electrons to the nonmetal atoms, forming ions;
(4) Atoms with an $ns^2np^1np^1np^0$ configuration (4 electrons in outer energy shell) have no pronounced tendency to lose or gain electrons in the formation of ions.

Although these generalizations are valid in enough cases to make them useful to the beginning chemistry student, exceptions do exist. Consequently, these statements should not be thought of as chemical law, but merely as memory aids for the cases most frequently encountered.

Electron Dot Formulas

To keep track of the electrons involved in electron-transfer reactions such as those discussed previously, a type of notation known as the *electron dot formula* is used. To write the electron dot formula of an atom or ion, the electrons in the outermost principal energy level are counted, and for each electron a dot is placed around the symbol for the element. The dots are paired so that they are consistent with the pairing of electrons in orbitals. For example, fluorine $(1s^2 2s^2 2p^5)$ is $\cdot \ddot{\text{F}} \colon$ (seven electrons in the outermost 2s and 2p orbitals).

The reaction between sodium and fluorine is written as follows:

$$\text{Na}\cdot + \colon \ddot{\text{F}} \cdot \longrightarrow \text{Na}^+ + \colon \ddot{\text{F}} \colon^-$$

Note that Na^+ has no dots around it, since there are no electrons in the shell corresponding to the outermost shell of the sodium atom (n = 3). The reaction between calcium and fluorine can be written using electron dot formulas as follows:

$$\text{Ca}\colon + 2\colon \ddot{\text{F}} \cdot \longrightarrow \colon \ddot{\text{F}} \colon^- + \text{Ca}^{2+} + \colon \ddot{\text{F}} \colon^-$$

Since the calcium ion has two electrons, it must react with two fluorine atoms, each of which takes one electron.

Properties of Ions

Ion Sizes

The sizes (radii) of atoms of practically all the elements can be measured by several experimental methods. Let us consider briefly the sizes of the ions that are produced by either electron gain or electron loss. The sizes of the ions are important because the strength of the forces holding ions together in ionic compounds depends on the sizes of the ions involved.

A sodium atom with its single 3s outermost electron has a radius of 1.86 A (A = Angstrom unit, 10^{-8} cm). One would expect that when this electron is removed (forming the Na^+ ion) the resulting ion would be smaller. This decrease in size results because there are now only 10 electrons attracted to a charge of +11 on the nucleus, and these electrons can be pulled in closer to the nucleus by this charge imbalance. This same type of phenomenon is observed for all metal ions. Figure 9–4 compares atomic and ionic sizes for seven elements.

The metal ions with multiple charges (Al^{3+}, Fe^{2+}, and so on) are much smaller than the corresponding metal atom.

Nonmetals that gain electrons to achieve closed-shell configurations increase in size during this process so that the negative ion is larger than the corresponding atom. Consider the information in Figure 9–4. This behavior is due to the addition of electrons to the orbitals of an atom without increasing the charge on the nucleus. The result is a crowding of the electrons due to a repulsion of filled orbitals to a point that expansion of the ion is the result.

Electrostatic Forces

Ionic bonding is due principally to electrostatic interaction between ions of opposite charge which leads to the stable, regular crystalline array of ions.

METALS

NONMETALS

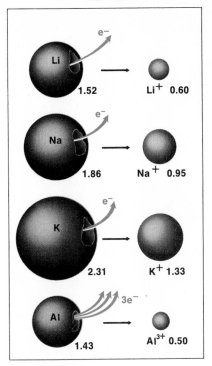

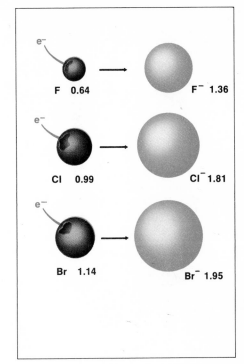

FIGURE 9-4   Relative sizes of selected atoms and ions. Numbers given are atomic or ionic radii in Angstrom units.

The geometrical array of ions is termed an *ionic lattice*, and the macroscopic substance is an ionic solid or salt (Figure 9–5).

Recall from Chapter 5 that Coulomb's law tells us that oppositely charged ions will attract each other with a force which depends on the charges and the distance between them. We can express this coulombic attraction as:

$$\text{Force} \begin{Bmatrix} \text{is} \\ \text{proportional} \\ \text{to} \end{Bmatrix} \frac{(\text{charge on positive ion}) \times (\text{charge on negative ion})}{(\text{distance of separation})^2}$$

or

$$F \propto \frac{(Q^+)(Q^-)}{r^2}$$

This expression leads us to expect that strong ionic bonds are formed when the ions possess high charges and are separated by small distances (which will occur for the smaller ions). The effects of ionic charge and size can be seen rather easily if we compare the melting points of several salts, all with similar crystal geometries. Let us compare the melting points of NaCl and KCl (the K^+ ion is larger than the Na^+ ion) and also those of CaO and BaO (the Ba^{2+} ion is larger than the Ca^{2+} ion). All four salts have a sodium chloride (rock salt) crystalline structure (see Figure 9–5). Table 9–2 shows that in both cases the salts with smaller positive ions have the higher melting points. Those with doubly charged ions have dramatically increased melting points.

159

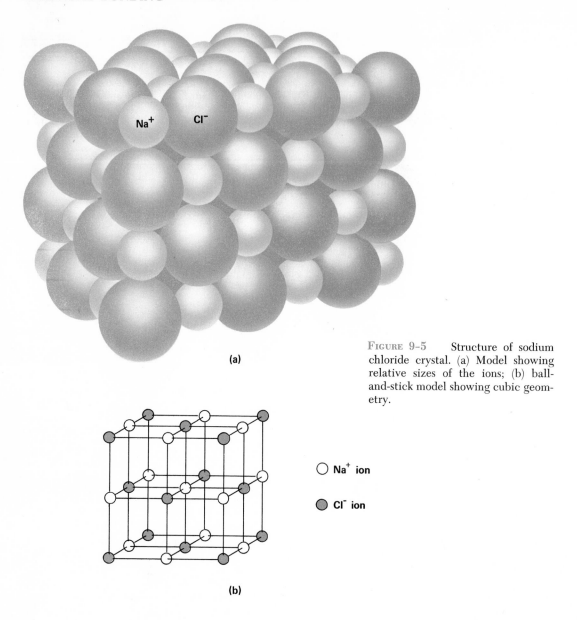

(a)

FIGURE 9–5   Structure of sodium chloride crystal. (a) Model showing relative sizes of the ions; (b) ball-and-stick model showing cubic geometry.

○ Na⁺ ion

● Cl⁻ ion

(b)

TABLE 9–2   THE EFFECTS OF IONIC CHARGE AND SIZE ON SOME SALTS WITH IDENTICAL STRUCTURES

		M⁺, RADIUS, A	MELTING POINT
Salts with singly charged ions	NaCl	0.95	801°C
	KCl	1.33	776°C
		M²⁺, RADIUS, A	MELTING POINT
Salts with doubly charged ions	CaO	0.99	2580°C
	BaO	1.35	1923°C

Ionic Lattice

The ionic bond is largely a property of ions in the lattice, with the kind of lattice also having something to do with the strength of the bonds. There are numerous kinds of ionic lattices; a common type is the sodium chloride lattice (Figure 9–5). Here, each Na^+ ion is surrounded by six Cl^- ions, and each Cl^- ion is in turn surrounded by six Na^+ ions. Each ion is attracted to all of the oppositely charged ions and repelled by all of the like charged ions in the lattice. Thus, there are no sodium chloride molecules in this solid.

One question arises at this point. If the oppositely charged ions attract one another, what keeps them from getting closer and closer to each other? This is prevented by the repulsion which arises when orbitals filled with electrons are moved very close together. There is a very considerable repulsion which prevents significant interpenetration of filled orbitals from different atoms into the filled orbitals of each other's space. As a consequence, each ion has an ionic radius which is reasonably constant and which represents the closest distance of approach of another ion's filled orbitals.

Other Properties of Ionic Compounds

Since the bonding between the ions in ionic compounds is strong, there is a tendency for them to be rather hard, crystalline solids with relatively high melting points. Since the electrons involved in bonding are localized on the individual ions and cannot move about in the lattice, solid ionic compounds are rather poor electrical conductors when compared to molten salts or metals. Molten ionic compounds contain free ions, since the melting process partially breaks down the structure that had existed in the lattice. These free ions can move in an electrical field and thus transport charge or an electric current from one place to another. Hence, molten salts are generally expected to be excellent conductors of electricity. Properties of ionic compounds are considered in greater detail in Chapter 11.

COVALENT BONDING

A second type of chemical bonding, which is found in molecular compounds and polymers, is covalent bonding. This type of bonding is distinctly different from ionic bonding in that electrons are shared between atoms rather than transferred from one atom to another.

The atoms that tend to bond covalently are those with relatively high ionization potentials. Combinations of these atoms contain no atom which can readily lose electrons. Hence, the atoms share electrons. Such substances, for example, as carbon dioxide (CO_2), carbon tetrachloride (CCl_4), ammonia (NH_3), hydrogen (H_2), chlorine (Cl_2), and water (H_2O) are all composed of molecules which are held together by covalent bonding.

Electron Sharing and Localization

The simplest example of covalent bonding is found in diatomic (two atoms) gaseous molecules such as H_2 (hydrogen), F_2 (fluorine), and Cl_2 (chlorine). A

161

hydrogen atom has an electronic configuration of $1s^1$. Comparing this with the electronic configuration of the noble gas nearest to hydrogen in the periodic table, helium (He, $1s^2$), we see that, if the hydrogen atom could somehow obtain one more electron, a very stable configuration could be reached (recall that the Li^+ ion with its $1s^2$ configuration is quite stable). For hydrogen, this configuration can be achieved by two hydrogen atoms sharing their single electrons. The electron dot formula for the resulting H_2 molecule is:

$$2H \cdot \longrightarrow H:H$$

Ionic bonding is not a reasonable possibility for the H—H bond since both hydrogen atoms have the same attraction for the electron pair, neither having the ability to completely remove the electron from the other.

For other atoms we may use the tendency to attain a closed-shell configuration (eight electrons) as a guide in writing electron dot formulas. Two fluorine atoms bound together at normal temperatures form a F_2 molecule. This bonding is due to each fluorine atom sharing its unpaired 2p electron.

$$2 :\overset{..}{\underset{..}{F}}\cdot \longrightarrow \quad :\overset{..}{\underset{..}{F}}:\overset{..}{\underset{..}{F}}:$$

$1s^22s^22p^5$ *Fluorine molecule*

Orbital Overlap

Since hydrogen gas is molecular, and since the interaction between the two atoms must be other than ionic, the question arises as to how one hydrogen atom can be bonded to another. We have already concluded, in the wave-mechanical model of the atom, that two electrons are compatible in an orbital of an atom if their spins are paired. It is reasonable to assume that the close approach of two unpaired electrons on different atoms will result in the formation of a covalent bond where both orbitals are made available to both of the paired electrons. Thus, the bond in the hydrogen molecule can be pictured as resulting from the overlap of two 1s orbitals, one from each atom, and the spin-pairing of the electrons involved (Figure 9–6).

Overlap of atomic orbitals can be visualized for all covalent bonds; and the greater the overlap, the more stable the bond. Since the types of orbitals which an atom has available for overlap are determined by wave mechanics, covalent bonding can be thought of as a logical consequence of the wave-mechanical model of the atom.

It is assumed that there is a stability associated with the pairing of two

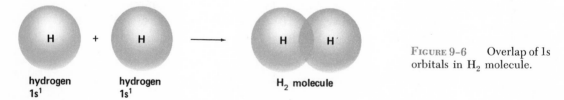

hydrogen
$1s^1$

hydrogen
$1s^1$

H_2 molecule

FIGURE 9–6   Overlap of 1s orbitals in H_2 molecule.

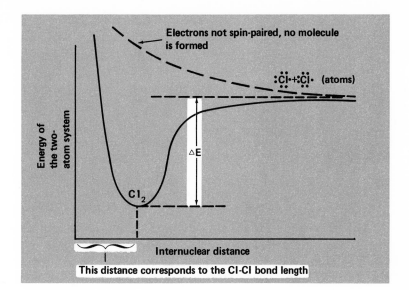

FIGURE 9-7   The relative stabilities of Cl_2 molecules and Cl atoms. The bond energy of the Cl-Cl bond is approximately equal to ΔE.

electrons of opposite spin from different atoms and the formation of a covalent bond. Figure 9–7 illustrates this for Cl atoms ($3s^2 3p^5$). The amount of energy released, ΔE, is approximately the bond energy, and is the same as the amount of energy necessary to break the bond and give gaseous atoms. In general, energy is released in the formation of covalent bonds, and energy is required to break such bonds. Notice in Figure 9–7 that any attempt to force the two atoms closer to each other than their equilibrium bond length, will require a large amount of energy, since there is a repulsion caused by the filled atomic orbitals of the two different atoms.

Single Bonds

Covalent bonds formed by the sharing of two electrons (a single election pair) between two atoms are called *single bonds*. Many molecules, both simple and complicated, contain one or more single covalent bonds. Let us consider some of these.

Hydrogen fluoride (HF) contains a single covalent bond between the two atoms. Fluorine ($2s^2 2p^5$) needs one electron to attain a closed-shell configuration as does hydrogen ($1s^1$). This could be accomplished by an overlap of the 1s hydrogen orbital with the half-filled 2p fluorine orbital. The result is shown in Figure 9–8.

FIGURE 9-8   Single bond formation in HF. (a) The electron dot representation; (b) the orbital overlap representation.

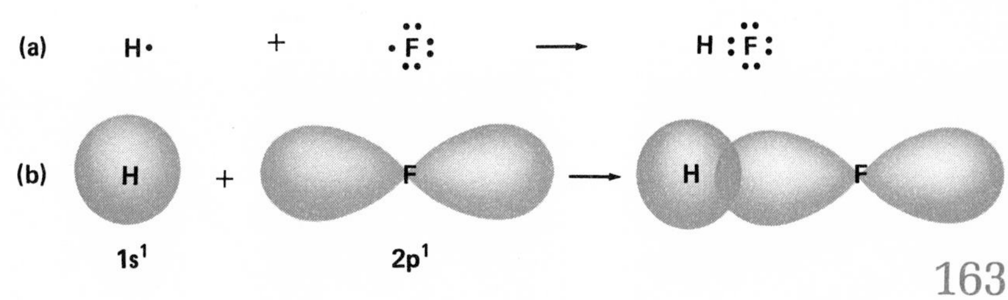

(a)   H·   +   ·F: ⟶   H:F:

(b)   H   +   F   ⟶   H   F
       1s¹   2p¹

163

An oxygen atom has a configuration $2s^2 2p^4$ and, therefore, needs two electrons to attain a closed-shell configuration. (We have already seen how oxygen atoms may gain two electrons and form O^{2-} ions). If two hydrogen atoms are present, there can be sharing of electrons between the atoms. The electron dot formula for the molecule that results, the water molecule, is given below. Oxygen is written as $:\ddot{O}\cdot$ because there are two unpaired electrons in the atom ($2p^4 = 2p^2 2p^1 2p^1$).

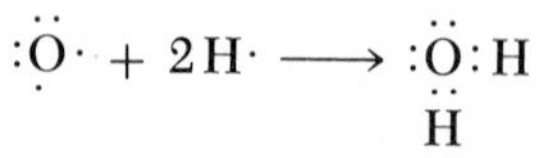

$$:\ddot{O}\cdot\ +\ 2\,H\cdot\ \longrightarrow\ :\ddot{O}\!:\!H$$
$$H$$
Water

Consider the nitrogen atom and its electronic configuration. Wave mechanics predict a configuration of $2s^2 2p^3 (2p^3 = 2p^1 2p^1 2p^1)$ with three unpaired electrons. The atom needs three electrons to attain a closed-shell configuration. One way this can be done is by forming three single bonds with some other atoms having one unpaired electron each. With hydrogen atoms, the result is ammonia (NH_3). The nitrogen atom is written as $\cdot\ddot{N}\cdot$ to show the three unpaired electrons.

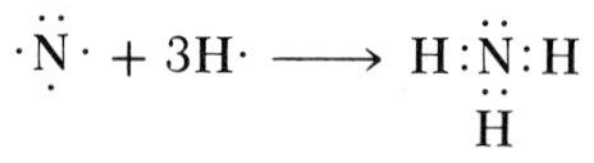

$$\cdot\ddot{N}\cdot\ +\ 3\,H\cdot\ \longrightarrow\ H\!:\!\ddot{N}\!:\!H$$
$$H$$
Ammonia

Note that fluorine, oxygen, and nitrogen all end up with an octet (8) of electrons in HF, H_2O, and NH_3. This leads to the generalization known as the *octet rule*, which predicts that second period elements on the periodic chart form compounds in such a way as to complete their octet of electrons. Even though this rule can often be applied usefully to elements with heavier atoms, it does not always apply since these atoms can have more than eight electrons in their outer shells.

In the examples just given we can see that for the atoms undergoing covalent bonding, the number of single bonds possible is the same as the number of unpaired electrons in the atom when it is ready to undergo bonding. This is generally true, as shown in the following examples:

H hydrogen, 1 unpaired electron $\longrightarrow$ 1 single bond
F fluorine, 1 unpaired electron $\longrightarrow$ 1 single bond
O oxygen, 2 unpaired electrons $\longrightarrow$ 2 single bonds
N nitrogen, 3 unpaired electrons $\longrightarrow$ 3 single bonds

Multiple Bonding

When an atom has less than seven electrons in its outermost shell ($ns^2 np^4$, for example), it can complete its octet in two ways. In the first way, it simply shares a single electron with each of several other atoms which can share a single electron. This leads to *single* covalent bonds. But the atom can also share two (or three) pairs of electrons with a single other atom. In this case there will be two (or three) bonds between these two atoms.

When two shared pairs of electrons join together two atoms, we speak of a *double bond*, and when three shared pairs are involved, the bond is called a *triple bond*. Examples of these bonds are found in many compounds of $O(2s^2 2p^4)$, $N(2s^2 2p^3)$, and $C(2s^2 2p^2)$ such as those shown in Figure 9–9.

Formula	Name	Electron Dot Structure
Double Bonds:		
CO_2	Carbon dioxide	$:\overset{..}{O}::C::\overset{..}{O}:$
C_2H_4	Ethylene	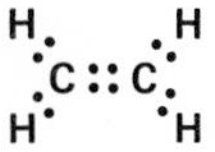 (electron dot structure of ethylene)
SO_3	Sulfur trioxide	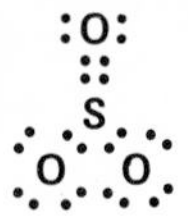 (electron dot structure of sulfur trioxide)
Triple Bonds:		
N_2	Nitrogen	$:N:::N:$
CO	Carbon monoxide	$:C:::O:$
C_2H_2	Acetylene	$H:C:::C:H$

FIGURE 9–9   Electron dot structures of some molecules containing multiple bonds.

As we can see from these structures, molecules may contain several types of bonds. Thus, ethylene (Figure 9–9) contains a double bond between the carbon atoms and single bonds between the hydrogen atoms and the carbon atoms. For convenience, an electron pair bond is often indicated by a dash as follows:

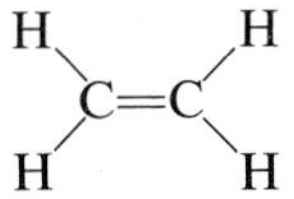

so that a single bond will be shown as H—H, a double bond as $H_2C=CH_2$, and a triple bond as N≡N. Note that in each of these cases the octet rule is obeyed; exceptional cases such as B_2H_6 can be cited, wherein the octet rule is not obeyed.

Each type of bond has characteristics which may be used to identify it. Triple bonds are shorter than double bonds, which in turn are shorter than single bonds. Bond energies normally increase with decreasing bond length due to greater orbital overlaps. Bond lengths and energies in some typical cases are shown in Table 9–3.

TABLE 9–3   SOME BOND LENGTHS AND BOND ENERGIES

Bond	C—C	C=C	C≡C	N—N	N=N	N≡N
Length, A	1.54	1.34	1.20	1.40	1.24	1.09
Bond energy (kcal/mole)	83	146	200	40	100	225

A = 10^{-8} cm; kcal/mole = thousands of calories necessary to break 6.02×10^{23} bonds.

165

POLAR BONDS

In a molecule like H_2 or F_2, where both atoms are alike, there is equal sharing of the electron pair. Where two unlike atoms are bonded together, however, the sharing of the electron pair is likely to be unequal, with the result that there is a negative and a positive portion of the bond. Such a bond is termed a *polar bond.* The negative pole is in the region of high electron density, and the positive pole is in the region of low electron density. Thus, in the molecule HF, the bonding pair of electrons is more under the control of the fluorine atom than of the hydrogen atom (Figure 9–10). This can also be expressed in terms of the centers of positive and negative charge. In the nonpolar hydrogen molecule, the center of negative charge is in the geometrical center of the molecule since there is a symmetrical distribution of negative charge about that center. The center of positive charge is located there also since the two hydrogen nuclei are equally spaced on either side. In contrast to the nonpolar hydrogen molecule, the polar HF molecule has two distinct centers of positive and negative polar structure (Figure 9–10).

A measure of the net ability of an atom to attract electrons to it in a chemical bond is called its *electronegativity*. Since the fluorine atom attracts the shared electrons in the H—F bond more than does the hydrogen atom, we say that the electronegativity of the fluorine atom is greater. Electronegativity is a function of nuclear charge and atomic size. Small atoms with large nuclear charges are the most electronegative. The trend of electronegativities in the periodic table is rather regular and follows an inverse trend in atomic size. Figure 9–11 shows these relationships.

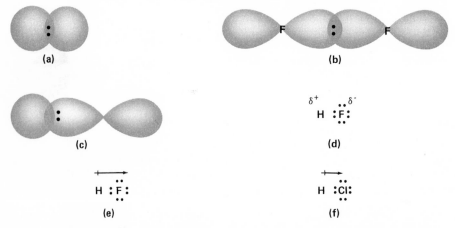

FIGURE 9–10   Polar bonds in HF and HCl. (a) Symmetrical distribution of electrons in H_2 results in the center of negative charge being identical with the center of positive charge. This is symbolized by the electron dots placed in the overlap area. (b) Overlap of p orbitals in F_2 also results in symmetrical distribution of charge. (c) Fluorine, being more electronegative than hydrogen, displaces electron pair towards the fluorine nucleus. Note the electron dots to the right of the overlap area which conveys the idea of polarity (separation of charge). (d) δ^+ (delta plus-fractional plus charge) and δ^- (delta negative-fractional negative charge) are used to indicate poles of charge. In (e) and (f) an arrow is used to indicate electron shift, the arrow having a plus tail to indicate partial positive charge on the hydrogen atom. Note that the longer arrow in HF structure indicates a greater degree of polarity than in HCl.

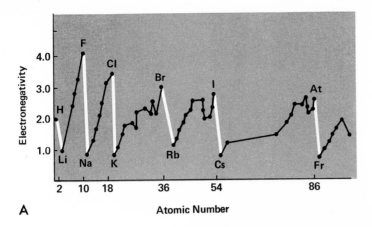

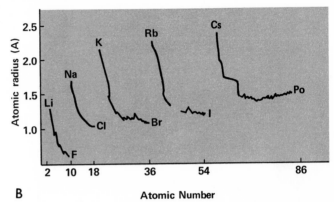

FIGURE 9-11   A. Electronegativities of the elements. B, Atomic sizes of the elements.

The electronegativity concept is useful in predicting the polarity of covalent bonds. Polarity is simply the state which results when the centers of positive and negative charge in a bond or a molecule are separated by some distance. Bonds joining different atoms are generally polar in that one of the atoms has distorted the electron distribution toward itself. Thus, polar covalent bonds occur in practically every molecule that has different kinds of atoms covalently bonded together. Additional examples of molecules with polar covalent bonds include NO, SO_2, CO_2, CCl_4, and H_2S.

HYDROGEN BONDING

For a series of molecular substances with similar structures, the boiling points ordinarily increase as the molecular weights increase. For example, the boiling points of fluorine (F_2), chlorine (Cl_2), bromine (Br_2), and iodine (I_2) increase with increasing molecular weight (Figure 9–12).

The general relationship between boiling points and molecular weights also holds for hydrogen chloride (HCl), hydrogen bromide (HBr), and hydrogen iodide (HI). However, the boiling point of hydrogen fluoride (HF), the lightest member of this series of compounds, is abnormally high (Figure 9–13). Irregularities similar to this are also found in other compounds in which hydrogen is bonded to fluorine, oxygen, and nitrogen. The increased interaction between molecules containing H—F, H—O, or H—N bonds is termed *hydrogen bonding*.

167

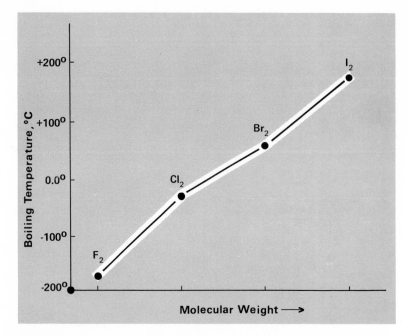

FIGURE 9-12 Boiling points of F_2, Cl_2, Br_2, and I_2, as a function of molecular weight. Temperature is given in °C.

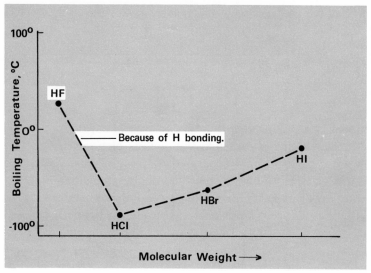

FIGURE 9-13 Boiling points of HF, HCl, HBr, and HI, plotted against molecular weights. Temperature is given in °C.

The explanation for hydrogen bonding is to be found in the extremely polar nature of the H—F, H—O, and H—N bonds, the polarity being due to the extreme electronegativity of these three elements. Consider the bonding in liquid HF. Since unlike-charged ends of these molecules (Figure 9-10) should attract each other, we expect HF molecules to be associated with one another. This association is illustrated in Figure 9-14. The increased association between HF molecules compared to that found in HCl offers a ready explanation for the unusually high boiling point of HF.

The structure of water provides another good example of hydrogen bonding. Water molecules are not linear but rather are angular with two nonbonding (unshared) electron pairs located toward one end of the molecule and the par-

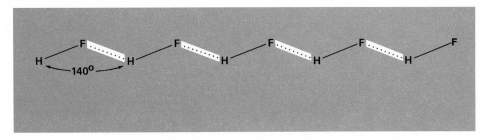

FIGURE 9-14   Hydrogen bonding in HF.

tially positive hydrogen atoms located toward the opposite end. The two polar bonds in this geometry result in a distinctly polar molecule.

$$\begin{array}{c} \delta^- \\ \overset{..}{\underset{..}{O}} \\ H \qquad H \\ \delta^+ \end{array}$$

In liquid and solid water where the molecules are close enough to interact, the hydrogen atoms on some of the water molecules tend to associate themselves with the nonbonding electrons on the oxygen atom of an adjacent water molecule (Figure 9–15). This is possible because of the small size of the hydrogen atom. The result of this association is called *hydrogen bonding* because the hydrogen atom acts as a sort of a bridge to hold two molecules together much like electron pairs hold atoms together in molecules, though the hydrogen bond is much weaker than an ordinary covalent bond.

The extent of hydrogen bonding in liquid water varies with temperature. In ice, hydrogen bonding is very extensive and almost all hydrogen atoms are involved in this sort of bonding. This results in a very open structure for ice (Figure 9–16). Consequently, at ordinary pressures ice is *less* dense than water. As ice is melted and the molecules gain more energy and move about, this bonding begins to break down. All the hydrogen bonds are not broken, however,

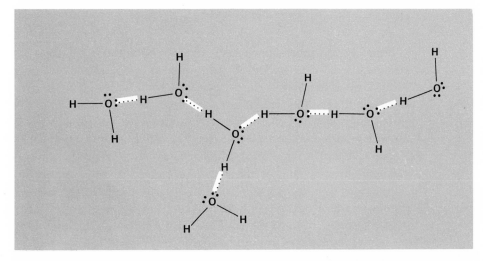

FIGURE 9-15   Hydrogen bonding in water.

169

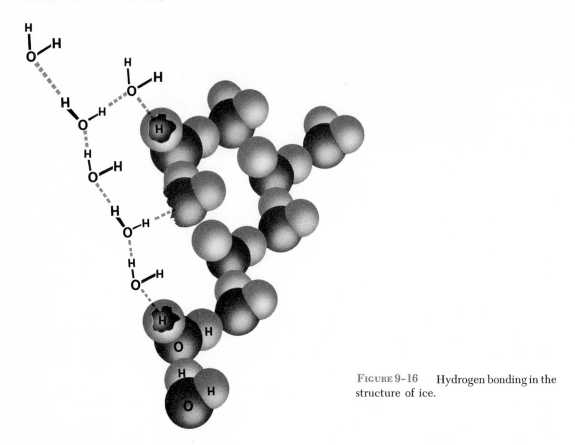

FIGURE 9-16 Hydrogen bonding in the structure of ice.

and large aggregates of water molecules exist in liquid water even near 100°C. As water is heated, thermal agitation tends to disrupt the hydrogen bonding until, in gaseous water, there is only a small fraction of the number of hydrogen bonds that are found in liquid or solid water.

VAN DER WAAL'S ATTRACTIONS

Since every substance known can exist in the solid state (even helium solidifies at $-272°C$ under 26 atmospheres pressure), we must assume that another type of attraction exists between atoms and molecules other than ionic and covalent bonding. In helium, the two 2s electrons are in a closed shell (nonbonding), and yet there is a slight attraction between two helium atoms. There are a large number of other cases in which weak attractions exist between molecules. These interactions, which are much weaker than ionic or covalent bonds, can arise from a variety of causes. One of these is the occurrence of instantaneous dipoles. An uneven electron distribution in an atom gives rise to a dipole in the atom, but this takes place on a very temporary basis. Such a dipole is termed instantaneous. These instantaneous dipoles can interact with each other, resulting in the attractive force (Figure 9–17).

This type of attraction is called Van der Waal's attraction, after the Dutch physicist who suggested its existence. These forces are responsible for many of

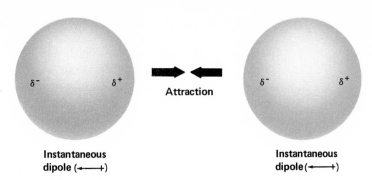

FIGURE 9-17   An illustration of Van der Waals' attraction. One instantaneous dipole interacts with another in a neighboring atom.

the forces in molecular crystals, in solid oxygen, in solid nitrogen, and in numerous other examples.

METALLIC BONDING ✕

Metals have some properties totally unlike those of other substances. For example, most metals are very good electrical conductors; they are shiny solids and have relatively high melting points (with a few notable exceptions such as mercury). Any theory concerning the bonding of metals must be consistent with these and other properties. Structural investigations of metals have led to the conclusion that metals are composed of very regular arrays (lattices) of metal ions in which the bonding electrons are very mobile (or delocalized). Figure 9–18 shows a diagram of a sodium lattice. The bonding electrons in a metal such as sodium are 3s electrons. These can be made to move rather easily through the lattice upon application of an electric field. In this way the metal acts as a conductor of electricity. As a consequence of this delocalization of bonding electrons, we cannot really write a satisfactory description of the bonding in a metal using any simple form of electron dot structure.

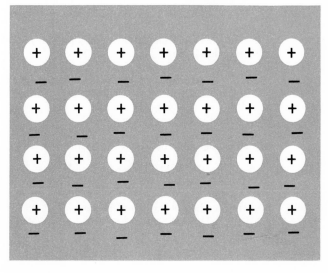

FIGURE 9-18   A cross-sectional view of the crystalline lattice in metallic sodium. *The bonding electrons are delocalized.*

171

CONCLUSION AND PREVIEW

Chemical bonding is a broad term used to describe the forces that hold atoms and molecules together. Consistent with the wave-mechanical picture of the atom, two primary causes of bonding have been presented: (1) the attraction between positive and negative charges and (2) the overlap of atomic orbitals in such a way that spin-paired electrons are shared between the atomic centers. Yet to be considered is the relationship between chemical bonding and molecular shapes; this will be the topic for study in Chapter 10.

QUESTIONS

1. Diamond (a form of carbon) has a melting point of 3500°C, whereas carbon monoxide (CO) has a melting point of −207°C. What does this tell about the bonding in these two substances?

2. Write the electronic configuration for the element potassium (atomic number 19). What will be the electronic configuration when a K^+ ion is formed?

3. Is Ca^{3+} a possible ion under normal chemical conditions? Why?

4. Write the symbols for the six elements with the highest ionization potential, selecting one from each period of the periodic table.

5. Select the electronic configurations which would be expected to lead to similar chemical behavior.

   (a) $1s^2 2s^2$
   (b) $1s^2 2s^2 2p^3$
   (c) $1s^2 2s^2 2p^6 3s^2 3p^6 4s^1$
   (d) $1s^2 2s^2 2p^6 3s^2$
   (e) $1s^2 2s^1$
   (f) $1s^2 2s^2 2p^6 3s^2 3p^3$

6. Fluorine (atomic number 9) has an electronic configuration of $1s^2 2s^2 2p^5$. How many electrons are in the outermost (valence) shell of a fluorine atom? How many electrons will be involved in chemical bonding?

7. Which salt would be expected to have the highest melting point, KCl or RbCl; MgO or BaO? Give a reason for your answer.

8. Write the electronic configuration for iodine (I, atomic number 53). How many covalent bonds should an iodine atom form?

9. Write the electron dot structures for the fluoride ion, F^-, the chloride ion, Cl^-, and the bromide ion, Br^-.

10. Draw the electron dot structure for water. Based on bonding theory, why is water's formula not H_3O?

11. Define the term *bond energy*.

12. Draw electron dot structures for the following molecules:

   (a) NF_3
   (b) CCl_4
   (c) C_2Cl_2
   (d) OF_2
   (e) H_2S
   (f) CO
   (g) N_2H_4
   (h) CH_3OH

13. The members of the nitrogen family, N, P, As, and Sb, form compounds with hydrogen, NH_3, PH_3, AsH_3, and SbH_3. The boiling points of these compounds are given:

SbH_3	−17°
AsH_3	−55°
PH_3	−87.4°
NH_3	−33.4°

   Comment on why NH_3 doesn't follow the downward trend of boiling points.

14. Match the substances listed below with the type of bonding responsible for holding units in the solid together.

Solid krypton (Kr)	Ionic
Ice	Covalent
Diamond	Metallic
CaF_2	Hydrogen bonding
Iron	Van der Waal's

15. Predict the general kind of chemical behavior (that is, loss, gain, or sharing of electrons) you would expect from atoms with the following electronic configurations:

    (a) $1s^2 2s^2 2p^6 3s^1$
    (b) $1s^2 2s^2 2p^5$
    (c) $1s^2 2s^2 2p^2$

16. Show how two fluorine atoms can form a bond with the overlap of their half-filled p orbitals.

17. Select the *polar* molecules from the following list and explain why they are polar:
    N_2, HCl, CO, and NO

18. How many bonds join the two atoms in the following?
    CN^-, Cl_2, S_2

19. Boron trichloride has the electron-dot formula: $\ddot{:}\overset{..}{Cl}:B:\overset{..}{Cl}\ddot{:}$. What does this tell you about the rule of eight even for Period II elements? $\overset{..}{:}\overset{..}{Cl}:$

20. Write the electronic configuration for the element gold (Au, atomic number 79). What ion of gold would you suggest which might have a degree of stability?

21. If covalent bonding is to white what ionic bonding is to black, polar covalent bonding would be represented by _____. On this scale, how could a more polar molecule be distinguished from one that is less polar?

22. Chromium (element No. 24) forms a stable ion in which the chromium has lost three electrons per atom. What does this tell you about the d electrons?

23. Use your chemical intuition and suggest a reaction that might occur between boron trichloride (problem 19 above) and ammonia, $H:\overset{..}{N}:H$.
    $\qquad\qquad\qquad\qquad\qquad\qquad\qquad\qquad\qquad\qquad\quad H$

SUGGESTIONS FOR FURTHER READING

Companion, A. L., *Chemical Bonding*, McGraw-Hill Book Company, New York, 1964.

Derjaguin, B. V., "The Force Between Molecules," *Scientific American*, Vol. 203, p. 47 (1960).

Eyring, Henry, and MuShik Jhon, "Significant Structure Theory of Water," *Chemistry*, Vol. 39, No. 9, p. 8 (1966).

Garfield, Eugene, Gabrielle S. Reavesz, and Nathin Rubin, "Fixed Valence and Molecular Formula Verification," *Chemistry*, Vol. 43, No. 9, p. 13 (1970).

Greenwood, N. N., "Chemical Bonds," *Education in Chemistry*, Vol. 4, p. 164 (1967).

Hochstrasser, R. M., *Behavior of Electrons in Atoms*, W. A. Benjamin, Inc., New York, 1964.

House, J. E., Jr., "Ionic Bonding in Solids," *Chemistry*, Vol. 43, No. 2, p. 18 (1970).

Kondratyev, V. N., *The Structure of Atoms and Molecules*, Dover Publications, Inc., New York, 1965.

Lagowski, J. J., *The Chemical Bond*, Houghton Mifflin Co., Boston, 1966.

Mellor, D. P., "The Noble Gases and Their Compounds," *Chemistry*, Vol. 41, No. 10, p. 16 (1968).

Pauling, L., *The Nature of the Chemical Bond*, Cornell University Press, Ithaca, New York, 1960.

Ryschkewitsch, G. E., *Chemical Bonding and the Geometry of Molecules*, Reinhold Publishing Corp., New York, 1963.

Sanderson, R. T., "Principles of Chemical Bonding," *Journal of Chemical Education*, Vol. 38, p. 382 (1961).

Webb, Valerie J., "Hydrogen Bond 'Special Agent'," *Chemistry*, Vol. 41, No. 6, p. 16 (1968).

MOLECULAR SHAPES

CHAPTER 10 ▬▬▬▬▬▬▬▬▬▬▬

INTRODUCTION

Although the wave-mechanical theory of the atom is very effective in explaining bonding and periodicity of the elements, it must also be capable of explaining related chemical phenomena, such as molecular shapes and associated properties, in order to be really useful. The shape (or geometry) of a molecule is determined by the arrangement of the atoms in the molecule with respect to each other; that is, by the angles between and the lengths of its covalent bonds. The angle formed by two intersecting lines drawn from the two nuclei of the attached atoms through the nucleus of the central atom is called the bond angle. Molecular shapes are established by many varied experimental techniques. A few of these methods will be described in support of some of the molecular structures presented in this chapter. Meanwhile, we shall see what molecular shapes are predicted by atomic theory and what modifications are needed for the theoretical and the experimental findings to agree.

MOLECULAR SHAPES PREDICTED BY ATOMIC ORBITALS

The starting point for a discussion of molecular shapes is the set of atomic orbitals, since these determine the preferred regions in space for electrons in the various atomic orbitals. Expected molecular geometries based on orbital overlap when s and p atomic orbitals are used to form bonds are shown in Figure 10–1.

Since an s orbital is spherically symmetrical, there is no preferred direction for the bonding with the other atom. Of course, with only one s orbital per valence shell, only one bond can be formed from this kind of orbital. Thus, lithium, with the electronic configuration $1s^2 2s^1$, can form only one bond.

There are three p orbitals per valence shell. The electrons in a set of p orbitals concentrate charge along the x, y, and z axes, which are at 90° angles to each other. Any two atoms bonded to a third atom by overlapping two p orbitals would be expected to form bonds enclosing an angle of 90°.

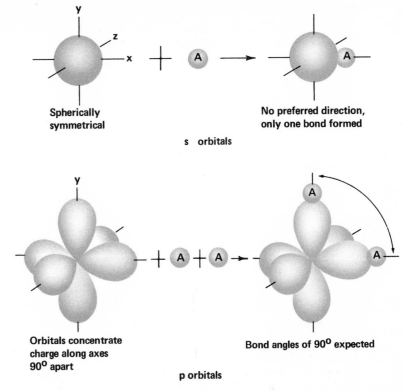

FIGURE 10-1 Predicted molecular shapes when s and p orbitals are used to form bonds. *A* represents any atom to be joined to the central atom.

Is this predicted bond angle actually found in molecules? Some molecules such as H_2S and H_2Se come close. However, instead of the expected bond angle of 90°, when s orbitals of two hydrogens overlap two of the p orbitals of S or Se, the angles are 92° for H_2S and 91° for H_2Se. It is rare for the bond angles of molecules to agree this closely with the angles predicted by combinations of simple atomic orbitals. In H_2O, for example, the bond angle is 104.5°.

But before yelling, "Down with the theory," we should realize its versatility and true value. With a slight but elegant adaptation, the wave-mechanical theory can be brought into agreement with experimentally derived molecular shapes of relatively simple molecules. The theory is then available for our plunge into the structural mysteries of complex molecules, enshrouded in complexity because of the vast number of atoms per molecule. To discard the theory now would be as tragic as finding a spaceship slightly off course and discarding it without making the small adjustment that will allow it to go on course into the great beyond.

The method of adjusting the pure atomic orbitals is called *hybridization*. This process allows bond angles to conform with experiment. Although hybridization is a mathematical process of the wave-mechanical theory, we can gain an insight into its methods and its usefulness through a qualitative examination of its features. The electron-pair repulsion theory complements hybridization because it visualizes the mathematical expertise of the hybridization method.

Both of these theories will be examined along with some of their applications and some of the experimental observations that prompted an on-course setting of the wave-mechanical theory.

EVIDENCE FOR THE TETRAHEDRAL CARBON ATOM

The molecular structures given in this chapter have been established by a variety of experimental methods. Often the evidence is similar to that used to establish the tetrahedral nature of compounds like CH_4, CCl_4, and CBr_4. For this reason, some of the arguments for the tetrahedral structure of these and similar compounds will be cited to illustrate some methods of establishing structures of molecules in general.

Inductive reasoning and data from several sources were combined many years ago to deduce a tetrahedral model for the bonds formed by carbon. For simplicity, methane, CH_4, will be used as an example to illustrate some of the lines of investigation although CCl_4, CBr_4, $C(CH_3)_4$, and similar compounds have the same basic geometry about the carbon atom and provide the same general kind of evidence.

When chlorine is made to react with methane, no particular C—H bond is more active than the other bonds. Consequently, when methyl chloride is formed, only one compound is formed (Figure 10–2). (There is only one melting point, one boiling point, one density, and so forth.) This means that structures that favor one bond breaking over another cannot be justified.

Methane has been shown experimentally to be nonpolar (that is, the overall centers of positive and negative charge of the molecule coincide). This means that the electrons are evenly distributed about the center of the molecule, and that the molecule has a symmetrical charge distribution. This is a good indication that all of the C—H bonds in methane are exactly alike in strength and length and, furthermore, that the bonds go out from the carbon atom in a symmetrical pattern.

Although a planar arrangement satisfies the equal bond length and strength requirements, it predicts too many compounds for the formula CH_2Cl_2. If the structure is planar, two different molecular structures would be permitted and, consequently, two different compounds would be predicted (Figure 10–3). A tetrahedral structure allows only one kind of molecule and, hence, one compound for CH_2Cl_2, which is actually the case.

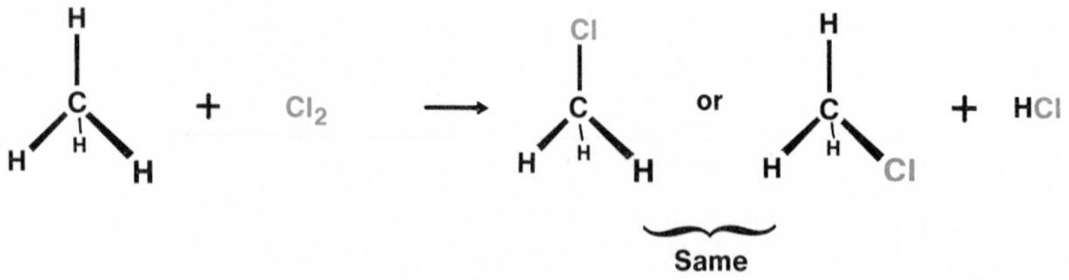

FIGURE 10-2   Reaction of methane with chlorine to form only one type of methyl chloride.

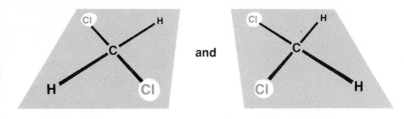

FIGURE 10–3 Possible structures for CH_2Cl_2 if the molecule were planar. There is only one compound for this formula; hence, the structure is not planar.

Probably the best source of evidence for establishing the position of atoms in a molecule as well as the length of bonds is x-ray crystallography. When a beam of x-rays of a uniform frequency is sent through a sample of materials, the pattern of the x-rays which reflects from the sample can be recorded on film and used to determine the positions of the atoms in a substance. The use of x-rays to determine the structure of matter is somewhat like using light to show the positions of several small mirrors in a box that you must view from the side (Figure 10–4). As the light beam passes into the box, the mirrors reflect the light, and the pattern found on the ceiling indicates the positions of the mirrors in the box. X-ray crystallography is not as simple as this illustration implies because the reflection does not indicate each atom, nor each plane of atoms, but rather the distance between planes of atoms in the crystal. By mathematical analysis, the pattern and arrangement of atoms in a crystal can then

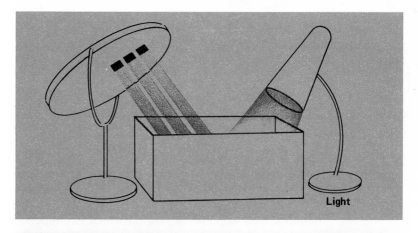

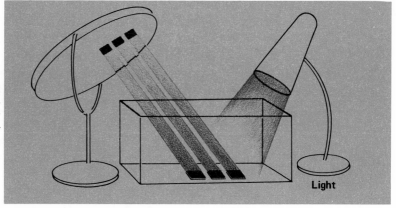

FIGURE 10–4 How light might be used to locate mirrors in a box. This is a very simplified illustration of how x-rays are used to indicate positions of atoms in molecules.

be determined. As a result of x-ray determinations, the structure of a substance like iodoform, CHI_3, is found to be that of a tetrahedron with all the C—I bond lengths equal and the I—C—I bond angles all close to 109.5°. X-rays also indicate the tetrahedral arrangement of the carbon atoms in diamond, where one carbon atom has four other carbon atoms bonded to it (Figure 10–5). This, incidentally, explains why diamond has such great hardness—it is interlaced by strong covalent bonds.

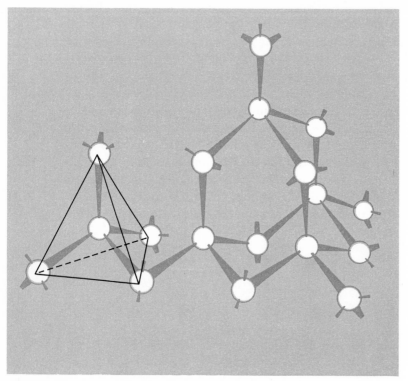

FIGURE 10–5   The tetrahedral arrangement of the carbon atoms in the diamond crystal. Each atom has four nearest neighbors, which are arranged about it at the corners of a regular tetrahedron.

Another indication of the similarity of strength and length of the bonds in methane arises from *infrared spectroscopy*. Two atoms bonded together in a molecule vibrate (move to and from each other) much like two balls on a spring (Figure 10–6). The frequency of each type of vibration for such a system depends on the mass of the balls and the attraction between them (strength of the spring). The greater the kinetic energy in the system, the greater the distance involved in the vibratory motion. However, the frequencies of vibration are independent of distance. According to Planck's theory introduced in Chapter 7, if the quantum of energy has the same frequency as one of the vibrations of the two atoms in

FIGURE 10–6   The vibration of two atoms bonded together can be compared to the vibration of two balls connected by a spring.

the bond, the atoms will absorb energy and vibrate with a greater distance between them.

In infrared spectroscopy, infrared energy of various frequencies (infrared energy is just beyond the visible region, with wavelengths of 0.0001 to 0.1 cm) is sent through the sample of material. The detector on the infrared instrument, which measures the intensity of infrared energy, will show that some energy has been absorbed at a particular wavelength and will portray the event as an absorbance peak on the trace being drawn on paper (Figure 10-7). The infrared spectrum of methane can be explained completely on the basis of the possible types of vibrations of a tetrahedral CH_4 molecule in which all four of the C—H bonds in methane are equivalent, so that this furnishes supporting evidence for such a structure. A regular tetrahedron is a four-sided geometric figure, and the four faces are triangles having equal areas.

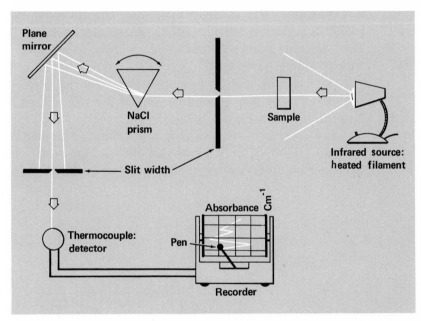

FIGURE 10-7   Outline of an infrared spectrograph. Absorbance peaks are shown on the recorder.

When all the other available evidence is combined, the tetrahedral structure of the carbon atom is indicated consistently, and it is now accepted as the correct arrangement when there are four groups bonded to a carbon atom. In the remainder of this book, many structures of compounds will be given without any attempt to relate the efforts that have been made to establish the proof of these structures. In general, the structures were established by methods and techniques similar to those discussed here.

HYBRIDIZATION

The tetrahedral structure of methane requires an adjustment of atomic theory. The isolated carbon atom in its ground state has the electron configuration

179

$1s^2 2s^2 2p_x^1 2p_y^1$ (the 2p electrons are unpaired). If the unpaired electrons in the 2p orbitals are used for bonding, this leads us to expect that carbon would form two bonds at an angle of 90°. Actually carbon forms four bonds, and each bond angle is 109°. This angle directs each bond from carbon toward each of the corners of a regular tetrahedron (Figures 10–8 and 10–9).

To explain these four equivalent bonds, we assume that prior to or during the formation of the bonds, four orbitals of the carbon atom are hybridized (or mixed) to form a new set of four equivalent orbitals. The carbon atom promotes one of the electrons from the 2s orbital to the empty $2p_z$ orbital to give the electronic arrangement $2s^1 2p_x^1 2p_y^1 2p_z^1$ for the valence shell electrons. These four orbitals are then combined mathematically according to specific rules to obtain four new equivalent hybrid orbitals. These are designated sp^3 orbitals since they are made from one s and three p orbitals. The new orbitals are directed toward the corners of a tetrahedron. The hybrid orbitals are shown graphically in Figure 10–8. The energy required for this transition is less than that which results from the formation of additional, stronger bonds with the hybrid orbitals. Thus, we have a more favorable energy change than would result from bonding with p orbitals alone.

Other sets of bonding orbitals with geometries very closely in accord with established molecular geometries can be derived in a similar fashion. Some of these are listed in Figure 10–9.

As examples of the use of hybrid orbitals, consider tetrahedral molecules such as CH_4, $SiCl_4$, and CCl_4. In all of these, the atomic orbitals of the central atom can be described as sp^3 hybrids. For molecules such as H_2O and NH_3,

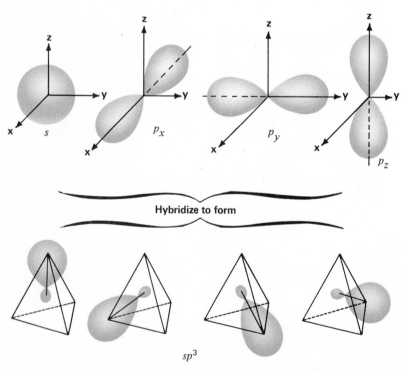

Hybridize to form

sp^3

FIGURE 10–8   Hybridization of s and p orbitals into sp^3 orbitals results in new spatial representations of the electron density patterns.

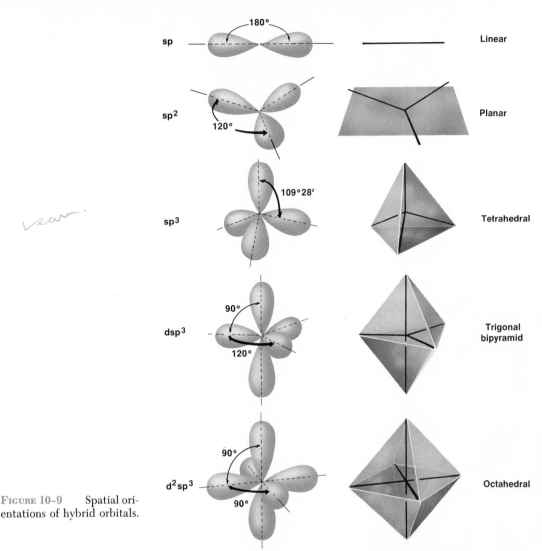

FIGURE 10-9   Spatial orientations of hybrid orbitals.

sp^3 hybrid orbitals can also be postulated if it is assumed that the nonbonding pairs of electrons (those on the O or N which are not involved in bonds to H), are hybridized along with the bonding electrons. The bond angles in these

Nonbonding pairs

H—Ö: N̈

    H   H H H

molecules would not be expected to be quite the same as normal tetrahedral bond angles. As shown in Figure 10–10, the bond angles in water and ammonia do close up somewhat.

ELECTRON-PAIR REPULSION THEORY

There is another, even simpler, notion which can be used to estimate bond angles in covalent molecules, based on the idea that electron pairs repel each other. In fact, this theory assumes that electron pairs around a central atom behave like a group of electrically charged ping-pong balls which are connected

181

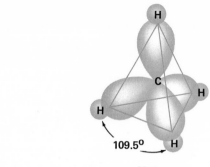

Methane

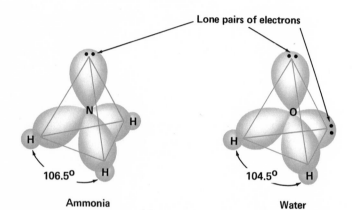

Lone pairs of electrons

Ammonia

Water

FIGURE 10-10 Molecules containing sp³ orbitals. Methane has only bonding pairs of valence electrons; water and ammonia have bonding and nonbonding pairs of valence electrons.

to a central point by strings. The balls would tend to be as far apart from each other as possible. This leads to the arrangements shown in Figure 10–11.

These ideas allow us to predict the shapes of molecules if the underlying electronic structure is known. Examples can be seen in BCl_3, $SiCl_4$, and SF_6.

BCl_3

Atomic B has $1s^2 2s^2 2p^1$ as its ground state electronic structure. One of the 2s electrons can be promoted to yield the configuration, $2s^1 2p_x^1 2p_y^1$. When these are paired up with the unpaired electrons of three chlorine atoms, $:\ddot{Cl}\cdot$, three electron-pair bonds are formed. This leads to the structure shown below:

$$\begin{array}{c} Cl \\ \cdot | \cdot \\ B \\ Cl \quad Cl \end{array} \quad 120°$$

(which actually is the structure found experimentally). The use of one s and two p orbitals to form sp^2 hybrid orbitals also leads to the same structure.

$SiCl_4$

Atomic Si has $1s^2 2s^2 2p^6 3s^2 3p_x^1 3p_y^1$ as its electronic structure. If one of the 3s electrons is promoted to the $3p_z$ orbital, four unpaired electrons will be available. When these are paired with the unpaired electrons of four chlorine atoms, $:\ddot{Cl}\cdot$, four electron-pair bonds are formed. These four pairs of electrons in the valence shell will repel each other to form the predicted tetrahedral structure.

$$\begin{array}{c} Cl \\ \cdot | \cdot \\ Si \\ \cdot | \cdot \\ Cl \\ Cl\text{-----------}Cl \end{array} \quad 109° \; 28'$$

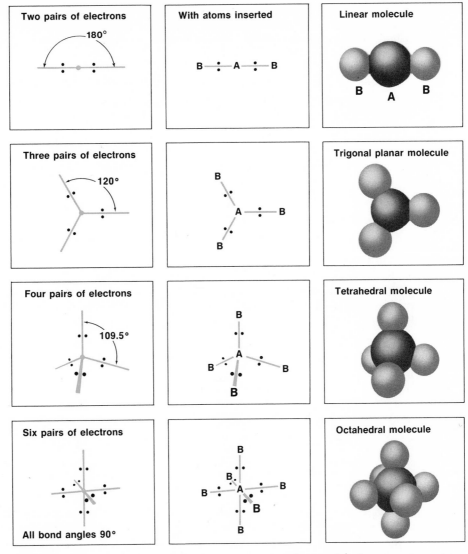

FIGURE 10-11   Arrangements of electron pairs according to the electron-pair repulsion theory.

The combination of one s and three p orbitals to form sp³ hybrids leads to the same structure. The actual structure is, indeed, tetrahedral.

SF_6

Here, the sulfur atom has the electronic configuration $1s^22s^22p^63s^23p^4$ prior to formation of the molecule. Because it has six fluorines bonded to it, the sulfur atom must use the $3s^2$ and $3p^4$ electrons to form bonding pairs. *Six* electron pairs in the valence shell lead to the prediction of an octahedral structure for the molecule, which is the structure found.

Since the sulfur forms six bonds, six orbitals must be involved. The 3s and 3p orbitals furnish four of these, and the sulfur must also use two of its empty 3d orbitals to obtain a set of sp^3d^2 hybrid orbitals to use in bonding. This also leads to an octahedral structure.

At this point you might ask how the presence of *nonbonding pairs* of electrons (in the outermost shell of the central atom) affect the disposition of the other electron pairs. In brief, we find that a pair of electrons occupies somewhat more volume when it is not involved in bonding.

Nonbonding electron pairs help to explain the structures of NH_3 and H_2O molecules discussed earlier and shown in Figure 10–10. It can be postulated, for example, that the NH_3 and H_2O molecules have bond angles (106.5° in NH_3 and 104.5° in H_2O) slightly less than the tetrahedral angle of 109°28′ for pure sp^3 hybrids because the lone, nonbonding pairs of electrons exert more repulsion than the bonding pairs do.

From the typical examples given here, it is obvious that both the electron-pair repulsion theory and the hybridization theory are consistent with experimental structures for molecules. The methods differ, however, in their approach. Hybridization utilizes the mathematical combination of s, p, and d atomic orbitals to form hybrid orbitals that are properly oriented to give the established bond angles. On the other hand, electron-pair repulsion theory assumes that electron pairs simply move as far away from each other as is possible while maintaining the proper bond length. The nonbonding pairs of electrons exert more repulsion than bonding pairs do.

MOLECULAR SHAPES OF ETHYLENE AND ACETYLENE— THE GEOMETRY AROUND DOUBLE AND TRIPLE BONDS

In Chapter 9, the electron dot structure of ethylene, C_2H_4, was written. It is shown again in Figure 10–12. Four electrons form a double bond between the two carbon atoms. Each carbon furnishes two unpaired electrons, which means two orbitals from each carbon are involved in the *double bond*. Ethylene is the simplest example of a large number of carbon compounds that have double bonds, and is discussed here as a typical example of a compound with a carbon-to-carbon double bond. This structure is important as a basis for a considerable portion of the carbon chemistry to be discussed in Chapter 14.

Experimentally, it is observed that ethylene is a planar molecule (all six atoms are in the same plane) and that all bond angles are close to 120°. Based on Figures 10–9 and 10–11, we would postulate sp^2 hybridization, which can come about in the following way. Each carbon atom can be excited to the electronic arrangement, $1s^2 2s^1 2p_x^1 2p_y^1 2p_z^1$. If the 2s and two of the p orbitals hybridize to form three sp^2 orbitals, single bonds can be formed by overlapping orbitals (Figure 10–12b). The second bond between the two carbon atoms is a lateral (or side) overlap of the two p orbitals not involved in the sp^2 hybridizations (Figure 10–12c and d). Since sp^2 hybrid orbitals are expected to give bond angles of approximately 120° and since this predicted model is planar, the theory successfully anticipates the bond angles and the flat structure of ethylene.

If two atoms are bonded by direct overlap of two s orbitals (as in H—H) or of an s orbital with a p orbital (as in HCl), the bond is called a sigma (σ)

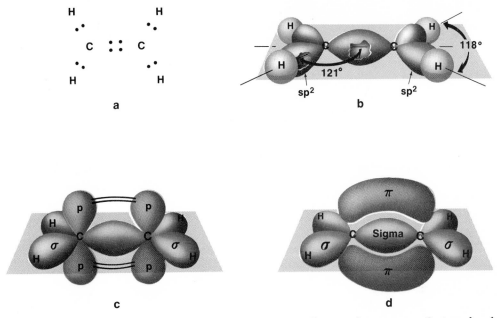

FIGURE 10-12 Structure and bonding in ethylene, C_2H_4. *A*, Electron-dot structure; *B*, sigma bond formation; *C*, assembly of p orbitals to form a π bond; *D*, representation of the sigma (σ) and pi (π) bonds.

bond (Figure 10–13). In the double bond of ethylene, two sp^2 hybrid orbitals (one from each carbon atom) overlap to form a sigma bond. The lateral overlap of p orbitals is called a pi (π) bond (Figure 10–13). The double bond in ethylene is composed of one sigma bond (sp^2—sp^2 end-on overlap) and one pi bond (p—p lateral overlap). The two CH_2 groups of the double bond in ethylene cannot rotate with respect to each other because the rotation would reduce the electron overlap of the pi bond.

Acetylene, C_2H_2, has a triple bond whose electron dot structure is shown in Figure 10–14a. Since the molecule is found experimentally to be linear (all four atoms in a line), the appropriate hybridization for the two carbon atoms would be sp hybridization. Overlap of these orbitals would occur. The other two bonds of the triple bond are formed by lateral overlap of two pairs of p orbitals (Figure 10–14b and c). Thus, the triple bond consists of one sigma bond and two pi bonds, and the predicted shape of the molecule agrees with the experimental facts.

CONCLUSION AND PREVIEW

The wave-mechanical theory has extended our knowledge of molecules. With minor modifications (such as hybridization), the theory is consistent with the established structures of molecules. When this capability is combined with its successes in accounting for bonding, periodicity, and spectra of the elements, it appears that this is a workable theory that relates macroscopic phenomena to the submicroscopic world. It is particularly amazing that a consistent, unified theory could emerge from the enormous number of complex facts about the

From two s orbitals

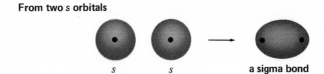

From two p orbitals (end-to-end overlap)

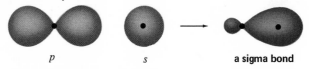

From an s and a p orbital

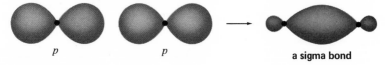

From two p orbitals (lateral overlap)

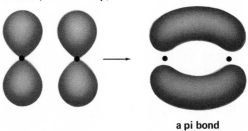

FIGURE 10-13   Ways of overlapping s and p orbitals to form sigma and pi bonds.

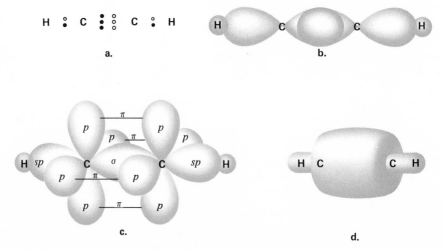

FIGURE 10-14   Structure and bonding of acetylene, C_2H_2. (a) Electron-dot structure, (b) sigma bond formation, (c) assembly of p orbitals to form two pi bonds, and (d) representation of the electron density pattern which is cylindrically symmetrical in the final molecule.

unseen submicroscopic world. Even with these great accomplishments, there are still facets of the wave-mechanical theory yet to be explored. Nevertheless, this survey of modern atomic theory has been developed and presented so that we can have a reliable theory to explain the phenomena encountered. In succeeding chapters we shall turn our attention to some of the chemical and physical properties of specific elements and compounds, using the wave-mechanical theory for an explanation of these properties.

QUESTIONS

1. Sketch separately the spatial distribution of electronic charge associated with an s, a p, and a d atomic orbital.

2. What is the meaning of the notation sp^3?

3. What is the geometry of bonding around a central atom having the following hybridization?

   (a) sp   (c) sp^3
   (b) sp^2 (d) d^2sp^3

4. What is the meaning of hybridization?

5. Account for the fact that carbon in methane, CH_4, forms four equivalent bonds, although the ground state electronic arrangement of an isolated carbon atom is $1s^2 2s^2 2p^2$.

6. For ethylene and acetylene

   (a) give the bond angles and molecular shape
   (b) account for these shapes theoretically.

7. What is a sigma bond; a pi bond?

8. If three orbitals are hybridized, how many new orbitals result?

9. Give the basic points of the valence-shell electron-pair repulsion theory.

10. Explain the 106.5° H—N—H bond angles of NH_3, the ammonia molecule.

11. List some methods used to determine molecular structures experimentally.

12. How many atomic positions must be specified in order to define a bond angle?

13. $BeCl_2$ is a molecule known to contain polar Be—Cl bonds, yet the $BeCl_2$ molecule is not polar. Explain.

14. Use the valence-shell electron-pair repulsion theory to predict the molecular shape of ethane, C_2H_6.

15. What happens to the third p orbital in sp^2 hybridization?

16. Suppose two of the sp^3 orbitals of a carbon were end-overlapped with two sp^3 orbitals of another carbon to form a double bond. If four hydrogens are bonded to the other four sp^3 orbitals of the carbon, what would be the H—C—H bond angles? Would this qualify as a possible orbital arrangement for ethylene on the basis of its consistency with the molecule being planar, its inability to twist about the carbon—carbon bond, and its bond angles of 120°?

17. In the *Suggestions for Further Reading*, the article "Structures of Noble Gas Compounds" discusses the structure of XeF_6. How many electron pairs would Xe have in its valence shell

when bonded to six fluorines? According to electron-pair repulsion theory, would this number of electron pairs predict an octahedral structure for the XeF_6 molecule?

18. There is an article in the *Suggestions for Further Reading* on nuclear magnetic resonance spectroscopy. Use this article, or another, as a basis for a report on how this powerful tool unravels molecular structure.

SUGGESTIONS FOR FURTHER READING

Bent, H. A., "The Tetrahedral Atom, Part I. Enter the Third Dimension," *Chemistry*, Vol. 39, No. 12, p. 8 (1966).

Bent, H. A., "The Tetrahedral Atom, Part II. Valence in Three Dimensions," *Chemistry*, Vol. 40, No. 1, p. 8 (1967).

Benfey, T., "Geometry and Chemical Bonding," *Chemistry*, Vol. 40, No. 5, p. 21 (1967).

Bragg, Sir Lawrence, "The Start of X-Ray Analysis," *Chemistry*, Vol. 40, No. 11, p. 8 (1967).

Brey, W., "Physical Methods for Determining Molecular Geometry," Reinhold Publishing Co., New York, 1965.

Eyring, H., and M. S. Jhon, "The Significant Structure Theory of Water," *Chemistry*, Vol. 39, No. 9, p. 8 (1966).

Gillespie, R. J., "The Valence Shell Electron Pair Repulsion Theory of Directed Valency," *Journal of Chemical Education*, Vol. 40, p. 295 (1963).

Griswold, E., "Chemical Bonding and Structure," Raytheon Education Co. (D. C. Heath), Lexington, Mass., 1968.

Jones, P. R., "Infrared Spectroscopy and Molecular Architecture," *Chemistry*, Vol. 38, No. 2, p. 5 (1965).

Lagowski, J. J., "Liquid Ammonia—A Unique Solvent," *Chemistry*, Vol. 41, No. 4, p. 10 (1968).

Luder, W. F., "Electron Repulsion Theory," *Chemistry*, Vol. 42, No. 6, p. 16 (1969).

Sheppard, W. J., "The Construction of Solid Tetrahedral and Octahedral Models," *Journal of Chemical Education*, Vol. 44, p. 683 (1967).

"Structures of Noble Gas Compounds," *Chemistry*, Vol. 39, No. 4, p. 17 (1966).

Wagner, J. J., "Nuclear Magnetic Resonance Spectroscopy—An Outline," *Chemistry*, Vol. 43, No. 3, p. 13 (1970).

Yamana, S., "An Easily Constructed Tetrahedron Model," *Journal of Chemical Education*, Vol. 45, p. 245 (1968).

PROTON (ACID-BASE) AND ELECTRON (OXIDATION-REDUCTION) TRANSFER REACTIONS

Two major classes of chemical reactions which have far-reaching applications in our lives are those of *proton transfer* and *electron transfer* between molecular or ionic species. Acids such as vinegar or lemon juice, and bases such as lime or baking soda are commonly encountered. Reactions of these acids with bases can be defined in terms of the transfer of protons from an acid to a base. Chemical oxidation and reduction are equally common. The rusting of iron, the burning of wood, and the bleaching of hair or fabrics are examples of oxidation, while the transformation of iron ore in a blast furnace to iron metal is a classic example of chemical reduction. Such redox (short for oxidation-reduction) reactions can be defined in terms of the transfer of electrons from one particle to another. It is evident that if we can understand how protons and electrons are transferred and the relative strengths of the forces involved in these transfers, we will be better able to control our environment.

Acid-base and redox reactions occur in the solid, liquid, or gaseous phases. However, the presentation will be made easier at this point if we begin our study with the liquid (solution) phase. This will require an understanding of some characteristics of the major types of solutions.

FORMATION OF SOLUTIONS

The process of solution is a familiar one. We can see sugar and salt dissolve in water, oil paints dissolve in turpentine, and grease dissolve in gasoline. Though they are less commonly observed, solid solutions such as metal alloys, and gaseous solutions like polluted air are equally possible. In a solution the substance present in greater amount is defined as the *solvent* and the one in smaller amount the *solute*. For example, in a glass of tea, water is the solvent and sugar, lemon juice,

189

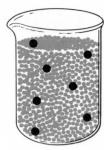

● **Sugar molecule**

· **Water molecule**

FIGURE 11-1   A schematic illustration at the molecular level of a sugar solution in water. Large circles represent the sugar molecules and the small circles water. Note that the size of the container and the size of the particles are not to scale.

and the tea itself are solutes. In this presentation of acid-base and redox reactions, most of the chemistry studied will be in water or aqueous solutions, where water is the solvent. Generally, one of the species exchanging protons or electrons is the solute. Complicating the study somewhat is the fact that solvent particles themselves can enter into both acid-base and redox reactions. Water molecules can and do exchange protons and electrons with solute particles under suitable conditions.

In theoretical terms a solution of sugar in water can be thought of in terms of an array of sugar molecules dispersed among the water molecules (Figure 11-1).

IONIC SOLUTIONS (ELECTROLYTES) AND MOLECULAR SOLUTIONS (NONELECTROLYTES)

When aqueous solutions are examined to see if they conduct electricity, we find that solutions fall into one of two categories: electrolytic solutions and

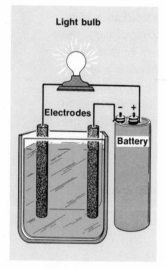

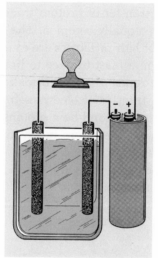

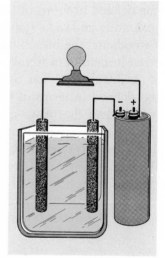

Light bulb

Electrodes

Battery

a. **Solution of table salt**
(an electrolytic solution)

b. **Solution of table sugar**
(a nonelectrolytic solution)

c. **Pure water**
(a nonelectrolyte)

FIGURE 11-2   A simple test for an electrolytic solution. In order for the light bulb to burn (a), electricity must flow from one part of the battery and return to the battery via the other part. To complete the circuit, the solution must conduct electricity. A solution of table salt, sodium chloride, results in a glowing light bulb. Hence, sodium chloride is an electrolyte. In b, the light bulb does not glow. Hence, table sugar is a nonelectrolyte. In c, it is evident that the solvent, water, does not qualify as an electrolyte since it does not conduct electricity in this test.

nonelectrolytic solutions. A simple apparatus such as that shown in Figure 11–2 can be used to determine into which classification a given solution falls.

The conductance of electrolytic solutions is readily explained, since the solute particles in such solutions are ions rather than molecules. Recall that sodium chloride crystals are composed of sodium ions, which are positively charged, and chloride ions, which are negatively charged. When sodium chloride dissolves in water, *ionic dissociation* occurs. The resulting solution (Figure 11–3a) is a

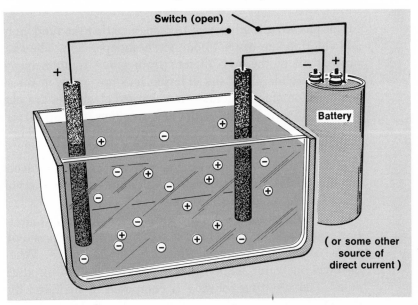

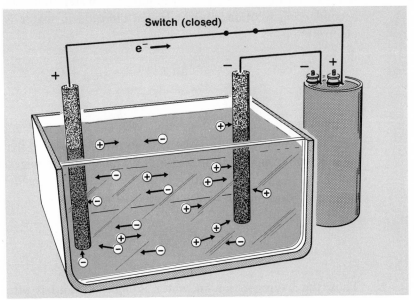

FIGURE 11-3   Conductance of electricity by ionic solution. *Upper,* Random distribution of hydrated ions in salt solution such that net charge at any point is zero. *Lower,* Negative electrode attracts positive ions (*cations*); positive electrode attracts negative ions (*anions*). If electrons are transferred from negative electrode to cation and from anion to positive electrode, the circuit is complete, and electricity will flow through the circuit.

191

constantly changing array of hydrated positive sodium ions and hydrated negative chloride ions dispersed in a water solvent. Of course, the solution as a whole is neutral since the total number of positive charges is equal to the total number of negative charges.

$$Na^+Cl^- \longrightarrow Na^+ + Cl^-$$
$$\text{(solid)} \qquad \text{(aqueous) (aqueous)}$$

In this picture it is important to recognize that the random motions of the sodium and chloride ions are not completely independent. The charges on the particles prevent all of the sodium ions from going spontaneously to one side of the container while all of the chloride ions are going to the other side. However, a net motion of ions occurs when charged electrodes are placed in an electrolyte solution (Figure 11–3b). If the negative ions give up electrons to one electrode while the positive ions receive electrons from the other electrode, a flow of electrons or electricity is produced.

Nonelectrolyte solutions composed of solvated molecules are insensitive to negatively and positively charged electrodes unless the voltage is so great that it breaks the molecules into ions.

Sometimes, ionic solutions arise when molecular materials dissolve in water. For example, hydrogen chloride, HCl, is a gas composed of diatomic molecules, each having one hydrogen and one chlorine atom. When hydrogen chloride dissolves in water an *ionization* reaction occurs, producing ions from molecules. The resulting solution is composed of hydrated hydrogen ions and hydrated chloride ions dispersed among the water molecules; consequently it is a conducting solution, and hydrogen chloride in water is properly termed an electrolyte.

$$HCl \longrightarrow H^+_{(aq)} + Cl^-_{(aq)}$$
$$\text{Molecules Yield} \qquad \text{Ions}$$

The hydrogen ion in aqueous systems is not free and unattached. Recall that water molecules are polar. A free proton (isolated hydrogen ion) could not exist in such a medium; it becomes attached to the negative end of one of the water dipoles. In fact the attraction of water dipoles for the polar HCl molecule probably brings about its ionization in the first place.

$$H^+ + H\!:\!\overset{\cdot\cdot}{\underset{\cdot\cdot}{O}}\!:\ \longrightarrow\ H\!:\!\overset{\cdot\cdot}{\underset{\cdot\cdot}{O}}\!:\!H^+$$
$$\qquad\quad H \qquad\qquad\quad H$$

Thus, the hydrogen ion in water is *hydrated* and is often referred to as the *hydronium* ion, H_3O^+ or $H^+(H_2O)$. When one considers hydrogen bonding between water molecules, it is very likely that other water molecules are attached to the molecule to which the proton is attached. The best representation we can give for the hydrogen ion in water then is $H^+(H_2O)_n$, where n is a large and constantly changing number, perhaps averaging about 4 or 5 in dilute solutions at room temperature.

CONCENTRATIONS OF SOLUTIONS

When sugar, sodium chloride, or alcohol dissolves in water, we can have either a concentrated or a dilute solution. Such a qualitative description of concentration is much less satisfactory and useful than a quantitative description which tells us just how much of a given material is dissolved in a specified volume. In chemical work we often express concentrations in moles of solute per liter of solution. Solutions of known concentration are prepared using volumetric flasks. These are glass vessels with the stems marked to indicate specific volumes, such as one liter. The procedure involves the steps shown in Figure 11–4.

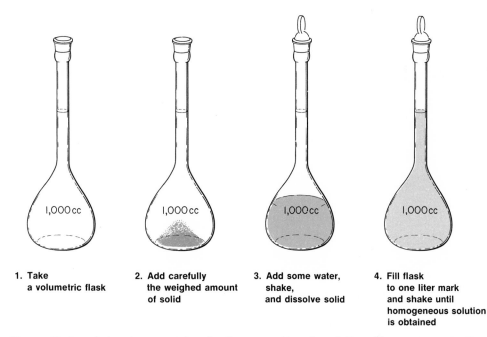

1. **Take**
   a volumetric flask

2. **Add carefully**
   the weighed amount
   of solid

3. **Add some water,**
   shake,
   and dissolve solid

4. **Fill flask**
   to one liter mark
   and shake until
   homogeneous solution
   is obtained

FIGURE 11-4   Laboratory procedure for the preparation of a solution of known concentration.

To show how concentrations are determined, let us consider a case where a 25 g sample of NaCl is carefully weighed, then transferred to a one-liter volumetric flask and dissolved in water. The next step is to add water to the flask until the solution has a total volume of one liter. In order to determine the concentration of such a solution, we need to know the number of moles of NaCl present. The formula weight of NaCl is $23 + 35.5$ or 58.5. We have 25 grams, so we have

$$\frac{25}{58.5} \text{ gram mole of NaCl per liter}$$

or 0.42 gram mole per liter. We usually indicate this as 0.42 M NaCl where M stands for gram moles per liter.

Suppose we have 86 grams of sugar ($C_{12}H_{22}O_{11}$, molecular weight 342) dissolved in a volume of 585 ml. What is the concentration of sugar in this solution? Since we have ($^{86}/_{342}$) gram mole of sugar or 0.24 gram mole of sugar, dissolved in $^{585}/_{1000}$ of a liter, the concentration of the sugar solution is

$$\frac{\dfrac{86}{342}\,\text{mole}}{\dfrac{585}{1000}\,\text{liter}} = 0.43 \text{ M sucrose.}$$

With this knowledge about solutions and their concentrations in mind, let us look at acid-base reactions occurring in solution.

ACIDS AND BASES

The terms "acid" and "base" have been used by chemists for several hundred years. These names were originally given to substances which showed certain properties. For example, acids have long been characterized as substances which are sour tasting, corrosive, and able to react with substances called bases. Bases, on the other hand, have a bitter taste, make the fingers or skin feel slippery on contact, and react with acids. As more and more information was collected on the properties of acids and bases, these simple definitions had to be refined. This has been carried to the point where current definitions of acids and bases are structural in nature.

PROTON TRANSFER AND NEUTRALIZATION

The definitions that we will use here are those first given by J. N. Brønsted and T. M. Lowry, in 1923.

Brønsted acid: A chemical species which can *donate* hydrogen ions (also called protons or H^+ ions) is an acid.
Brønsted base: A chemical species which can *accept* hydrogen ions is a base.

To illustrate these definitions we again consider the reaction between gaseous hydrogen chloride (HCl) and water:

$$\text{HCl(gas)} + \text{H}_2\text{O} \longrightarrow \text{H}_3\text{O}^+ + \text{Cl}^-$$

Acid	*Base*	*Hydronium ion*	*Chloride ion*
		(an acid)	*(a weak base)*

Hydrochloric acid

Examination of the above reaction shows that the HCl molecule has donated a proton (H^+ ion) to the water molecule. This behavior is understandable in terms of the electronic structures of the reacting molecules.

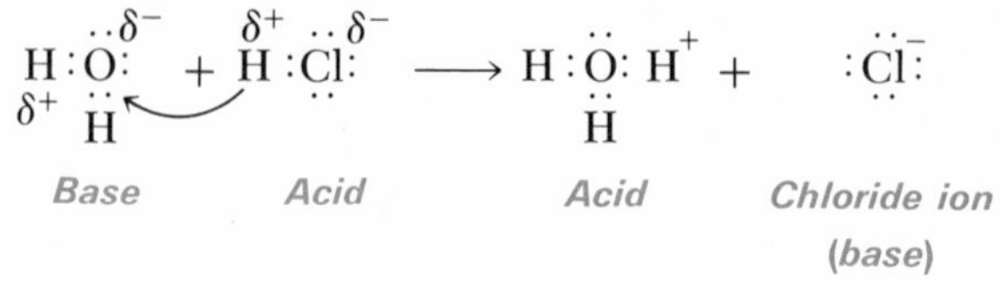

Base	*Acid*	*Acid*	*Chloride ion*
			(base)

194

This reaction is reversible, but when equilibrium is established, essentially all the HCl has been converted to H_3O^+ and Cl^-. A concentrated (~ 12 M) solution of hydrogen chloride in water is mostly a solution of H_3O^+ ions (hydronium ions) and Cl^- (chloride) ions.

If the ionic solid sodium oxide, Na_2O, is dissolved in water, a vigorous reaction occurs producing a solution containing sodium ions (Na^+) and hydroxide ions (OH^-). In this process the oxide ion (O^{2-}) reacts with water to form the hydroxide ion. In this, as in other such aqueous reactions, it is understood that the ions are hydrated (i.e., water molecules are bonded to them on a transitory basis).

$$:\overset{..}{\underset{..}{O}}:^{2-} \ + \ H:\overset{..}{\underset{\underset{\displaystyle H}{|}}{O}}: \ \longrightarrow \ 2:\overset{..}{\underset{..}{O}}:H^-$$

Base   Acid   Hydroxide ion

There are many other bases that take a proton from a water molecule in this way, for example:

$$NH_3 \ + \ H_2O \ \rightleftharpoons \ NH_4^+ \ + \ OH^-$$

Ammonia Acid   Acid Base
(base)

$$CN^- \ + \ H_2O \ \rightleftharpoons \ HCN \ + \ OH^-$$

Cyanide Acid   Acid Base
(base)

According to the Brønsted–Lowry definition, water acts as an acid in these reactions and donates a proton to the other molecule or ion which acts as a base. A species such as water that can either donate or accept protons is called *amphiprotic*. The existence of amphiprotic species implies that acid-base reactions possess a reciprocal nature; an acid and a base react to form another acid (to which a proton has just been added) and another base (from which a proton has just been removed). Because water is the most commonly used solvent, it is also the most usual reference compound for acid-base reactions.

In principle, at least, one water molecule can transfer a proton to another water molecule.

$$2H_2O \ \rightleftharpoons \ H_3O^+ \ + \ OH^-$$

This reaction takes place to only a very small extent, as is indicated by our arrows of unequal length. Since H_3O^+ and OH^- are produced in equal amounts when only water is present, pure water is neither acidic nor basic, but is described as neutral.

A chemical species in water solution is commonly spoken of as an acid if it donates protons to water and increases the concentration of H_3O^+ or $H^+_{(aq)}$. Similarly, a base in water solution is commonly described as a compound whose addition to water will increase the concentration of OH^-. Since water is not the only possible solvent, these concepts are too narrow for general scientific use; they have been extended by the definitions given above, which focus on the essential feature of such acid-base behavior—that is, the donation or acceptance of a proton (H^+) in a reaction.

It has long been known that where acids react with bases the properties of both species disappear. The process involved is called *neutralization*. To get a more precise picture of acid-base neutralization reactions, let us consider what happens when a solution of hydrochloric acid is mixed with a solution of sodium hydroxide. The hydrochloric acid contains H_3O^+ and Cl^- ions; the sodium hydroxide solution contains Na^+ and OH^- ions. When these two solutions are mixed, a reaction occurs between H_3O^+ and OH^-.

$$Na^+ + \underset{\text{Acid}}{Cl^- + H_3O^+} + \underset{\text{Base}}{OH^-} \longrightarrow H_2O + H_2O + Na^+ + Cl^-$$

If we have an equal number of H_3O^+ and OH^- ions, they will react to produce a neutral solution, with the hydronium ions (H_3O^+) donating their protons to the hydroxide ions (OH^-), forming molecules of water. Such reactions are called *neutralization* reactions because the acids and bases neutralize each other's properties. If we have more H_3O^+ ions than OH^- ions, the extra H_3O^+ will make the resulting solution *acidic*. If we have more OH^- ions than H_3O^+ ions, only a fraction of the OH^- ions will be neutralized, and the extra OH^- ions will make the resulting solution *basic*.

THE STRENGTHS OF ACIDS AND BASES

Because the water molecule is itself a weak base, the strongest acid (that is, the best proton donor) that can exist in water is the hydronium ion (H_3O^+). If *strong acids* such as perchloric acid ($HClO_4$) or nitric acid (HNO_3) are added to water they *donate all of their protons to water*, and the resultant solutions contain H_3O^+ and ClO_4^-, or H_3O^+ and NO_3^-. All of these reactions are reversible, at least in principle, but when strong acids are dissolved in water the equilibria that result favor the reaction products to a very great extent.

Reactants   Products

$$\underset{\text{Acid}}{HNO_3} + \underset{\text{Base}}{H_2O} \longrightarrow H_3O^+ + NO_3^-$$

$$\underset{\text{Acid}}{H_2SO_4} + \underset{\text{Base}}{H_2O} \longrightarrow \underset{\text{Acid}}{H_3O^+} + \underset{\text{Base}}{HSO_4^-}$$

Since there are a large number of ions present, the solution conducts electricity well; HNO_3 and H_2SO_4 are termed strong electrolytes.

Not all acids lose protons as readily to water as do nitric acid and sulfuric acid. Some acid anions, because of their structures, are capable of competing with water for the proton being exchanged. The result of this competition is the establishment of an equilibrium between neutral acid molecules and hydronium ions in water solution.

Acetic acid, the acid found in vinegar, is one of these weak acids. The molecular structure for acetic acid ($HC_2H_3O_2$) is as follows:

Acidic hydrogen

The hydrogen atom bonded to the oxygen in the molecule is the only one that is donated to a base in water solution; for that reason it is designated an acidic hydrogen. When acetic acid is dissolved in water, some ions are produced. However, most of the acetic acid molecules do not donate protons to water molecules. The result is an equilibrium mixture containing a few H_3O^+ and $C_2H_3O_2^-$ ions; neutral acetic acid molecules are more abundant than acetate ions in the solution. The reaction here is:

$$HC_2H_3O_2 + H_2O \rightleftarrows H_3O^+ + C_2H_3O_2^-$$

Acetic Acid   Water   Hydronium ion   Acetate ion

The relatively few ions in an acetic acid solution do not conduct electricity very effectively; consequently, acetic acid is a weak electrolyte.

Another way of looking at this reaction is to realize that in the equilibrium mixture there are two bases: the water molecule H_2O, and the acetate ion $C_2H_3O_2^-$. Since the reverse reaction is more important in determining the position of equilibrium, the acetate ion must be a stronger base than the water molecule. Table 11–1 gives some common acids and bases ranked according to their relative strengths.

TABLE 11-1   RELATIVE STRENGTHS OF SOME ACIDS AND BASES

INCREASING ACID STRENGTH						INCREASING BASE STRENGTH
perchloric acid,	$HClO_4$	+	$H_2O \rightleftarrows H_3O^+$	+	ClO_4^-	
hydrochloric acid,	HCl	+	$H_2O \rightleftarrows H_3O^+$	+	Cl^-	
nitric acid,	HNO_3	+	$H_2O \rightleftarrows H_3O^+$	+	NO_3^-	
hydronium ion,	H_3O^+	+	$H_2O \rightleftarrows H_3O^+$	+	H_2O	
hydrofluoric acid,	HF	+	$H_2O \rightleftarrows H_3O^+$	+	F^-	
acetic acid,	$HC_2H_3O_2$	+	$H_2O \rightleftarrows H_3O^+$	+	$C_2H_3O_2^-$	
water,	H_2O	+	$H_2O \rightleftarrows H_3O^+$	+	OH^-	
hydroxide ion,	OH^-	+	$H_2O \rightleftarrows H_3O^+$	+	O^{2-}	

THE pH SCALE

Because water solutions of dilute acids and bases are used so extensively, it is convenient to have a simple way of designating the acidity or basicity of these solutions. The pH scale was devised for this purpose; it furnishes a number which describes the acidity of a solution. The pH is defined as follows: $pH = -\log[H_3O^+]$. The pH scale, for practical purposes, runs from 0 to 14. A pH of 7 indicates a neutral solution like pure water, a pH below 7 indicates an acid solution, and a pH above 7 indicates a basic solution.

The pH number is related to the concentration of the hydrogen ions in an aqueous solution expressed as a power of 10. Actually pH is the negative of the power. A hydrogen ion concentration of 0.00001 M or 1×10^{-5} has a pH of 5. To have a pH of 7 for pure water, the hydrogen ion concentration would have to be 1×10^{-7} M.

To understand this we will need to explore the very slight acid-base reaction   197 between water molecules:

$$H_2O + H_2O \rightleftharpoons H_3O^+ + OH^-$$

The reaction is an ionization reaction producing ions which should cause water to conduct electricity. But, we have said that water is a nonelectrolyte. Actually, with very sensitive electrical equipment, pure water can be shown to conduct electricity *very slightly* due to the very small amount of ionization. Laboratory measurements reveal that 0.0000001 or 1×10^{-7} mole per liter of water ionize at 25°C. Pure water contains 55 moles of water per liter of volume. Consequently, we can see that the actual amount of the water that ionizes is very small compared to the total amount of water present. Pure water, then, which is defined to be neutral, contains 1×10^{-7} mole per liter of hydrogen ions (or hydronium ions), and its pH is 7. In pure water the concentrations of H_3O^+ and OH^- ions must be equal since each time an H_3O^+ ion is produced in the ionization reaction an OH^- ion is also produced. Therefore, in pure water the concentration of the OH^- ion is also 1×10^{-7} mole per liter.

Figure 11–5 graphically displays the relationship between the pH number and the concentration of the hydrogen ion. It also gives the approximate pH values for common solutions.

A close examination of Figure 11–5 reveals that basic solutions such as ammonia water have hydrogen ion concentrations less than that of pure water.

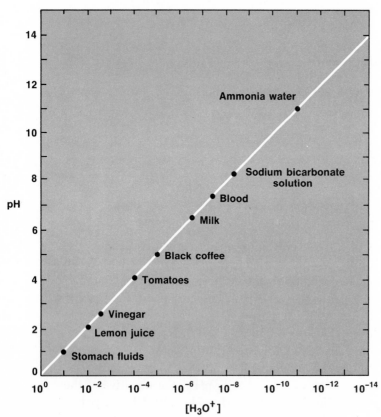

FIGURE 11-5 A plot of pH versus hydrogen in concentration, $[H_3O^+]$. Note that the pH *increases* as the $[H_3O^+]$ decreases. The pH values of some common fluids are given for reference.

Note that the pH of a typical sample of ammonia water is 11. This is equivalent to a hydrogen ion concentration of 1×10^{-11} or 0.00000000001 mole per liter. You may wonder how it is possible for ammonia water to have fewer hydrogen ions than an equal volume of water. The answer to this puzzle lies in the fact that this ionization reaction, like all chemical reactions, will reach a state of chemical equilibrium. In the following paragraphs an explanation is given, but a facility with exponential notation and logs is required of the reader.

The usual quantitative definition of pH is:

$$pH = -\log[H_3O^+]\,°$$

where $[H_3O^+]$ is the hydrogen ion concentration of the solution in moles per liter. A 0.1 M solution of HCl has $[H_3O^+] = 0.1$, because HCl ionizes completely in water:

$$HCl + H_2O \longrightarrow H_3O^+ + Cl^-$$

For this solution, the pH is given by the expression:

$$pH = -\log[H_3O^+] = -\log(0.1)$$

The log of 0.1 is -1, so

$$pH = -(-1) = +1$$

For neutral water we find $[H_3O^+] = [OH^-] = 10^{-7}$ mole liter, so its pH is:

$$pH = -\log[H_3O^+] = -\log(10^{-7}) = -(-7) = +7$$

As you can see, solutions with pH values *less* than 7 are acidic. What about basic solutions? What happens to the hydrogen ion concentration in a water solution which is also 0.1 M in NaOH? First we have to recall the equilibrium which occurs in *all* aqueous solutions:

$$2H_2O \rightleftharpoons H_3O^+ + OH^-$$

It has been shown experimentally that the product of the hydrogen ion concentration and hydroxide ion concentration is the same constant value for all aqueous solutions at the same temperature.

$$[H_3O^+][OH^-] = a \text{ constant}$$

This constant is called the ion product of water, is given the symbol K_W, and has the experimental value of about 10^{-14} at 25°C.

$$K_W = [H_3O^+][OH^-] = 10^{-14}$$

° The logarithm of a number is the exponent or power to which 10 must be raised to equal the number. The log of 10^{-7} is -7; the log of 4×10^{-7} is -6.4 because $10^{-6.4}$ equals 4×10^{-7}.

Since the product is always 10^{-14} at $25°C$, an acid with a high $[H_3O^+]$ will have a low $[OH^-]$, and a basic solution with a high $[OH^-]$ will have a low $[H_3O^+]$. The adjustment of the concentrations that ensures the constant product comes through shifting the water equilibrium. For example, if sufficient acid is added to water to provide a hydrogen ion concentration of 0.1 M, enough of the hydrogen ions react with the hydroxide ions present in water to lower the $[OH^-]$ from its value of 10^{-7} M in neutral water to the much smaller concentration of about 10^{-13} M.

Returning to our 0.1 M NaOH solution, we know that this compound produces Na^+ and OH^- ions in solution and that it is completely dissociated. From this we can say:

$$NaOH + H_2O \xrightarrow[100\%]{} Na^+_{(aq)} + OH^-_{(aq)}$$

$$[OH^-] = 0.1 \text{ M}$$

The $[H_3O^+]$ value can now be obtained because we know that

$$[H_3O^+][OH^-] = 10^{-14}, \text{ so}$$

$$[H_3O^+](0.1) = 10^{-14}, \text{ and}$$

$$[H_3O^+] = 10^{-13} \text{ M}$$

The pH of this solution is then:

$$pH = -\log[H_3O^+] = -\log(10^{-13}) = -(-13) = +13$$

SALTS

PREPARATION OF SALTS

The chemical compounds known as *salts* play a vital role in nature, in animal growth and life, and in the manufacture of various chemicals for human use. They are formed as the products of an acid-base neutralization, as in the example:

$$(K^+ + OH^-) + (H_3O^+ + Cl^-) \longrightarrow \rightleftharpoons K^+ + Cl^- + 2\ H_2O$$

Potassium hydroxide   Hydrochloric acid
in water   in water

Crystallize out
the salt
KCl (solid)

Salts contain species held together by *ionic bonding* (see Chapter 9). Potassium chloride, for example, is composed of an equal number of K^+ ions and Cl^- ions arranged in definite positions with respect to one another in a lattice (see Figure 9–5). This lattice when extended in three dimensions includes vast numbers of ions which make up a crystal. The salt crystal must be electrically neutral; it can have neither an excess nor a deficiency of positive or negative charge.

Let us imagine that we have at our disposal the ions listed below, and let us see what salts could result.

IONS		POSSIBLE SALTS	SALT NAME
Na^+	sodium	NaCl	sodium chloride
Ca^{2+}	calcium	$NaNO_3$	sodium nitrate
Cl^-	chloride	Na_2SO_4	sodium sulfate
NO_3^-	nitrate	$NaC_2H_3O_2$	sodium acetate
SO_4^{2-}	sulfate	$CaCl_2$	calcium chloride
$C_2H_3O_2^-$	acetate	$Ca(NO_3)_2$	calcium nitrate
		$Ca(C_2H_3O_2)_2$	calcium acetate
		$CaSO_4$	calcium sulfate

In the examples just given, notice that in order to attain an electrically neutral lattice, it is necessary to balance the charges of the ions. A sodium (Na^+) ion requires just one chloride ion (Cl^-), and the NaCl lattice contains an equal number of Na^+ and Cl^- ions. A sulfate ion (SO_4^{2-}) with two negative charges must have its negative charge balanced by two positive charges. This may be done by using two Na^+ ions:

$$2Na^+ + SO_4^{2-} \longrightarrow Na_2SO_4$$

or one Ca^{2+} ion:

$$Ca^{2+} + SO_4^{2-} \longrightarrow CaSO_4$$

It is possible to form many solid salts by mixing water solutions of different soluble salts with each other. For example, both lead acetate and sodium chloride are soluble in water. If we prepare solutions of these salts and then mix the solutions, we find the insoluble salt, lead chloride, precipitates from the mixture.

$$\underbrace{2Na^+ + 2Cl^-}_{\substack{\text{Sodium chloride} \\ \text{in solution}}} + \underbrace{Pb^{2+} + 2C_2H_3O_2^-}_{\substack{\text{Lead acetate in} \\ \text{solution}}} \longrightarrow \underbrace{PbCl_2}_{\substack{\text{Solid} \\ \text{lead} \\ \text{chloride}}} + \underbrace{2Na^+ + 2C_2H_3O_2^-}_{\substack{\text{Sodium acetate in} \\ \text{solution}}}$$

SALTS IN SOLUTION

An important property of salts is their ability to dissolve in suitable solvents to yield solutions. A salt solution is a homogeneous mixture of solvent molecules and ions (frequently bonded to solvent molecules). The amount of a salt that will dissolve in a given quantity of solvent tells us the salt's solubility in that solvent. The preparation of lead chloride shown above is an example of a reaction made possible by the differences in solubilities of different salts in the same solvent. It is because of these differences in solubilities that we can make roads out of calcium carbonate (limestone), which is insoluble in water, but not calcium chloride, which is water soluble.

Because of the amount of water on the earth and the importance of water as a major component of living matter, the study of aqueous (water) solutions of salts is important in understanding ourselves and the changing earth about us. Consider what happens at the ionic level when a sodium chloride crystal is placed in contact with water. We know that sodium chloride is soluble in water. This means that most, if not all, of the potential energy in the crystal lattice is somehow overcome in the solution process.

The surface of the salt crystal appears calm when the crystal is placed in water, but on the ionic level there is a great deal of agitation. Water molecules are trying to pull the ions of the salt away from the lattice and into solution. Water molecules have sufficient polarity to interact strongly with the ions and tie them up. Once this takes place the ion is removed from the crystal and the crystal lattice now has a gap in it where the ion was removed. Ions that are tied up by solvent molecules are termed solvated ions and the process of ion-solvent interaction is called solvation. But just how are these ions solvated? What causes this solvation?

The answers lie in the structure of the water molecule and its ability to interact with ions. As we saw in Chapters 8 and 9, water is a *polar* molecule. Because of its polar nature, the water molecule is ideally suited to interact with ions. The negative end of the molecule is attracted to positive ions, and the positive end is attracted to negative ions. As a result, several water molecules will interact with each ion. Figure 11–6 shows several water molecules solvating a Na^+ ion. The positive ends of the water molecule will tend to interact with a negative ion; this is shown for the Cl^- ion in Figure 11–6.

We have learned earlier that energy is required to break chemical bonds, or conversely, energy is released when a bond is formed. The solvation process we have just described involves the breaking of ionic bonds and the formation of ion-solvent molecule bonds. The energy required to separate all of the ions of a salt is a measure of the *lattice energy*. Lattice energies are relatively large, but they are partially compensated for by the energy liberated in the solvation of the ions called the *solvation energy*.

When the dissolving of a salt such as calcium chloride ($CaCl_2$) causes the solution to become warmer, the solvation energy must be greater than the lattice energy. On the other hand, the dissolving of a salt like ammonium chloride (NH_4Cl) causes the solution to become cooler because the lattice energy is greater than the solvation energy, so energy must be taken from the system to dissolve the salt. Some salts like NaCl dissolve without much change in temperature because the lattice energy approximately equals the solvation energy. Salts like AgCl have a much greater lattice energy than solvation energy and are, therefore, practically insoluble in water. These energy changes are illustrated in the graphs of Figure 11–7.

All salts have what may be termed "solubility limits" for a given solvent: at a given temperature, a certain number of grams of salt, and *no more*, will dissolve in 100 grams of solvent. A solution that contains all the dissolved salt that it can hold is termed a "saturated solution." One might ask just why this type of solubility limit is found in nature. The reason for this can be understood if we remember that the oppositely charged ions of the salt in solution actually attract one another. If one crowds the solution with solvated ions to too great an extent and "ties up too many solvent molecules," the ions will begin re-forming the crystal lattice. This is solvation in reverse. When undissolved salt is in contact with a saturated solution of that salt, there is a dynamic equilibrium

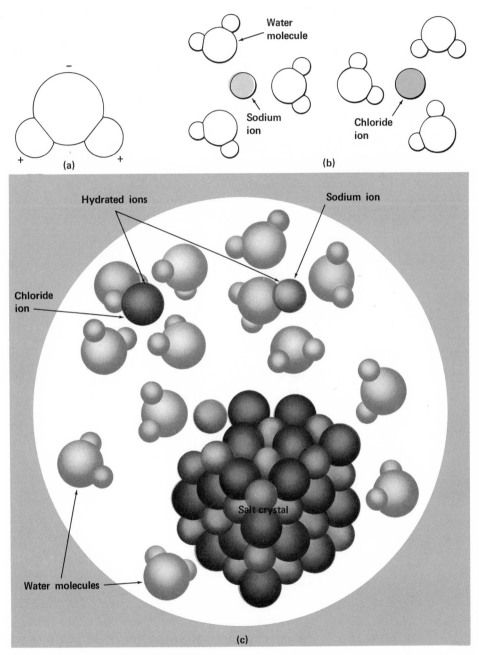

FIGURE 11-6 Dissolution of sodium chloride in water. (a) Geometry of water molecule, a polar molecule. (b) Solvation of sodium and chloride ions due to interaction (bonding) between these ions and water molecules. (c) Dissolution occurs as collisions between water molecules and crystal ions result in the removal of the crystal ion. In the process the ion becomes completely solvated.

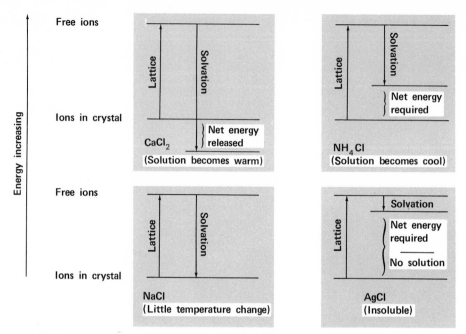

FIGURE 11-7   Energy relationships associated with the solution of salts. Free ions are the ions of the salt separated from each other.

established with the salt crystal being broken down at one point while being formed in another. The effect of temperature on the solubility of some common salts is shown in Figure 11–8.

USES OF SALTS

Salts are used for so many different applications that it would be impossible to list them all. We can, however, look at the uses of several salts and see just how versatile they are.

Sodium Chloride (Table Salt)

This is the best-known salt and perhaps the most important to man. Physiologically, mammals need this salt for an ionic balance in the bloodstream and for the production of hydrochloric acid in the stomach to aid in the digestion of food. Very early in man's history, sodium chloride was used in barter, and wars have actually been fought for control of salt deposits. Commonly, sodium chloride is used for medicinal purposes, such as in water solutions, to wash wounds and to treat burns.

Nitrate Salts

Salts containing the nitrate ion (NO_3^-), such as sodium nitrate, ($NaNO_3$), potassium nitrate (KNO_3), and ammonium nitrate (NH_4NO_3), find special uses in fertilizers and in the manufacture of explosives. Sodium nitrate and potassium

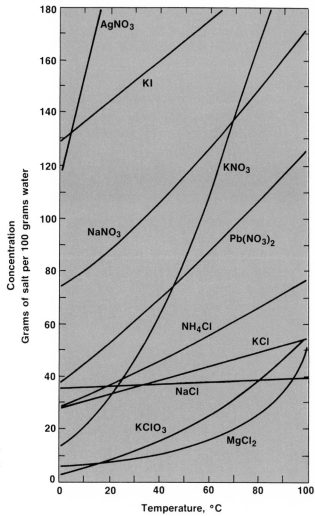

FigURE 11-8   The effect of temperature
on the solubility of some common salts in
water.

nitrate occur in natural deposits while ammonium nitrate can be prepared by
a reaction involving ammonia (NH_3) and nitric acid.

$$NH_3 + HNO_3 \rightleftharpoons NH_4NO_3$$

Base   Acid   Ammonium
nitrate salt

Explosions are possible when mixtures of nitrate salts and combustible
materials are mixed. Gunpowder contains KNO_3. Ammonium nitrate will explode
by itself under certain conditions, and a number of disasters have occurred when
large amounts of this salt have exploded in warehouses and in the holds of
ships in a harbor. Under more controlled conditions, NH_4NO_3 decomposes to
yield dinitrogen oxide (N_2O) and water. Dinitrogen oxide is called laughing gas
and is used as an anesthetic in certain forms of surgery.

$$NH_4NO_3 \xrightarrow{Heat} N_2O + 2H_2O$$

Dinitrogen
oxide
(laughing gas)

205

This gas also has a limited solubility in milk and cream and finds a use in canned whipped cream. When the valve is opened, the cream and gas escape together, and upon reaching the atmosphere the gas expands, thus "whipping" the cream.

Sodium Hydrogen Carbonate

Sodium hydrogen carbonate, also called sodium bicarbonate ($NaHCO_3$), has the common name, baking soda. The negative ion of the salt, the hydrogen carbonate ion, is a base and is capable of reacting with water in an acid-base reaction.

$$HCO_3^- + H_2O \rightleftharpoons H_2CO_3 + OH^-$$

Base   Acid   Acid   Base

When this salt is placed in a mixture containing water, such as biscuit batter, this reaction takes place to a slight extent. The product of the reaction, carbonic acid (H_2CO_3), is quite unstable and tends to decompose under the influence of heat, yielding the products CO_2 and water. The decomposition of the H_2CO_3 causes the reaction of bicarbonate with water to take place to a greater extent. In effect, the equilibrium is shifted to the products side, and the overall reaction is essentially a complete conversion of the bicarbonate ion to CO_2 and water. The reaction is:

$$HCO_3^- + H_2O \longrightarrow \begin{cases} H_2CO_3 \xrightarrow{heat} H_2O + CO_2 \\ + OH^- \qquad\qquad\qquad (Gas) \end{cases}$$

The CO_2 gas which is produced in the reaction causes the biscuit batter to rise. Baking powder contains a weak acid which shifts the equilibrium to the right. Self-rising flour contains a small amount of baking soda and a weak acid to produce a controlled amount of CO_2 gas as the mixture is heated in the oven.

The HCO_3^- ion can also act as a weak acid:

$$HCO_3^- + OH^- \rightleftharpoons HOH + CO_3^{2-}$$

Acid   Base   Acid   Base

This explains its use as a neutralizer of spilled acids and bases in the laboratory.

Barium Sulfate

This salt may be prepared by taking advantage of its insolubility in water. If a solution of barium chloride ($BaCl_2$) is mixed with sodium sulfate solution (Na_2SO_4), insoluble $BaSO_4$ precipitates:

$$(Ba^{2+} + 2Cl^-) + (2Na^+ + SO_4^{2-}) \longrightarrow BaSO_4 \text{ (Solid)} + (2Na^+ + 2Cl^-)$$

In the diagnosis of stomach ulcers and intestinal disorders x-rays are often used. Since these organs are normally transparent to x-rays, a suitable material must be placed in the organ to render it partially opaque. Barium is a good element for this since it is essentially opaque toward the type of x-rays used in medicine. Barium ions, however, are quite poisonous and cannot be taken internally. The answer lies in the extreme insolubility of $BaSO_4$. A syrup containing $BaSO_4$ is swallowed to introduce the barium into the digestive system. There is no danger of barium poisoning when using this method, and after diagnosis the salt passes from the body essentially unaltered.

ELECTRON TRANSFER OXIDATION-REDUCTION

A very common type of reaction is one in which the control of valence shell electrons passes from one species to another. These reactions are called oxidation-reduction reactions. Examples include:

$$2Na + Cl_2 \longrightarrow 2NaCl\,(Na^+ + Cl^-)$$
$$2H_2 + O_2 \longrightarrow 2H_2O$$
$$2Fe + 3Br_2 \longrightarrow 2FeBr_3$$

Because the degree of control of atoms over their valence shell electrons is not a very accurately defined concept, we take the elements in the starting compounds as standards and compare these to the atoms in the compound formed as products. If the atom has either lost electrons or lost some control over the electrons, we say it has been *oxidized*. If it has gained electrons or gained a greater degree of control over the valence shell electrons, we say it has been *reduced*. In the reactions above, the sodium, hydrogen, and iron atoms have all lost electrons, so these elements have been oxidized. The species which have caused this change (Cl_2, O_2, and Br_2) are *oxidizing agents*. Since these molecules have picked up electrons, they have been reduced, and the species which caused this change (Na, H_2, and Fe) are *reducing agents*.

The term oxidation has its historical background in the fact that the first oxidizing agent whose chemical behavior was thoroughly studied was oxygen. The phenomenon was then named after the element. As the understanding of chemical reactions deepened, it became apparent that the reaction of a metal with oxygen was very similar to its reaction with fluorine, chlorine, or bromine. Thus, in each of the following reactions:

$$4Fe + 3O_2 \longrightarrow 2Fe_2O_3$$
$$2Fe + 3F_2 \longrightarrow 2FeF_3$$
$$2Fe + 3Cl_2 \longrightarrow 2FeCl_3$$

the iron loses valence electrons to the other reactant. After this similarity was noted, the concept of oxidation was generalized to cover all situations in which an atom, ion, or molecule lost electrons. In the same manner the concept of reduction is now used to cover all situations in which an atom, ion, or molecule gains electrons.

The formation of a salt by the direct combination of the elements is an example of an oxidation-reduction reaction. When sodium reacts with chlorine to form sodium chloride, the following reactions take place.

$$Na \longrightarrow Na^+ + e^- \qquad \textit{(Oxidation of sodium)}$$
$$Cl_2 + 2e^- \longrightarrow 2Cl^- \qquad \textit{(Reduction of chlorine)}$$

Since two electrons are gained per chlorine molecule, two sodium atoms must be oxidized for every Cl_2 molecule reduced; the chemical reaction is

$$2Na + Cl_2 \longrightarrow 2Na^+ + 2Cl^-$$

207

In the above reaction, chlorine is the oxidizing agent, the species that brings about the oxidation of the sodium. In a similar sense the sodium is termed the reducing agent since it brings about the reduction of the chlorine.

REDUCTION OF METALS

One of the most practical applications of the oxidation-reduction principle is the winning of metals from their ores. The majority of metals are found in nature in compounds; that is, the metals are in an oxidized state. In order to obtain the metal, it must be reduced. In addition to electricity there are a variety of chemicals which can supply the electrons for the reduction process. Some of the ways that metals can be reduced from their ores are given in Table 11–2.

TABLE 11-2   SOME WAYS USED TO WIN METALS FROM THEIR ORES BY REDUCTION

METAL	OCCURRENCE	A REDUCTION PROCESS	USES OF METAL
Cu	Cu_2S, chalcocite	Air blown through melted ore $$Cu_2S + O_2 \longrightarrow 2\ Cu + SO_2\uparrow$$ $(Cu^+ + e^- \longrightarrow Cu)$	Electrical wiring, boilers, pipes, brass (Cu 85%, Zn), bronze (Cu 90%, Sn)
Na	NaCl, rock salt	Electrolysis of fused chloride $$2\ NaCl \longrightarrow 2\ Na + Cl_2$$ $(Na^+ + e^- \longrightarrow Na)$	Coolant in nuclear reactors
Mg	$MgCl_2$, seawater	Electrolysis of fused chloride $$MgCl_2 \longrightarrow Mg + Cl_2$$ $(Mg^{2+} + 2\ e^- \longrightarrow Mg)$	Light alloys duralumin (.5% Mg, Al), dorrmetal H (90.7% Mg, Al, Zn, Mn), flares, some flash bulbs
Ca	$CaCO_3$, chalk, limestone, marble	Thermal reduction for very pure Ca $$CaCO_3 + 2\ HCl \longrightarrow CaCl_2 + H_2O + CO_2$$ $$3\ CaCl_2 + 2\ Al \xrightarrow{heat} 3\ Ca + 2\ AlCl_3$$ $(Ca^{2+} + 2\ e^- \longrightarrow Ca)$	Bearing metal alloys (0.7% Ca, Pb, Na, Li)
Al	$Al_2O_3 \cdot H_2O$, bauxite	Electrolysis in fused cryolite, Na_3AlF_6, at 800–900°C. $$2\ Al_2O_3 \longrightarrow 4\ Al + 3\ O_2$$ $(Al^{3+} + 3\ e^- \longrightarrow Al)$	Packaging, airplane parts, alloys, roofing, siding
Ag	Ag_2S, argentite	Reduction by a more active metal $$Ag_2S + 4\ CN^- \rightleftharpoons 2\ Ag(CN)_2^- + S^{2-}$$ $$2\ Ag(CN)_2^- + Zn \longrightarrow Zn(CN)_4^{2-} + 2\ Ag\downarrow$$ $(Ag^+ + e^- \longrightarrow Ag)$	Jewelry, electrical wiring
Zn	ZnS, zinc blende	Roasting in air with carbon $$ZnS + 2\ O_2 + C \longrightarrow Zn + SO_2\uparrow + CO_2\uparrow$$ $(Zn^{2+} + 2\ e^- \longrightarrow Zn)$	Galvanizing iron sheet
Hg	HgS, cinnabar	Roasting in air $$HgS + O_2 \xrightarrow{heat} Hg\uparrow + SO_2\uparrow$$ $(Hg^{2+} + 2\ e^- \longrightarrow Hg)$	Electrodes, arc rectifiers
Fe	Fe_2O_3, haematite	Reduction by carbon monoxide $$Fe_2O_3 + 3\ CO \longrightarrow 3\ CO_2 + 2\ Fe$$ $(Fe^{3+} + 3\ e^- \longrightarrow Fe)$	Alloyed with C (0.1 to 1.5%) to make steels; (stainless steel has 15% Cr, Fe, C)

ELECTROLYSIS

Several metals are either won from their ores or purified by electrolysis as noted in Table 11–2. *Electrolysis* is the term used to designate a chemical reaction carried out by supplying electrical energy.

Figure 11–9 outlines the principal parts of an electrolysis apparatus. Electrical contact between the external circuit and the solution is obtained by means of electrodes, which are often made of graphite or metal. The electrode at which electrons enter the cell is termed the *cathode,* and the electrode at which they leave is the *anode.* The battery produces a direct current of electrons which flow toward one electrode (the cathode) and away from the other electrode (the anode). The cathode is defined as the electrode at which reduction takes place; the anode as the electrode at which oxidation takes place. A salt solution can easily be observed to conduct electricity if a light bulb is part of the circuit. When the switch is closed, the positive ions in solution migrate toward the cathode. Soon, evidence of a chemical reaction can be seen at the electrodes. Depending on the substances present in the solution, gases may be evolved or metals formed at the electrodes. The ions that migrate to the electrodes are not necessarily the species undergoing reaction at the electrodes because sometimes the solvent undergoes reaction more easily. Whatever happens, the chemical reactions that are taking place at the cathode and anode are due to electrons being forced through the solution. The chemical reaction at the cathode is one that uses up electrons (reduction). At the anode, electrons are taken from species in solution, so the chemical reaction at the anode is one that gives up electrons (oxidation).

Figure 11–9 is a specific illustration of the electrolysis of a copper sulfate solution. Such an electrolysis can be used either to plate an object with a layer of pure copper or to purify an impure sample of copper metal; copper metal

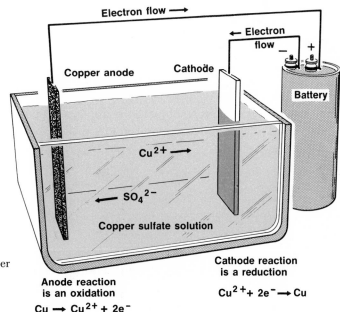

FIGURE 11–9   Electrolysis of a copper sulfate solution.

Electron flow ⟶

← Electron flow

Copper anode   Cathode

Battery

Cu^{2+} ⟶

← SO_4^{2-}

Copper sulfate solution

Anode reaction is an oxidation
$Cu \rightarrow Cu^{2+} + 2e^-$

Cathode reaction is a reduction
$Cu^{2+} + 2e^- \rightarrow Cu$

is transferred from the positive electrode to the negative electrode. If the positive electrode is impure copper to be purified, electrolysis deposits the copper as very pure copper on the negative electrode. If one desires to plate an object with copper, he has only to render the surface conducting and make it the negative electrode; it will become coated with copper, with the copper coating growing thicker as the electrolysis is continued. If the object is a metal, it will conduct electricity by itself. If it is a nonmetal, its surface can be lightly dusted with graphite powder to render it conducting.

Now let us examine how the electrolysis transfers the copper from the positive electrode to the negative electrode. Electrons flow out of the negative terminal of the battery, through the wire, and into the negative electrode. Somehow this negative charge must be used up at the surface of the electrode.

How? Consider what happens when the electrons build up on the negative electrode (cathode). Since the positive copper ions are nearby, they will be attracted to the surface and take electrons from it. Thus, the Cu^{2+} ions are reduced:

$$Cu^{2+} + 2e^- \longrightarrow Cu$$

In a similar way the negative sulfate ions migrate to the positive electrode (anode). However, it turns out that it is easier to get electrons from the copper metal of the electrode than it is from the sulfate ions. As each copper atom gives up two electrons, it passes into solution:

$$Cu \longrightarrow Cu^{2+} + 2e^-$$

In effect, then, the copper of the positive electrode (anode) passes into solution; the copper ions in solution migrate to the negative electrode (cathode) and plate out as copper metal. The sulfate ions move in such a way as to keep the solution neutral. Large amounts of copper are purified in this way each year. Silver and gold plating can be done in a similar fashion.

RELATIVE STRENGTHS OF OXIDIZING AND REDUCING AGENTS

When a piece of metallic zinc is placed in a solution containing blue copper ions (Cu^{2+}), an oxidation-reduction reaction occurs:

$$Zn + Cu^{2+} \longrightarrow Zn^{2+} + Cu$$

Evidence for this reaction is the deposit of copper on the zinc and the gradual decrease in the intensity of the blue color of the solution indicating that the Cu^{2+} ions are being removed.

The oxidation of zinc by copper ions can be thought of as a competition between zinc ions (Zn^{2+}) and copper ions (Cu^{2+}) for the two electrons. Since the reaction proceeds almost to completion, the Cu^{2+} ions obviously win out in the competition.

The *activity* of a metal is a measure of its tendency to lose electrons. Zinc is a more active metal than copper on the basis of this experiment. This means that given an equal opportunity, the first reaction will take place to a greater extent:

$$Zn \longrightarrow Zn^{2+} + 2e^-$$
$$Cu \longrightarrow Cu^{2+} + 2e^-$$

In terms of electronic theory, it can be said that a copper atom has a greater attraction for its outermost electrons than does a zinc atom.

Experiments of this type with various pairs of metals and other reducing agents yield an activity series of the elements which ranks the various oxidizing and reducing agents according to their *strengths* or *tendency* for the electron transfer to take place. An iron nail will be partly dissolved in a solution of a copper salt containing Cu^{2+} ions, with copper being deposited on the nail that remains. From this, it is determined that iron, like zinc, is more active than copper. The reaction that takes place is:

$$Fe + Cu^{2+} \longrightarrow Fe^{2+} + Cu.$$

Now, which is more active, zinc or iron? This question can be answered by placing an iron nail in a solution containing Zn^{2+} ions and, in a separate container, a strip of zinc in a solution containing Fe^{2+} ions. It is found that the zinc strip is eated away in the solution containing Fe^{2+} ions. The reaction is

$$Zn + Fe^{2+} \longrightarrow Fe + Zn^{2+}$$

Nothing happens to the iron nail in the solution of Zn^{2+} ions. We deduce that Zn loses electrons more readily than Fe.

Such an activity series can be extended to include other metals and even nonmetals as well. The concentrations of the ions in solution and other factors often must be considered for accurate work, but for our purposes these will be ignored. Table 11–3 gives an activity series of some oxidizing and reducing agents.

TABLE 11–3   RELATIVE STRENGTHS OF SOME OXIDIZING AND REDUCING AGENTS

	REDUCING AGENTS			OXIDIZING AGENTS	
INCREASING STRENGTH OF REDUCING AGENTS	Na	$\rightleftharpoons$	$1\ e^-$	+	Na^+
	Ca	$\rightleftharpoons$	$2\ e^-$	+	Ca^{2+}
	Mg	$\rightleftharpoons$	$2\ e^-$	+	Mg^{2+}
	Zn	$\rightleftharpoons$	$2\ e^-$	+	Zn^{2+}
	Fe	$\rightleftharpoons$	$2\ e^-$	+	Fe^{2+}
	H_2	$\rightleftharpoons$	$2\ e^-$	+	$2\ H^+$
	Cu	$\rightleftharpoons$	$2\ e^-$	+	Cu^{2+}
	Fe^{+2}	$\rightleftharpoons$	$1\ e^-$	+	Fe^{3+}
	Ag	$\rightleftharpoons$	$1\ e^-$	+	Ag^+
	$2\ H_2O$	$\rightleftharpoons$	$4\ e^-$	+	$4\ H^+ + O_2$
	$2\ F^-$	$\rightleftharpoons$	$2\ e^-$	+	F_2

INCREASING STRENGTH OF OXIDIZING AGENTS

An application of these ideas can be seen in the use of *cathodic protection* to reduce corrosion. For this, the metal to be protected, say an iron pipeline, is connected to a more active metal, such as magnesium. In moist earth, an electrolytic cell is set up in which the magnesium is oxidized and transfers electrons to the iron. This keeps the iron from being oxidized by the air or any other reagent which might otherwise remove electrons from it.

The value of Table 11–3 is that it allows us to predict the feasibility of reactions involving the species in it. Thus, a reducing agent in the table is able to form any species below it from the oxidized form of that species. For example, sodium can form calcium metal from calcium ions:

$$2Na + Ca^{2+} \longrightarrow Ca + 2Na^+$$

and also zinc from zinc ions:

$$2Na + Zn^{2+} \longrightarrow Zn + 2Na^+$$

Zinc can reduce silver ions:

$$Zn + 2Ag^+ \longrightarrow 2Ag + Zn^{2+}$$

but not calcium ions:

$$Zn + Ca^{2+} \longrightarrow\!\!\!\times\!\!\!\longrightarrow \textbf{\textit{No reaction}}$$

The series arranges oxidizing agents in the order of their effectiveness also. Fluorine, F_2, can oxidize water, silver, ferrous ion, or any species above F^- in the listing. Cupric ion, Cu^{2+}, can oxidize H_2, Fe, Zn, and the metals above Cu in the table, because it has a greater tendency to take on electrons than the ions that are formed. Thus:

$$Cu^{++} + Fe \longrightarrow Fe^{++} + Cu$$
$$Cu^{++} + Mg \longrightarrow Mg^{++} + Cu$$

but not

$$Cu^{++} + Ag \longrightarrow\!\!\!\times\!\!\!\longrightarrow \textbf{\textit{No reaction}}$$

BATTERIES

One of the most useful applications of oxidation-reduction reactions is in the production of electrical energy. A device that produces an electron flow (current) is called an electrical cell. Although a series of cells is a battery, we shall use the term battery, exclusively. Consider the reaction between zinc and copper ions that was previously discussed. If zinc is placed in a solution containing Cu^{2+} ions, the electron transfer takes place between the zinc metal and the copper ions, and the energy liberated simply causes a slight heating of the solution and the zinc strip.

212

If the zinc can be separated from the copper solution, and the two connected in such a way to allow current flow, the reaction proceeds, but now the electrons are transferred through the connecting wires. Figure 11–10 shows a battery that can be constructed to make use of the energy evolved in the reaction of Zn with Cu^{2+}.

The anode reaction is the oxidation of zinc to Zn^{2+} ions.

$$Zn \longrightarrow Zn^{2+} + 2e^-$$

The electrons flow from the Zn electrode through the connecting wire, light the lamp in the circuit, and then flow into the copper cathode where reduction of Cu^{2+} ions takes place:

$$Cu^{2+} + 2e^- \longrightarrow Cu \ \textit{(Metal)}$$

The copper is deposited on the copper cathode.

This flow of electrons (negative charge) from the anode to the cathode compartment in the battery must be neutralized electrically. A "salt bridge" is provided which connects the two compartments. Negative ions (SO_4^{2-}) flow through this bridge readily since it is constructed of a salt solution such as K_2SO_4 in a glass tube with porous plugs at either end. During the operation of the battery, a number of negative charges just equal to the number of electrons used

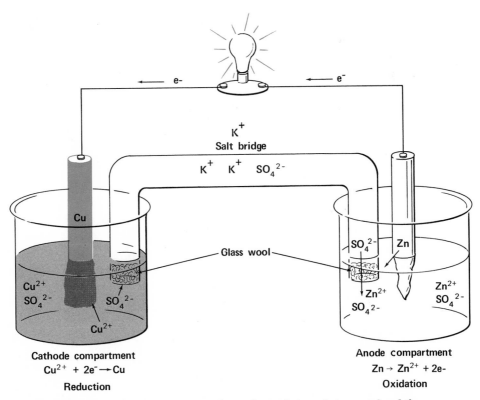

FIGURE 11–10   A simple battery involving the oxidation of zinc metal and the reduction of Cu^{2+} ions.

TABLE 11-4   CHARACTERISTICS OF SOME BATTERIES

SYSTEM	ANODE (OXIDATION)	CATHODE (REDUCTION)	ELECTROLYTE	TYPICAL OPERATING VOLTAGE PER CELL
Dry cell	Zn	MnO_2	NH_4Cl—$ZnCl_2$	0.9—1.4 volts
Edison storage	Fe	Ni oxides	KOH	1.2—1.4
Cadmium	Cd	Ni oxides	KOH	1.1—1.3
	Cd	AgO	KOH	1.0—1.1
Lead storage	Pb	PbO_2	H_2SO_4	1.95—2.05
Mercury cell (used in transistor radios)	Zn(Hg)	HgO	KOH—ZnO	1.30

at the cathode will pass through the bridge to the Zn^{2+} solution. Likewise, the copper solution is kept neutral by negative SO_4^{2-} ions flowing into the salt bridge and positive K^+ ions flowing into the solution. The reaction of zinc with Cu^{2+} continues until the battery runs down; that is, until equilibrium is attained.

Many different oxidation-reduction combinations are used in a suitable battery to produce electrons. A few of the more popular ones are listed in Table 11–4.

CONCLUSION AND PREVIEW

The general principles examined in this chapter can be extended to explain the principal types of reactions of simple inorganic compounds. There are other examples of chemical equilibria, acid-base behavior, and oxidation-reduction reactions with far more complex molecules, but the basic features of these processes are the same as with very simple compounds. It will be found that an understanding of these phenomena provides a firm basis for extrapolation and analogies to explain similar behavior in complicated organic or biochemical molecules.

QUESTIONS

1. Define acid-base reactions in terms of protons, and oxidation-reduction reactions in terms of electrons.

2. Define the solute and solvent in (a) coffee, (b) a 5 per cent alcohol in water solution, (c) a 5 per cent water in alcohol solution, and (d) a solution of 50 per cent alcohol and 50 per cent water.

3. What is the one test that all aqueous electrolytes must pass?

4. Give an example of ionic dissociation. Give an example of ionization. What is the difference between the two?

5. What is the difference between the hydrogen ion and the hydronium ion?

6. NaCl has a formula weight of 58.5. If 58.5 g of this salt is dissolved in enough water to make a liter of solution, what is the molarity of the solution? What would be the molarity if 2 liters of solution were made with the same amount of salt? 0.5 liter?

7. Write the equation for a chemical reaction in which water acts as a Brønsted acid; as a Brønsted base.

8. A solution of hydrochloric acid is electrolyzed. The products are hydrogen at the cathode and chlorine at the anode. Write the reactions occurring at each electrode and tell which ions move toward the cathode and which toward the anode.

9. Classify each of the following as acids or bases, using the Brønsted-Lowry definitions: H_2SO_4, CO_3^{2-}, Cl^-, HCO_3^-, O^{2-}, H_2O.

10. Predict the formulas of salts formed between the following pairs of ions:

$$Na^+ \text{ and } SO_4^{2-}$$
$$Ca^{2+} \text{ and } I^-$$
$$Mg^{2+} \text{ and } NO_3^-$$
$$Ca^{2+} \text{ and } PO_4^{3-}$$
$$K^+ \text{ and } Br^-$$

11. Moist baking soda is often put on acid burns. Why? Write an equation for the reaction assuming the acid to be hydrochloric (HCl).

12. Using the table of relative strengths of acids and bases, predict whether the products or reactants would be favored in the following reactions:

$$HClO_4 + C_2H_3O_2^- \rightleftharpoons HC_2H_3O_2 + ClO_4^-$$
$$HF + OH^- \rightleftharpoons H_2O + F^-$$
$$O^{2-} + HNO_3 \rightleftharpoons OH^- + NO_3^-$$

13. Using the table giving the relative oxidizing and reducing powers, predict whether or not the following reactions would be expected to proceed.

$$Ag + Na^+ \rightleftharpoons Ag^+ + Na$$
$$H_2 + Cu^{2+} \rightleftharpoons Cu + 2H^+$$
$$F_2 + 2Fe^{2+} \rightleftharpoons 2Fe^{3+} + 2F^-$$
$$Mg + 2Ag^+ \rightleftharpoons 2Ag + Mg^{2+}$$

14. A certain salt, MX, is found to raise the temperature when dissolved in water. Which is the larger quantity, the solvation energy or the lattice energy? Explain.

15. Describe what happens when an ionic solid dissolves in water.

16. What ions are present in water solutions of the following salts: Na_2SO_4, $CaBr_2$, $Mg(NO_3)_2$?

NOTE: The following questions require a thorough understanding of the mathematical relationships presented in this chapter.

17. What is the molar concentration of a solution prepared by dissolving 90 grams of acetic acid, CH_3COOH, in a liter of solution?

18. What is the pH of a 0.001 M solution of HCl?

19. What is the hydrogen ion concentration in a solution with a pH of 8?

20. What is the hydroxide ion concentration in a solution with a pH of 5?

21. Analysis of 20 ml of a solution shows that it contains 1.0 gram of NaCl. What is the molarity of this solution?

22. A solution is prepared by mixing 500 ml of 2 M NaOH with 500 ml of 2 M HCl. What is its pH?

23. Show by calculation that pure water is 55.6 M.

24. How many grams of the salt will be required to make each of the following solutions?

    (a) 20 ml of 0.1 M KCl
    (b) 250 ml of 0.3 M $AgNO_3$
    (c) 2 1 of 4.0 M Na_2CO_3
    (d) 400 ml of 0.23 M $Fe(NO_3)_3 \cdot 9H_2O$

25. A saturated solution of KCl, at 20°C, contains 34.7 grams per liter. What is the molarity of a solution which is prepared when 100 ml of this concentrated solution is taken and diluted up to one liter with distilled water?

26. What is the molar concentration of water in a solution which contains 48 grams of water in sufficient ethyl alcohol to bring the total volume up to one liter?

27. When water trickles through limestone ($CaCO_3$), the water dissolves some of it. Much drinking water is obtained from such sources. If you drink 4 liters of water a day, all of which is saturated with calcium carbonate (0.00153 gram of $CaCO_3$ per 100 ml at 25°C), how many moles of calcium carbonate will you ingest from this source in the course of a week? If the recommended daily adult intake of calcium is 800 mg per day (as Ca^{2+}), is it feasible to satisfy your calcium requirement by drinking such water?

28. A 0.01 M solution of sodium hydroxide is combined with an equal volume of 0.005 M hydrochloric acid. What is the pH of the resultant solution?

SUGGESTIONS FOR FURTHER READING

Eyring, H., and MuShik, J., "Significant Structure Theory of Water," *Chemistry*, Vol. 39, No. 9, p. 8 (1966).

Haggin, J., "New Method of Iron Production," *Chemistry*, Vol. 39, No. 5, p. 24 (1966).

Manufacturing Chemists' Association, "Ammonia, Industry Profile," *Chemistry*, Vol. 39, No. 6, p. 14 (1966).

Morris, D. L., "Brønsted-Lowry Acid-Base Theory—A Brief Survey, *Chemistry*, Vol. 43, No. 3, p. 18 (1970).

Newman, D. S., "Fused Salts," *Chemistry*, Vol. 39, No. 4, p. 7 (1966).

Schaar, B. E., "Chance Favors the Prepared Mind. VI. Nobel's Dynamite," *Chemistry*, Vol. 39, No. 6, p. 19 (1966).

Schmuckler, J. S., and Mogue, P. H.: "Dilute Solutions of Strong Acids: The Effect of Water on pH," *Chemistry*, Vol. 42, No. 9, p. 14 (1969).

Steele, D., "The Chemistry of the Metallic Elements," Pergamon Press, New York, 1966.

SCIENCE AND TECHNOLOGY AS A NEW PHILOSOPHY

CHAPTER 12

A PHILOSOPHICAL AND HISTORICAL BACKGROUND

The various factors which led to the systematic development of science and its associated technologies in the Western World are both numerous and complex. Rather than attempt to develop an overall "complete" view, we will list only a few of the empirical results. The topic as a whole is the subject of active investigation and controversy by scholars. One of the key points seems to be the development of printing techniques which enormously increased the availability of the accumulated knowledge of mankind. Another seems to be the continuous refusal to accept authority as the ultimate judge of truth in certain areas; this indeed still seems to be one of the major differences between mathematics and the sciences and other areas of human knowledge. No one ever states, as proof of the validity of a piece of scientific information, that so-and-so said this and therefore it must be true. This skeptical attitude is to be found throughout all nations at all times; however, it was only in Western Europe that such a frame of mind came to be cultivated systematically and ultimately formed the basis of certain intellectual organizations whose primary purpose was the search for and propagation of a kind of impersonal knowledge.

The most important aspects of this type of knowledge are the ways in which it changes. With the passage of time, it becomes more extensive, more accurate, more broadly disseminated, *and*, most important of all, more concisely summarized in terms of generalizations or their equivalent: mathematical equations giving the quantitative relationships connecting a set of properties.

Because the physical sciences deal with our environment, this growth of knowledge leads to an ever increasing understanding of the physical world and the ways it can be manipulated to obtain desired ends. This manipulation of our environment can attain a high degree of sophistication. Presently, modern technology has two types of roots. The one derives from the practical knowledge of craftsmen and frequently has been transmitted orally or by apprenticeships. The other is in the sciences and has been transmitted by publications of various sorts and formal education. The dual roots tend to fuse in practice and reinforce each other. The older method of oral tradition had its ups and downs, as is especially obvious in certain areas where knowledge of techniques has died out (e.g., the early Middle Ages in Europe as contrasted with the Roman Empire). The newer method of transmission by publication tends to make such knowledge available to all who can read and who have the required training for understanding.

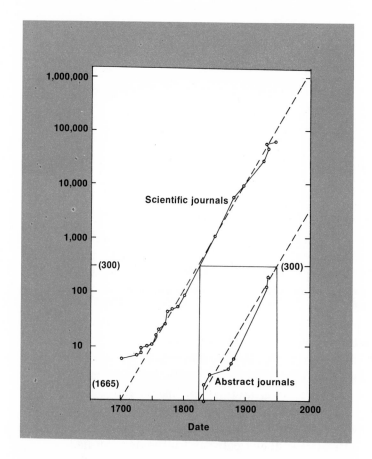

FIGURE 12-1   Rate of increase in number of scientific journals since 1665. (Reproduced with permission of Yale University Press.)

As time has passed, the effectiveness of this new method of collecting and extending knowledge has become more apparent to the ordinary citizen. It has also become accepted by scholars in more disciplines and has led to the extension of this method to new areas of knowledge which previously had not been the subject of systematic study.

At this point one might well ask for a definition of science or the scientific method. While it is difficult to give such definitions which are acceptable to all, a close approximation may be achieved. First of all, "science" consists of that which has been collected by the use of the scientific method. The term "scientific method" refers more to a mental attitude than to a procedure. The attitude is one which requires the strictest intellectual honesty in the collection

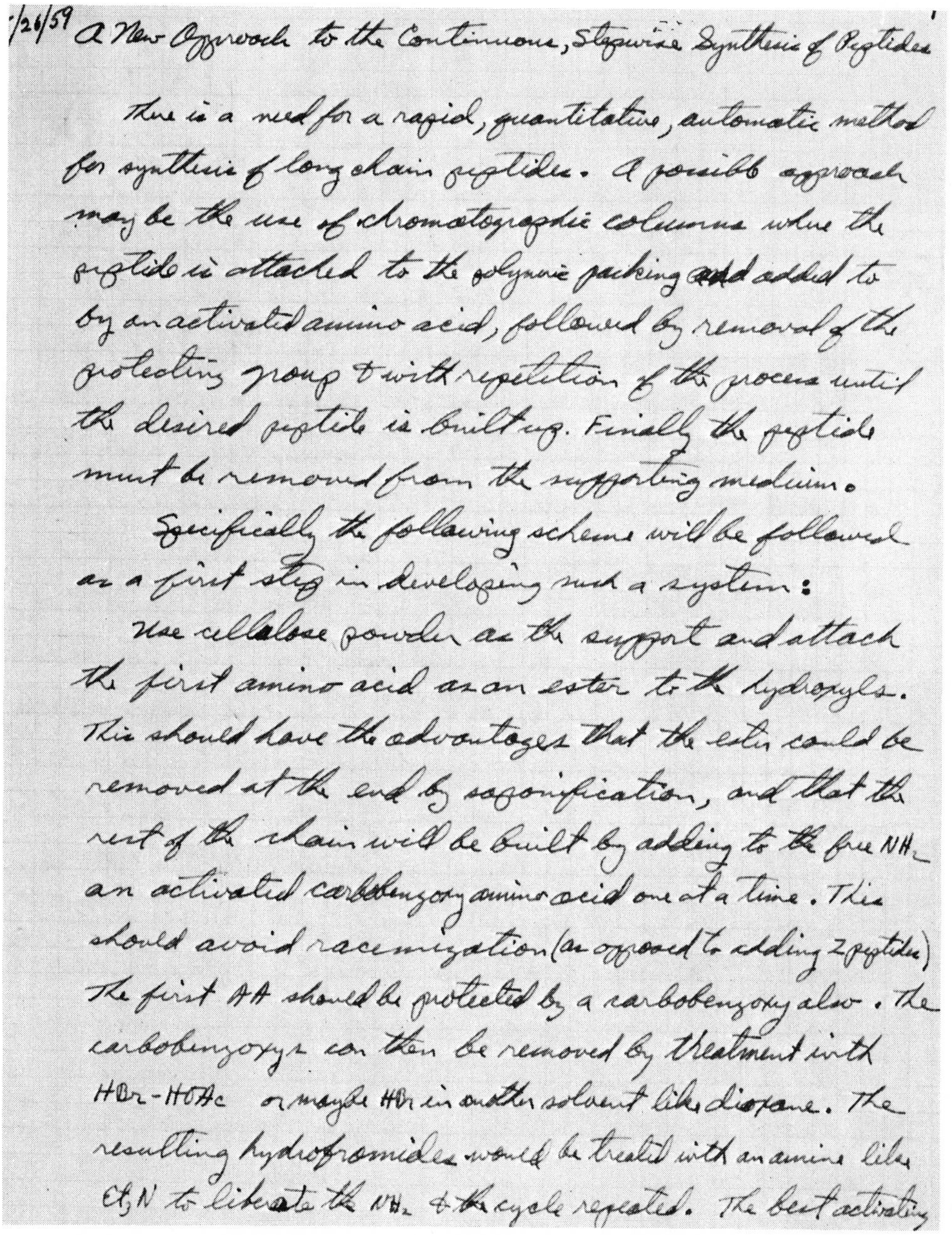

7/26/59 *A New Approach to the Continuous, Stepwise Synthesis of Peptides*

There is a need for a rapid, quantitative, automatic method for synthesis of long chain peptides. A possible approach may be the use of chromatographic columns where the peptide is attached to the polymeric packing and added to by an activated amino acid, followed by removal of the protecting group & with repetition of the process until the desired peptide is built up. Finally the peptide must be removed from the supporting medium.

Specifically the following scheme will be followed as a first step in developing such a system:

Use cellulose powder as the support and attach the first amino acid as an ester to the hydroxyls. This should have the advantages that the ester could be removed at the end by saponification, and that the rest of the chain will be built by adding to the free NH_2 an activated carbobenzoxy amino acid one at a time. This should avoid racemization (as opposed to adding 2 peptides) The first AA should be protected by a carbobenzoxy also. The carbobenzoxys can then be removed by treatment with $HBr-HOAc$ or maybe HBr in another solvent like dioxane. The resulting hydrobromides would be treated with an amine like Et_3N to liberate the NH_2 & the cycle repeated. The best activating

FIGURE 12–2   A page from the notes of Dr. Bruce Merrifield (Professor of Biochemistry at Rockefeller University), the designer of a protein-making machine. Progress in science is made through disciplined human thought, taking advantage of previous knowledge and useful theories. See Figure 17–26. (From Chemical and Engineering News, August 2, 1971.)

219

of data and, hopefully, in the arrangement of this data in a pattern that reveals the underlying basis of the observed behavior. The data normally must be collected under conditions which can be reproduced anywhere in the world, so that new data can be obtained to confirm or to refute the old (geologists and astronomers are excused from these rigid requirements). The results obtained are thus independent of differences in language, culture, religion, or economic status in the various observers and represent a unique type of truth.

Over the last 200 years this accumulated knowledge has been put to use on a very extensive scale in Europe and in those areas of the world which have had the means to follow the example set by England, which underwent this "industrial revolution" first. The result has been the development of a society largely dependent upon and supported by a technology which is itself undergoing constant change. The first consequence of this technology has been to increase the *rate* at which things can be produced. This in turn has continually changed the occupational patterns of millions of human beings, and has brought forcefully to mind the persistence of *change* in our pattern of life.

FIGURE 12-3   Name some discoveries that led to this change in transportation. How many of them are chemical in nature? In this text we will discuss some changes in transportation.

These changes have influenced profoundly the way in which people think about their material wants and the ways in which they can be satisfied. For example, there seems to be little argument with the statement, "If the number of human beings on the earth could be stabilized, a much higher standard of living could prevail over most of the earth." A statement such as this would have been greeted with widespread derision 500 years ago. Today, people *expect* the continuation of technological innovation to solve many of their own and the world's problems. This attitude has given rise to the phrase, "the revolution of rising expectations," to describe such a state of mind.

TECHNOLOGY: ITS TRIUMPHS AND PROBLEMS

Almost as soon as the industrial revolution began in England, the public realized that technological progress brought with it a series of problems. The first to be noticed was the necessity for progress to be accompanied by changing patterns of employment.

It is obvious that if a machine makes as much thread as 100 men can make,

FUMIFUGIUM:

O R,

The Inconvenience of the AER,

A N D

SMOAKE of LONDON

DISSIPATED.

TOGETHER

With ſome REMEDIES humbly propoſed

By J. E. Eſq;

To His Sacred MAJESTIE,

A N D

To the PARLIAMENT now Aſſembled.

Publiſhed by His Majeſties Command.

Lucret. l. 5.

Carbonumque gravis vis, atque odor inſinuatur
Quam facile in cerebrum?——

LONDON:

Printed by W. GODBID, for GABRIEL BEDEL, and THOMAS
COLLINS; and are to be ſold at their Shop at the Middle
Temple Gate, neer Temple Bar. M.DC.LXI.

Re-printed for B. WHITE, at Horace's Head, in Fleet-ſtreet.
M DCC LXXII.

FIGURE 12-4   Title page from J. Evelyn, F.R.S., *The Smoake of London.* The Latin quotation is from the Roman poet Lucretius (97–53 B.C.). It may be translated, "How easily the heavy potency of carbons and odors sneaks into the brain!" (Courtesy A. E. Gunther and the University Press, Oxford.)

the men are released to do other work. The 100 men, however, do not look on this as an advantage, especially if they are settled in their place of employment with their families. The new opportunities that result from such a machine are rarely of benefit directly to the men displaced. The wealth of their country is increased since there are now 100 men able to do other work. However, the initial reaction of the men in 19th century England was to riot and break up the machinery.

The increased use of fuels of all sorts, especially the introduction of coal and coke into metallurgical plants, led to widespread problems with air pollution which were recognized and discussed over 100 years ago.

The most important point of these results is the realization that technological progress is always obtained at some cost, and the cost may not be obvious at the outset.

A very important technological development was recognized as necessary in 1890 by Sir William Crooks, who addressed the British Association for the Advancement of Science on the problem of the fixed nitrogen supply (that is, nitrogen in a chemical form plants are capable of using in growth). At that time, scientists recognized that nitrogen compounds were necessary in fertilizers and that the future food supply of mankind would be determined by the amount of nitrogen compounds which could be made available for this purpose. The source of these nitrogen supplies was then limited to rapidly depleting supplies of guano in Peru and to sodium nitrate in Chile. It was realized that when these were exhausted, widespread famine would result *unless* an alternative supply could be developed. This problem was recognized first by English scientists as a potentially acute one because by the 1890's England had become very dependent upon *imported food supplies*.

Widespread interest in this problem led to research on a number of chemical reactions by which nitrogen can be obtained from the relatively inexhaustible supply present in the air. Air is 21 per cent oxygen (O_2) and 79 per cent nitrogen (N_2). The nitrogen in the air is present as the rather unreactive molecule N_2, and in this form it can be used as a source of other nitrogen compounds only by a few kinds of bacteria. Some is also transformed into NO by lightning, and when this is washed into the soil by rain it can be utilized by plants. Needless

FIGURE 12-5   An early German ammonia plant. (From "The Realm of Chemistry," Econ-Verlag Gmb H, Düsseldorf, 1965.)

FIGURE 12-6   Man usually has the capability of making very large quantities of materials that he finds useful. Sometimes mass production precedes a complete understanding of the consequences. In this industrial process, ribbons of soap are produced and are subsequently converted into flakes or bars. (Courtesy of the Procter and Gamble Company.)

to say, the amount of nitrogen transformed into chemical compounds useful to plants by these processes is quite limited.

Several chemical reactions were developed to form useful compounds from atmospheric nitrogen, but the best known and most widely used one has an ironic history. While England was interested in nitrogen for fertilizers, Germany was interested in nitrogen for explosives. The German General Staff realized that the British Navy could blockade German ports and cut them off from the sources of nitrogen compounds in South America. As a consequence, when a German chemist named Fritz Haber showed the potential of an industrial process in which nitrogen reacts with hydrogen in the presence of a suitable catalyst to form ammonia, the German General Staff was quite interested and furnished support through the German chemical industry for the study of the reaction and the development of industrial plants based on it. The first such plant was in operation by 1911, and by 1914 such plants were being built very rapidly by Germany.

When the First World War broke out in August, 1914, many people thought that a shortage of explosives based on nitrogen compounds would force the war to end within a year. Unfortunately, by this time the nitrogen fixing industry in Germany was capable of supplying the needed compounds in large amounts. This process thus prolonged the war considerably and resulted in an enormous increase in mortality. Subsequently, the ammonia process has been used on a huge scale to prepare fertilizers and now is largely responsible for the fact that the earth can support a population of four billion. Ammonia production by this process exceeds 30,000 tons per day in the United States alone.

The same type of problem seems to arise from the development of many technological processes. The control of nuclear energy brings with it the ability to make nuclear explosives. The development of rapid and convenient means of transportation such as the automobile and the airplane also brings forth new weapons of war and air pollution problems. Man, however, must learn to control his technology in such a manner as to maximize its benefits and minimize its

223

disadvantages. These problems arise with *all* technological developments, even the most primitive. The discovery of the techniques necessary to the manufacture of iron led first to the development of new weapons (swords) by their discoverers, the Hittites, who then proceeded to conquer their neighbors and lead the first successful invasion of Egypt (ca. 1550 B.C.).

TECHNOLOGY AND THE HUMAN ENVIRONMENT

The growth in large scale technology has a very large number of effects, both direct and indirect, on the human environment. The examination of a few of these shows just how complex these consequences can be.

An obvious case is the development of atomic energy. When the incredibly large amounts of energy which could be released by nuclear reactions were first recognized, the development of such energy sources capable of providing energy for mankind on a scale previously thought to be impossible was placed on a top priority basis. After some nuclear reactors had been built and actually placed in operation, it was evident that their operation was accompanied by some

FIGURE 12-7   Land burial trench at the Oak Ridge National Laboratory reservation. Each day's accumulation of waste containers is buried by 3 or more feet of earth. (From *Radioactive Wastes*. Courtesy of the U.S. Atomic Energy Commission, Washington, D.C.)

TABLE 12-1   ESTIMATED ANNUAL CROP LOSS DUE TO PESTS
AND DISEASE (PER CENT OF TOTAL U.S. CROP DESTROYED)°

CROP	INSECTS	WEEDS	DISEASE
Corn	12	10	12
Rice	4	17	7
Wheat	6	12	14
Potatoes	14	3	19
Cotton	19	8	12

° From Scientific Aspects of Pest Control, Natural Academy of Science Publication No. 1402.

serious potential risks to their human users. The first was the danger of some potential disaster such as the explosion of a boiler, with the consequent dispersal of radioactive material. The second danger was in the generation of radioactive products as the nuclear reaction proceeded. The uranium used in such reactors was transformed into a wide variety of fission products which made the operation of the pile more difficult as time went on. The re-purification of the uranium could be accomplished chemically, but what was to be done with the radioactive wastes generated in the process? Obviously they could not be dumped into a sewer since many of them have relatively long half-lives. Present practice calls for solid radioactive wastes to be placed in vaults and buried at carefully chosen

FIGURE 12-8   The concentration of DDT in birds of prey, such as hawks, results in poor egg shell formation along with other abnormalities in the reproductive cycle.

225

FIGURE 12-9   (From "The Washington Star." Cartoon by Gib Crockett.)

sites. Liquid wastes are placed in underground storage tanks. The problem which faces us here is obtaining the benefits of the technology with the minimum disruption of our own environment.

The same type of problem arises whenever we introduce a specific chemical compound into our environment to accomplish one single thing. The compound is often capable of a variety of actions and can lead to consequences undreamed of by those who introduce it. Many such examples can be found in the fields of drugs and insecticides. Thalidomide was designed to be an effective tranquilizer, and for this purpose it is an unqualified success. Unfortunately, when taken by pregnant women, it prevents the normal development of the children they bear.

Man must always combat insects in his struggle for a food supply, and the most effective aid he now receives in this struggle is from insecticides. Unfortunately, many insecticides are rather unspecific poisons.

Arsenic compounds were used on a very large scale as insecticides, but a realization of their great toxicity for man has led to their replacement by other compounds equally effective but far less toxic to man. However, many of the compounds which are most effective in protecting crops for man's use are also very toxic to man and capable of causing death when improperly handled

FIGURE 12-10   Human thought versus chemicals. Will man ever completely control chemicals—or will they control him? (Courtesy of Varian, Palo Alto, California.)

or ingested. Even those insecticides which are not obviously harmful to man often have side effects which make them unattractive. DDT, for example, is very effective for the control of a wide variety of insects and, as far as can be ascertained, is nontoxic for men. Because of its widespread use and the fact that it is accumulated in fish and animal fats, it is found to concentrate enormously in birds of prey (eagles, pelicans, and ospreys), whose diet is principally fish, and to disrupt their reproductive cycles. A side effect such as this has led to extensive agitation to replace DDT with other compounds which are equally effective as insecticides yet free from this particular side effect. It is possible, however, that the compounds used to replace DDT will have other side effects which will not become obvious until after they have been used for some time. It would seem that the use of any insecticide will have a considerable effect upon the bird population of an area, if only through the changes it introduces into the food supply of the birds.

Chemistry, however, is not the only science which can furnish technological processes for the control of insects. There is a wide variety of biological processes which can be used for the same purposes. One is to furnish assistance to the biological species which destroy insects, such as other parasitical insects. Another

227

is the introduction of large numbers of sterile insects. When these mate with normal insects, no offspring are produced; the number of a type of undesirable insect can be reduced considerably by this process, at least temporarily. These examples are given to emphasize the fact that there are usually several very different kinds of processes which can be used to solve any given practical problem. In the future these will be assessed, at least in part, on their ability to leave the environment undamaged as well as on their monetary cost.

TECHNOLOGICAL DEVELOPMENT AND ITS ENVIRONMENTAL CONSEQUENCES

A very good example of a highly desirable technological development which has consequences that are obviously not so desirable is seen in the development of fertilizers. Most men agree that an abundant food supply is good. Most would admit, too, that every crop harvested from a field removes essential nutrients from that field: nitrogen, phosphorus, potassium, and so forth. By replacing these lost elements with fertilizer, we can restore or enhance the amount of food we obtain from the field. The increased yields obtained with increased application of fertilizers have been well established and commonly held to be desirable. It is also well established that increased fertilization of a field *increases* the concentration of these essential nutrients in the rain water which runs off such land.

The change in the mineral concentration in the runoff water is capable of causing drastic changes in the rivers and lakes into which it drains. By increasing the amount of minerals available, we have greatly stimulated the growth of surface algae in rivers and lakes, and as these grow they choke out other forms of life. The growth of such slimes makes the water much less capable

FIGURE 12-11   This pond was fertilized with chemicals to produce an excessive growth of algae. Part *A* shows the water covered with the plant growth: Part *B* shows the decay that follows when the concentrated form of life cannot be sustained. (Courtesy of Dr. D. L. Brockway, Federal Water Quality Administration.)

of supporting its normal population of fish, and ultimately these die off. The algae also make the water more difficult to purify for drinking purposes. After a time the increased food supply is paid for by a general disruption of the biology of the waterways which drain an area. This disruption has occurred in many areas in the United States where the movement of water through lakes and rivers is slow.

Other examples of this same kind of process can be seen in the development of energy sources, the use of antibiotics, weed killers, and, in fact, any kind of process which releases chemical compounds into the environment in sufficient quantities to cause an appreciable percentage increase in its composition in the environment. This holds for the hydrocarbons released from automobile engines and for the carbon dioxide and sulfur dioxide released by electric power plants. The change which these increased concentrations of various compounds cause cannot always be estimated on the basis of experiments covering only a short period of time since some of the effects may be very long range ones which build up slowly.

IS TECHNOLOGY ESCAPING CONTROL?

It is quite easy to imagine a situation in which a technological process can introduce into our environment drastic and irreversible changes which set a chain of events into action before we can stop them. This is especially easy to visualize in the case of changes which might be provided by the continuous increase in the carbon dioxide content of our atmosphere. This is caused by the increased use of "fossil" fuels, such as coal and petroleum, to furnish power for energy generation and transportation. What will be the long range effects, if any, from this steadily increasing amount of carbon dioxide in our atmosphere? At the present time, no one really knows.

There are other situations where the consequences of technology seem more obviously to be moving beyond the control of man. The ability to build nuclear weapons is now spreading quite rapidly and soon may be well within the power of all but the smallest nations. How are these to be controlled? What mechanisms can be developed to prevent mankind from destroying itself in an atomic holocaust triggered by some insignificant local dispute? This is obviously a very urgent problem.

The application of technology to the problems of warfare has already produced some frightening developments, including chemical and biological warfare agents whose discovery has closely followed the extension of human knowledge in these sciences. The same kind of knowledge that allows more effective drugs and insecticides to be synthesized facilitates the synthesis of more effective agents for gas warfare. The understanding of cellular behavior that allows us to produce new varieties of high yielding grains also can be used to develop new techniques of biological warfare.

Obviously, a distinction must be made between scientific knowledge which can be used for good or evil purposes and the types of technology that are

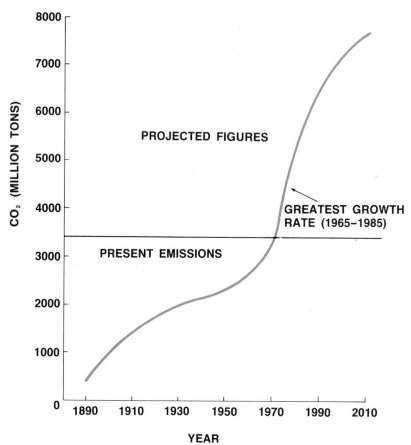

INDUSTRIAL
CO₂ EMISSIONS TO U.S. ATMOSPHERE
1890–2000

PROJECTED FIGURES

GREATEST GROWTH
RATE (1965–1985)

PRESENT EMISSIONS

CO₂ (MILLION TONS)

YEAR

FIGURE 12-12 CO_2 industrial emissions in millions of tons per year.

developed specifically for destructive purposes. The kind of emotional thinking which puts all technological developments under some kind of moral ban can lead only back to a new Dark Age. Most of the materials of the world can be used for a variety of purposes, which are ethical to varying degrees. In mankind's upward struggle he must always use his intelligence as well as his emotions if he is to succeed. He must be able to select the good developments from the indifferent ones and from the bad ones. This has *always* been a problem for men and probably always will be. In such a situation, ignorance can be catastrophic; only by a study of such processes and an evaluation of their probable consequences can rational selections be made. In this case it makes more sense to ask if man is being given control over things beyond his understanding than to ask "Is Technology Escaping Control?" Technology is always *initiated* by men and is under some sort of actual or potential control by them.

There is still another manner in which fears arise over the developing course of technology and man's ability to control it. The vast majority of mankind has no knowledge of the scientific principles upon which technology is based and

230

accordingly must accept the opinions of others on technological matters. This dependence upon experts, in turn, arouses fears of these experts and the damage they can do either by error or by evil intent. This fear, in some people, is an unreasoning, blind, driving force which leads them to condemn all technology. The only antidote to such fear is an understanding of technology based on study. Mankind's perennial enemy has been ignorance and the prejudice it generates. Human progress has always been the result of activities of that small percentage of people who accept neither the ignorance nor the popular prejudices of their fellow human beings.

In all fairness, it must be noted that the benefits of technology far outweigh its disadvantages, and also that its transformation of the human condition is still in its infancy. There is, even now, an enormous number of practical problems facing mankind which can be solved only by new scientific discoveries and the technological advances which they will make possible. It is the responsibility of all of us to learn about the sciences and to develop an understanding of the possible ways in which this knowledge may be developed to the advantage of mankind. In these learning processes, we will discover an avenue to responsible participation in decision making, an understanding of technological advances, and a thorough enjoyment of our investigations.

QUESTIONS

1. Select a law or generalization from a book on chemistry and trace down the supporting evidence. Do the same from a book on economics or sociology.

2. Cite three technological advances, give their direct benefits, and list a problem arising from each.

3. List 10 technological advances since 1940 which affect your life.

4. Name two specific changes that were made in the fall of 1970 to decrease the amount of air pollution caused by automobiles.

5. Try to think of a reason not to dump liquid chemical wastes in abandoned mine shafts.

6. Distinguish between science and technology.

7. Discuss three of the most important practical problems facing mankind, in terms of how science and technology have helped to bring them about and how they are being solved.

8. Look up the number of deaths in the U. S. due to malaria for the past century. Did a "break" occur? When?

9. Is there any human evidence of technological advances resulting in the decline of civilization? Consider an article in the *Journal of Occupational Medicine*, 7:53–60 (1965).

10. Over coffee one morning, a friend states, "The results of chemical technology have all been harmful to mankind!" He goes on to list smog, water pollution, DDT, chemical and biological warfare agents, and so forth. Could you balance the argument by listing some advances in this area which have been for the general good of man?

SUGGESTIONS FOR FURTHER READING

Commoner, B. (Ed.), "Science and Survival," Viking Press, New York, 1966.
Helfrich, H. W., Jr. (Ed.), "The Environmental Crisis," Yale University Press, New Haven, 1970.

Kranzberg, M., and C. W. Pursell, Jr. (Eds.), "Technology in Western Civilization," Oxford University Press, New York, 1967. (In 2 volumes; a comprehensive collection of well-illustrated essays.)

Novich, S., "A New Pollution Problem," *Environment*, Vol. 11, No. 4, p. 2 (1969).

U.S. Atomic Energy Commission, "Radioactive Wastes," 1969.

Vavoulis, A., and A. Colver (Eds.), "Science and Society—Collected Essays," Holden-Day, Inc., San Francisco, 1966.

Wagner, Richard H., "Environment and Man," W. W. Norton, Inc., New York, 1971.

Woodwell, G. M., C. F. Wooster, Jr., and D. A. Isaacson, "DDT Residues in an East Coast Esturary: A Case of Biological Concentration of a Persistent Insecticide," *Science*, Vol. 156, p. 821 (1967).

USEFUL MATERIALS FROM THE EARTH, SEA, AND AIR

CHAPTER 13

INTRODUCTION

A large number of the things we use in everyday life result from chemical reactions which are carried out on an industrial scale. They include iron and steel products, fabricated items of aluminum, magnesium, copper, glass, ceramics, and cement. Also included are fertilizers, detergents, chemicals used for purifying water, and numerous chemicals used to alter our surroundings. The purpose of this chapter is to discuss some of the chemistry involved in preparing these products from natural resources found in the earth, sea, and air.

Industrial chemistry should be and probably will be limited by two relatively new demands. Not only must it produce a useful arrangement of atoms, but it must also provide for the return of the unused atoms to nature in a desirable form—that is, the eventual recycling of the atoms. A reasonable economic package must place enough demand on the desired arrangement of atoms to provide for negating the undesired arrangements.

METALS AND THEIR PREPARATION

Metals occur in the crust of the earth mostly as compounds, though some of the less active ones such as copper, silver, and gold can be found also as free elements. Because of extensive and continuing geological processes which have formed the crust of the earth over the last billion years, the distribution of elements is anything but uniform. In order to obtain a metal from the crust of the earth, one must first find an ore, which is defined simply as a mineral from which the element can be extracted economically. Color Plate I pictures some natural metals and ores.

Some elements which are actually not particularly abundant in the earth's crust are nevertheless very familiar to us because they tend to occur in very

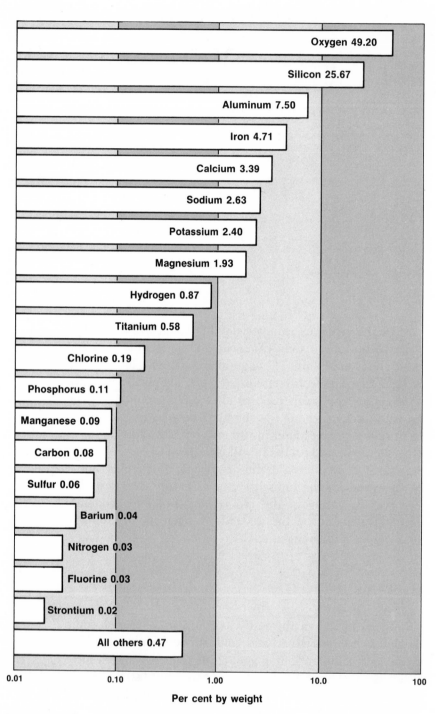

Oxygen 49.20
Silicon 25.67
Aluminum 7.50
Iron 4.71
Calcium 3.39
Sodium 2.63
Potassium 2.40
Magnesium 1.93
Hydrogen 0.87
Titanium 0.58
Chlorine 0.19
Phosphorus 0.11
Manganese 0.09
Carbon 0.08
Sulfur 0.06
Barium 0.04
Nitrogen 0.03
Fluorine 0.03
Strontium 0.02
All others 0.47

0.01   0.10   1.00   10.0   100

Per cent by weight

FIGURE 13-1   Abundance of the elements in the earth's crust (per cent by weight).

concentrated localized deposits from which they can easily be extracted. Examples of these are lead, copper, and tin, none of which is among the more abundant elements in the crust of the earth (Figure 13–1). Other elements that actually form a much larger percentage of the crust of the earth are almost unknown to us because concentrated deposits of their ores are less commonly found. An example is titanium, the tenth most abundant element in the crust of the earth (Table 13–1).

TABLE 13-1   SOME COMMON ORES

METAL	CHEMICAL FORMULA OF COMPOUND OF THE ELEMENT	NAME OF ORE
Aluminum	$Al_2O_3 \cdot xH_2O$	bauxite
Copper	Cu_2S	chalcocite
Zinc	ZnS	sphalerite
	$ZnCO_3$	smithsonite
Iron	Fe_2O_3	hematite
	Fe_3O_4	magnetite
Manganese	MnO_2	pyrolusite
Chromium	$FeO \cdot Cr_2O_3$	chromite
Calcium	$CaCO_3$	limestone
Lead	PbS	galena

The preparation of metals from their ores involves chemical reduction, (Chapter 11). Indeed, the concept of oxidation and reduction developed from metallurgical operations. As presented earlier, iron in iron oxide is in the form of Fe^{3+}. If we *reduce* Fe^{3+} ions to Fe atoms, we must find a source of electrons. (Recall that oxidation is the loss of electrons and that reduction is the gain of electrons.) Sometimes the desired metal is in solution (e.g., magnesium in the sea), where it exists in the oxidized form Mg^{2+} ions. To obtain magnesium from the sea, we must add electrons to these ions (reduction) to produce magnesium atoms.

IRON

The sources of most of the world's iron are large deposits of the iron oxides in Minnesota, Sweden, France, Venezuela, Russia, Australia, and England. In nature these oxides frequently are mixed with impurities, so the production of iron usually incorporates steps to remove such impurities. Iron ores are reduced to the metal by using carbon, in the form of coke, as the reducing agent.

The reduction of the iron ore is carried out in a blast furnace (Figure 13–2). The solid material fed into the top of the blast furnace consists of a mixture of an oxide of iron (Fe_2O_3), coke (C), and limestone ($CaCO_3$). A blast of heated air is forced into the furnace near the bottom. The reactions which occur within the blast furnace are:

$$2C \;+\; O_2 \;\longrightarrow\; 2CO + heat$$
Carbon   Oxygen   Carbon monoxide

$$Fe_2O_3 \;+\; 3CO \;\longrightarrow\; 2Fe + 3CO_2 + heat$$
Iron oxide   Carbon monoxide   Iron   Carbon dioxide

235

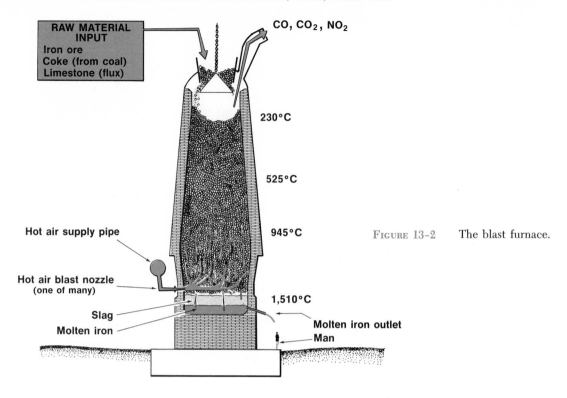

FIGURE 13-2   The blast furnace.

Limestone is added because iron ores usually contain silica (SiO_2) as an impurity. The limestone reacts as follows:

$$CaCO_3 \longrightarrow CaO + CO_2$$

Calcium carbonate   Calcium oxide   Carbon dioxide

$$CaO + SiO_2 \longrightarrow CaSiO_3$$

Calcium oxide   Silicon dioxide   Calcium silicate

The calcium silicate, or slag, has a melting point low enough to allow it to exist as a liquid in the furnace. Consequently, as the blast furnace operates, two molten layers collect in the bottom. The lower, denser layer is mostly liquid iron which contains a fair amount of dissolved carbon and often smaller amounts of other impurities. The upper, lighter layer is primarily molten calcium silicate with some impurities. From time to time the furnace is tapped at the bottom and the molten iron drawn off. Another outlet somewhat higher in the blast furnace base can be opened to remove the liquid slag.

The iron that is obtained from the blast furnace contains too much carbon for most uses, and so is made into steel. Steel is an alloy of iron with a relatively small amount of carbon (less than 1.5 per cent); it may also contain other metals. In order to convert iron to steel, the carbon content must be reduced; this is done by burning out the excess carbon with oxygen.

There are several processes by which the excess carbon is removed; a recent development that has been very widely adopted is the basic oxygen process (Figure 13–4). In this process pure oxygen is blown through molten iron by means

236

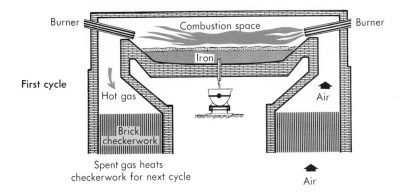

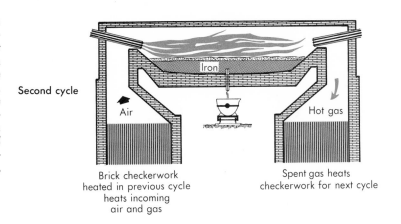

FIGURE 13-3 The open hearth furnace for the conversion of iron to steel. Air and fuel are blown in first in one direction and then the other in order that heat, which is costly, might be stored in the brick checkerwork. (From Lee, Van Orden and Ragsdale: *General and Organic Chemistry*. Philadelphia, W. B. Saunders Co., 1971.)

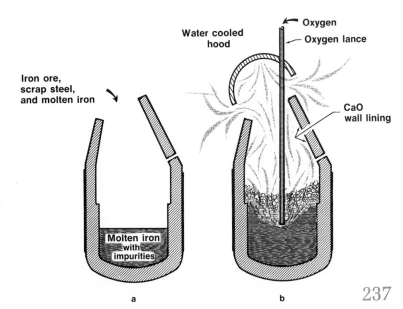

FIGURE 13-4 The basic oxygen process furnace. Much of the steel manufactured today is refined by blowing oxygen through a furnace charged with ore, scrap, and molten iron.

237

of a refractory tube which is pushed below the surface of the iron. At elevated temperatures, the dissolved carbon reacts very rapidly with the oxygen to give gaseous carbon monoxide or dioxide, which then escapes. At the same time the lining of the container, which is made of a basic material with a high melting point (e.g., MgO or CaO), reacts with acidic impurities present, such as silicon dioxide:

$$CaO \ + \ SiO_2 \longrightarrow CaSiO_3$$

Calcium   Silicon   Calcium
oxide   dioxide   silicate

After the carbon content has been reduced to a suitable level, the molten steel is formed into desired shapes. During the processing it is subjected to carefully controlled heat treatment to insure that the steel has the desired mechanical properties. This is necessary in order to control the crystalline form of the steel, which determines its pliability, toughness, and other properties. The total world production of iron and steel is now in excess of one billion tons per year.

CORROSION, RUSTING, AND STAINLESS STEEL

Many metals undergo corrosion when exposed to moist air over a long period of time. Typically, corrosion is a reaction with the oxygen and water of the air which transforms a metal into its oxide or hydroxide. Corrosion of iron is called rusting and leads to the transformation of iron into rust ($Fe_2O_3 \cdot xH_2O$). The initial reaction is the oxidation of iron to ferrous hydroxide:

$$2Fe + O_2 + 2H_2O \longrightarrow \quad 2Fe(OH)_2$$

Moist air   *Ferrous hydroxide*
                       iron (II) hydroxide

Iron (II) hydroxide is itself subject to further oxidation in moist air to give iron (III) hydroxide:

$$4Fe(OH)_2 + O_2 + 2H_2O \longrightarrow 4Fe(OH)_3 \text{ (or } Fe_2O_3 \cdot 3H_2O)$$

Moist air   *Rust*

The iron (III) hydroxide loses water readily to form iron (III) oxides with variable amounts of water. The rusting process transforms iron metal back into a compound very similar in composition to its ore and thus undoes all the effort expended in obtaining the metal. The replacement of rusted objects costs several billion dollars a year in the United States alone.

Rusting also occurs when we make an iron object into an electrochemical cell or battery, often without realizing it. This happens when part of the iron surface becomes wet in contact with air (Figure 13–5). The oxygen in the air will dissolve in the water and remove electrons from the iron:

$$O_2 + 2H_2O + 4e^- \longrightarrow 4OH^-$$

Corrosion and rusting are obviously very serious problems which have occupied the attention of scientists for years. There are many ways of preventing

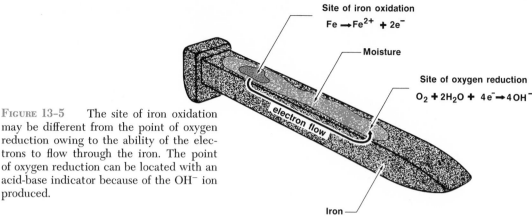

Site of iron oxidation
$$Fe \longrightarrow Fe^{2+} + 2e^-$$

Moisture

Site of oxygen reduction
$$O_2 + 2H_2O + 4e^- \longrightarrow 4OH^-$$

electron flow

Iron

FIGURE 13-5   The site of iron oxidation may be different from the point of oxygen reduction owing to the ability of the electrons to flow through the iron. The point of oxygen reduction can be located with an acid-base indicator because of the OH⁻ ion produced.

or reducing corrosion of a metal object; three of these are: (1) protective coatings, (2) cathodic protection, and (3) alloying (stainless steel is a corrosion-resistant alloy of iron).

Protective coatings are applied to prevent the access of atmospheric oxygen to the iron surface. Such coatings may be of paint, enamel, grease, or another more resistant metal such as chromium. This method is successful as long as cracks or holes do not develop in the coating. Galvanized iron contains a surface coating of the more active metal zinc, which forms an oxide that tends to be hard and impervious to the air (see below).

Cathodic protection is a method by which the iron is connected to a more reactive metal, such as magnesium, so that the reactive metal will be oxidized rather than the iron (Figure 13–6). In such a system, the oxygen is reduced at the steel surface via the reaction:

$$O_2 + 2H_2O + 4e \longrightarrow 4OH^-$$

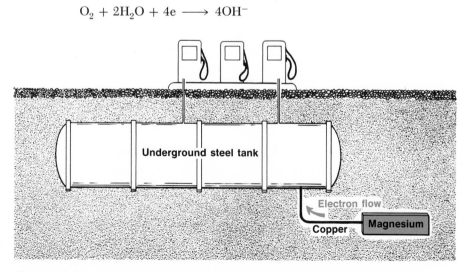

Underground steel tank

Electron flow

Copper   Magnesium

FIGURE 13-6   Cathodic protection. If magnesium is connected to the steel tank to be protected, the magnesium is more easily oxidized than the iron or copper connecting wire. The magnesium serves as the anode. Hence, the cathode is protected with no points of oxidation occurring on its surface. The anode is the electrode where oxidation occurs; reduction occurs at the cathode.

239

but the more reactive metal (magnesium) is the source of the electrons:

$$Mg \longrightarrow Mg^{2+} + 2e^-$$

As a result, the magnesium is corroded rather than the iron. This method is used to protect iron or steel objects buried in the ground, such as pipelines. Of course, protection ends when the magnesium is depleted.

The terms *cathode* and *anode* are often confusing to the student. The cathode is always the electrode where reduction occurs, and the anode is the electrode where oxidation occurs. Cathodic protection is so named because the iron to be protected serves as the cathode, while the more active metal becomes the anode. Hence, at the iron cathode oxidation cannot occur.

Stainless steels are alloys of iron which contain other metals, such as nickel, chromium, or cobalt. They are made by melting iron and the alloying elements together in an electric furnace. The resulting alloys are resistant to corrosion. In the presence of oxygen they form a very thin, tough, and impervious adherent layer of metal oxide on their surfaces. The layer is *so* thin it is essentially transparent (the luster of the metal is retained). This protects the underlying metal from further contact with the oxygen of the air and renders the objects "stainless," or very resistant to corrosion under normal circumstances. A stainless steel alloy which is widely used is the so-called 18-8: 18 per cent chromium, 8 per cent nickel, and the rest iron. Stainless steels are used in household items; larger amounts are used in the construction of industrial plants which handle great quantities of hydrochloric acid and other highly corrosive materials.

ALUMINUM

Seven and one-half per cent of the crust of the earth is aluminum, in the form of Al^{3+} ions. However, because of the difficulty of reducing Al^{3+} to Al, only recently has man learned to isolate and use this abundant element. Aluminum metal is soft and has a low density. Many of its alloys, however, are quite strong. Hence, it is an excellent choice when a light weight, strong metal is required. In structural aluminum, the high chemical reactivity of the element is offset by the fact that a transparent, hard film of aluminum oxide, Al_2O_3, forms over the surface, protecting it from further oxidation:

$$4Al + 3O_2 \longrightarrow 2Al_2O_3$$

The principal ore of aluminum is bauxite, a hydrated aluminum oxide, $Al_2O_3 \cdot xH_2O$. Because impurities in the ore have undesirable effects on the properties of aluminum, these must be removed, generally by the purification of the ore. This is accomplished with the Bayer process, which is based upon the fact that aluminum oxide and aluminum hydroxide are weak acids. The principal impurities are iron oxides, which are not soluble in sodium hydroxide solution. The Bayer process separates these by treating the mixture with a sodium hydroxide solution, which dissolves the aluminum and leaves the iron:

240

$$Al_2O_3 \cdot xH_2O(\text{solid}) + Fe_2O_3(\text{solid}) \xrightarrow[\text{solution}]{NaOH} Al(OH)_4^- + Na^+ + Fe_2O_3(\text{solid})$$

The mixture is filtered; the $Al(OH)_3$ is then carefully precipitated out of the clear solution by adding carbon dioxide (an acid), thus making the solution less basic:

$$CO_2 + Al(OH)_4^- \longrightarrow Al(OH)_3\downarrow + HCO_3^-$$

The aluminum hydroxide is then heated to transform it into pure anhydrous aluminum oxide:

$$2Al(OH)_3 \xrightarrow{\text{heat}} Al_2O_3 + 3H_2O$$

Aluminum metal is obtained from the purified oxide by electrolysis in molten cryolite (Fig. 13–7). Cryolite, Na_3AlF_6, has a melting point of 1006°C; the molten compound dissolves considerable amounts of aluminum oxide, which in turn lowers its melting point. This mixture of cryolite and aluminum oxide is electrolyzed in a cell with carbon anodes and a carbon cell lining that serves as the cathode on which aluminum is deposited. As the operation of the cell proceeds, the molten aluminum sinks to the bottom of the cell. From time to time the cell is tapped and the molten aluminum is run off into molds.

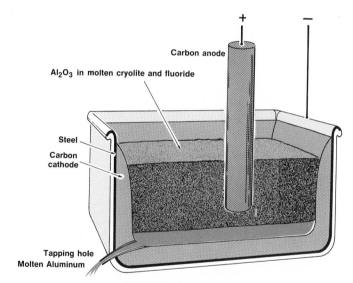

FIGURE 13-7   Schematic drawing of a furnace for producing aluminum by electrolysis of a melt of Al_2O_3 in Na_3AlF_6 and AlF_3. The molten aluminum collects in the carbon cathode container.

Aluminum is used both as a structural metal and as an electrical conductor in high voltage transmission lines. It competes with copper as an electrical conductor because of its lower cost. The lower cost allows larger diameter aluminum wires to be used to offset the fact that an aluminum wire has a lower electrical conductivity than a copper wire of the same diameter. At the present time the world production of aluminum is about 10 million tons per year. Over 500,000 tons of aluminum went into food and beverage cans in the USA in 1971.

COPPER

Although copper metal occurs in the free state in some parts of the world, the supply available from such sources is quite insufficient for the world's need.

The majority of the copper obtained today is from various copper sulfide ores, most of which must be concentrated prior to the chemical processes which produce the metal. These minerals include $CuFeS_2$ (chalcopyrite), Cu_2S (chalcocite), and CuS (covellite). Because the copper content of these ores is around 1 to 2 per cent, the powdered ore is first concentrated by the flotation process (Figure 13–8).

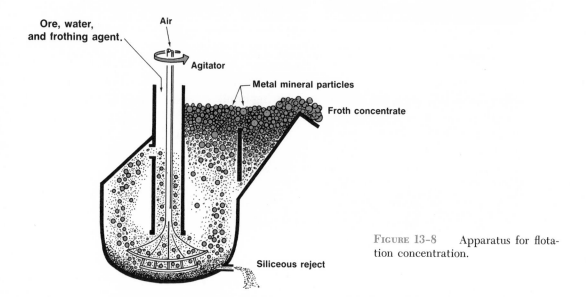

FIGURE 13-8   Apparatus for flotation concentration.

In the flotation process, the powdered ore is mixed with water and a frothing agent such as pine oil. A stream of air is blown through to produce froth. The gangue in the ore, which is composed of sand, rock, and clay, is easily wet by the water and sinks to the bottom of the container. In contrast, a copper sulfide particle is hydrophobic—it is not wet by the water. The copper sulfide particle becomes coated with oil and is carried to the top of the container in the froth. The froth is removed continuously, and the copper sulfide minerals are recovered from it.

The preparation of copper metal from a copper sulfide ore involves roasting in air to oxidize some of the copper sulfide and any iron sulfide present:

$$2Cu_2S + 3O_2 \longrightarrow 2Cu_2O + 2SO_2\uparrow$$

$$2FeS + 3O_2 \longrightarrow 2FeO + 2SO_2\uparrow$$

Subsequently the mixture is heated to a higher temperature, and some copper is produced by the reaction:

$$Cu_2S + 2Cu_2O \longrightarrow 6Cu + SO_2\uparrow$$

The iron oxides then form a slag. The product of this operation is a *matte*, a mixture of copper metal and sulfides of copper, iron, other ore constituents, and slag. The molten matte is heated in a converter with silica materials. When

242

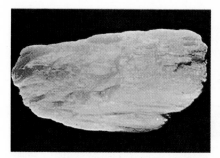

Wollastonite—$CaSiO_3$, a native calcium silicate from metamorphic rock.

Pyrite—FeS_2, iron disulfide.

Talc—$3MgO \cdot 4SiO_2 \cdot H_2O$, hydrated magnesium silicate.

Limonite—$2Fe_2O_3 \cdot 3H_2O$, hydrous ferric oxide.

Calcite—$CaCO_3$, calcium carbonate.

Selenite—$CaSO_4 \cdot 2H_2O$, waferlike crystal cluster.

Barytes—$BaSO_4$, blue and yellow crystals, on purple hematite.

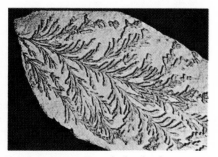

Pyrolusite—MnO_2, dark dendrites on sandstone.

Plate I   *Photos courtesy of Pfizer Inc.*

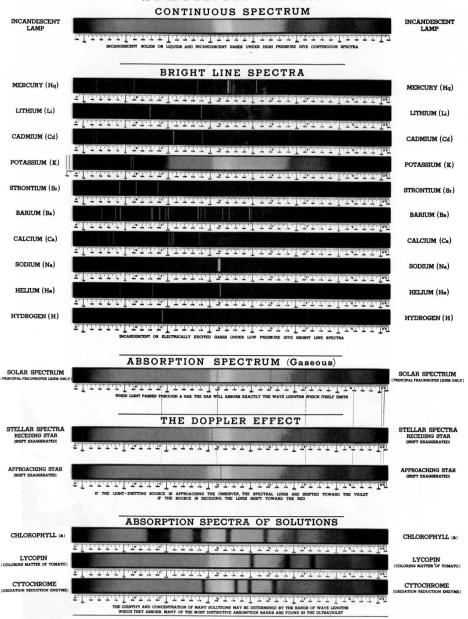

SPECTRUM CHART
CONTINUOUS SPECTRUM

INCANDESCENT LAMP

INCANDESCENT LAMP

INCANDESCENT SOLIDS OR LIQUIDS AND INCANDESCENT GASES UNDER HIGH PRESSURE GIVE CONTINUOUS SPECTRA

BRIGHT LINE SPECTRA

MERCURY (Hg)

LITHIUM (Li)

CADMIUM (Cd)

POTASSIUM (K)

STRONTIUM (Sr)

BARIUM (Ba)

CALCIUM (Ca)

SODIUM (Na)

HELIUM (He)

HYDROGEN (H)

INCANDESCENT OR ELECTRICALLY EXCITED GASES UNDER LOW PRESSURE GIVE BRIGHT LINE SPECTRA

ABSORPTION SPECTRUM (Gaseous)

SOLAR SPECTRUM
(PRINCIPAL FRAUNHOFER LINES ONLY)

SOLAR SPECTRUM
(PRINCIPAL FRAUNHOFER LINES ONLY)

WHEN LIGHT PASSES THROUGH A GAS, THE GAS WILL ABSORB EXACTLY THE WAVE LENGTHS WHICH ITSELF EMITS

THE DOPPLER EFFECT

STELLAR SPECTRA
RECEDING STAR
(SHIFT EXAGGERATED)

STELLAR SPECTRA
RECEDING STAR
(SHIFT EXAGGERATED)

APPROACHING STAR
(SHIFT EXAGGERATED)

APPROACHING STAR
(SHIFT EXAGGERATED)

IF THE LIGHT-EMITTING SOURCE IS APPROACHING THE OBSERVER, THE SPECTRAL LINES ARE SHIFTED TOWARD THE VIOLET
IF THE SOURCE IS RECEDING, THE LINES SHIFT TOWARD THE RED

ABSORPTION SPECTRA OF SOLUTIONS

CHLOROPHYLL (a)

CHLOROPHYLL (a)

LYCOPIN
(COLORING MATTER OF TOMATO)

LYCOPIN
(COLORING MATTER OF TOMATO)

CYTOCHROME
(OXIDATION REDUCTION ENZYME)

CYTOCHROME
(OXIDATION REDUCTION ENZYME)

THE IDENTITY AND CONCENTRATION OF MANY SOLUTIONS MAY BE DETERMINED BY THE BANDS OF WAVE LENGTHS
WHICH THEY ABSORB. MANY OF THE MOST DISTINCTIVE ABSORPTION BANDS ARE FOUND IN THE ULTRAVIOLET

Some typical spectra in the visible region. An emission spectrum (Continuous and Bright Line) represents all of the light emitted by the source of light while an absorption spectrum (Solar Spectrum) represents the light from the source minus the light absorbed by an intervening absorbing medium. Spectral lines from sources moving toward or away from the observer are shifted to shorter or longer wave lengths respectively, the Doppler effect. Wave lengths for spectral lines are given in angstrom units (one-hundredth million of a centimeter). (Courtesy of Sargent-Welch Scientific Company, Skokie, Illinois)

Plate II

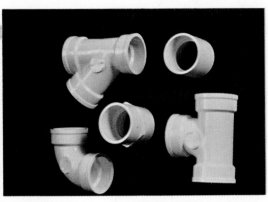

Plumbing made of plastic (polyvinyl chloride).

Courtesy of the Olin Corporation.

Teflon cookware.

Courtesy of the Du Pont Company.

Harmon Killebrew can "powder" a baseball but is no match for a sheet of plexiglas.

Courtesy of the Rohm and Haas Company.

Plate III

Industrial wastes pollute natural waters (Cumberland River, Nashville, Tennessee).

A foaming problem at the outlet of a sewer system (Cumberland River, Nashville, Tennessee).

Heavy growth of algae in the recreational area of a middle-Tennessee lake.

Photos by Dr. David J. Wilson, Vanderbilt University.

Plate IV

Air pollution in New York.

Industrial pollution in Rockwood, Tennessee.

Plate V

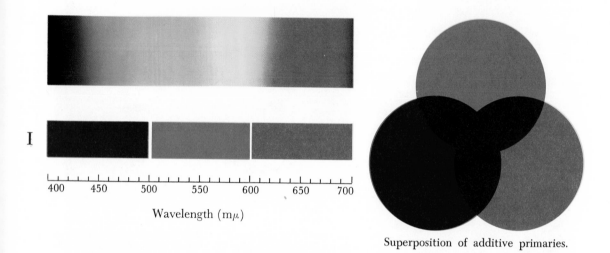

I

Superposition of additive primaries.

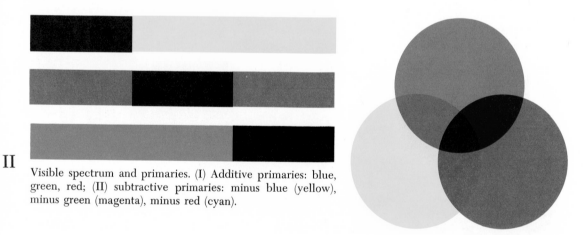

Wavelength (mμ)

II

Visible spectrum and primaries. (I) Additive primaries: blue, green, red; (II) subtractive primaries: minus blue (yellow), minus green (magenta), minus red (cyan).

Superposition of subtractive primaries.

From "A Chemist's View of Color Photography," by Arnold Weissberger, *American Scientist*, 58(6), 1970; courtesy of Eastman Kodak Company, Rochester, N. Y.

Plate VI

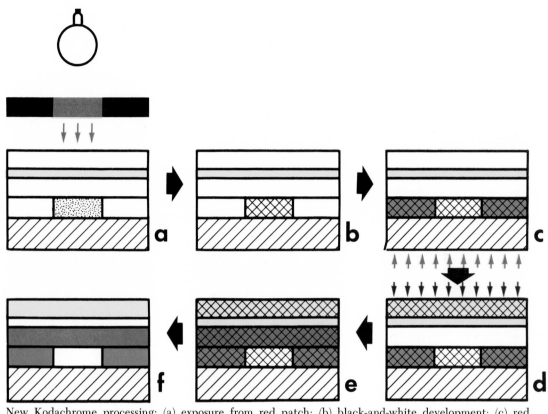

New Kodachrome processing: (a) exposure from red patch; (b) black-and-white development; (c) red exposure (through base) and cyan development using newly exposed silver halide; (d) blue exposure (from top) and yellow development using newly exposed silver halide; (e) fogging magenta development using residual silver halide (which is left only in green-sensitive layer); (f) silver and yellow filter layer removed by bleach and fix.

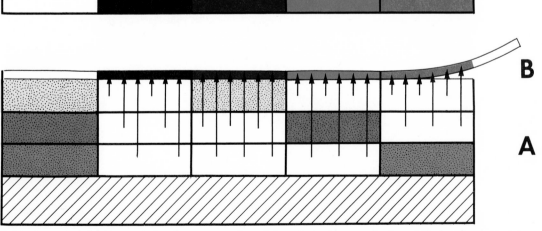

Polacolor principle. Layer A has been exposed and dyes have been made immobile by development. Mobile dyes have been transferred to receiving sheet B by contact.

From "A Chemist's View of Color Photography," by Arnold Weissberger, *American Scientist*, 58(6), 1970; courtesy of Eastman Kodak Company, Rochester, N. Y.

Plate VII

air is blown through the molten material in the converter, two reaction sequences occur. First, the iron is converted to a slag:

$$2FeS + 3O_2 \longrightarrow 2FeO + 2SO_2$$
$$FeO + SiO_2 \longrightarrow FeSiO_3$$
(Molten slag)

The remaining copper sulfide is converted to copper metal:

$$2Cu_2S + 3O_2 \longrightarrow 2Cu_2O + 2SO_2$$
$$Cu_2S + 2Cu_2O \longrightarrow 6Cu + SO_2$$

The copper produced in this manner is crude or "blister" copper and is purified electrolytically.

In the electrolytic purification of copper, the crude copper is first cast into anodes; these are placed in a water solution of copper sulfate and sulfuric acid. The cathodes are made of pure copper. As electrolysis proceeds, copper is oxidized at the anode, moves through the solution as Cu^{2+} ions and is deposited on the cathode (Chapter 11). The voltage of the cell is regulated so that more active impurities (such as iron) are left in the solution and less active ones are not oxidized at all. These less active impurities include gold and silver, and they collect as "anode slime," an insoluble residue. The anode slime is subsequently worked up to recover these rarer metals.

The copper produced by the electrolytic cell is 99.95 per cent pure and is suitable for use as an electrical conductor. Copper for this purpose must be pure because very small amounts of impurities, such as arsenic, considerably reduce the electrical conductivity of copper.

MAGNESIUM

Magnesium, with a density of 1.74 grams per cubic centimeter, is the lightest structural metal in common use. It is a relatively active metal chemically because it loses electrons easily. For this reason it is most often used in alloys designed for light weight and great strength. Magnesium "ores" include sea water, which has a magnesium concentration of 0.13 per cent, and dolomite, a mineral with the composition $CaCO_3 \cdot MgCO_3$. Because there are six million tons of magnesium present as Mg^{2+} salts in every cubic mile of sea water, the sea can furnish an almost limitless amount of this element.

The recovery of magnesium from sea water (Figure 13–9) begins with the precipitation of magnesium hydroxide by the addition of lime to sea water:

$$CaO + H_2O \longrightarrow Ca^{2+} + 2OH^-$$
$$Mg^{2+} + 2OH^- \longrightarrow Mg(OH)_2$$

The magnesium hydroxide is removed by filtration and then neutralized with hydrochloric acid to form the chloride:

$$Mg(OH)_2 + 2H^+ + 2Cl^- \rightleftharpoons Mg^{2+} + 2Cl^- + 2H_2O$$

243

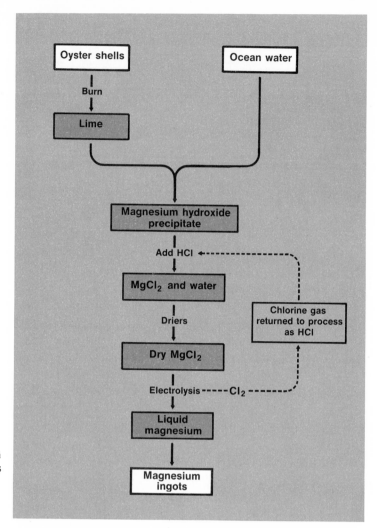

FIGURE 13-9 Flow diagram showing how magnesium metal is produced from sea water.

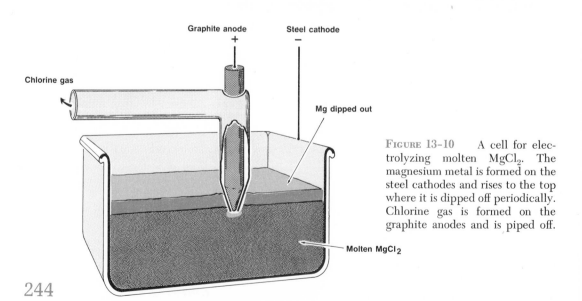

FIGURE 13-10 A cell for electrolyzing molten $MgCl_2$. The magnesium metal is formed on the steel cathodes and rises to the top where it is dipped off periodically. Chlorine gas is formed on the graphite anodes and is piped off.

The water is evaporated; this is followed by the electrolysis of molten magnesium chloride in a huge steel pot that serves as the cathode (Figure 13–10). Graphite bars serve as the anodes.

$$\textit{Electrolysis}$$
$$Mg^{2+} + 2Cl^- \xrightarrow{e^-} Mg + Cl_2\uparrow$$
$$(\textit{Melted})$$

At the cathode:
$$Mg^{2+} + 2e^- \longrightarrow Mg$$

At the anode:
$$2Cl^- \longrightarrow Cl_2\uparrow + 2e^-$$

As the melted magnesium forms, it floats to the surface and is periodically removed. The chlorine is recovered and reacted with air and natural gas to form hydrochloric acid for use in dissolving the magnesium hydroxide:

$$4Cl_2 + 2CH_4 + O_2 \longrightarrow 2CO + 8HCl$$

The lime used to precipitate the magnesium as the hydroxide is obtained by heating limestone or oyster shells:

$$CaCO_3 \xrightarrow{heat} CaO + CO_2$$

The total world production of magnesium is only about 250,000 tons per year, although it is potentially available on a larger scale.

THE KINGDOM OF SILICON: SILICATES, GLASS, CERAMICS, SILICONES, AND CEMENT

While the carbon-carbon bonds of organic compounds serve as the backbone for the enormous number of organic compounds, a different kind of linkage serves the same purpose in a large number of important *inorganic* structures. This is the —Si—O— linkage which is found in silica itself (sand) and silicates, which are basic constituents of glasses, ceramics, and cement.

SILICA

SiO_2 occurs naturally in large amounts as sand or more rarely in much larger crystals (quartz) (Figure 13–11). It has a melting point of 1710°C. If the melted material is cooled, a *glass* is usually obtained. Crystalline quartz consists of a polymeric structure in which each silicon (Si) atom is bonded tetrahedrally to four oxygen atoms (Figure 13–12a), and each oxygen atom is bonded to two silicon atoms. The bonding thus extends throughout the crystal (Figure 13–12b). When silica is melted, some of the bonds are broken and the units move with respect to each other. When the liquid is cooled, the re-formation of the original solid entails a reorganization which is hard to achieve because of the difficulty the groups experience in moving. The high viscosity

245

QUARTZ CRYSTALS

FIGURE 13-11   Quartz. Courtesy of McGraw-Hill.

liquid structure is thus partially preserved on cooling to give the characteristic feature of a *glass*, which is an apparently solid material (pseudo-solid) with some of the randomness in structure characteristic of a liquid. This unique structure accounts for one of the typical properties of a glass: it breaks irregularly rather than splitting along a plane like a crystal.

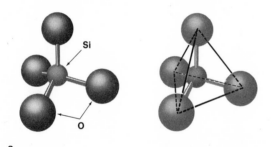

Si

O

a

FIGURE 13-12   (a) Tetrahedral structure of silicon and oxygen in silicates. (b) Chain of tetrahedra showing that an oxygen is common at each point of contact between tetrahedra.

b

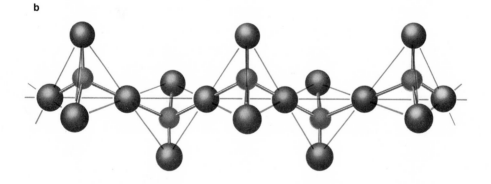

GLASS

The difference between a crystalline solid and a glass can be more easily appreciated by use of a *two*-dimensional analogy. If we consider a two-dimensional model, G_2O_3, in which each G atom is bonded to three oxygen atoms and each oxygen to two G atoms, we have a regular crystalline structure with the arrangement shown in Figure 13-13.

246

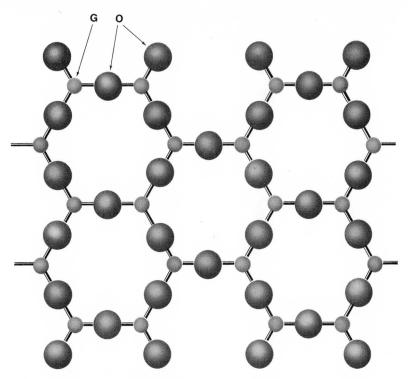

FIGURE 13-13   Regular crystalline structure for SiO_2 in two dimensions. In two dimensions the ratios of atoms would be 2:3; hence, G_2O_3.

If the substance having this structure is melted, some of the G—O bonds are broken and new ones are formed. In the liquid we do not have the long-range order that we have in the crystalline solid. When the substance is cooled to form a glass, some of the irregular atomic arrangements are frozen in. Each G atom is still bonded to two oxygens and each oxygen bonded to two G atoms; but we have an irregular structure such as shown in Figure 13-14.

Glass has an analogous irregular structure in *three* dimensions. This structure can be modified by the addition of molten metal oxides because the metal ions form ionic (nondirectional) bonds with oxygen atoms that had previously been bonded rigidly to G atoms. The resultant mixture has a lower melting point and viscosity than pure SiO_2 (melting point = 1710°C).

By the addition of metal oxides the melting temperature of glass is reduced to about 700°C. The oxides used to effect this change are sodium oxide (added as Na_2CO_3) and calcium oxide (added as $CaCO_3$). Because SiO_2 is an acidic oxide, it can displace CO_2 from sodium carbonate and calcium carbonate. For example,

$$CaCO_3 + SiO_2 \longrightarrow CaSiO_3 + CO_2$$

$$Na_2CO_3 + SiO_2 \longrightarrow Na_2SiO_3 + CO_2$$

The ions of sodium and calcium occupy spaces in the lattice, breaking some of the Si—O bonds. The soda-lime glass produced in this way will be clear and colorless only if the purity of the ingredients has been carefully controlled.   247

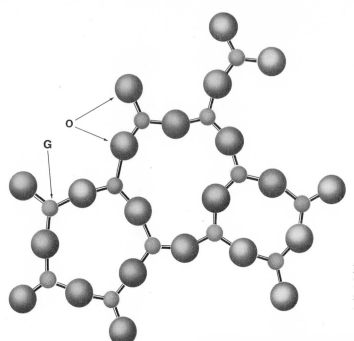

FIGURE 13–14   Irregular crystalline structure for SiO_2 in two dimensions. In the two dimensional analogy the formula would be G_2O_3.

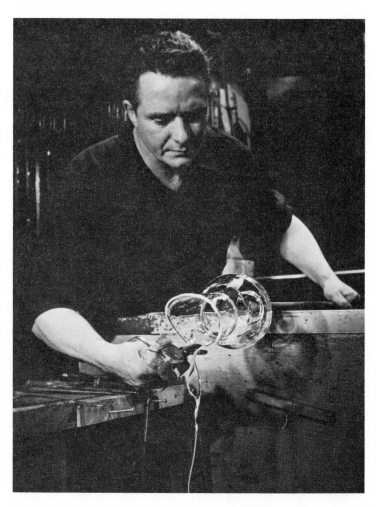

FIGURE 13–15 Craftsman working with molten glass. (Courtesy of the Corning Glass Company.)

TABLE 13-2   SPECIAL GLASSES

SPECIAL ADDITION OR COMPOSITION	DESIRED PROPERTY
1. Large amounts of PbO with silica and Na_2CO_3	crystal glass
2. SiO_2, B_2O_3, and small amounts of Al_2O_3 (borosilicate glass)	small thermal expansion "Pyrex," "Kimex," etc.
3. One part SiO_2 and four parts PbO	stops (absorbs) large amounts of x-rays and γ-rays— "lead glass"
4. Large concentration of CdO	neutron absorbing glass
5. High concentration of pure As_2O_3	transparent to infrared radiation

If too much iron oxide is present, the glass will be green; other metal oxides produce other colors. The materials for the glass are melted together in a gas or oil fired furnace; as they react, bubbles of CO_2 are evolved. In order to insure that the glass is homogeneous, it is heated to about 1500°C; at this temperature its viscosity is low, and the bubbles of gas trapped in it are allowed to escape. The mixture is cooled somewhat and drawn off to be shaped into bottles, sheets, or other forms (Figure 13–15).

It is possible to incorporate a wide variety of materials into glass for special purposes. Table 13–2 gives some examples.

CERAMICS

Ceramic materials have been made since well before the dawn of recorded history. They are generally materials fashioned at room temperature from clay or other natural earths and then permanently hardened by heat. Clays with a wide variety of properties are found in a considerable range of ceramic materials, from bricks to table china. The techniques developed with natural clay have been applied to a wide range of other inorganic materials in recent years. The result has been a considerable increase in the kinds of ceramic materials available. One can now obtain ceramic magnets as well as ceramics suitable for rocket nozzles—both were developed from mixtures of inorganic oxides by the use of ceramic technology, which includes as an indispensable process the heating of the materials to make them hard and resistant to wear.

The three basic ingredients of common pottery are silicate minerals: clay, sand, and feldspar. The term *clay* includes materials with a wide range of chemical compositions which are produced from the weathering of granite and gneiss rocks. Feldspars are aluminosilicates containing potassium, sodium, and other ions in addition to silicon and oxygen. An approximation of the weathering process can be made if we write feldspar as a mixture of oxides: $K_2O \cdot Al_2O_3 \cdot 6SiO_2$; this includes, among other reactions, the reaction of the mineral with water containing dissolved carbon dioxide to form clay.

$$K_2O \cdot Al_2O_3 \cdot 6SiO_2 + 2H_2O + CO_2 \longrightarrow Al_2O_3 \cdot 2SiO_2 \cdot 2H_2O + 4SiO_2 + 2K^+ + CO_3^{2-}$$

A clay mineral

The essential feature of the clay mineral is that it occurs in the form of extremely minute platelets which, when wet, are plastic and can easily be shaped. When dry they are rigid; if heated to an elevated temperature they become permanently rigid and are no longer subject to easy dispersion in water. When these clays are mixed with feldspars and silica, heating them produces a mixture of crystals held together by a matrix of glass-like material. The clays can be used by themselves to make bricks, flowerpots, and clay pipe, but finer quality ceramic materials contain purified clays and other ingredients in carefully controlled proportions. The clay is mixed with potter's flint (a form of silica) and feldspar in various proportions. The clay is used to make the mixture pliable; the silica decreases the amount of shrinkage which occurs after drying and firing in a kiln (Figure 13–16); the feldspar lowers the temperature needed for adequate firing. The feldspar acts as a glass-like material to hold the grains of clay and flint together. It is usually necessary to dry clay before firing it; otherwise the

FIGURE 13-16   Fired pottery. Different colors and surface textures can be achieved by fusing the coating material with the clay itself. The coating ceases to be a "coat" and becomes a part of the whole, the entire structure being held together by covalent bonds. (Courtesy of the Robinson-Ransbottom Pottery Company.)

rapid loss of water from the surface and the slower loss from the interior of the object will cause cracks in the clay.

Natural clays are generally extremely complex mixtures. If these are used in ceramics without treatment, the finished materials have a color and physical properties characteristic of the impurities present. The first pieces of fine oriental chinaware arrived in Europe during the late Middle Ages, and European potters envied and admired the obviously superior product. This led to the beginning of systematic studies on the effect of composition on the nature of the ceramic produced and to a keen appreciation of the role of the purity of the clay in determining the color and potential decorative development of the piece.

Alchemists made notable achievements in this area. One of these, Johann Friedrich Bottger, worked from about 1705 to 1719 for King Augustus of Saxony, who kept him almost as a prisoner. The king hoped to gain power from the alchemist's discoveries. Bottger succeeded in developing several novel ceramic materials, of which the most important was the first white glazed porcelain made in Europe (in 1709). Bottger devoted the rest of his life to the perfection of the manufacture and decoration of this material, in which he enjoyed considerable success. The china was made in Meissen and was both glazed and vitrified. The glazing was accomplished by coating the pieces with a material which melted and produced an impermeable layer on the surface. Vitrification was produced by firing the clay at a temperature sufficient to melt a portion of the material and, in effect, produce an impermeable glass which held the remaining particles together.

In the past few decades new ceramic materials have been developed and used on an increasingly wide scale. Nearly pure alumina (Al_2O_3) and zirconia (ZrO_2) are now used as bases for ceramic materials which are excellent electrical or thermal insulators. Magnetic ceramics, or ferrites, containing iron compounds are used as memory elements in computers.

In recent years a new class of materials, the glass ceramics, has been discovered; these have unusual but very valuable properties. Normally glass breaks because once a crack starts, there is nothing to stop it from spreading. It was discovered that if glass is treated by heating until a very large number of tiny crystals has developed in it, the resulting material, when cooled, is much more resistant to breaking than normal glass. The process has to be controlled carefully to obtain the desired properties. The materials produced in this way are generally opaque and are used for cooking utensils and kitchen ware. They include materials marketed under the name "Pyroceram" (Figure 13–17). The initial manufacturing process is similar to that of other glass objects, but once they have been formed into their final shapes, they are heat treated to develop their special properties.

SILICONES

Because silicon and carbon are in the same family of the periodic table, chemists have investigated the similarities and differences of these elements. The most outstanding characteristic of carbon is its ability to bond to itself to form compounds that are quite stable to hydrolysis, moderate heat, and a variety of

251

FIGURE 13-17   Cookware
made of Pyroceram. (Courtesy of
Corning Glass Company.)

chemical reagents. The study of silicon soon revealed that although silicon can
form bonds to itself, these bonds are normally subject to rapid splitting in the
presence of water:

$$-\underset{|}{\overset{|}{Si}}-\underset{|}{\overset{|}{Si}}- + 2H_2O \longrightarrow 2-\underset{|}{\overset{|}{Si}}-OH + H_2\uparrow$$

This type of reaction delayed the commercial development of many silicon
compounds. Eventually chemists discovered a way to use this hydrolytic reaction
to form a special kind of compound which subsequently found wide use as the
basis for high temperature lubricants and polymers. These compounds are the
silicones (Figure 13–18).

In order to understand the submicroscopic structure of the silicones we
return to the tetrahedral structure of methane:

$$\underset{\underset{\textit{Methane}}{H}}{\overset{H}{\underset{|}{\overset{|}{C}}}}\ _{H}\ ^{H}$$

It is not surprising to find that the corresponding silicon compound, silane, is also tetrahedral:

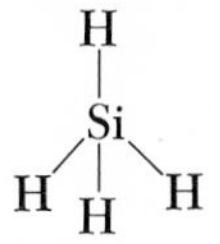

$$\begin{array}{c} H \\ | \\ Si \\ H\, |\, H \\ H \end{array}$$

Silane

Methyl chloride, CH_3Cl, reacts with silicon over a copper catalyst to form a substituted silane:

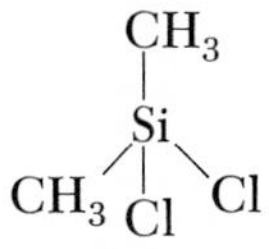

$$\begin{array}{c} CH_3 \\ | \\ Si \\ CH_3\, |\, Cl \\ Cl \end{array}$$

Dimethyldichlorosilane

The reaction is:

$$2CH_3Cl + Si \xrightarrow[\text{catalyst}]{Cu} (CH_3)_2SiCl_2$$

In contact with water the substituted silane readily hydrolyzes to form first a dihydroxy compound, in which the chlorine atoms are replaced by hydroxyl groups:

$$(CH_3)_2SiCl_2 + H_2O \longrightarrow (CH_3)_2Si(OH)_2 + 2HCl$$

The dihydroxy compound, in turn, is unstable and reacts with itself to eliminate a water molecule:

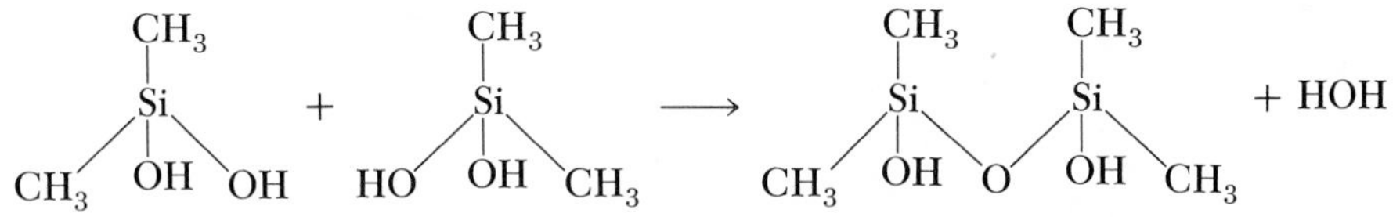

$$\begin{array}{c} CH_3 \\ | \\ Si \\ CH_3 \quad OH \quad OH \end{array} + \begin{array}{c} CH_3 \\ | \\ Si \\ HO \quad OH \quad CH_3 \end{array} \longrightarrow \begin{array}{c} CH_3 \qquad CH_3 \\ | \qquad | \\ Si \qquad Si \\ CH_3 \quad OH \quad O \quad OH \quad CH_3 \end{array} + HOH$$

However, the reaction does not stop here. Note that two OH groups remain in the dimer structure. The reaction continues to form a polymer with a molecular weight which runs into the hundreds or even thousands of atomic weight units. The result is a silicone with the structure (the open bonds on either end indicate indefinite length):

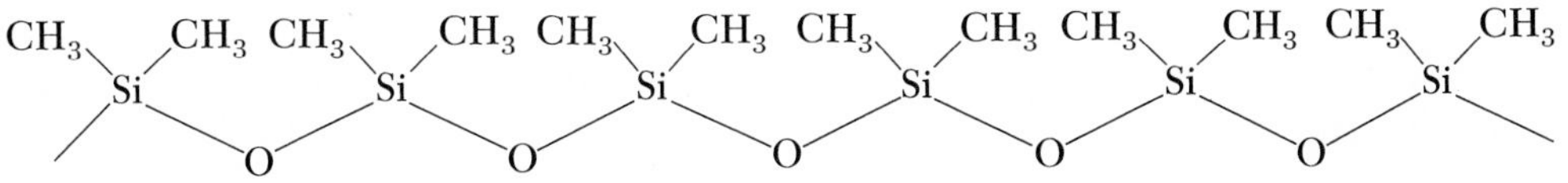

$$\begin{array}{c} CH_3 \quad CH_3 \; CH_3 \quad CH_3 \; CH_3 \quad CH_3 \; CH_3 \quad CH_3 \; CH_3 \quad CH_3 \; CH_3 \quad CH_3 \\ Si \qquad Si \qquad Si \qquad Si \qquad Si \qquad Si \\ O \qquad O \qquad O \qquad O \qquad O \end{array}$$

In addition to being more stable at high temperatures than hydrocarbon oils, silicone oils having the above structure decrease less in viscosity when cooled. The increase in viscosity for a typical silicone oil cooled from 38° to $-37°$ is sevenfold. A hydrocarbon oil with the same viscosity at 38° would increase in viscosity by a factor of 1800 when cooled to $-37°$.

The properties of silicones are varied by altering their basic structure. This is accomplished in two ways: (1) by using other hydrocarbon groups, such as ethyl (C_2H_5) or propyl (C_3H_7), to replace the methyl groups, and (2) by forming chemical bonds between the silicone chains. Silicone rubber is composed of very

253

FIGURE 13–18   Examples of the use of silicone in the space program. Soles of lunar boots worn by the Apollo astronauts are made of high-strength silicone rubber. A silicone compound is also used for the air-tight seal of the lunar module hatch from which Astronaut Edwin E. Aldrin, Jr., has just emerged in this photo of the first manned landing on the moon on July 20, 1969.

high molecular weight units that are bridged together with ethylene groups:

$$
\begin{array}{c}
\text{———Chain 1———Si———Chain 1———} \\
\text{H———C———H} \\
\text{H———C———H} \\
\text{———Chain 2———Si———Chain 2———}
\end{array}
$$

Over 3,000,000 pounds of silicone rubber are produced in the United States each year; approximately 1100 pounds go into every jet airliner. The first footprints on the moon were made with silicone rubber boots, the rubber readily withstanding the extreme temperature on the lunar surface.

"Silly Putty," a silicone widely distributed as a toy, is intermediate between silicone oils and silicone rubber. It is an interesting material with elastic properties on sudden deformation, but its elasticity is quickly overcome by its ability to flow like a liquid when allowed to stand.

CEMENTS

A cement is a material used to bind other materials together. Portland cement contains calcium, iron, aluminum, silicon, and oxygen in varying pro-

A

B

FIGURE 13-19   A, A cement kiln. Note the rollers on the supports which allow the giant cylinder to rotate. As the kiln turns, the powder moves down the cylinder because it is at a slight angle. Intense heat is produced by the combustion of gaseous fuels. The powder loses volatile materials as it moves along and the finished product is discharged from the lower end. B, The first kiln used for making cement in the United States was constructed by David Saylor over a century ago at Coplay, Pennsylvania. The kiln still stands as pictured here. (Courtesy of the Portland Cement Northwestern States Company, Mason City, Iowa.)

portions. It has a structure somewhat similar to that described earlier for glass, except that in cement some of the silicon atoms have been replaced by aluminum atoms. Cement reacts in the presence of water to form a hydrated colloid of large surface area which subsequently undergoes recrystallization and reaction to bond to itself and to bricks or stone. Cement is made by roasting a powdered mixture of calcium carbonate (limestone or chalk), silica (sand), or aluminosilicate mineral (kaolin, clay, or shale) and iron oxide. The roasting is carried out at a temperature of up to 870°C in a rotating kiln (Figure 13-19). As the materials pass through the kiln, they lose water and carbon dioxide and ultimately form a "clinker," in which the materials are partially fused together. The "clinker" is then ground to a very fine powder after the addition of a small amount of calcium sulfate (gypsum). The composition of Portland cement is 60 to 67 per cent CaO, 17 to 25 per cent SiO_2, 3 to 8 per cent Al_2O_3, up to 6 per cent Fe_2O_3, and small amounts of magnesium oxide, magnesium sulfate, and potassium and sodium oxides.

The reactions which occur during the setting of cement are quite complex and involve the reaction of the various constituents with water and, subsequently at the surface, with the carbon dioxide in air. The initial reaction of cement with water gives a sticky gel which results from the hydrolysis of the calcium silicates. This sticks to itself and to the other particles (sand, crushed stone, or gravel). The gel has a very large surface area and is responsible for the strength of concrete. The setting process also involves the formation of small densely interlocked crystals after the initial solidification of the wet mass. This continues for a long time after the initial setting and increases the compressive strength of the cement. Water is required since the setting reactions involve hydration. For this reason freshly poured concrete is kept moist for several days. Over 400,000,000 tons of cement are manufactured each year, most of which is used to make concrete. Concrete, like many other materials containing Si—O bonds, is highly noncompressible but lacks tensile strength. If concrete is to be used where it is subject to tension, it must be reinforced with steel.

SOME IMPORTANT COMPOUNDS

A number of useful inorganic materials can be prepared on a large scale from naturally occurring materials such as air, water, and minerals.

We will examine the following substances in some detail:

sulfuric acid (H_2SO_4),
ammonia (NH_3),
hydrochloric acid (muriatic acid) (HCl),
sodium hydroxide (lye) (NaOH),
chlorine (Cl_2),
sodium hydrogen carbonate (baking soda) ($NaHCO_3$),
sodium carbonate (washing soda) (Na_2CO_3),
phosphoric acid (H_3PO_4).

Sulfuric Acid, H_2SO_4

Sulfur in underground mineral deposits is brought to the surface by the Frasch process (Figure 13-20), which utilizes the fact that sulfur can be melted

256

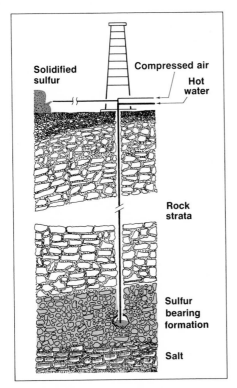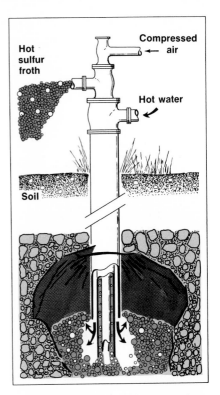

FIGURE 13-20   Diagram of the mining of sulfur by Frasch process which uses melting and pressure. Three concentric pipes are directed into the sulfur deposit. Compressed air and hot water are sent down two of the pipes and molten sulfur is forced up the third pipe.

by superheated steam. The molten sulfur is raised to the surface of the earth by means of compressed air and is then allowed to cool in large vats.

The conversion of sulfur to sulfuric acid is carried out by means of a four-step process called the contact process (Figure 13–21). In the first step the sulfur is burned in air to give mostly sulfur dioxide.

$$S + O_2 \longrightarrow SO_2(g)$$

The conversion of the gaseous SO_2 to SO_3 is then achieved by passing it over a hot catalytically active surface, such as platinum or vanadium pentoxide:

$$2SO_2 + O_2 \xrightarrow{\text{Catalyst}} 2SO_3(g)$$

Although SO_3 can be converted directly into H_2SO_4 by passing it into water, the enormous amount of heat released in the reaction causes the formation of a stable fog of H_2SO_4. This is avoided by passing the SO_3 into H_2SO_4:

$$SO_3 + \underset{\textit{Sulfuric acid}}{H_2SO_4} \longrightarrow \underset{\textit{Pyrosulfuric acid}}{H_2S_2O_7}$$

and then diluting the $H_2S_2O_7$ with water:

$$H_2S_2O_7 + H_2O \longrightarrow 2H_2SO_4$$

By this method several millions of tons of sulfuric acid are prepared each year.

257

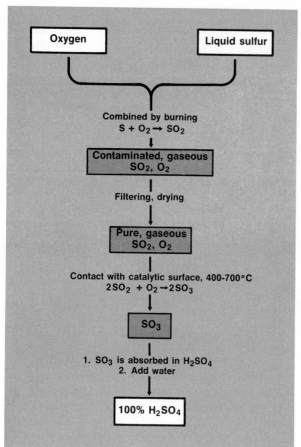

FIGURE 13-21   Flow diagram of contact process for producing sulfuric acid.

The acid is used in huge quantities in the manufacture of fertilizers, in the petroleum industry, and in the production of steel. Also it plays an important role in the manufacture of organic dyes, plastics, and drugs, among many other products. The cost of sulfuric acid, about 1¢ per pound, has not changed much in 300 years, a tribute to man's improving technology.

AMMONIA, NH₃

For many years chemists looked upon the vast amounts of elemental nitrogen in the atmosphere and wondered how it could be turned into useful compounds. The key lay in mixing nitrogen gas and hydrogen gas and using platinum metal as a catalyst. This nineteenth-century discovery produced a very useful product, ammonia. The reaction is reversible; platinum is used as a catalyst.

$$N_2(g) + 3H_2(g) \xrightarrow{Pt} 2NH_3(g) + Heat$$

The manipulation of this reaction was studied in great detail by the German chemist F. Haber, who developed a high pressure process for the synthesis. The mixture of nitrogen and hydrogen is heated under pressure and passes over the

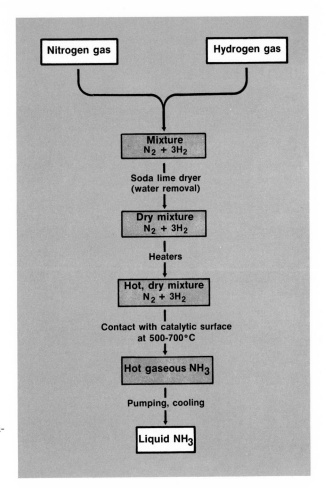

FIGURE 13-22   The Haber process for synthesizing ammonia.

catalyst. The ammonia can then be condensed out of the mixture because of its relatively high boiling point (Figure 13–22).

Most other industrial nitrogen compounds are prepared from ammonia because it is readily available and low in cost. An example is *nitric acid*, HNO_3. This is made by burning ammonia over a platinum catalyst to obtain NO, nitric oxide (or mononitrogen oxide):

$$4NH_3 + 5O_2 \xrightarrow{Pt} 4NO + 6H_2O$$

The NO reacts readily with O_2 from the air to form NO_2:

$$2NO + O_2 \longrightarrow 2NO_2$$

This, in turn, reacts with water to give nitric acid and NO, which is recycled:

$$H_2O + 3NO_2(g) \longrightarrow 2HNO_3 + NO(g)$$

Ammonia is used in a water solution as a fertilizer, or it may be transformed into salts for this purpose. For example:

$$2NH_3 + H_2SO_4 \longrightarrow \quad (NH_4)_2SO_4$$

Ammonium sulfate

259

SODIUM HYDROXIDE (NaOH), CHLORINE (Cl₂), AND HYDROGEN CHLORIDE (HCl)

Sodium hydroxide, hydrogen, and chlorine are prepared simultaneously by the electrolysis of a concentrated solution of sodium chloride in water (Figure 13–23). There are several variations in the basic process which improve its efficiency and the purity of the products. The basic reactions, which occur at non-reactive solid electrodes, are:

$$\textit{Anode: } 2Cl^- \longrightarrow Cl_2(g) + 2e^- \qquad \textit{(Oxidation)}$$
$$\textit{Cathode: } 2H_2O + 2e^- \longrightarrow 2OH^- + H_2 \qquad \textit{(Reduction)}$$

The reaction replaces the Cl^- of the NaCl with OH^-.

Purer sodium hydroxide is produced when the process is carried out using a mercury cathode. In this instance Na^+ is reduced to Na metal which dissolves in the mercury electrode:

$$\textit{Cathode: } Na^+ + e^- \xrightarrow{Hg} Na(Hg)$$

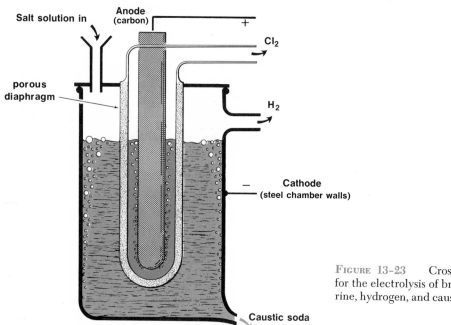

FIGURE 13–23   Cross section of cell for the electrolysis of brine to yield chlorine, hydrogen, and caustic soda (NaOH).

The sodium-mercury amalgam* is a liquid. It is removed from the cell continuously and reacted with pure water:

$$2Na(Hg) + 2H_2O \longrightarrow 2Na^+ + 2OH^- + H_2(g)$$

The chlorine gas is collected from the anode compartment, compressed into tanks, and sold for further use. The hydrogen gas is also collected and sold. An

———————————

* An amalgam is an alloy of metal with mercury; it can be either liquid or solid.

important reaction between hydrogen and chlorine is the production of hydrogen chloride gas and hydrochloric acid.

$$H_2(g) + Cl_2(g) \longrightarrow 2HCl(g)$$

This process can be used to prepare hydrogen chloride which is quite pure. Hydrochloric acid is a solution of hydrogen chloride in water. The use of mercury in this process is a potential hazard as mercury contamination of streams results when the waste solutions from this process are thrown out.

Because electrical energy is quite expensive, an electrochemical process is economical only if all the products are sold. In recent years the demand for chlorine has grown very rapidly while the demand for sodium hydroxide has not. Consequently it became necessary to devise some way of disposing of relatively large amounts of sodium hydroxide. This has been accomplished, in part, by transforming it into sodium hydrogen carbonate:

$$Na^+ + OH^- + CO_2(g) \longrightarrow Na^+ + HCO_3^- \longrightarrow NaHCO_3$$

Because of the surplus of NaOH, this process has largely replaced older processes for the manufacture of $NaHCO_3$.

An alternate method for making HCl involves heating NaCl and H_2SO_4. The reaction occurs in two stages and can be driven to completion by heating:

$$NaCl + H_2SO_4 \longrightarrow HCl(g) + NaHSO_4$$
$$NaCl + NaHSO_4 \longrightarrow HCl(g) + Na_2SO_4$$

In principle, chlorine can be prepared by a number of reactions in which chloride is oxidized to Cl_2:

$$2Cl^- \longrightarrow Cl_2(g) + 2e^-$$

A number of oxidizing agents, including O_2 and hydrogen peroxide, H_2O_2, are suitable for this purpose.

PHOSPHORIC ACID, H_3PO_4

Phosphoric acid or its salts are used in the manufacture of a large number of materials encountered in our daily lives. These include baking powder, carbonated beverages, detergents, fertilizers, and fire-resistant textiles. There are several processes available for the manufacture of phosphoric acid, but the basic raw material is usually the calcium phosphate which occurs naturally in apatite minerals of the general formula $Ca_5X(PO_4)_3$, where X may be F^-, Cl^-, or OH^-.

An electric furnace is used to heat a mixture of the phosphate ore $[Ca_5F(PO_4)_3]$, silica (SiO_2), and coke (C). At an elevated temperature a reaction occurs in which elemental phosphorus vapor (P_4) is produced:

$$4Ca_5F(PO_4)_3 + 18SiO_2 + 15C \longrightarrow 18CaSiO_3 + 2CaF_2 + 15CO_2\uparrow + 3P_4\uparrow$$

261

The phosphorus is then condensed from the gaseous vapor and purified. Because of the low melting point of phosphorus (44.1°C), it can easily be stored and handled as a liquid if it is protected from air, in which it ignites spontaneously.

Elemental phosphorus is transformed into phorphoric acid by oxidation with air to give P_4O_{10}:

$$P_4(gas) + 5O_2 \longrightarrow P_4O_{10}(gas)$$

which is then hydrated by absorption into hot phosphoric acid containing about 10 per cent water:

$$P_4O_{10} + 6H_2O \xrightarrow{H_3PO_4} 4H_3PO_4$$

Arsenic, when present in the original ore, is carried through to the phosphoric acid produced, at which point it can be precipitated by treatment with H_2S. This is important if the phosphoric acid is to be used in the manufacture of food products.

QUESTIONS

1. Name three metals that you would expect to find free in nature. Name three that you would not.

2. What is the primary reducing agent in the production of iron from its ore?

3. Why is CaO necessary for the production of iron in a blast furnace?

4. What is the difference between iron and steel?

5. Both iron and magnesium will oxidize in the air. Why is the oxidation of iron a much greater problem than the oxidation of magnesium?

6. How is it possible that oxidation and reduction can occur at different points on a piece of iron?

7. Give examples of three ways in which metals can be protected from corrosion.

8. Describe the solution used in the commercial cell for the electrolytic reduction of aluminum.

9. What is "blister" copper?

10. Why is it so important to purify electrolytically industrial quantities of copper to a level above 99.9 per cent pure?

11. Oyster shells are used as a source of what chemical in the production of magnesium from sea water? What is the role of this chemical in the process?

12. What is the difference between quartz and sand?

13. Explain how the structures of glass and a liquid are similar.

14. At the molecular level, compare a hydrocarbon oil with a silicone oil. Also, make a comparison at the experimental level.

15. Describe the contact process for the production of sulfuric acid.

16. Contrast the electrolysis of molten NaCl to the electrolysis of an aqueous solution of NaCl.

17. Give reactions involved in the preparation of

$Ca(OH)_2$ from $CaCO_3$
H_2SO_4 from S
H_3PO_4 from P_4
Na_2CO_3 from $NaHCO_3$

18. Explain what is meant by the term "flotation." Why is this an important process?

19. In the production of copper from its sulfide ores, large amounts of SO_2 gas are formed as a by-product. What useful product can be prepared from this material? What are the consequences of allowing the gas to escape into the atmosphere? See Chapter 21.

20. Name four items in everyday use which contain (a) silicon, (b) aluminum, (c) calcium.

21. Write the reaction for the formation of a silicone from $(C_2H_5)_2SiCl_2$.

22. A typical lime-soda glass has a composition reported as 70 per cent SiO_2, 15 per cent Na_2O, and 10 per cent CaO. What weights of sand (SiO_2), sodium carbonate (Na_2CO_3), and calcium carbonate ($CaCO_3$) must be melted together to make this glass? The carbonates are decomposed by heat to evolve carbon dioxide gas.

23. What volume of chlorine gas (at STP) is obtained when 500 grams of sodium chloride is electrolyzed into sodium hydroxide?

24. What is the maximum weight (in pounds) of magnesium that can be obtained from 1000 pounds of sea water?

SUGGESTIONS FOR FURTHER READING

Andrade, J., "Materials Science and Engineering—A Modern Multidiscipline," *Chemistry*, Vol. 43, No. 4, p. 13 (1970).
Brown, G. H., "Liquid Crystals," *Chemistry*, Vol. 40, No. 9, p. 10 (1967).
Chandler, M., "Ceramics in the Modern World," Doubleday & Co., Inc., Garden City, New York, 1968.
"Encyclopedia of Chemical Technology," Second Edition, Interscience Publishers, New York, 1963.
Farber, E., "Oxygen—The Element With Two Faces," *Chemistry*, Vol. 39, No. 5, p. 17 (1966).
"Industry Profile," *Chemistry*, Vol. 39, No. 6, pp. 17–18 (June, 1966).
"McGraw-Hill Encyclopedia of Science and Technology," McGraw-Hill Book Co., New York, 1960.
Maloney, F. J. T., "Glass in the Modern World," Doubleday & Co., Inc., Garden City, New York, 1968.
Schaar, B. E., "Dental Amalgams," *Chemistry*, Vol. 40, No. 8, p. 21 (1967).
Stone, J. K., "Oxygen in Steel Making," *Scientific American*, Vol. 218, No. 4, p. 24 (1968).

THE UBIQUITOUS CARBON ATOM— AN INTRODUCTION TO ORGANIC CHEMISTRY

CHAPTER 14 ═══════════════════════════════

The importance of compounds of the element carbon to life on earth cannot be overestimated. Consider what the world would be like if all the carbon and carbon compounds were suddenly removed. The result would be somewhat like the barren surface of the moon! Many of the little everyday things often taken for granted would be quite impossible without the versatile element carbon. In an ordinary pencil for example, the "lead" in the pencil (made from graphite, an elementary form of carbon), the wood, the rubber in the eraser, and the paint on the surface are all carbon or carbon compounds. The paper in this book, the cloth in its cover, and the glue holding it together are also made of carbon compounds. All of the clothes one wears, including the leather in shoes, would not exist. If carbon compounds were removed from the human body, there would be nothing left except water and a small residue of minerals, and the same is true of all forms of living matter. Fossil fuels, foods, and most drugs are essentially made of carbon compounds. In addition, many carbon compounds such as plastics and detergents, which are not directly connected with the life processes, play a vital role in one's life.

There are nearly two million different carbon compounds that have been studied and described in the chemical literature, with thousands of new ones being reported every year. Although there are 89 other naturally occurring elements, the number of known carbon compounds is many times greater than that of the known compounds which contain no carbon. The very large and important branch of chemistry devoted to the study of carbon compounds is called *organic chemistry*. The name "organic" is actually a relic of the past, when chemical compounds produced from once-living matter were called "organic" and all other compounds were called "inorganic."

THEORETICAL BASIS FOR THE LARGE NUMBER OF ORGANIC COMPOUNDS

The enormous number of organic compounds has intrigued chemists for over a hundred years. The atomic theory, as developed earlier for all atoms, describes a structure for the carbon atom which explains this multiplicity of carbon compounds. The peculiar structure of this atom allows it *to form covalent bonds with other carbon atoms in a seemingly endless array of possible combinations.* A simple organic molecule may contain a single carbon—carbon bond, whereas a complex one may contain literally thousands of such bonds. A few other elements are capable of forming stable bonds between like atoms. These include such elements as nitrogen, N_2, oxygen, O_2, and sulfur, S_8, to name a few. But only S, Sn, Si, and P can form long chain molecules, and none of these can approach carbon in the ability to do this.

An additional factor in the large number of carbon compounds lies in the *stability* of carbon chains. The carbon chains are not normally subject to attack by water or, at ordinary temperatures, by oxygen. Chains formed by atoms of other elements undergo reaction with either water or oxygen, or both, much more easily than do carbon chains.

A further reason for the large number of organic compounds is the ability of a given number of atoms to combine in more than one molecular pattern and, hence, produce more than one compound. Such compounds, each of which has molecules containing the same number and kinds of atoms, but arranged differently relative to each other, are called *isomers*. For example, the molecular structure represented by A—B—C is different from the molecular structure A—C—B, as is C—A—B; these three species are isomers. If one considers the number of possible ways the digits one through nine can be ordered to make nine-digit numbers, he can begin to imagine how a single group of atoms could possibly form hundreds of different molecules. Carbon, with its ability to bond to other carbon atoms, is especially well suited to form isomers.

A final factor explaining the large number of organic compounds is the ability of the carbon atom to form strong covalent bonds with numerous other atoms, such as nitrogen, oxygen, sulfur, and the halogens. As a result, there are large classes of organic compounds, each of which has a *functional group*, a particular combination of atoms, which appears in each member of that class. For example, organic acids have a carboxyl group attached to another carbon atom.

$$\left(-C\diagup_{\diagdown OH}^{\nearrow O}\right) \textit{ Carboxyl group}$$

The remainder of this chapter will be devoted to a closer examination of these three basic reasons for the great number of organic compounds. Some interesting properties of a few of these compounds will be discussed.

CHAINS OF CARBON ATOMS— THE HYDROCARBONS

Only two elements, hydrogen and carbon, are required to explain the existence of literally thousands of compounds known as hydrocarbons. In Chapter 8 the electron structure for carbon was given.

265

$$\text{C} \qquad 1s^2 2s^2 2p^1 2p^1 2p^0$$

As was shown earlier, when the 2s and 2p orbitals are hybridized, a tetrahedral arrangement of bonds occurs if a carbon atom is bonded to four other atoms (Figure 14–1 and 14–2). The direct, head-on overlap of one of the sp³ hybrid orbitals in carbon with a 1s orbital of hydrogen produces a carbon-hydrogen *sigma bond*. Methane (CH₄), the simplest of the hydrocarbons, has four of these bonds (Figure 14–2).

A tetrahedron

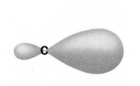

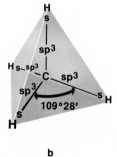

A single sp³ hybrid orbital

FIGURE 14-1   Orbital configuration for sp³ hybrid orbitals (back lobes omitted in tetrahedral drawing).

Four sp³ hybrid orbitals

Carbon has an intermediate electronegativity value (a measure of an atom's attraction for an electron pair in a covalent bond). This intermediate value is not large enough to enable a carbon atom to completely remove electrons from metals, but it is sufficient to keep even the most electronegative atom, fluorine, from completely removing an electron from carbon. As a result, carbon atoms tend not to form ionic bonds but rather to share electrons in the formation of covalent bonds. In addition to the tendency for carbon to form covalent bonds with many other atoms, there is also a remarkable inclination for carbon atoms to form relatively strong covalent bonds with each other.

If two carbon atoms are close enough to overlap their sp³ orbitals, a sigma bond will result. The remaining six orbitals form single covalent bonds with six hydrogens in *ethane* (C₂H₆). The structure of ethane is illustrated in two dimensions by the following formulas:

$$
\begin{array}{cc}
\text{H} \ \ \text{H} & \qquad \text{H} \ \ \text{H} \\
\text{H} \!:\!\! \overset{..}{\underset{..}{\text{C}}} \!:\!\! \overset{..}{\underset{..}{\text{C}}} \!:\! \text{H} \quad \text{or} & \text{H}\!-\!\!\overset{|}{\underset{|}{\text{C}}}\!-\!\!\overset{|}{\underset{|}{\text{C}}}\!-\!\text{H} \\
\text{H} \ \ \text{H} & \qquad \text{H} \ \ \text{H}
\end{array}
$$

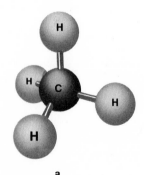

FIGURE 14-2   Methane. (*a*) Ball-and-stick model showing tetrahedral structure. (*b*) Geometry of regular tetrahedron. (*c*) Model of methane, CH₄, showing relative size of atoms in relationship to interatomic distances.

a   b   c

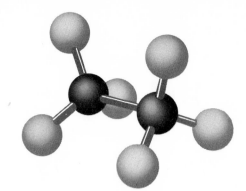

FIGURE 14-3   Ball-and-stick model of ethane.

Ethane

The geometric shape of an ethane molecule is indicated in Figure 14–3.

Ethane is composed of two CH_3 groups (methyl groups, from methane) joined by a single carbon—carbon sigma bond. Since this sigma bond is symmetrical about the bond axis (the line drawn between the carbon atoms), there is only a little interference when one methyl group spins around the bond axis relative to the other methyl group. In Figure 14–4 it is evident that the hydrogen atoms can be lined up, in a staggered relationship, or in any intermediate position. The eclipsed (or lined-up) form is actually a slightly high energy form due to atomic repulsions; the staggered form, in which repulsions are less important, is a slightly lower energy form.

By applying what we have learned, it is a simple matter to extend the concept of carbon—carbon bonding to a three-carbon molecule such as *propane* (C_3H_8).

$$
\begin{array}{ccc}
\text{H} & \text{H} & \text{H} \\
\text{H}:\text{C}:\text{C}:\text{C}:\text{H} & \text{or} & \text{H}-\text{C}-\text{C}-\text{C}-\text{H} \\
\text{H} & \text{H} & \text{H}
\end{array}
$$

In Figure 14–5, note that the three carbon atoms in propane do not lie in a straight line because of the tetrahedral bonding about each carbon atom.

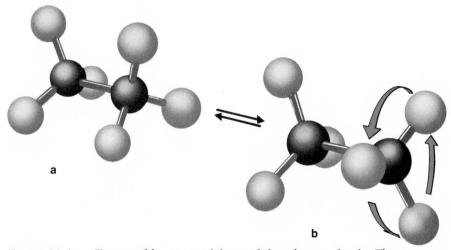

a

b

FIGURE 14-4   Two possible rotational forms of the ethane molecule. The hydrogen atoms in the methyl groups may be in an eclipsed position (*a*), staggered (*b*), or in any intermediate position.

267

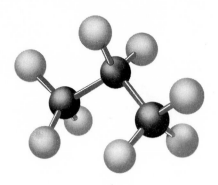

FIGURE 14-5   Ball-and-stick model of propane.

Propane

It is apparent that these bonding concepts can be extended to a four-carbon molecule and to a limitless number of even larger hydrocarbon molecules. Actually, many such compounds are known; some, such as natural rubber, are known to contain over a thousand carbon atoms in a chain.

STRUCTURAL ISOMERS

When given the task of writing the electron dot formula for butane, C_4H_{10}, the student will soon discover that there are two possible structures.

	n-Butane	Isobutane
Melting point	−138.3°C	−160°C
Boiling point	−0.5°C	−12°C
Density	0.601 g/ml	0.557 g/ml

The two formulas represent two distinctly different compounds. Both are well known, each with its own particular set of properties. It must be concluded then, that molecular formulas such as C_4H_{10} are sometimes ambiguous, and that structural formulas are necessary. If no carbon atom is attached to more than two other carbon atoms, the carbon chain is said to be a *straight chain* structure. Actually, as shown in Figure 14–6, the carbon chain is bent (109° 28′) at each carbon atom, but it is called a straight chain because the carbon atoms are bonded together in succession one after the other. The student might note that many molecular shapes are possible for n-butane because of the possible rotational motion about the single bonds. These arrangements (called conformations) do not constitute different molecules; rather, there tends to be a mixture of all possible shapes at a given temperature due to the ease of bond rotation.

If one carbon atom is bonded to either three or four other carbon atoms in a molecule, the molecule is said to have a *branched chain*. Isobutane is an example of a branched-chain hydrocarbon (Figure 14–6). Isobutane and n-butane are called *structural isomers* because both molecules contain exactly the same number and kinds of atoms, C_4H_{10}, but the molecules are structured (put together) differently. Structural isomerism is somewhat like a child building many different structures with the same collection of building blocks.

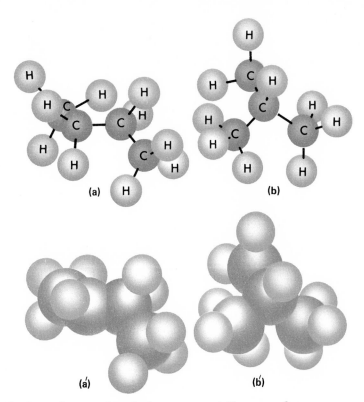

FIGURE 14-6 (a) and (a') Normal butane. (b) and (b') Isobutane.

(a) (b)

(a) (b)

The two butanes (and all hydrocarbon molecules) are essentially nonpolar since the C—C bonds are nonpolar, and the slightly polar C—H bonds are symmetrically arranged to cancel each other out. The forces holding these molecules together in the liquid, therefore, are Van der Waal's attractions, which depend upon the surface area of a molecule and the closeness of approach of the molecules to each other. In general, a branched-chain isomer has a lower boiling point than a straight-chain isomer since the branched-chain isomer does not permit intermolecular distances as short as those of a straight chain and has less surface area and, therefore, less intermolecular attraction. Melting points of isomers generally do not follow the same pattern since they depend upon the ease with which the molecules fit into a crystalline lattice, but they can be easily explained and predicted on this basis.

It is important to distinguish between different bond rotational arrangements (conformations) and structural isomers (configurations). To change from one rotational arrangement to another, only motion about a bond is required. However, to change from one structural isomer to another (for example, from isobutane to n-butane), it is necessary to break bonds and to form new ones.

Consider the isomeric pentanes with the formula, C_5H_{12}. There are three of these:

	n-Pentane	Isopentane	Neopentane
Melting point	−130°C	−160°C	−20°C
Boiling point	36°C	28°C	9.5°C
Density	0.626 g/ml	0.620 g/ml	0.614 g/ml

269

Note that the three isomers are predicted by bonding theory, and, furthermore, the theory predicts no other possible isomers for C_5H_{12} since there are no other ways to unite the 17 atoms in keeping with the atomic orbitals available (octet rule for Period II). These three isomers of pentane are well known, and no others have ever been found.

Table 14–1 gives the number of isomers predicted by theory for some larger molecular formulas, starting with C_6H_{14}. Every predicted isomer, *and no more*, has been isolated and identified for the C_6, C_7, and C_8 groups. However, not all of the C_{15}'s and C_{20}'s have been produced, but there is sufficient belief in the theory to presume that if enough time and effort were spent, all of the isomers could eventually be produced. We conclude that *structural isomerism* helps to explain the vast number of carbon compounds.

TABLE 14-1   STRUCTURAL ISOMERS OF SOME HYDROCARBONS

FORMULA	ISOMERS PREDICTED	FOUND
C_6H_{14}	5	5
C_7H_{16}	9	9
C_8H_{18}	18	18
$C_{15}H_{32}$	4347	—
$C_{20}H_{42}$	366,319	—

Structural isomers can also exist in molecules with double and triple carbon—carbon bonds. For example, two of the four isomers of C_4H_8, 1-butene and *trans*-2-butene, have the following structures:

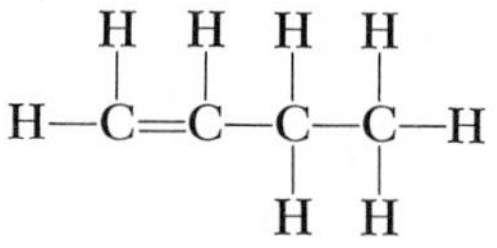

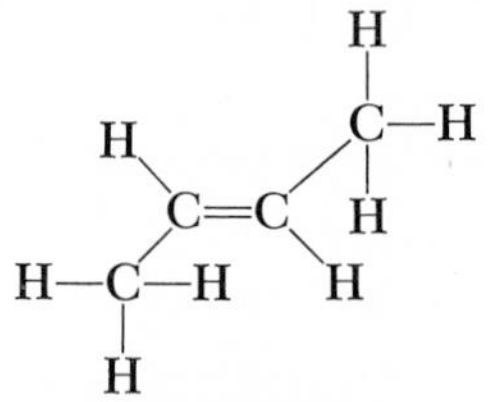

	1-Butene	trans-2-Butene
Melting point	−185.4°C	−106.0°C
Boiling point	−6.3°C	1.0°C
Density (at −20°C)	0.641 g/ml	0.649 g/ml

The number placed before the name butene indicates the position number of the double bond. *Trans*-2-butene is an example of a geometrical isomer which results from double bonds within a molecule. These will be considered later. Note that the properties of 1-butene and 2-butene definitely indicate two *different* compounds. 1-Butyne and 2-butyne illustrate structural isomerism due to the positioning of triple bonds.

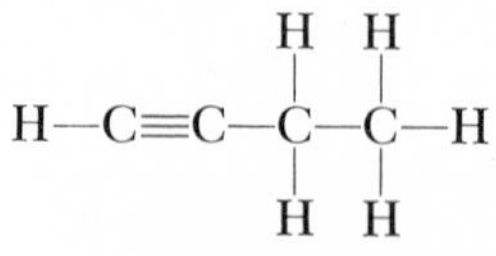

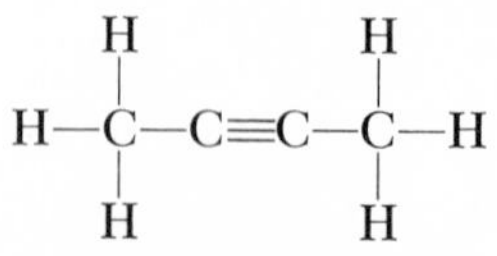

	1-Butyne	2-Butyne
Melting point	−122.5°C	−32.3°C
Boiling point	8.1°C	27°C
Density	0.678 g/ml	0.691 g/ml

270

Hydrocarbons containing a double bond are isomers with single bonded ring-type structures. For example, in addition to the isomers given before for C_4H_8, another isomer exists which has a single bond ring structure.

$$
\begin{array}{cc}
\text{H} & \text{H} \\
| & | \\
\text{H—C—C—H} \\
| & | \\
\text{H—C—C—H} \\
| & | \\
\text{H} & \text{H}
\end{array}
$$

	Cyclobutane
Melting point	*−50°C*
Boiling point	*13°C*
Density	*0.703 g/ml*

Cyclobutane has a much higher boiling point than the other two compounds because its structure allows closer intermolecular distances and greater intermolecular attractions.

NOMENCLATURE

With so many organic compounds, a system of common names quickly fails due to the shortage of names. It is evident that a system of nomenclature is needed that makes use of numbers as well as names. Much attention has been given to this problem, and several international conventions have been held to work out a satisfactory system that can be used throughout the world. The International Union of Pure and Applied Chemistry has given its approval to a very elaborate nomenclature system (IUPAC System), and this system is now in general use.

A few of these IUPAC names will be needed, and an appreciation for the basic simplicity of the approach in naming organic compounds is desirable. Inscrutable names, such as some of those encountered on medicine bottles, are converted to systematic names that are descriptive of the molecules involved. The names of a few of the hydrocarbons and the idea of the numbered prefix are presented in Table 14–2, and additional points in nomenclature will be presented in later sections.

Often students become a bit confused when writing and interpreting structural formulas for organic compounds because they fail to realize that there are many equivalent ways of writing a single structural formula. For example,

$$
\begin{array}{c}
\text{H H H H} \\
| \; | \; | \; | \\
\text{H—C—C—C—C—H} \\
| \; | \; | \; | \\
\text{H H H H}
\end{array}
\quad \text{can be written as}
$$

can be written as

$$
\begin{array}{c}
\text{H   H} \\
| \quad\quad | \\
\text{H—C——C—H} \\
| \quad\quad | \\
\text{H—C—H H} \\
| \\
\text{H—C—H} \\
| \\
\text{H}
\end{array}
\quad \text{or} \quad
\begin{array}{c}
\text{H   H} \\
| \quad\quad | \\
\text{H—C————C—H} \\
| \quad\quad\quad | \\
\text{H—C—H H—C—H} \\
| \quad\quad\quad | \\
\text{H   H}
\end{array}
$$

For branched-chain hydrocarbons it becomes necessary to name submolecular groups. The $—CH_3$ group is called the methyl group; this name is derived from methane by dropping the -ane and adding -yl. Any of the 10 hydrocarbons listed in Table 14–2 could give rise to a similar group. For example, the propyl

TABLE 14-2   THE FIRST TEN STRAIGHT-CHAIN HYDROCARBONS

NAME	FORMULA	STRUCTURAL FORMULA

Methane	CH_4	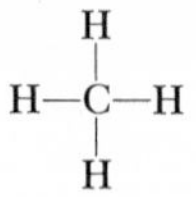 <pre> H
		    |
		H—C—H
		    |
		    H</pre>

Ethane	C_2H_6	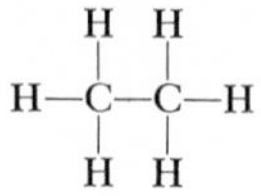 <pre> H H
		    | |
		H—C—C—H
		    | |
		    H H</pre>

Propane	C_3H_8	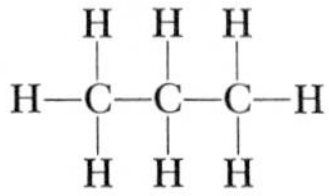 <pre> H H H
		    | | |
		H—C—C—C—H
		    | | |
		    H H H</pre>

n-Butane	C_4H_{10}	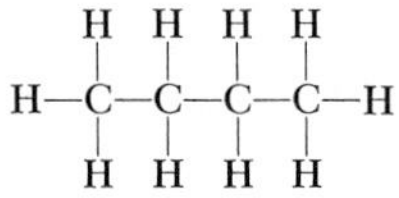 <pre> H H H H
		    | | | |
		H—C—C—C—C—H
		    | | | |
		    H H H H</pre>

n-Pentane	C_5H_{12}	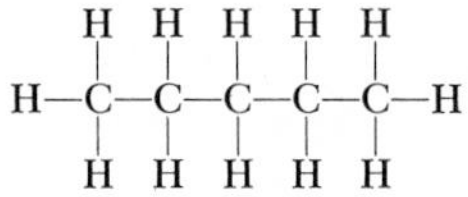 <pre> H H H H H
		    | | | | |
		H—C—C—C—C—C—H
		    | | | | |
		    H H H H H</pre>

n-Hexane	C_6H_{14}	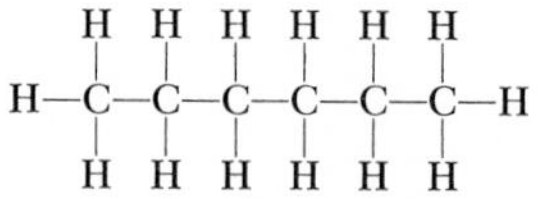 <pre> H H H H H H
		    | | | | | |
		H—C—C—C—C—C—C—H
		    | | | | | |
		    H H H H H H</pre>

n-Heptane	C_7H_{16}	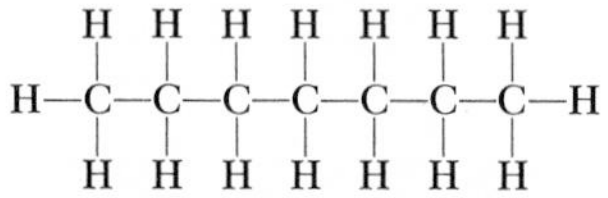 <pre> H H H H H H H
		    | | | | | | |
		H—C—C—C—C—C—C—C—H
		    | | | | | | |
		    H H H H H H H</pre>

n-Octane	C_8H_{18}	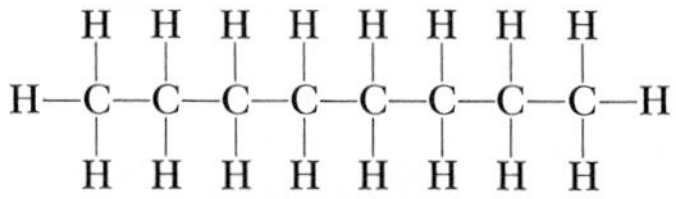 <pre> H H H H H H H H
		    | | | | | | | |
		H—C—C—C—C—C—C—C—C—H
		    | | | | | | | |
		    H H H H H H H H</pre>

n-Nonane	C_9H_{20}	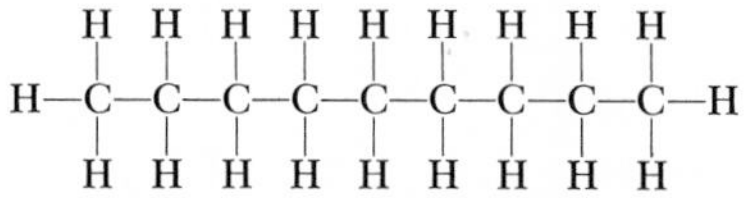 <pre> H H H H H H H H H
		    | | | | | | | | |
		H—C—C—C—C—C—C—C—C—C—H
		    | | | | | | | | |
		    H H H H H H H H H</pre>

n-Decane	$C_{10}H_{22}$	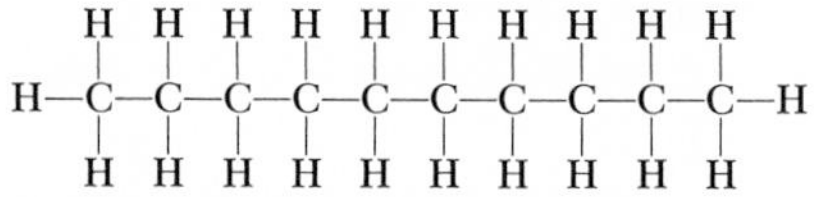 <pre> H H H H H H H H H H
		    | | | | | | | | | |
		H—C—C—C—C—C—C—C—C—C—C—H
		    | | | | | | | | | |
		    H H H H H H H H H H</pre>

group would be $-C_3H_7$. In order to illustrate the use of the group names, consider this formula:

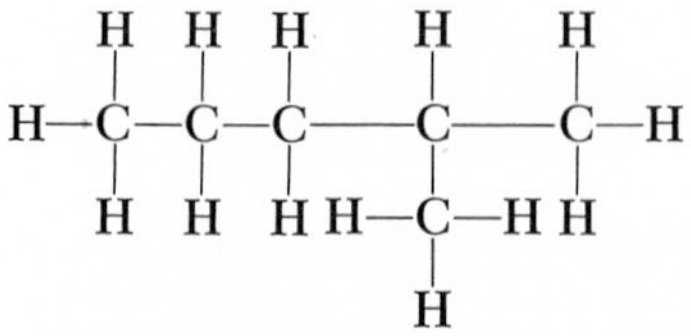

<pre>   H H H   H   H
    | | | | |
H—C—C—C———C———C—H
    | | | H—C—H H
    H H H |
              H</pre>

The longest carbon chain in the molecule is five carbon atoms long, hence, the root name is pentane. Furthermore, it is a methylpentane (written as one word) because a methyl group is attached to the pentane structure. In addition, a number is needed because the methyl group could be bonded to either the second or third carbon atom.

3-Methylpentane   *2-Methylpentane*

Note that 2-methylpentane is the same as 4-methylpentane since the latter would be the same molecule turned around; the accepted rule requires numbering from the end of the carbon chain that will result in the smallest numbers. Therefore, 2-methylpentane is the correct name.

Any number of substituted groups can be handled in this same fashion. Consider the formula for 4,4-diethyl-3,5,6-tetramethyloctane.

If a double bond appears in a hydrocarbon, the root name, which indicates the number of carbon atoms, must be modified to reflect the double bond structure and its position. Changing -ane to -ene indicates the presence of the double bond, and a number is used to indicate its position. For example:

$CH_2{=}CH_2$ *Ethene[1]*
(common name: ethylene)

$CH_2{=}CHCH_3$ *Propene[1]*

$CH_2{=}CHCH_2CH_3$ *1-Butene*

$CH_3CH{=}CHCH_3$ *2-Butene*

$CH_3{-}\overset{\displaystyle CH_3}{C}{=}CHCH_2CH_3$ *2-Methyl-2-pentene*

If a triple bond is present the -ane is changed to -yne. Examples:

$H{-}C{\equiv}C{-}H$ *Ethyne*
(common name: acetylene)

$CH_3CH_2C{\equiv}CH$ *1-Butyne*

A point of difficulty often arises because the use of common names persists. Consider, for example:

$CH_3{-}\overset{\displaystyle CH_3}{\underset{\displaystyle H}{C}}{-}CH_3$

The correct name is methylpropane. However, this compound is more often referred to by its common name, isobutane. Only under special circumstances does the multiplicity of names for a single compound create confusion.

[1]No number is necessary for ethene or propene because there is only one possible position for the double bond.

273

OPTICAL ISOMERS

In the preceding section, it was pointed out that structural isomers sometimes differ because there is more than one way in which a given set of atoms can be held together. However, it is possible for some sets of atoms to form two isomeric molecules, both of which have the same set of bonds as well as the same set of atoms. This special case of structural isomerism is called *optical isomerism.*

Optical isomerism is possible when a molecular structure is asymmetric (without symmetry). One common example of an asymmetric molecule is one containing a tetrahedral carbon atom bonded to four *different* atoms or groups of atoms. Such a carbon atom is called an *asymmetric* carbon atom. Consider the molecule CBrClIH.

A study of Figure 14–7 shows that there are two ways to arrange the four different atoms in the tetrahedral positions about the central carbon atom. These result in two nonsuperposable mirror image molecules which are optical isomers.

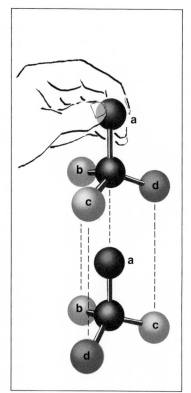
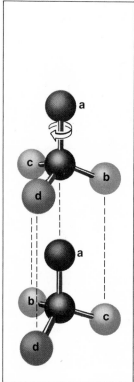

FIGURE 14-7   Optical isomers. Four different atoms, or groups of atoms, are bonded to tetrahedral center atoms so that the upper isomeric form cannot be turned in any way, thus exactly substituting for the lower structure. These are nonsuperimposable mirror images.

There are a few examples of nonsuperposable mirror images in the macroscopic world. Consider right- and left-hand gloves for instance. They are mirror images of one another and are nonsuperposable. In Figure 14–8, this mirror image relationship is shown for isomeric forms of alanine, the molecules of which contain a tetrahedral carbon atom surrounded by an amine group ($-NH_2$), a methyl group ($-CH_3$), an acid group ($-COOH$), and a hydrogen atom. Note

274

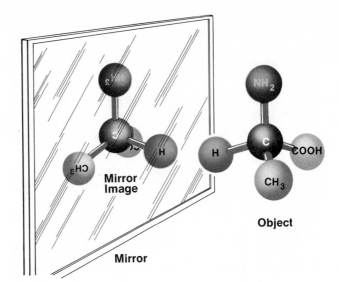

FIGURE 14-8   Optical isomers of the amino acid, alanine.

$$H_2N - \overset{\displaystyle H}{\underset{\displaystyle CH_3}{\overset{|}{\underset{|}{C}}}} - COOH.$$

The D-form is the mirror image of the L-form.

in Figure 14–8 that the *carbon atoms* in the methyl and acid groups are not asymmetric since these atoms are not bonded to four different groups.

The properties of optical isomers are almost identical. Different compounds whose molecules are mirror images of one another have the same melting point, the same boiling point, the same density, and will be the same in many other physical and chemical properties. However, they always differ in one physical property; they rotate the plane of *polarized* light in opposite directions. According to the wave theory of light, a light wave traveling through space vibrates at right angles to its path (Figure 14–9). A group of such rays traveling together vibrate in random directions, all of which are at right angles to the path of travel. If such a group of waves is passed through a polarizing crystal, such as Iceland spar (a form of $CaCO_3$), the emerging waves along the incoming axis will vibrate in only one plane, perpendicular to the light path. Such light is said to be *plane polarized*. When plane polarized light is passed through a solution of D-lactic acid, the light is still polarized but the plane of vibration is rotated somewhat in a clockwise direction. If the other lactic acid isomer is substituted (L-lactic acid), just the opposite rotation of the light is obtained.

Optical isomers can also differ in chemical properties. Sometimes one optical isomer will have a chemical effect not shown by its counterpart. The hormone adrenalin is one of a pair of optical isomers and its structure can be represented as:

275

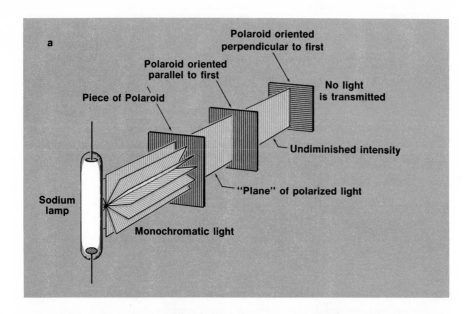

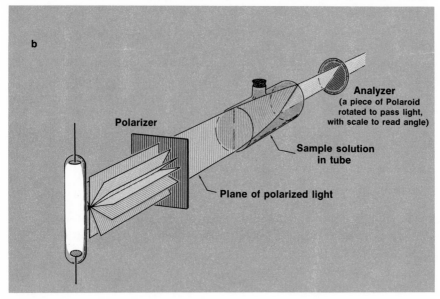

FIGURE 14-9   Rotation of plane polarized light by an optical isomer. (*a*) A sodium lamp provides a yellow light (monochromatic). The original beam is nonpolarized; it vibrates in all directions at right angles to its path. After passing a Polaroid filter, the light is vibrating in only one direction. This polarized light will pass another Polaroid filter if lined up properly but will not pass the third Polaroid filter if it is at right angles to the other two. Hence, the direction of the Polaroid filters determine the direction of polarization. (*b*) The plane of polarized light is rotated by a solution of an optically active isomer. The analyzer can be a second Polaroid filter which can be rotated to find the angle for maximum transmission of light. If the solution rotates the plane of polarized light, the analyzer will not be at the same angle as the polarizer for maximum transmission.

where C* designates the asymmetric carbon atom. Only the isomer that rotates plane polarized light to the left is effective in starting a heart that has stopped beating momentarily or in giving a person unusual strength during times of great emotional stress. It is also interesting to note that during the contraction of muscles the body produces only the L-form of lactic acid and not the D-form. The concentration of this lactic acid in the blood is associated with the feeling of tiredness, and a period of rest is necessary to reduce the concentration of this chemical.

Large organic molecules may have many asymmetric carbon atoms within the same molecule. At each such carbon atom there exists the possibility of *two* arrangements of the molecule. The total number of possible molecules, then, increases exponentially with the number of asymmetric centers. With two asymmetric carbon atoms there would be 2^2 or four possible structures; for three, there would be 2^3 or eight possible structures. It should be emphasized that each of the eight isomers could be made from the *same* set of atoms with the *same* set of chemical bonds. Dextrose, a simple blood sugar, contains four asymmetric carbon atoms per molecule. Thus, there are 2^4 or 16 isomers in the family of compounds to which dextrose belongs. Obviously, then, the concept of optical isomers is significant in explaining the vast number of carbon compounds.

$$
\begin{array}{c}
O \\
\parallel \\
C-H \\
| \\
H-C-OH \\
| \\
HO-C-H \\
| \\
H-C-OH \\
| \\
H-C-OH \\
| \\
HO-C-H \\
| \\
H
\end{array}
$$

Dextrose

GEOMETRICAL ISOMERS

As we noted in our discussion of molecular shapes, the geometry of bonds around a carbon atom involved in a carbon—carbon double bond is not tetrahedral. Rather, the three bond axes, one along the double bond and one each along the two single bonds, all lie in the same plane. The bond angles are each approximately 120 degrees. This geometry and the lack of rotation about a double bond gives rise to another type of isomerism, called *geometric isomerism*.

Consider the compound ethylene, C_2H_4. The spatial relationship between the atoms in a molecule of ethylene can be adequately represented with all of the six atoms lying in the same plane.

$$
\begin{array}{ccc}
H & \overset{121°}{} & H \\
& \diagdown \diagup & \\
118°\;\diagup & C = C & \diagdown \\
H & & H
\end{array}
$$

If one of the hydrogen atoms bonded to each of the carbon atoms is replaced by chlorine, the result is *1,2-dichloroethene*, CHCl=CHCl (the 1 and 2 in the 1,2-dichloroethene indicate that the two chlorine atoms are attached to different carbon atoms). Laboratory experiments demonstrate that there are two compounds with this same formula. This experimental conclusion is in agreement with an understanding of the molecular geometries; the two chlorine atoms could be close together or far apart. If they are close together, the isomer is called the *cis* isomer; and if far apart, the *trans* isomer. Note that the two isomeric compounds have significant differences in their properties.

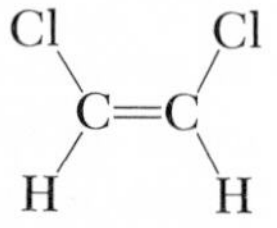

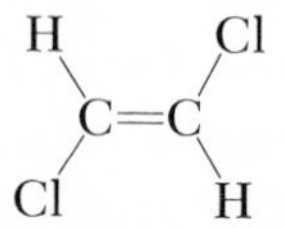

	cis-*structure* 1,2-*dichloroethene*	trans-*structure* 1,2-*dichloroethene*
Melting point	−80.5°C	−50°C
Boiling point	60.3°C	47.5°C
Density	1.284 g/ml	1.256 g/ml

The reason for the *rigidity* of the carbon—carbon double bond can readily be explained in terms of orbital theory. It was pointed out in Figure 11–9 that in a double bond, four electrons are shared between the bonded atoms. In each of the two carbon atoms, the 2s orbital along with two of the three p orbitals are hybridized to form three sp² hybrid orbitals. These three orbitals lie in a plane, at angles of 120 degrees and each is capable of forming a single covalent bond (sigma bond) with head-on orbital overlap. The third p orbital, not involved in the hybridization, extends above and below (at right angles to) the plane of the three sp² orbitals (Figure 14–10). If the two carbon atoms of ethylene overlap sp² hybrid orbitals to form a sigma bond, there is one rotational position about

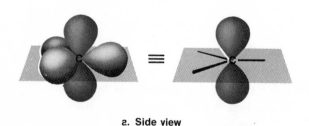

2. Side view

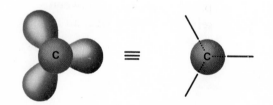

278

b. Top view

FIGURE 14-10   (a) Geometry of three sp² hybrid orbitals and one p orbital about a carbon atom. The two illustrations in (a) are equivalent (≡). (b) Looking down on the single p orbital, one can see that the three sp² orbitals extend to the corners of an equilateral triangle.

this bond in which the p orbitals (one on each carbon) are lined up with each other (Figure 14–11). When this occurs, a side-to-side overlap of the p orbitals is possible both above and below the plane of the other bonds. The side-to-side overlap results in a π bond, which is somewhat weaker than the sigma bond. It is evident that the carbon—carbon double bond would have to break the π bond in order to be free to rotate (Figure 14–11).

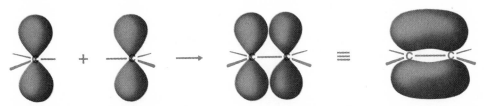

FIGURE 14-11   Double bond formation between carbon atoms. Direct overlap of sp^2 orbitals result in a sigma bond. At the proper angle adjacent p orbitals overlap to form π bond. The double bond is made up of one sigma bond and one π bond.

As a general rule, *trans* isomers are higher melting than *cis* isomers due to the greater ease with which the *trans* molecules can fit into the lattice and form strong intermolecular bonds.

Often, when there is a carbon—carbon double bond in an organic molecule, the possibility exists for *cis* and *trans* isomers. Sometimes a number of such bonds can be found in the same molecule, giving rise to numerous isomeric compounds.

FUNCTIONAL GROUPS

As pointed out earlier, carbon also forms covalent bonds with a number of other elements. As a result, certain groups of atoms called *functional groups* appear over and over again in different organic compounds. Consider, for example, the —OH group, called the hydroxyl group. A large number of molecules containing an —OH group have properties which are characteristic of a class of organic compounds called *alcohols*.

The —OH group has the same combining power as an atom of hydrogen. This means that, in any of the hydrocarbons considered thus far, any of the hydrogen atoms could be replaced by an —OH group to form an alcohol. A single hydrocarbon molecule can give rise to a number of alcohols if there are different isomeric positions for the —OH group. Three different alcohols result when a hydrogen atom is replaced by an —OH group in n-pentane, depending on which hydrogen atom is replaced (Table 14–3).

When one or more functional groups appear in a molecule, the IUPAC name reveals the group name. For example, the name of an alcohol will use the root of the name of the analogous hydrocarbon to indicate the number of carbon atoms and the suffix -ol to denote an alcohol. As before, a number is used to indicate the position of the alcohol group.

Some major functional groups that we will consider in following chapters are listed in Table 14–4. The symbol —R is usually a hydrocarbon group such as methyl (—CH$_3$) or ethyl (—C$_2$H$_5$).

279

TABLE 14-3   ALCOHOLS DERIVED FROM PENTANE (C_5H_{12})

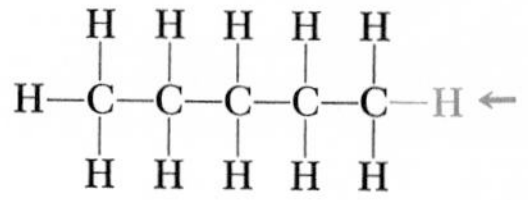

H—C—C—C—C—C—H ←   gives   H—C—C—C—C—C—O—H

Substitute an —OH for an end-hydrogen

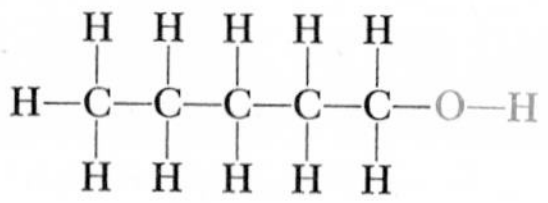

1-Pentanol

H—C—C—C—C—C—H   gives   2-Pentanol

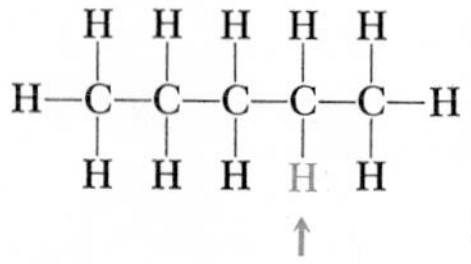

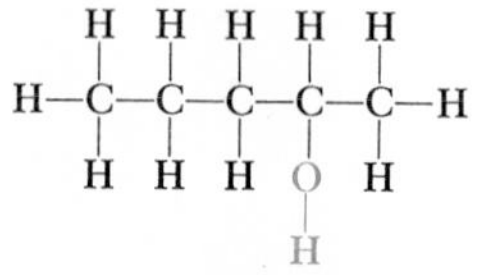

Substitute an —OH for a 2-carbon hydrogen

H—C—C—C—C—C—H   gives   3-Pentanol

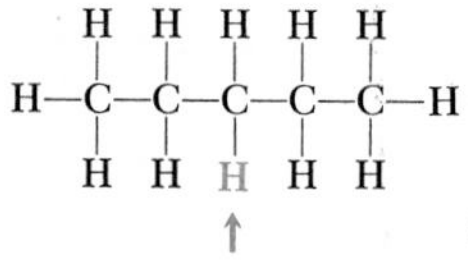

Substitute an —OH for a 3-carbon hydrogen

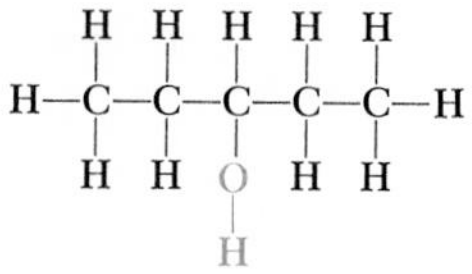

TABLE 14-4   SOME FUNCTIONAL GROUPS FOUND IN ORGANIC COMPOUNDS

GROUP	NAME	TYPICAL COMPOUND	COMPOUND NAME	COMMON USE
R—OH	Alcohol (hydroxyl)	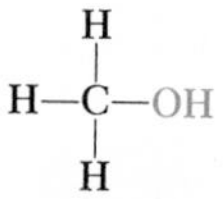 H—C—OH	Methanol (wood alcohol)	Solvent
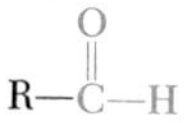 R—C—H (=O)	Aldehyde	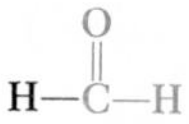 H—C—H (=O)	Methanal (Formaldehyde)	Preservative
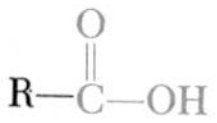 R—C—OH (=O)	Acid (carboxyl)	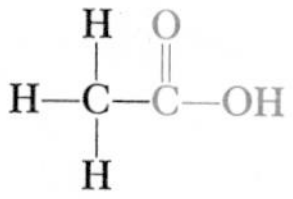 H—C—C—OH (=O)	Ethanoic Acid (Acetic acid)	Vinegar
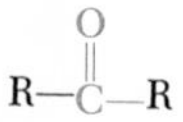 R—C—R (=O)	Ketone	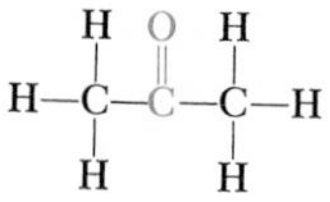 H—C—C—C—H (=O)	Propanone (Acetone)	Solvent
R—O—R	Ether	C_2H_5—O—C_2H_5	Diethyl ether (Ethyl ether)	Anesthetic
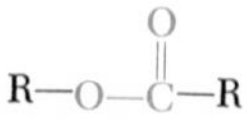 R—O—C—R (=O)	Ester	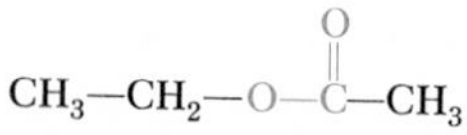 CH_3—CH_2—O—C—CH_3 (=O)	Ethyl ethanoate (Ethyl acetate)	Solvent in fingernail polish
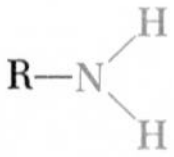 R—N(H)(H)	Amine	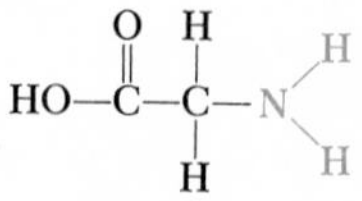 HO—C—C—N(H)(H) (=O)	2-Aminoethanoic acid (Glycine)	Amino acid found in proteins

AROMATIC COMPOUNDS

All of the hydrocarbons we have discussed up to this point have localized electronic structures; that is, the bonding electrons are essentially fixed between two atomic centers as in C—C or C=C bonds. For a large group of organic compounds known as *aromatic* compounds, this type of complete electron localization is not found. Rather, these compounds have delocalized electrons which are spread over the entire molecule, a behavior which leads to some interesting chemical properties.

The simplest aromatic compound is *benzene* (C_6H_6). The molecular structure of benzene is that of a ring of carbon atoms in a plane. The bonds between these carbon atoms are shorter than single bonds, but longer than double bonds. Since the measured bond angles are 120°, this implies sp^2 hybridization in the carbon atoms.

One way to account for the bonding and structure of the benzene molecule is to use two of the three sp^2 hybrid orbitals of each carbon to form *sigma* bonds between the carbon atoms and to involve the six remaining p orbitals of the six carbon atoms in the formation of π bonds. These π bonds are not exactly the same as those discussed earlier for double and triple bonds because there is a ring system of p orbitals in an aromatic molecule. Each p orbital overlaps with the p orbitals on both neighboring carbon atoms. This causes the π bonds to be less well fixed (or localized) between alternating pairs of p orbitals. Figure 14–12 illustrates this π bonding. The π bonds are averaged rather evenly around

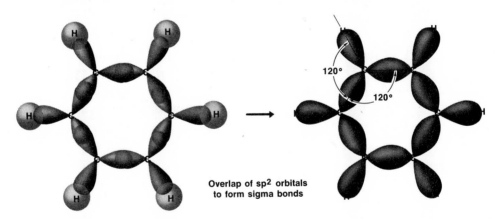

120°

120°

Overlap of sp2 orbitals
to form sigma bonds

The p orbitals on carbon are omitted

σ-skeleton only

FIGURE 14–12   Bonding in an aromatic compound, benzene, C_6H_6.

Lateral overlap of p orbitals to form pi bonds

the ring; we say they are *delocalized*. This explains why the carbon—carbon bond in aromatic molecules is intermediate between the single and double bonds. Various symbols have been used to represent the benzene ring, but the symbol at the right below is perhaps the most used. The hydrogen atoms are not shown.

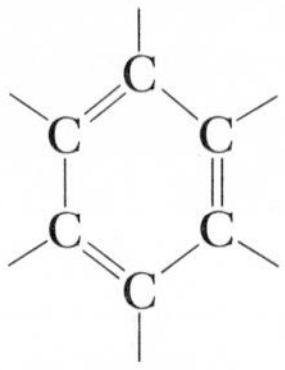

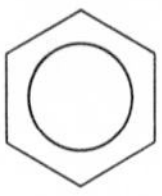

There are a large number of aromatic compounds that can be derived from benzene by replacing one or more of the hydrogen atoms about the ring. These molecules exhibit a wide variety of properties and differ greatly in their chemical reactivity. Some interesting examples of isomerism are also possible. Consider the three different compounds with the formula C_8H_{10} found in some coal tars. There are several names given to these isomers:

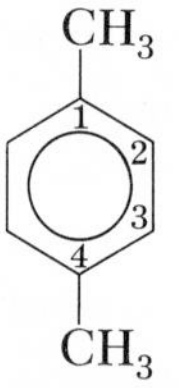

1,4-dimethylbenzene (*para-xylene*) m.p. 13.2°C   *1,3-dimethylbenzene* (*meta-xylene*) m.p. −47.4°C   *1,2-dimethylbenzene* (*ortho-xylene*) m.p. −29°C

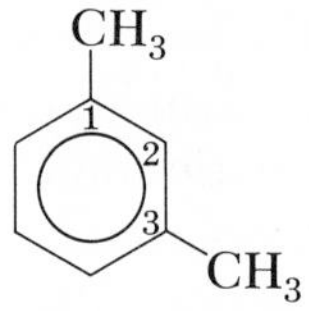

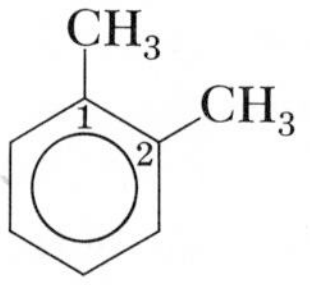

Each of these isomers has two methyl groups substituted for hydrogen atoms on the ring. The prefixes, *para-*, *meta-*, and *ortho-* are used if there are only two groups on the benzene ring.

If more than two groups occur, a number system is most useful. Consider the following compounds:

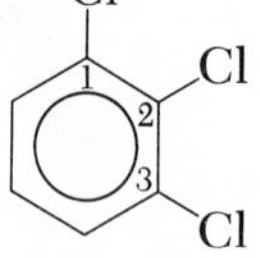

1,2,3-Trichlorobenzene 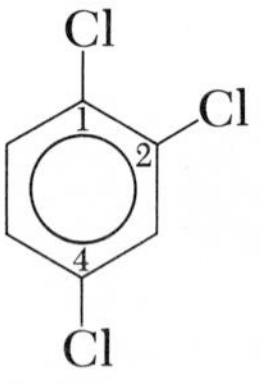   *1,2,4-Trichlorobenzene* 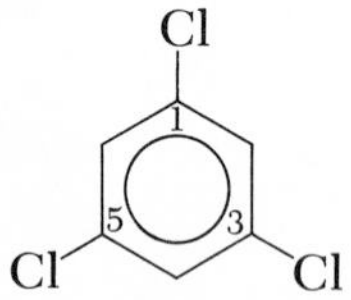   *1,3,5-Trichlorobenzene*

There are no other ways of drawing these three isomers and only three tri-chlorobenzenes have been isolated in the laboratory. We shall look at some of the reactions of aromatic compounds in this and the next chapter.

SOURCES OF ORGANIC COMPOUNDS

Carbon compounds come from numerous sources. They are separated by chemical and physical methods, such as distillation and extraction, and from

282

natural materials which were once alive or which resulted from living organisms; alternatively they may be prepared in the laboratory from inorganic chemicals such as carbon monoxide and water or from other organic compounds. The largest sources of naturally occurring organic compounds are the fossil fuels, coal and petroleum, and vegetable matter such as wood, plant stalks, and so forth. A series of distillations (fractional distillation) separate petroleum ether, gasoline, kerosene, oils, and asphalt from petroleum (Table 14–5); each of these materials can be further separated into a multitude of individual compounds.

TABLE 14-5   PETROLEUM FRACTIONS

FRACTION	COMPOSITION	DISTILLATION RANGE
Natural gas	C_1—C_4	Less than 20°
Liquefied petroleum	C_5—C_6	20—60°
Gasoline	C_4—C_8	40—200°
Kerosene	C_{10}—C_{16}	175—275°
Fuel oil, diesel oil	C_{15}—C_{20}	250—400°
Lubricating oils	C_{18}—C_{22}	Above 300°
Asphalt or	C_{20} and above	Non-volatile
petroleum coke	(complicated structures)	

By heating coal or wood in the absence of air (destructive distillation), coal or wood tar can be separated from the coke or charcoal (carbon). A ton of a typical soft coal produces about 140 lbs of coal tar, which is about one-half pitch, the rest being composed of chemicals such as naphthalene, benzene, phenol, cresols, toluene, and xylenes. The amount of benzene in a 140 lb coal tar sample is not great (about 2 lbs); yet, in 1966, 110 million gallons of benzene were produced from coal tar.

The enormous increase in man's use of organic compounds in the last century forces us to recognize that the sources of these materials are limited and subject to eventual exhaustion. Thus, petroleum sources from which we draw so much of our energy and which serve as starting materials for the synthesis of industrial organic compounds, cannot be expected to last beyond 50 to 100 years at the most. The transformation of many of the natural sources of organic compounds into disposable convenience items which soon end up in a garbage dump represents an extremely short-sighted use of irreplaceable natural resources. Most of us realize that we must begin ultimately to recycle a much larger percentage of the material we use in our daily lives, but few seem ready to begin effective efforts in this direction now. It seems very probable that the next decade will force us to change many of our casual attitudes toward the utilization of our natural sources of organic compounds.

SYNTHETIC ORGANIC COMPOUNDS

The naturally occurring organic compounds are not only important in and of themselves, but they also serve as starting materials for making numerous organic compounds that do not occur in nature. Modification of existing mole-

283

cules can take the form of rearranging atoms within molecules, making giant molecules from small ones (polymerization), making small molecules from large ones (cracking), substituting one functional group for another, and similar changes. Plastics, synthetic fibers, and drugs are important organic compounds that result from these kinds of reactions.

The synthesis of new compounds is seldom done by trial and error; the synthetic chemist often has as an aid in planning a synthesis some idea of the step by step structural changes the reactants undergo in becoming products. A description of the step by step structural changes is the *mechanism* of a reaction. Regardless of the reaction, all mechanisms have some things in common. For example, every mechanism must have the reactant molecules colliding with proper orientation (turned correctly in relationship to each other) and with sufficient energy to react. In addition, most organic reaction mechanisms obey the *principle of minimum structural change,* which means that most of the structure of the reacting species goes through the reaction unchanged. Only a few specific bonds are broken or made. Every mechanism usually involves breaking the weakest bond first unless the attacking species is specifically reactive toward a group containing stronger bonds. A specific mechanism may contain some unique features, but all mechanisms adhere to these general principles.

Let us briefly look at some of the more important types of organic reactions and their mechanisms and see how they can lead the organic chemist to create new compounds.

FREE RADICAL SUBSTITUTION OF SIMPLE HYDROCARBONS

Under the proper conditions it is possible to substitute a halogen (most often Cl or Br) atom for a hydrogen atom in simple hydrocarbon molecules.

$$-\overset{|}{\underset{|}{C}}-H + Cl_2 \longrightarrow -\overset{|}{\underset{|}{C}}-Cl + HCl$$

This is an important reaction since it is useful in preparing molecules that are much more reactive (and hence, more valuable in other synthetic processes) than the hydrocarbons from which they came. A typical reaction might involve the chlorination of ethane at 25°C in the presence of light:

$$H-\overset{\overset{\displaystyle H}{|}}{\underset{\underset{\displaystyle H}{|}}{C}}-\overset{\overset{\displaystyle H}{|}}{\underset{\underset{\displaystyle H}{|}}{C}}-H + Cl_2 \xrightarrow[25°]{light} H-\overset{\overset{\displaystyle H}{|}}{\underset{\underset{\displaystyle H}{|}}{C}}-\overset{\overset{\displaystyle H}{|}}{\underset{\underset{\displaystyle H}{|}}{C}}-Cl + HCl$$

Chloroethane

The mechanism of this type of reaction is an interesting one. It involves a series of reaction steps and chemical species called *free radicals* that have a single unpaired electron.

Step 1. The chlorine molecule is broken apart into two chlorine atoms (free radicals) by a photon of light:

$$Cl_2 \xrightarrow{light} 2Cl\cdot$$

(Other outer shell electrons omitted)

Step 2. The chlorine free radicals, being very reactive species, attack the ethane molecules creating new (ethyl) free radicals:

$$H-\overset{\overset{\displaystyle H}{|}}{\underset{\underset{\displaystyle H}{|}}{C}}-\overset{\overset{\displaystyle H}{|}}{\underset{\underset{\displaystyle H}{|}}{C}}-H + Cl\cdot \longrightarrow H-\overset{\overset{\displaystyle H}{|}}{\underset{\underset{\displaystyle H}{|}}{C}}-\overset{\overset{\displaystyle H}{|}}{\underset{\underset{\displaystyle H}{|}}{C}}\cdot + HCl$$

Step 3. The ethyl radicals in turn can react with chlorine molecules to produce product:

$$H-\overset{\overset{\displaystyle H}{|}}{\underset{\underset{\displaystyle H}{|}}{C}}-\overset{\overset{\displaystyle H}{|}}{\underset{\underset{\displaystyle H}{|}}{C}}\cdot + Cl_2 \longrightarrow H-\overset{\overset{\displaystyle H}{|}}{\underset{\underset{\displaystyle H}{|}}{C}}-\overset{\overset{\displaystyle H}{|}}{\underset{\underset{\displaystyle H}{|}}{C}}-Cl + Cl\cdot$$

Reactions (2 and 3) continue until all the molecules have reacted and the process is terminated.

Step 4. Termination of the reaction can take place in a number of ways other than using up all the reactants:

$$Cl\cdot + Cl\cdot \longrightarrow Cl_2$$

$$H-\overset{\overset{\displaystyle H}{|}}{\underset{\underset{\displaystyle H}{|}}{C}}-\overset{\overset{\displaystyle H}{|}}{\underset{\underset{\displaystyle H}{|}}{C}}\cdot + Cl\cdot \longrightarrow H-\overset{\overset{\displaystyle H}{|}}{\underset{\underset{\displaystyle H}{|}}{C}}-\overset{\overset{\displaystyle H}{|}}{\underset{\underset{\displaystyle H}{|}}{C}}-Cl$$

$$\underset{\text{Reactants}}{HC-\overset{\overset{\displaystyle H}{|}}{\underset{\underset{\displaystyle H}{|}}{C}}\cdot + \cdot\overset{\overset{\displaystyle H}{|}}{\underset{\underset{\displaystyle H}{|}}{C}}-CH} \longrightarrow \underset{\text{Product}}{HC-\overset{\overset{\displaystyle H}{|}}{\underset{\underset{\displaystyle H}{|}}{C}}-\overset{\overset{\displaystyle H}{|}}{\underset{\underset{\displaystyle H}{|}}{C}}-CH}$$

This free radical mechanism for halogenation of hydrocarbons is also a prototype for the polymerization reactions which we shall study in Chapter 16. There we shall see that many useful products result from reactions of this type.

ADDITION REACTIONS

Halogens react with organic molecules containing double or triple carbon—carbon bonds in a manner different from that just discussed. For example, chlorine reacts with ethylene to produce 1,2-dichloroethane. In effect, this is

$$\underset{\text{Ethylene}}{H-\overset{\overset{\displaystyle H}{|}}{C}=\overset{\overset{\displaystyle H}{|}}{C}-H + Cl_2} \longrightarrow \underset{\text{1,2-Dichloroethane}}{H-\overset{\overset{\displaystyle H}{|}}{\underset{\underset{\displaystyle Cl}{|}}{C}}-\overset{\overset{\displaystyle H}{|}}{\underset{\underset{\displaystyle Cl}{|}}{C}}-H}$$

an *addition* reaction since two chlorine atoms have been added to the molecule. The mechanism for this reaction makes use of a special property of the C=C bond. Since there are four electrons localized between the two carbon atoms, they can interact with a chlorine molecule and cause its electron distribution to be distorted. This electron distortion is called *polarization* and can result in the breaking of the Cl—Cl bond and forming a bond between a carbon and a chlorine atom.

285

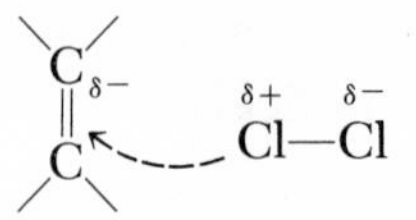

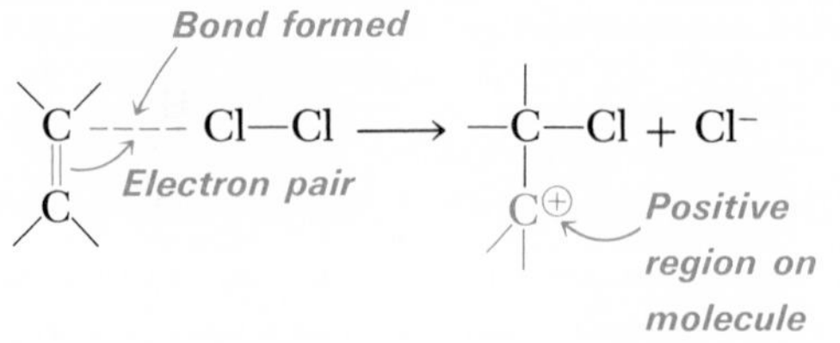

Bond formed

Electron pair

$$-\overset{|}{\underset{|}{C}}-Cl + Cl^-$$

Positive region on molecule

The resulting positive pole on the molecule then reacts with the negative chloride ion:

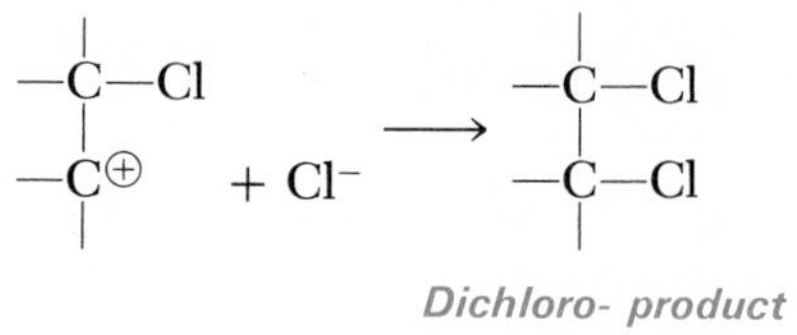

Dichloro- product

SUBSTITUTION REACTIONS OF AROMATIC COMPOUNDS

When benzene reacts with bromine in the presence of iron, the product which results is bromobenzene, a substitution product in which a hydrogen atom has been replaced by a bromine atom. On the surface, this reaction appears

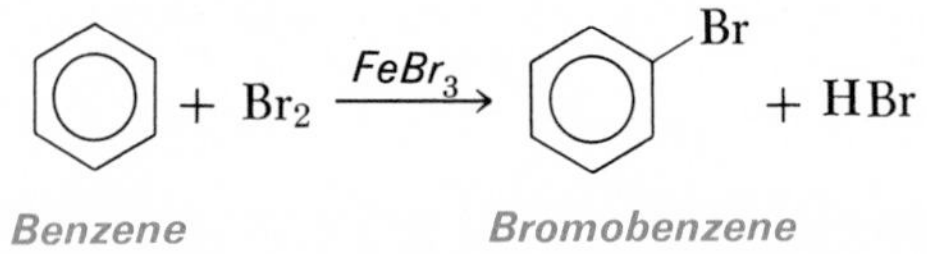

Benzene   *Bromobenzene*

to be like the substitution reactions discussed for simple hydrocarbons. A difference exists, however, because the conditions are not those that give free radicals. This reaction is actually more like the addition reactions just described.

Bromine oxidizes some of the iron present to produce $FeBr_3$, which in turn reacts with bromine to give $FeBr_4^-$ and Br^+ ions:

$$Br_2 + FeBr_3 \rightleftharpoons FeBr_4^- + Br^+$$

The Br^+ species then attacks the electron-rich π electron orbitals on the benzene

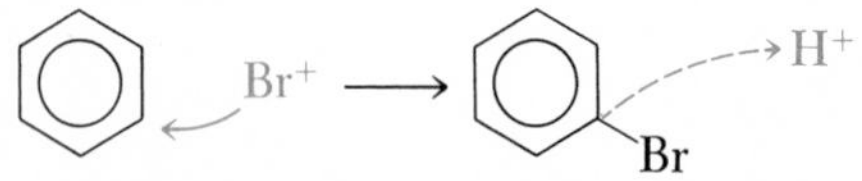

ring. Hydrogen leaves the ring as H^+ ion and reacts with the $FeBr_4^-$ giving HBr:

$$H^+ + FeBr_4^- \longrightarrow HBr + FeBr_3$$

CONCLUSION AND PREVIEW

No wonder carbon is ubiquitous; there are over two million recorded compounds containing carbon. These millions of carbon compounds are due to the following effects:

1. The ability of carbon to form covalent bonds to other carbon atoms almost without limit.
2. The ability of a given number of carbon atoms to combine in more than one molecular pattern.
3. The ability of carbon to form stable covalent bonds to a large number of other atoms.
4. The persistence of carbon chains in the presence of reagents such as oxygen and water which usually destroy chains of other atoms.

Structural differences explain how two compounds can have the same chemical composition by weight, and yet have different physical and chemical properties. Some order can be brought out of chaos with an understanding of isomerism and a systematic approach to the possible molecular structures.

In the next chapter we shall take a closer look at some of the applications of organic chemistry and we shall see how molecular structure is important in determining properties of polymers and other compounds.

In Chapter 17, we shall look at some important organic compounds associated with the basic life processes. There we will see some examples of the chemistry of functional groups.

QUESTIONS

1. *Saturated hydrocarbons* are so named because they have the maximum amount of hydrogen present for a given amount of carbon. The saturated hydrocarbons have the general formula C_nH_{2n+2} where n is a whole number. What are the names and formulas of the first four members of this series of compounds?

2. Using diagrams and bonding theory, explain why nearly free rotation is allowed around a carbon—carbon single bond but not around a carbon—carbon double bond.

3. Using the periodic chart and electron-dot formulas, illustrate a sigma bond in a compound other than those given in this chapter.

4. Draw the structural formula for each of the five isomeric hexanes, C_6H_{14}.

5. Write the structural formulas for: (a) 2-methylbutane, (b) ethylpentane, (c) 4,4-dimethyl-5-ethyloctane, (d) methylbutane, (e) 2-methyl-2-hexene, (f) 3-methyl-3-hexanol.

6. Give the IUPAC name for:

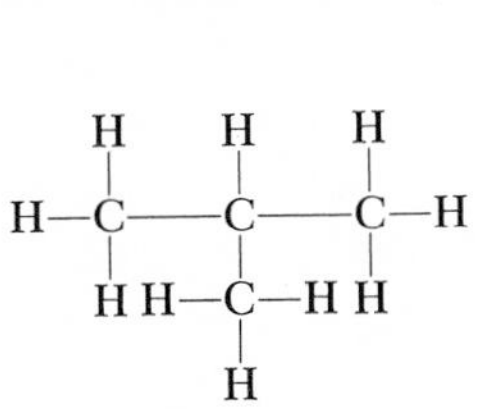

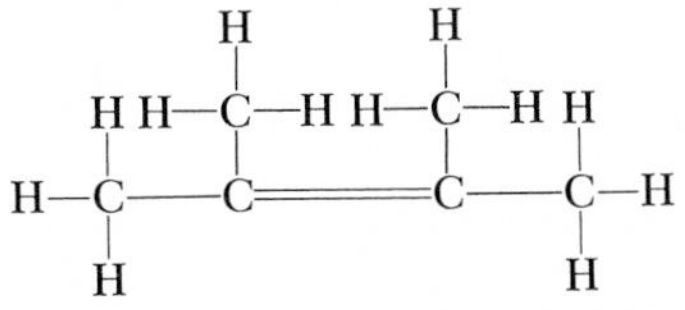

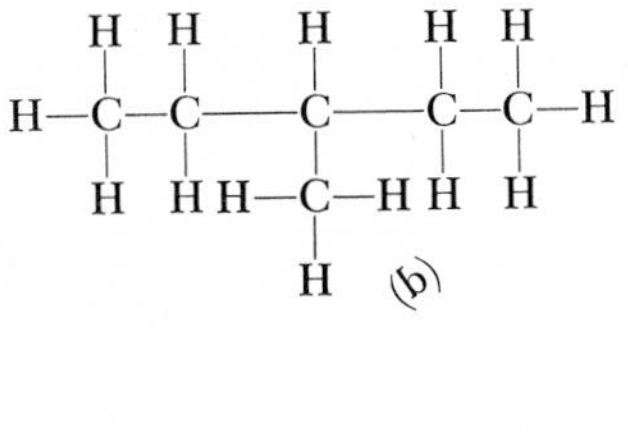

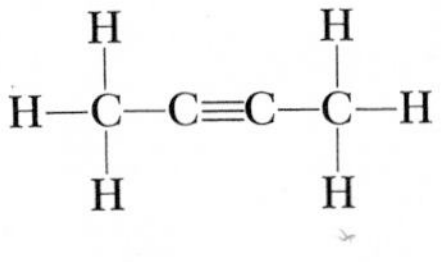

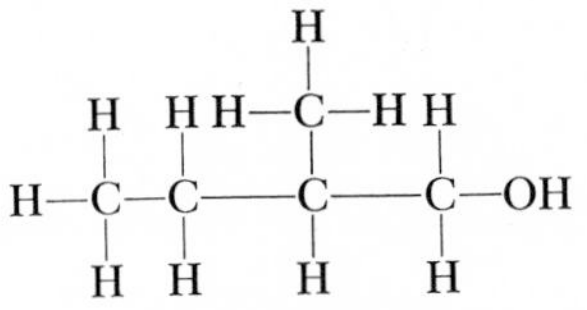

7. If cyclohexane has the formula C_6H_{12}, and if it has *no* double bonds, what would you suggest as its structural formula?

8. How can optical isomers be distinguished from each other?

9. What kind of symmetry does

does not have? Explain the term: asymmetric carbon atom.

10. If a solution of the D optical isomer of an optical pair rotates plane polarized light clockwise, and the L isomer does the opposite, what do you think an equimolar mixture of the two would do?

11. How many optical isomers can there be for a molecular structure containing eight asymmetric carbon atoms?

12. Explain the statement: The formation of a π bond between two carbon atoms does not release as much energy as the formation of a sigma bond, yet more energy is released in the formation of a carbon—carbon double bond than in the formation of a carbon—carbon single bond.

13. Alanine contains two functional groups; what are they?

14. Which do you think would be better for use for medicinal purposes, pure adrenalin obtained from natural products (found in nature) or the pure compound as synthesized in the laboratory? Why?

15. Ethyl alcohol, $H-\overset{H}{\underset{H}{C}}-\overset{H}{\underset{H}{C}}-OH$, can be oxidized to acetic acid, $H-\overset{H}{\underset{H}{C}}-C\overset{O}{\underset{OH}{}}$. How does this illustrate the *principle of minimum structural change?*

16. Methane reacts with chlorine at room temperature in the presence of light to form chloromethane. Write the mechanistic steps for this reaction. Name the species involved.

17. What structural feature characterizes aromatic compounds?

18. Indicate the functional groups present in the following molecules:

(a) $CH_3CH_2CH_2COOH$

(b) $CH_3CH_2NH_2$

(c) $CH_3\underset{NH_2}{CH}-CH_2CH_2COOH$

(d) $CH_3-\underset{OH}{CH}-CH_2COOH$

(e) $CH_3-\underset{O}{\overset{\|}{C}}-CH_2CH_2COOH$

(f) $CH_3-\underset{NH_2}{CH}-CH_2OH$

19. Name the following compounds: $CH_3OCH_2CH_3$, $CH_3CH_2CH_2COOH$, $CH_3CH_2CH_2-C\equiv CH$, $CH_3CH_2CH_2NH_2$.

288   20. Write structural formulas for butanoic acid, aminomethane, 2-butanol, and 3-aminopentane.

21. A compound contains 81.8% C and 18.2% H, and has a molecular weight of 44. Can you write a structure for it?

22. Two isomeric compounds, A and B, are found to have molecular weights of 46 and 52.2% C, 13% H, and 34.8% O. Compound A is found to give every evidence of hydrogen bonding to itself in the liquid state while compound B does not. Suggest structural formulas for A and B on the basis of these data.

23. Give an example of:

    (a) an alkane
    (b) an amine
    (c) a carboxylic acid
    (d) an ether
    (e) an ester
    (f) an alkene
    (g) an alkyne
    (h) an alcohol
    (i) a ketone

24. Write two structural formulas for compounds which can have each of the molecular formulas listed:

    (a) $C_5H_{12}O$
    (b) C_3H_6O
    (c) $C_5H_{10}O_2$

25. Draw the *cis* and *trans* isomers for:

    (a) 1,2-dibromoethene
    (b) 1-bromo-2-chloroethene

SUGGESTIONS FOR FURTHER READING

Dodge, B. S., "Louis Pasteur's Looking Glass World," *Chemistry*, Vol. 41, No. 2, p. 16 (1968).
Mills, G. A., "Ubiquitous Hydrocarbons," *Chemistry*, Vol. 44, No. 2, p. 8; No. 2, p. 12 (1971).
Orchin, M., "Determining the Number of Isomers From A Structural Formula," *Chemistry*, Vol. 42, No. 5, p. 8 (1969).
Roberts, J. D., "Organic Chemical Reactions," *Scientific American*, Vol. 197, p. 117 (1957).
von Tamelen, E. E., "Benzene, The Story of Its Formula, 1865–1965," *Chemistry*, Vol. 38, No. 1, p. 6 (1965).
Westheimer, F. H., "The Structural Theory of Organic Chemistry," *Chemistry*, Vol. 38, No. 6, p. 13 (1965), and Vol. 38, No. 7, p. 10 (1965).

SOME APPLICATIONS
OF ORGANIC CHEMISTRY

CHAPTER 15

It is difficult to overestimate the importance of carbon chemistry to man. Every month several hundred new organic compounds are prepared. A few of these new compounds become important as medicines, plastics, textiles, solvents, food additives, cosmetics, or some other consumer product. A precious few may provide an important clue to the mechanism of a fundamental chemical reaction in the human body. Most, however, become laboratory curiosities and for the present, at least, have no practical application.

The preparation of new and different compounds through chemical reaction is called *organic synthesis*. The million or so organic compounds now known and characterized have been synthesized in the laboratories of the world during the past 150 years. Prior to 1828, it was widely believed that chemical compounds synthesized by living matter could not be made without the living matter—a "vital force" was necessary for the synthesis. In 1828, a young German chemist, Friedrich Wöhler, destroyed the vital force myth and opened the door to the myriad organic synthesis that we now know. Wöhler heated a solution of silver cyanate and ammonium chloride, neither of which had been derived

Frederic Wöhler (1800–1882) was Professor of Chemistry at the University of Berlin and then at Göttingen. His preparation of the organic compound urea from ammonium cyanate, an inorganic source, did much to overturn the theory that organic compounds must be prepared in living organisms. One of the first to study the properties of aluminum, he discovered the element beryllium and is known for many other outstanding contributions to chemistry.

from any living substance. From these he prepared urea, a major waste product found in urine.

$$\text{AgCNO} \quad + \quad \text{NH}_4\text{Cl} \quad \longrightarrow \quad \text{AgCl} \quad + \quad \text{NH}_4\text{CNO}$$

Silver cyanate   Ammonium   Silver chloride   Ammonium
chloride   (Precipitate)   cyanate

$$\text{NH}_4\text{CNO} \quad \longrightarrow \quad \text{H}_2\text{N}\overset{\displaystyle O}{\overset{\|}{\text{C}}}\text{NH}_2$$

Ammonium cyanate   Urea

The inevitable questioning and probing of Wöhler's work could not destroy the fact that urea had been prepared from materials not related to living materials. The notion of a mysterious vital force declined as other chemists began to synthesize more and more organic chemicals without the aid of a living system. Soon it was shown that chemistry could do more than imitate the products of living tissue; it could form unique materials of its own.

Advances in understanding the structure of organic compounds (see Chapter 14) gave organic synthesis a tremendous boost. Knowing the structure of compounds, the organic chemist could predict by analogy with simpler molecules what reactions might take place when organic reagents were used. The time-consuming trial and error technique of synthesis became less fashionable. Very elegant and reliable schemes of synthesis could now be constructed.

In this chapter emphasis will be given to some important organic compounds that are not only useful in themselves but are necessary for the synthesis of many other organic compounds. The dependence of synthesis upon a knowledge of structure will be pointed out from time to time.

SOME USES OF HYDROCARBONS

Complex mixtures of hydrocarbons, compounds containing only carbon and hydrogen, occur in enormous quantities in nature as petroleum and natural gas (Chapter 14). Many other organic compounds are prepared from these materials after they are separated into their constituents. From the simplest hydrocarbon, methane, come such diverse consumer products as plastic dishes, acrylic fibers, vinyl paints, and neoprene rubber, and such industrial products as Teflon and cattle feed.

Natural gas consists primarily of low molecular weight hydrocarbons. It is predominantly methane (CH_4), but ethane (C_2H_6), propane (C_3H_8), and butane (C_4H_{10}) are also present. This mixture is conveyed in long pipelines from the areas in which it occurs to cities where it is used as a fuel. Propane and butane, which can be liquefied by the use of moderate pressures, are the principal constituents of bottled gas, which is used in rural or isolated communities. Because propane is more volatile than butane, the composition of bottled gas is adjusted to contain more propane in colder climates and more butane in warmer ones.

The lower molecular weight hydrocarbons can also be prepared from higher molecular weight hydrocarbons by a process known as *cracking*. In this process, the gaseous hydrocarbon is passed over a catalyst at elevated temperatures and pressures and is broken down into smaller molecular weight fragments. The cracking of n-butane is a simple example.

$$CH_3CH_2CH_2CH_3 \xrightarrow[600°]{catalyst} CH_2=CH_2 + CH_2=CH_2 + H_2$$

These small molecules are most useful in the chemical synthesis of other products. If large hydrocarbon molecules in heavy petroleum oils are cracked, molecules in the C_5 to C_{10} range are produced. Molecules in this range are useful as fuels.

Gasoline is actually a mixture of low molecular weight liquid hydrocarbons which evaporate easily. Vaporized gasoline, combined with air, is used to provide energy in internal combustion engines. This mixture must burn evenly and rapidly if the operation of the engine is to be smooth. When the burning of the gasoline-air mixture is too rapid or irregular, preignition occurs in the combustion chamber, resulting in a small explosion which is heard as a "knock" in the engine. This knocking can spell disaster to the owner of an automobile since it will eventually lead to the breakdown of the internal parts of the automobile's engine. A great deal of research has been directed toward reducing the knocking of gasoline, and it has been discovered that this can be avoided by two procedures. The first is to use only hydrocarbons with certain structures which are known to burn satisfactorily in an automobile engine. The second is to add "antiknock" agents to a more motley mixture of hydrocarbons to achieve the same effect. The best known antiknock additive is tetraethyl lead, $Pb(C_2H_5)_4$. Tetraethyl lead vapor is extremely toxic and gasolines containing it ordinarily contain a dye to indicate its presence. Gasoline containing this compound is known as "ethyl" gasoline. Antiknock additives in general act to slow down free radical reactions (Chapter 14) that take place in the combustion chamber of an engine. By slowing these reactions down, the gasoline-air mixture can burn more smoothly, producing more power with a reduced likelihood of damaging the engine. When tetraethyl lead is used in gasolines, it is also necessary to add 1,2-dibromoethane, $BrCH_2—CH_2Br$, to assist in the removal of the lead from the engine. The burning of a mixture of these compounds produces lead bromide, $PbBr_2$, which is eliminated as a vapor in the engine exhaust. We will consider the effects of the presence of lead in gasoline in more detail in Chapter 21 on air pollution.

An arbitrary scale for rating the relative knocking properties of gasolines has been developed. Normal heptane knocks considerably and is assigned an octane rating of 0.

$$CH_3CH_2CH_2CH_2CH_2CH_2CH_3 \text{ octane rating} = 0$$

while isooctane (2,2,4-trimethylpentane) is far superior in this respect and is assigned an octane rating of 100.

$$CH_3—\underset{\underset{CH_3}{|}}{\overset{\overset{CH_3}{|}}{C}}—CH_2—\underset{\underset{H}{|}}{\overset{\overset{CH_3}{|}}{C}}—CH_3 \text{ octane rating} = 100$$

To determine the octane rating of a gasoline, it is used in a standard engine and its knocking properties are recorded. This is compared to the behavior of mixtures of n-heptane and isooctane, and the percentage of isooctane in the mixture with identical knocking properties is called the octane rating of the gasoline. Thus, if a gasoline has the same knocking characteristics as a mixture of 9 per cent n-heptane and 91 per cent isooctane, it is assigned an octane rating of 91. This corresponds to a regular grade of gasoline. Since the octane rating scale was established, fuels have been developed which are superior to isooctane, so the scale has been extended well above 100. A primary source of the present high octane gasolines is a chemical process known as *catalytic reforming*. Under the influence of certain catalysts, such as finely divided platinum, straight-chain hydrocarbons with low octane numbers can be re-formed into their branched-chain isomers, which have higher octane numbers.

$$CH_3CH_2CH_2CH_2CH_3 \xrightarrow[\text{Heat}]{\textit{Platinum}} CH_3CH_2\underset{\overset{|}{CH_3}}{CH}CH_3$$

n-Pentane   *2-Methylbutane*

SOME IMPORTANT COMPOUNDS OF CARBON, HYDROGEN, AND OXYGEN

Carbon, hydrogen, and oxygen can be combined to form an enormous number of compounds. As we have seen in Chapter 14, considerable order is introduced into the study of these compounds when they are divided into classes on the basis of the functional groups they contain. The *alcohols*, the *organic acids* and their derivatives (compounds that can be made from them), and the *esters* and *soaps* are very important compounds of carbon, hydrogen, and oxygen since they find such wide application in our everyday lives.

ALCOHOLS

When a hydroxyl (—OH) group is attached to a nonaromatic carbon skeleton (an R-group), the resulting R—OH molecule has properties common to a class of compounds called alcohols. Formulas and names for some important molecules of this type are given in Table 15–1.

Methanol (Methyl Alcohol)

Methanol was originally called wood alcohol since it was obtained by the destructive distillation of wood. It is the simplest of all alcohols and has the formula CH_3OH. In the older method for the production of wood alcohol, hardwoods such as beech, hickory, maple, or birch are heated in the absence of air in a retort (Figure 15–1). Methanol, which is 92 to 95 per cent pure, can be obtained by fractional distillation of the liquid which results.

In 1923, the price of wood alcohol in the United States was 88¢ per gallon. 293

TABLE 15-1   SOME IMPORTANT ALCOHOLS

FORMULA	IUPAC NAME	COMMON NAME
$\begin{array}{c} H \\ | \\ H{-}C{-}OH \\ | \\ H \end{array}$	Methanol	Methyl alcohol (wood alcohol)
$\begin{array}{c} H\quad H \\ |\quad | \\ H{-}C{-}C{-}OH \\ |\quad | \\ H\quad H \end{array}$	Ethanol	Ethyl alcohol (grain alcohol)
$\begin{array}{c} H\quad H\quad H \\ |\quad |\quad | \\ H{-}C{-}C{-}C{-}OH \\ |\quad |\quad | \\ H\quad H\quad H \end{array}$	1-Propanol	n-Propyl alcohol
$\begin{array}{c} H\quad H\quad H \\ |\quad |\quad | \\ H{-}C{-}C{-}C{-}H \\ |\quad |\quad | \\ H\quad O\quad H \\ \quad | \\ \quad H \end{array}$	2-Propanol	Isopropyl alcohol (rubbing alcohol)
$\begin{array}{c} H \\ | \\ H{-}C{-}OH \\ | \\ H{-}C{-}OH \\ | \\ H \end{array}$	1,2-Ethanediol	Ethylene glycol (antifreeze)
$\begin{array}{c} H \\ | \\ H{-}C{-}OH \\ | \\ H{-}C{-}OH \\ | \\ H{-}C{-}OH \\ | \\ H \end{array}$	1,2,3-Propanetriol	Glycerol (glycerin)

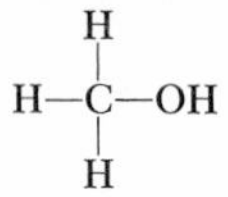

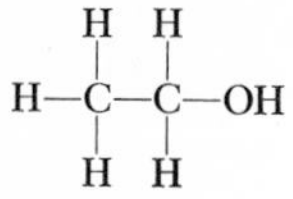

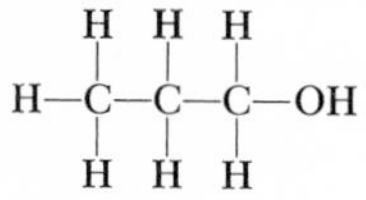

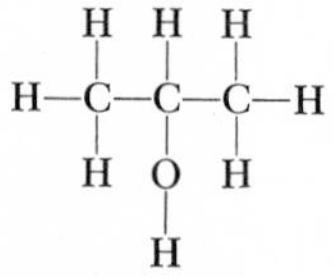

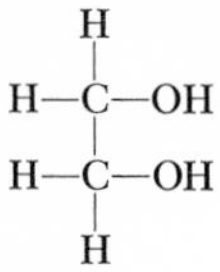

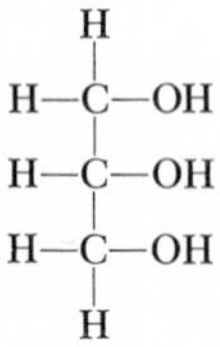

In that year German chemists discovered how to produce this useful compound synthetically. Methanol is formed when carbon monoxide, CO, and hydrogen are heated at a pressure of 200 to 300 atmospheres over a catalyst of mixed oxides (90% ZnO—10% Cr_2O_3).

$$CO + 2H_2 \xrightarrow[300°C]{ZnO-Cr_2O_3} CH_3OH$$

As a result of this synthetic process, German industrialists were able to sell pure methanol at 20¢ per gallon. Even a high tariff was not able to save the wood distillers in their outdated operations. The synthetic product dominated the market in the United States.

The production of synthetic methanol in the United States rose to over 3.4 billion pounds in 1967. About one-half of this is used in the production of formaldehyde (used in plastics), 30 per cent in the production of other chemicals, and smaller amounts used for jet fuels, antifreeze mixtures, solvents, and as a denaturant (a poison added to ethyl alcohol to make it unfit for beverages). Methanol is a *deadly poison*; it causes blindness in less than lethal doses. Many deaths and injuries have resulted when this alcohol was mistakingly substituted for ethyl alcohol in beverages.

294

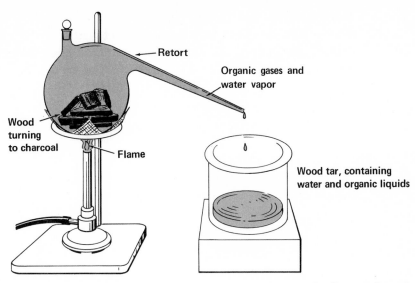

FIGURE 15-1   Destructive distillation of wood.

Ethanol

Ethanol (ethyl alcohol) is called grain alcohol because it can be fractionally distilled from the fermented mash made from corn, rice, barley, or other grains which are sources of carbohydrates (Chapter 17). Fermentation is a breakdown of complex organic molecules, such as carbohydrates, brought about by means of enzymes. Enzymes are complex organic molecules produced by living cells that act as catalysts (Chapter 18). If the enzyme diastase is mixed with ground grain and water and the mixture is allowed to stand at 40°C for a period of time, the starch in the grain will be changed into the sugar maltose. (The more detailed chemistry of carbohydrates is discussed in Chapter 17).

$$2(C_6H_{10}O_5)_n \quad + nH_2O \longrightarrow \quad nC_{12}H_{22}O_{11}$$
Starch (a carbohydrate)   *Maltose (a sugar)*

The subscript n in the formula for starch indicates that starch is made up of many $C_6H_{10}O_5$ units. Brewers call the resulting mixture of maltose and water the wort. The wort is diluted and mixed with yeast and held at a temperature of 30°C for 40 to 60 hours. The living yeast cells secrete two enzymes, maltase and zymase. The maltase causes the sugar, maltose, to decompose into a simple sugar, glucose.

$$C_{12}H_{22}O_{11} + H_2O \longrightarrow 2C_6H_{12}O_6$$
Maltose   *Glucose*

The glucose, in turn, is converted by zymase to alcohol and carbon dioxide.

$$C_6H_{12}O_6 \longrightarrow 2CO_2 + 2C_2H_5OH$$
Glucose   *Ethanol*

A solution of 95 per cent ethyl alcohol and 5 per cent water can be recovered from the mash by fractional distillation.

295

Synthetic ethyl alcohol is produced on a large scale for industrial use. The direct hydration of ethylene by addition of water (Chapter 14) accounts for more than 80 per cent of all ethanol production. Using ethylene obtained from the cracking of petroleum, and a large excess of water, the reaction which occurs is:

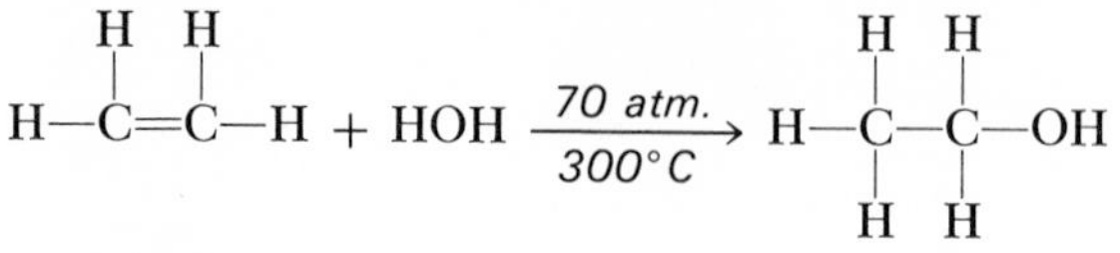

$$H-\underset{\underset{H}{|}}{\overset{\overset{H}{|}}{C}}=\underset{\underset{H}{|}}{\overset{\overset{H}{|}}{C}}-H + HOH \xrightarrow[300°C]{70\ atm.} H-\underset{\underset{H}{|}}{\overset{\overset{H}{|}}{C}}-\underset{\underset{H}{|}}{\overset{\overset{H}{|}}{C}}-OH$$

Pure ethyl alcohol is 200 proof (exactly twice the per cent). Apart from the alcoholic beverage industry, ethyl alcohol is used widely in solvents and in the preparation of chloroform, ether, and many other organic compounds.

Propanols (Propyl Alcohols)

When one considers the possible structures for propyl alcohol, it is apparent that two isomers are possible.

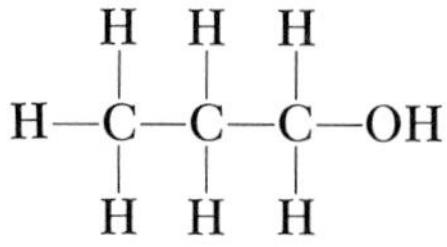

1-Propanol
(n-propyl alcohol)

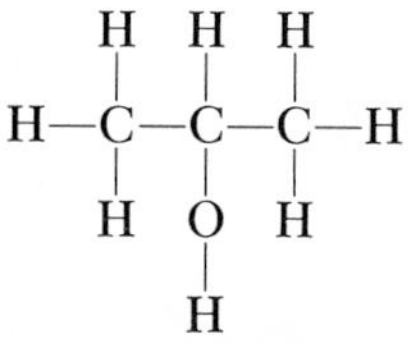

2-Propanol
(isopropyl alcohol)

Of the two propanols, 1-propanol is the more expensive; it is prepared by the oxidation of simple hydrocarbons. It finds uses as a solvent and as a raw material in the manufacture of other organic compounds.

The hydration of propylene yields 2-propanol (isopropyl alcohol) which is sold as rubbing alcohol since it has a greater germicidal activity than the other simple alcohols.

Ethylene Glycol and Glycerol (Glycerin)

More than one alcohol group (—OH) can be present in a single molecule. Ethylene glycol, the base of permanent antifreeze, and glycerin are examples of such compounds.

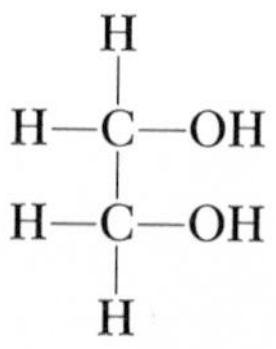

Ethylene glycol
(1,2-ethanediol)

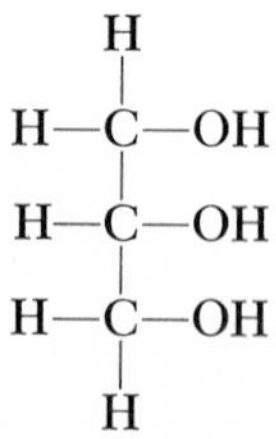

Glycerol (Glycerin)
(1,2,3-propanetriol)

Glycerol has many uses in the manufacture of drugs and cosmetics, in the production of nitroglycerin and numerous other chemicals. Perhaps the most important compounds of glycerol are its natural esters (fats and oils) which we shall discuss later in this chapter.

Hydrogen Bonding in Alcohols

The physical properties of water, methanol, ethanol, the propanols, ethylene glycol, and glycerol offer another interesting example of the effects of *hydrogen bonding* between molecules in liquids. For a more complete discussion of hydrogen bonding, see Chapter 9. In Table 15–2 the boiling points for these compounds are listed.

TABLE 15–2	BOILING POINTS FOR SOME —OH COMPOUNDS	
Water	HOH	100°C
Methanol	CH_3OH	64.6°
Ethanol	CH_3CH_2OH	78.5°
1-Propanol	$CH_3CH_2CH_2OH$	97.2°
2-Propanol	$CH_3CHOHCH_3$	82.3°
Ethylene glycol	CH_2OHCH_2OH	198°
Glycerol	$CH_2OHCHOHCH_2OH$	290°

Since boiling involves breaking the bonds between liquid molecules as they pass into the gas phase, a higher boiling point indicates stronger intermolecular forces holding the molecules together. Another factor is also present: as the molecules become larger, higher boiling points result, since more energy is required to separate the longer chain molecules from the liquid into the gaseous phase, owing in part to the larger Van der Waals attraction. A graph showing the boiling points of the normal alcohols (straight carbon chains with the —OH group on an end carbon) as a function of chain length is given in Figure 15–2.

Methyl alcohol, like water, has an —OH group and some hydrogen bonding is to be expected, as shown in Figure 15–3. Hydrogen bonding explains why methyl alcohol (molecular weight = 32) is a liquid, whereas propane (C_3H_8, molecular weight = 44), an even heavier molecule, is a gas at room temperature. Methyl alcohol has only one hydrogen through which it can hydrogen bond, while

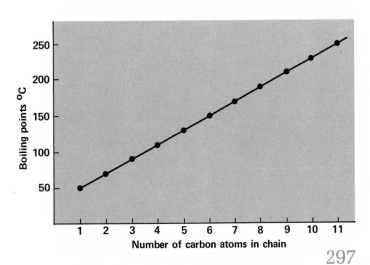

FIGURE 15-2 Boiling points of straight chain alcohols (—OH group on an end carbon).

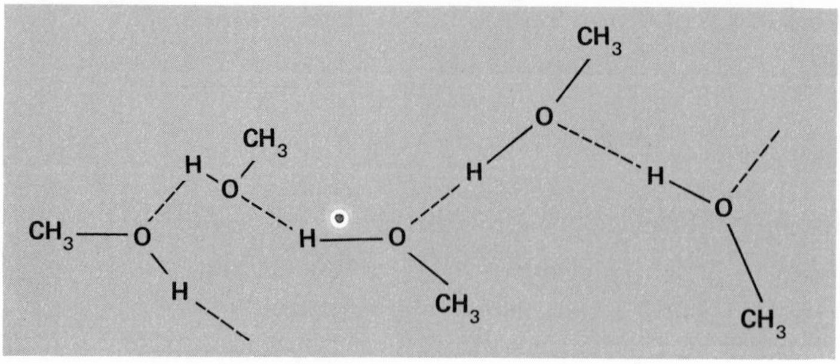

FIGURE 15-3 Hydrogen bonding in methanol.

water can hydrogen bond from either of its two hydrogen atoms. Thus, water, with more extensive intermolecular bonding, has the higher boiling point even though it is made up of lighter molecules.

Both methyl and ethyl alcohol can be used as antifreeze, but there is one serious disadvantage in their use. They tend to fractionally distill out of the coolant at the temperatures of a hot gasoline engine. Protection against freezing is then lost over a period of time. Ethylene glycol is equally effective (molecule for molecule) in lowering the freezing point of water, and its high boiling point (198°C) makes it a permanent type antifreeze. This property makes ethylene glycol more desirable even though it takes almost twice as much ethylene glycol by weight than methanol for the same amount of protection of a car's cooling system. Ethylene glycol with suitable additives to protect the radiator system is sold under a number of brand names. The higher boiling point of ethylene glycol is readily explained in terms of the two —OH groups per molecule and the enhanced possibility for hydrogen bonding. Glycerol, with three —OH groups per molecule, has an even higher boiling point, as well as a very high viscosity (resistance to flow).

ORGANIC ACIDS

In Chapter 11 an acid was defined as a species that has a tendency to donate protons. We will now consider a group of organic compounds which are proton donors, the carboxylic acids containing the carboxyl group, $-C\begin{smallmatrix}O\\\\OH\end{smallmatrix}$. The electronegative character of the C=O group tends to drain electron density away from the region between the oxygen and hydrogen atom, giving the hydrogen atom acidic character. The strength of an organic acid depends on the group which is attached to the carboxyl group. If the attached group has a tendency to pull electrons away from the carboxyl group, the result is a stronger acid. For example, trifluoroacetic acid is a much stronger acid than acetic acid.

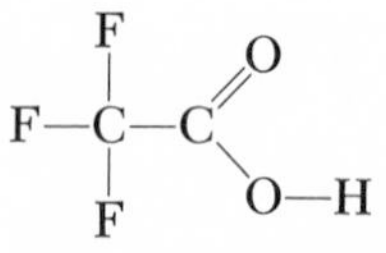

Trifluoroacetic acid 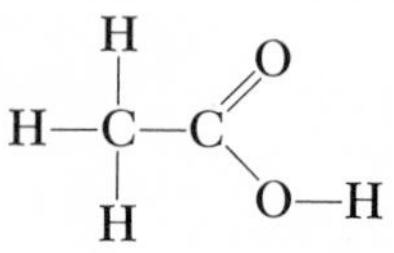 *Acetic acid*

The electronegative fluorine atoms withdraw electron density from the region of the carboxyl group and facilitate the loss of the hydrogen ion. However, trifluoroacetic acid is still weaker than a strong mineral acid such as sulfuric acid.

298

Formic Acid

The simplest organic acid is formic acid, in which the carboxyl group is attached directly to a hydrogen atom.

$$H-C\underset{O-H}{\overset{O}{\diagup}}$$

Formic acid

This acid is found in ants and other insects and is part of the irritant that produces itching and swelling after a bite. The sodium salt of formic acid is readily prepared by heating carbon monoxide (CO) with sodium hydroxide (NaOH).

$$CO + NaOH \xrightarrow[\text{6-10 min.}]{200°} HCOO^-Na^+$$

Sodium formate

If the resulting salt is mixed with a mineral acid, formic acid can be distilled from the mixture.

$$HCOO^-Na^+ + H_3O^+ + Cl^- \longrightarrow HCOOH + Na^+Cl^- + H_2O$$

| Sodium formate | Hydrochloric acid | Formic acid | Sodium chloride |

Acetic Acid

This acid is the most widely used of the organic acids. It is found in vinegar, an aqueous solution containing 4 to 5 per cent acetic acid. Flavor and colors are imparted to vinegars by the constituents of the alcoholic solutions from which they are made. Ethyl alcohol in the presence of certain bacteria and air is oxidized to acetic acid.

$$CH_3CH_2OH + O_2 \xrightarrow{Bacteria} CH_3COOH + H_2O$$

Ethyl alcohol   *Oxygen*   *Acetic acid*   *Water*

The bacteria, called mother of vinegar, forms a slimy growth in a vinegar solution. The growth of bacteria can sometimes be observed in a bottle of commercially prepared vinegar after it has been opened to the air.

Acetic acid is an important starting material for making other chemicals and is a convenient acidic material when a cheap organic acid is needed.

Fatty Acids

A fatty acid is made up of molecules containing a carboxyl group attached to a long hydrocarbon chain. The chains often contain only single carbon—carbon bonds, but may contain carbon—carbon double bonds as well. Two examples are stearic acid and palmitic acid.

$$CH_3CH_2CH_2CH_2CH_2CH_2CH_2CH_2CH_2CH_2CH_2CH_2CH_2CH_2CH_2CH_2CH_2C\underset{OH}{\overset{O}{\diagup}}$$

Stearic acid, CH$_3$—(CH$_2$)$_{16}$—COOH

$$CH_3CH_2CH_2CH_2CH_2CH_2CH_2CH_2CH_2CH_2CH_2CH_2CH_2CH_2CH_2C\underset{OH}{\overset{O}{\diagup}}$$

Palmitic acid, CH$_3$—(CH$_2$)$_{14}$—COOH

299

Stearic acid is obtained by the hydrolysis (reaction with water) of animal fat; palmitic acid results from the hydrolysis of palm oil, a liquid fat. These two fatty acids are especially important in the manufacture of soaps.

ESTERS

In the presence of certain strong mineral acids, organic acids react with alcohols to form a class of compounds called esters. For example, when ethyl alcohol is mixed with acetic acid in the presence of sulfuric acid, ethyl acetate is formed. This reaction is a dehydration in which sulfuric acid acts as a catalyst. Ethyl acetate is a common solvent and is often used as fingernail polish remover.

$$CH_3-CH_2-O-[H + HO]-\underset{\underset{O}{\|}}{C}-CH_3 \xrightarrow{H_2SO_4} CH_3-CH_2-O-\underset{\underset{O}{\|}}{C}-CH_3 + H_2O$$

Some of the odors of common fruits are due to the presence of mixtures of volatile esters (Table 15–3). In contrast, higher molecular weight esters often have a distinctly unpleasant odor.

TABLE 15-3   SOME ALCOHOLS, ACIDS, AND THEIR ESTERS

ALCOHOL	ACID	ESTER	ODOR OF THE ESTER
$CH_3CHCH_2CH_2OH$ | CH_3 Isopentyl alcohol	CH_3COOH Acetic acid	$CH_3CHCH_2CH_2-O-\underset{\underset{O}{\|}}{C}-CH_3$ | CH_3 Isopentyl acetate	Banana
$CH_3CHCH_2CH_2OH$ | CH_3 Isopentyl alcohol	$CH_3CH_2CH_2CH_2COOH$ n-Valeric acid	$CH_3CHCH_2CH_2-O-\underset{\underset{O}{\|}}{C}-CH_2CH_2CH_2CH_3$ | CH_3 Isopentyl n-valerate	Apple
$CH_3CH_2CH_2CH_2OH$ n-Butyl alcohol	$CH_3CH_2CH_2COOH$ n-Butyric acid	$CH_3CH_2CH_2CH_2-O-\underset{\underset{O}{\|}}{C}-CH_2CH_2CH_3$ Butyl n-butyrate	Pineapple
CH_3CHCH_2OH | CH_3 Isobutyl alcohol	CH_3CH_2COOH Propionic acid	$CH_3CHCH_2-O-\underset{\underset{O}{\|}}{C}-CH_2CH_3$ | CH_3 Isobutyl propionate	Rum

FATS, OILS, AND SOAPS

Fats and oils are esters of glycerol (glycerin) and a fatty acid. R, R', and R'' stand for the hydrocarbon chains in the following equation.

The term *fat* is usually reserved for solids (butter, lard, tallow) and oil for liquids (castor, olive, linseed, tung, and so forth). Since fats and oils are products of living organisms, the biochemist is naturally interested in them, and uses the term *lipid* to apply to fats, oils, and other fat soluble compounds.

$$
\begin{array}{c}
CH_2\text{—}OH \\
| \\
CH\text{—}OH \\
| \\
CH_2\text{—}OH
\end{array}
\;+\;
\begin{array}{c}
\overset{\displaystyle O}{\underset{\|}{}}\\
HO\text{—}C\text{—}R \\
HO\text{—}C\text{—}R' \\
HO\text{—}C\text{—}R''
\end{array}
\;\rightleftharpoons\;
\begin{array}{c}
CH_2\text{—}O\text{—}C\text{—}R \\
| \\
CH\text{—}O\text{—}C\text{—}R' \\
| \\
CH_2\text{—}O\text{—}C\text{—}R''
\end{array}
\;+\; 3\,H_2O
$$

Glycerol (one molecule)   Fatty acid (3 molecules, which may or may not be the same)   A fat or oil molecule   3 molecules of water

Saturation (all single bonds with maximum hydrogen content) in the carbon chain of the fatty acids is usually found in solid or semi-solid fats, whereas unsaturated fatty acids (one or more double bonds) are usually found in oils. Hydrogen can be catalytically added to the double bonds of an oil to convert it into a semi-solid fat. For example, liquid soybean and other vegetable oils are hydrogenated to produce cooking fats and margarine.

Consumers in Europe and North America have historically valued butter as a source of fat. As the population increased the advantages of a substitute for butter became apparent, and efforts to prepare such a product began about a hundred years ago. One problem which arose was the fact that common fats are almost all *animal* products with very pronounced tastes of their own. Analogous compounds from vegetable oils, which have mixed flavors, were generally *unsaturated* and consequently *oils*. A solid fat could be made from the much cheaper vegetable oils if an inexpensive way could be discovered to add hydrogen across their double bonds. After extensive experiments, many catalysts were found, of which finely divided nickel is among the most effective. The nature of the process can be illustrated by the reaction:

$$
\begin{array}{c}
H_2C\text{—}O\text{—}C\text{—}(CH_2)_7\text{—}CH{=}CH(CH_2)_7CH_3 \\
| \\
HC\text{—}O\text{—}C\text{—}(CH_2)_7\text{—}CH{=}CH(CH_2)_7CH_3 \\
| \\
H_2C\text{—}O\text{—}C\text{—}(CH_2)_7\text{—}CH{=}CH(CH_2)_7CH_3
\end{array}
\;\xrightarrow[\sim 200\,°C]{H_2,\;Ni}\;
\begin{array}{c}
H_2C\text{—}O\text{—}C\text{—}(CH_2)_7CH_2\text{—}CH_2\text{—}(CH_2)_7CH_3 \\
| \\
HC\text{—}O\text{—}C\text{—}(CH_2)_7\text{—}CH_2\text{—}CH_2\text{—}(CH_2)_7CH_3 \\
| \\
H_2C\text{—}O\text{—}C\text{—}(CH_2)_7\text{—}CH_2\text{—}CH_2\text{—}(CH_2)_7CH_3
\end{array}
$$

Triolein, a liquid fat (oil)   Tristearin, a solid fat

Oils which are commonly subjected to this process include those obtained by pressing cottonseed, peanuts, corn germ, soya bean, coconut, and safflower seeds. In recent years, as it became apparent that hydrogenation destroys essential fatty acids, soft, partially hydrogenated products have been placed on the market. For the three essential fatty acids, the best sources are safflower oil for linoleic, soya bean oil for linolenic, and peanut oil for arachidonic, and these have been incorporated in many margarines, as their labels testify.

There is considerable interest in the relationship between saturated fats in the diet and heart disease. It is known that an increase in solid fats increases the concentration of the complicated biochemical cholesterol in the blood

301

stream. Since it is thought by some physicians that a rise in cholesterol is associated with hardening of the arteries, there has been much interest in replacing solid fats with liquid oils in the diet.

Naturally occurring fats and oils can be hydrolyzed in strongly basic solutions to form glycerol and salts of the fatty acids. Such hydrolysis reactions are called *saponification* reactions because the sodium or potassium salts of the fatty acids formed are *soaps*. Pioneers prepared their soap by boiling animal fat with an alkaline solution obtained from the ashes of hard wood. The resulting soap could be "salted out" by adding sodium chloride and making use of the fact that soap is less soluble in a salt solution than in water.

$$CH_3(CH_2)_{16}COO-CH_2$$
$$CH_3(CH_2)_{16}COO-CH + 3NaOH \longrightarrow 3CH_3(CH_2)_{16}COO^-Na^+ + HO-CH$$
$$CH_3(CH_2)_{16}COO-CH_2$$

Stearic acid ester	Sodium Stearate	Glycerol
(Glycerol tristearate)	(A soap)	

$HO-CH_2$

The cleansing action of soap can be explained in terms of its molecular structure. Material that is water soluble can be readily removed from the skin or a surface by simply washing with an excess of water. To remove a sticky sugar syrup from one's hands, the sugar is dissolved in water and rinsed away. Many times the material to be removed is oily and water will merely run over the surface of the oil. Since the skin has natural oils, even substances, such as ordinary dirt, which are not oily themselves, can cover the skin in a greasy layer. The cohesive forces (forces between like molecules tending to hold them together) within the water layer are too large to allow the oil and water to intermingle (Figure 15–4). When present in an oil-water system, soap molecules such as sodium stearate,

$$CH_3CH_2CH_2CH_2CH_2CH_2CH_2CH_2CH_2CH_2CH_2CH_2CH_2CH_2CH_2CH_2CH_2C\begin{matrix} O \\ \diagdown \\ O^-Na^+ \end{matrix}$$

will move to the interface between the two liquids. The carbon chain, which is a nonpolar structure, will readily mix with the nonpolar grease molecules, whereas the highly polar —COO⁻Na⁺ group enters the water layer (Figure 15–4b). The soap molecule will then tend to lie across the oil-water interface. The grease is broken up into small droplets, each surrounded by hydrated soap molecules (Figure 15–4c). The surrounded oil droplets cannot come together again since the exterior of each is covered with —COO⁻Na⁺ groups which strongly interact with the surrounding water. If enough soap and water are available, the oil will be swept away forming a clean and water-wet surface.

SYNTHETIC DETERGENTS (SYNDETS)

If Ca^{2+} or Mg^{2+} ions are present in water, an ordinary soap will precipitate as an insoluble salt and the water is said to be hard. The ring on the bath tub and the scum in the washer are visible signs of this precipitate. Only after all these interfering ions have been precipitated can the added soap cause the water

FIGURE 15-4   The cleansing action of soap. (*a*) A piece of glass coated with grease inserted in water gives evidence for the strong adhesion between water and glass at 1, 2 and 3. The water curves up against the pull of gravity to wet the glass. The relatively weak adhesion between oil and water is indicated at 4 by the curvature of the water away from the grease against the force tending to level the water. (*b*) A soap molecule, having oil soluble and water soluble ends, will orient at an oil-water interface such that the hydrocarbon chain is in the oil (with molecules that are electrically similar-nonpolar) and the COO⁻Na⁺ group is in the water (highly charged polar groups interacting electrically). (*c*) In an idealized molecular view, a grease particle, 1, is surrounded by soap molecules which in turn are strongly attracted to the water. At 2 another droplet is about to break away. At 3 the grease and clean glass interact before the water moves between.

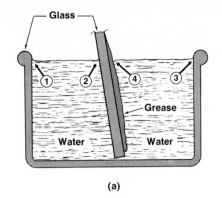

(a)

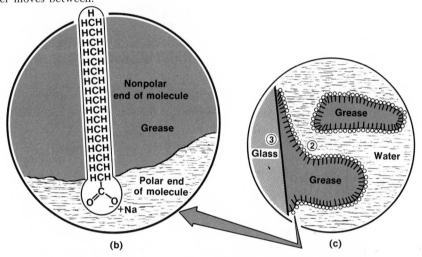

(b)   (c)

to form suds. Water softeners can remove the hard water ions, replacing them with ions such as Na⁺ which do not interfere with the ordinary soap's action, but this is often an expensive process. Synthetic detergents (syndets) are similar to soaps in that they are composed of molecules having water-soluble and oil-soluble ends, but syndets have an advantage over soaps in hard water in that they do not form insoluble precipitates with Ca^{2+} or Mg^{2+}. A typical syndet molecule with a water-soluble $-SO_3^-Na^+$ group and an oil-soluble hydrocarbon group is shown below.

$$CH_3(CH_2)_{11}OH \xrightarrow{H_2SO_4} CH_3(CH_2)_{11}OSO_3H \xrightarrow{NaOH} CH_3(CH_2)_{11}OSO_3^- Na^+$$

Lauryl alcohol   *Lauryl hydrogen sulfate*   *Sodium lauryl sulfate*

The most widely used syndets are sodium salts of substituted benzene-sulfonic acids. Normally, lone hydrocarbon chains, designated —R, are attached to the benzene ring by substitution reactions (Chapter 14).

Benzene-sulfonic Acid

303

Syndets are effective cleansing agents, but they pose problems in water treatment and purification. By careful synthesis of these molecules, chemists have been able to produce syndets which, once "used," can be broken down into simpler molecules by the action of bacteria. Since about 1965, these biodegradable syndets have been used almost exclusively in the United States (Chapter 20).

USEFUL PRODUCTS FROM
AROMATIC ORGANIC REACTIONS

A large number of chemists are engaged in the synthesis of organic compounds. In educational and industrial laboratories throughout the world they prepare new and different compounds on a small scale, expand the scale to pilot plant operation, or set up and manage large scale industrial processes.

The thousands of chemical changes required to synthesize organic compounds have a few characteristics in common.

1. Many chemical changes are usually required to synthesize a single organic compound. Each chemical change is called a step, and each step produces an intermediate compound that is used in the next step. The final step produces the desired end product.

2. Normally only one functional group undergoes change in each step. The rest of the molecule remains intact and unchanged. This is known as the principle of minimum structural change.

3. From one principal starting material can come many diverse products. The kind of product depends upon the reactants and the conditions imposed.

4. The more steps in the synthesis, the less the per cent yield of final product. The starting material and each intermediate are only partially converted to the next intermediate because of equilibrium considerations, side reactions, or both, which convert some of the starting material into undesirable products. If an intermediate product must be removed and purified before proceeding to the next step, some additional material is lost. The principal purification methods are recrystallization and extraction for solids and distillation for liquids.

These principles will be illustrated by the preparation of chlorobenzene from a natural product, benzene, and then the use of chlorobenzene to prepare such diverse and useful products as aspirin, oil of wintergreen, sulfa drugs, and an organic dye.

In the previous chapter we saw how bromobenzene could be prepared from benzene in the presence of a suitable catalyst, such as iron(III) bromide, $FeBr_3$. Benzene undergoes the same kind of a reaction with chlorine to produce chlorobenzene.

benzene   Chlorobenzene

Chlorobenzene is manufactured on an enormous scale and is widely used in industrial processes to prepare other organic compounds.

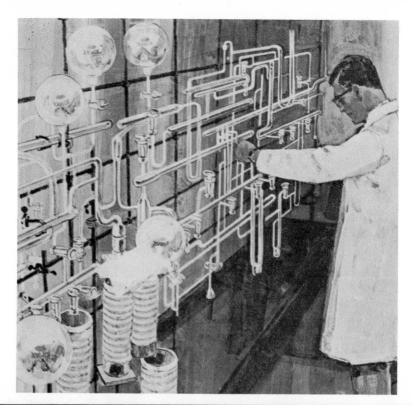

FIGURE 15-5   Organic syntheses begin on a small scale in the laboratory (*top*), but graduate to a very large scale if they become commerically important. The top view shows a typical laboratory bench. The bottom view shows an automatic, modern chemical plant. In both cases the chemical reactions are the same; the scope is different. (*bottom*, courtesy of the Du Pont Company, Textile Fiber Department.)

305

ASPIRIN, OIL OF WINTERGREEN
AND SALICYLAMIDE

The series of steps required to synthesize aspirin, oil of wintergreen, and salicylamide from chlorobenzene is outlined in Figure 15–6. Each structure represents the beginning or end of a step in the synthesis. Aspirin and oil of wintergreen require five steps; salicylamide requires six. Only the principal organic substance is given as the product for each step. Other products such as NaCl (a coproduct with phenol) and water (a coproduct with sodium phenolate) are sometimes important in the synthesis because they have to be removed to avoid interference with subsequent steps. However, in giving a broad outline of the synthetic process, the coproducts are generally omitted; only those products made in a previous step and then required for subsequent steps are included. In Figure 15–6 the step by step structural changes can be followed by noting groups in color. Conditions and additional reactants for each step are written over the arrow. These conventions will be used to summarize the organic syntheses presented in this chapter.

Some intermediates are useful compounds in their own right. Phenol, commonly called carbolic acid, is used to prepare plastics such as Bakelite, drugs, dyes, and other compounds. Phenol also has medicinal application as a topical anesthetic for some types of lesions and in the treatment of mange and colic in animals. Methyl salicylate, or oil of wintergreen, is used as a flavoring agent in addition to being an intermediate in the synthesis of salicylamide, a compound with many of the properties of aspirin but reportedly less toxic.

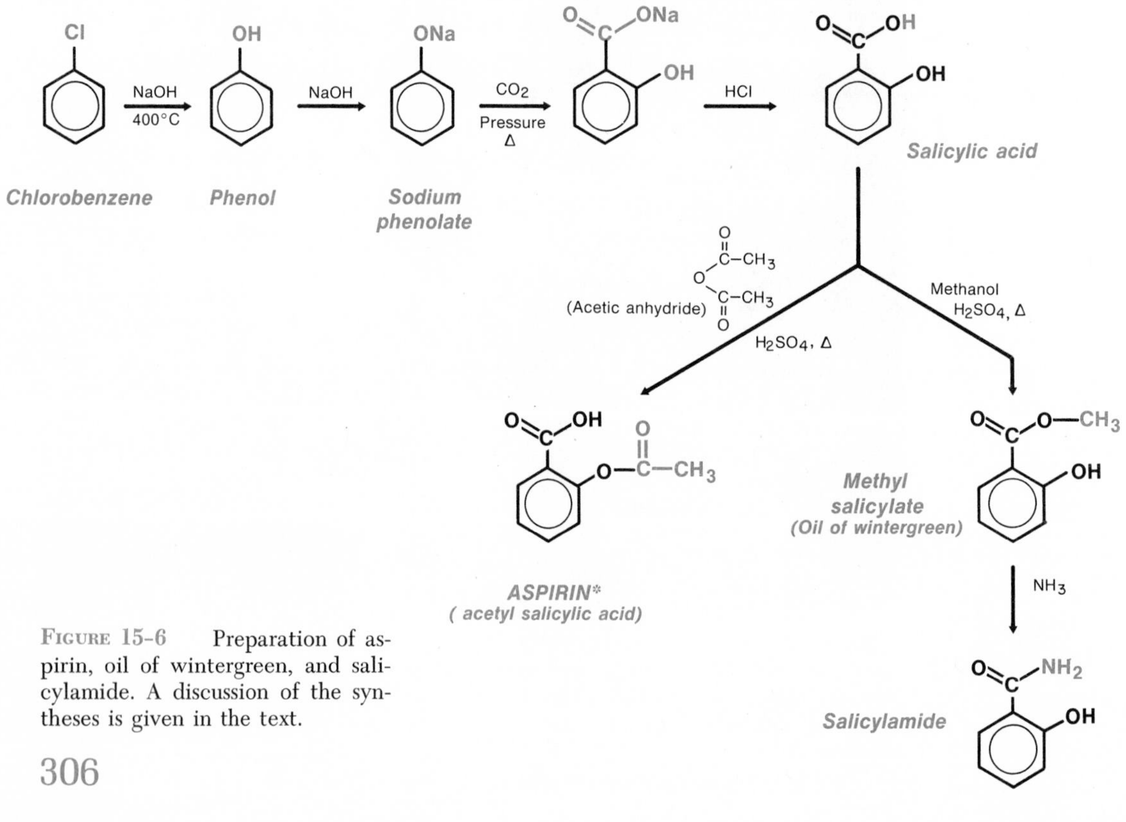

FIGURE 15–6   Preparation of aspirin, oil of wintergreen, and salicylamide. A discussion of the syntheses is given in the text.

Several of the reactions in these syntheses are already familiar from discussions in previous chapters. The conversion of phenol into sodium phenolate is an acid-base reaction. Phenol, with an acidic hydrogen in the hydroxyl group, reacts with a base, sodium hydroxide, to give a salt, sodium phenolate, and water. The reaction of salicyclic acid to form oil of wintergreen is an esterification. The organic acid reacts with an alcohol in the presence of a strong mineral acid to produce an ester, methyl salicylate, and water.

An important group of useful products from aromatic organic reactions is comprised of the sulfa drugs. Sulfa drugs represent a group of compounds discovered in a conscious search for materials which could control bacterial infections in living animals. In 1904, the German chemist, Paul Ehrlich, realized that infectious diseases could be conquered if toxic chemicals could be found that attacked parasitic organisms rather than host cells. Ehrlich achieved some success toward his goal—the use of dyes against African sleeping sickness and arsenic compounds against syphilis. The mass outbreak of typhus during World War I, the loss of many wounded due to secondary bacterial infection, and the great influenza epidemic of 1917–1918 prepared the medical world for discoveries that might eradicate infectious diseases.

After experimenting with several drugs, Gerhard Domagk, a pathologist in the I. G. Farbenindustrie Laboratories in Germany, found, in 1935, that prontosil, a coloring matter or dye, was active against bacterial infection in mice. Prontosil as such is not effective in killing bacteria, but it hydrolyzes into sulfanilamide, which is effective.

This discovery led to the synthesis and testing of a large number of related compounds in the search for drugs which were more effective or less toxic to the infected animal. By 1964, more than 5000 sulfa drugs had been prepared and tested. Some of the more effective ones are listed in Table 15–4. Sulfanilamide was first prepared by Schroeter in 1906; Fritz Mietzsch and Joseph Klarer, chemists, and Gerhard Domagk, a pathologist, are credited with the synthesis of several other sulfonamides (many of these were later known as sulfa drugs) in 1931.

TABLE 15-4

PRODUCTION OF SULFA DRUGS AND ANTIBIOTICS PER YEAR IN THOUSANDS OF POUNDS

	1942	1943	1946	1952	1956	1966	1968
Sulfa Drugs	5,435	10,006	5,104	5,786	3,817	5,450	4,794
Antibiotics	0	0	38	1,487	1,967	9,652	

Sulfa drugs inhibit the growth of bacteria by preventing the synthesis of the essential vitamin, folic acid. In this process, the compound p-aminobenzoic acid

p-Aminobenzoic acid

307

is incorporated into a much larger molecule. The close structural similarity of sulfanilamide and p-aminobenzoic acid permits sulfanilamide to be incorporated into the enzymatic system instead of p-aminobenzoic acid. By bonding tightly, sulfanilamide shuts off the production of the essential folic acid, and the bacteria die. Not all bacteria are susceptible to sulfa drugs. However, the drugs are effective on streptococci, staphylococci, many gram-negative and gram-positive bacteria, and protozoa such as coccidia.

During World War II sulfa drugs, along with some of their relatives, the antibiotics, were used with particular effectiveness to prevent infection. Penicillin and other antibiotics, streptomycin, bacitracin, chloromycetin, aureomycin, and terramycin have partially replaced the sulfa drugs as agents against infection

TABLE 15-5   A FEW OF THE MORE EFFECTIVE SULFA DRUGS

Name	Formula
Sulfanilamide	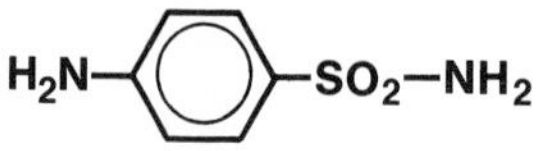
Sulfapyridine	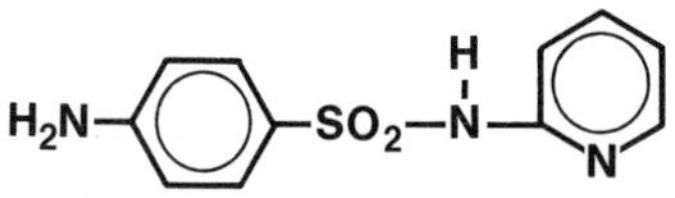
Sulfadiazine	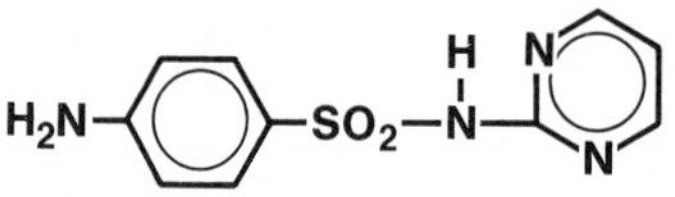
Sulfathiazole	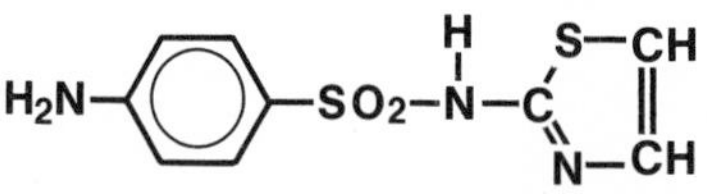
Sulfaguanidine	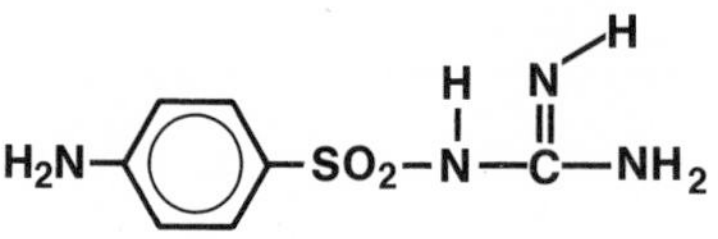
4,4'-diamino-diphenylsulfone	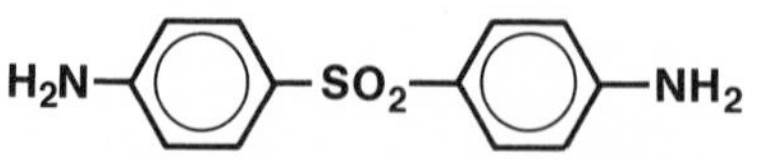
Prontosil	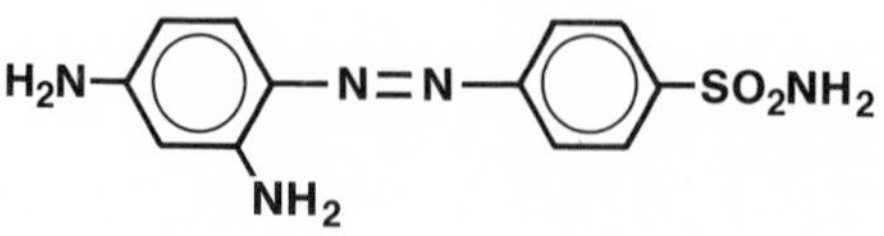

(Table 15–5). Antibiotics will be discussed in detail in Chapter 22. The yearly production of about 5 million pounds of sulfa drugs is explained by the relatively low cost of sulfa drugs, their widespread use in veterinary medicine, their general effectiveness, and their superiority over antibiotics in treating some diseases such as infectious diseases of the urinary tract.

Many sulfa drugs are derivatives of sulfanilamide, which is prepared from chlorobenzene as summarized in Figure 15–7. This synthesis, beginning with acetanilide, is commonly repeated in school laboratories. Great care must be exercised in handling the chlorosulfonic acid, ClSO₃H, since it is very corrosive and can cause severe burns. The synthesis employs an often-used technique in organic synthesis. The amine group in aniline is first "blocked" with an acetyl group prior to adding the chlorosulfonic acid. Blocking of the —NH₂ group is necessary to prevent its reaction with chlorosulfonic acid. With the NH₂ blocked, the acid reacts at the para position of the ring, substituting a —SO₂Cl group for a hydrogen; the hydrogen reacts with an OH of the chlorosulfonic acid to form a molecule of water as a coproduct. The technique of blocking one group while another is attacked is necessary in many organic syntheses.

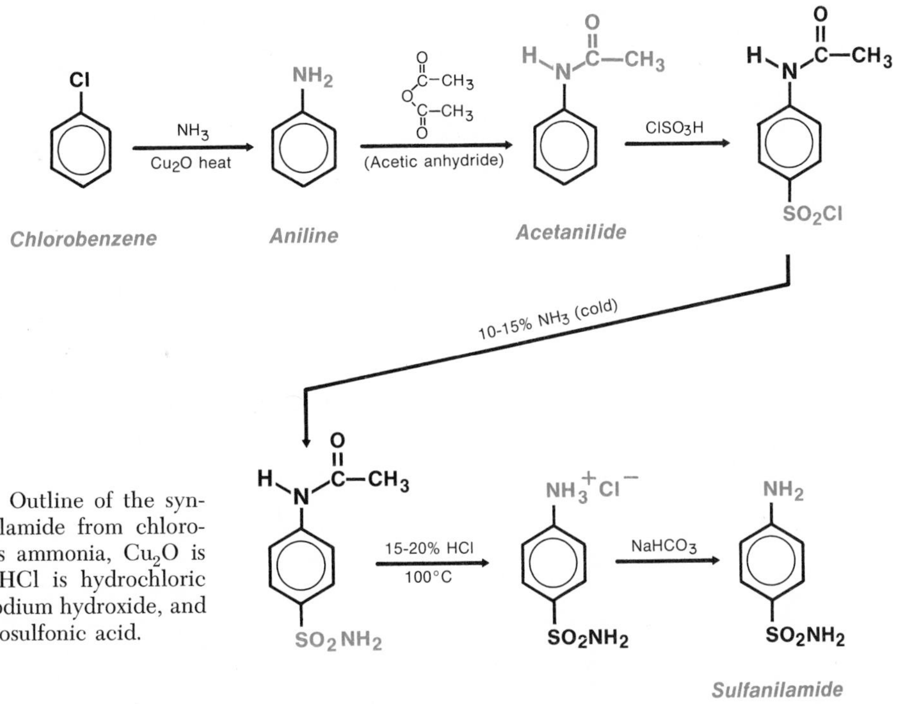

FIGURE 15-7 Outline of the synthesis of sulfanilamide from chlorobenzene. NH₃ is ammonia, Cu₂O is cuprous oxide, HCl is hydrochloric acid, NaOH is sodium hydroxide, and ClSO₃H is chlorosulfonic acid.

ORGANIC DYES

Until the middle of the last century all of the dyes available to man came from natural sources. Most of these were vegetable extracts of one sort or another; a few were animal products. The range of colors was limited as was the utility of the dyes. If a natural dye did not have the chemical groupings necessary to react with the chemical groupings of a particular fabric, the fabric could not be dyed with that material.

The history of natural dyes is very interesting. Egyptian mummies have been found wrapped in cloth dyed with juice from the madder plant. Alexander the Great is supposed to have deceived the Persians into thinking that his army was badly wounded by splashing his soldiers with a red dye, probably madder juice. The dye in madder juice is the compound, alizarin. Dark blue indigo dye has been known for over 4000 years. When the Romans invaded England, they found it inhabited by the Picts, who both tattooed and painted themselves with indigo. The word *Briton* is apparently the Latin form of a Celtic word meaning painted men. A legend recorded on coins of Tyre attests that Hercules discovered Tyrian purple (royal purple) when his dog bit a snail which stained the dog's jaws purple. Mark Anthony's flight from the crucial naval battle of Actium was especially conspicuous because he fled in Cleopatra's barge, which had sails of royal purple. The value of dyes in the ancient world was indicated by the fact that Cortez extorted cochineal, a crimson dyestuff, as well as gold from the Aztecs.

The first useful synthetic dye was prepared in 1856 by William Perkin, a young English chemist. Only 18 years old at the time, Perkin prepared mauve in his small home laboratory by the oxidation of aniline. Charmed by the purple color, he tested it as a dye and found that it resisted both sunshine and soap. Perkin called the dye aniline purple, but it became known popularly as mauve. Perkin's discovery clearly showed the possibilities of synthetic dyes, and within a few years a large number of new dyes were developed. Well over 100,000 dyes of various sorts have been prepared, and quite often new ones must be developed before new polymeric fibers can be dyed. Of those developed and tested, perhaps 2000 or so are in production at various chemical plants throughout the world.

The basic problem in the synthesis of a dye is to prepare a molecule which will absorb some wavelengths from white light and reflect back a desired color. Certain atomic groupings must be present in aromatic dyes to absorb light and produce colored compounds. These groups are called chromophores (Greek: *chroma* = color + *phoros* = bearer); some are listed in Table 15–6.

The absorption of certain wavelengths of light by the double bonds in chromophores causes the coloration in these compounds. The electrons in a double bond can absorb certain small quanta of energy and be promoted to a higher energy level. The particular wavelengths absorbed are extracted from the incoming light, and the reflected light is the color of the combined unabsorbed wavelengths (Figure 15–8). If only one double bond is present in a molecule, the electrons are held too tightly, and high-energy ultraviolet light

TABLE 15-6   CHROMOPHORES

—N=N—	Azo group	C=C	Ethylenic group
C=O	Carbonyl group	—N(=O)O	Nitro group
C=S	Thiocarbonyl group		
Quinoid group	Quinoid group	—C=C—C=C—	Conjugated group

310

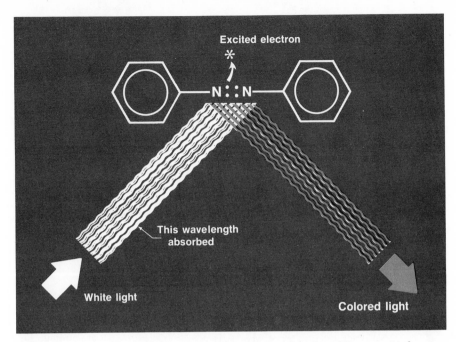

FIGURE 15-8   An explanation for the color of aromatic dyes. Electrons in the conjugated system absorb certain wavelengths of light and reflect the rest. The assortment of wavelengths reflected determine the color of the dye. A wavelength is absorbed only if that amount of energy will take an electron to an existing higher energy level.

is necessary to excite them. If, however, a chromophore is bonded to an aromatic system or to another conjugated system (alternating double and single bonds which lead to a pi system of electrons), the electrons move more freely, excite more easily, and absorb in the lower energy, visible light range of the spectrum. For example, CH_3—N=N—CH_3 has only a double bond, is not conjugated, and absorbs ultraviolet light at 3400 Angstroms. This compound is colorless in visible light. When the azo group is attached to two benzene molecules, two "racetracks" of electrons, the energy required to excite an electron to a higher-energy orbital is decreased; bluish light is absorbed. Azobenzene, ⬡—N=N—⬡, is orange and absorbs light in the visible region at 4450 Angstroms.

In addition to chromophores, dyes must also contain reactive groups that allow them to form bonds with the fiber molecule. These groups, when attached to the conjugated system, often increase the wavelength at which the dye absorbs light. These anchoring and color intensifying groups are called auxochromes (Latin: *auxilium* = aid); some are listed in Table 15-7.

Dyes are synthesized by incorporating appropriate combinations of chro-

TABLE 15-7   AUXOCHROMES

—OH Phenolic hydroxyl group
—NH_2 Amino group
—COOH Carboxyl group
—SO_3H Sulfonic acid group

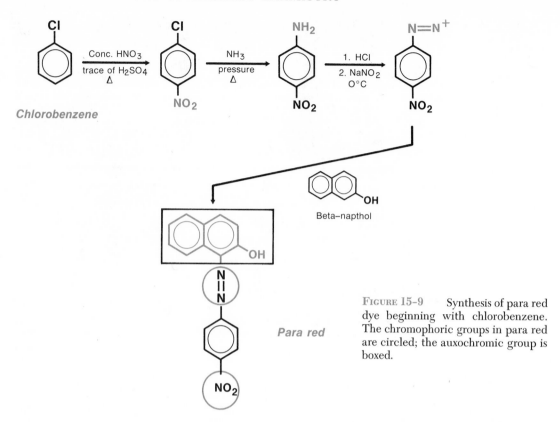

Beta–napthol

FIGURE 15-9   Synthesis of para red dye beginning with chlorobenzene. The chromophoric groups in para red are circled; the auxochromic group is boxed.

mophores and auxochromes into a molecule. An example of dye synthesis is the preparation of a red dye, para red, from chlorobenzene (Figure 15–9). Note in the structure of para red that the azo and nitro chromophoric groups and the phenolic hydroxyl auxochromic group are bonded to the conjugated ring systems, the structural pattern necessary to produce an aromatic dye.

Para red is used to dye cotton by a technique known as ingrain dyeing. In this technique the last step of the synthesis is carried out directly on the fiber rather than by synthesizing the dye first and then trying to apply it to the fiber. This is an effective dyeing technique because smaller molecules can penetrate crevices in the fiber which might be very difficult for a large molecule to enter. After the synthesis, the large molecules are trapped where they are formed, like a ship in a bottle. Besides the physical entrapment of the dye molecule, cotton holds onto para red by means of hydrogen bonds. In addition to some other methods of dyeing, this technique is illustrated in Figure 15–10.

Direct dyeing is a chemical reaction in which a reactive part of the dye molecule, usually an acidic or basic group, reacts directly with some group on the fiber itself. Wool and silk are particularly easy to dye in this way since they are composed of proteins (Chapter 17) with available amino (—NH₂) and carboxyl (—COOH) groups. The amino groups, because they are basic, will react with and bind to the fiber any dye that has an acidic group. Likewise the carboxyl groups, because they are acidic, will react with and bind any dye that has a basic group. Picric acid will dye wool and silk because it has an acidic phenolic hydrogen which reacts with the amino groups of wool and silk, as shown in Figure 15–10.

When a fiber is relatively unreactive to dyes, it can often be dyed by vat dyeing. In this process the cloth is immersed in a vat of the soluble dye, and the dye penetrates into the inner crevices of the fiber. While imbedded in the fibers, the dye is rendered insoluble to prevent it from being easily washed from the cloth. This method is particularly useful for dyeing cotton; the chemistry of this technique is illustrated with indigo in Figure 15–10.

Some dyes do not adhere very well to a fiber unless it is first treated with a substance known as a mordant (Latin: *mordere* = to bite). Mordants interact with both the fiber and the dye, forming a link between them. Metal oxides are often used as mordants, and anyone who has tried to remove an iron rust stain from cotton can testify to the strong adherance between the two. Reactive centers in a dye, suitable for causing a reaction with mordants, are the acidic carboxyl and phenolic hydroxyl groups. Alizarin with phenolic hydroxyl groups will dye cotton that has previously been treated with a metallic oxide, as shown in Figure 15–10. The mordant frequently changes the color of the dye since it reacts chemically with it. For example, alizarin is blood-red, but it gives violet colors with iron mordants and brownish red colors with chromium mordants.

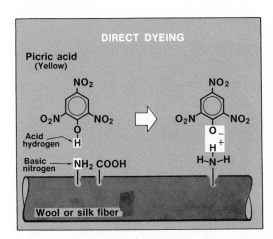

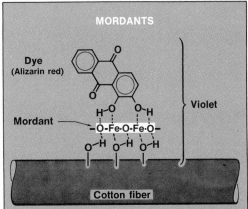

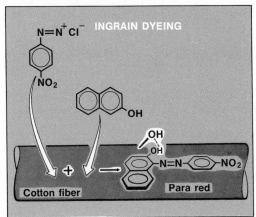

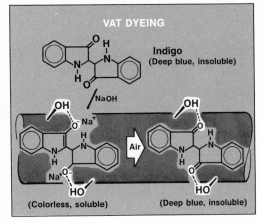

FIGURE 15–10   Some examples of some methods of dyeing cloth. Details are given in the text.

313

CONCLUSION AND PREVIEW

The many applications of organic chemistry are largely an outgrowth of our understanding of the types of reactions undergone by relatively simple molecules. The major sources of organic chemicals used on a large scale are petroleum and natural gas. Ethyl alcohol, acetic acid, and some other simple compounds are prepared by the fermentation of carbohydrates. Soaps and synthetic detergents are examples of useful compounds consisting of long chains of carbon atoms with functional groups on one end. The variety of useful products which can be prepared from benzene and its derivatives is illustrated by the synthesis of phenol, aspirin, sulfanilamide, and para red. Chemists are working steadily to find new, useful compounds and shorter, more efficient synthetic methods.

The incorporation of color producing (chromophore) groups in organic molecules yields dyes when the molecules also contain groups that form bonds with fiber molecules (auxochromes). In the next chapter we will consider another widely used class of organic compounds—the polymers.

QUESTIONS

1. Would you expect $CH_3CH_2CH_2CH_2CH_2CH_2CH_2CH_2CH_3$ to be an important useful constituent of bottled gas? Explain your answer. Would it be useful in gasoline? Explain.

2. What products would you expect to be formed in the reaction of the following:

   (a) methanol and acetic acid
   (b) l-propanol and stearic acid
   (c) ethylene glycol and acetic acid

3. Explain how hydrogen bonding could play a significant role in fixing the boiling point of acetic acid.

4. Write structural formulas for the possible alcohols with the composition C_4H_9OH.

5. Draw a structural formula for each of the following:

   (a) an alcohol
   (b) an organic acid
   (c) an ester
   (d) glycerin

6. What is meant by

   (a) proof rating of an alcohol
   (b) octane rating of a gasoline
   (c) denatured alcohol

7. Beginning with petroleum, outline the steps and write out the chemical equations for the production of ethyl acetate.

8. How does a soap cleanse?

9. Discuss the fundamental characteristics of the following:

   (a) synthetic detergents
   (b) soaps

10. Which would you expect to boil at a higher temperature, $CH_3CH_2CH_2CH_3$ or CH_3CH_2OH? Why?

314

11. Why will the addition of strong alkali aid in unstopping a greasy sink drain?

12. Draw structural formulas for

    (a) aspirin
    (b) oil of wintergreen
    (c) phenol
    (d) aniline

13. What chemical reactions can be used to distinguish between

    (a) C_2H_5OH and CH_3COOH
    (b) CH_3NH_3 and CH_3OH
    (c) $CH_3COOC_2H_5$ and CH_3COOH

14. Tell how you could prepare the following compounds:

    (a) chlorobenzene
    (b) urea
    (c) acetanilide

15. Describe the types of binding between dye and wool fibers when dyed by the direct dyeing method.

16. Suppose in the synthesis of aspirin from chlorobenzene each step produces a 50 per cent yield. What fraction of an original amount of starting material would actually remain as a final product?

17. Describe three ways a dye is held on a fiber.

18. A gaseous compound which is 85.6 per cent C and 14.4 per cent H by weight reacts with water to form a new compound which has the composition of 52.1 per cent C, 34.8 per cent O, and 13.1 per cent H. What would be the properties of this new compound?

19. Antifreezes work on a principle that the freezing point of a solution is lowered owing to the presence of dissolved molecules. It is the number of dissolved molecules relative to the number of molecules of solvent which is of primary importance. Why does it take 62.07 grams of ethylene glycol and only 32.04 grams of methanol to lower the freezing point of a sample of water by a given amount?

20. Why should some synthetic detergents not be utilized by bacteria in surface waters while fatty acid soaps are easily broken down into simpler molecules?

21. Pick a product of organic synthesis which has widespread use and is generally considered harmful to the environment; determine its raw materials and synthesis steps. Are there any useful products produced along the way to the final product?

22. What volume of 1 M NaOH is needed to transform 1 gram of acetic acid into sodium acetate?

SUGGESTIONS FOR FURTHER READING

Campaigne, E., "Wöhler and the Overthrow of Vitalism," *Journal of Chemical Education*, Vol. 32, No. 8, p. 403 (1955).

Collier, H. O., "Aspirin," *Scientific American*, Vol. 209, No. 5, p. 97 (1963).

Ferguson, L. N., "Hydrogen Bonding and the Physical Properties of Substances," *Journal of Chemical Education*, Vol. 33, No. 6, p. 267 (1956).

Greenbery, L. A., "Alcohol in the Body," *Scientific American*, Vol. 186, No. 6, p. 86 (1953).

Juster, N. J., "Color and Chemical Constitution," *Journal of Chemical Education*, Vol. 39, No. 11, p. 596 (1962).

Rossini, F. D., "Hydrocarbons in Petroleum," *Journal of Chemical Education*, Vol. 37, No. 11, p. 554 (1960).

Sates, M., "Analgesic Drugs," *Scientific American*, Vol. 215, No. 5, p. 131 (1966).

Schaar, B. E., "Aniline Dyes," *Chemistry*, Vol. 39, No. 1, p. 12 (1966).

MAN-MADE GIANT MOLECULES—THE SYNTHETIC POLYMERS

CHAPTER 16 ━━━━━━━━━━━━━━━━

It is impossible for most people to get through a day without using a dozen or more materials which were completely unknown 30 to 50 years ago. Many of these materials are plastics of one sort or another. Examples include plastic dishes and cups, combs, automobile steering wheels and seat covers, telephones, pens, plastic bags for foods and wastes, plastic pipes and fittings, plastic water-dispersed paints, false eyelashes and wigs, a wide range of synthetic fibers (suits and stockings), glues, and flooring materials. In fact these materials are so widely used that they are usually taken for granted. We rarely wonder where they come from or how man has achieved the ability to produce materials of this sort for various special uses. All these materials are composed of *giant molecules*. The purpose of this chapter is to examine some of the chemistry and chemical technology that lie behind this flood of indispensable plastic objects.

WHAT ARE GIANT MOLECULES?

We recognize that a molecule is a grouping of atoms held together by chemical bonds of the covalent type. Most of the molecules studied in this chapter are derived from small organic molecules containing less than two dozen atoms per molecule. Some examples are shown in Figure 16–1.

Giant molecules are formed by the bonding together of smaller molecules in very large numbers. They do this by interacting to form covalent bonds to each other, thus forming a giant molecule. An example is styrene, which can polymerize (react with itself) as shown in Figure 16–2.

The basic feature of these molecules is that chemical bonds join together hundreds of atoms into a single unit. Plastics consist of aggregates of molecules with very high molecular weights; most of their properties are determined by this fact. By the 1880's, techniques had been developed that enabled chemists

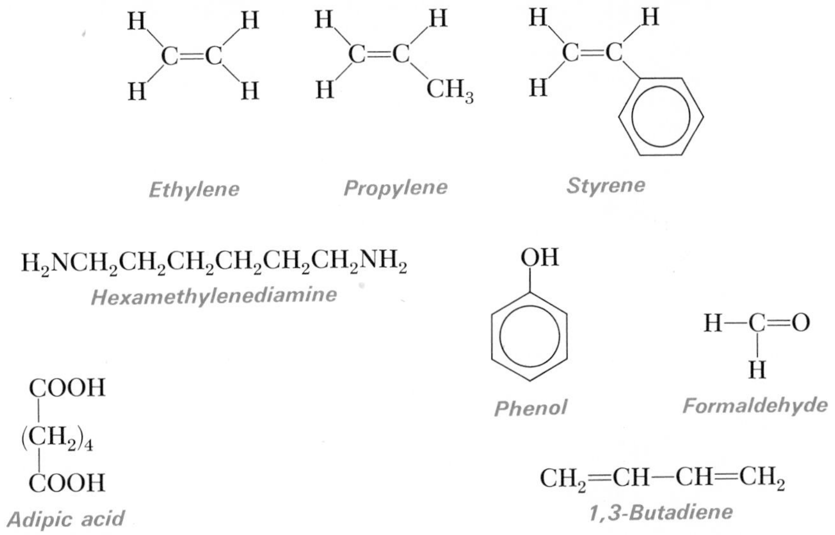

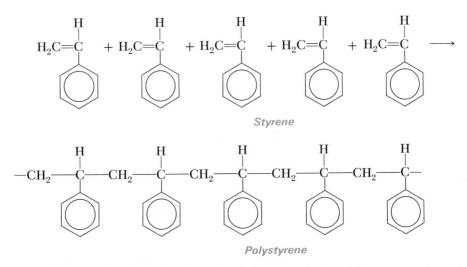

Ethylene   Propylene   Styrene

H₂NCH₂CH₂CH₂CH₂CH₂CH₂NH₂
Hexamethylenediamine

COOH
|
(CH₂)₄
|
COOH
Adipic acid

OH

Phenol

H—C=O
|
H
Formaldehyde

$CH_2=CH-CH=CH_2$
1,3-Butadiene

FIGURE 16–1   A few examples of small molecules used in the synthesis of giant molecules.

to determine the approximate molecular weights of rubber, cellulose, proteins, and some other materials of very high molecular weights. One of these methods, involving the change in the freezing point of water, is illustrated in Figure 16–3. The basic idea is quite simple: the freezing point of water is depressed if other molecules are dissolved in it, and the depression of the freezing point is directly proportional to the number of molecules (or moles) of solute added, as long as their total concentration is not too high. One mole of any solute dissolved in 1000 grams of water depresses the freezing point by 1.86°C. The results of studies using various methods for the determination of molecular weight have led to values of about 12,000 for rubber; cellulose-like materials such as starch have molecular weights of 40,000; and the molecular weights of proteins range from

Styrene

Polystyrene

FIGURE 16–2   A portion of a giant molecule, polystyrene, formed by addition of styrene monomer units.

317

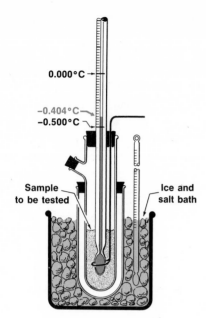

0.000°C →

−0.404°C
−0.500°C

Sample
to be tested

Ice and
salt bath

FIGURE 16–3   Molecular weights of solutes can be determined by the lowering of the freezing point below that of the pure solvent. In this example, water, which freezes at 0°C, dissolves a solute, and the resulting solution freezes at −0.404°C. A salt-ice bath is used in this instance to cool the sample.

a thousand up to a million. These are very large molecules compared to water (molecular weight = 18), ethylene (molecular weight = 28), or even sucrose (sugar) (molecular weight = 342).

Many chemists were reluctant to accept the concept of giant molecules, but in the 1920's, one persistent German chemist, H. Staudinger, championed the idea and introduced a new term, *macromolecule*, for these giant molecules. Staudinger devised experiments that yielded accurate molecular weights, and, in addition, he synthesized "model compounds" to test his theory. One of his first model compounds was prepared from styrene.

$$H_2C{=}CH$$

Styrene

Under the proper conditions styrene molecules undergo a *polymerization* reaction to yield polystyrene, a giant molecule. The word *polymer* means "many parts" (Greek: *poly* = many, *meros* = parts). The styrene molecule is a monomer (Greek: *mono* = one), the recurring unit in the giant molecule (Figure 16–2).

The macromolecule polystyrene is represented as a long chain of monomer units bonded to each other. The polymer chain is not an endless one; some polystyrenes made by Staudinger were found to have molecular weights of about 600,000, corresponding to a chain of about 12,000 atoms. The polymer chain can be indicated as

$$R{-}CH_2{-}CH{-}\!\!\left(CH_2{-}CH\right)_n CH_2{-}CH{-}R$$

318

where R represents some terminal group, such as an impurity, and n is a large number.

Polystyrene is a clear, hard, water-white solid at room temperature. Since it can be molded easily at 250°C, the term *plastic* has become associated with it and similar materials. Polystyrene has so many useful properties that its commercial production, which began in Germany in 1929, today exceeds 2.5 billion pounds per year.

There are two broad categories of plastics. One, when heated *repeatedly*, will soften and flow; when it is cooled, it hardens. Materials which undergo such changes when heated and cooled are called *thermoplastics*. Polystyrene is one example of these common plastics. The other category consists of materials which are plastic when first heated, but when heated further, they form a set of interlocking bonds which makes them harden as they are cooled. When reheated, they cannot be softened and re-formed without extensive degradation. These materials are called *thermosetting plastics* and include materials such as a phenol-formaldehyde polymer (Bakelite), which is described later in this chapter. Thermosetting plastics are used extensively in glues and special applications but have been displaced from many of their previous uses by the more easily handled thermoplastics.

POLYMERIZATION PROCESSES

In order to get a better understanding of polymers it is necessary to look at representative examples of the different types of polymerization processes. These different types of polymerization are (1) addition polymerization reactions, in which small molecules with double bonds react to form bonds to each other, losing double bonds in the process; (2) condensation polymerization reactions, in which two small molecules react to form bonds to each other, splitting out a smaller molecule such as water; and (3) rearrangement polymerization reactions, in which two different molecules form bonds with each other by rearranging their bonding patterns. Generally one of the reacting species in such a process contains a double bond or a cyclic structure whose rearrangement permits the construction of a polymer.

Since addition polymerization is presently the most important industrial polymerization, we will examine several such reactions in detail. It must be remembered, however, that changes in technology can rapidly alter the relative attractiveness of two different industrial processes which give the same product. As a result, some of the reactions which are now carried out on a large scale may be replaced by more efficient processes in the near future.

ADDITION POLYMERIZATION

In the previous section it was noted that some polymers, such as polystyrene, are made by adding monomer to monomer to form a polymer chain of great length. Perhaps the easiest addition reactions to understand chemically are those

319

involving monomers containing double bonds. The simplest monomer of this group is ethylene, C_2H_4.

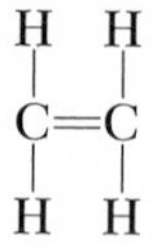

$$\begin{array}{cc} H & H \\ | & | \\ C & = C \\ | & | \\ H & H \end{array}$$

Ethylene

When ethylene is heated under pressure in the presence of oxygen, polymers with molecular weights of about 30,000 are formed. This polymer is a long chain hydrocarbon (Figure 16–4).

$$nCH_2 = CH_2 \xrightarrow[pressure]{O_2, \ heat} \cdots -CH_2-CH_2-CH_2-CH_2-CH_2-CH_2 \cdots$$

FIGURE 16–4   A portion of the polyethylene polymer.

In order to enter into reaction, the double bond of ethylene must be broken. This forms reactive sites at either end:

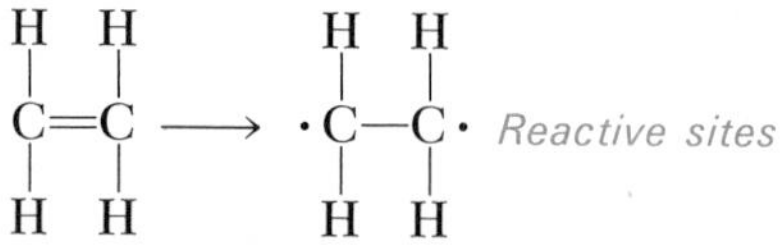

$$\begin{array}{cc} H & H \\ | & | \\ C = C \\ | & | \\ H & H \end{array} \longrightarrow \begin{array}{cc} H & H \\ | & | \\ \cdot C - C \cdot \\ | & | \\ H & H \end{array} \ \textit{Reactive sites}$$

The partial breaking of the double bond (also called initiation) can be accomplished by physical means such as heat, ultraviolet light, x-rays, and high energy electrons. The initiation of the polymerization reaction can also be accomplished with chemical initiators such as organic peroxides. These initiators, which are very unstable, break down to yield free radicals which react readily with the molecule containing the double bond. Benzoyl peroxide is one commonly used free radical initiator.

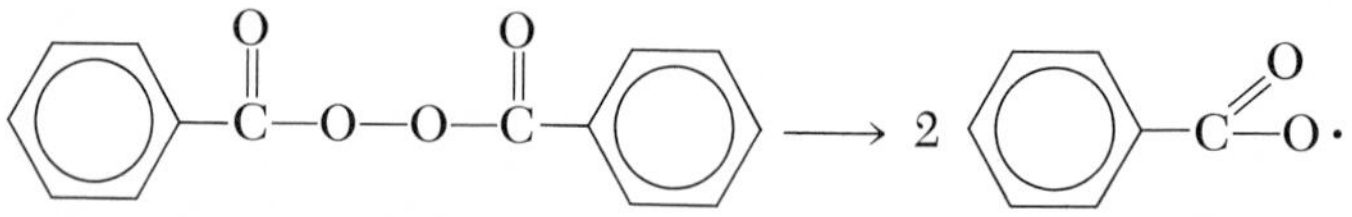

Benzoyl peroxide   *(A free radical, R·)*

It readily decomposes to form a free radical (R·):

$$\text{Peroxide} \longrightarrow R\cdot$$

$$R\cdot + CH_2 = CH_2 \longrightarrow R - CH_2 - CH_2 \cdot$$
Ethylene   *(Another free radical)*

The polymer grows as the free radical reacts with other ethylene molecules to form a long hydrocarbon chain:

$$R - CH_2 - CH_2 \cdot + CH_2 = CH_2 \longrightarrow R - CH_2 - CH_2 - CH_2 - CH_2 \cdot$$

$$R - CH_2 - CH_2 - CH_2 - CH_2 \cdot + CH_2 = CH_2 \longrightarrow$$

$$R - CH_2 - CH_2 - CH_2 - CH_2 - CH_2 - CH_2 \cdot$$

Some time after the chain has begun to form, *termination* of the polymerization process occurs. Since several initiation reactions were started at about the same time, after a little while many long chains exist in the reaction mixture. Occasionally, two long chains may meet and link up their reactive sites.

$$R-CH_2-CH_2-(CH_2-CH_2)_n-CH_2-CH_2 \cdot + \cdot CH_2-CH_2-(CH_2-CH_2)_m-R$$

$$\longrightarrow R-CH_2-CH_2-(CH_2-CH_2)_n-CH_2-CH_2-CH_2-CH_2-(CH_2-CH_2)_m-R$$

(In this example n and m are large, and probably different, numbers.)

In addition to this process, the initiator free radicals that were not used up in the initiation process begin to terminate the build-up of certain polymer chains.

$$R-(CH_2-CH_2)_n-CH_2-CH_2 \cdot + \cdot R \longrightarrow R-(CH_2-CH_2)_n-CH_2-CH_2-R$$

Polyethylenes formed under various pressures and catalytic conditions have different physical properties owing to different molecular structures. For example, the catalyst, chromium oxide, yields almost exclusively the linear polyethylene shown previously. If ethylene is heated to 230°C at a pressure of 200 atm, irregular branches result. Under these conditions, free radicals undoubtedly attack the chain at random positions, thus causing the irregular branching.

$$-CH_2-CH_2-CH_2-CH_2 \longrightarrow -CH_2-CH-CH_2-CH_2- + H\cdot$$

$$RCH_2CH_2\cdot \nearrow$$

$$\begin{array}{c} CH_2 \\ | \\ CH_2 \\ | \\ R \end{array}$$

The molecules in linear polyethylene can line up with one another very easily, yielding a tough, high-density compound which is useful in making toys, bottles, and so forth. The polyethylene with irregular branches is less dense, more flexible, and not nearly as tough as the linear polymer since the molecules are generally farther apart and their arrangement is not as precisely ordered. This material is used for squeeze bottles and other similar applications. The structural differences show up in the infrared spectra of the polymers, and we can confirm our hypothesis about cross-linking from such spectra.

Addition polymerization is interesting in terms of *bond breaking* and *bond making*. Consider that when two atoms joined by a sigma and a pi bond in ethylene react to form a polymer, both atoms in the end form only sigma bonds. Since more energy is released when a sigma bond is formed than is lost when a pi bond is disrupted, a result of the polymerization of ethylene is the release of heat. This can be a problem when the reaction is carried out on a large scale since too much heat can cause a runaway reaction and an explosion. To offset this problem, it is necessary to cool the reaction vessel with water.

The problem of heat generation during addition polymerization reactions of other monomers can be dealt with in a similar fashion. For example, vinyl chloride

$$\begin{array}{cc} H & H \\ | & | \\ C & = C \\ | & | \\ H & Cl \end{array}$$

is polymerized as a suspension in water (suspension polymerization); the water, because of its large heat capacity and proximity with the droplets of vinyl chloride, can easily take away the heat. The reaction vessels are huge 4000 gallon glass-lined steel pots. Since vinyl chloride is quite volatile (b.p. = −13.6°), the pressure in the vessel is relatively high until the reaction is near completion. When the pressure drops to a certain value, the reaction is stopped and the unreacted monomer is recycled. The polyvinyl chloride (PVC) is then centrifuged and washed. This product is usually less pure than polyethylene since the washing fails to remove all impurities.

One disadvantage to suspension polymerization is the time involved—about eight hours. This is because of the immiscibility of vinyl chloride and water, which means that the droplets are a little too large to mix effectively with the initiator. The problem is partially overcome by stabilizing the suspension of the monomer and water by the addition of a soap. The organic end of the polymer dissolves in the organic phase, and its polar end seeks the water phase. As a result, small droplets of the organic phase are emulsified in the water phase and stabilized by the soap molecules at the interface (emulsion polymerization). Such an emulsion can be polymerized much more rapidly.

Polyvinyl chloride which is to be used as an electrical insulator is generally not made by emulsion polymerization because the soap, which remains in the polymer as an impurity, decreases the insulating ability of the polymer. However, vinyl acetate

$$
\begin{array}{c}
\quad\quad\quad\quad\ \ \overset{\displaystyle O}{\overset{\displaystyle \|}{}} \\
H \quad O-C-CH_3 \\
|\quad\ | \\
C=C \\
|\quad\ | \\
H \quad H
\end{array}
$$

is generally polymerized by emulsion polymerization since most of the polyvinyl acetate made is used in making emulsion (latex or water-base) paints. At the end of the polymerization process the material is almost ready for use; only a plasticizer—a material which is added to make the polymer film more flexible—and a pigment need to be added. When the paint is applied to a surface, the water evaporates and leaves a polymer film containing the pigment and plasticizer.

There is a large group of derivatives of ethylene which undergo addition polymerization via other processes similar to those presented. Table 16–1 summarizes some pertinent information on these materials.

COPOLYMERIZATION

After examining Table 16–1, one might well wonder what would happen if a mixture of two monomers were polymerized. This type of reaction has been studied in detail and the products are called *copolymers*. If we polymerize pure monomer A, we get a homopolymer, poly A:

—AAAAAAAAA—

TABLE 16-1   ETHYLENE DERIVATIVES WHICH UNDERGO ADDITION POLYMERIZATION

FORMULA	MONOMER NAME	POLYMER NAME	USES
$CH_2{=}CH_2$	Ethylene	Polyethylene	coats, milk cartons, wire insulation, bread boxes
$CH_2{=}CHCl$	Vinyl chloride	Polyvinyl chloride (PVC)	as a vinyl acetate copolymer in phonograph records, credit cards, rain wear
$CH_2{=}CH$ $\|$ C_6H_5	Vinyl benzene (Styrene)	Polystyrene	combs
$CH_2{=}CH$ $\|$ $O{-}C{-}CH_3$ $\|\|$ O	Vinyl acetate	Polyvinylacetate	latex paint
$CH_2{=}CH$ $\|$ CN	Acrylonitrile	Polyacrylonitrile (PAN)	rug fibers
$CH_2{=}CH{-}CH{=}CH_2$	Divinyl (1,3-Butadiene)	BUNA rubbers	tires and hoses
$CH_2{=}C{-}CH_3$ $\|$ $O{-}C{-}CH_3$ $\|\|$ O	Methyl methacrylate	Polymethyl methacrylate (Plexiglas, Lucrite)	transparent objects, lightweight "pipes"
$CF_2{=}CF_2$	Tetrafluoroethylene	Polytetrafluoroethylene (TFE) (Teflon)	Insulation, bearings, nonstick fry pan surfaces

Likewise, if pure monomer B is polymerized, we get a homopolymer, poly B:

—BBBBBBBBBB—

If the monomers A and B are mixed and then polymerized, we get a copolymer, of which the following are examples:

—ABBABAAABB—

—AABABABABB—

—BABABBAABA—

In such polymers the order of the units is often completely random, but the properties of the copolymer will be determined by the ratio of A to B. An example involving vinyl acetate and vinyl chloride would be as follows:

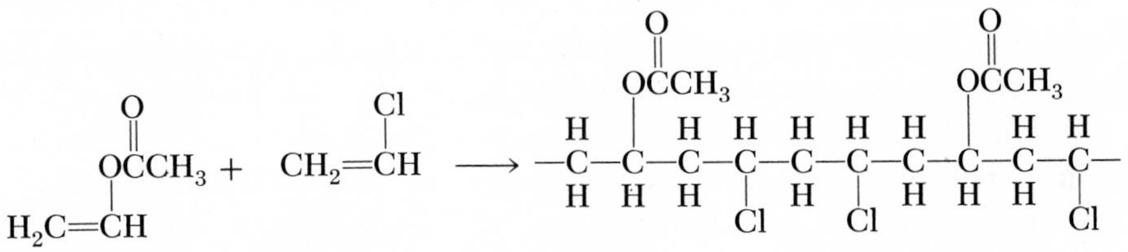

Vinyl acetate   *Vinyl chloride*   *Copolymer*

323

A copolymer can have useful properties that are different from and often superior to those of the polymers made of its pure constituents. An example is the synthetic rubber known as SBR, which is a styrene-butadiene copolymer with a ratio of styrene to butadiene of about 1 to 3:

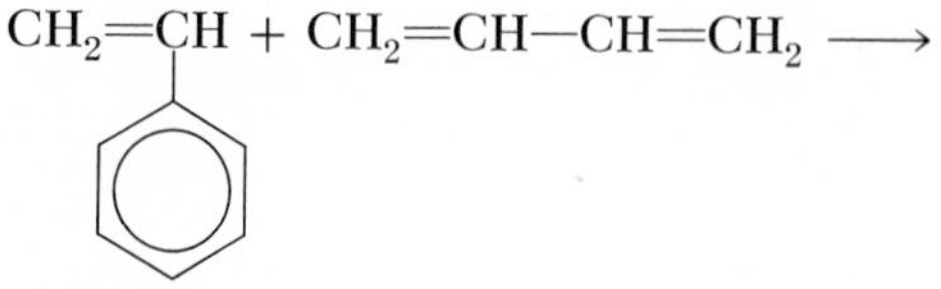

$$CH_2{=}CH + CH_2{=}CH{-}CH{=}CH_2 \longrightarrow$$

Styrene   *Butadiene*

$$-CH_2{-}CH{=}CH{-}CH_2{-}CH_2{-}\overset{H}{C}{-}CH_2{-}CH{=}CH{-}CH_2{-}CH_2{-}CH{=}CH{-}CH_2{-}$$

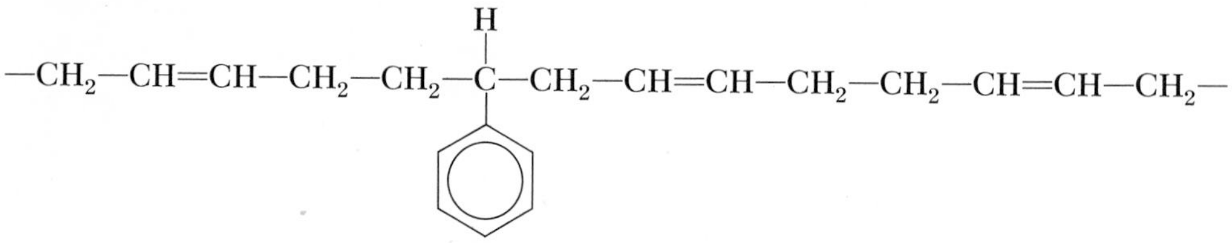

SBR copolymer (Styrene-butadiene rubber)

The double bonds in the copolymer are used in the vulcanization process (discussed later) to help form bonds between adjacent copolymer chains. About 2 million tons of SBR is used annually in the manufacture of tires.

CONDENSATION POLYMERIZATION

A chemical reaction in which two molecules react by splitting out or eliminating a smaller molecule is called a condensation reaction. For example, acetic acid and aniline form an amide by the elimination of water:

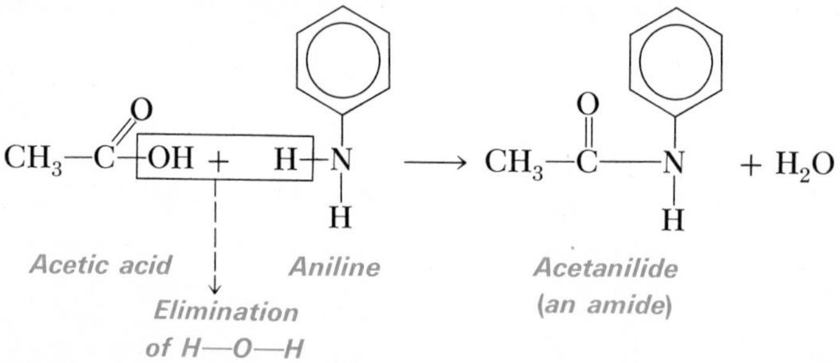

$$CH_3{-}C{-}OH + H{-}N \longrightarrow CH_3{-}C{-}N + H_2O$$

Acetic acid   *Aniline*   *Acetanilide (an amide)*

Elimination of H—O—H

This important type of chemical reaction does not depend upon the presence of a double bond in the reacting molecules. Rather, it requires the presence of two kinds of functional groups on two different molecules. Reaction results in the formation of a bond between the molecules (condensation), accompanied by the splitting out of a small molecule such as water. If each reacting molecule has two functional groups, both of which can react, it is then possible for condensation reactions to lead to a polymer. If we take a molecule with two carboxyl groups, such as terephthalic acid, and another molecule with two alcohol groups, such as ethylene glycol, each molecule can react at both ends. Reacting a molecule of terephthalic acid with a dialcohol such as ethylene glycol initially produces an ester molecule with an acid group on one end and an alcohol group on the other.

324

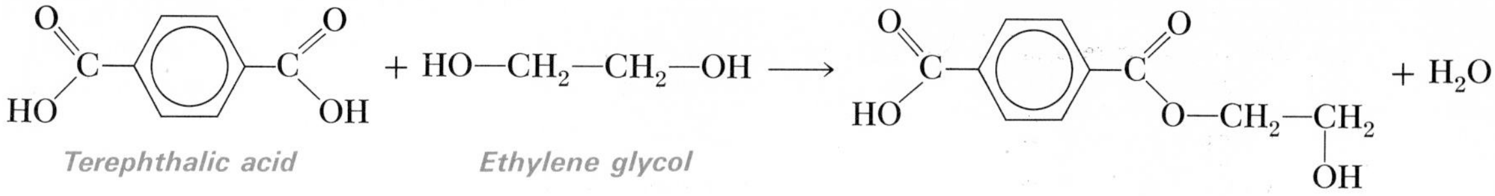

Terephthalic acid   Ethylene glycol

(An ester)

Subsequently, the remaining acid group can react with another alcohol group, and the alcohol group can react with another acid molecule. The process continues until an extremely large polymer molecule, known as a *polyester*, is produced; it has a molecular weight of about 10,000 to 20,000.

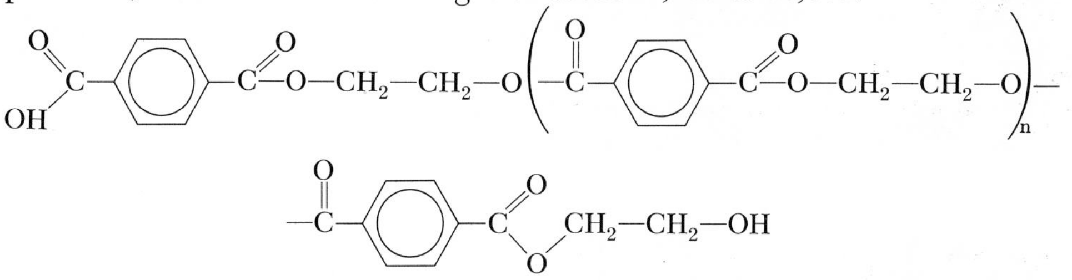

Poly(ethylene glycol terephthalate)

Poly(ethylene glycol terephthalate) is used in making textile fibers marketed under such names as "Dacron" and "Terylene," and films such as "Mylar." The film material has unusual strength and can be rolled into sheets one thirteenth the thickness of a human hair. In film form, this polyester is often used as a base for magnetic recording tape and for frozen food packaging.

In addition to the many polyesters made from difunctional acids and alcohols, some very useful polymers can be prepared from trifunctional alcohols and difunctional acids. For example, glycerol contains three alcohol groups, and orthophthalic acid two acid groups. When reacted together, the presence of the unreacted carboxyl and alcohol groups makes a *cross-linking* of two polymer chains possible.

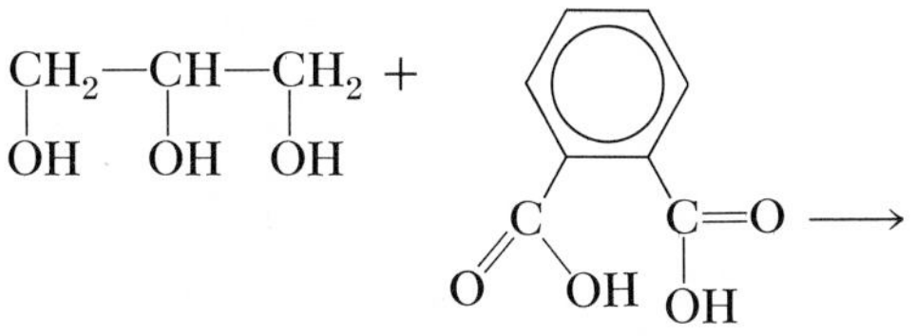

Gylcerol   Orthophthalic acid

The generalized pattern for cross-linking is shown in Figure 16–5.

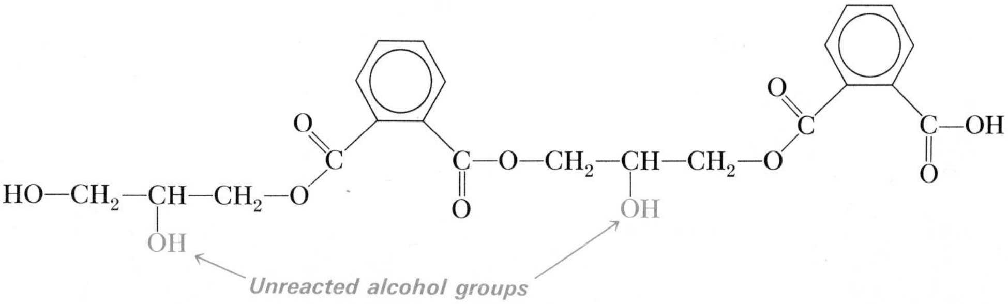

Unreacted alcohol groups

Cross-linked polymers of this type are known as *alkyd* resins and find uses in enamels and in the preparation of false teeth. Cross-linking polymer chains causes a drastic reduction in their solubilities.

325

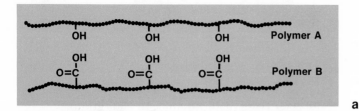

a

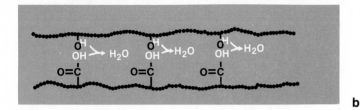

b

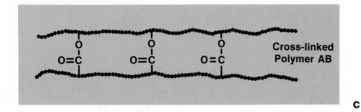

c

FIGURE 16-5   Cross-linking of two polymer chains.

POLYAMIDES (NYLONS)

Earlier we saw how an acid and an amine can condense to form an amide and eliminate a water molecule. Reactions of this type yield a group of polymers which perhaps have had a greater impact upon society than any other type. These are the polyamides, or *nylons*.

In 1928, the Du Pont Company embarked upon a program of basic research headed by Dr. Wallace Carothers, who had been hired from the Harvard University faculty. His research interests were high molecular weight compounds, such as rubber, proteins, resins, and the reaction mechanisms that produced these compounds. Carother's research group discovered many polymers that were suitable for commercial fibers. In February, 1935, the culmination of this research was a product known as nylon 66, prepared from adipic acid and hexamethylene diamine:

$$\underset{\text{HO}}{\overset{\text{O}}{\|}}{C}-(CH_2)_4-C\underset{\text{OH}}{\overset{\text{O}}{\|}} \quad + \quad H_2N-(CH_2)_6-NH_2 \longrightarrow$$

Adipic acid   *Hexamethylene diamine*

$$\underset{\text{HO}}{\overset{\text{O}}{\|}}{C}-(CH_2)_4-\boxed{\underset{\text{H}}{\overset{\text{O}}{\|}}{C}-N}-(CH_2)_6-\boxed{\underset{\text{H}}{\overset{\text{O}}{\|}}{N}-C}-(CH_2)_4-\boxed{\underset{\text{H}}{\overset{\text{O}}{\|}}{C}-N}-(CH_2)_6-\ +\ xH_2O$$

Nylon 66

This material could easily be extended into fibers that were stronger than natural fibers and more chemically inert. The discovery of nylon jolted the American textile industry at almost precisely the right time. Natural fibers were not

326

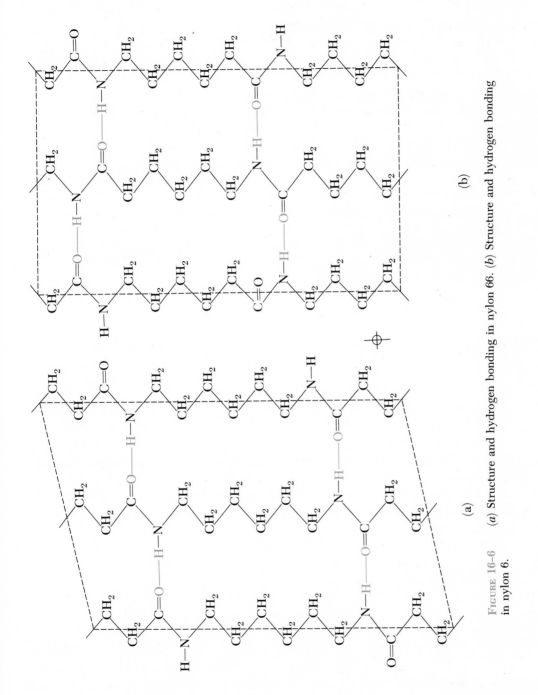

FIGURE 16-6   (a) Structure and hydrogen bonding in nylon 66. (b) Structure and hydrogen bonding in nylon 6.

327

meeting the needs of twentieth century Americans. Silk was not durable and was very expensive, wool was scratchy, linen crushed easily, and cotton did not lend itself to high fashion. As women's hem-lines rose in the mid-1930's, silk stockings were in great demand, but they were very expensive. Nylon changed all that almost overnight. It could be woven into the sheer hosiery women wanted, and it was much more durable than silk. The first public sale of nylon hose took place in Wilmington, Delaware (the hometown of Du Pont's main office), on October 24, 1939. They were so popular that they had to be rationed. World War II caused all commercial use of nylon to be abandoned until 1945. Not until 1952 was the nylon industry able to meet the demands of the hosiery industry and to release nylon for other uses as a fiber and as a thermoplastic.

Many nylons have been prepared and tried on the consumer market, but two, nylon 66 and nylon 6, have been most successful. Nylon 6 is prepared from caprolactam, which comes from aminocaproic acid. Notice how aminocaproic acid resembles a mixture of adipic acid and hexamethylenediamimine in that it contains an amine group on one end of the molecule and an acid group on the other end.

$$H_2N-(CH_2)_5-C\overset{O}{\underset{OH}{\diagdown}} \xrightarrow{-H_2O} (CH_2)_5 \diagup \overset{N-H}{\underset{C=O}{\diagdown}} \xrightarrow{polymerization} \textbf{\textit{Nylon 6}}$$

Aminocaproic acid   *Caprolactam (An intermediate)*

The basic structural linkage in nylon 6, the $-\overset{H}{\underset{}{N}}-\overset{}{\underset{O}{C}}-$ bond (amide bond), occurs every five carbon atoms and gives rise to a slightly more regular structure (Figure 16–6a) than that of nylon 66, in which six carbons intervene between two amide bonds, then four carbons, then six, and so on (Figure 16–6b).

Figure 16–6a and b illustrates another facet of the structure of *nylons—hydrogen bonding*. This type of bonding explains why the nylons make such good fibers. In order to have good tensile strength, the chains of atoms in a polymer should be able to attract one another but not so strongly that the plastic cannot be initially extended to form the fibers. Ordinary covalent chemical bonds linking the chains together would be too strong. Hydrogen bonds, with a strength about one-tenth that of an ordinary covalent bond, link the chains in the desired manner. We will see later that this type of bonding is also of great importance in protein structures.

PHENOL-FORMALDEHYDE RESINS

In 1871, a German chemist, Adolf von Bayer, mixed phenol and formaldehyde and observed that their reaction led to the formation of a hard, porous gray mass—a resin:

$$\text{OH} \quad \bigcirc \quad + \quad H-\overset{H}{\underset{}{C}}=O \quad \longrightarrow \quad \textit{Resin}$$

Phenol   *Formaldehyde*

Since this material had no properties that interested von Bayer and was impossible to characterize exactly, he simply reported his findings and went on to other reactions which he considered more useful. However, the Belgian chemist, Leo Baekeland (who later became an American citizen), saw in this resin a material with considerable potential. Baekeland found ways to improve the yields and properties of the phenol-formaldehyde resin, which he patented under the name Bakelite. His primary contribution was a method of interrupting the condensation to yield a workable material which could be mixed with a filler and "cured" in a hot mold until it had thermoset. In simplified form the reaction involved is a condensation, and eventually many cross-linking bonds are formed:

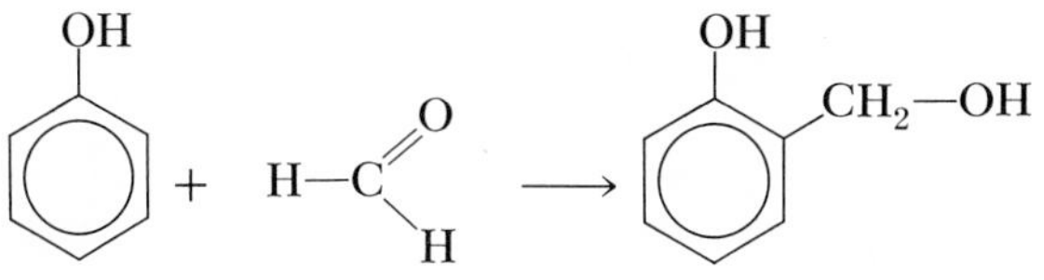

Phenol *Formaldehyde* *A phenol alcohol*

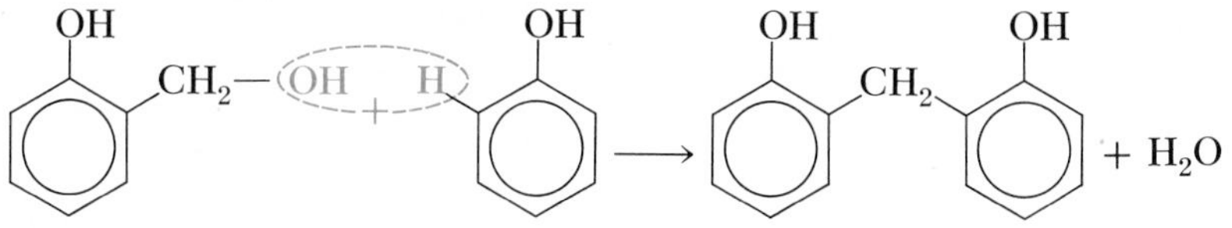

Further reactions take place until a resinous material occurs as the end product (Figure 16–7).

A polymer of this type is a *thermosetting* plastic. That is, once it has been heat cured, it cannot be softened without degradation. Such polymers have many

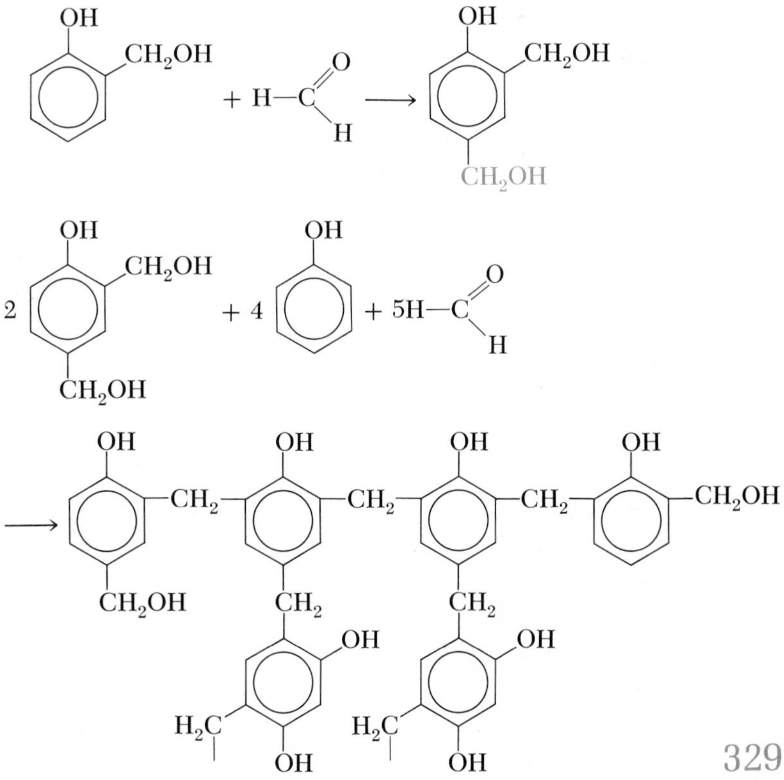

FIGURE 16–7 Condensation of phenol and formaldehyde to form a phenolic plastic.

desirable properties and are low in cost. Their hardness and insulating properties make them excellent as moldings for radios and other containers for electrical circuits. In addition, they can be machined quite easily. Indeed, thermosetting polymers gave a great impetus to the use of synthetic polymers, and only a handful of synthetics have matched these resins for economy and general usefulness.

REARRANGEMENT POLYMERIZATION

Some molecules polymerize by rearrangement reactions to yield very useful products. Molecules containing the isocyanate group (—NCO), for example, will react with almost any other molecule containing an active hydrogen atom (such as in an —OH or —NH$_2$ group) in a rearrangement process. An example is the reaction of hexamethylene diisocyanate and butanediol. The urethane linkage

$$\left(-\underset{\underset{H}{|}}{N}-\underset{\underset{O}{\|}}{C}-O- \right)$$

is a rearrangement of the same atoms in the isocyanate and alcohol groups and is similar to the amide bond in nylons.

$$OCN-(CH_2)_6-NCO + \boxed{H}O(CH_2)_4O\boxed{H} \longrightarrow$$

Hexamethylene diisocyanate   *Butanediol*

$$OCN-(CH_2)_6-\underset{\underset{H}{|}}{N}-\underset{\underset{O}{\|}}{C}-O-(CH_2)_4OH$$

Product molecule

In addition, using difunctional molecules gives rise to a polymer chain—a polyurethane.

$$-(CH_2)_6-\underset{\underset{H}{|}}{N}-\underset{\underset{O}{\|}}{C}-O-(CH_2)_4-O-\underset{\underset{O}{\|}}{C}-\underset{\underset{H}{|}}{N}-(CH_2)_6-\underset{\underset{H}{|}}{N}-\underset{\underset{O}{\|}}{C}-O-$$

A polyurethane

A typical polyurethane is structurally similar to a polyamide (nylon), and in Europe has found uses similar to those of nylon in this country. Polyurethanes have viscosities and melting points that make them useful for foam applications. Foamed polyurethanes are known as "foam rubber" and "foamed plastics" when copolymerized with substances such as polystyrene.

TAILOR-MADE MOLECULES

When the structure of polypropylene is drawn out to illustrate the three-dimensionality of the carbon-carbon bonds, it appears that there are three unique structures (Figure 16–8).

In the first structure, called *isotactic*, all the methyl (—CH$_3$) groups are in identical positions along the polymer chain. When the methyl groups alternate directions away from the chain, a *syndiotactic* arrangement results; and, of course, there is the possibility of a random, or *atactic*, arrangement. Polypropylenes of these three types were actually prepared and named by Professor

330

(a)

(b)

(c)

FIGURE 16-8   Three different structures of polypropylene. Each structure imparts different properties to the plastic. See text for discussion.

Giulio Natta in Italy in 1955. Using a novel catalyst, first used by Professor Karl Ziegler in Germany to increase the yields of straight chain, high density, polyethylene, Natta was the first to be able to control the growth phase of a polymerization reaction. This type of control is called *stereochemical control*. As a

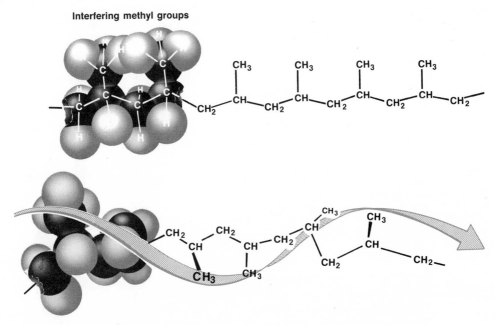

Interfering methyl groups

FIGURE 16-9   (a) Isotactic polypropylene. The bulky methyl groups (1, 2, 3 . .) are close to each other in this arrangement. (b) To eliminate crowding of the methyl groups on one side of the polymer chain, the chain flexes to produce a helical arrangement.

result of this discovery chemists were not only able to choose the monomer (formula) and the extent of polymerization (molecular weight), but also the fine structural features of the polymer chain itself.

Natta's three polypropylenes are different in ways that can be related to their structures. The isotactic material actually has a helical chain structure, owing to bulkiness of the methyl groups interacting with each other (Figure 16–9), which allows the chains to approach each other closely. Isotactic polypropylene melts at 170°C and can easily be formed into fibers. The atactic polypropylene chains cannot approach as closely because of the randomness of the methyl groups; this material is amorphous and rubbery.

ELASTOMERS

A very interesting application of stereochemical control over polymerization has been in the manufacture of synthetic rubber. Natural rubber has the composition $(C_5H_8)_n$ and when decomposed in the absence of oxygen, yields isoprene.

$$CH_2{=}\overset{\overset{\displaystyle CH_3}{|}}{C}{-}CH{=}CH_2$$

Isoprene

Natural rubber occurs as latex (a suspension of rubber particles in water) that oozes from rubber trees when they are cut. When the rubber particles are precipitated from the latex, a gummy mass is obtained that is elastic and water repellent but also very sticky, especially when warm. In 1839, after 10 years' work on this material, Charles Goodyear discovered that heating it with sulfur produced a material that was no longer sticky, but still elastic, water repellent, and resilient. A material of this sort is called an elastomer. Vulcanized rubber, as Goodyear called his product, contains short chains of sulfur atoms bonded between the polymer chains of the natural rubber. The sulfur chains help to align the polymer chains so that the material does not undergo a permanent change when stretched, but springs back to its original shape and size when the stress is removed (Figure 16–10).

In later years chemists searched for ways to make a synthetic rubber so that the nation would not be completely dependent on natural rubber during

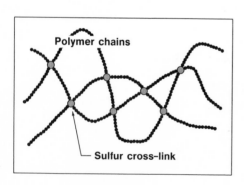

a. Before stretching

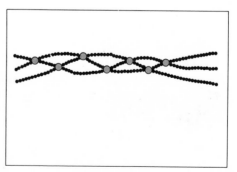

b. Stretched

FIGURE 16–10   Stretched vulcanized rubber retains its elasticity.

emergencies, as during the first years of World War II. In the mid-1920's, German chemists polymerized butadiene

$$CH_2{=}CH{-}CH{=}CH_2$$
Butadiene

(structurally similar to isoprene) to produce Buna rubber,

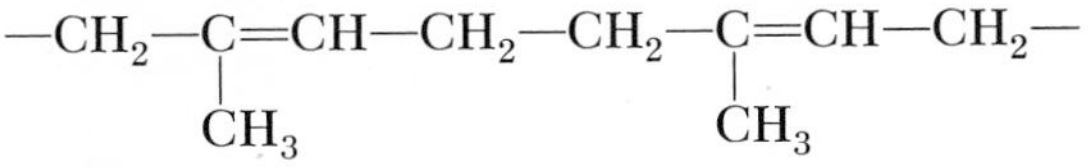

$$-CH_2{-}\underset{\underset{CH_3}{|}}{C}{=}CH{-}CH_2{-}CH_2{-}\underset{\underset{CH_3}{|}}{C}{=}CH{-}CH_2{-}$$

so named because it was made from butadiene (Bu-) catalyzed by sodium (-Na). Later a different copolymer rubber, Buna-S, a 75:25 copolymer of butadiene and styrene, was prepared. This material is still manufactured and known as SBR (styrene-butadiene rubber). However, none of these materials was as good as natural rubber for use in automobile tires. In the United States, Carothers, just prior to his discovery of nylon, discovered chloroprene

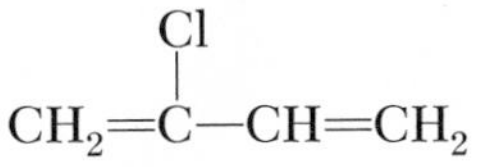

$$CH_2{=}\underset{\underset{Cl}{|}}{C}{-}CH{=}CH_2$$
Chloroprene

and polymerized it to make polychloroprene (neoprene).

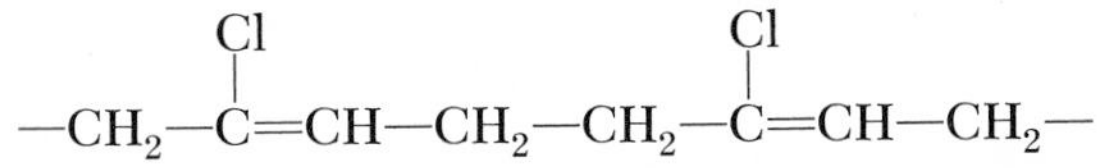

$$-CH_2{-}\underset{\underset{Cl}{|}}{C}{=}CH{-}CH_2{-}CH_2{-}\underset{\underset{Cl}{|}}{C}{=}CH{-}CH_2{-}$$
Polychloroprene

Polychloroprene has some superior properties owing to the presence of the chlorine atom and is still widely used today as an oil resistant rubber.

The behavior of polyisoprene, it was later learned, is due to the stereoregularity within the polymer chain. If we write the formula for polyisoprene as

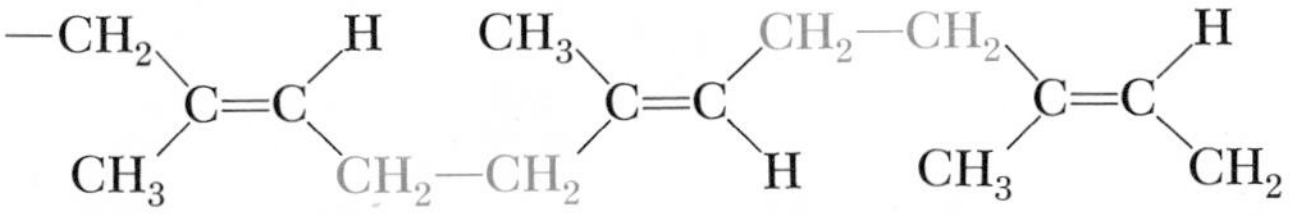

Poly-trans-isoprene (the —CH$_2$—CH$_2$— groups are trans)

we cannot appreciate the fact that *cis* and *trans* arrangements can exist about the nonrotating double bond (Chapter 14).

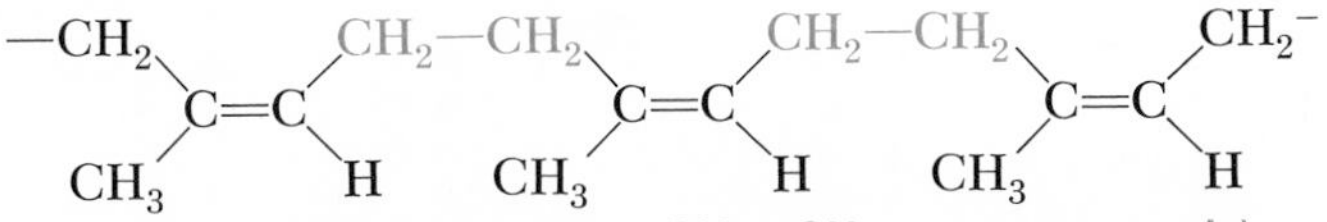

Poly-cis-isoprene (the —CH$_2$—CH$_2$— groups are cis)

Natural rubber is poly-*cis*-isoprene. However, the *trans* material also occurs in nature, in the leaves and bark of the sapotacea tree, and is known as *gutta-percha*. It is used as a thermoplastic for golf ball covers, electrical insulation, and other such applications. Without an appropriate catalyst, polymerization of isoprene yields a solid that is like neither rubber nor gutta-percha.

In 1955, chemists at Goodyear and Firestone discovered, almost simultaneously, how to use stereoregulating catalysts similar to those developed by

333

Figure 16-11   Ready to take the cure, these "green" truck tires at Goodyear's Danville, Virginia, plant will assume their familiar "doughnut" shape when they are cured, or vulcanized. This involves placing the tires in the huge molds, just behind the technician, and subjecting them to heat and pressure. Note the familiar cured product in the open molds. Goodyear estimated that the industry shipped 28 million truck tires to motor vehicle manufacturers and operators during 1971. (Photo through the courtesy of the Goodyear Tire and Rubber Company, Akron, Ohio.)

Ziegler to prepare poly-*cis*-isoprene. This material was, therefore, natural rubber. Today, synthetic poly-*cis*-isoprene can be manufactured cheaply and is used when natural rubber is in short supply.

POLYMER ADDITIVES—TOWARD AN END USE

Few plastics produced today find end uses without some kind of modification. Of those that do, the most important ones are *polystyrene*, which can be molded into partition boxes and refrigerator boxes, and *poly(methyl methacrylate)*, which is extruded into rod

Methyl methacrylate   *Poly(methyl methacrylate)*

and cast into sheet. In this section we shall examine briefly some of the ways plastics can be modified to make them more useful. It should be remembered that most often the reason a given plastic is chosen from all those available is that it has properties which represent a desirable compromise among all considerations.

FOAMING AGENTS

Most people are familiar with foamed plastics in the form of polystyrene foam (drinking cups, picnic coolers, and so on) or polyurethane foam (foam rubber). To produce useful foamed materials chemical engineers add to a liquid plastic, or its ingredients, a suitable foaming agent. The foaming agent must be capable of producing gas bubbles within the plastic mass when the plastic is not too soft or too hard (Figure 16–12). Two methods are commonly used.

Hydrocarbons such as pentane or neopentane are termed *physical foaming*

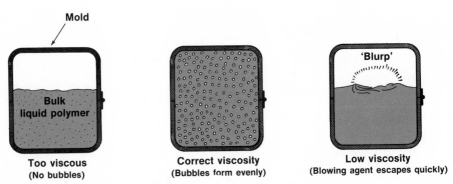

FIGURE 16–12   The preparation of foamed plastics requires that blowing conditions be adjusted carefully.

335

agents because they can foam a plastic simply by "boiling" to produce gas spaces. Polystyrene is foamed by dissolving pentane in liquid polystyrene under pressure. When this mixture is extruded from an outlet to atmospheric pressure, the pentane volatilizes, leaving the polystyrene full of small holes. The polystyrene is quickly solidified by cooling. Planks, boards, or logs of foamed styrene can be made in this way. To produce the small polystyrene beads that are compressed into shaped objects, such as picnic coolers, nucleating agents—small, solid grains—are added to the mixture; the polymer then builds up on them.

Chemical foaming agents produce a gas by chemical reaction, as their name implies. Some of these materials are rather exotic chemicals. Azo-bisformamide (ABFA), also known as dehydrobisurea DBU, is one of the most popular chemical foaming agents.

$$H_2N-\underset{\underset{O}{\|}}{C}-N{=}N-\underset{\underset{O}{\|}}{C}-NH_2$$

ABFA

Others, however, are quite simple chemically, such as carbon dioxide, which is used in making foamed polyurethane. The CO_2 is generated when a simple isocyanate is reacted with water.

$$R-NCO \ + \ H_2O \ \longrightarrow \ R-\underset{\underset{H}{|}}{N}-H \ + \ CO_2\uparrow$$

Isocyanate

There are two types of polyurethane foams, flexible and nonflexible. Both types can be prepared by carrying out simultaneously the polymerization and

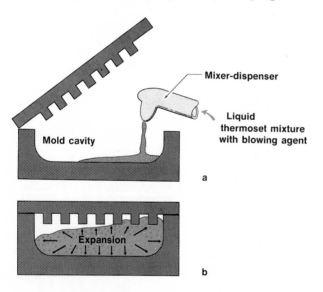

Mixer-dispenser

Liquid
thermoset mixture
with blowing agent

Mold cavity

a

Expansion

b

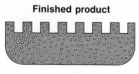

Finished product

FIGURE 16-13   Mixing and molding a thermoset foam.

c

$$OCN\!-\!\overset{\displaystyle\bigcirc}{}\!-\!NCO \; + \; HO\!-\!CH_2\!-\!\underset{\overset{\displaystyle|}{OH}}{CH}\!-\!CH_2\!-\!OH$$

$$\overset{\displaystyle N}{\underset{\displaystyle O}{\overset{\displaystyle C}{}}}$$

$$\longrightarrow \; -\!\overset{O}{\overset{\|}{C}}\!-\!\overset{H}{\overset{|}{N}}\!-\!\overset{\displaystyle\bigcirc}{}\!-\!\overset{H}{\overset{|}{N}}\!-\!\overset{O}{\overset{\|}{C}}\!-\!O\!-\!CH_2\!-\!\underset{\overset{\displaystyle|}{O}}{CH}\!-\!CH_2\!-\!O-$$

$$N\!-\!H$$
$$C\!=\!O$$
$$O$$
$$-\!O\!-\!CH_2\!-\!CH\!-\!CH_2\!-\!O-$$

FIGURE 16–14   Preparation of a type of nonflexible polyurethane foam.

foaming processes. Often this is done in a mold so that the product is produced directly (Figure 16–13). Flexible polyurethane foam, such as that found in foam pillows, contains little cross-linking but is a linear type made from a di-isocyanate (OCN—R—NCO) and diols (HO—R—OH). Polyisocyanates and polyols are used to prepare the nonflexible foamed polyurethane. Extensive cross-linking results in a thermosetting polymer.

PLASTICIZERS

Many times a plastic such as polyethylene or polyvinylchloride turns out to be too stiff for its intended application. Chemists have found that the addition of certain molecules, called plasticizers, can render a polymer more flexible. The structure of a polymer largely determines its flexibility. The polymer chains are intertwined and somewhat aligned in the solid state, and it is the Van der Waal's attractions that hold the polymer chains together. Although these forces are relatively weak compared to chemical bonds, they are still strong enough

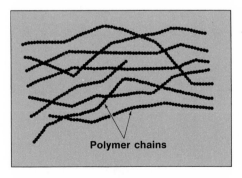

Rigid

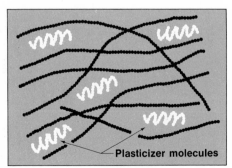

Less rigid

FIGURE 16–15   Schematic representation of how a plasticizer works.

337

to prevent adequate flexing of the solid in many instances. Certain molecules, such as di-(2-ethylhexyl) phthalate (DOP) and di-(2-ethylhexyl) adipate (DOA), fit in between the polymer chain and weaken the Van der Waal's attractions, producing dramatic increases in flexibility (Figure 16–15).

In 1968, 1.35 billion pounds of plasticizers were used in plastic formulation. The extent of plastic food wraps, for example, would be much more limited without DOA, which has FDA approval.

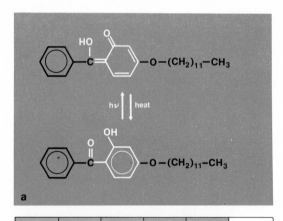

DOP   *DOA*

STABILIZERS

Most plastics used in outside locations, such as signs, tarpaulins, indoor-outdoor carpeting, auto seat covers, and toys, must be protected from sunlight

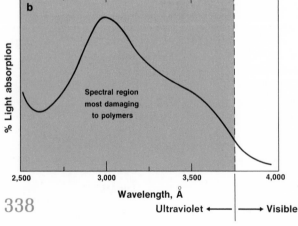

a

b

% Light absorption

Spectral region most damaging to polymers

2,500   3,000   3,500   4,000

Wavelength, Å

Ultraviolet ←——→ Visible

FIGURE 16–16 (*a*) A possible mechanism by which a molecule can absorb ultraviolet radiation and thus protect a plastic. This particular molecule is almost 100 per cent efficient in the process and finds wide use in plastics. (*b*) Typical ultraviolet absorption by a 2-hydroxybenzophenone-type compound used to protect plastics from sunlight.

in the 2900 to 4000 Angstrom (or ultraviolet) regions because photons in this spectral region have sufficient energy to break some chemical bonds found in organic molecules. Because of their structures, certain organic molecules are specific ultraviolet absorbers. These may be added directly to a plastic (about 0.1 per cent by weight) that is to be subjected to sunlight. Molecules that absorb ultraviolet light liberate energy as heat, thus preventing broken bonds in the polymer structure (Figure 16–16). Unfortunately, plastic formulations intended for outdoor use often do not contain sufficient ultraviolet stabilizers to prevent decay (Figure 16–17).

FIGURE 16–17   Sunlight (mostly ultraviolet) damages most plastics such as polypropylene webbing. The results are shorter life and higher costs to the consumer. Plastics containing ultraviolet-absorbing chemicals have a longer outdoor life.

In addition to aging due to photochemical damage, many polymers undergo rapid air oxidation if left unprotected. One of the most susceptible is polypropylene, which contains a hydrogen atom attached to a tertiary carbon atom (i.e., a hydrogen atom bonded to a carbon atom which is bonded to three other carbon atoms):

$$-CH_2-\overset{\displaystyle CH_3}{\underset{\displaystyle H}{C}}-CH_2-$$

Tertiary carbon

Polypropylene

Experiments on simpler molecules indicate that the hydrogens on a tertiary carbon atom can be abstracted in an oxidation process more easily then those on a secondary carbon atom, such as found in the more stable polyethylene:

$$-\overset{\displaystyle H}{\underset{\displaystyle H}{C}}-\overset{\displaystyle H}{\underset{\displaystyle H}{C}}-\overset{\displaystyle H}{\underset{\displaystyle H}{C}}-\overset{\displaystyle H}{\underset{\displaystyle H}{C}}-$$

Secondary carbons

Polyethylene

339

Some compounds act as antioxidants by inhibiting free radical oxidation. The oxidation of a polymer chain may be represented as follows:

$$R + \text{Initiator} \longrightarrow \text{Free radical, R·}$$
$$|$$
$$H$$
$$R· + O_2 \longrightarrow ROO·$$
$$ROO· + RH \longrightarrow ROOH + R·$$
$$\text{etc.}$$

Antioxidants, A—H, in which the H is readily oxidized, intercept these free radical mechanisms as they proceed in the solid:

$$ROO· + A\text{—}H \longrightarrow \text{Inert products}$$

PROCESSING METHODS

In order to make full use of a polymer's useful properties, the material must generally be processed into some desired form. Foaming, which results in many useful plastic articles, has already been mentioned.

Many thermoplastics can be extruded into filaments which have sufficient strength and flexibility that they are used as fibers. Among these are nylon and polypropylene. In a typical extrusion process (Figure 16–18) the molten plastic is compressed and made to flow from specially shaped holes in a die. When nylon is extruded, the molten polymer is heated to 285°C, which requires that it be protected from air oxidation by pure nitrogen. The nylon filaments are produced from the die at rates of about 2500 feet per minute.

Many plastic articles are molded into shape by a variety of techniques. For example, polyvinylchloride plastic tubing can be made by a continuous extrusion process.

The techniques of molding individual objects from plastics depend on the type of product desired. Everyone has seen plastic objects and wondered how

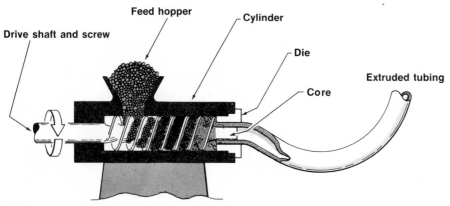

FIGURE 16-18   Screw-type extrusion.

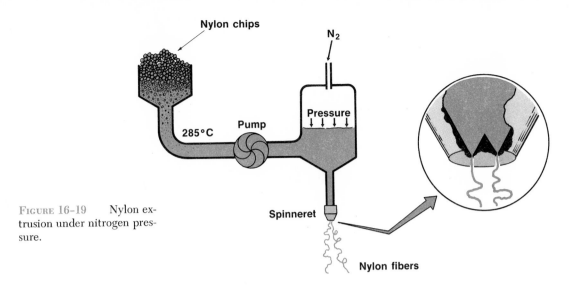

FIGURE 16-19   Nylon extrusion under nitrogen pressure.

they were made. Hollow toy balls and similar toys, perhaps the most interesting, are made by rotation molding. The two-part die (Figure 16–20) is placed in a gimbel which allows rotation about two perpendicular axes. The polymer is in a liquid form when it is added to the die. Rotation causes the polymer to coat the inner surface of the die; cooling sets the thermoplastic.

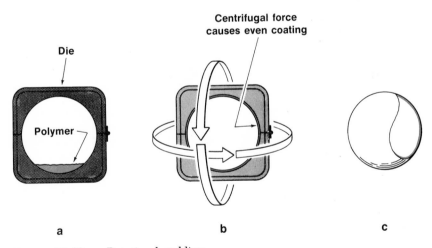

FIGURE 16-20   Rotational molding.

Blow molding is the process used to manufacture plastic bottles, artificial display fruit, and many toys. All of these items have relatively thin walls. To make a blow-molded article, a short plastic tube is extruded into the mold halves as they close, cutting off the top and bottom of the tube. Air is then injected into the tube until the plastic surface just fills the contours of the mold. (Figure 16–21). The mold opens, releasing the article. This is a very high speed process, some machines producing up to 24,000 bottles per hour and requiring over a ton of plastic raw material each hour.

341

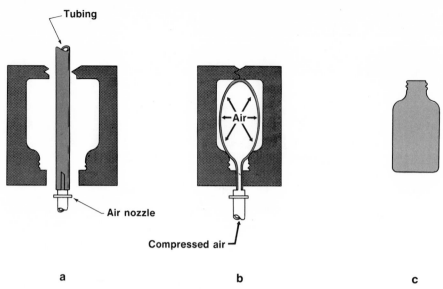

FIGURE 16-21   Blow molding of a hollow plastic article.

Injection molding is used to make many intricately shaped plastic articles such as telephone cases, small appliance cases, waste baskets, and refrigerator bowls. In this high-speed process, the plastic is softened just as it enters the mold by pressure applied by a screw (Figure 16-22). In the mold the plastic cools and solidifies. Almost all known thermoplastics can be shaped by injection molding.

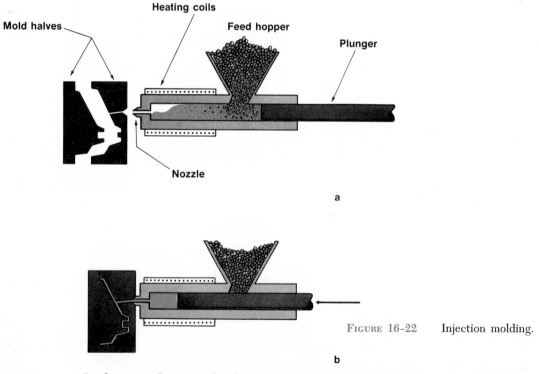

FIGURE 16-22   Injection molding.

In the manufacture of lightweight polystyrene products made from blown beads, mentioned earlier, a process known as expandable bead molding is used.

If blowing agent is added in excess, the blowing process can be stopped just prior to using all the agent. If the beads are then placed in a mold and heated, the plastic softens and fuses into a desired shape (Figure 16–23). This process is used to make ice chests and other foamed polystyrene products.

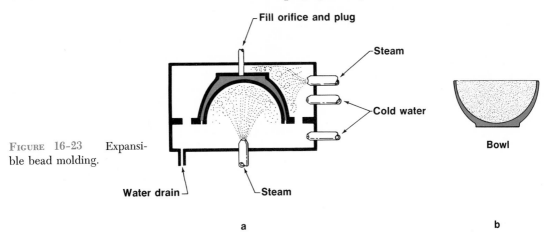

Fill orifice and plug

Steam

Cold water

FIGURE 16-23   Expansible bead molding.

Water drain

Steam

Bowl

a

b

THE FUTURE OF POLYMERS

As we have seen in this chapter, the development and use of synthetic polymers is quite recent. Polyethylene, for example, was discovered by accident in 1933; yet by 1969, its production in the United States amounted to billions of pounds. Chemists are constantly synthesizing new polymers and finding applications for them. The space age has brought with it the need for polymers, especially in electronics and as special coatings, that can withstand high temperatures without breaking down. Among these compounds are the polyimides, prepared from the polycondensation of a diacidanhydride and a diamine. Some of these polymers have very high service temperatures (Figure 16–24).

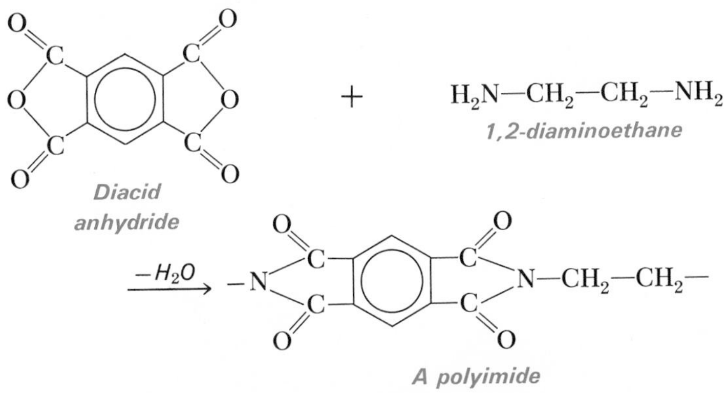

FIGURE 16-24   Preparation of a polyimides, plastics that can withstand high temperatures without breaking down.

Diacid anhydride

$+$ $H_2N-CH_2-CH_2-NH_2$

1,2-diaminoethane

$\xrightarrow{-H_2O}$

A polyimide

Because polymers are used so extensively in the world today, the problem of waste disposal is inevitable. Engineers have envisioned plants in which solid wastes from cities would undergo first a magnetic separation to remove iron and steel objects, then a ballistic separation, based on density since glass and aluminum objects are more dense than plastics. The plastics thus separated could be treated in two ways. If suitable separation methods could be developed, thermo-

343

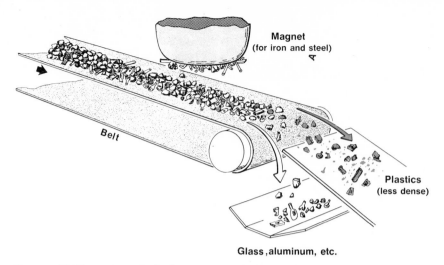

Magnet
(for iron and steel)

Belt

Plastics
(less dense)

Glass, aluminum, etc.

FIGURE 16-25   A method of separation of plastics from other wastes prior to recycling the plastics for reuse.

plastics could be reprocessed into new items (e.g., if all the nylon could be separated from polystyrene). Thermosetting plastics could not be treated this way, however, because breaking the cross-linking would cause complete molecular degradation. If separation and reuse were not feasible, combustion units built near cities could actually use plastics as fuels since they are mostly carbon and hydrogen; but some plastics contain elements which could create massive pollution if released into the atmosphere. An example would be polyvinylchloride, which on burning yields hydrogen chloride, a very corrosive gas. However, the combustion products could be recycled as raw materials for other chemical syntheses.

Hopefully, these and similar problems will be solved to almost everyone's satisfaction as man begins to understand fully his place on this planet.

QUESTIONS

1. Explain how polymers could be prepared from

(a) $CH_3-\underset{\underset{H}{|}}{C}=\underset{\underset{H}{|}}{C}-CH_3$

(b) $HO-\underset{\overset{\parallel}{O}}{C}-CH_2-CH_2-\underset{\overset{\parallel}{O}}{C}-OH$

(c) $\underset{\underset{OH}{|}}{CH_2}-\underset{\underset{OH}{|}}{CH}-\underset{\underset{OH}{|}}{CH_2}$

(d) $H_2N-CH_2-\langle\bigcirc\rangle-CH_2-NH_2$

2. Beginning with petroleum, outline the steps involved in the preparation of polyethylene.

3. Draw the monomer structural formula for

    (a) polyethylene
    (b) neoprene
    (c) polystyrene
    (d) Teflon

344

4. How many moles of ethylene gas are required to manufacture a 100 gram polyethylene bottle?

5. What are the monomers used to prepare the following polymers?

$-CH_2-CH_2-CH_2-CH_2-CH_2-CH_2-CH_2-CH_2-CH_2-$

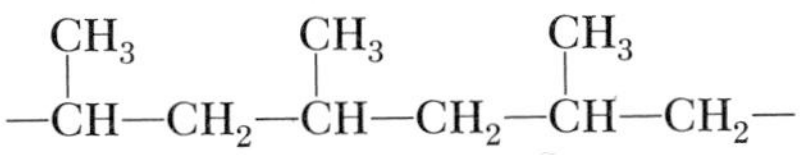

(c) 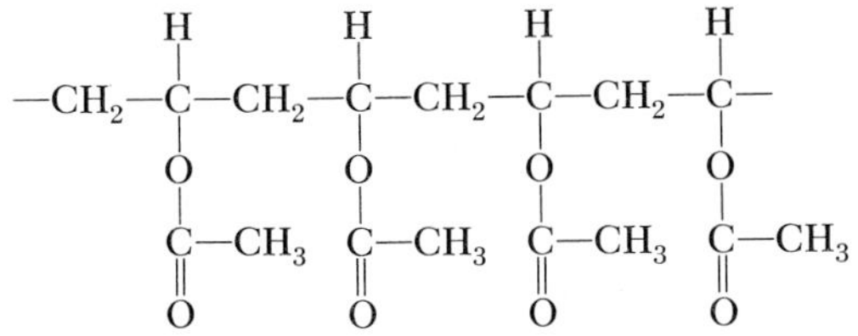

6. Write equations showing the formation of polymers by the reaction of the following pairs of molecules:

(a) 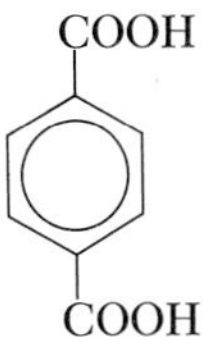 and $HOCH_2CH_2OH$

(b) $HOOC-CH_2-CH_2-COOH$ and $H_2NCH_2CH_2NH_2$

(c) 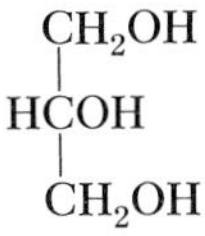 and 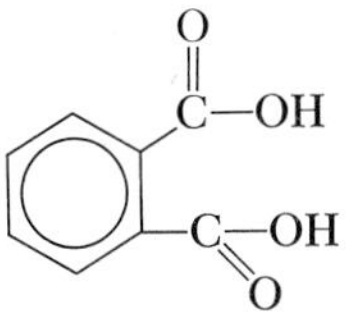

7. A specimen of polystyrene is found to consist of molecules with an average molecular weight of 12,000 a.w.u. What is the average number of styrene monomers in each unit?

8. You are given two specimens of plastic, A and B, to identify. One is known to be nylon, and the other poly(methyl methacrylate). Analysis of A shows it to contain C, H, and O, while B contains C, H, O, and N. What are A and B?

9. What was the first synthetic polymer?

10. What structural features are necessary in a molecule for it to undergo addition polymerization?

11. What is meant by the term *macromolecule*?

12. Give an example of a thermoplastic and a thermosetting plastic and compare their properties.

13. Orlon has a polymeric chain structure of

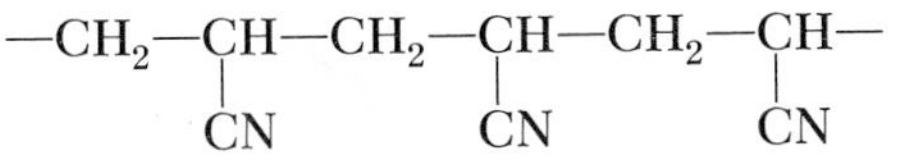

What is the orlon monomer?

14. Illustrate how the presence of too much peroxide initiator could cause an addition polymerization to stop prematurely.

15. What is a copolymer? Give an example of one.

16. What feature do all condensation polymerization reactions have in common?

17. Give an example of the possibilities that exist if a trifunctional acid reacts with a difunctional alcohol.

18. What are the starting materials for nylon 66?

19. Hydrogen bonding is possible in nylon. Show how.

20. Name an additional hazard to firemen when a building containing large quantities of PVC water pipe burns.

21. What structural features account for the differences between flexible and firm polyurethane foams?

22. Explain why a polymer with extensive cross-linking could not readily be extruded into a flexible fiber.

23. How does vulcanizing render rubber a very good elastomer?

24. Explain how a plasticizer can render a polymer more flexible.

SUGGESTIONS FOR FURTHER READING

Billmeyer, F. W. Jr., "Measuring the Weight of Giant Molecules," *Chemistry*, Vol. 39, No. 3, p. 8 (1966).

Frauer, A. H., "High Temperature Plastics," *Scientific American*, Vol. 221, No. 1, p. 96 (1969).

Kaufman, Morris, "Giant Molecules," Doubleday, New York, 1968.

Mark, H. F., "The Nature of Polymeric Materials," *Scientific American*, Vol. 217, No. 3, p. 149 (1967).

Mark, H. F., "Giant Molecules," Time Incorporated, New York, 1966.

Natta, G., "How Giant Molecules are Made," *Scientific American*, Vol. 197, No. 3, p. 98 (1957). (1957).

"Nylon: The First 25 Years," The Du Pont Company, Wilmington, 1963.

BIOCHEMISTRY—
BASIC STRUCTURES

CHAPTER 17

INTRODUCTION

The goal of biochemistry is to develop a chemically based understanding of living cells of all types. This includes the determination of the kinds of atoms which are present, the investigation of how they are joined together to form the larger structural units present in cells, and the study of the chemical reactions by which living cells furnish the energy required for the life processes of growth, movement, and reproduction. Furthermore, it seeks an understanding of how basic chemical structures and processes are combined to produce the complicated and highly specialized cells found in most plants and animals.

In this chapter only a very few aspects of biochemistry will be examined. These are: (1) carbohydrates, (2) amino acids and proteins, (3) nucleic acids, (4) enzymes, and (5) synthetic preparation of giant biochemical molecules.

CARBOHYDRATES

Carbohydrates are composed of the three elements carbon, hydrogen, and oxygen. Three structural groups are prevalent in carbohydrates: alcohol (—OH),

aldehyde $\left(-\overset{\overset{\displaystyle O}{\displaystyle \|}}{C}-\right)$, and ketone $\left(-\overset{\overset{\displaystyle O}{\displaystyle \|}}{C}H\right)$. The carbohydrates can be classified into three main groups: monosaccharides (Latin: *saccharum*, sugar), oligosaccharides, and polysaccharides. Monosaccharides are simple sugars that cannot be broken down into smaller units by mild acid hydrolysis. Hydrolysis of a molecule of an oligosaccharide yields two to six molecules of a simple sugar; complete hydrolysis of a polysaccharide produces many monosaccharide units.

Carbohydrates are synthesized by plants from water and atmospheric carbon dioxide. The process is called photosynthesis since it is a synthetic reaction that occurs when energized by light. It is an endothermic process; consequently, carbohydrates are energy-rich compounds. These compounds serve as important

347

sources of energy for the metabolic processes of plants and animals. Glucose, $C_6H_{12}O_6$, along with some of the other simple sugars, are quick energy sources for the cell. Polysaccharides, such as starch, store large amounts of energy that can be used by the cell only when the complex unit is broken down into mono-saccharides.

Some complex carbohydrates are also used by cells for structural purposes. Cellulose, for example, partially accounts for the structural properties of wood.

MONOSACCHARIDES

Approximately 70 monosaccharides are known; 20 of these simple sugars occur naturally. Unlike many organic compounds these sugars are very soluble in water owing to the numerous —OH groups present which can form hydrogen bonds with water.

The most common simple sugar is D-glucose. The monosaccharide requires three structures for its adequate representation (Figure 17–1).

Structure (a), in which the carbon atoms are numbered for later reference, depicts the "straight-chain" structure with the aldehyde group (—CHO) in position 1. The properties of a water solution of D-glucose cannot be explained by this structure alone. Most of the molecules at any given time exist in ring form, structures (b) and (c), which results when carbon 1 bonds to carbon 5 through an oxygen atom. Both ring structures are possible since the OH group on carbon 1 may form in such a way to point either along the plane of the molecule or out of the plane. It should be emphasized that a solution of D-glucose contains a mixture of three forms in a dynamic state of equilibrium. The equilibrium is shifted towards the ring forms; very limited amounts of the open-chain form are ever present.

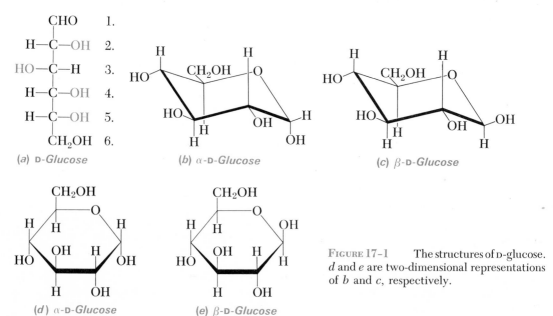

(a) D-*Glucose*

(b) α-D-*Glucose*

(c) β-D-*Glucose*

(d) α-D-*Glucose*

(e) β-D-*Glucose*

FIGURE 17–1   The structures of D-glucose. *d* and *e* are two-dimensional representations of *b* and *c*, respectively.

348

Examination of structure (a) in Figure 17–1 shows that four of the carbon atoms (2 through 5) are asymmetric (four different groups attached to each carbon). Consequently, there are two possible structures about each asymmetric carbon (Chapter 14). Structures like (a), a two-dimensional projection of the three-dimensional molecule, indicate either right- or left-hand geometry by placing the OH group on either the right or left side of the carbon chain. Since there are four asymmetric carbon atoms, there are a total of 16 (that is, 2^4) isomers in this group of simple sugars.

Because it is very sweet, D-glucose is used in the manufacture of candy and in commercial baking. This simple sugar, also called dextrose, grape sugar, and blood sugar, is found widely in fruits, vegetables, blood, and tissue fluids. A solution of D-glucose is fed intravenously when a readily available source of energy is needed to preserve life. As will be discussed later, many polysaccharides including starch are composed of glucose units and serve as a source of this important chemical upon hydrolysis of their complex structures.

Another very important monosaccharide is D-fructose. Its structure is given in Figure 17–2.

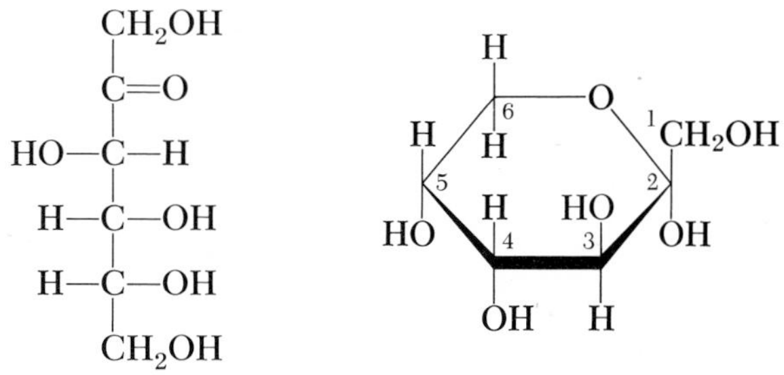

FIGURE 17–2   The structures of D-fructose. The β-ring structure differs from the α-ring structure in that the CH_2OH groups are in reversed positions on carbon 2.

(a) Ketone structure

(b) α-Ring structure

OLIGOSACCHARIDES

The most important oligosaccharides are the disaccharides (two simple sugar units per molecule). Examples include the following widely used sugars:

sucrose (from sugar cane or sugar beets), which consists of a glucose unit and a fructose unit,

maltose (from starch), which consists of two glucose units, and

lactose (from milk), which consists of a glucose unit and a galactose (an isomer of glucose) unit.

The formula for these disaccharides, $C_{12}H_{22}O_{11}$, is not simply the sum of two monosaccharides, $C_6H_{12}O_6 + C_6H_{12}O_6$. A water molecule must be added (hydrolysis) to obtain the monosaccharides. The structures for sucrose, maltose, and lactose, along with the hydrolysis reactions, are given in Figure 17–3.

The disaccharides are important as foods; sucrose is produced in a very high state of purity on an enormous scale. The annual production of this food amounts to over 40,000,000 tons per year. Originally produced in India and Persia, sucrose is now used universally as a sweetener. About 40 per cent of the world sucrose

349

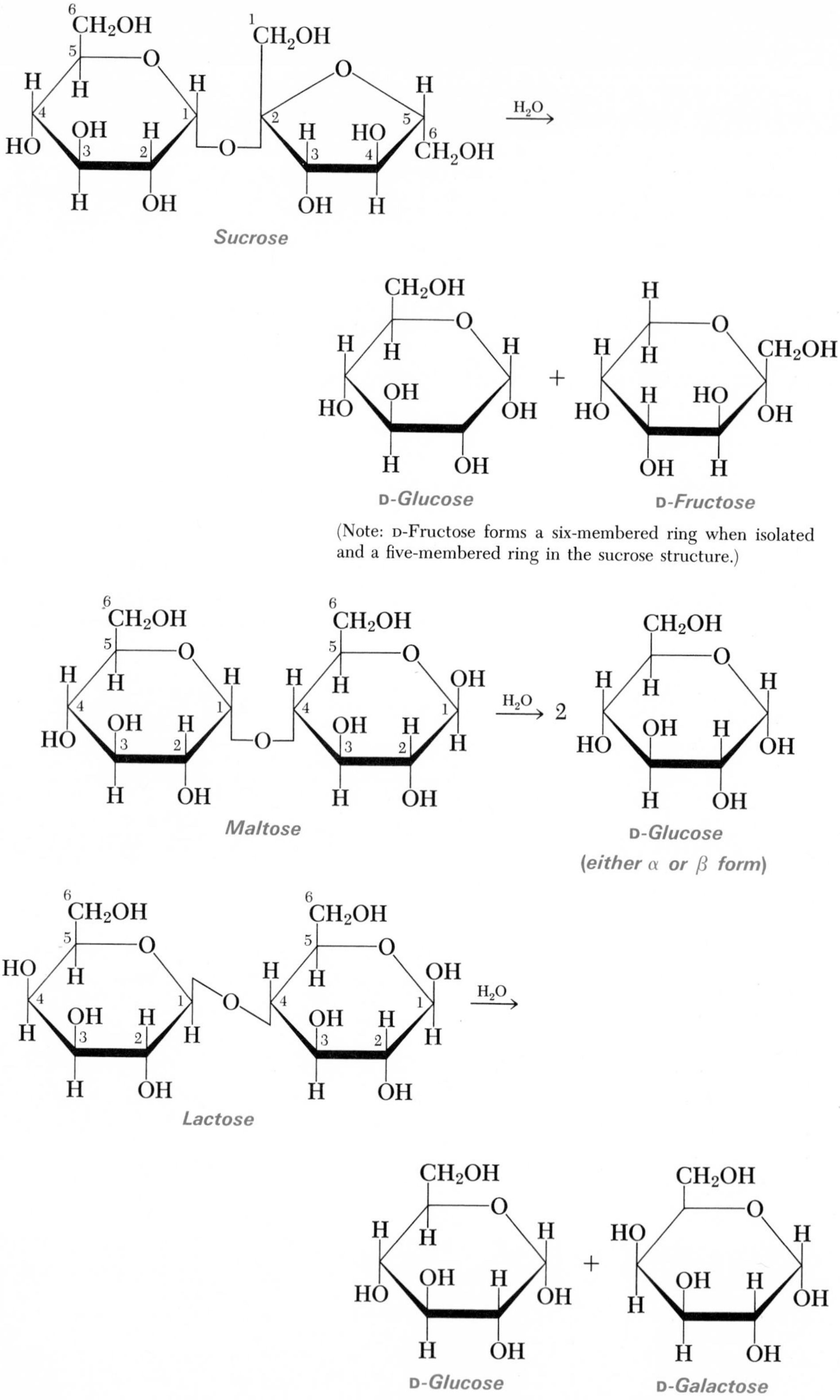
Sucrose
D-Glucose
D-Fructose
(Note: D-Fructose forms a six-membered ring when isolated and a five-membered ring in the sucrose structure.)
Maltose
D-Glucose
(either α or β form)
Lactose
D-Glucose
D-Galactose

FIGURE 17-3 Hydrolysis of disaccharides (sucrose, maltose, and lactose).

production comes from sugar beets and 60 per cent from sugar cane. Sugar provides a high caloric value (1794 kcal per pound); it is also used as a preservative in jams, jellies, and other foods.

POLYSACCHARIDES

There is an almost limitless number of possible structures in which monosaccharide units (monosaccharide molecules minus one water molecule at each bond between units) can be combined. Molecular weights are known to go above 1,000,000. Apparently nature has been very selective in that only a few of the many different monosaccharide units are found in polysaccharides.

Starches and Glycogen

Starch is found in plants in protein-covered granules. These granules are disrupted by heat, and a part of the starch content is soluble in hot water. Soluble starch is *amylose* and constitutes 10 to 20 per cent of most natural starches; the remainder is *amylopectin*. Amylose gives the familiar blue-black starch test with iodine solutions; amylopectin turns red in contact with iodine.

Structurally amylose is a straight chain of D-glucose units, each one bonded to the next, just as the two units are bonded in maltose (Figure 17–3). While there are amylose structures in excess of 1000 glucose units, molecular weight studies indicate the average "molecule" contains fewer than 500 units. The structure of amylose is illustrated in Figure 17–4.

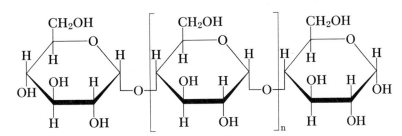

FIGURE 17–4   Amylose structure. n represents a large number of D-glucose units; the type of polysaccharide determines the value of n.

Amylopectin is made up of branched D-glucose units (Figure 17–5). Its molecular weight is generally higher than that for amylose. Partial hydrolysis of amylopectin yields mixtures called *dextrins*. Complete hydrolysis, of course, yields D-glucose. Dextrins are used as food additives, mucilages, paste, and in finishes for paper and fabrics.

Glycogen serves as an energy reservoir in animals as does starch in plants. Glycogen has essentially the same structure as amylopectin (branched chains of glucose units) except that it has an even more highly branched structure.

351

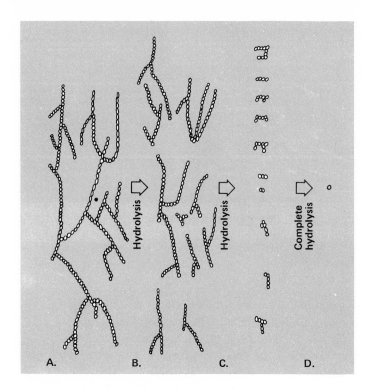

FIGURE 17-5   A, Partial amylopectin structure; B, dextrins from
incomplete hydrolysis of A; C,
oligosaccharides from hydrolysis of
dextrins; D, final hydrolysis product: D-glucose. Each circle represents a glucose unit.

Cellulose

Cellulose is the most abundant polysaccharide in nature. Like amylose, it
is composed of D-glucose units. The difference between the structure of cellulose
and that of amylose lies in the bonding between the D-glucose units. In cellulose,
all of the glucose units are in the β-ring form in contrast to the α-ring form
in amylose. Review the ring forms in Figure 17–1 and compare the structures
in Figures 17–4 and 17–6.

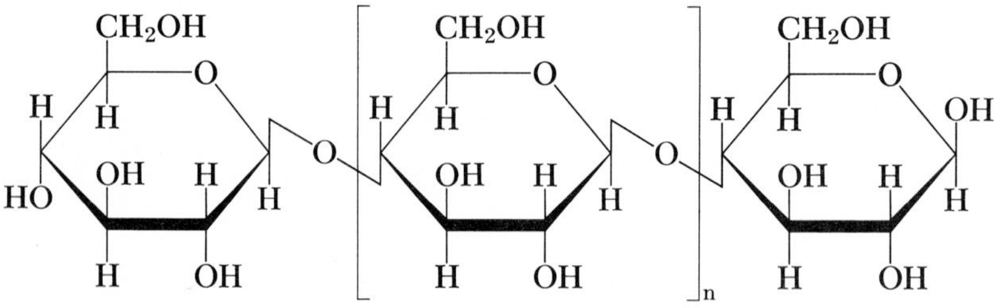

FIGURE 17-6   The structure of cellulose. The value of n may go as high as
10,000 in the natural product.

The different structures of starch and cellulose account for their difference
in digestibility. Human beings and carnivorous animals do not have the necessary
enzymes (biochemical catalysts) to break down the cellulose structure as do
numerous microorganisms. Cellulose is readily hydrolyzed to D-glucose in the
laboratory by heating a suspension of the polysaccharide in the presence of a
strong acid. Unfortunately, at present this is not an economically feasible solution
to man's growing need for an adequate food supply.

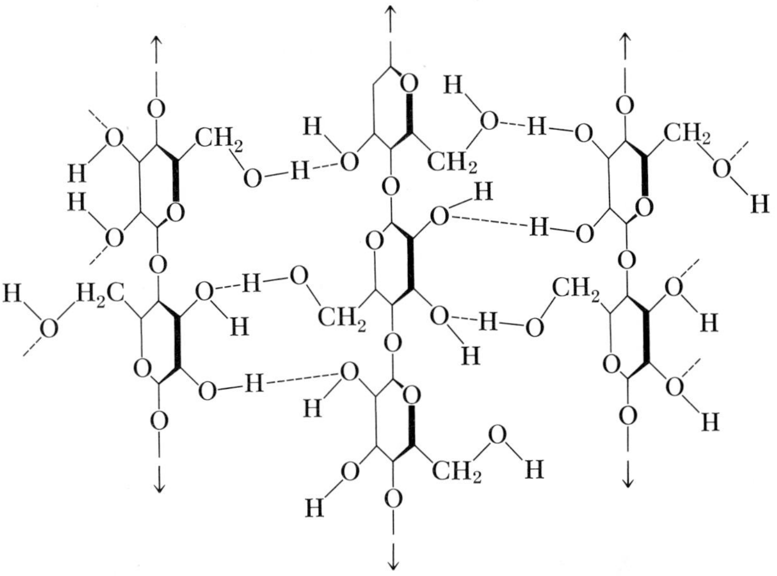

FIGURE 17-7   The properties of cotton, about 98 per cent cellulose, can be explained in terms of this submicroscopic structure. A small group of cellulose "molecules," each with from 2000 to 9000 units of D-glucose, are held together in an approximately parallel fashion by hydrogen bonding (····). When several of these *chain bundles* cling together in a relatively vast network of hydrogen bonds, a *microfibril* results; the microfibril is the smallest microscopic unit that can be seen. The macroscopic *fibril* is a collection of numerous microfibrils. The absorbent nature of cotton is readily explained in terms of the smaller water molecules being held by hydrogen bonding in the numerous capillaries that exist between the cellulose chains.

Paper, rayon, cellophane, and cotton are essentially cellulose. Figure 17–7 relates the structure of cotton fiber to the chains of glucose units at the submicroscopic level.

PROTEINS, AMINO ACIDS, AND THE PEPTIDE LINKAGE

Proteins occur in all the major regions of living cells. These compounds serve a wide variety of functions including motion of the organism, defense mechanism against foreign chemicals, metabolic regulation of cellular processes, and cell structure. The close relationship between protein structures and life was first noted by the German chemist, G. T. Mulder, in 1835. He named these compounds proteins from the Greek *proteios*, meaning first, indicating this to be the starting point in the chemical understanding of life.

Proteins are macromolecules with molecular weights ranging from 5000 to several million. Like the polysaccharides, these macrostructures are composed of recurring submicroscopic units, the fundamental units in the case of proteins being *amino acids*. Proteins and amino acids are made primarily from four elements: carbon, oxygen, hydrogen, and nitrogen. Other elements occur in trace amounts, the one most often encountered being sulfur.

353

AMINO ACIDS

The complete hydrolysis of a typical protein yields a mixture of about 20 amino acids. Some proteins lack one or more of these acids, others have small amounts of other amino acids characteristic of a given protein, but the 20 given in Table 17–1 dominate. In a few instances, one amino acid will constitute a major fraction of a protein (the protein in silk, for example, is 44 per cent glycine), but this is not common.

TABLE 17-1   COMMON AMINO ACIDS

All of the amino acids except proline and hydroxyproline have the general formula:

$$R-\underset{\underset{NH_2}{|}}{\overset{\overset{H}{|}}{C}}-C\overset{O}{\underset{OH}{}}$$

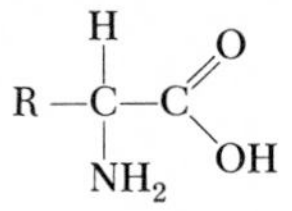

in which R is the characteristic group for each acid. The R groups are as follows.

1. Glycine —H
2. Alanine —CH_3
3. Serine —CH_2OH
4. Cysteine —CH_2SH
5. Cystine —CH_2—S—S—CH_2—

6. Threonine° —CH—CH_3 with OH below

7. Valine° CH_3—CH—CH_3

8. Leucine° —CH_2—CH—CH_3 with CH_3 below

9. Isoleucine° —CH with CH_3 and CH_2—CH_3

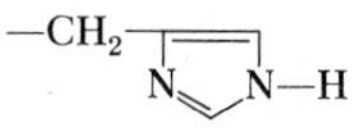

The structures for the other two are:

19. Proline H_2C——CH_2 / H_2C—N—$CHCO_2H$ / H

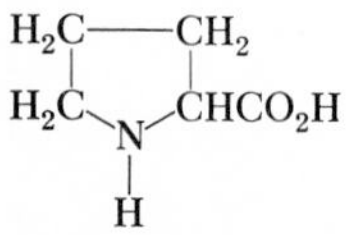

10. Methionine° —CH_2—CH_2—S—CH_3
11. Aspartic acid —CH_2CO_2H
12. Glutamic acid —CH_2—CH_2—CO_2H
13. Lysine° —CH_2—CH_2—CH_2—CH_2—NH_2
14. Arginine —CH_2—CH_2—CH_2—$NH\overset{\overset{NH}{||}}{C}NH_2$
15. Phenylalanine° —CH_2—⬡
16. Tyrosine —CH_2—⬡—OH
17. Tryptophan° —CH_2—(indole ring)
18. Histidine —CH_2—(imidazole ring N N—H)

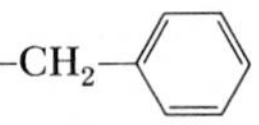

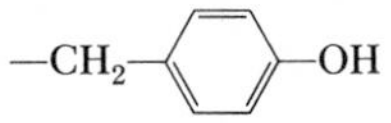

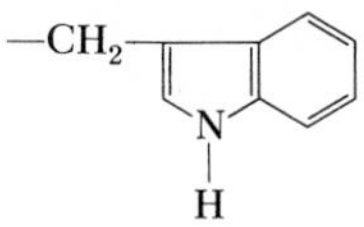

20. Hydroxyproline HOHC——CH_2 / H_2C—N—$CHCO_2H$ / H

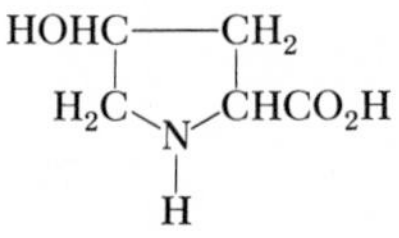

° Essential amino acids

As the name suggests, amino acids contain an amine group (—NH_2) and an acid (carboxyl) group (—COOH). In all of the amino acids listed in Table 17–1, the amine group and the acid group are bonded to the same carbon atom. Of these acids, 18 have the general formula:

$$R-\underset{\underset{NH_2}{|}}{\overset{\overset{H}{|}}{C}}-C\overset{O}{\underset{OH}{}}$$

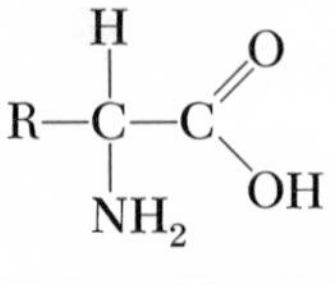

where R is a characteristic group for each amino acid. The simplest amino acid is *glycine,* in which R is a hydrogen atom:

$$H-\overset{\overset{\displaystyle H}{|}}{\underset{\underset{\displaystyle NH_2}{|}}{C}}-C\overset{\displaystyle O}{\underset{\displaystyle OH}{\diagup}}$$

Glycine

The human body is capable of synthesizing some amino acids needed for protein structures, but it is unable to provide others at a rate necessary for normal growth and development. The latter are designated *essential* amino acids and must be ingested in the food supply. The *nonessential* amino acids are just as necessary as the essential amino acids for life but can be made from any protein or nonprotein food supply. The essential amino acids are indicated in Table 17–1.

THE PEPTIDE LINKAGE

Amino acid units are linked together in protein structures by the peptide linkage. This same linkage was illustrated in nylon 66, in which an acid and an amine were condensed to form the polymer and the peptide bond. As it applies to proteins, this type of chemical bond can be understood in terms of a hypothetical reaction, which is a gross oversimplification of natural protein synthesis. If the acid group of one glycine molecule reacts with the basic amine group of another, the two are joined through the peptide linkage, and one molecule of water is eliminated for each link formed.

$$H-N-\overset{\overset{\displaystyle H}{|}}{\underset{\underset{\displaystyle H}{|}}{C}}-\overset{\overset{\displaystyle O}{||}}{C}-OH \; + \; H-N-\overset{\overset{\displaystyle H}{|}}{\underset{\underset{\displaystyle H}{|}}{C}}-\overset{\overset{\displaystyle O}{||}}{C}-OH \longrightarrow$$

Glycine   *Glycine*

Peptide linkage

$$H-N-\overset{\overset{\displaystyle H}{|}}{\underset{\underset{\displaystyle H}{|}}{C}}-\overset{\overset{\displaystyle O}{||}}{C}-N-\overset{\overset{\displaystyle H}{|}}{\underset{\underset{\displaystyle H}{|}}{C}}-\overset{\overset{\displaystyle O}{||}}{C}-OH \; + \; HOH$$

Glycylglycine

If this hypothetical reaction is carried out with two different amino acids, glycine and alanine, two different *dipeptides* are possible.

Peptide linkages

$$H-N-\overset{\overset{\displaystyle H}{|}}{\underset{\underset{\displaystyle H}{|}}{C}}-\overset{\overset{\displaystyle H}{|}}{\underset{\underset{\displaystyle H}{|}}{C}}-N-\overset{\overset{\displaystyle H}{|}}{\underset{\underset{\displaystyle CH_3}{|}}{C}}-\overset{\overset{\displaystyle O}{||}}{C}-OH$$

Glycylalanine   *Alanylglycine*

355

Twenty-four *tetra*peptides are possible if four amino acids (for example, glycine, Gly; alanine, Ala; serine, Ser; and cystine, Cy) are linked in all possible combinations. They are:

Gly-Ala-Ser-Cy	Ala-Gly-Ser-Cy	Ser-Ala-Gly-Cy	Cy-Ala-Gly-Ser
Gly-Ala-Cy-Ser	Ala-Gly-Cy-Ser	Ser-Ala-Cy-Gly	Cy-Ala-Ser-Gly
Gly-Ser-Ala-Cy	Ala-Ser-Gly-Cy	Ser-Gly-Ala-Cy	Cy-Gly-Ala-Ser
Gly-Ser-Cy-Ala	Ala-Ser-Cy-Gly	Ser-Gly-Cy-Ala	Cy-Gly-Ser-Ala
Gly-Cy-Ser-Ala	Ala-Cy-Gly-Ser	Ser-Cy-Ala-Gly	Cy-Ser-Ala-Gly
Gly-Cy-Ala-Ser	Ala-Cy-Ser-Gly	Ser-Cy-Gly-Ala	Cy-Ser-Gly-Ala

If 17 different amino acids are used, the sequences alone would make 3.56×10^{14} uniquely different 17-unit molecules.° Although there are numerous protein structures in nature, these represent an extremely small fraction of the possible structures. Of all the many different proteins which could possibly be made from a set of amino acids, a living cell will make only a relatively small, select number.

PROTEIN STRUCTURES

The *primary structure* of a protein indicates only the order of the sequence of amino acid units in the *polypeptide* chain. Since the single bonds in the chain allow free rotation around the bond, it is reasonable to assume that there is an almost infinite number of possible conformations. Because of interactions, such

Linus Pauling (1901–   ): A scientist of great versatility and accomplishment. His interests have included the determination of the molecular structures of crystals by x-ray diffraction and theories of the chemical bond. His work on the structure of spiral polypeptide chains led to the Nobel Prize in 1954.

as hydrogen bonding, between atoms in the same chain, certain conformations, called *secondary structures*, are favored. In 1954, Linus Pauling received the Nobel Prize for his work on secondary protein structure. Along with R. B. Corey, he suggested the two secondary structures discussed in the following paragraphs.

° If the amino acids are all different, the number of arrangements is n! (read n factorial). For five different amino acids, the number of different arrangements is 5! (or $5 \times 4 \times 3 \times 2 \times 1 = 120$).

Polyglycine is the synthetic protein made entirely of the amino acid glycine. In polyglycine the hydrogen attached to the nitrogen atom and the oxygen bonded to the carbon are both well suited to engage in hydrogen bonding. In the two stable conformations of polyglycine, maximum advantage is taken of the hydrogen bonding available. In one conformation, the hydrogen bonding is between adjacent chains of the polypeptide; in the other, hydrogen bonding occurs between atoms within the same chain.

Figure 17–8 illustrates a sheetlike structure in which row after row of the polypeptide are joined by hydrogen bonding. Note that all the oxygen and nitrogen atoms are involved in hydrogen bonding. Most of the properties of silk can be explained in terms of this type of structure for fibroin, the protein of silk, which contains 44 per cent glycine.

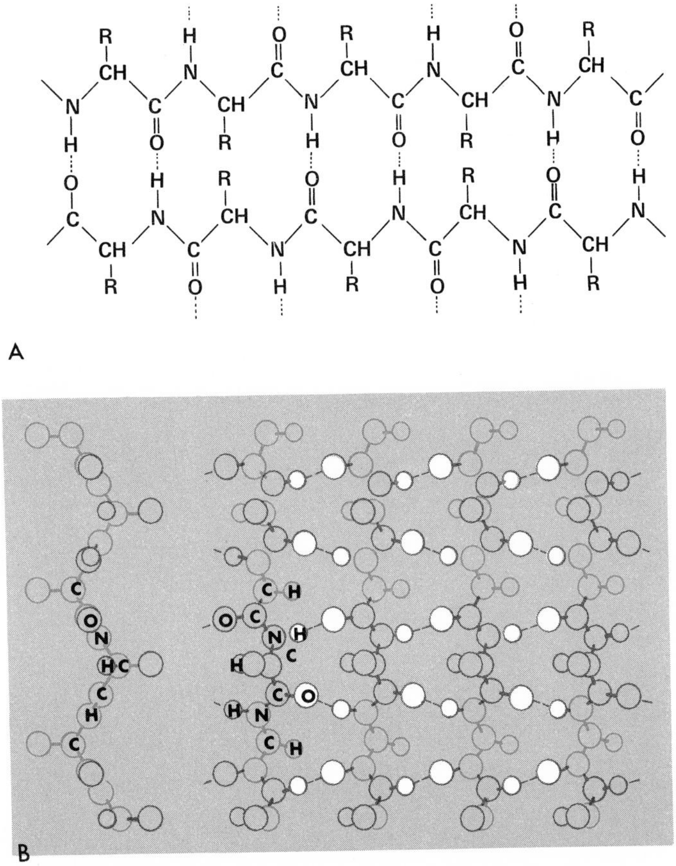

A

B

FIGURE 17–8   Sheet structure for polypeptide. In A, the two-dimensional drawing emphasizes that all of the oxygen and nitrogen atoms are involved in hydrogen bonding for the most stable structure. B illustrates the bonding in perspective showing that the sheet is not flat; rather, it is sometimes called a pleated sheet structure.

Hydrogen bonding is possible within a single polypeptide chain if the secondary structure is helical (Figure 17–9). Bond angles and bond lengths are such that the nitrogen atom forms hydrogen bonds with the oxygen atom in the third amino acid unit down the chain (Figure 17–10).

Collagen is the principal fibrous protein in mammalian tissue. It has remarkable tensile strength which makes it useful in structuring bones, tendons, teeth, and cartilage. Three polypeptide chains, each of which is twisted into a left-handed helix, are twisted into a right-handed super helix to form an

357

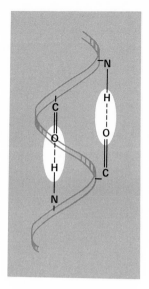

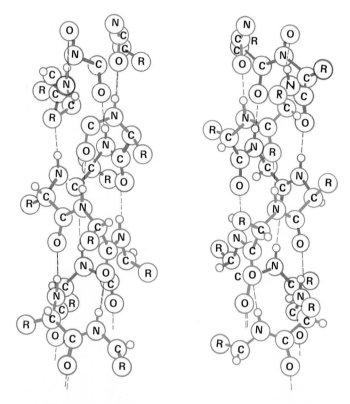

FIGURE 17-9   Helix structure for a polypeptide in which each oxygen atom can be hydrogen bonded to a nitrogen atom in the third amino acid unit down the chain.

FIGURE 17-10   Drawing of the helix structure for polypeptides suggested by Pauling and Corey. Both left-handed and right-handed helices are possible. R is the characteristic group for each amino acid as shown in Table 17-1.

extremely strong fibril, as shown in Figure 17–11. A bundle of such fibrils forms the macroscopic protein.

FIGURE 17-11   The structure of collagen.

The structure of collagen illustrates a third level of protein structure, *tertiary* structure. The primary structure is the sequence of amino acids in the protein;

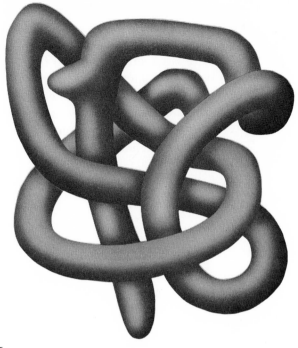

FIGURE 17-12   Typical folded structure of the helix in a globular protein.

A

FIGURE 17-13   *A*, The structure of heme; *B*, a model, from two views, of the hemoglobin structure. The heme structures are indicated by disks. (Courtesy of M. F. Perutz and *Science*, 140:863, 1963.)

B

the secondary structure is the helical form of the protein chain; and the tertiary structure is the twisted or folded form of the helix. Another tertiary structure is illustrated by globular proteins. In these structures, the helix chain (secondary structure) is folded and twisted into a definite geometric pattern. This pattern may be held in place by one or more of several different kinds of chemical bonds, depending on the particular functional groups in the amino acids involved (Table 17–1). Figure 17–12 illustrates the folded structure of a typical globular protein. Abnormal hemoglobin structures are unable to transfer oxygen in the blood if the wrong amino acid is in a given position in the polypeptide structure (Figure 17–13).

ENZYMES

Living organisms practice an impressive economy in the use of their molecular resources. Much of this efficiency is accomplished by the use of small amounts of very powerful catalysts (materials which accelerate reaction rates) which dominate the chemistry of protoplasm. A very large group of biochemicals, the *enzymes*, offer an excellent example of this type of catalysis. Enzymes are polypeptides.

Most biochemical reactions do not readily occur in the absence of cellular materials. For example, a mixture of amino acids does not readily form polypeptides, nor will starch in water spontaneously decompose into glucose. However, both of these reactions, along with numerous others, occur readily under normal cell conditions. Obviously, then, there are catalysts in protoplasm which facilitate very complex biochemical reactions at body temperature of 37°C.

There appears to be a different enzyme for almost every biochemical reaction; the reaction occurs only in the presence of this enzyme. Maltase is an enzyme that catalyzes the hydrolysis of maltose into two molecules of D-glucose. This is the only known function of maltase, and no other enzyme can substitute for it in this catalytic activity.

The explanation for the specific activity of enzymes is to be found in molecular shapes. Enzymes are globular proteins with very definite tertiary protein structures (Figure 17–12). The highly specific action of maltase can be explained if its globular structure accurately accommodates a maltose molecule. When the two units come together, strain is placed on the bonds holding the two simple sugar units together. As a result water is allowed to enter, and hydrolysis occurs. Sucrose cannot be hydrolyzed by maltase because of the different geometry involved. Another enzyme, sucrase, hydrolyzes sucrose effectively.

Some enzymes can be crystallized in solid form and can perform their catalytic activity outside the cell. Yeast cells are not necessary for fermentation; only the enzymes from yeast are needed (Chapter 15).

NUCLEIC ACIDS

Like the polysaccharides and the polypeptides, the *nucleic acids* are high molecular weight substances, with molecular weights up to several million.

Nucleic acids are found in all living cells, with the exception of the red blood cells of mammals, and the structures of these compounds are believed to be directly related not only to the characteristics of the individual cell but of the gross organism itself. The almost infinite variety of possible structures for nucleic acids allow information in coded form to be recorded in molecular complexes in somewhat similar fashion to the complex of language symbols used to convey ideas in this book. Such stored information controls the inherited characteristics of the next generation as well as many of the ongoing life processes of the organism.

Hydrolysis of nucleic acids yields one of two simple sugars, phosphoric acid (H_3PO_4), and a group of nitrogen compounds that have basic (alkaline) properties. Based on these hydrolysis products, the nucleic acids can be classified as either those that contain the sugar D-*deoxyribose*, or those that contain D-*ribose*. The former are called *deoxyribonucleic acids* (DNA) and the latter *ribonucleic acids* (RNA). DNA is found primarily in the nucleus of the cell, whereas RNA is found mainly in the cytoplasm outside of the nucleus.

RIBOSE AND DEOXYRIBOSE

The structures for the two sugars in nucleic acids are shown in Figure 17–14. The names and formulas for the basic nitrogen compounds in nucleic acids are given in Figure 17–15.

FIGURE 17–14 The structure of α-D-ribose and α-2-deoxy-D-ribose. In the IUPAC names given, α indicates the one of two ring-forms possible; D distinguishes the isomers that rotate plane polarized light in opposite directions, and the 2 indicates the carbon to which no oxygen is attached in the second sugar.

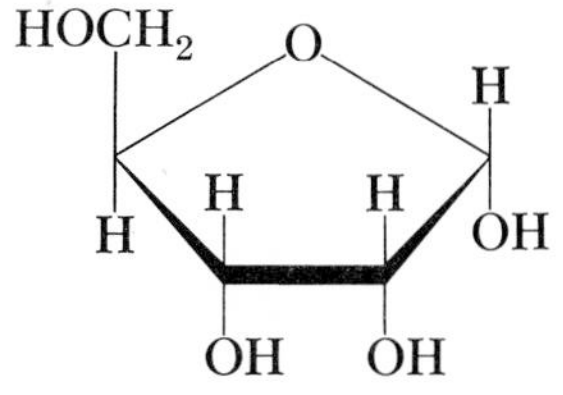

α-D-*ribose*

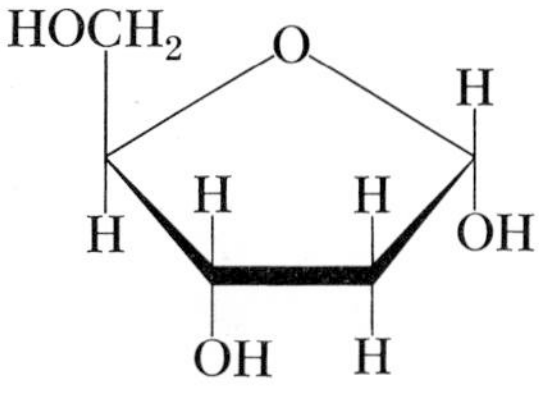

α-2-*deoxy*-D-*ribose*

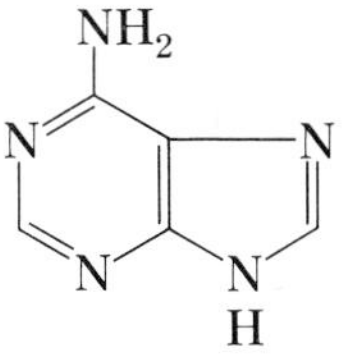

Adenine

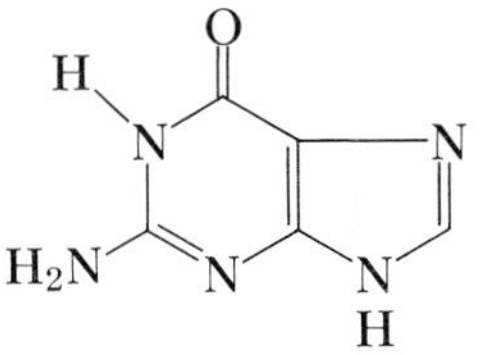

Guanine

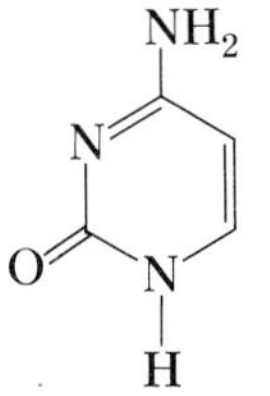

Cytosine

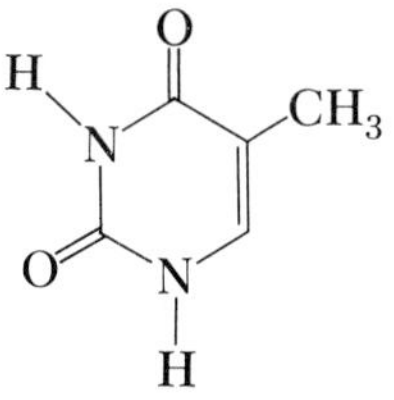

Thymine

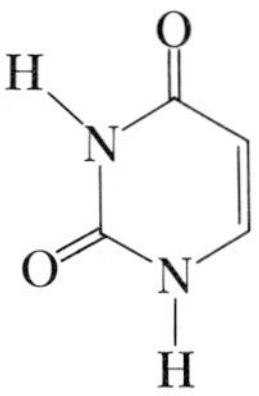

Uracil

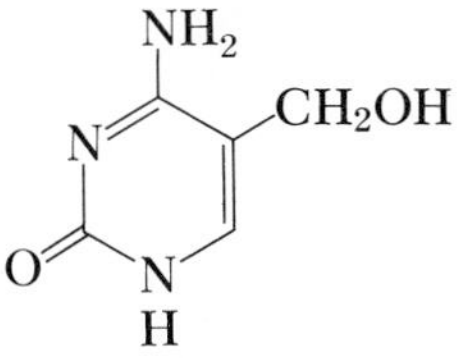

5-Hydroxymethylcytosine

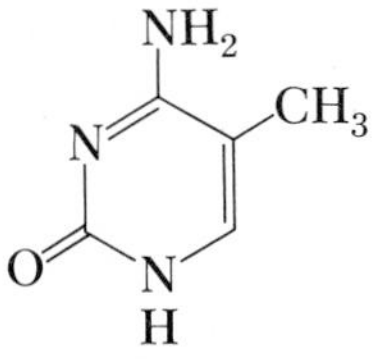

5-Methylcytosine

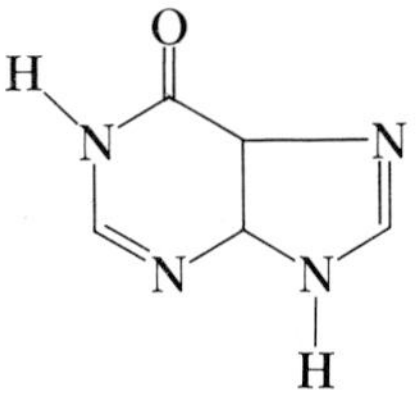

Hypoxanthine

FIGURE 17–15   Nitrogenous bases obtained from the hydrolysis of nucleic acids.   361

NUCLEOTIDES

Incomplete hydrolysis of DNA or RNA yields *nucleotides*. These substances contain a simple sugar unit, one of the nitrogenous base units, and one or two units of phosphoric acid. An example of the nucleotide structure is illustrated by inosinic acid, Figure 17–16.

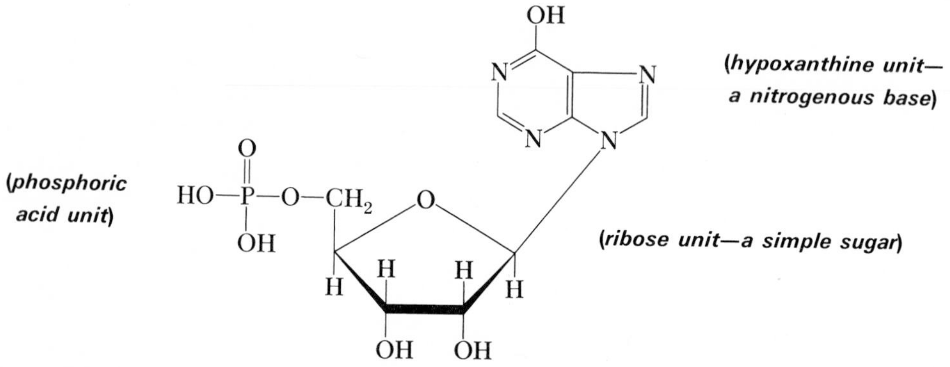

FIGURE 17–16   Inosinic acid (a nucleotide). If other bases are substituted for hypoxanthine, a number of nucleotides are possible for each of the two sugars. There is ample evidence that the nucleotides found in both DNA and RNA have the general structure indicated for inosinic acid.

POLYNUCLEOTIDES

In addition to the mononucleotides, partial hydrolysis of DNA or RNA yields oligonucleotides which have a few nucleotide units in their molecular structure. The structure for a trinucleotide is illustrated in Figure 17–17. Obviously, a large number of oligonucleotides are possible when one considers the choice of base structures and the different sequence possibilities for the chain of nucleotides.

DNA and RNA are polynucleotides. The number of possible structures for these molecules, which have molecular weights as high as a few million, appears to be almost limitless. Since DNA is a major part of the chromosome material in the nucleus of a cell, it seems reasonable to assume that the organism's characteristics are coded in the DNA structure. It has been estimated that there are over two million different species of organisms. If each individual requires a different DNA structure, there are ample combinations of nucleotides for each individual to be unique. It is now believed that some kinds of RNA transfer the information coded in the DNA structure to control the chemistry in the cytoplasmic region of the cell.

SECONDARY STRUCTURE OF DNA AND RNA

In 1953, two scientists, Watson and Crick, proposed a secondary structure for DNA that has since gained wide acceptance. Figure 17–18 illustrates the

362

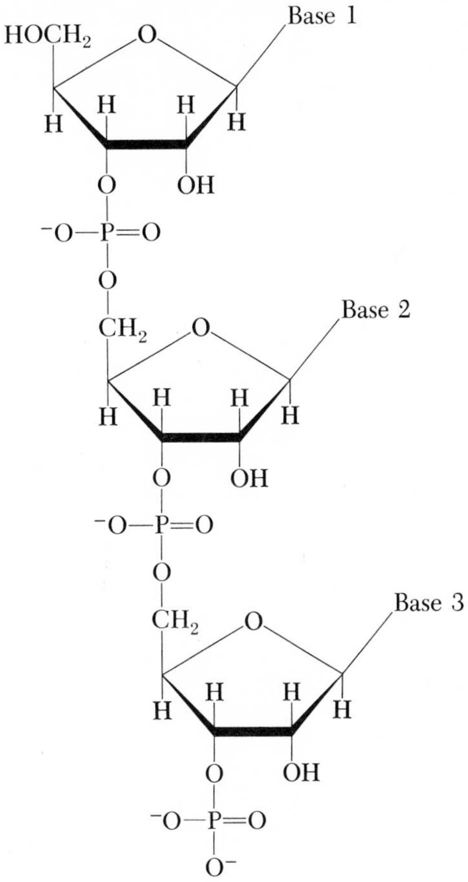

FIGURE 17-17   Bonding structure of a trinucleotide. Bases 1, 2, and 3 represent any of the nitrogenous bases obtained in the hydrolysis of DNA and RNA. The primary structure of both DNA and RNA is an extension of this structure to produce molecular weights as high as a few million.

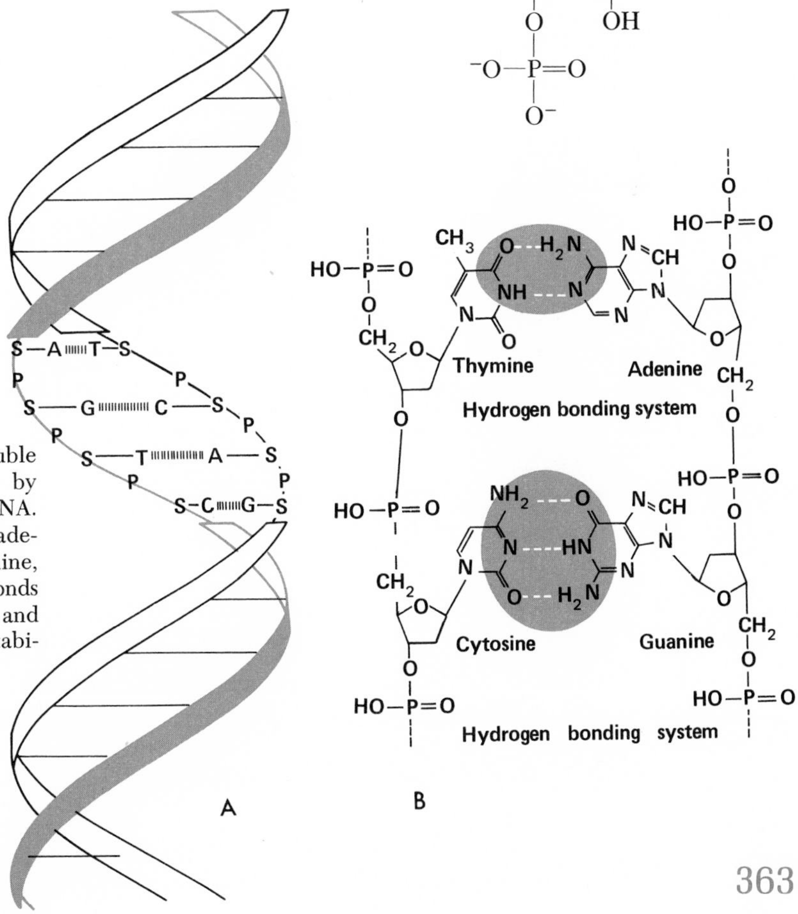

FIGURE 17-18   A, Double helix structure proposed by Watson and Crick for DNA. S-sugar, P-phosphate, A-adenine, T-thymine, G-guanine, C-cytosine. B, Hydrogen bonds in the thymine—adenine and cytosine—guanine pairs stabilize the double helix.

proposed structure in which two polynucleotides are arranged in a double helix stabilized by hydrogen bonding between the adjoining base groups. RNA is generally a single strand of helical polynucleotide.

SYNTHESIS OF LIVING SYSTEMS

In his quest for a molecular understanding of living systems, man must finally put his theories to the ultimate test of synthesis. If he can synthesize some of the complex molecules described above and see them successfully participate in the life processes, he can be reassured that he is on the right track. It should be emphasized that the biochemist is presently working at the molecular level, and as yet only a relatively few of the giant molecules have been characterized. The syntheses in a living cell are far too complex for our present methods to duplicate. However, in spite of the enormity of this undertaking, remarkable strides have been made recently, and interest in current research in this area is intense.

FIGURE 17–19   F. H. C. Crick (1916–   ) (right) and J. D. Watson (1928–   ) (left), working in the Cavendish Laboratory at Cambridge, built scale models of the double helical structure of DNA based on the x-ray data of M. H. F. Wilkins. Knowing distances and angles between atoms, they compared the task to the working of a three-dimensional jigsaw puzzle. Watson, Crick, and Wilkins received the Nobel Prize in 1962 for their work relating to the structure of DNA.

VIRUS STRUCTURE

A virus is a parasitic chemical complex that can reproduce only when it has invaded a host cell. It has the ability to disrupt the life processes of the host cell and order the cell contents therein to reproduce the virus structure. The isolated virus unit has neither the enzymes nor the smaller molecules necessary to reproduce itself alone.

A virus is a polynucleotide surrounded by a layer of protein. One virus that has been studied in detail is the tobacco mosaic virus, illustrated in Figure 17–20.

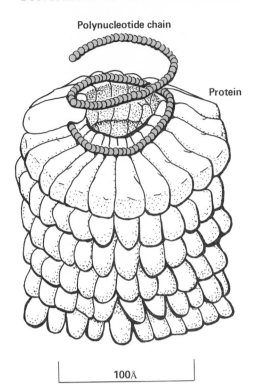

Polynucleotide chain

Protein

100Å

FIGURE 17-20   The structure of the tobacco mosaic virus. The polynucleotide chain is coiled, and there is one protein unit for each three nucleotide units. Part of the polynucleotide chain is exposed for clarity. This structure is based on X-ray studies.

REPLICATION OF DNA

Almost all of the cells of one organism contain the same chromosome structure in their nuclei. This structure remains constant regardless of whether the cell is starving or has an ample supply of food materials. Each organism begins life as a single cell with this same chromosome structure; in sexual reproduction one-half of this structure comes from each parent. These well known biological facts, along with recent discoveries concerning polynucleotide structures, lead to the conclusion that the DNA structure is faithfully copied during normal cell division (mitosis—both strands) and is only partly copied in cell division producing reproductive cells (meiosis—only one strand).

A prominent theory of DNA replication, based on verifiable experimental facts, suggests that the double helix of the DNA structure unwinds and each half of the structure serves as a template or pattern to reproduce the other half from the molecular environment (Figure 17–21).

RNA AND PROTEIN SYNTHESIS

Three major types of RNA have been identified. They are messenger RNA (mRNA), transfer RNA (tRNA), and ribosomal RNA (rRNA). Each has a characteristic molecular weight and base composition. Messenger RNA's are generally the largest, with molecular weights between 25,000 and one million. They contain from 75 to 3000 mononucleotide units. Transfer RNA's have molecular

365

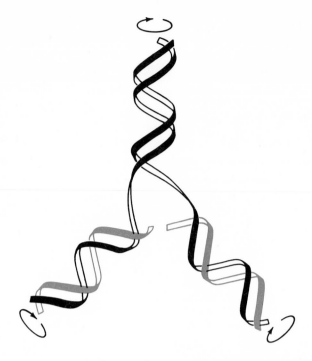

FIGURE 17-21   Replication of DNA structure. When the double helix of DNA (black) unwinds, each half serves as a template on which to assemble subunits (color) from the environment.

weights in the range of 23,000 to 30,000 and contain 75 to 90 mononucleotide units. Ribosomal RNA's make up as much as 80 per cent of the total cell RNA and have molecular weights between those of mRNA's and tRNA's.

All RNA's consist of a single polyribonucleotide strand. Most RNA is found in the cytoplasm and ribosomes of the cell (Figure 17–22), but in liver cells as much as 11 per cent (largely mRNA) of the total cell RNA is found in the nucleus. Besides having different molecular weights, the three types of RNA appear to differ in function.

Messenger RNA is formed in the nucleus during the process of transcription, during which the sequence of bases in one strand of the chromosomal DNA serves as the template for monoribonucleotides to order themselves into a single strand of mRNA (Figure 17–23). The bases of the mRNA strand complement those of the DNA strand. A pair of complementary bases is so structured that each one fits together and forms a hydrogen bond. Messenger RNA contains only the four bases adenine (A), guanine (G), cytosine (C), and uracil (U). DNA contains principally the four bases adenine (A), guanine (G), cytosine (C), and thymine (T). The base pairs are as follows:

DNA	mRNA
A	U
G	C
C	G
T	A

This means that every place a DNA has an adenine base (A), the mRNA will transcribe a uracil base (U), and so on, provided the necessary enzymes are present.

After transcription, mRNA passes from the nucleus of the cell to a ribosome,

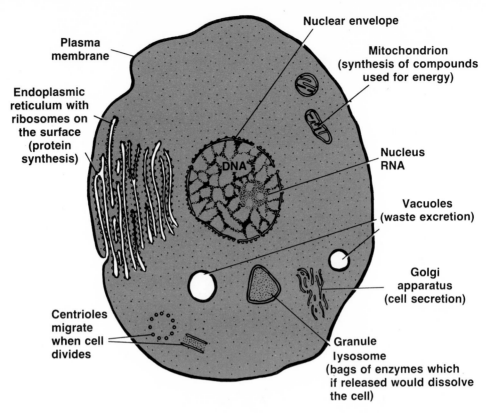

Plasma membrane

Nuclear envelope

Endoplasmic reticulum with ribosomes on the surface (protein synthesis)

Mitochondrion (synthesis of compounds used for energy)

DNA

Nucleus RNA

Vacuoles (waste excretion)

Golgi apparatus (cell secretion)

Centrioles migrate when cell divides

Granule lysosome (bags of enzymes which if released would dissolve the cell)

FIGURE 17–22   Diagrammatic generalized cell to show the relationships between the various components of the cell.

where it serves as the template for the sequential ordering of amino acids during protein synthesis. As its name implies, messenger RNA contains the sequence message for ordering amino acids into proteins. Each of the thousands of different proteins synthesized by cells is coded by a specific mRNA or segment of a mRNA molecule.

Transfer RNA's act as carriers of specific amino acids during protein synthesis on a ribosome. Each of the 20 amino acids found in proteins has at least one corresponding tRNA, and some have multiple tRNA's. For example, there are five distinctly different tRNA molecules specifically for the transfer of the amino acid, leucine, in cells of the bacterium, *Escherichia coli*. At one end of a tRNA molecule is a trinucleotide base sequence that fits a trinucleotide base sequence on mRNA. At the other end of a tRNA molecule is a specific base sequence of three terminal nucleotides—CCA—with a hydroxyl group exposed on the terminal adenine nucleotide group. This hydroxyl group reacts with a specific amino acid by an esterification reaction and with the aid of enzymes.

$$(Mononucleotides)_{75-90} \ \text{CCA—OH} + \text{HO}\overset{\displaystyle O}{\overset{\displaystyle \|}{\text{C}}}\text{CH(NH}_2)\text{R} \longrightarrow$$

tRNA   Amino acid

$$(Mononucleotides)_{75-90}\text{CCA—O}\overset{\displaystyle O}{\overset{\displaystyle \|}{\text{C}}}\text{CH(NH}_2)\text{R} + \text{H}_2\text{O}$$

tRNA-amino acid

367

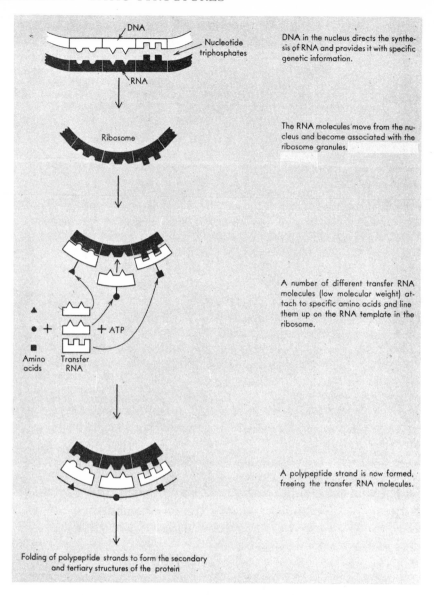

DNA in the nucleus directs the synthesis of RNA and provides it with specific genetic information.

The RNA molecules move from the nucleus and become associated with the ribosome granules.

A number of different transfer RNA molecules (low molecular weight) attach to specific amino acids and line them up on the RNA template in the ribosome.

A polypeptide strand is now formed, freeing the transfer RNA molecules.

FIGURE 17-23   A schematic illustration of the role of DNA and RNA in protein synthesis. (From McElroy, W. D., "Cell Physiology and Biochemistry," second edition, Prentice-Hall, Inc., Englewood Cliffs, N.J., 1964.)

The tRNA's line up on the mRNA by means of hydrogen bonding, and the amino acids carried by the tRNA's are transferred enzymatically to the end of a growing peptide chain.

Although ribosomal RNA makes up a large fraction of total cellular RNA, its function is not yet clear. It contains the four major bases A, G, C, and U; but like tRNA, rRNA contains a few other bases, including methylated derivatives of the major bases.

SYNTHETIC PROTEIN

A mixture of amino acids will react to form polypeptide structures outside of a living cell. Only relatively simple catalytic agents are required, and the

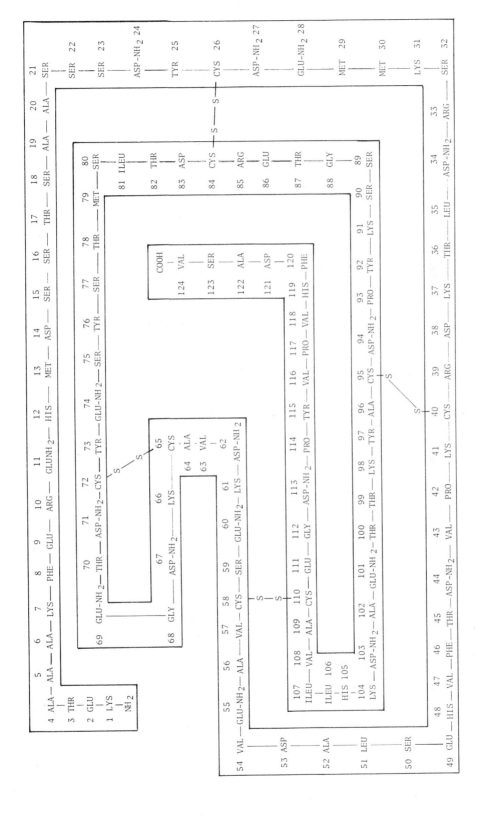

FIGURE 17-24   The structure of ribonuclease A, an enzyme. This protein structure is composed of 124 amino acid units. The structure is partially explained by sulfur-sulfur bonds between some of the acid units—for example, between 40 and 95.

complex enzymatic environment of cytoplasm is not necessary. However, a mixture of only a few amino acids will result in a multitude of different protein structures.

Methods too complex to be detailed here have been devised to construct protein structures with a desired sequence of amino acids. In order to obtain the desired bond between an amino acid and the peptide already constructed, it is necessary to block all the functional groups except those which are to undergo the peptide reaction. With the chemical blocking groups in place, the particular amino acid is added and the peptide bond is made. This is followed by removal of the blocking groups. To build a polypeptide with 20 amino acid units, this complicated process would have to be repeated 19 times. As the peptide chain grows, it becomes increasingly difficult to carry out the chemical operations without disturbing the bonds previously formed.

In spite of the difficulties, the customized synthesis of a prescribed protein has progressed steadily. In 1953, Vincent du Vigneaud synthesized a hormone from 9 amino acids. For his work he received a Nobel Prize in 1955. Other, more lengthy molecules have been synthesized similarly by starting with the individual amino acids and adding them in proper sequence to make the desired protein. Notable hormone syntheses were β-corticotropin with 39 amino acid units, and insulin, with 58 units. In 1969, two teams of researchers synthesized the first enzyme, ribonuclease (Figure 17–24), to be assembled outside the living cell from individual amino acids. Ribonuclease contains 124 amino acid units. One team was led by Rockefeller University's Robert B. Merrifield and Bernd Gutte, and the other team at Merck Sharp & Dohme research laboratories was led by Robert G. Denkewalter and Ralph F. Hirschmann.

Merrifield's method has had wider application. In this automated technique, an insoluble solid support, polystyrene, acts as an anchor for the peptide chain during the synthesis (Figure 17–25). The first amino acid is firmly bonded to a small polystyrene bead, and each of the other amino acids is then added one at a time in a stepwise manner. The synthesis of ribonuclease required 369 chemical reactions and 11,931 steps in a continuous operation on a machine developed for this purpose.

In 1970, a group led by Choah Hao Li at the University of California used Merrifield's method to synthesize human growth hormone (HGH) (Figure 17–26). HGH has 188 amino acid units and a molecular weight of about 21,500. This hormone is produced naturally by the front lobe of the pituitary gland, a pea-sized body located at the base of the brain. In humans, it has a dual function—it controls milk formation and it regulates many aspects of growth. In childhood, excess secretions of HGH can cause giantism; too little HGH causes dwarfism.

These synthetic proteins are identical in every respect to the natural products. They perform the same functions and given the same results on analysis. The limits of the present methods are unknown; in time even longer molecules are expected to be duplicated.

SYNTHETIC NUCLEIC ACIDS

Progress in the synthesis of polynucleotides has been difficult, principally because of the difficulties involved in determining the proper blocking groups. Progress, although slow, is being made.

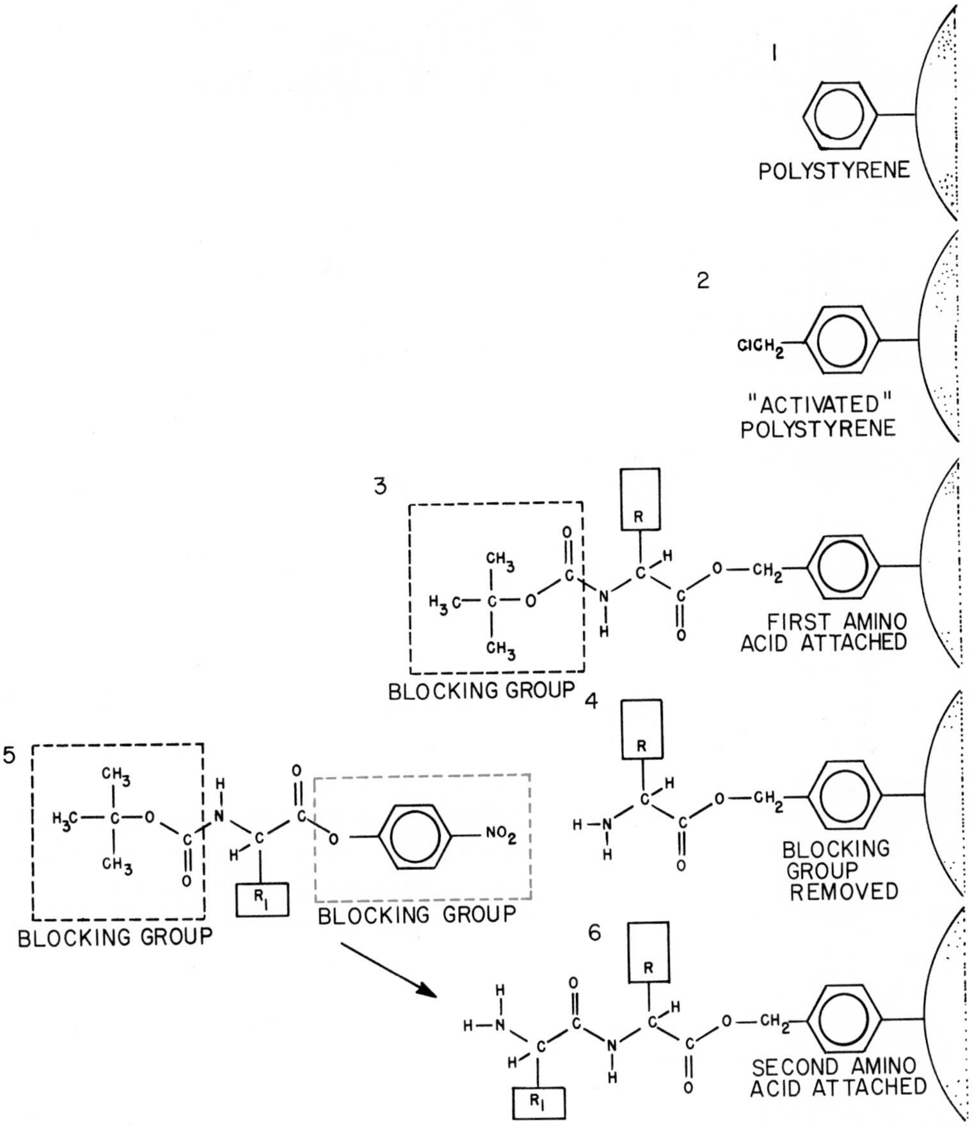

FIGURE 17-25   Solid phase method of synthesizing a protein is carried out stepwise from the carboxyl end toward the amino end of the peptide. An aromatic ring of the polystyrene (1) is activated by attaching a chloromethyl group (2). The first amino acid (black), protected by a butyloxycarbonyl (Boc) group (black box), is coupled to the site (3) by a benzyl ester bond and is then deprotected (4). Subsequent amino acid units are supplied in one of two activated forms; a second unit is shown in one of these forms, the nitrophenyl ester of the amino acid (5). The ester (colored box) is eliminated as the second unit couples to the first. Then the second unit is deprotected, leaving a dipeptide (6). These processes are repeated to lengthen the peptide chain.

371

FIGURE 17–26   Dr. Li (center) checks experiment with Dr. Richard Noble (left) and Dr. Donald Yanashiro (right). This group was the first to synthesize the HGH polypeptide. (From *Chemical and Engineering News*, January 11, 1971.)

In 1959, Arthur Kornberg synthesized a DNA-type polynucleotide, for which he received a Nobel Prize. He used natural enzymes as templates to arrange the nucleotides in the order of a desired polynucleotide. His product was not biologically active. In 1965, Sol Spiegelman synthesized the polynucleotide portion of an RNA virus. This polynucleotide was biologically active and reproduced itself readily when introduced into living cells. In 1967, Mehran Goulian and Kornberg synthesized a fully infectious virus of the more complicated DNA type.

In 1970, Gobind Khorana synthesized a complete, double-stranded, 77-nucleotide gene. He, too, used natural enzymes to join previously synthesized, short, single-stranded polynucleotides into the double-stranded gene.

Work is under way to synthesize other genes such as the tyrosine-suppressor transfer-RNA found in the bacteria *Escherichia coli*. This work, as have all previous syntheses of polynucleotides, relies on natural enzymes and extracts from natural systems to order the nucleotides. The real breakthrough of using strictly chemical means to align and bond the nucleotides in a desired sequence is on the horizon. This goal may have already been reached by the time you read this material since research in this area is moving very rapidly.

If man can construct DNA, can he then control the genetic code? Genes are the submicroscopic, theoretical bodies proposed by early geneticists to explain the transmission of characteristics from parents to progeny. It was thought that genes composed the chromosomes, which are large enough to be observed through the microscope as the central figures in cell division. It is now generally believed that DNA structures carry the message of the genes; hence,

DNA contains the *genetic code*. If man can construct DNA, he could very well alter its structure and thereby control the genetic code.

A *mutation* occurs when an individual characteristic appears that has not been inherited but has been passed along as an inherited factor to the next generation. A mutation can readily be accounted for in terms of the DNA genetic code; that is, some force alters the nucleotide structure in a reproductive cell. Some sources of energy, such as gamma radiation, are known to produce mutations. This is entirely reasonable because certain kinds of energy can disrupt some bonds, which will re-form in another sequence.

If man can control the genetic code, can he control hereditary diseases such as sickle cell anemia, gout, some forms of diabetes, or mental retardation? If the understanding of detailed DNA structure and the enzymatic activity in building these structures continues to grow, it is reasonable to believe that some detailed relationships between structure and gross properties will emerge. If this happens, it may be possible to build compounds which, when introduced into living cells, can combat or block inherited characteristics.

CONCLUSION AND PREVIEW

A very few of the many possible areas of biochemistry have been explained in this chapter. In each of these areas we can see the application of chemical principles and molecular structures in the development of an understanding of some processes in living cells. Our knowledge in these areas has increased steadily, and there is every reason to suspect that this increase will continue because of its importance to the understanding of ourselves.

In the next chapter, we shall consider some important biochemical reactions that provide living cells with energy. After this, we shall examine some chemicals that are toxic to the human body.

QUESTIONS

1. Show the structure of the product which would be obtained if two alanine molecules (Table 17–1) react to form a dipeptide.

2. What is an essential amino acid?

3. The ketone structure of D-fructose has three asymmetric carbon atoms per molecule. How many isomers result from these asymmetric centers?

4. What polysaccharide yields only D-glucose upon complete hydrolysis? What disaccharide yields the same hydrolysis product?

5. What is the difference between the starch, amylopectin, and the "animal starch," glycogen?

6. What is the chief function of glycogen in animal tissue?

7. Explain the basic difference between starch, amylose, and cellulose.

8. What functional groups are always present in each molecule of an amino acid?

9. Give the name and formula for the amino acid simplest in structure. What natural product has a high percentage of this amino acid?

373

10. If three amino acids form all of the possible tripeptides, how many would there be?

11. What is the meaning of the terms: *primary, secondary,* and *tertiary structures of proteins?*

12. What three molecular units are a part of nucleotides?

13. Based on the structure in Figure 17–15, explain the meaning of the prefix *deoxy-* in deoxyribonucleic acid.

14. How many trinucleotides with the structure indicated in Figure 17–18 could be made with the nitrogenous bases listed in Figure 17–16?

15. What are the basic differences between DNA and RNA structures?

16. What stabilizing forces hold the double helix together in the secondary DNA structure proposed by Watson and Crick?

17. What is the meaning of the term *enzyme specificity?*

18. What two molecular structures are present in viruses?

19. Does a strand of DNA actually duplicate itself base for base in the formation of a strand of messenger RNA? Explain.

20. Distinguish between the three types of RNA as to molecular weight, number of amino acid units per molecule, and function.

21. (a) Describe the general method of synthesizing DNA *in vitro* (in a test tube) at present.
    (b) Why would the synthesis of a polynucleotide from the individual phosphoric acid, sugar, and nitrogenous bases be a breakthrough in controlling the genetic code?

22. Discuss the feasibility of solving sociological and psychological problems via religion, DNA alteration, chemical suppressants, political pressures, and/or persuasive dialogue.

23. Check the recent issues of *Science* or other scientific news publications to update the work done on synthesis of proteins and polynucleotides.

SUGGESTIONS FOR FURTHER READING

"A Step Toward Synthetic Life," *Chemistry,* Vol. 41, No. 2, p. 27 (1968).

"Anatomy of a Virus," *Chemistry,* Vol. 39, No. 4, p. 20 (1966).

Battista, O. A., "Sugar—The Chemical with a Thousand Uses," *Chemistry,* Vol. 38, No. 4, p. 12 (1965).

"Bonding Habits of DNA Bases," *Chemistry,* Vol. 41, No. 8, p. 34 (1968).

Crick, F., "The Genetic Code: III," *Scientific American,* p. 55, October (1966).

Davies, D. R., "X-ray Diffraction and Nucleic Acids," *Chemistry,* Vol. 40, No. 2, p. 8 (1967).

Fraenkel-Conrat, H., "The Genetic Code of a Virus," *Scientific American,* p. 46, October (1964).

Hofmann, K., Khorana, H., and Spiegelman, S., "The Synthesis of Living Systems," *Chemical and Engineering News,* Vol. 45, August 7, p. 144 (1967).

"Interlocking Rings of DNA," *Chemistry,* Vol. 41, No. 2, p. 26 (1968).

"Is DNA the Master Molecule," *Chemistry,* Vol. 41, No. 6, p. 23 (1968).

Mazur, A., and Harrow, B., *Biochemistry: A Brief Course,* W. B. Saunders Company, Philadelphia, 1968.

Merrifield, R. B., "The Automated Synthesis of Proteins," *Scientific American,* Vol. 218, No. 3, p. 56 (1968).

Pauling, L., Corey, R., and Hayward, R., "The Structure of Protein Molecules," *Scientific American,* p. 51, July (1954).

"Portrait of a Gene," *Chemistry,* Vol. 42, No. 8, p. 20 (1969).

Raw, I., "Enzymes, How They Operate," *Chemistry,* Vol. 40, No. 6, p. 8 (1967).

"Trouble on the DNA Front," *Chemistry,* Vol. 43, No. 9, p. 24 (1970).

"Viruses as Invaders of Living Cells," *Chemistry,* Vol. 41, No. 5, p. 29 (1968).

Watson, J. D., *The Double Helix,* Atheneum, New York, 1968.

Zimmerman, J., "First Synthesis of an Enzyme, Ribonuclease," *Chemistry,* Vol. 42, No. 4, p. 21 (1969).

BIOCHEMICAL PROCESSES

━━━━━━━━━━ **CHAPTER 18**

INTRODUCTION

In Chapter 17 we examined the structures of some molecules that are important in biochemical systems. We will now look at some of the reactions by which such molecules are constructed, serve their function, and are subsequently destroyed. There are three aspects of biochemical systems which should be kept in mind in this study. First, the contents of a living cell are in a dynamic state; the molecules are constantly being synthesized and degraded. However, the healthy living cell is characterized by a "steady state" condition in which the rates of buildup and breakdown are nearly the same at any time. Second, biochemical processes are very general in that the same basic chemistry is employed by a wide variety of cells. For example, the same types of reactions utilized to obtain energy from a compound in man also occur in simple unicellular organisms. Third, biochemical systems are composed of literally thousands of different kinds of molecules. Out of such an apparent chaos comes an ordered array of reactions that supports all life forms. In view of this complexity, it is evident that an elementary treatment of what is presently known about biochemical reactions will have to deal only with selected highlights.

BIOCHEMICAL ENERGY AND ATP

Not only do the structures of biochemicals determine the gross structures of organisms, they also control the basic flow of energy that is constantly needed to support life.

In biochemical processes energy is required to drive the process to yield the desired products. This energy comes from the ultimate energy source in our environment: *the sun*. The biochemist, then, is interested in how the energy of sunlight can be stored in various compounds, which can in turn supply energy to feed the life processes.

375

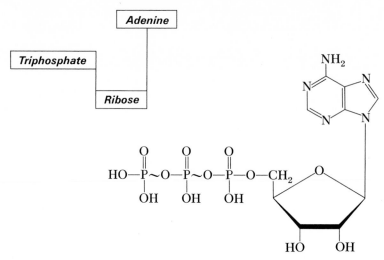

FIGURE 18-1   Molecular structure of adenosine triphosphate (ATP). Note
the similarity between ATP and the basic unit of nucleic acid, the nucleotide,
Chapter 17.

The energy obtained by the oxidation of foods is mostly used to synthesize
the key molecule, adenosine triphosphate, which is often written ATP (Figure
18–1).

ATP is the basic reservoir of energy in the living cell. This molecule is the
immediate source of energy in muscular contraction, and the energy released
by it allows many energy-requiring biochemical reactions to occur.

ATP molecules each contain two so-called high energy phosphate bonds.
These are marked by wiggle bonds ($\sim$) in Figure 18–1. In the presence of a
suitable catalyst, ATP will undergo a three-step hydrolysis. (Recall that hydrolysis
is a reaction in which a chemical bond is broken and water is added—H to
one part and OH to the other—at the point of rupture.) The hydrolysis of ATP
to adenosine diphosphate (ADP) and phosphoric acid releases about 12 kilocalories per mole. The second hydrolysis of ADP to adenosine monophosphate

(AMP), also produces about 12 kilocalories of energy per mole. Finally, the hydrolysis of AMP to adenosine, which involves a low energy bond, releases only about 2.5 kilocalories per mole.

$$\boxed{\text{ADP}} + \text{HOH} \xrightarrow{\text{catalyst}}$$

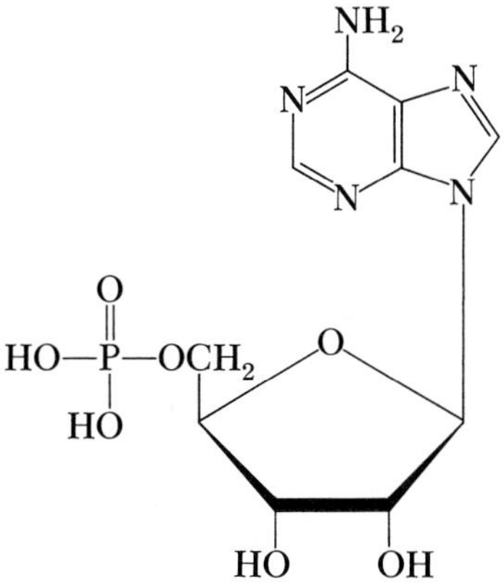

$$+ H_3PO_4 + 12 \text{ Kcal}$$
(approx.)

Adenosine monophosphate
(AMP)

$$\boxed{\text{AMP}} + \text{HOH} \xrightarrow{\text{catalyst}}$$

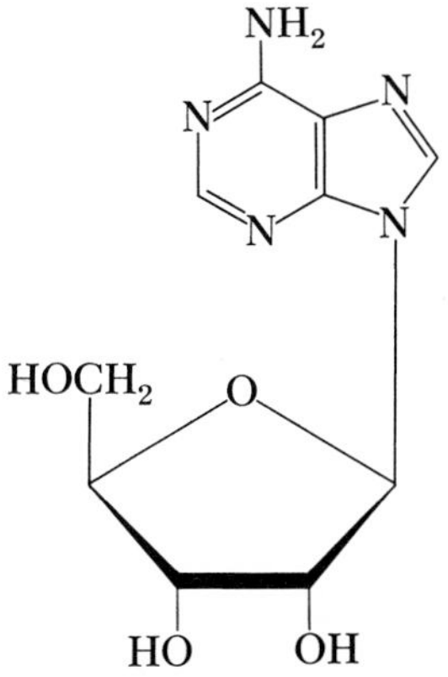

$$+ H_3PO_4 + 2.5 \text{ Kcal}$$
(approx.)

Adenosine

Studies of the energetics of chemical reactions show that not all the energy released is available to initiate other chemical change. The energy figures for these three reactions are given in this context only to show the relatively large amount of energy associated with the hydrolysis of the first two phosphate units in the ATP structure. A more meaningful figure is the energy change that is available to initiate other chemical change; this is termed the change in *free* energy. A change in free energy is symbolized by ΔG. For the three hydrolytic reactions of ATP the approximate free energy changes are:

ATP hydrolysis: $\Delta G = -7.4$ Kcal per mole
ADP hydrolysis: $\Delta G = -6.8$ Kcal per mole
AMP hydrolysis: $\Delta G = -2.2$ Kcal per mole

(Note: The negative sign means that energy is evolved.)
These free energy figures indicate the amount of chemical potential energy 377

in the phosphate bonds in the ATP molecule. Such energy-producing reactions provide the energy necessary for energy-requiring biochemical reactions to occur, thus supplying the energy necessary for the basic life processes.

INTERMEDIATE ROLE OF ATP (COUPLED REACTIONS)

The human body consumes large amounts of glucose and oxygen and subsequently excretes large amounts of carbon dioxide and water. The sugar is oxidized in the body to carbon dioxide and water. A major source of energy to support human life comes from this oxidative process, the loss in free energy being 686 kilocalories per mole of glucose:

$$C_6H_{12}O_6 + 6O_2 \longrightarrow 6CO_2 + 6H_2O$$

$$\Delta G = -686 \text{ Kcal per mole}$$

The body also ingests many amino acids, or their equivalent protein structures, and builds them into peptide and protein structures characteristic of the individual. The formation of the peptide bond, unlike the oxidation of glucose, absorbs energy, the net free energy change being about 0.45 kilocalories per mole of bonds, depending on the particular structure involved:

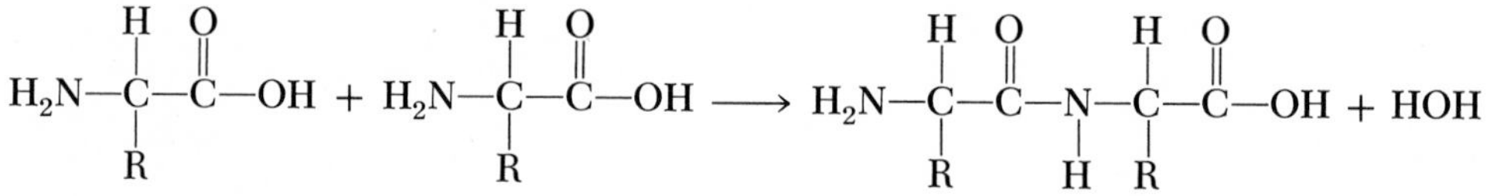

Amino acid (1)   Amino acid (2)   A dipeptide

$$\Delta G = +0.45 \text{ Kcal per mole}$$

The preceding discussion might seem to imply a rather uncomplicated process. It might seem obvious to suggest that the amino acids acquire the energy directly from the oxidation of glucose and use it to form the peptide structure. However, a very simple experiment shows that this is not the case. If oxygen, glucose, and amino acids are mixed together in water, the glucose and oxygen are consumed, and carbon dioxide bubbles out of the mixture; but the amino acids remain essentially unchanged. In addition, energy is released from the mixture in the form of heat. Evidently there is no way for the energy from the oxidation of glucose to be utilized by the amino acids to form peptides.

The problem of transferring energy from the one reaction to the other is solved in an organism by the presence of ADP, inorganic phosphate (P), and enzymes. The process can be illustrated by a specific chemical reaction that occurs during the anaerobic metabolism of glucose. One step of the overall reaction is the following transformation:

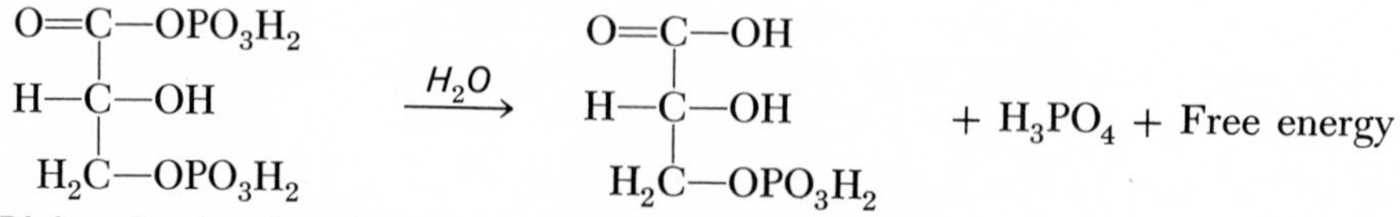

1,3-Diphosphoglyceric acid   3-Phosphoglyceric acid

If this reaction is carried out in the presence of a suitable enzyme (phospho-glyceryl kinase), magnesium ion, and ADP, it follows a different course, and the energy released is used to form ATP from ADP:

$$\left.\begin{array}{l} \text{1,3-Diphosphoglyceric acid} \\ +\text{ADP} \\ +\text{Enzyme} + \text{Mg}^{2+} \end{array}\right\} \longrightarrow \left\{\begin{array}{l} \text{3-Phosphoglyceric acid} \\ +\text{ATP} \\ +\text{Enzyme} + \text{Mg}^{2+} \end{array}\right.$$

This type of reaction scheme, in which the energy of one reaction is used to drive another reaction, is called a set of *coupled reactions*. Without ATP or a similar compound to store the energy, the formation of products merely leads to the release and loss of energy. Cells utilize coupled reactions to transform the compounds from which they get their energy. As a result, they manufacture ATP, whose energy can then be used to drive other processes or reactions necessary to the life of the cell. The transformation of simple molecules by means of coupled reactions usually requires a large number of steps, many of which can be used to generate ATP. The aerobic oxidation of glucose actually proceeds by a sequence of over a dozen individual steps to give the overall reaction

$$\text{Glucose} + 38\text{Phosphate} + 38\text{ADP} + 6\text{O}_2 \longrightarrow 6\text{CO}_2 + 44\text{H}_2\text{O} + 38\text{ATP}$$

ENZYMES

An enzyme is a biochemical catalyst. Like other catalysts, a given enzyme facilitates a reaction that would not occur, or occur only very slowly, under a given set of conditions. As you can determine by experiment, glucose does not readily burn in air. If, however, some cigarette ashes, or other catalysts, are placed on its surface, combustion can be initiated easily with a match. Further-more, once the sugar burns, it liberates much more energy than was required to initiate the reaction:

$$\underset{\text{Sugar}}{\text{C}_6\text{H}_{12}\text{O}_6} + \underset{\text{Oxygen}}{6\text{O}_2} \longrightarrow \underset{\substack{\text{Carbon} \\ \text{dioxide}}}{6\text{CO}_2} + \underset{\text{Water}}{6\text{H}_2\text{O}};\ \Delta\text{G} = -686\ \text{Kcal per mole}$$

The energy required to get the reaction started is called the *activation energy*. If an enzyme can lower the activation energy to the average kinetic energy of the molecules at the temperature of a living cell (or in the laboratory system), the reaction can proceed rapidly at that temperature. Glucose, to continue the example cited earlier, is oxidized rapidly and efficiently at ordinary temperatures in the presence of biochemical catalysts. To be sure, the oxidation of glucose in a living cell requires many enzymes and many steps, but the point to remem-ber is that the enzymatic catalysis produces the same final result as the combus-tion at elevated temperature, namely carbon dioxide, water, and 686 kilocalories of usable energy per mole of sugar. Figure 18–2 illustrates the concepts of

379

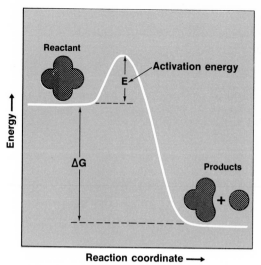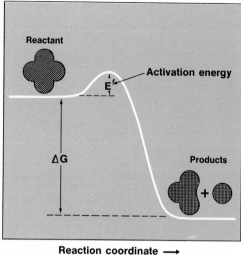

a. No enzyme present   b. Enzyme present

FIGURE 18–2   Enzyme Effect on Activation Energy. The vertical coordinate represents increasing energy and the horizontal one the course of the reaction in going from reactants to products. For energy-producing reactions the reactant molecule is at a higher energy than the product molecule, as illustrated in *a*. The difference between these energies is the net free energy change of the reaction. However, it is necessary for the reactant molecule to "get over" the energy barrier (acquire the activation energy, E) in going from reactant to product. Note that the activation energy is given back along with the free energy. The enzyme lowers the activation energy, E', as illustrated in *b*, while the free energy change remains the same. The net effect is to obtain the free energy of the reaction with a smaller expenditure of activation energy.

activation energy, the energy available from an energy-producing reaction, and the reduction of the activation energy by an enzyme.

Enzymes are remarkable catalysts in that they are highly specific for a given reaction. There are literally hundreds of them in the cell medium, each one doing a very specific catalytic job.

How does an enzyme work? How can it lower the activation energy and be so specific for a given reaction at the same time? While a definitive answer cannot be given at this time (the matter is presently under intensive research), a "lock-and-key" analogy has been a fruitful approach to the problem. Just as a key can separate a padlock into two parts and subsequently remain unchanged ready to unlock other identical locks, so the enzyme makes possible a molecular change (Figure 18–3). With enough energy the lock could be separated without the key, and with enough energy the molecular alteration could occur without the enzyme.

Sometimes an enzyme requires another molecule, called a coenzyme, in order to function as a catalyst. This situation is rather similar to the lock-and-key hypothesis, as can be seen in Figure 18–4. (In the figure, the term *apoenzyme* refers to the protein portion of the enzyme; the *active site* refers to the protein part which is directly engaged in the process of catalysis.)

A chemical illustration will demonstrate these points. The peptide glycyl-glycine is hydrolyzed very slowly by water to give two molecules of glycine:

380

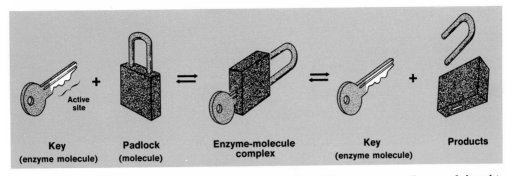

FIGURE 18-3   Lock-and-Key theory for enzymatic catalysis. While it is generally agreed that this analogy is an oversimplification, it does make one very important point: the enzyme makes a difficult job easy by reducing the energy required to get the job started. It also suggests that the enzyme has a particular structure at an active site which will allow it to work only for certain molecules, similar to a key that fits the shape of a particular keyhole.

$$\underset{\text{Glycylglycine}}{\overset{\text{H}\quad\text{O}}{\underset{|}{\text{HC}}-\overset{\text{O}}{\underset{|}{\text{C}}}-\underset{|}{\text{N}}-\text{CH}_2-\overset{\text{O}}{\underset{\text{OH}}{\text{C}}}} + \text{H}_2\text{O} \longrightarrow \underset{\text{Glycine}}{2\text{CH}_2-\overset{\text{O}}{\underset{\text{OH}}{\text{C}}}}$$

Figure 18-5 illustrates how this reaction can be facilitated by an enzyme. The reactant, or the molecule to be changed (in this case glycylglycine), is referred to as the substrate. The substrate forms an intermediate compound with the enzyme, and as a result the substrate is activated for further reaction. The

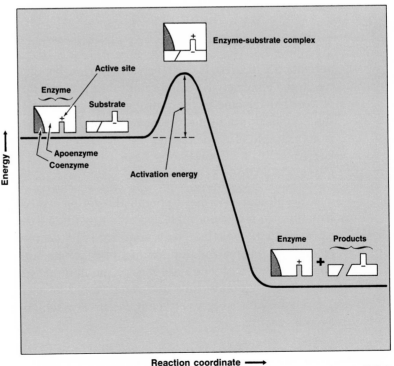

FIGURE 18-4   A reaction utilizing an enzyme which requires a coenzyme. Here the essential feature is that the combination of enzyme plus coenzyme allows the reaction to proceed with a lower activation energy. Apoenzyme is the name given to the enzyme structure that combines with the coenzyme.

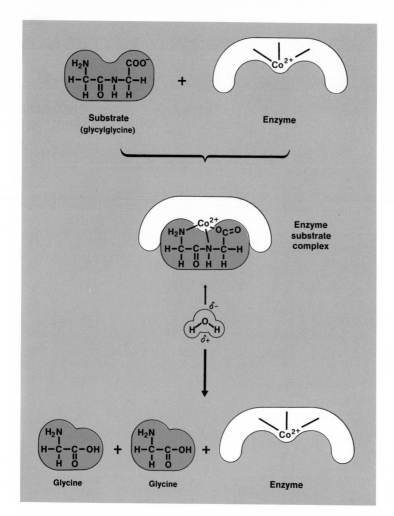

FIGURE 18-5   Operation of an enzyme. The substrate molecule is bonded to the enzyme (glycylglycine dipeptidase) through chemical bonding, the negative oxygen and the nitrogen atoms binding to the positive cobalt ion in the enzyme. The bonding of the substrate makes it more susceptible to attack by water. Hydrolysis occurs and the glycine molecules are released by the enzyme which is then ready to play its catalytic role again.

activation can arise via extensive hydrogen bonding, interaction with a metal ion in the enzyme, or a number of other processes. In this case the coordination of the glycylglycine to the positive charge of a cobalt ion (Co^{2+}) makes it more susceptible to attack by the negative end of a water molecule. The Co^{2+} ion is an example of a coenzyme.

PHOTOSYNTHESIS

Recall that a synthetic chemical reaction involves the buildup of a chemical structure from starting materials. Photosynthesis, then, is an appropriate name for a synthetic reaction that employs light energy as a driving force in the reaction. While many reactions might be included under this definition, the name photosynthesis has been reserved for the natural synthesis of carbohydrates and sugars from carbon dioxide and water, using solar energy to power the process.

Photosynthesis is a very complex process which belies the relatively simple overall reaction:

$$6CO_2 + 6H_2O + 686\ Kcal \longrightarrow C_6H_{12}O_6 + 6O_2$$

| Carbon dioxide | Water | Energy (Sunlight) | Glucose | Oxygen |

This equation indicates that carbon dioxide is reduced to sugar (reduction is the gain of electrons; see Chapter 11):

$$6CO_2 + 24H^+ + 24e^- \longrightarrow C_6H_{12}O_6 + 6H_2O$$

and that water is oxidized (oxidation is the loss of electrons):

$$12H_2O \longrightarrow 6O_2 + 24H^+ + 24e^-$$

Note that these two half reactions, the one reduction and the other oxidation, give the overall reaction when added together. Note also that the 24 hydrogen ions and the 24 electrons cancel.

This process is not completely understood. However, some aspects are presented here as part of our examination of the energy flow through biochemical systems.

Photosynthesis is generally considered in terms of the *light reaction,* one that can occur only in the presence of light energy, and the *dark reaction,* which can occur in the dark but feeds on the high energy structures produced in the light reaction. Actually, both the light and dark reactions are a series of reactions, all occurring simultaneously in the green plant cell. As you might suspect, the light reaction is unique to green plants while the dark reaction is characteristic of both plant and animal cells.

THE LIGHT REACTION

Photosynthesis is initiated by a quantum of light energy. The green plant contains certain pigments that readily absorb light in the visible region of the spectrum. The most important of these are the chlorophylls, *chlorophyll a* and *chlorophyll b* (Figure 18–6).

Chlorophyll a   Chlorophyll b

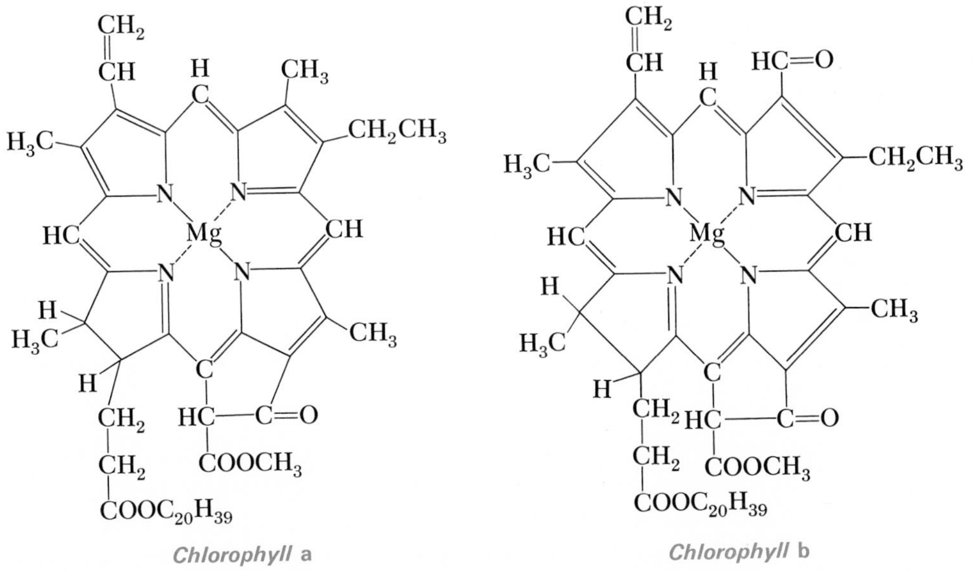

FIGURE 18–6   The molecular structure of chlorophyll a and chlorophyll b. Note that both are magnesium complexes and that they have complex ring systems which contain many π electrons.

383

Note that both are magnesium complexes and that they have complex ring systems which contain many π electrons (Chapter 14). Such ring systems usually absorb light in the visible region of the spectrum. For example, chlorophyll is green because it absorbs light in the violet region (about 4000 Angstroms) and the red region (about 6500 Angstroms) and allows the green light in between those wavelengths to be reflected, or transmitted.

When a chlorophyll unit absorbs photons of light, some of its electrons are elevated to such high energy levels that they can be transferred easily to other molecular units. There is actually a sequence of such units. One such agent is the oxidized form of nicotinamide adenine dinucleotide phosphate ($NADP^+$):

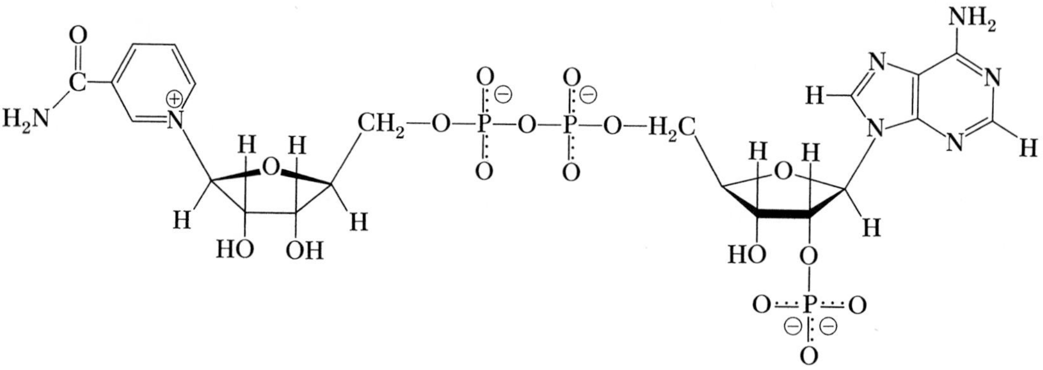

Nicotinamide adenine dinucleotide phosphate
(NADP)

Note that *dinucleotide* is an appropriate name for this compound since it is composed of two nucleotides (Chapter 17) joined together between the two phosphate groups. For our purposes this dinucleotide will be considered in its plus-one cationic form and represented by $NADP^+$. When a high energy electron from chlorophyll is acquired by a molecule of $NADP^+$, the reduction product is a molecule of NADPH:

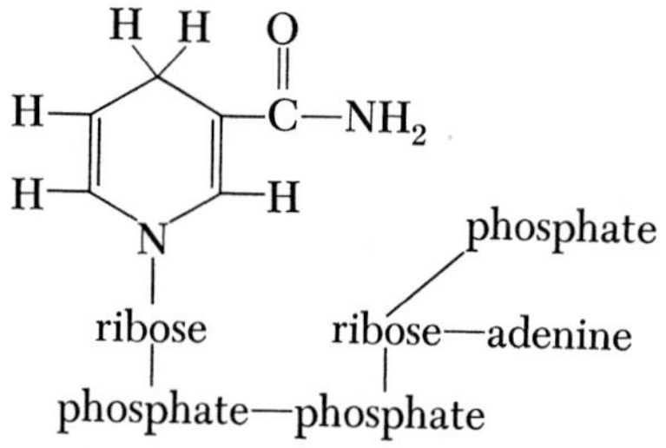

Reduced adenine dinucleotide phosphate
(NADPH)

The reduced form of $NADP^+$, NADPH, is now a reducing agent. The reducing power can now be passed to another oxidizing agent similar to $NADP^+$. In the process NADPH would be oxidized back to $NADP^+$, which would then be ready to serve in the same capacity again. While the details of the elaborate electron transport systems are presently unknown, one point is firmly established at this point: the reducing power decreases as it is passed along from one compound to another. It is important to remember that it is this reducing power that is

384

coupled to the ADP-ATP reaction. This is the energy that is necessary to form the last bond in the ATP structure:

$$\text{ADP} + \text{Phosphate} + \text{Energy} \longrightarrow \text{ATP}$$

At this point we have the energy necessary to run biochemical systems in the ATP structure, energy traced directly from the radiant energy of the sun. If the plant cell is given the minerals along with carbon dioxide and water, it, or subsequent living cells, can use the energy in ATP to help carry out the complex biochemical reactions that take place in a living cell.

Thus far little mention has been made of the electrochemical charge balance and the source of oxygen in photosynthesis, both of which are important parts of the light reaction. Water, in the presence of chloroplasts (plant cell bodies containing chlorphyll) and light, is decomposed to oxygen, hydrogen ions, and electrons:

$$2H_2O \longrightarrow 4H^+ + 4e^- + O_2$$

The hydrogen ions are available for other reactions, such as the one for $NADP^+$ discussed previously, and oxygen is liberated from the plant cell. To summarize the light reaction: radiant (light) energy decomposes water, providing a source of electrons and protons; the oxygen is the by-product, while other photons of light furnish energy through several intermediate reactions to form ATP.

THE DARK REACTION

The dark reaction responsible for the conversion of gaseous carbon dioxide to nongaseous compounds, a process known as *fixation*, was discovered by Calvin. Calvin studied the uptake of radioactive carbon in carbon dioxide by plant cell chloroplasts when they are illuminated by light for very short periods of time. He illuminated the plants for definite periods of time and then analyzed the plant cells to determine which compounds contained the most radioactive carbon. As the time periods were reduced, more of the radioactive carbon was found in those compounds into which it had been initially incorporated and less in compounds which had been formed in subsequent reactions. Thus, after only five seconds' illumination, radioactive carbon is found in the compound 3-phosphoglyceric acid:

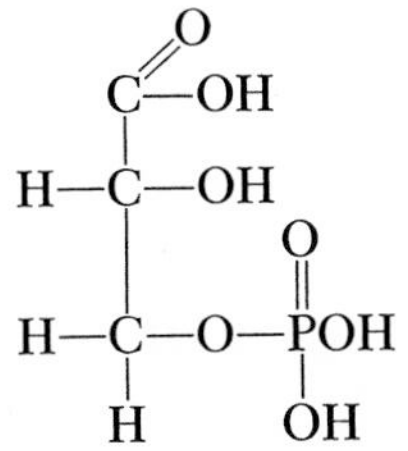

3-Phosphoglyceric acid

This compound is apparently formed *in the initial reaction* in which CO_2 from the air reacts with some molecule present in the plant. Calvin discovered that

385

the key was the reaction of atmospheric CO_2 with ribulose 1,5-diphosphate to give two molecules of 3-phosphoglycerate:

$$
\begin{array}{ccc}
\left.\begin{array}{c}
{}^*CO_2 \\
\textit{Carbon dioxide} \\
+ \\
CH_2OPO_3^= \\
| \\
C{=}O \\
| \\
HCOH \\
| \\
HCOH \\
| \\
CH_2OPO_3^=
\end{array}\right\}
& \longrightarrow &
\begin{array}{c}
{}^\circ COO^- \\
| \\
HCOH \\
| \\
CH_2OPO_3^= \\
+ \\
COO^- \\
| \\
HCOH \\
| \\
CH_2OPO_3^= \\
\textit{3-Phosphoglycerate}
\end{array}
\end{array}
$$

Ribulose 1,5-diphosphate

$^\circ C$ indicates the fate of radioactive carbon as determined by Calvin's experiments.

The 3-phosphoglycerate is then transformed into other carbohydrates in reactions which regenerate ribulose 1,5-diphosphate for further uptake of more atmospheric CO_2. The energy needed to carry out these reactions is furnished by the NADPH and ATP generated from the light reaction.

Figure 18–7 presents the cyclic character of the dark reaction of photosynthesis. Note that carbon dioxide enters the cycle at the upper left and that sugars are removed at the lower right.

The energy needed to carry out the dark reactions is furnished by the high energy compounds produced in the light reaction. After examining the gross breakdown of food (digestion), we shall turn our attention to how living cells can tap various energy sources, remake the active intermediate ATP, and thus obtain power for the ongoing activities of the cell.

DIGESTION

From a chemical point of view, digestion is the breakdown of ingested foods through hydrolysis. The products of these hydrolytic reactions are relatively small molecules that can be absorbed through the walls of the alimentary canal into the body fluids where they are used for metabolic processes. The hydrolytic reactions of digestion are catalyzed by enzymes, there being a specific enzyme for each hydrolysis. The hydrolysis of carbohydrates ultimately yields simple sugars, proteins yield amino acids, and fats yield fatty acids. The activities of some of the more important enzymes are described in Table 18–1.

CARBOHYDRATE DIGESTION AND ABSORPTION

The principal forms of carbohydrates in our food are (1) high molecular weight polymers such as starch (glycogen), (2) disaccharides such as sucrose, and (3) simple sugars such as glucose and fructose. The first enzyme capable of facilitating the splitting of polysaccharides and of acting on ingested food is furnished by the saliva and is named salivary amylase, or ptyalin. Its action produces limited amounts of the disaccharide maltose from starch or glycogen.

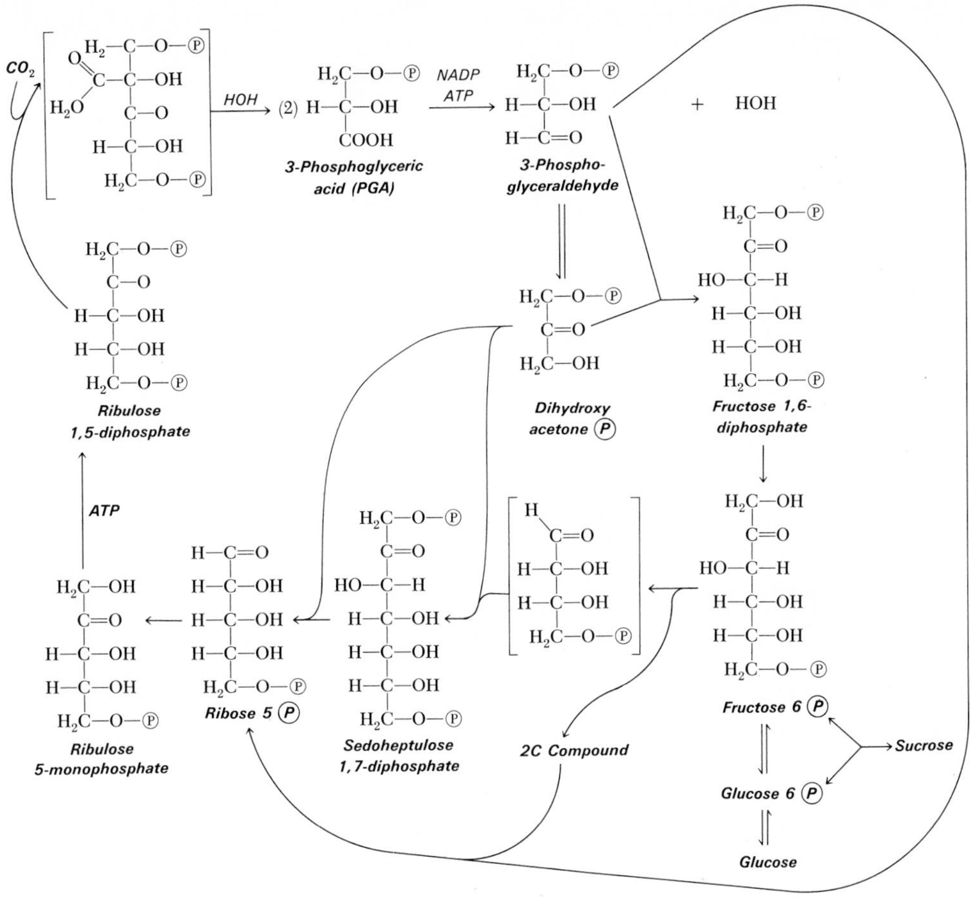

CO_2 H_2O

3-Phosphoglyceric acid (PGA)

3-Phosphoglyceraldehyde

NADP ATP

+ HOH

Ribulose 1,5-diphosphate

Dihydroxy acetone (P)

Fructose 1,6-diphosphate

ATP

Ribulose 5-monophosphate

Ribose 5 (P)

Sedoheptulose 1,7-diphosphate

2C Compound

Fructose 6 (P)

Glucose 6 (P)

Sucrose

Glucose

FIGURE 18-7 Abbreviated version of the dark reaction of photosynthesis. The energy needed to carry out the dark reactions is furnished from the high energy compounds produced in the light reaction.

Ptyalin is inactivated by the high acidity in the stomach, so its activity stops there. The stomach furnishes no enzymes which can catalyze the splitting of carbohydrate polymers.

When food passes from the stomach, its acidity is neutralized by a secretion of the pancreas, which also contains enzymes (Table 18–1) that can facilitate the splitting of some of the polysaccharides to maltose, the transformation of maltose to glucose, and the catalysis of hydrolytic reactions. The final result is a mixture of such simple sugars as glucose, fructose, and galactose. The net result of these processes is a supply of simple sugars, which are then absorbed into the blood stream where the control of blood sugar is regulated by the hormone insulin. If the sugar level is too high, the polysaccharide glycogen is produced in the liver; if the blood sugar is low, the stored glycogen is hydrolyzed.

LIPID DIGESTION AND ABSORPTION

The term *lipid* is used to describe a group of compounds which include fats and oils and other substances whose solubility characteristics are similar to

387

TABLE 18-1   PRINCIPAL DIGESTIVE ENZYMES

ENZYME	SOURCE	SUBSTRATE	PRODUCTS	OPTIMAL pH
Ptyalin	Salivary glands	Starch	Smaller carbohydrate polymers (minor physiologic role)	6–7
Pepsin	Chief cells of stomach	Protein	Polypeptides	1.6–2.4
Gastric lipase	Stomach	Fat	Glycerides, fatty acids (minor physiologic role)	—
Enterokinase	Duodenal mucosa	Trypsinogen	Trypsin	—
Trypsin	Exocrine pancreas	Denatured proteins and polypeptides	Small polypeptides (also activates chymotrypsinogen to chymotrypsin)	8.0
Chymotrypsin		Proteins and polypeptides	Small polypeptides	8.0
Nucleases		Nucleic acids	Nucleotides	—
Carboxypeptidases		Polypeptides	Smaller polypeptides°	—
Pancreatic lipase		Fat	Glycerides, fatty acids, glycerol	8.0
Pancreatic amylase		Starch	Maltose units	6.7–7.0
Aminopeptidases	Intestinal glands	Polypeptides	Smaller polypeptides†	8.0
Dipeptidase		Dipeptide	Amino acids	—
Maltase		Maltose	Hexoses	5.0–7.0
Lactase		Lactose	(glucose, galactose	5.8–6.2
Sucrase		Sucrose	and fructose)	5.0–7.0
Nucleotidase		Nucleotides	Nucleosides, phosphoric acid	—
Nucleosidase		Nucleosides	Purine or pyrimidine base, pentose	—
Intestinal lipase		Fat	Glycerides, fatty acids and glycerol	8.0

° Removal of C-terminal amino acid
† Removal of N-terminal amino acid

those of fats and oils. These compounds are not all structurally related, and we will consider only the triglycerides in this brief discussion. A typical triglyceride is palmitooleostearin. Its structure is shown along with the hydrolytic products—fatty acids and glycerol—in the following reaction:

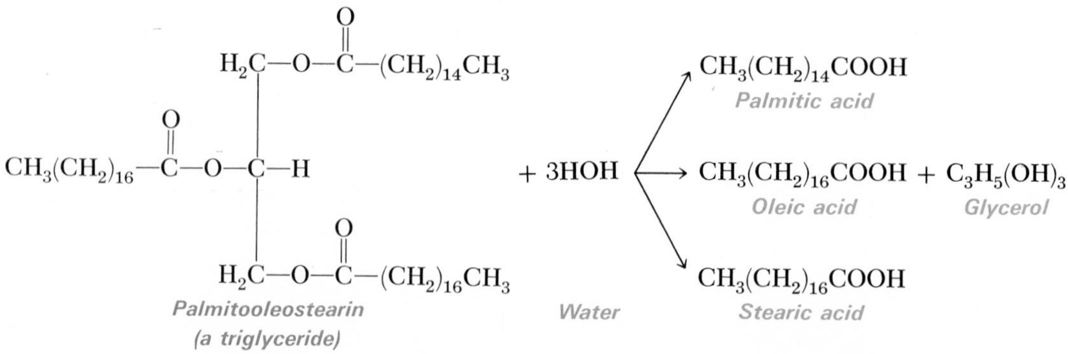

Palmitooleostearin (a triglyceride)   + 3HOH   Water

$CH_3(CH_2)_{14}COOH$ Palmitic acid

$CH_3(CH_2)_{16}COOH + C_3H_5(OH)_3$ Oleic acid   Glycerol

$CH_3(CH_2)_{16}COOH$ Stearic acid

The triglycerides, which can be either solid or liquid (Chapter 17), are not digested in either the mouth or stomach. The enzyme lipase, which is secreted by the pancreas, facilitates hydrolysis of the ester linkages in the triglyceride in the small intestine. However, this is difficult because the enzyme is water soluble, and the triglyceride is likely to be in an oil globule that is insoluble in water. The smaller the oil droplets are, the easier it is for the hydrolysis to occur. Ideally the oil would exist in an emulsion (in tiny droplets), as in salad dressing. The bile fluid from the liver contains salts that help emulsify the oil, and thereby greatly facilitates the digestion of fats.

PROTEIN DIGESTION AND ABSORPTION

The hydrolysis of proteins begins in the stomach and continues in the small intestine. Several different types of enzymes are known to be involved. These enzymatic systems must be controlled very carefully, for they have the potential of digesting the walls of the stomach and intestines. A number of these enzymes are secreted in an inactive form. For example, pepsin, which is secreted in the stomach, is first present in a form termed prepepsin. The molecular weight of prepepsin is 42,600. Prepepsin, in the presence of the acid of the stomach, is broken by still another enzyme to pepsin. The molecular weight of pepsin is 34,500. It is reasonable to believe that prepepsin, the enzyme that breaks it down, and the concentrated acid normally have no effect on the stomach wall. After mixing, pepsin is formed, and this enzyme would have considerable action on the stomach protein had it been formed under the mucous lining, a lining which is constantly sloughing off like the outer skin.

Pepsin facilitates the breakdown of only about 10 per cent of the bonds in a typical protein, leaving polypeptides with molecular weights from 600 up to 3000. In the small intestine, hydrolysis is completed to small peptides which are absorbed through the intestinal wall.

Some protein enzymes are sold commercially. Meat tenderizers are protein materials that speed up partial digestion of meat. Enzymes are used as stain removers in detergents, although their effect on the skin is open to some question. Related enzymes are also used to free the lens of the eye prior to cataract surgery.

GLUCOSE METABOLISM

The biochemical process by which living cells obtain energy from glucose and similar structures is a long one. The *initial* sequence of reactions can follow two courses; one of these does not use oxygen, while the other one does. When this first sequence takes place in the absence of air, it is called *anaerobic glycolysis* and results in the formation of lactic acid. This process furnishes only a relatively small amount of energy. The overall reaction can be represented by the equation

$$C_6H_{12}O_6 + 2ADP + 2P_i \longrightarrow 2CH_3\overset{\overset{\displaystyle H}{|}}{\underset{\underset{\displaystyle OH}{|}}{C}}\overset{\displaystyle O}{\underset{}{C}}{-}OH + 2ATP$$

Glucose   *Lactic acid*

where P_i represents inorganic phosphate such as $H_2PO_4^-$.

While this process furnishes only 16 kilocalories per mole of chemically useful energy (in the form of ATP), compared to the over 691 kilocalories per mole theoretically obtainable from the complete combustion of glucose, it is an especially useful process because it can occur in the complete absence of oxygen. This means that it can furnish the energy required for very rapid short spurts of muscular effort when oxygen cannot be supplied rapidly enough to carry out other energy-furnishing processes.

A considerably larger amount of energy can be obtained when glucose is metabolized in the presence of air, which is called *aerobic glycolysis;* in this reaction the product is pyruvic acid:

$$C_6H_{12}O_6 + O_2 + 6ADP + 6P_i \longrightarrow 2CH_3C-C\overset{O}{\underset{O}{\diagdown}}OH + 6ATP + 2H_2O$$

Glucose

Pyruvic acid

In this process 48 kilocalories per mole of glucose is obtained in a useful form. This process utilizes about 35 per cent of the total energy released in the overall reaction.

$$C_6H_{12}O_6 + O_2 \longrightarrow 2CH_3C-C\overset{O}{\underset{O}{\diagdown}}OH + 2H_2O + 141 \text{ Kcal per mole}$$

This illustrates an important fact: *only a portion of the energy theoretically available from a chemical reaction can be utilized by the cells which carry out the reaction.* The reaction sequence by which glucose is converted into pyruvic acid is complicated and can be found in biochemistry textbooks.

THE KREBS CYCLE

After the pyruvate has been formed through glycolysis, it enters into a cyclic series of reactions (the Krebs cycle), the net result of which is the formation of carbon dioxide and more ATP. In fact, there are 15 molecules of ATP formed from each pyruvate ion, or 30 from each glucose molecule, since, as indicated in the preceding equation, two pyruvate ions are formed from each glucose molecule. Figure 18–8 presents a brief summary of the Krebs cycle.

A complete discussion of the chemistry of this cycle is beyond the scope of this presentation. However, three points are worth remembering: (1) pyruvate from sugar is consumed—note that it enters from the top in Figure 18–8; (2) carbon dioxide is produced—note that it leaves the cycle in two places; (3) coupled reactions, indicated by the curved arrows, allow the energy in pyruvate to be used to synthesize ATP.

THE METABOLISM OF FATS

The energy needed by animal cells comes from the oxidation of carbohydrates, fats, and proteins. Both fats and carbohydrates are stored by the body

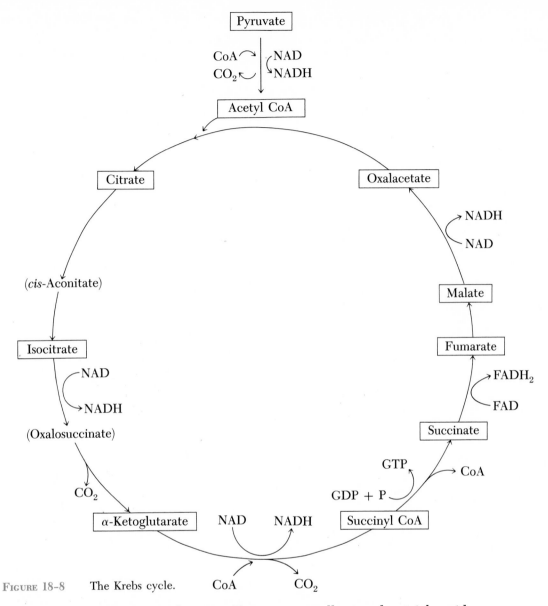

FIGURE 18-8   The Krebs cycle.

to serve as a future source of energy. Fats are generally stored as triglycerides, although both absorption and oxidation generally proceed through the free fatty acid, $R-CH_2COOH$, where R is a long hydrocarbon chain.

The key to the oxidation of fatty acids was discovered in 1904, by Knoop, who used fatty acids carrying phenyl groups to characterize the process. The phenyl groups in these compounds are not metabolized in the body, so they can be used as a kind of tracer. When the first compound in this series, benzoic acid, was fed to animals, hippuric acid was isolated from the urine:

391

The second compound fed to animals was phenylacetic acid; phenylacetylglycine was found in the urine:

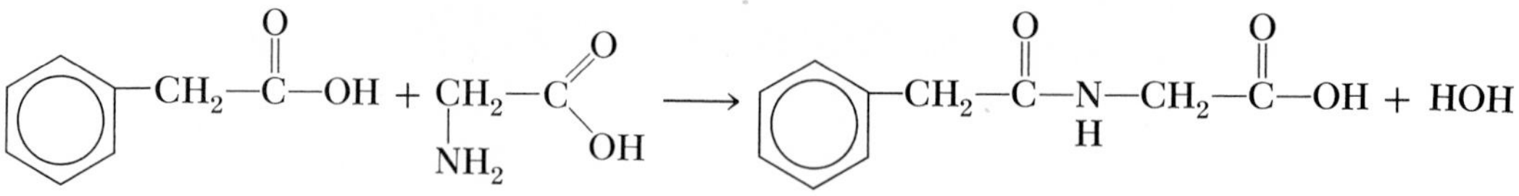

Phenylacetic acid + Glycine → Phenylacetylglycine + HOH

Phenylacetic acid   *Glycine*   *Phenylacetylglycine*

The experiment was extended to continuously longer fatty acids, and it was found that when these fatty acids, which have phenyl groups on their terminal carbon atoms, are ingested, either hippuric acid *or* phenylacetylglycine is obtained in the urine as a product:

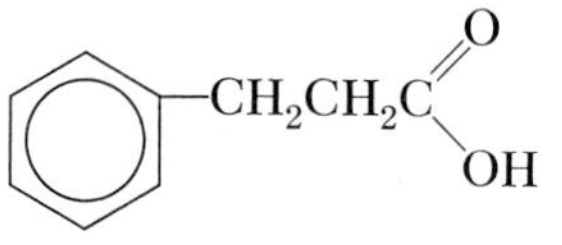

[phenyl—CH₂CH₂C(=O)OH]   *produces* 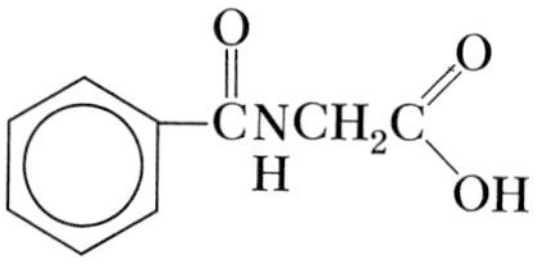   [*Hippuric acid*]

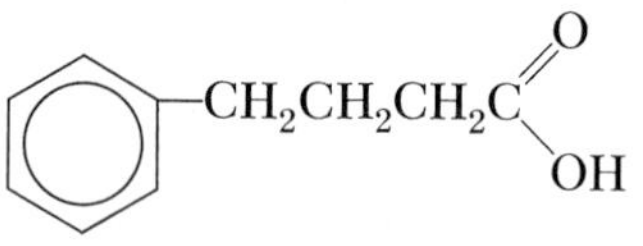

[phenyl—CH₂CH₂CH₂C(=O)OH]   *produces*   [*Phenylacetylglycine*]

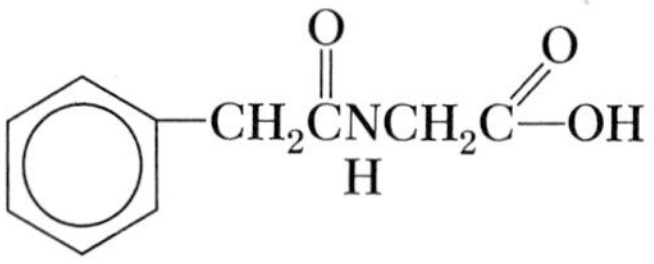

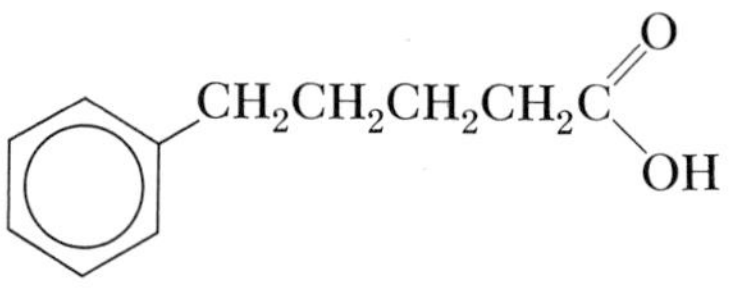

[phenyl—CH₂CH₂CH₂CH₂C(=O)OH]   *produces* 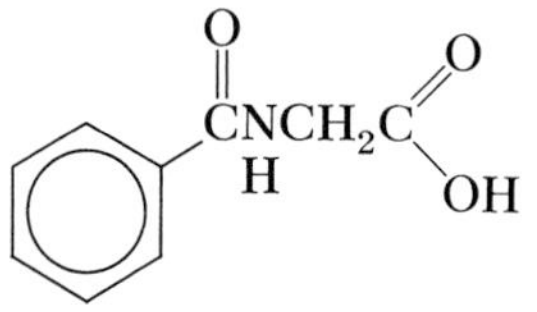   [*Hippuric acid*]

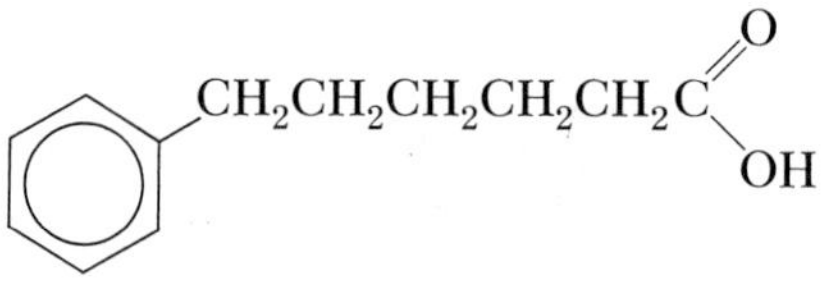

[phenyl—CH₂CH₂CH₂CH₂CH₂C(=O)OH]   *produces* 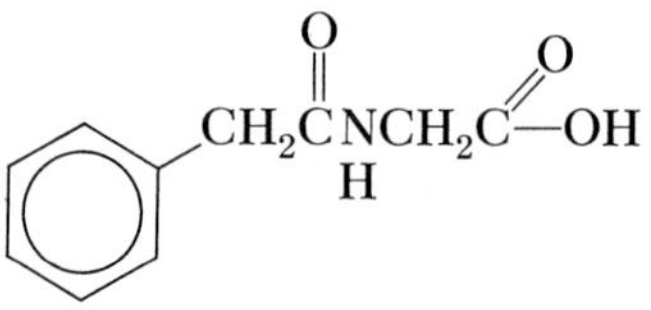   [*Phenylacetylglycine*]

To understand this puzzle two observations must be carefully noted. First, the compounds with an odd number of carbon atoms in the chain yielded hippuric acid and those with an even number of carbon atoms in the chain yielded phenylacetylglycine. Second, the only difference between the hippuric acid molecule and the phenylacetylglycine molecule is a —CH₂— group next to the benzene ring in phenylacetylglycine. Knoop knew from his experience in organic chemistry that the most likely point of attack for the oxidation of these molecules is the carbon atom two removed from the COOH group. This carbon is termed the β *carbon* and the one next to the COOH group the α *carbon*. If oxidation occurs at the β carbon, the result is the loss of two carbons from the chain; the β carbon ends up in another COOH group.

Consider the structure:

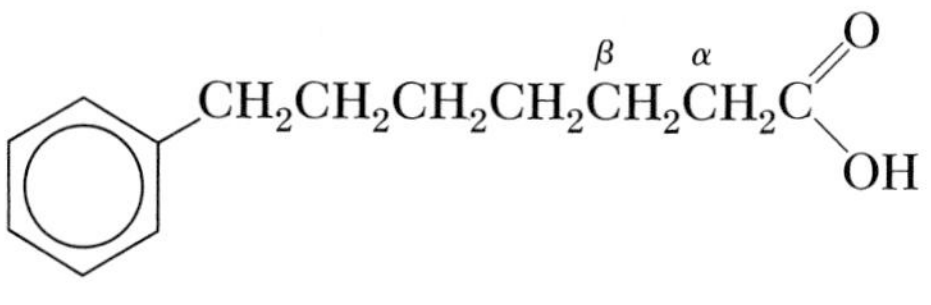

The first oxidation step would result in:

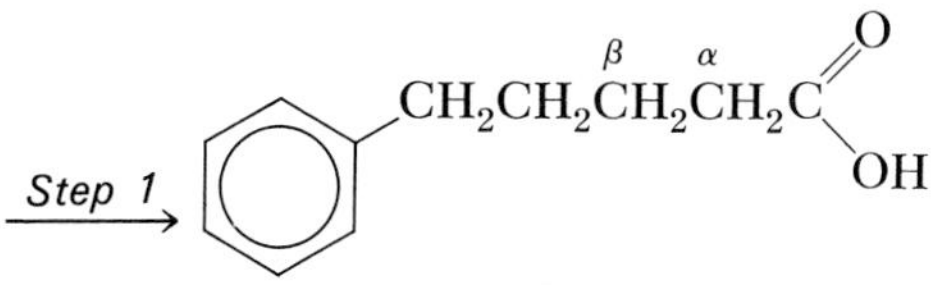

The second would result in:

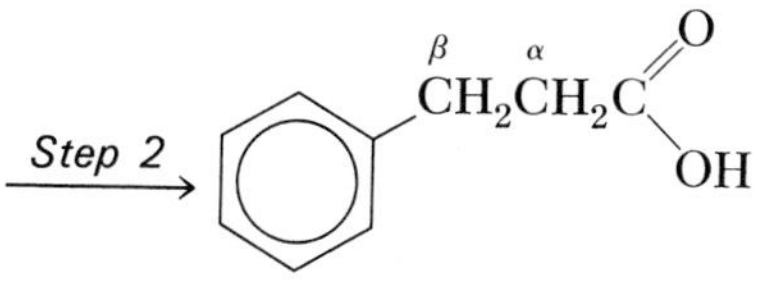

The third oxidation step would result in:

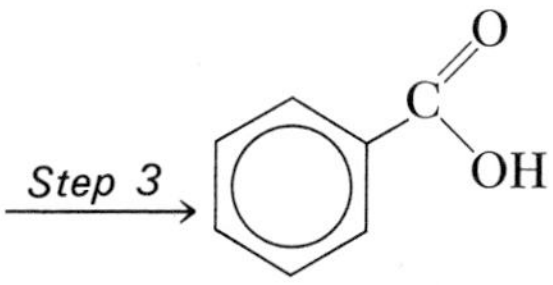

Benzoic acid

which reacts with glycine to form:

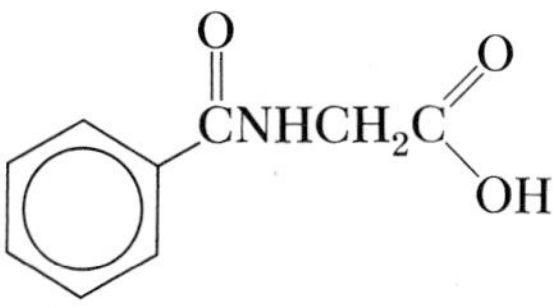

Hippuric acid

If the carbon chain has an even number of carbon atoms, the side chain oxidation stops with:

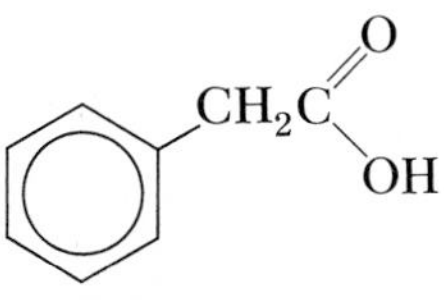

which would react with glycine to form:

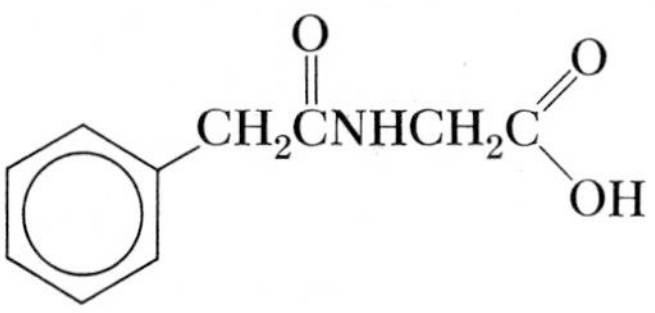

Phenylacetylglycine

393

Knoop felt this to be proof that fats were oxidized in two carbon "bites." Subsequent studies have confirmed the main outline of Knoop's theory and, furthermore, they have shown that the two carbons removed are incorporated into acetyl coenzyme A (acetyl CoA) as the acetyl group (Figure 18–9).

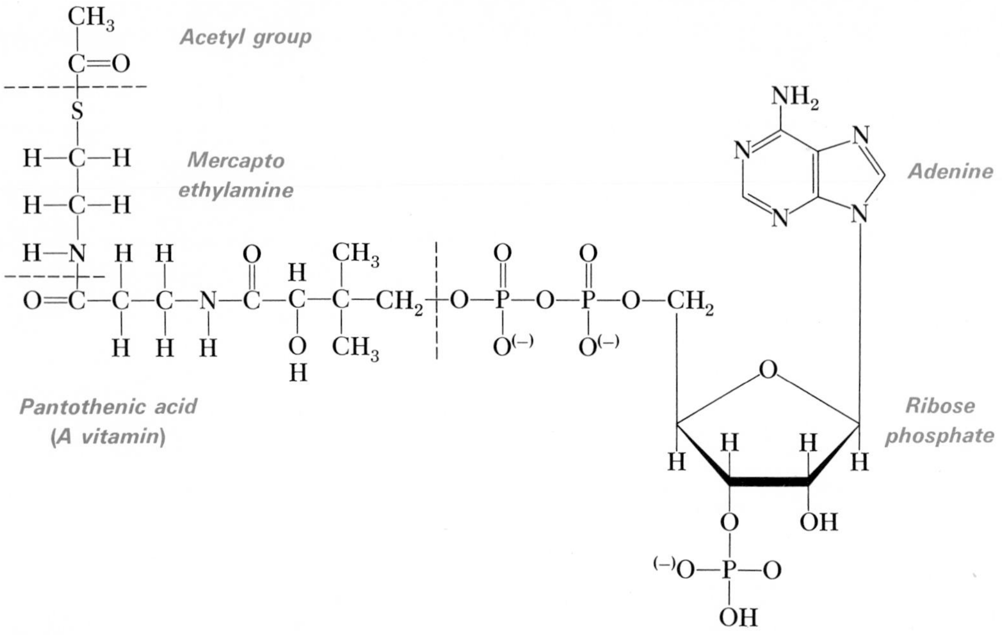

FIGURE 18-9   The structure of acetyl coenzyme A (CoA). An acetyl group is shown attached to the CoA to give acetyl CoA.

The acetyl coenzyme A can now transport the acetyl group to the Krebs cycle from which the carbons will exit as carbon dioxide. Again, the energy of fats is used in the production of high energy intermediates such as ATP.

It is worth noting in passing that pantothenic acid, a vitamin, is necessary to the structure of acetyl coenzyme A. Humans must obtain this vitamin in their foods (Figure 18–9). Common sources include egg yolk, liver, salmon, and peanuts.

NATURAL PROTEIN SYNTHESIS

The proteins of the body are being replaced and resynthesized continuously from the amino acids available to the body, many of which are obtained from the diet. The amino acids and proteins in the body can be considered as constituents of a "nitrogen pool"; additions to and losses from the pool are shown in Figure 18–10.

The use of isotopically labeled amino acids has made possible studies on the average lifetimes of amino acids as constituents in proteins—that is, the time it takes the body to replace a protein in a tissue. For a process that must be extremely complex, it is very rapid. Only minutes after radioactive amino acids are injected into animals, radioactive protein can be found. Although all the proteins in the body are continually being replaced, the rates of replacement

394

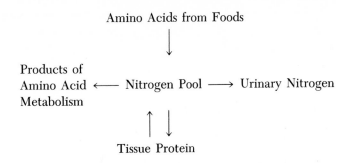

Amino Acids from Foods

Products of
Amino Acid ⟵ Nitrogen Pool ⟶ Urinary Nitrogen
Metabolism

FIGURE 18-10   The nitrogen pool.

Tissue Protein

have been found to vary. Half of the proteins in the liver and plasma are replaced in *six days*. The time is much longer for muscle proteins, about 180 days, and replacement of protein in other tissues, such as bone collagen, takes even longer.

Recall that each organism has its own kind of protein. The number of possible arrangements of 20 amino acid units is more than the number of atoms in the known universe, yet proteins characteristic of a given organism can be synthesized in a matter of a few minutes. It should come as no surprise, then, that a vast amount of research has been devoted to this problem in recent years. Although it is thought that the general scheme of protein synthesis is now understood, you should realize that many of the details are still to be worked out.

As stated in Chapter 17, it is the DNA structure in the cell nucleus that holds the code for protein synthesis. Messenger RNA, like all forms of RNA, is synthesized in the cell nucleus. Figure 18–11 shows how a DNA unit containing the coded information partially unwinds to serve as a pattern, or template, for RNA synthesis. The roles of messenger RNA and transfer RNA in protein synthesis are illustrated again and summarized in Figure 18–12.

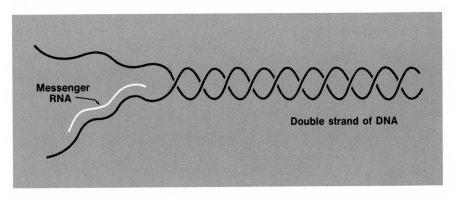

Messenger
RNA

Double strand of DNA

FIGURE 18-11   Messenger RNA is formed on the partially unwound DNA structure in the cell nucleus.

ATP is important in protein synthesis. Now that this compound has been introduced, a discussion of its role in making protein will supplement the discussion presented in Chapter 17. In the first step of protein synthesis, a molecule of ATP activates an amino acid to form a unit that generally can be termed adenine-ribose-triphosphate-amino acid:

$$\text{ATP} + \begin{array}{c}\text{Amino}\\ \text{acid}\end{array} \longrightarrow \begin{array}{c}\text{ATP-amino acid--}\\ \text{activated species}\end{array}$$

395

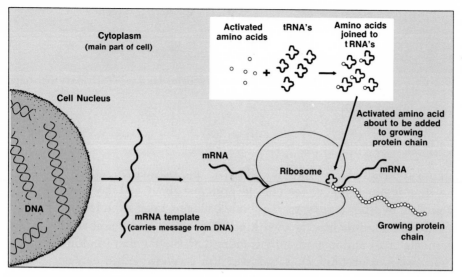

FIGURE 18-12   How proteins are synthesized. Information in DNA structure is built into messenger RNA template. Messenger RNA is then bonded to the ribosomal RNA, and the two together direct the sequence of amino acid addition, the amino acid units being brought to the site by transfer RNA units.

This activated species is then able to attach the amino acid unit to a molecule of transfer RNA (Figure 18–13). The transfer RNA and its amino acid migrate to the ribosome where the amino acid is given up in the formation of a polypeptide. The transfer RNA is then free to migrate back and repeat the process. A different transfer RNA is required for each amino acid transported; if a polypeptide is being made up of 12 different amino acids, at least 12 different kinds of transfer RNA are required. All these reactions are generally catalyzed by enzymes.

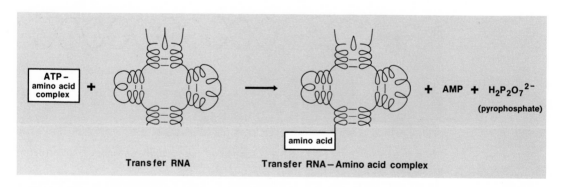

FIGURE 18-13   Bonding of activated amino acid to transfer RNA.

Messenger RNA is used only once, or at most a few times, before it is depolymerized. While this may seem to be a terrible waste, it allows the cell to produce different proteins on very short notice. As conditions change, a different type of messenger RNA comes from the nucleus, a different protein is made, and the cell adequately responds to a changing environment.

CONCLUSION

After biological macromolecules have been synthesized, they are incorporated into the appropriate part of the cell or structure, often being held in a particular conformation by hydrogen bonds. In any cell the components are in a dynamic relationship with their environment, constantly being replaced at varying rates by new molecules of exactly the same structure. The methods by which the macromolecules are incorporated into the larger functioning parts of a cell are not completely known.

We have examined the way in which the energy of ATP is obtained and then used for chemical synthesis, but we have not looked at other important functions of this molecule, such as the furnishing of energy for muscle movement or locomotion. The living cell has an ability to transform the chemical energy of ATP almost directly into motion.

It can be seen, from what has been examined in this chapter, that our knowledge of biochemical systems, of both the structures and processes in cells, is becoming continually more detailed and sophisticated. This knowledge has been built up in a systematic fashion from detailed studies on living systems and the molecules obtained from them. There is every reason to believe that these studies will lead to a continuing expansion of our knowledge of the science of life and the manner in which living systems obey basic scientific laws.

QUESTIONS

1. Write a basic equation for the digestion of: (a) starch to a disaccharide; (b) a disaccharide to a simple sugar; (c) a protein to amino acids; (d) a triglyceride to fatty acids.

2. List the different kinds of RNA that are employed in protein synthesis and briefly describe the role played by each.

3. What is the metal in chlorophyll? Recall a similar metal complex from Chapter 17.

4. If you were to "feed" radioactive carbon dioxide to a green plant, what would be the first radioactive carbon compound formed? Who made this discovery?

5. Give the structure of ATP and point out the region of the molecule that contains bonds the hydrolysis of which is involved in coupled reactions.

6. What is meant by coupled reactions? Give an example. Why do the energetics of biochemical systems make coupled reactions necessary?

7. Give two characteristics of all enzymes.

8. Using the energy diagrams, explain the concept of activation energy for a chemical reaction and show the effect of an enzyme on the activation energy.

9. Point out three important similarities of the lock-and-key analogy to enzymatic activity.

10. What is a coenzyme and why are they sometimes necessary?

11. What type of compound first absorbs light energy in photosynthesis? Give an example.

12. In photosynthesis why is it partially correct to say that light is an oxidizing agent?

397

13. What are the two major divisions in photosynthesis? Express in words what is accomplished in each.

14. What is the source of oxygen in photosynthesis?

15. What part of photosynthesis could even take place in an animal cell?

16. Since chlorophyll loses electrons because of light, it must subsequently gain electrons from somewhere. Where do they come from?

17. Some people like to say that the contents of the alimentary canal are not "in the body." Can you explain this statement?

18. If protein digestion is facilitated by enzymes, and these enzymes are produced in body organs made of proteins, explain why the enzymes do not cause rapid digestion of the organs themselves?

19. What is pyruvic acid? Why is it so important in getting energy from sugars?

20. Explain how it was learned that fatty acids are broken down by units containing two carbon atoms.

21. What do you think would be the result if the body was deprived of its supply of pantothenic acid? To what general class of compounds does pantothenic acid belong?

22. If a certain muscle requires 10 calories for contraction and obtains this ultimately by the hydrolysis of ATP to ADP, what is the minimum number of molecules of ATP that are needed to furnish the energy for such a contraction process?

23. What other fatty acids would you expect to find as intermediates in the oxidation of caproic acid?

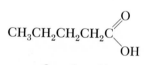

Caproic acid

24. What volume of O_2 gas (measured under standard conditions of temperature and pressure) is released to the atmosphere for each gram of CO_2 transformed to glucose by photosynthesis?

25. What is the basic nature of the digestion processes for large molecules?

26. What happens if an amino acid is needed for protein synthesis and the amino acid can neither be made by the body nor obtained from the diet? Does the modern theory of protein synthesis include an explanation of the role of essential amino acids? (These are amino acids which cannot be manufactured by the human body; they must be obtained in the diet.)

SUGGESTIONS FOR FURTHER READING

General

Baldwin, E., "The Nature of Biochemistry," Second Edition, Cambridge University Press, Cambridge, England, 1967.

"Bio-Organic Chemistry," Readings From *Scientific American*, W. H. Freeman and Company, San Francisco, 1968.

Harrison, K., "Guidebook to Biochemistry," Second Edition, Cambridge University Press, Cambridge, England, 1965.

Lehninger, A. L., "Bioenergetics," W. A. Benjamin Inc., N. Y., 1965.

Mazur, A., and Harrow, B., "Biochemistry: A Brief Course," W. B. Saunders Co., Philadelphia, 1968.

Carbohydrate Metabolism

Horecker, R. L., "Pathways of Carbohydrate Metabolism and Their Biological Significance," J. Chem. Ed. *42*, p. 244 (1965).

Oesper, P., "Error and Trial: The Story of the Oxidative Reaction of Glycolysis," J. Chem. Ed. *45*, p. 607 (1968).

Photosynthesis

Rabinowitch, E. I., and Govindjec, "The Role of Chloroplasts in Photosynthesis," Scientific American *213:* 1, p. 74 (1965).

Hydrogen Bonding

McClellan, A. L., "Significance of Hydrogen Bonds in Biological Structures," J. Chem. Ed. *44*, p. 547 (1967).

Nutrition

Bogert, L. J., Briggs, G. M., and Calloway, D. H., "Nutrition And Physical Fitness," Eighth Edition, W. B. Saunders Co., Philadelphia, 1966.

Enzymes

Locke, D. M., "Enzymes—The Agents of Life," Crown Publishers, New York, 1969.

TOXIC SUBSTANCES IN MAN'S ENVIRONMENT

CHAPTER 19 ━━━━━━━━━━━━━━━━━

INTRODUCTION

In this chapter and the two that follow, we shall consider some of the toxic substances in our environment. *Toxic* refers to substances which cause harm to man in a chemical way. The emphasis will vary from chapter to chapter owing to the many aspects of the subject. In this chapter we shall categorize toxic substances—poisons—principally by their mechanism of action. In Chapter 20, dealing with water pollution, examples of these same substances will be discussed again, but the emphasis will be shifted to their effects on environment. A similar organization will be found in Chapter 21, on air pollution. In both of these chapters, the discussion will concentrate on methods of solving these problems.

TOXIC SUBSTANCES

The human body is an incredibly complex system of chemical compounds. Life is dependent upon the assembly of just the right amounts of many different elements. The body has many mechanisms to regulate its intake of materials, but these operate only within limits. When presented with an excessive amount of a single element or compound, toxic reactions can be expected. As we shall see, there is an upper limit to the amount of any compound which the body can accommodate; this is true even of those normally considered as harmless. Everyone can remember the child who ate too much candy (glucose) and got a stomach ache.

Sometimes chemicals, present in small amounts, upset the biochemical reactions of the body, and we usually limit the term *toxic substances* to these materials. Lethal doses are customarily expressed in milligrams (mg) of substance per kilogram (kg) weight of the subject. For example, the cyanide ion (CN^-) is generally fatal to humans in a dose of 1 mg of CN^- per kg of body weight. For a 200-pound person, 0.0032 ounce of cyanide is a lethal dose. Examples of somewhat less toxic substances and the probable lethal doses for an average person are:

Morphine	1–50 mg per kg
Aspirin	50–500 mg per kg
Methyl alcohol	0.5–5 g per kg
Ethyl alcohol	5–15 g per kg

There is a wide variety of ways in which toxic substances may act; but before we examine these, let us see how we can obtain a quantitative measure of toxicity. This is done by introducing various dosages of substances to be tested into laboratory animals (such as rats). That dosage which would be lethal to 50 per cent of a large number of the animals under controlled conditions is called the LD_{50} (lethal dosage$_{50\%}$) and is reported in milligrams of poison per kilogram of body weight. Thus if a statistical analysis of data on a large population of rats showed that a dosage of 1 mg per kg was lethal to 50 per cent of the population tested, the LD_{50} for this poison would be 1 mg per kg. Obviously, metabolic variations and other differences between species will produce different LD_{50} values for a given poison in different animals. For this reason such data cannot be extrapolated to humans with any assurance, but it is safe to assume that a substance with a low LD_{50} value for several animal species will also be quite toxic to humans.

Toxic substances can be classified into several catagories which are descriptive of the way in which they disrupt the chemistry of the body. Thus the modes of action of toxic substances can be classified as *corrosive, metabolic, neurotoxic, mutagenic,* and *carcinogenic.* These describe types of biochemical processes, which will serve as the basis of our discussion. Before beginning, however, we should distinguish between two terms used to describe poisoning: chronic and acute. *Chronic poisoning* occurs when the dose is sufficient to cause moderate, persistent symptoms. *Acute poisoning,* caused by a larger dose, is potentially lethal in the absence of suitable treatment.

CORROSIVE POISONS

Toxic substances that react locally on tissues are corrosive poisons. Examples include strong acids and alkalies and many oxidants such as those found in laundry products, which can destroy tissues. Sulfuric acid (found in auto batteries) and hydrochloric acid (also called muriatic acid when it is used for cleaning purposes) are very dangerous corrosive poisons. Death has resulted from the swallowing of 1 ounce of concentrated (98 per cent) sulfuric acid, and much smaller amounts can cause extensive damage and severe pain.

These materials act by first dehydrating cellular structures; this is caused by the high charge density on the acid protons. After the cell dies, its protein structures are destroyed. Acids catalyze the splitting of the peptide bonds by water (hydrolysis) to form a carboxyl group and an amine:

$$R - \overset{\overset{\displaystyle O}{\|}}{C} - \overset{\overset{\displaystyle H}{|}}{N} - R + H_2O \xrightarrow[\text{from acid}]{H^+} R - \overset{\overset{\displaystyle O}{\|}}{C} - OH \ + \ H \overset{\overset{\displaystyle H}{|}}{N} - R$$

Peptide link　　　　　　　　　　*Carboxyl group*　　*Amine*

In the early stages of this process there will be a large proportion of larger fragments present. Subsequently as more bonds are broken, smaller and smaller fragments result, leading to the ultimate disintegration of the tissue.

Some poisons act by undergoing chemical reaction to produce corrosive poisons. Phosgene, the deadly gas used during World War I, is an example. When inhaled, it is hydrolyzed in the lungs to hydrochloric acid, which causes pulmonary edema (a collection of fluid in the lungs) owing to the dehydrating effect of the strong acid on tissues. The victim dies of suffocation because oxygen cannot be absorbed effectively by damaged tissues.

$$\underset{Cl}{\overset{\overset{\displaystyle O}{\|}}{\underset{}{C}}}\,\underset{Cl}{} + H_2O \longrightarrow \quad 2HCl \quad + \quad CO_2$$

Phosgene　　　　　　　*Hydrochloric*　*Carbon*
　　　　　　　　　　　　　acid　　*dioxide*

Sodium hydroxide, NaOH (lye—a component of drain cleaners), is a very strongly alkaline, or basic, substance that can be just as corrosive to tissue as strong acids. The hydroxide ion which it furnishes also catalyzes the splitting of peptide linkages:

$$R - \overset{\overset{\displaystyle O}{\|}}{C} - \overset{\overset{\displaystyle H}{|}}{N} - R + H_2O \xrightarrow[\text{from base}]{OH^-} R - \overset{\overset{\displaystyle O}{\|}}{C} - OH + H \overset{\overset{\displaystyle H}{|}}{N} - R$$

Both acids and bases, as well as other types of corrosive poisons, continue their action until they are consumed in chemical reactions.

Plants thought to contain corrosive poisons are poison ivy and poison sumac (Figure 19–1). Since not all persons are susceptible to poison ivy, in all likelihood its action is also allergenic in nature. Some of the toxic components of these plants have been identified. The toxic characteristics of poison ivy are usually attributed to four substances, all related chemically to phenol (carbolic acid), of which the principal one is urushiol.

Only about half of the individuals who come in contact with the plants seem to be affected, and there is no known cure for the poison. Since it takes about 15 minutes for the poison to penetrate the skin, a quick and thorough washing with soap and water or trisodium phosphate (to neutralize the acidic urishol) is highly desirable.

Some corrosive poisons destroy tissue by oxidizing it. This is characteristic of substances such as ozone, nitrogen dioxide, and possibly iodine, which destroy enzymes by oxidizing their functional groups. Specific groups such as the —SH and —S—S— groups in the enzyme, are believed to be converted by oxidation to nonfunctioning groups; alternatively, the oxidizing agents may break chemical bonds in the enzyme, leading to its inactivation.

402

A summary of some common corrosive poisons is presented in Table 19–1.

TABLE 19-1   SOME CORROSIVE POISONS

SUBSTANCE	FORMULA	TOXIC ACTION	POSSIBLE CONTACT
Hydrochloric acid	HCl	Acid hydrolysis	Concrete floor cleaner
Sulfuric acid	H_2SO_4	Acid hydrolysis, dehydrates tissue—oxidizes tissue	Auto batteries
Phosgene	ClCOCl	Acid hydrolysis	Combustion of chlorine—containing plastics (PVC or Saran)
Sodium hydroxide	NaOH	Base hydrolysis	Lye, drain cleaners
Trisodium phosphate	Na_3PO_4	Base hydrolysis	Detergents, household cleaners
Sodium perborate	$NaBO_3 \cdot H_2O$	Base hydrolysis—oxidizing agent	Laundry detergents, denture cleaners
Ozone	O_3	Oxidizing agent	Air, electric motors
Nitrogen dioxide	NO_2	Oxidizing agent	Polluted air
Iodine	I_2	Oxidizing agent	Antiseptic
Hypochlorite ion	OCl^-	Oxidizing agent	Bleach
Peroxide ion	O_2^{2-}	Oxidizing agent	Bleach
Oxalic acid	$H_2C_2O_4$	Reducing agent, precipitates Ca^{2+}	Bleach, ink eradicator, leather tanning, rhubarb, spinach, tea
Sulfite ion	SO_3^{2-}	Reducing agent	Bleach

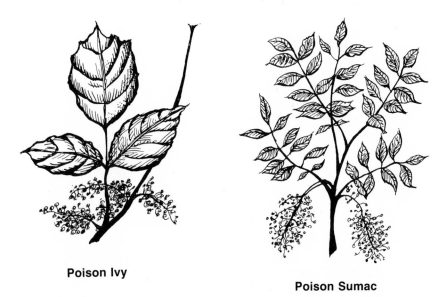

Poison Ivy

Poison Sumac

FIGURE 19-1   Poison ivy and poison sumac. The toxic nature of poison ivy is usually attributed to four substances, all related chemically to phenol (carbolic acid), of which the principal one is urushiol.

403

METABOLIC POISONS

These substances are more subtle than the tissue-destroying corrosive poisons. In fact, many of them do their work without actually indicating their presence until it is too late. Metabolic poisons can cause illness or death by interfering with some vital biochemical mechanism to such an extent that it ceases to function or is prevented from functioning efficiently.

Carbon Monoxide

The interference of carbon monoxide with extracellular oxygen transport is one of the best understood processes of metabolic poisoning. As early as 1895, J. Haldane, an English biologist, noted that carbon monoxide deprives body cells of oxygen, but only much later did researchers find that carbon monoxide, like oxygen, combines with hemoglobin:

$$O_2 + \text{hemoglobin} \longrightarrow \text{oxyhemoglobin}$$
$$CO + \text{hemoglobin} \longrightarrow \text{carboxyhemoglobin}$$

Laboratory tests show that carbon monoxide reacts with hemoglobin to give a compound which is much more stable than the compound of hemoglobin and oxygen. In addition, the presence of carboxyhemoglobin interferes with the normal oxygen release by oxyhemoglobin (the normal oxygen carrier in the blood). Thus while carbon monoxide ties up some hemoglobin, it blocks still more from supplying oxygen to the cells.

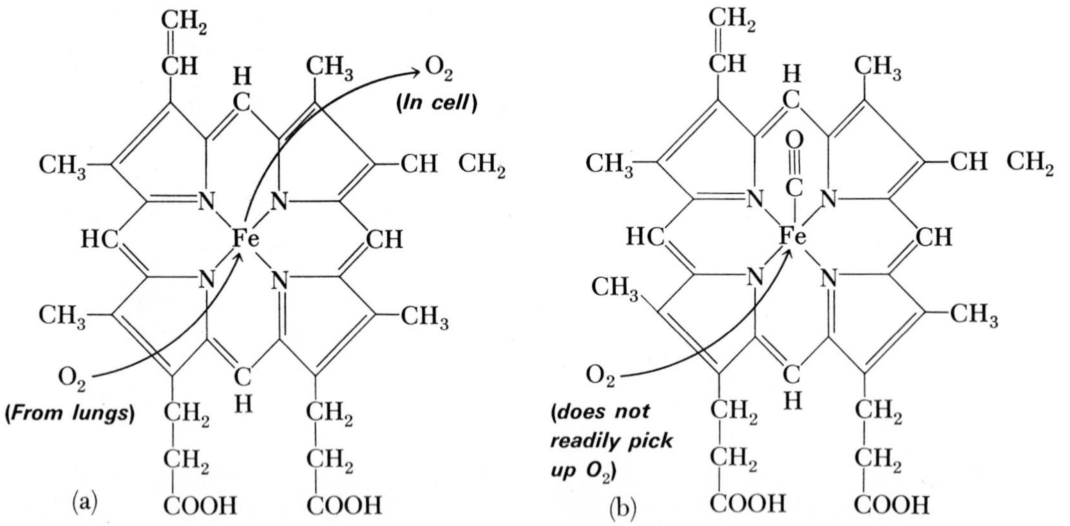

FIGURE 19-2   The interference of the hemoglobin oxygen transport mechanism. (a) illustrated by carbon monoxide bonding tightly to heme (b).

An organic material which undergoes incomplete combustion will always liberate carbon monoxide. Sources include auto exhausts, smoldering leaves, and charcoal burners. The most important initial factor in carbon monoxide poisoning

404

is the concentration of the carbon monoxide. Of course, the number of breaths per minute also determines the amount a subject receives. Breathing an atmosphere which is 0.1 per cent carbon monoxide for four hours converts approximately 60 per cent of the hemoglobin of an average adult to carboxyhemoglobin (Table 19–2).

TABLE 19-2   CONCENTRATION OF CO IN ATMOSPHERE
VERSUS PERCENTAGE OF HEMOGLOBIN (Hb) SATURATED°

CO concentration in air	0.01%	0.02%	0.10%	1.0%
Percentage of hemoglobin molecules saturated with CO†	17	20	60	90

° A few hours of breathing time is assumed.
† Normal human blood contains up to 5% carboxyhemoglobin (COHb).

Because the reaction is reversible, time is required to achieve the maximum conversion to carboxyhemoglobin, during which the CO competes with large numbers of oxygen molecules. For example, in air that is 0.1 per cent CO, oxygen molecules outnumber CO molecules 200 to 1. This numerical imbalance helps to counteract the greater combining power of CO with hemoglobin. The reaction is reversible and is in competition with the oxygen-hemoglobin reaction, consequently, as soon as the victim is exposed to fresh air (provided he is still breathing), the carboxyhemoglobin (COHb) reaction is reversed, owing to the greater concentration of oxygen:

$$COHb + O_2 \rightleftharpoons HbO_2 + CO$$

For this reason carbon monoxide is not a cumulative poison. It should be pointed out, however, that permanent damage can be sustained by an individual if certain vital cells (e.g., brain cells) are deprived of oxygen for too long.

Individuals differ in their tolerance of carbon monoxide, but generally those with anemia and a low reserve of hemoglobin (such as children) are more susceptible to its effects.

Cyanide

The cyanide ion (CN^-) is the toxic agent in cyanide salts such as the sodium cyanide, used in electroplating. Since the cyanide ion is a relatively strong base, it reacts easily with many acids (weak and strong) to form the volatile hydrocyanic acid (HCN). This reaction occurs readily when a cyanide salt is mixed with even weak acids such as acetic acid:

$$CH_3COOH + NaCN \rightleftharpoons HCN + CH_3COO^- + Na^+$$

Acetic acid   *Hydrocyanic acid*

HCN boils at a relatively low temperature (26°C), and is a gas at temperatures slightly above room temperature. It is often used as a fumigant in storage bins and holds of ships since it is toxic to most forms of life and can penetrate into tiny openings, even into insect eggs.

405

Natural sources of cyanide ions include the seeds of the cherry, plum, peach, apple, and apricot fruits. Hydrocyanic acid is produced from the seeds by hydrolysis of certain compounds, such as amygdalin, contained in them:

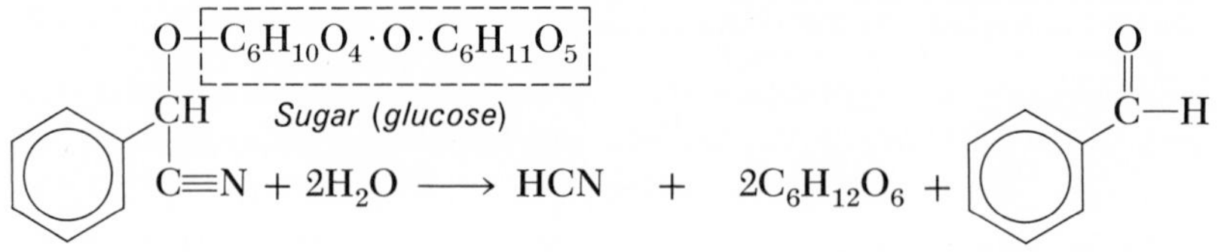

| Amygdalin | Hydrogen cyanide | Glucose | Benzaldehyde |

$$O-[C_6H_{10}O_4 \cdot O \cdot C_6H_{11}O_5]$$
$$CH \quad Sugar\ (glucose)$$
$$C{\equiv}N + 2H_2O \longrightarrow HCN + 2C_6H_{12}O_6 +$$

The cyanide is not toxic as long as it is tied up in the amygdalin. Presumably if enough apple or peach seeds were hydrolyzed in warm acid, sufficient HCN would result to cause considerable danger. There are a few recorded instances of humans being poisoned by eating large numbers of apple seeds. It appears that amygdalin is not confined to the seeds; amounts as high as 66 mg per 100g have been reported in peach leaves.

Cyanide is one of the most rapidly working poisons. Lethal doses taken orally act in minutes. Cyanide poisons by asphyxiation, as does carbon monoxide; but the mechanism of cyanide poisoning has a different twist (Figure 19–3). Instead of preventing the cells from getting oxygen, cyanide interferes with *oxidative enzymes*, cytochrome oxidase, for example. Oxidases are metalloprotein enzymes, usually containing iron or copper. They catalyze the oxidation of metabolites such as glucose:

$$\text{Metabolite (H)}_2 + \tfrac{1}{2}\,O_2 \xrightarrow{\text{\textit{Oxidase}}} \text{Oxidized metabolite} + H_2O + \text{Energy}$$

The iron atom in the cytochrome oxidase alternates between the Fe^{2+} and Fe^{3+} states to provide electrons for the reduction of O_2 and regains electrons from other steps in the process.

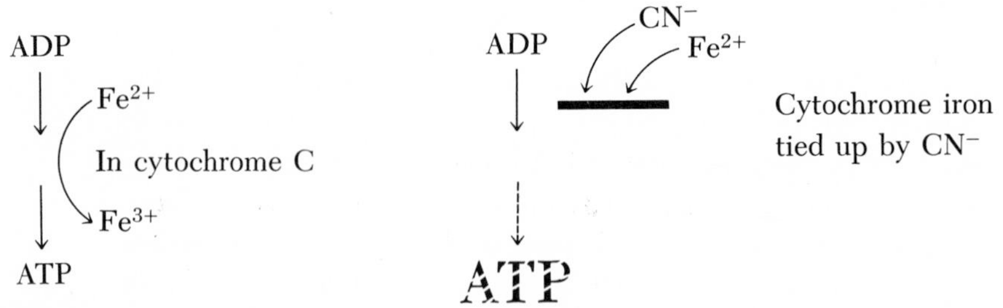

FIGURE 19–3   The mechanism of cyanide (CN⁻) poisoning. Cyanide tightly binds to the enzyme cytochrome C, an iron compound, thus blocking the vital ADP-ATP reaction in cells.

Cyanide forms complexes with the metal ions of the oxidases and renders them incapable of reducing oxygen or oxidizing the metabolite.

406

$$\text{Cytochrome oxidase (Fe)} + CN^- \longrightarrow \text{cytochrome oxidase (Fe)} \cdots CN^-$$

In essence the electrons of iron or copper are "frozen"—they cannot participate in the oxidation-reduction processes. Plenty of oxygen gets to the cells, but the mechanism by which the oxygen is used in the biochemistry of life is stopped. Hence, the cell dies, and if this occurs fast enough in the vital centers, the victim dies.

The body has a mechanism for ridding itself of cyanide. The cyanide-oxidative enzyme reaction is reversible, and other enzymes, such as rhodanase, exist in almost all cells that can convert cyanide to relatively harmless thiocyanate; for example:

$$CN^- + \underset{\textit{(Thiosulfate)}}{S_2O_3^{2-}} \xrightarrow{\textit{rhodanase}} \underset{\textit{(Thiocyanate)}}{SCN^-} + SO_3^{2-}$$

This mechanism is not as effective in protecting a cyanide-poison victim as it might appear since there is only a limited amount of thiosulfate available in the body at a given time.

Fluoroacetic Acid

Another metabolic poison is fluoroacetic acid. This molecule contains the very stable C—F bond, the same type found in fluorocarbons such as Teflon.

$$\begin{array}{ccc} F & O & \\ | & \| & \\ H—C—C—O—H \\ | & & \\ H & & \end{array}$$

Fluoroacetic acid

Nature has used the synthesis of this molecule as part of a defense mechanism for certain plants. Native to South Africa, the gilfbaar plant contains lethal quantities of fluoroacetic acid. Cattle that eat these leaves usually sicken and die.

Sodium fluoroacetate, the sodium salt of this acid (alias compound 1080), is a potent rodenticide. Because it is odorless and tasteless, it is especially dangerous, and its sale in this country is strictly regulated by law.

Fluoroacetate is toxic because the body uses it to synthesize fluorocitric acid, and fluorocitric acid subsequently produced blocks the Krebs cycle (Figure 19-4). The C—F linkage apparently ties up the enzyme aconitase, thus preventing it from converting citrate to isocitrate.

This is an instance in which the poison is sufficiently similar to the substance (the substrate) which normally reacts with the enzyme that it provides effective competition for the active sites on the enzyme. If the poison has sufficient affinity for the active groups of the enzyme, it occupies the active sites on the enzyme in much the same manner that the normal substrate occupies the same sites. If the enzyme is part of a chain of enzymatic reactions governing a physiological mechanism, that mechanism is blocked when further reaction on the poison molecule is impossible. The blocking of the Krebs cycle by fluorocitrate is a typical example of this. The fluorocitrate competes with citrate for the active sites on aconitase, thereby inhibiting the enzyme and blocking the Krebs cycle. If fluoroacetate is not present in excessive amounts, its action can be reversed simply by increasing the concentration of available citrate.

407

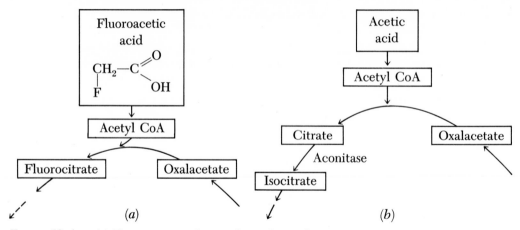

FIGURE 19-4   (a) Fluoroacetic acid is synthesized into fluorocitrate which then forms a stable bond with the enzyme aconitase. This blocks the normal Krebs cycle, a portion of which is shown (b).

Heavy Metals

Perhaps the most common of all the metabolic poisons are the heavy metals. These include such common elements as lead and mercury as well as many less common ones such as cadmium, chromium, and thallium. In this group we will also include the infamous poison, *arsenic*, which is really not a metal but metal-like in many of its properties, including its toxic action.

This classic homicidal poison occurs in small amounts in many foods because it is a constituent of some garden insecticides and pesticides (Table 19–3); it

TABLE 19-3   SOME ARSENIC-CONTAINING INSECTICIDES

NAME	FORMULA
Lead arsenate	$Pb_3(AsO_4)_2$
Calcium arsenate	$Ca_3(AsO_4)_2$
Paris green (copper acetoarsenite)	$(CuO)_3As_2O_3 \cdot Cu(C_2H_3O_2)_2$

can also come from the soil. Because arsenic compounds are still used occasionally to spray fruits and vegetables, minute quantities of arsenic occur in many foods as an impurity. The Federal Food and Drug Administration (FDA) limits the allowable amount of arsenic in foods to 0.15 milligram per pound of food, and this apparently causes no harm. Several drugs contain arsenic, but these compounds, such as arsphenamine, which has found some use in treating syphilis, contain covalently bonded arsenic.

$$H_2N \qquad\qquad NH_2$$

$$HO-\!\!\!\!\bigcirc\!\!\!\!-As\!=\!As-\!\!\!\!\bigcirc\!\!\!\!-OH$$

Arsphenamine

Metals owe their toxicity primarily to their ability to react with and inhibit sulfhydryl (—S—H) enzyme systems such as those involved in the production of cellular energy. For example, glutathione (a tripeptide of glutamic acid, cysteine, and glycine) occurs in most tissues and illustrates the interaction of a metal with sulfhydryl groups. The metal replaces the hydrogen on the sulfhydryl group (Figure 19–5).

$$2 \text{ Glutathione} + \text{Metal ion (M}^{2+}) \longrightarrow \text{M (Glutathione)}_2 + 2\text{H}^+$$

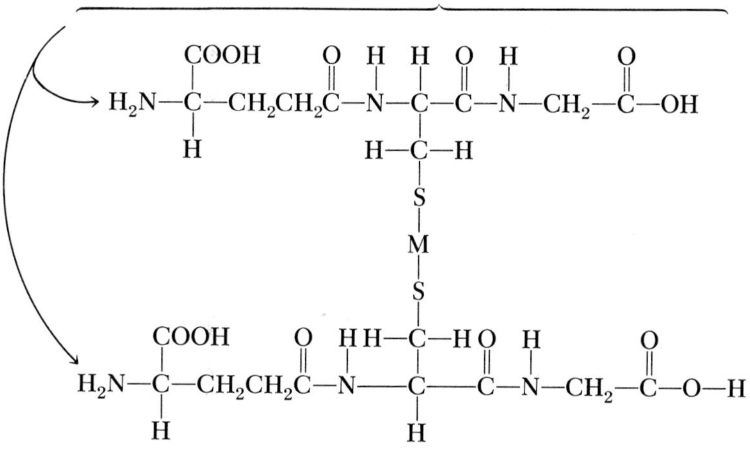

Glutathione

FIGURE 19–5   Glutathione reaction with a metal.

The typical forms in which toxic arsenic compounds are encountered are the inorganic ions, such as arsenate (AsO_4^{3-}) and arsenite (AsO_3^{3-}). The reaction of an arsenite, such as sodium arsenite, with sulfhydryl groups results in a complex in which the arsenic unites with two sulfhydryl groups, which may be on the same molecule.

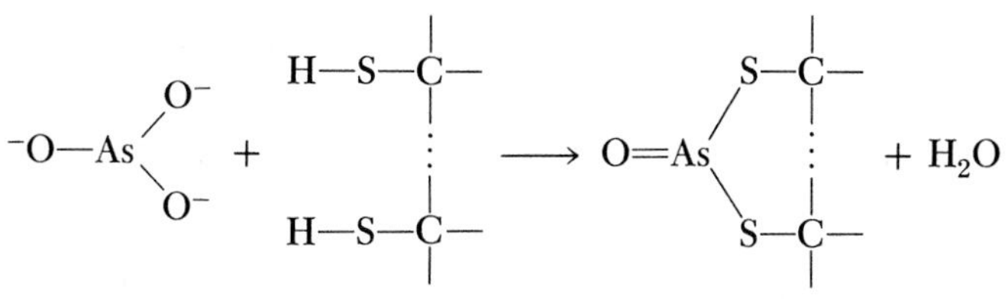

Arsenite   Sulfhydryl   Arsenic
                groups   complex

The problem of developing a compound to counteract the arsenic-containing gas, *Lewisite* (used in World War I), led to an understanding of how arsenic acts as a poison and subsequently to the development of an antidote.

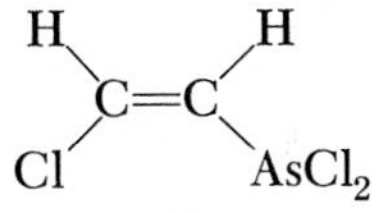

Lewisite

Once the sulfhydryl interaction was understood, a number of British scientists set out to find a suitable compound that contained highly reactive sulfhydryl

409

groups to combine with the arsenic and render the poison ineffective. The result of this research was a compound now known as British Anti-Lewisite (BAL).

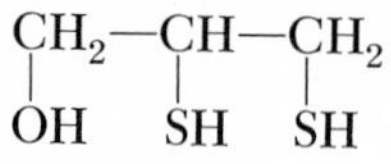

$$CH_2 - CH - CH_2$$
$$\;\;|\qquad|\qquad|$$
$$OH\quad SH\quad SH$$

BAL

The BAL bonds to the metal at several sites and forms a *chelate* (Greek: *chela*, claw), a term applied to a reacting agent that envelopes a species of metal ion. BAL is one of a large number of compounds that can act as *chelating agents* for metals (Figure 19–6).

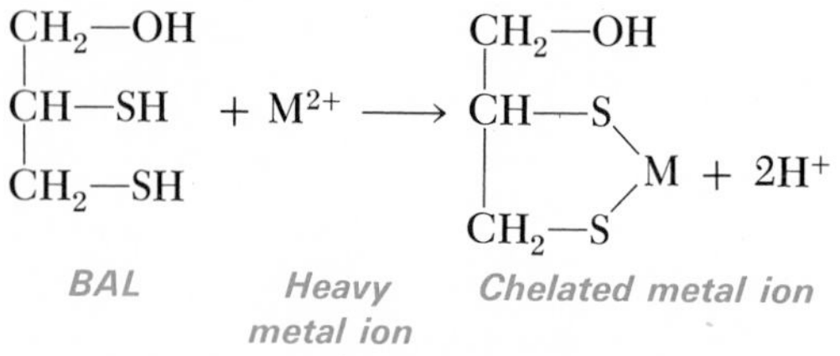

$$
\begin{array}{ccc}
CH_2\!-\!OH & & CH_2\!-\!OH \\
| & & | \\
CH\!-\!SH & +\; M^{2+} \longrightarrow & CH\!-\!S \\
| & & \qquad\qquad\;\;\diagdown \\
CH_2\!-\!SH & & \qquad\qquad\qquad M + 2H^+ \\
& & \qquad\qquad\;\;\diagup \\
& & CH_2\!-\!S
\end{array}
$$

BAL   Heavy   Chelated metal ion
            metal ion

FIGURE 19-6   BAL chelation of heavy metal ion.

With the arsenic or other heavy metal tied up, the sulfhydryl groups in vital enzymes are freed and can resume their normal functions. BAL is a standard therapeutic item in a hospital's poison emergency center and is used routinely to treat arsenic and mercury poisoning.

Mercury deserves some special attention because it has a rather peculiar fascination for some people, especially children, who love to touch it, coat coins with it, and play with it (Figure 19–7). It is poisonous, and to make matters worse, mercury and its salts are cumulative in the body. This means the body has no quick means of ridding itself of this element and there tends to be a buildup of the toxic effects leading to *chronic* poisoning.

While mercury is rather unreactive compared to other metals, it is quite volatile and easily absorbed through the skin. In the body the metal atoms are oxidized to Hg_2^{2+} [mercury (I) ion] and Hg^{2+} [mercury (II) ion]. Many compounds of both Hg_2^{2+} and Hg^{2+} ions are known to be toxic. A vivid description of the psychic changes produced in an individual by mercury poisoning can be seen in the Mad Hatter, a character in Lewis Caroll's "Alice in Wonderland." An old practice in the fur felt hat industry involved the use of mercuric nitrate, $Hg(NO_3)_2$, to stiffen the felt. This not only accounted for the Mad Hatter's odd behavior, but also gave the workers in the hat factories chronic mercury poisoning, with symptoms known as "hatter's shakes."

Today mercury poisoning is a potential hazard to those working with or near this metal or its salts, such as the dentist (who uses it in making amalgams for fillings), various medical and scientific laboratory personnel (who routinely use mercury compounds or mercury pressure gauges), and some agricultural workers (who frequently employ mercury salts as fungicides).

Lead is another widely encountered metal poison. The body's method of handling lead provides an interesting example of a "metal equilibrium" (Figure

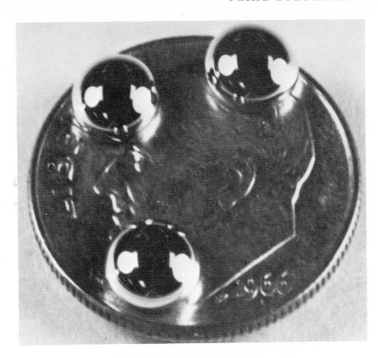

FIGURE 19-7 The first step to coating a coin with mercury. Children love to coat coins with metallic mercury—a very dangerous practice since mercury is easily inhaled and also passes through the skin.

19–8). Lead often occurs in foods (100 to 300 μg per kg), beverages (20 to 30 μg per liter), public water supplies (100 μg per liter—from lead sealed pipes), and even air (2.5 μg per cubic meter—from lead compounds in auto exhausts). With this many sources and contacts per day, it is obvious that the body must be able to rid itself of this poison, otherwise everyone would have died long ago of lead poisoning! The average human can rid himself of about two milligrams of lead a day through the kidneys and intestinal tract; the daily intake is normally less than this. However, if intake exceeds this amount, accumulation and storage results. In the body, lead not only resides in soft tissues, but is also deposited in bone. In the bones lead acts on the bone marrow, while in tissues it behaves like other heavy metal poisons, mercury and arsenic, for example. Lead, like mercury and arsenic, can also affect the central nervous system.

Lead salts, unless they are very insoluble, are always toxic, and their toxicity is directly related to the salt's solubility. One common covalent lead compound, tetraethyl lead, $Pb(C_2H_5)_4$, a component of most antiknock gasolines, is different from other metal compounds in that it is readily absorbed through the skin. Even metallic lead can be absorbed through the skin; cases of lead poisoning have resulted from repeated handling of lead foil, bullets, and other lead objects.

One of the truly tragic aspects of lead poisoning is the fact that even though lead-pigmented paints have not been used in this country for interior painting during the past 30 years, children are still poisoned by lead from old paint. Health experts estimate that up to 225,000 children become ill from lead poisoning each year, with many experiencing mental retardation or other neurological problems. The reason for this is twofold. Lead-based paints still cover the walls of many older dwellings. Coupled with this is the fact that many children in poverty-stricken areas are ill-fed and anemic. These children develop a peculiar appetite trait called *pica*, and among the items that satisfy their cravings are pieces of

411

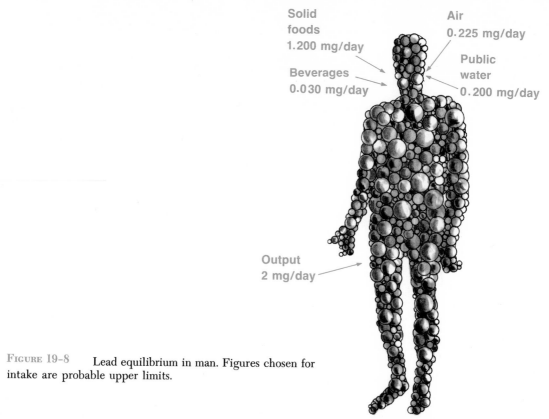

Solid
foods
1.200 mg/day

Air
0.225 mg/day

Beverages
0.030 mg/day

Public
water
0.200 mg/day

Output
2 mg/day

FIGURE 19-8   Lead equilibrium in man. Figures chosen for
intake are probable upper limits.

flaking paint, which may contain lead. In 1969, about 200 children in the United
States alone died of lead poisoning and untold thousands suffered permanent
damage.

In order to remove lead from the human body toxicologists have come upon
another chemical which is a very effective chelating agent—ethylenediaminetetraacetic acid, also called EDTA (Figure 19-9).

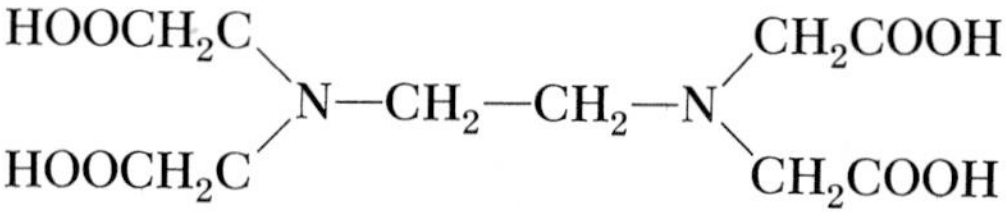

$$\text{HOOCH}_2\text{C} \diagdown \atop \text{HOOCH}_2\text{C} \diagup \text{N}-\text{CH}_2-\text{CH}_2-\text{N} \diagup^{\text{CH}_2\text{COOH}}_{\diagdown\text{CH}_2\text{COOH}}$$

EDTA
(Ethylenediaminetetraacetic acid)

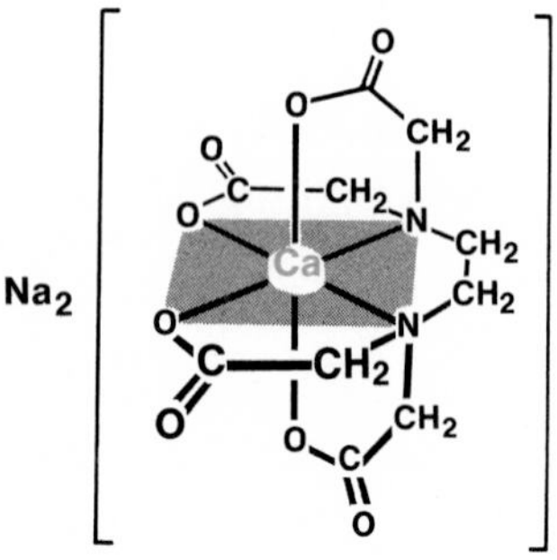

FIGURE 19-9   A calcium EDTA complex.

Na₂

412

The calcium disodium salt of EDTA is used in the treatment of lead poisoning because EDTA by itself would remove too much of the body's essential calcium supply.

In solution EDTA has a greater tendency to complex with lead (Pb^{2+}) than with calcium (Ca^{2+}). The lead chelate is then excreted in the urine. The reaction is

$$[CaEDTA]^{2-} + Pb^{2+} \longrightarrow [PbEDTA]^{2-} + Ca^{2+}$$

NEUROTOXINS

Some metabolic poisons are known to limit their action to the nervous system. These include poisons such as strychnine and curare (the South American Indian's dart poison) as well as the dreaded nerve gases developed for chemical warfare. The exact modes of action of most neurotoxins are not known for certain, but investigations have advanced our knowledge of the action of some. (Concurrently, these investigations have provided insight into the working of the nervous system.)

A nerve impulse or stimulus is actually transmitted along a nerve fiber by electrical impulses. The nerve fiber connects with either another nerve fiber or with some other cell (such as a gland or cardiac, smooth, or skeletal muscle) capable of being stimulated by the nerve impulse (Figure 19–10). Neurotoxins often act at the point where two nerve fibers come together, which is called a synapse. The impulse is transmitted across the synapse or to the organ by chemical action. When the impulse reaches the end of certain nerves, a small quantity of acetylcholine

$$CH_3\overset{O}{\overset{\|}{C}}-OCH_2-CH_2-\overset{CH_3}{\overset{|}{\underset{CH_3}{N^+}}}-CH_3 \quad OH^-$$

Acetylcholine

is liberated; this activates a receptor on an adjacent nerve or organ (Figure 19–11). The acetylcholine is thought to activate a nerve ending by changing the permeability of the nerve cell membrane. The means by which it does this is not clear, but it may be related to an ability to dissociate fat-protein complexes or to penetrate the surface films of fats. Such effects can be brought about by as little as 10^{-6} mole of acetylcholine and could alter the permeability of a cell by allowing ions to cross the cell membrane more freely.

To enable the receptor to receive further impulses, the enzyme cholinesterase breaks down acetylcholine into acetic acid and choline (Figure 19–11).

$$CH_3\underset{O}{\overset{\|}{C}}-OCH_2CH_2-\overset{CH_3}{\overset{|}{\underset{CH_3}{N^+}}}-CH_3, OH^- + H_2O \xrightarrow{\textit{Cholinesterase}} CH_3\overset{O}{\overset{\|}{C}}-OH + HOCH_2CH_2\overset{CH_3}{\overset{|}{\underset{CH_3}{N^+}}}-CH_3, OH^-$$

Acetylcholine   *Water*   *Acetic acid*   *Choline*

413

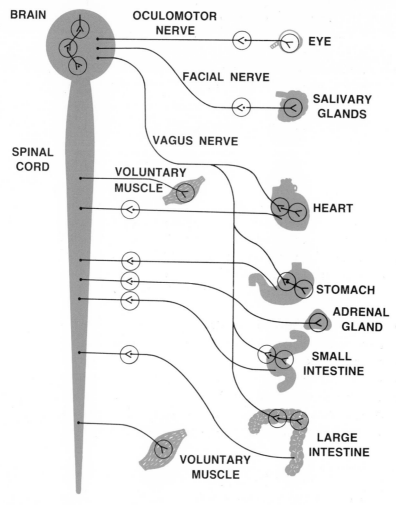

FIGURE 19-10   "Cholinergic" nerves, which transmit impulses by means
of acetylcholine, include nerves controlling both voluntary and involuntary
activities. Exceptions are parts of the "sympathetic" nervous system that
utilize norepinephrine instead of acetylcholine. Sites of acetylcholine secretion are circled in color; poisons that disrupt the acetylcholine cycle
can interrupt the body's communications at any of these points. The role
of acetylcholine in the brain is uncertain, as is indicated by the broken
circles.

In the presence of potassium and magnesium ions, other enzymes, such as
acetylase, resynthesize new acetylcholine from the acetic acid and the choline
within the afferent (incoming) nerve ending:

$$\text{Acetic Acid} + \text{Choline} \xrightarrow{\textit{Acetylase}} \text{Acetylcholine} + \text{H}_2\text{O}$$

The new acetylcholine is available for transmitting another impulse across the
gap.

Neurotoxins can affect the transmission of nerve impulses at nerve endings
in a variety of ways. The *anticholinesterase poisons* prevent the breakdown of
acetylcholine by deactivating cholinesterase. These are usually structurally
analogous to acetylcholine, so they bond to the enzyme cholinesterase and

414

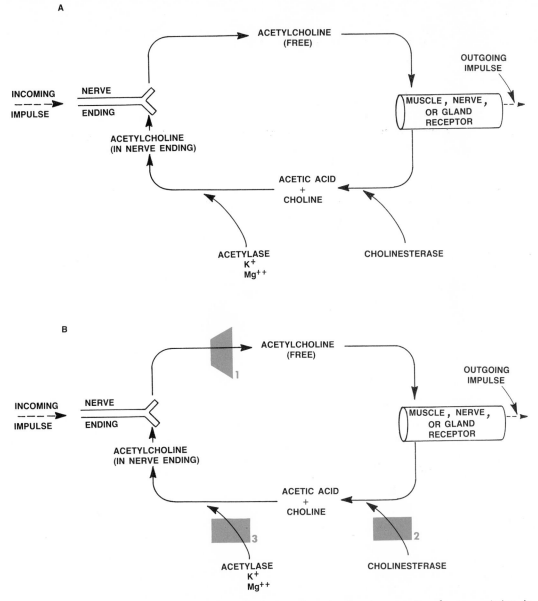

FIGURE 19-11   The acetylcholine cycle, a fundamental mechanism in nerve impulse transmission, is often affected by many poisons. An impulse reaching a nerve ending in the normal cycle (*A*) liberates acetylcholine, which then stimulates a receptor. To enable the receptor to receive further impulses, the enzyme *cholinesterase* breaks down acetylcholine into acetic acid and choline; other enzymes resynthesize these into more acetylcholine. (*B*) Botulinus and dinoflagellate toxins inhibit the synthesis, or the release, of acetylcholine (*1*). The "anticholinesterase" poisons inactivate cholinesterase, and therefore prevent the breakdown of acetylcholine (*2*). Curare and atropine desensitize the receptor to the chemical stimulus (*3*).

deactivate it (Figure 19–12). In effect, the cholinesterase molecules bound by the poison are held so effectively that the restoration of proper nerve function must await the manufacture of new cholinesterase. In the meantime, the excess acetylcholine overstimulates nerves, glands, and muscles, producing irregular heart rhythms, convulsions, and death. Many of the organic phosphates which

415

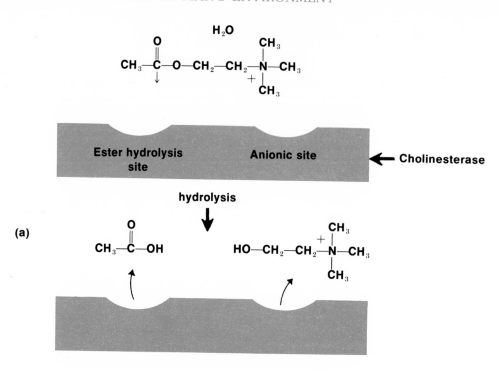

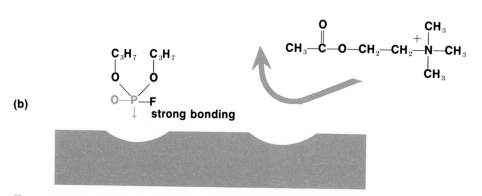

FIGURE 19-12   (a) The mechanism of cholinesterase breakdown of acetylcholine; (b) The tie-up of cholinesterase by an anticholinesterase poison like the nerve gas DFP blocks the normal hydrolysis of acetylcholine since the molecule cannot bind to the enzyme.

are widely used as insecticides are metabolized in the body to produce anticholinesterase poisons. For this reason they should be treated with extreme care. Some poisonous mushrooms also contain an anticholinesterase poison. Figure 19-13 illustrates the structure of some of these substances.

Neurotoxins, such as atropine and curare, are able to occupy the receptor sites on nerve endings of organs which are normally occupied by the impulse-

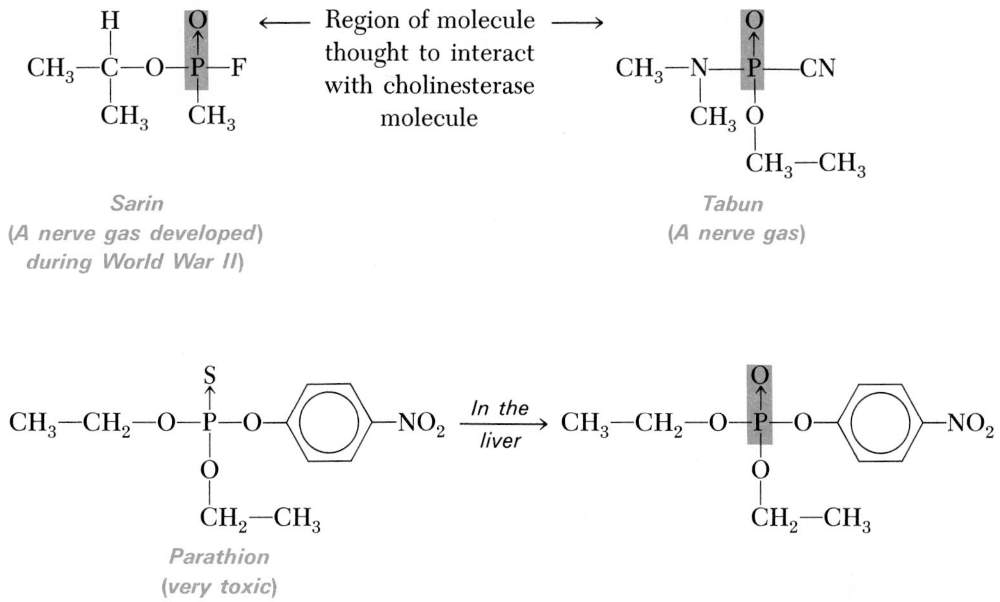

Sarin
(*A nerve gas developed*
during World War II)

Tabun
(*A nerve gas*)

Parathion
(*very toxic*)

FIGURE 19-13   Some anticholinesterase poisons. The potent insecticide Parathion is converted in the liver to a molecule much like the nerve gases (colored area).

carrying acetylcholine. When atropine or curare occupies the receptor site, no stimulus is transmitted to the organ. Acetylcholine in excess causes a slowing of the heart beat, a decrease in blood pressure, and excessive saliva, whereas atropine and curare produce excessive thirst and dryness of the mouth and throat, a rapid heart beat, and an increase in blood pressure. The normal responses to acetylcholine activation are absent, and the opposite responses occur when there is sufficient atropine to block the receptor sites.

Neurotoxins of this kind can be extremely useful in medicine. For example, atropine is used to dilate the pupil of the eye to facilitate examination of its interior. Applied to the skin, atropine sulfate, or other atropine salts, relieve pain by deactivating sensory nerve endings on the skin. Atropine is also used as antidote for anticholinesterase poisons. Curare, used by South American Indians in poison darts, was brought to Europe by Sir Walter Raleigh in 1595. It was purified in 1865, and its structure was determined in 1935. It has long been used as a muscle relaxant.

A well known, natural organic compound that blocks receptor sites, similar to curare and atropine, is nicotine. This powerful poison causes stimulation and then depression of the central nervous system. The probable lethal dose for a 70-kilogram man is less than 0.3 gram. It is interesting to note that pure nicotine was first extracted from tobacco and its toxic action observed *after* tobacco smoking was established as an acceptable habit.

Some other natural products which affect the central nervous system and can be neurotoxic in comparatively small amounts are listed in Table 19-4.

417

NAME	NORMAL CONTACT	LETHAL DOSE (g/70 kg man)	FORMULA
Atropine	Dilation of pupil of the eye	0.1 g	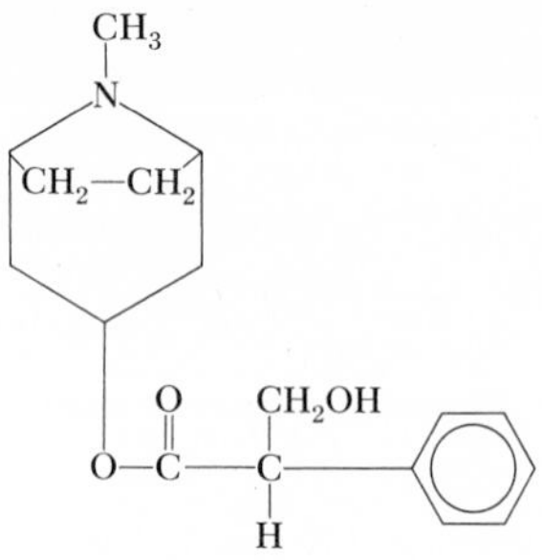
Curare	Muscle relaxant		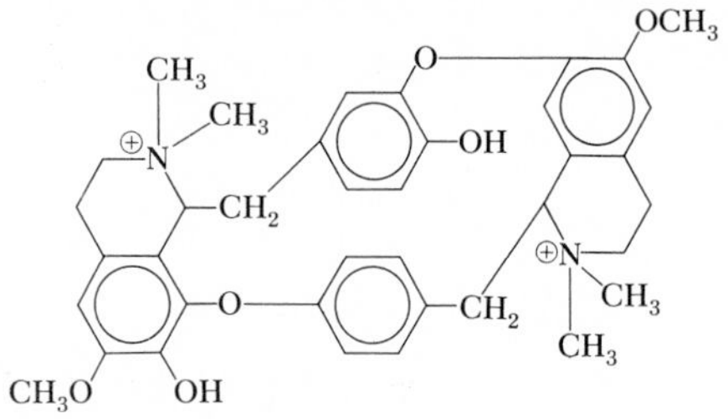
Nicotine	Tobacco insecticide	0.3 g	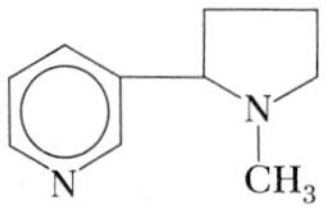
Caffeine	Coffee, tea, cola drinks		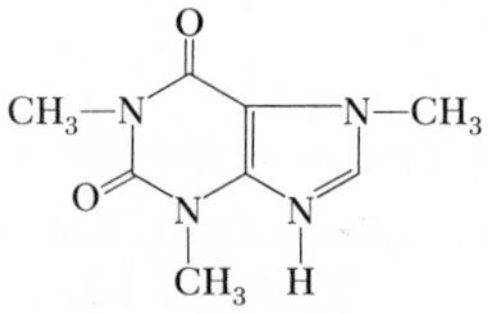
Morphine	Opium— pain killer		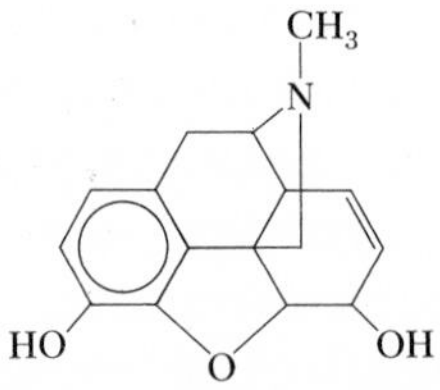
Codeine	Opium— pain killer	0.3 g	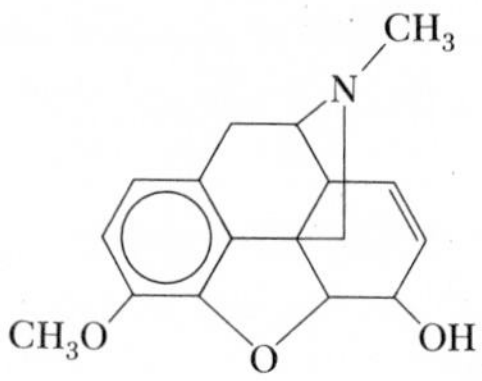
Cocaine	Leaves of *Erythroxylon coca* in South America	1 g	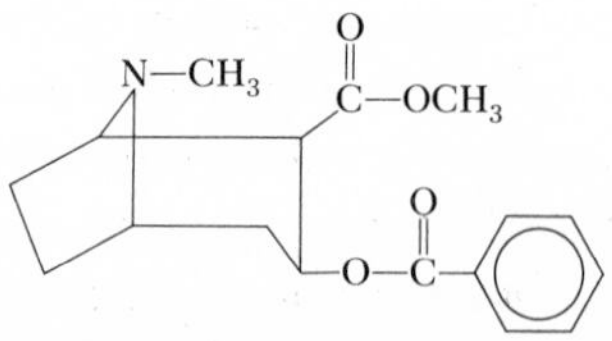

° *Alkaloid* is broadly defined as a physiologically active compound found in plants and containing amino nitrogen atoms (consequently it has basic properties). The nitrogen atom, or atoms, are frequently found as part of rings.

MUTAGENS

Mutagens are chemicals capable of altering the genes and chromosomes sufficiently to cause transmissible abnormalities in offspring. Chemically mutagens alter the structures of DNA and RNA, which compose the genes (and, in turn, the chromosomes) that transmit the traits of parent to offspring. Mature sex, or germinal, cells of man normally have 23 chromosomes; body, or somatic, cells have 23 *pairs* of chromosomes.

Although many chemicals are under suspicion because of their mutagenic effects on laboratory animals, it should be emphasized that no one has yet shown conclusively that any chemical induces mutations in the germinal cells of man. Part of the difficulty of determining the effects of mutagenic chemicals in man is the extreme rarity of mutation. A specific genetic disorder may occur as infrequently as only once in 10,000 to 100,000 births. Therefore, to obtain meaningful, statistical data, a carefully controlled study of the entire population of the United States would be required. In addition, the very long time between generations presents great difficulties, and there is also the problem of tracing a medical disorder to a single, specific chemical out of the tens of thousands of chemicals with which man comes in contact.

If there is no direct evidence for specific chemical mutagenic effects in man, why, then, the interest in the subject? The possibility of a deranged, deformed human race is frightening; the chance for an improved human body is hopeful; and the evidence for chemical mutation in plants and lower animals is established. A wide variety of chemicals is known to alter chromosomes and to produce mutations in rats, worms, bacteria, fruit flies, and other plants and animals. Some of these are listed in Table 19–5.

The horrible Thalidomide disaster in 1961 focused worldwide attention on chemically induced birth defects. Thalidomide, a tranquilizer and sleeping pill, caused gross deformities (flipper-like arms, shortened arms, no arms or legs, and other defects) in children whose mothers used this drug during the first two months of pregnancy. The use of this drug resulted in more than 4000 surviving malformed babies in West Germany, more than 1000 in Great Britain, and about 20 in the United States. With shattering impact, this incident demonstrated that a compound can appear to be remarkably safe on the basis of animal studies (so safe, in fact, that Thalidomide was sold in West Germany without prescription) and yet cause catastrophic effects in humans. While the tragedy focused attention on chemical mutagens, Thalidomide presumably does not cause genetic damage in the germinal cells, and is really not mutagenic. Rather, Thalidomide, when taken by a woman during early pregnancy, causes direct injury to the developing embryo. Even though Thalidomide is generally believed not to be mutagenic, the disaster spotlighted the serious potential hazards of supposedly nontoxic chemicals in our environment.

Experimental work on the chemical basis of the mutagenic effects of nitrous acid (HNO_2) has been very revealing. Repeated studies have shown that nitrous acid is a potent mutagen in bacteria, viruses, molds, and other organisms. In 1953, at Columbia University, Dr. Stephen Zamenhof demonstrated experimentally that nitrous acid attacks DNA. Specifically, nitrous acid reacts with the adenine, guanine, and cytosine bases of DNA by removing the amino group

TABLE 19-5   MUTAGENIC SUBSTANCES AS INDICATED BY
EXPERIMENTAL STUDIES ON PLANTS AND ANIMALS

SUBSTANCE	EXPERIMENTAL RESULTS
Aflatoxin (from mold, *Aspergillus flavus*)	Mutations in bacteria, viruses, fungi, parasitic wasps, human cell cultures, mice
Benzo(α)pyrene (from cigarette and coal smoke)	Mutations in mice
Caffeine	Chromosome changes in bacteria, fungi, onion root tips, fruit flies, human tissue cultures
Captan (a fungicide)	Mutagenic in bacteria and molds; chromosome breaks in rats and human tissue cultures
Dimethyl sulfate (used extensively in chemical industry to methylate amines, phenols, and other compounds)	Methylates DNA base guanine; potent mutagent in bacteria, viruses, fungi, higher plants, fruit flies
LSD (lysergic acid diethylamide)	Chromosome breaks in somatic cells of rats, mice, hamsters, white blood cells of humans and monkeys
Maleic hydrazide (plant growth inhibitor; Trade names Slo-Gro, MH-30)	Chromosome breaks in many plants and in cultured mouse cells
Mustard gas (dichlorodiethyl sulfide)	Mutations in fruit flies
Nitrous acid (HNO_2)	Mutations in bacteria, viruses, fungi
Ozone (O_3)	Chromosome breaks in root cells of broadleaf plants
Solvents in glue (glue sniffing) (toluene, acetone, hexane, cyclohexane, ethyl acetate)	4% more human white blood cells showed breaks and abnormalities (6% versus 2% normal)
TEM (triethylenemelamine) (anticancer drug, insect chemosterilants)	Mutagenic in fruit flies, mice

of each of these compounds. The eliminated group is replaced by an oxygen atom (Figure 19–14). The changing of the bases may garble a part of DNA's genetic message, and in the next replication of DNA the new base may not form a base pair with the proper nucleotide base.

For example, adenine (A) typically forms a base pair with thymine (T) (Figure 19–15). However, when adenine is changed to hypoxanthine, the new compound forms a base pair with cytosine (C). In the second replication, the cytosine forms its usual base pair with guanine (G). Thus, where an adenine-thymine (A-T) base pair existed originally, a guanine-cytosine (G-C) pair now exists. The result is an alteration in the DNA's genetic coding, so that a different protein could be formed later.

Do all of these findings mean that nitrous acid is mutagenic in humans? Not necessarily. We do know that sodium nitrite has been widely used as a preservative, color enhancer, or color fixative in meat and fish products for at least the past 30 years. It is currently used in such foods as frankfurters, bacon, smoked ham, deviled ham, bologna, Vienna sausage, smoked salmon, and smoked shad. The sodium nitrite is converted to nitrous acid by hydrochloric acid in

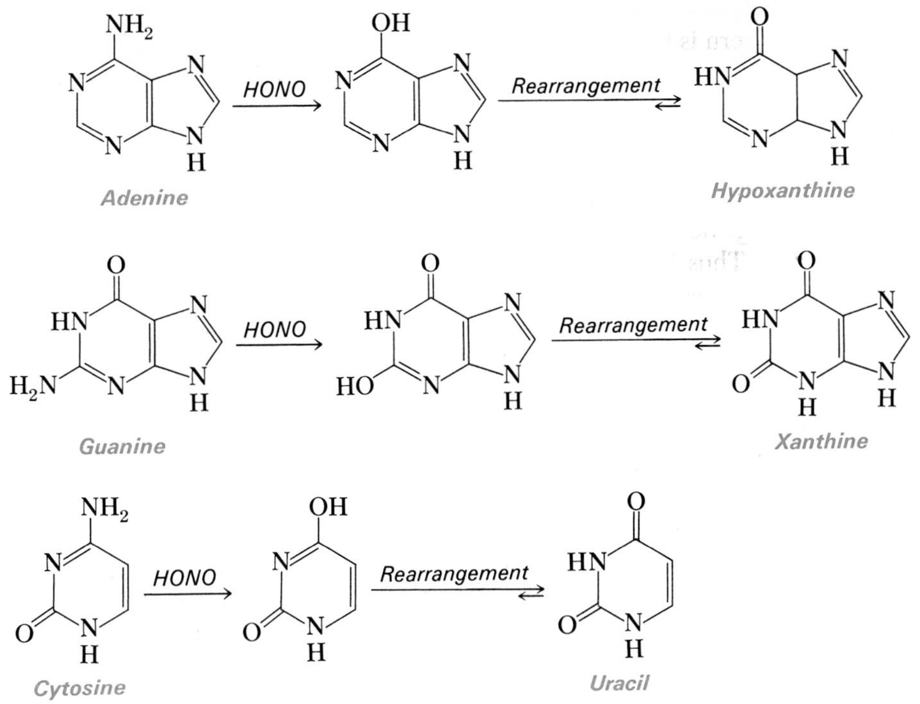

FIGURE 19-14   Reaction of nitrous acid with nitrogenous bases of DNA. Nitrogen gas (N_2) and water are also products of each reaction.

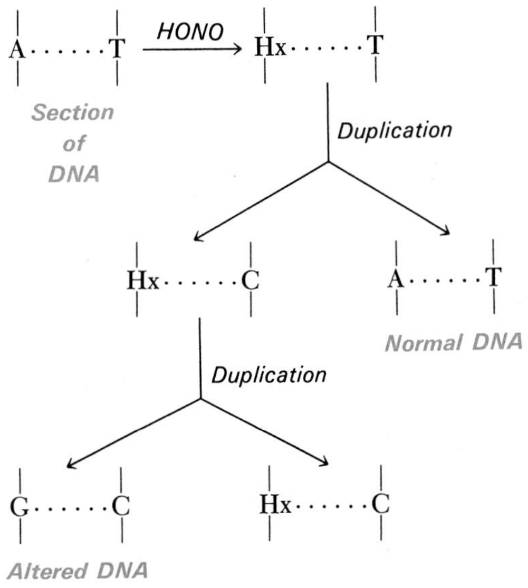

FIGURE 19-15   Alteration of DNA genetic code by base pairing nitrous acid-converted nitrogenous bases. The bases are adenine (A), cytosine (C), guanine (G), hypoxanthine (Hx), and thymine (T).

421

the human stomach:

$$NaNO_2 + HCl \longrightarrow HNO_2 + NaCl$$

Some scientists theorize that the mutagenic effects of nitrous acid in lower organisms are sufficiently ominous to suggest strongly that the use of sodium nitrite in foods be severely curtailed, if not completely banned. A number of European countries already restrict the use of sodium nitrite in foods. The concern is that this compound, after being converted in the body to nitrous acid, may cause mutation in somatic cells (and possibly in germinal cells) and thus could possibly produce cancer in the human stomach. Other scientists doubt that nitrous acid is present in germinal cells and, therefore, seriously question whether this compound could be a cause of genetically produced birth defects in man. The uncertainty of extrapolating results obtained in animal studies to human beings hovers over the mutagenic substances.

Thus far research has concentrated on the action of chemicals in causing mutations in bacteria viruses, molds, fruit flies, mice, rats, human white blood cells, and so on. Perhaps in the next 10 to 20 years it will be demonstrated that these chemicals can produce transmissible alternation of chromosomes in human germinal cells. Meanwhile, many scientists are pressing for a more vigorous research effort to expand our knowledge of chemically induced mutations and of their potentially harmful effects in man. One intriguing theory that will surely invoke experimental examination is the belief that some compounds cause cancer because they are first and foremost mutagenic. The supporting evidence at present is still extremely inconclusive.

Carcinogens

Carcinogens are chemicals that cause cancer. Cancer is an abnormal growth condition in an organism that manifests itself in at least three ways. The rate of cell growth (that is, the rate of cellular multiplication) in cancerous tissue differs from the rate in normal tissue. Cancerous cells may divide more rapidly or more slowly than normal cells. Cancerous cells spread to other tissues; they know no bounds. Normal liver cells divide and remain a part of the liver. Cancerous liver cells may leave the liver and be found, for example, in the lung. Most cancer cells show partial or complete loss of specialized functions. Although located in the liver, cancer cells no longer perform the functions of the liver.

Cancer of epithelial tissue is called a *carcinoma*, whereas cancer in the connective tissues is classified as a *sarcoma*. Cancer from one tissue can spread to other tissues as *metastases*, which, in man, are found in many locations such as in the lung, liver, bone, brain, and kidney. An abnormal growth is classified as cancerous, or *malignant*, when microscopic examination shows that it is invading neighboring tissues or when metastases are found. An abnormal growth is *benign* if it is localized at the original site.

Attempts to determine the cause of cancer have evolved from early studies in which the disease was linked to a person's occupation. If was first noticed in 1775 that persons employed as chimney sweeps in England had a higher rate of skin cancer than the general population. It was not until 1933 that benzo(α)pyrene, $C_{20}H_{12}$, (a 5-ringed aromatic hydrocarbon) was isolated from coal dust and shown to be carcinogenic. In 1895, the German physician Rehn noted

Benzo(α)pyrene

three cases of bladder cancer, not in a random population, but in employees of a factory which manufactured dye intermediates in the Rhine Valley. Rehn attributed these cancers to his patients' occupation. These and other cases confirmed that at times as many as 30 years passed between the time of the initial employment and the occurrence of bladder cancer. The principal product of these factories was aniline. While aniline was first thought to be the carcino-

Aniline   *2-Naphthylamine*

genic agent, it was later shown to be noncarcinogenic. It was not until 1937 that continuous, long-term treatment with 2-naphthylamine, one of the suspected dye intermediates, in dosages of up to 0.5 g per day produced bladder cancer in dogs. Since then other dye intermediates have been shown to be carcinogenic.

A vast amount of research effort has verified the carcinogenic behavior of a large number of diverse chemicals. Some of these are listed in Table 19–6. This research has led to the formulation of a few generalizations concerning the relationship between chemicals and cancer.

For example, carcinogenic effects on lower animals are extrapolated to man. The mouse has come to be the classic animal for studies of carcinogenicity. Strains of inbred mice and rats have been developed which are uniform and show a standard response.

Some carcinogens are relatively nontoxic in a single, large dose, but may be quite toxic, often increasingly so, when administered continuously. Thus, much patience, time, and money must be expended in carcinogenic studies. The development of a sarcoma in humans, from the activation of the first cell to the clinical manifestation of the cancer, takes from 20 to 30 years. With life expectancy of an average person in the United States now set at about 70 years, it is not surprising that the number of deaths due to cancer is increasing.

Cancer does not occur with the same frequency in all parts of the world. Breast cancer occurs with lower frequency in Japan than in the United States or Europe. Cancer of the stomach, especially in males, is more common in Japan

423

TABLE 19-6 A Sample of Carcinogenic Compounds[°]

Compound	Use or Source	Formula of a Representative Compound
Carbon tetrachloride	Dry cleaning agent, solvent	CCl_4
Dioxane	Solvent in chemical industry, cosmetics, glues, deodorants	
Cyclamates	Artificial sweeteners	
Aromatic amines		
2-Naphthylamine	Optical bleaching agent	
Benzidine	Polymer formation, manufacture of dyes, stain for tissues, rubber compounding	
N-2-Fluorenylacetamide	Herbicide	
Azo dyes		
N, N-Dimethyl-p-phenyl-azoaniline	Yellow dye (once used as a food color)	
Nitrosoamines		
N-Ethyl-N-nitroso-n-butylamine	Insecticide, gasoline and lubricant additive	
Polynuclear compounds		
Benzo(α)pyrene	Cigarette and coal smoke	
Aflatoxin	Mold on peanuts and other plants (*Aspergillus flavus*)	

[°] Most of the evidence is based on studies with mice.

than in the United States. Cancer of the liver is not widespread in the Western Hemisphere but accounts for a high proportion of the cancers in the Bantu in Africa and in certain populations in the Far East. The widely publicized incidence of lung cancer is higher in the industrialized world and is increasing at an appreciable rate.

Some compounds cause cancer at the point of contact: benzo(α)pyrene, for example, which has already been mentioned. Other compounds cause cancer in an area remote from the point of contact. The liver is particularly susceptible to such compounds. Since the original compound does not cause cancer on contact, some other compound made from it must be the cause of cancer. For example, it appears that the substitution of a NOH group for a N—H group in an aromatic amine derivative produces at least one of the active intermediates for carcinogenic amines. If R denotes a two or three ring aromatic system, then the process can be represented as follows:

$$\underset{\substack{\text{Inactive} \\ \text{on contact}}}{\overset{\text{H}}{\underset{|}{\text{RNCOCH}_3}}} \longrightarrow \underset{\substack{\text{Active on} \\ \text{contact}}}{\overset{\text{OH}}{\underset{|}{\text{RNCOCH}_3}}} \longrightarrow \underset{\substack{\text{Other unknown} \\ \text{intermediates}}}{\underline{\text{RX?}} \rightarrow \underline{\text{RY?}}} + \text{tissue} \longrightarrow \text{Tumor cell}$$

As indicated by the variety of chemicals in Table 19–6 many molecular structures produce cancer. This has hindered the isolation of many carcinogenic substances. Even closely related structures can produce totally different effects. The 2-naphthylamine mentioned earlier is carcinogenic, but repeated testing gives negative results for 1-naphthylamine.

1-Naphthylamine
(Noncarcinogenic)

2-Naphthylamine
(Carcinogenic)

For some types of cancer there are discrete and distinct periods which ultimately result in cancer. These may be identified as the initiation period, the development period, and the progression period. A single, minute dose of a carcinogenic polynuclear aromatic hydrocarbon, such as benzo(α)pyrene, applied to the skin of mice produces the permanent change of a normal cell to a tumor cell. This is the initiation step. No noticeable reaction occurs unless further treatment is made. If the area is painted repeatedly with noncarcinogenic croton oil, even up to one year later, carcinomas appear (Figure 19–16). This is the development period. Additional fundamental alterations in the nature of the cells occur during the progression period. If there is no initiator, there are no tumors. If there is initiator but no promoter, there are no tumors. If the initiator is

425

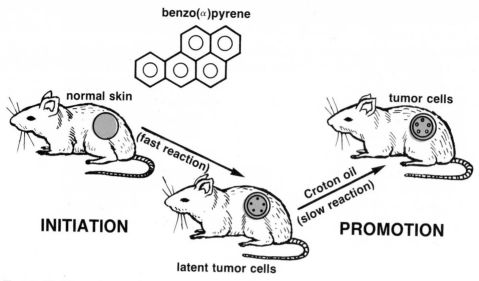

FIGURE 19-16   A second chemical can promote tumor growth in mice after an initiation period. Treatment of mice with croton oil produces no tumor nor does treatment with small quantities of benzo (α) pyrene alone.

followed by repeated doses of promoter, tumors appear. This seems to indicate that cancer cannot be contracted from chemicals unless repeated doses are administered or applied.

Just how do these toxic substances work? Cancer might be caused if the carcinogen combines with growth control proteins, rendering them inactive. During the normal growth process the cells divide and the organism grows to a point and stops. Cancer is abnormal in that cells continue to divide and portions of the organism continue to grow. One or more proteins are thought to be present in each cell with the specific duty of preventing replication of DNA and cell division. Virtually all of the carcinogens bind firmly to proteins, but so do some similar compounds which are noncarcinogenic. The specific growth proteins involved are not yet known for any of the carcinogens, despite considerable efforts to find them.

Another theory suggests that carcinogens react with and alter nucleic acids so the proteins ultimately formed on the messenger RNA are sufficiently different to alter the cell's function and growth rate. The carcinogen may be included in the DNA or RNA strands by covalent bonding or it may be entangled in the helix by a clathrate-type complex. The carcinogenic compounds nitrosodimethylamine and mustard gas have been shown to react with nucleic acids.

While researchers collect data in their laboratories and speculate on the theoretical structural causes of cancer, we can studiously avoid compounds known to cause cancer in man.

HALLUCINOGENS

When not used under professional supervision, hallucinogens produce temporary changes in perception, thought, and mood. These substances include mescaline, lysergic acid diethylamide (LSD), cannabiol (the active component of marijuana), and a broadening field of more than 50 other substances. They are included in this chapter as a separate section because they can be toxic in comparatively small doses and because most of them are known to function in the body in more than one way. Also, they deserve a special section because of the current interest in them.

Several characteristics of some of the more famous hallucinogens are given in Table 19–7. Each of these substances is capable of disturbing the mind and producing bizarre and even colored interpretations of visual and other external stimuli. Mescaline is one of the oldest known hallucinogens, having been isolated from the peyote plant in 1896 by Heffter. Indeed, as early as 1560 the Mexican Indians who ate or drank the peyote experienced "terrible or ludicrous visions; the inebriation lasting for two or three days and then disappearing." Today there are many substances that produce these experiences; the most powerful one known is LSD. Our brief discussion will be restricted to this compound.

TABLE 19–7   SOME HALLUCINOGENIC CHEMICALS

CHEMICAL	ALIASES	NOTES	ACTIVE COMPOUND	FORMULA
LSD	Acid Crackers The chief The hawk	Very powerful hallucinogen	D-Lysergic acid diethylamide	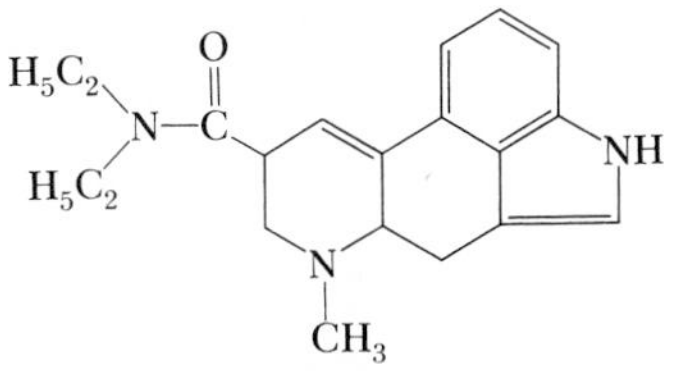
Mescaline		Extracted from mescal buttons from peyote cactus in South and Central America	3,4,5-Trimethoxy-phenethylamine	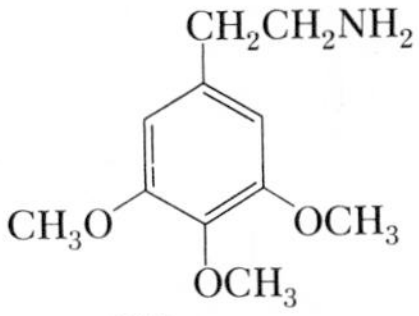
Marijuana	Grass Pot Reefers Locoweed Hash Mary Jane	Leaves of the 5-leaf-per-frond marijuana plant (Cannabis sativa)	3,4-Trans-tetra-hydrocannabinol	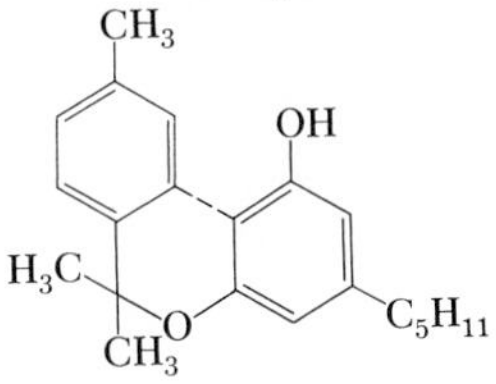
Amphetamines	Pep pills Speed Bennies Ups	Eighteen times more active than mescaline	2,4,5-Trimeth-oxyamphetamine (for example)	

Literally hundreds of scientists are doing research on the effects of hallucinogens, including LSD. They are probing the effects on chromosome alteration, nerve and brain function, tissue damage, psychological changes, and a host of other physiological and psychological effects. Scientists are not near a consensus on the toxic effects that LSD can cause. Collecting valid data is very difficult; many users of LSD overestimate the quantity of the drug they have used, the purity is often in question, and some take other drugs in addition. Even if the

427

purity of the LSD is known, the results are difficult to interpret. For example, a study in which mice were given 0.05 to 1.0 micrograms of LSD on the seventh day of pregnancy showed a 5 per cent incidence of badly deformed mouse embryos. Another study, probably equally valid and meticulous, showed no unusual fetal damage to rat embryos when 1.5 to 300 micrograms of LSD was administered during the fourth or fifth day of pregnancy. Babies with depressed skulls were aborted from two mothers who had taken LSD during the early weeks of pregnancy. Two other women, also LSD users, delivered full term, apparently normal babies. The debate over LSD's dangers is continuing and is not likely to be resolved soon.

There are, however, some dangers that are well documented. These drugs destroy one's sense of judgment. Such things as height, heat, or even a moving truck may seem to hold no danger for the person under the influence of a hallucinogen. Excessive or prolonged use of LSD can cause a person to "freak out" and possibly sustain permanent brain damage. After one "trip," a user can experience another "trip" some time later, unexpectedly, without taking any more drug.

Out of the darkness of LSD-induced suicides and brain damage has come some new understanding of how the brain works. It is an interesting coincidence that soon after LSD was found localized in areas of the brain responsible for eliciting man's deep-seated emotional reactions, the compounds that are probably responsible for transmitting the impulse across synapses in the brain were discovered. Serotonin and norepinephrine are thought to act in a manner similar

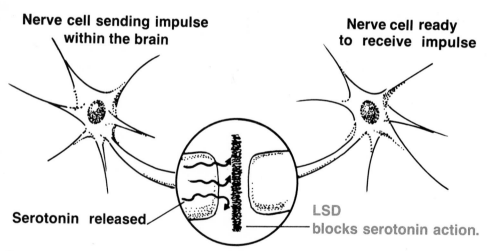

Nerve cell sending impulse within the brain

Nerve cell ready to receive impulse

Serotonin released

LSD blocks serotonin action.

FIGURE 19–17   LSD blocks the flow of serotonin from one brain nerve cell to another.

to acetylcholine, which was discussed earlier. These compounds carry the message from the end of one neuron to the end of another across the synapse. Somewhat later Dilworth Wooley discovered that LSD blocks serotonin action (Figure 19–17). This has led to an interesting theory to explain the LSD trip.

The theory of the hallucinogenic mechanism is, roughly, as follows: the LSD releases in some chemical way those emotional experiences that are generally hidden away, or chemically stored in the lower midbrain and the brain stem. Serotonin or norepinephrine, or both, inhibit the escape of these experiences into consciousness, as they turn off certain excitations. LSD interacts with serotonin or norepinephrine at the synapse and nullifies their blocking effect. Thus, these stored, previously rejected thoughts are allowed to enter the conscious part of the brain where the imaginary trip occurs.

This theory is built on the experimental fact that LSD blocks the action of serotonin; but there are other theories as well as some unexplained phenomena involved. Psychiatry is fraught with theories of why LSD causes an uplifting experience in some people and a frightening one in others. There are theories about dosage, purity of the compound, psychiatric state of the tripper, and conditions of administration. If there is anything close to a consensus, it is that the more controlled and relaxed the surroundings, the "better" the trip will be; even this, however, is debatable. Many trips, even under medical supervision, have not been satisfactory.

ALCOHOLS

Some well known toxic effects are found with the alcohols, but a complete chemical explanation has eluded scientists so far.

Methyl alcohol (methanol or wood alcohol) is highly poisonous, and unlike the other simple alcohols, it is, in effect, a cumulative poison in humans. It has a specific toxic effect on the optic nerve, causing blindness with large doses. After its rapid uptake by the body, oxidation occurs in which the alcohol is first converted into formaldehyde and then to formic acid, which is eliminated in the urine.

$$CH_3OH \xrightarrow{\text{Oxidation}} H-\overset{\overset{\displaystyle O}{\|}}{C}-H \xrightarrow{\text{Oxidation}} H\overset{\overset{\displaystyle O}{\|}}{C}OH$$

Methyl alcohol   *Formaldehyde*   *Formic acid*

This is a slow process and, for this reason, daily exposure to methyl alcohol can cause an extremely dangerous buildup of the alcohol in the body. The toxic effect on the optic nerve is thought to be caused by the oxidative products.

Ethyl alcohol (ethanol, grain alcohol) is found in alcoholic beverages, yet it is toxic like the other simple alcohols and is quantitatively absorbed by the gastrointestinal tract. About 58 per cent of a dose is absorbed in 30 minutes, 88 per cent in 1 hour, and 93 per cent in 90 minutes. Over 90 per cent of the ethyl alcohol is then slowly oxidized to acetic acid, carbon dioxide, and water, mainly in the liver.

The intoxicated person's staggering gait, stupor, and nausea are caused by the presence of acetaldehyde, but the chemical reactions involved are not fully understood. The compound disulfiram (*Antabuse*) is sometimes given as a treatment for chronic alcoholism because it blocks the oxidative steps beyond acetaldehyde. The accumulation of acetaldehyde causes nausea, vomiting, blurred vision, and confusion. This is supposed to encourage the partaker to avoid this severe sickness by avoiding alcohol. Interestingly, this drug was discovered by

429

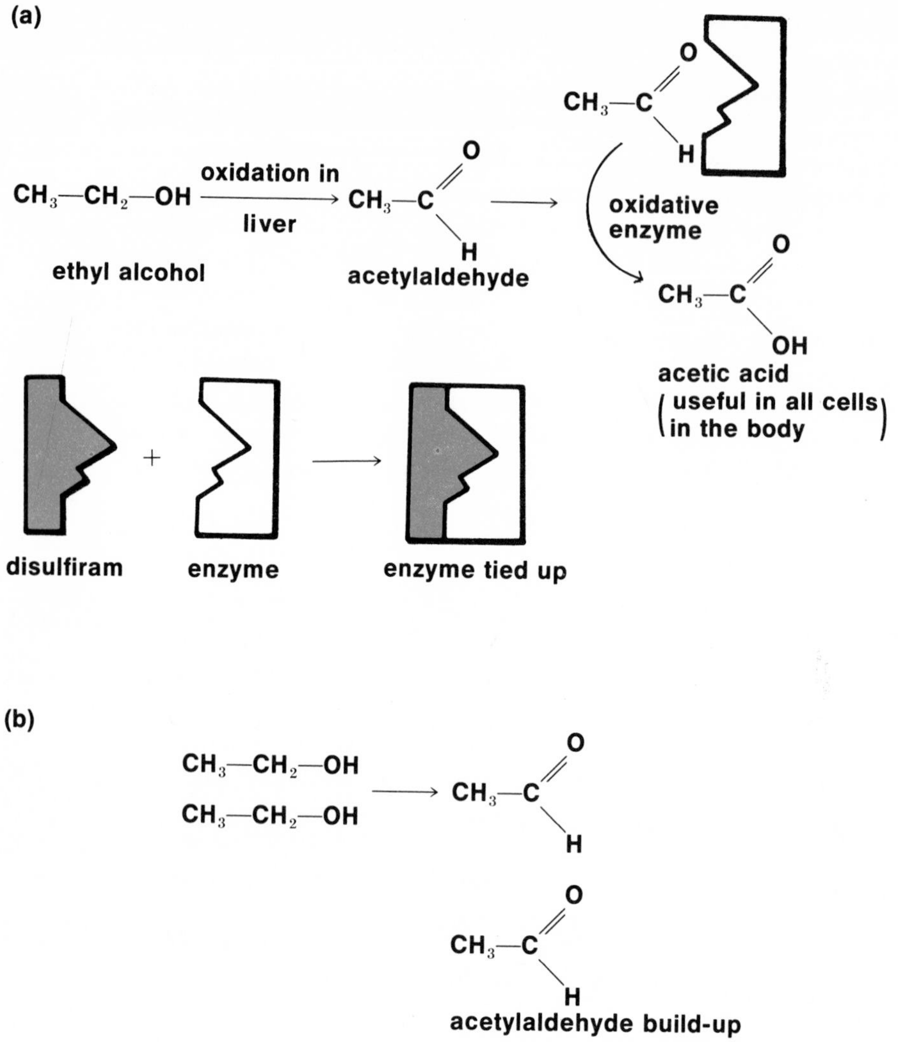

(a)

(b)

FIGURE 19-18   The blocking of ethyl alcohol oxidation by disulfiram (Antabuse). When the normal oxidative process (*a*) is stopped, acetylaldehyde builds up in the body (*b*).

two researchers who took a dose for another purpose and got violently ill that evening at a cocktail party.

DETOXIFICATION PROCESSES IN THE BODY

As we have just seen, the human body possesses a number of ways of getting rid of small amounts of most poisons, and this is the reason so few poisons are truly cumulative. These processes involve excretion in many cases, but they can also involve chemical transformations that change the poison into a less toxic substance which can be eliminated more easily. Such chemical transformation is called detoxification. The presence of some chemicals induces an increase in the concentration of enzymes capable of effecting chemical changes on some toxic molecules. Ultimately the less toxic material is removed by both excretion and several different chemical processes.

Phenol is an example of this. It is removed from the body by at least three routes: direct excretion, transformation into phenol sulfate which is then excreted, and transformation into phenol glucuronic acid which is excreted:

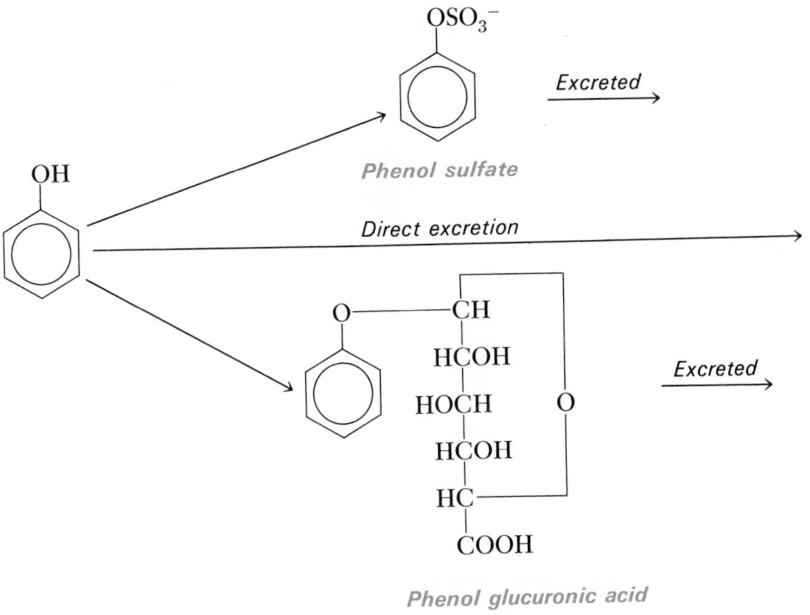

FIGURE 19-19   Removal of phenol from the body.

Other processes, including oxidation and reduction, are also possible. Thus aromatic nitro compounds are reduced to the amines:

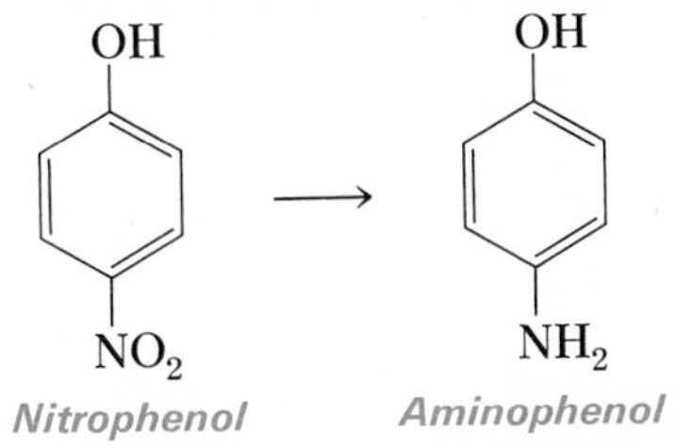

Aniline is also excreted as aminophenol. It is detoxified by an oxidative process.

In many processes the toxic material is combined with an amino acid or other acid to give a less toxic material which is eliminated:

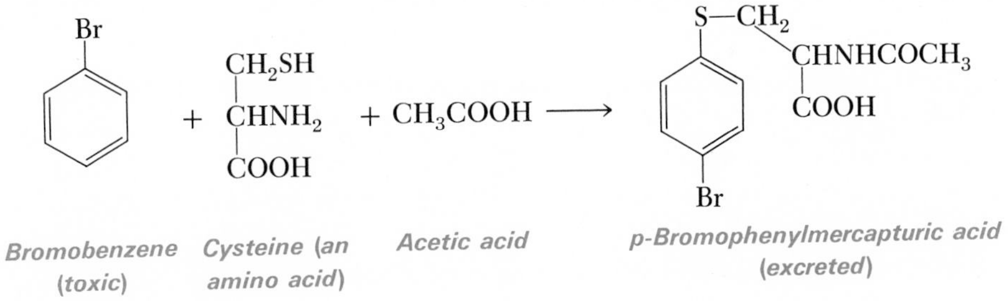

431

Although the body does possess such defenses against small amounts of toxic materials, the detoxification processes usually cannot detoxify large amounts of a poison. Since the ability of the body to cope with poisons is limited, you should never foolishly expose yourself to any toxic material. In some cases the intermediates produced in the detoxification reactions are actually more toxic than the initial molecule, even though they are more readily excreted.

SUMMARY AND CONCLUSION

The chemical bases for many toxic effects are known. For others, only the effect is known, but the basic processes involved are not. In this chapter, we have chosen some common toxic substances and have given the chemical reactions and structural interactions that are believed to produce the toxic effects. In a broad sense, toxic substances either destroy some important biochemical via hydrolysis, oxidation, or reduction, or interfere with the molecule's function by combining with it. The importance of enzymes in the body has been re-emphasized by a study of toxic substances. Toxic substances generally inhibit enzymes, and, as if in retaliation, enzymes are often effective in rendering toxic substances nontoxic. An understanding of the chemical basis of toxicity assists in the development of treatment in a rational chemical manner.

QUESTIONS

1. Give an example of a toxic substance that is toxic as a result of
   (a) hydrolysis
   (b) disguising itself as another compound
   (c) attack on an enzyme

2. True or False. Explain your answer concisely.
   (a) Thalidomide is a known mutagen.
   (b) The corrosive action of poison ivy is well understood.
   (c) There are no known chemical compounds that cannot be toxic under some circumstances.

3. The repeated application of a fused ring hydrocarbon such as benzo(α)pyrene fails to produce a tumor in mice. Does this mean this compound is definitely noncarcinogenic? Give a reason for your answer.

4. Describe the chemical mechanism by which the following substances show their toxic effects.
   (a) fluoroacetic acid
   (b) phosgene
   (c) curare

5. Should any laws and regulations be placed on the use of any of the following?
   (a) LSD
   (b) marijuana
   (c) cyclamates
   (d) ethanol

   Explain your answers.

432   6. Discuss some of the pros and cons of testing toxic substances on animals.

7. Give word chemical reactions for

    (a) action of NaOH on tissue (2 reactions)
    (b) detoxification of phenol
    (c) reaction of EDTA and lead ion

8. What questions do you think need to be answered before the action of ethyl alcohol is understood?

9. Assume a normal diet has the quantity of lead in a given quantity of food as stated in the text. What would a person's total food intake of lead be per day?

10. Look up corresponding data for mercury and make a calculation similar to that in Question 9.

11. How would the body rid itself of an excess of sodium benzoate (a food preservative)? Hint: consider the reactions on pages 391 through 394 of Chapter 18.

SUGGESTIONS FOR FURTHER READING

Adams, E., "Poisons," *Scientific American,* Vol. 201, p. 76 (1959).

Brookes, V. J., and Jacobs, M. B., "Poisons," Second Edition, D. Van Nostrand Co., Inc., Princeton, N.J., 1958.

Chisolm, J. J., Jr., "Lead Poisoning," *Scientific American,* Vol. 224, No. 2, p. 15 (1971).

"Environmental Lead and Public Health," U.S. Environmental Protection Agency, Air. Pollution Control Office, Research Triangle Park, N.C., 1971.

Goldwater, L. J., "Mercury in the Environment," *Scientific American,* Vol. 224, No. 5, p. 15 (1971).

Hoffer, A., and Osmond, H., "The Hallucinogens," Academic Press, New York, 1967.

Loomis, T. A., "Essentials of Toxicology," Lea and Febiger, Philadelphia, 1968.

McGrady, P., "The Savage Cell; A Report on Cancer and Cancer Research," Basic Books, New York, 1964.

Maxwell, K. E., "Chemicals and Life," Dickenson Publishing Co., Inc., Belmont, Calif., 1970.

Robinson, T., "Alkaloids," *Scientific American,* Vol. 201, No. 1, p. 113 (1959).

Sanders, H. J., "Chemical Mutagens," *Chemical and Engineering News,* Vol. 47, No. 21, p. 50, (1969), and Vol. 47, No. 23, p. 54 (1969).

Sanders, H. J., "Allergy—A Protective Mechanism Out of Control," *Chemical and Engineering News,* Vol. 48, No. 20, p. 84 (1970).

Weisburger, J. H., and Weisburger, E. K., "Chemicals as Causes of Cancer," *Chemical and Engineering News,* Vol. 44, No. 7, p. 124 (1966).

WATER:
ITS USE AND MISUSE

CHAPTER 20 ━━━━━━━━━━

Water is certainly one of the most important and unusual of all the chemical compounds. Its intimate participation in life processes marks it as indispensable to all forms of life known to man. The rapidly expanding scope of human life, with its associated industrial development, is increasing the demand for water supplies so rapidly that future needs may easily be underestimated. The time has passed when man can take an abundant water supply for granted.

It is obvious that over the centuries to modern times the world water supply has remained essentially constant. Pollution often diminishes the water supply to less than comfortable levels in regions in which it would otherwise be quite adequate.

Water pollution arises from two main sources. First, water, being an unusually good solvent, readily dissolves objectionable materials when available. Second, and by far the most serious source, are the activities of man; mercury poisoning in Lake Erie and raw sewage effluents are but two examples.

Not all the chemical knowledge needed to understand water pollution problems and their control is at hand. However, much is known, and research on the chemical problems of polluted water is intense. More important, the technological know-how, much of which is patterned after natural purification processes, is available to solve the water pollution problem. Certainly, purifying polluted water will raise the cost. A wise observer of modern man has stated, "Man will pay for pure water, one way or the other. Either he will pay affordable costs of purification or he will pay the catastrophic costs of an increasingly sick environment."

WATER REUSE

It is estimated that 4350 billion gallons of rain (snow and ice) fall on the contiguous United States each day. Of this amount, 3100 billion gallons return to the atmosphere by evaporation and transpiration. The discharge to the sea and underground reserves amounts to 800 billion gallons daily, leaving 400 billion gallons of surface water each day for personal or commercial use. The 48 states withdrew from natural sources 40 billion gallons per day in 1900, 325 billion gallons per day in 1960, and it is estimated that the demand will be at least 900 billion gallons per day by the year 2000. It is evident then that the *reuse* of surface water is the only way that human needs for water can be met in the future.

The reuse of surface water is not new. The fact that rivers serve as a source of water for many cities as well as receivers for their sewage and industrial waste effluents means that a series of cities on the banks of a river will use some of the same water over and over again. If man chooses not to purify his water after use or, in some cases, does not allow nature time to do the job, the river becomes progressively more polluted downstream. The polluted water may contain excessive *heat,* dissolved *chemicals,* and an undesirable population of *microorganisms,* along with *silt.* However, it should be noted that what is pollution to one user may not be to the next. The industrial user who needs water for cooling may primarily be concerned with heat content. The director of the municipal water plant is likely to be concerned with microorganisms, chemicals, and then heat, in that order.

Since the reuse of water is necessary, and since the purification of used water before reuse is often necessary, the question arises as to where the purification should occur. A delay to the point immediately prior to reuse actually leads to many chemical problems and great difficulties. In addition, this kind of action reduces most surface waters to open sewers.

Most will conclude that users must purify their used waters before returning them to rivers, lakes, and even the oceans to maintain a natural ecological balance. Society will have a rich bonus if it can expand its ability to use purified water several times before returning it to nature. An increasing number of industrial users are learning to use treated sewage for their purposes, prior to future purification and eventual return to the river. Much chemistry is involved in the purification and reuse of water. Obviously, the abatement of water pollution calls for some understanding of the chemical problems involved.

THE CHEMISTRY OF WATER POLLUTION

There was a time when polluted water could be thought of in terms of natural silt and contaminations associated with the natural wastes of animals and humans. As man's use of water has increased and diversified, pollution has become distinctly more chemical in nature. The U.S. Public Health Service has classified water pollutants into eight broad categories, listed in Table 20–1.

A consideration of each of the pollutants in Table 20–1 involves their chemical aspects. Some of these chemical problems will be presented after we have looked at nature's purification processes and the pollutants it is able to control.

NATURAL WATER PURIFICATION

Water is a natural resource which, within limitations, is continuously renewed. The familiar hydrologic or water cycle (Figure 20–1) offers a number

435

TABLE 20-1   CLASSES OF WATER POLLUTANTS, WITH SOME EXAMPLES

1. Oxygen-demanding wastes	plant and animal material
2. Infectious agents	bacteria and viruses
3. Plant nutrients	fertilizers, such as nitrates and phosphates
4. Organic chemicals	pesticides, such as DDT, detergent molecules
5. Other minerals and chemicals	acids from coal mine drainage, inorganic chemicals such as iron from steel plants
6. Sediment from land erosion	clay silt on stream bed may reduce or even destroy life forms living at the solid-liquid interface
7. Radioactive substances	waste products from mining and processing of radioactive material, radioactive isotopes after use
8. Heat from industry	cooling water used in steam generation of electricity

of opportunities for nature to purify its water. The world-wide *distillation* process results in rain water containing only traces of nonvolatile impurities, along with gases dissolved from the air. *Crystallization* of ice from ocean salt water results in relatively pure water in the form of ice floes. *Aeration* of ground water as it trickles over rock surfaces, as in a rapidly running brook, allows volatile impurities, previously dissolved from mineral deposits, to be released to the air. *Sedimentation* of solid particles occurs in slow-moving streams and lakes. *Filtration* of water through sand and other permeable solids rids it of suspended matter, such as silt. Next, and of very great importance, are the *oxidation processes*. Essentially all naturally occurring organic materials—plant and animal tissue, as well as their waste materials—are changed through a complicated series of oxidation steps in surface waters to simple molecules common to the environment. Finally, another process that has been used by nature is dilution. Most, if not all, pollutants found in nature are rendered harmless if reduced below certain levels of concentration by dilution with pure water.

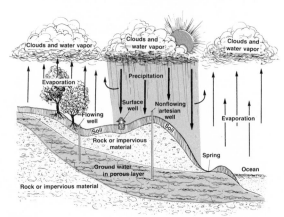

FIGURE 20-1   The water cycle in nature.

Before the advent of the exploding human population and the industrial revolution, natural purification processes were quite adequate to provide ample water of very high purity in all but the arid regions; even in these regions pure water was available from time to time. Nature's purification processes can be thought of as massive but somewhat delicate. In many instances the activities of man push the natural purification processes beyond their limit, and polluted water accumulates.

A simple example of nature's inability to handle man-made pollution is in dragging gravel from stream beds. This excavation leaves large amounts of suspended matter in the water. For miles downstream from a source of this pollutant, aquatic life is essentially destroyed. Eventually, the solid matter settles, and normal life can be found again in the stream.

A more complex example, one for which there is not nearly so much reason to hope for the eventual solution by natural purification, is the degradability of organic materials. A *biodegradable* substance is composed of molecules which are broken down to simpler ones in the natural environment. For example, cellulose suspended in water will eventually be converted to carbon dioxide and water.

$$(C_6H_{12}O_6)_x \xrightarrow[\text{Biochemical catalysts}]{\text{Oxidizing agents}} CO_2 + H_2O$$

Cellulose   Carbon   Water
                               dioxide

The overall oxidation process is a complicated one, involving microorganisms and their enzymatic secretions. Many reaction steps are involved; a detailed study of all the chemistry involved here is yet to be made. Some organic molecules, notably some of those synthetically produced, are not easily biodegradable; these molecules simply stay in the natural waters or are absorbed by life forms and remain for long periods of time.

Detergents offer an interesting contrast in biodegradability. Recall that detergents are composed of long organic molecules, one end of which is soluble in water and the other in oil. Branched-chained detergent molecules, such as

$$CH_3-CH-CH_2-CH-CH_2-CH-CH_2-CH-\bigcirc-SO_3^{\ominus}Na^{\oplus}$$
$$\quad\;\; | \qquad\quad | \qquad\quad | \qquad\quad |$$
$$\quad\;\; CH_3 \qquad CH_3 \qquad CH_3 \qquad CH_3$$

are not readily biodegradable. The enzymes of most of the bacteria present do not catalyze this decomposition. When such detergents were widely used, foaming problems occurred in sewers, streams, and, in extreme cases, in the purified fresh water supply (Figure 20–2). With the subsequent advent of straight-chain detergents, such as

$$CH_3CH_2CH_2CH_2CH_2CH_2CH_2CH_2CH_2CH_2CH_2CH_2OSO_3^{\ominus}Na^{\oplus}$$

biodegradability is readily achieved.

Man now has the ability to make chemically stable materials, such as DDT, and radioisotopes, such as strontium-90, in such large quantities that they cannot effectively be dealt with by dilution. If such materials are judged to be harmful

437

FIGURE 20-2   Foam in natural waters. A high concentration of organic molecules dissolved in natural water causes foam because of the lowered surface tension of the water. If the water is badly polluted, the foam may exist miles from the pollution source. (From Singer, S. F.: Federal interest in estuarine zones builds. Environmental Science & Technology, 3:2, 1969.)

to the natural environment, man must choose not to make or use them or must assist nature in dealing with these species. No one really knows what catastrophic effects would be caused by the sinking of a ship loaded with DDT or the release of strontium-90 that would be produced in a nuclear war.

Even nature's pure rain water is in jeopardy in isolated instances. In areas in which heavy concentrations of automobile fumes collect, poisonous lead compounds have been found in rain water in concentrations many times higher than that allowed in drinking water. The concentration of the lead can be correlated with the concentration of exhaust fumes in the air. Fortunately, lead does not long remain in water, since it generally forms insoluble compounds.

It is evident that man is polluting and will continue to pollute water beyond nature's ability to purify it. Consequently, we must stop the pollution or develop purification techniques to correct it. Fortunately, even if we do not understand thoroughly all of the chemistry involved, we can imitate natural purification processes and purify most, if not all, of our polluted waters.

BIOCHEMICAL OXYGEN DEMAND

The oxidation processes of nature in the purification of water deserve special attention. After many uses, water has an increased concentration of organic

438

material. Even in the natural state, life forms associated with natural waters are constantly putting organic debris into the water. Thus all natural bodies of water supporting life have a *biochemical oxygen demand* (BOD), the amount of oxygen required to oxidize the organic material to simple inorganic molecules. The oxygen is required by microorganisms, such as many forms of bacteria, to utilize organic matter as food in their metabolic reactions. The waste waters of modern society often have very high concentrations of organic material, which have a very large biochemical oxygen demand. In extreme cases, more oxygen is required than is available from the environment. Ultimately, given near normal conditions and enough time, the microorganisms will convert huge quantities of organic matter into the following chemical end-products.

$$\text{Organic carbon} \longrightarrow CO_2$$
$$\text{Organic hydrogen} \longrightarrow H_2O$$
$$\text{Organic oxygen} \longrightarrow H_2O$$
$$\text{Organic nitrogen} \longrightarrow NO_3^-$$

In the extreme case, when the BOD is higher than the oxygen available, radical changes occur in the chemistry and biology of the system. Septic conditions result; fish and other fresh water aquatic life can no longer survive. The aerobic bacteria (those that require oxygen for the decomposition processes) die. As a result of the death of these life forms, even more lifeless organic matter results and the BOD soars. Nature, however, has a back-up system for such conditions. A whole new set of microorganisms (anaerobic bacteria) takes over; these organisms take oxygen from oxygen-containing compounds to convert organic matter to CO_2 and water. Organic nitrogen is converted to elemental nitrogen by these bacteria. Given enough time, enough oxygen may become available, and aerobic oxidation with aquatic life will return.

A stream containing 10 parts per million (ppm) by weight (just 0.001 per cent)* of an organic pollutant from sewage, the formula of which can be represented by $C_6H_{10}O_5$, will contain 0.01 g of this material per liter.

$$?g = 1 \text{ liter of water}$$

$$?g = 1 \text{ liter} \times \frac{1000 \text{ ml}}{l} \times \frac{1 \text{ g}}{ml} = 1000 \text{ g}$$

0.001 per cent of this is the pollutant:

0.001 per cent of 1000 g = $(0.00001)(1000 \text{ g}) = 0.010$ g

To transform this pollutant to CO_2 and H_2O, the bacteria present use oxygen as described by the equation:

$$C_6H_{10}O_5 + 6O_2 \longrightarrow 6CO_2 + 5H_2O$$

$$\begin{array}{cc} \textit{Relative} & \textit{Relative} \\ \textit{weight} & \textit{weight} \\ \textit{162} & \textit{192} \end{array}$$

---

* Parts per million is a concentration unit often used by the chemist. It means, just as the name implies, so many parts for every million (10^6) parts. Ten ppm by weight means 10 parts of impurity for every 10^6 parts of the entire sample; so $\frac{10}{10^6} \times 100 = 10^{-3}$ per cent or 0.001 per cent by weight.

The 0.01 g of pollutant requires 0.012 g of dissolved oxygen.

$$?\text{g oxygen} = 0.010 \text{ g pollutant} \times \frac{192 \text{ g oxygen}}{162 \text{ g pollutant}} = 0.012 \text{ g oxygen}$$

At 68°F the solubility of oxygen in water under normal atmospheric conditions is 0.0094 g of oxygen per liter.

Since the BOD (0.012 g per liter) is greater than the equilibrium concentration of dissolved oxygen (0.009 g per liter), as the bacteria utilize the dissolved oxygen in this stream, the oxygen concentration of the water will soon drop too low to sustain any form of fish life (Figure 20–3). Only if the water is flowing very vigorously in a shallow stream (this facilitates the absorption of more oxygen from the air via aeration), are the life forms common to fresh water likely to survive.

FIGURE 20-3   Fish kills can be caused by the lack of a substance necessary for life, such as oxygen, or the presence of toxic materials that interfere with the life processes. A heavy concentration of organic matter in a stream may depress the oxygen concentration below that required to support fish life. (From Tidwell, P.: Anti-pollution: A fish management priority. The Tennessee Conservationist, 37:4, 1971.)

In the study of polluted water, it is important to know how much oxygen a given sample of polluted water will require. Simple laboratory measurements give a very good estimate. For example, a known volume of the sewage is diluted with a known volume of standardized sodium chloride solution of known oxygen content. This mixture is then held at 20°C for five days in a closed bottle. At the end of this time the amount of dissolved oxygen remaining in the sample is determined, and the amount of oxygen which has been consumed is taken to be the biochemical oxygen demand.

THERMAL POLLUTION

Thermal pollution results when water is used for cooling purposes and in the process has its own temperature raised. Water has a very high *heat capacity* (the heat required to raise a unit of weight of water one degree) and a high *heat of vaporization* (the heat required to change a unit of weight of water to gaseous water); this combination makes water an ideal cooling fluid for thermal power stations, nuclear energy generators, and industrial plants. The extent to which thermal pollution can occur is illustrated by the Thames River, the center of an industrial complex in England. The yearly average temperature in this river, corrected for changes in climatic conditions, rose from 53°F in 1930 to 60°F in 1950. The heat released into this river in 1930 was at the average rate of 550 megawatts. In 1950 the rate became 3700 megawatts. The river, even with various means of cooling (evaporation, conduction, and discharge to the sea), simply could not dissipate the heat as fast as it has been fed in by man. Table 20–2 lists the contributors to this heat inflow.

TABLE 20-2   HEAT CONTRIBUTORS TO INCREASED HEAT CONTENT OF THAMES IN 1950

	PERCENT
Fossil fuel power stations	75
Industrial effluents	6
Sewage effluents	9
Fresh water discharge	6
Biochemical activity	4

The most obvious result of thermal pollution is to make the water less efficient for further cooling applications. Far more important, however, are the biological and biochemical implications.

The oxygen content of water in contact with air is dependent on the temperature of the water, and increases as the water temperature is decreased (Table 20–3). Also, the rate at which water dissolves oxygen is directly proportional to the *difference* between the actual concentration of oxygen present and the equilibrium value. This is extremely fortunate, since it means the rate of solution of oxygen increases sharply as the oxygen is used up. Thus, the rate of absorption of oxygen from the air is greatly facilitated in a shallow, cold mountain stream compared to the case of a deep lake behind a dam in a warmer river.

The increased temperature of the Thames River also increased the oxygen deficit by four per cent over what it otherwise would have been. However, the

FIGURE 20-4   Thermal pollution. Plant effluent causes the Calumet River to "steam." Such heat discharges can change the temperature of a natural body of water enough to vastly alter the ecological balances. (From Chemical and Engineering News ["Technology"], Feb. 24, 1969.)

TABLE 20-3   SOLUBILITY OF OXYGEN IN WATER AT DIFFERENT TEMPERATURES

TEMPERATURE (°C)	SOLUBILITY (PARTS PER MILLION)
25	8.4
15	10.2
5	12.5

TABLE 20-4   A PROJECTION OF THE THERMAL POLLUTION PROBLEM IN THE UNITED STATES°

YEAR	PER CAPITA ELECTRICAL ENERGY CONSUMPTION IN U.S. (48 STATES, ACTUAL OR PROJECTED)	COOLING WATER NEEDS (BILLION GALLONS PER DAY)
1950	2,000 Kwh	
1968	6,500 Kwh	
1980	11,500 Kwh	
2000		200†
		450†

° Notes: (a) Annual surface runoff is 1250 billion gallons per day.
(b) Generation of electrical energy is doubling every 10 years.
(c) 500 new power stations will be needed by the year 2000, at the present growth rate.

† Projected.

biochemical result of this factor alone could not be determined, since the river was and would have been anaerobic as a result of other pollutants in 1950. Even though we readily admit there is much about the consequences of thermal pollution that we do not know, two conclusions seem obvious: (1) thermal pollution aggravates the problem of oxygen supply, and (2) a significant rise in the temperature of a stream can drastically change or even destroy entire biological populations.

Solutions to thermal pollution involve cooling the water in evaporation towers or storage lakes before returning it to the natural body of water. Cycling the water for reuse after cooling has obvious advantages.

SEWAGE

Sewage can and does vary widely in chemical composition from community to community. A comprehensive treatment of sewage chemistry would encompass all of the major topics in water pollution. We shall limit ourselves here to the following definition: sewage is a very dilute suspension or solution of organic substances in water, the organic material being approximately 0.05 per cent of the total weight. If only animal wastes are involved, sewage is biodegradable. Many industrial wastes and household products are not. We shall reserve for following sections particular pollutants, such as fertilizers or industrial wastes.

Recall from your study of organic chemistry that organic compounds are composed of the elements carbon, hydrogen, oxygen, nitrogen, phosphorus, sulfur, and occasionally other elements. The overall view of sewage purification is quite simple; the organic molecules are oxidized to simple inorganic molecules or ions.

$$\text{Organic Compounds} \xrightarrow{\textit{Oxidation}} \text{CO}_2, \text{H}_2\text{O}, \text{NO}_2^-, \text{NO}_3^- \\ \text{HCO}_3^-, \text{PO}_4^{3-}, \text{SO}_4^{2-}$$

Again in nature, it appears to be a very simple process, if we do not overload the system. In part, sewage simply is allowed to flow into natural waters, and there nature provides enzymes (catalysts from microorganisms) and the oxygen to complete the oxidation process. However, the fate of Lake Erie and numerous rivers has shown that this sewage disposal system will just not work if too much sewage is present.

One might hope that the modern biochemist would understand all of the enzymatic steps in the chemical degradation of these organic molecules and hence could improve on this chemistry so that quick and sudden decomposition can be achieved. This is not the case. The problem is a formidable one. Organic molecules are made up of atoms held together by covalent bonds which are relatively strong. Their decomposition then is a painfully slow, step-by-step catalytic process. Thus far, the best that man has been able to do is of a technological, not chemical, nature. By withholding the sewage from natural streams and enriching the bacterial concentration (hence, enzymatic activity) along with oxygen, the decomposition time is shortened. Sterilization of the excessive bacterial concentrations results in a purified sewage compatible with natural waters.

443

Some details of this and other modern developments in water purification are given later in this chapter.

Because sewage contains material which has some value as a fertilizer, it is possible to dispose of it via sewage farms in which it is used to irrigate and fertilize food crops. This method is widely used in Europe and Asia and is quite effective both in disposing of the sewage and in keeping the streams free from pollution. The enormous prejudice in the United States against food grown in this manner has prevented its use here, but it would seem to be a method with many advantages.

FERTILIZERS

Contamination of water by fertilizers leads to very undesirable effects. In a large measure, this contamination results from the phosphate (PO_4^{3-}) and nitrate (NO_3^-) present in fertilizers. It is generally believed that phosphate and nitrate encourage the growth of large amounts of algae. It is certain that nitrate in sufficient concentration is toxic to most higher organisms.

Limnology is the study of physical, biochemical, and biological aspects of fresh-water systems. One of the conclusions of limnology is that massive but relatively delicate balances exist in such systems. The growth of a limited kind and quantity of algae is good for a lake; such growth is a source of needed oxygen via photosynthesis:

$$6CO_2 + 6H_2O \xrightarrow[Catalysts]{Sunlight} C_6H_{12}O_6 + 6O_2$$

Carbon   Water   Sugar
dioxide

However, in waters that contain excessive amounts of phosphate and nitrate leached from fertilized farm land, algae growths are sometimes so massive that they tend to choke out desirable life forms. Such "blooms" of algae may lead to eutrophication. In this state, the oxygen-producing algae actually produce an oxygen-deficient environment, especially at lower depths. This is because of the massive quantity of dead plant material quickly produced by the algae; the BOD rises sharply.

Recent research studies by Wyandotte Chemical Company suggest that the phosphate and nitrate are not the primary causes of algae blooms. While the issue is not yet settled to the satisfaction of the scientific community, this research suggests that the excessive organic pollutant is the primary cause of algae blooms. Bodies of water in which the phosphate concentrations were kept abnormally low—less than 10 parts per billion—produced algae blooms. It was reasoned that the carbon dioxide at higher-than-normal levels resulting from the bacterial oxidation of organic material allowed the excessive algae growth. This "defense" of higher-than-normal phosphate concentrations in our natural waters has to be judged, at best, not entirely satisfactory since the limnological problem is so complex and our present knowledge is so limited.

In the meantime, W. R. Grace Company, at the end of a ten-year project, has announced a process that would reduce the great bulk of phosphate from

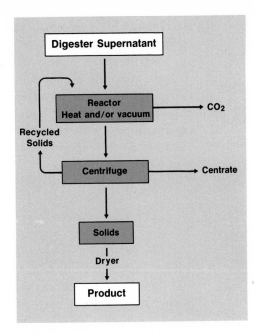

FIGURE 20-5 W. R. Grace process for removing phosphates from sewage.

municipal sewage. The process (Figure 20-5) can be accomplished in most cases without any addition of chemicals. The Grace process involves heating the sewage to 70°C. The solution becomes alkaline as a result of the decomposition of ammonium bicarbonate and other chemicals.

$$NH_4HCO_3 \longrightarrow NH_3 + H_2O + CO_2$$
Ammonium   *Ammonia* *Water*   *Carbon*
bicarbonate   *dioxide*

The ammonia causes the solution to be basic (alkaline) and the carbon dioxide is removed with a vacuum. Phosphate is then precipitated as magnesium ammonium phosphate.

$$NH_3 + H_2O \longrightarrow NH_4^+ + OH^-$$
Ammonia   *Water*   *Ammonium* *Hydroxide*
                         ion   *ion*

$$Mg^{2+} + NH_4^+ + PO_4^{3-} \longrightarrow MgNH_4PO_4$$
Magnesium *Ammonium* *Phosphate*   *Magnesium ammonium*
ion   *ion*   *ion*   *phosphate*

The solid product is centrifuged and dried; it is a marketable product. The process also reduces nitrogen, BOD, and COD (chemical oxygen demand) levels.

The presence of nitrate in water can be quite dangerous, and normal procedures used in purifying water for municipal supplies do not remove it. When the concentration of nitrate is over a few parts per million (ppm), it acts as a poison and destroys the ability of human hemoglobin to carry oxygen, as a result of the reduction of nitrate (NO_3^-) to nitrite NO_2^-. The nitrite interferes with the transport of oxygen by hemoglobin and leads to cyanosis (a condition in which the surface of the body becomes blue because of a lack of oxygen in

445

the blood). This is especially dangerous to infants, who are abnormally susceptible to poisoning (and death) via nitrate in their drinking water.

Fertilizers are made from simple chemicals, such as sodium nitrate, $NaNO_3$, that are quite soluble in water. During and after a hard rain the surface run-off carries much of the fertilizer away. On the other hand, natural fertilizers are not water soluble, but release nitrate, phosphate, and metal ions at a rate commensurate with the rate of uptake by the plants. New fertilizer products are now on the market that imitate nature in releasing their soil nutrients at a slower rate.

DETERGENTS (Syndets)

The detergent industry in the United States has annual sales of some $1.5 billion. Large amounts of these products are discharged into our sewer systems and then into our natural waters. Serious environmental problems can be expected if these waste materials are not biodegradable or if the degraded molecules or ions have a deleterious effect. The switch from branched-chained detergents to linear-chained molecules (see previous discussion) has apparently solved the biodegradability problem. Given sufficient time, these molecules are completely broken down by bacterial-enzymatic action. However, the decomposition products are not as yet known to be wholesome for the environment.

Detergents contain *detergent builders;* these are additives that make the surfactant (detergent) molecules more efficient and effective. In some detergents, as much as 40 per cent of the total weight is in the form of detergent builders. A common detergent builder is sodium tripolyphosphate.

$$3Na^+, \left[\begin{array}{c} \overset{O}{\underset{||}{}} \overset{O}{\underset{}{}} \\ P \\ O \quad\quad O \\ O=P \quad\quad P=O \\ O \quad O \quad O \end{array} \right]^{3-}$$

Many polyphosphate structures exist, and a commercial preparation is likely to contain several. At least one positive action of detergent builders is to tie up metal ions such as Fe^{3+} that cause hard water.

$$Fe^{3+} + (PO_3)_3^{3-} \longrightarrow Fe(PO_3)_3$$
$$\textit{Complex}$$

Because of the use of phosphorus in detergent builders more than 500 million pounds of this element end up yearly in our waste waters in the form of phosphate. In 1967, the average American detergent contained 9.4 per cent by weight of the element phosphorus.

Replacements have been suggested for phosphates in detergent builders. These include nitrilotriacetic acid (NTA),

$$\begin{array}{c} \text{CH}_2\text{COOH} \\ \text{N—CH}_2\text{COOH} \\ \text{CH}_2\text{COOH} \end{array}$$

NTA

organic polyelectrolytes (organic structures with a relatively large negative charge per unit), and sodium citrate.

Since NTA is readily biodegradable, detergent makers thought it to be the answer until it was found that NTA solubilizes heavy metals, producing toxic effects for various life forms. The degradability of the organic polyelectrolytes is not yet established, and sodium citrate, though readily degradable, is not as good a detergent builder as are the phosphates. A multimillion dollar industry awaits the development of a suitable chemical.

A word might be added about enzymatic presoaks (an $80 million dollar a year market) for stain removal. The enzymes are added to assist catalytically in the breakdown of the chemical bonds (degradation) within the stain material, thus aiding in the breakup and washing process. These products make a very small, if not trivial, contribution to cleanliness, and at the same time more than double the phosphate pollution. The enzyme presoak is essentially a detergent mixture and in most cases contains a higher percentage of phosphate builders than nonenzymatic detergents: four leading brands contained 17.4, 17.5, 16.0 and 15.3 per cent phosphorus, compared to the 9.4 per cent figure for detergents in general. In addition, it is not surprising to find that residues of the enzymes in clothing cause some skin disorders.

PESTICIDES

The use of synthetic insecticides increased enormously on a world-wide basis after World War II. As a result, insecticides such as DDT have found their way into lakes and rivers. There is a great variety of pesticides, and their use frequently leads to severe damage to other forms of animal life, such as fish and birds. The widespread use of pesticides was originally justified by the need for more food for a growing human population. However, it has to be admitted that the toxic reactions and peculiar biological side-effects of many of the pesticides were not thoroughly studied or understood prior to their widespread use.

A good case in point is DDT. This insecticide, which had not been shown to be toxic to man in doses up to those received by factory workers involved in its manufacture (400 times the average dosage over years of exposure), does have peculiar biological consequences. The structure of DDT is such that it is

DDT

not metabolized (broken down) very rapidly by animals or fish; it is deposited and stored in the fatty tissues. The biological half-life of DDT is about eight years; that is, it takes about eight years for an animal to metabolize one-half of an amount it assimilates. The enzymes are just not present to catalyze the breakup of this molecule. If ingestion continues at a steady rate, it is evident that DDT will build up within the animal over a period of time. For many animals this is not a problem, but for some predators, such as eagles and ospreys, which feed on other animals and fish, the consequences are disastrous. The DDT in the fish eaten by such birds is concentrated in the bird's body, which attempts to metabolize the large amounts by an alteration in its normal metabolic pattern. This alteration involves the use of molecules which normally regulate the calcium metabolism of the bird and are vital to its ability to lay strong eggs. When these molecules are diverted to their new use, they are chemically modified and are no longer available for the egg-making process. As a consequence, the eggs the bird does lay are easily damaged, and their survival rate decreases drastically. This process has led to the nearly complete extinction of eagles and ospreys in some parts of the United States where formerly they were numerous.

Dieldrin

Heptachlor

Not only does DDT have a long biological half-life, but it also is not readily biodegradable, and there is a resultant buildup of this molecule in aquatic water. DDT and other insecticides such as *dieldrin* and *heptachlor* are referred to as *persistent pesticides*. Substitutions of other substances with biodegradable structures are now made where possible. The compound chlordan is an example of just such a substitution. It is interesting to note that the structural differences between heptachlor (persistent) and chlordan (short-lived) is relatively slight (look at the chlorine atom on the lower five-membered ring).

Chlordan

There are many other insecticides which are actually much more toxic to man than is DDT. These include inorganic materials based on arsenic compounds, as well as a wide variety of phosphorus derivatives based on structures of the type

$$\begin{matrix} R & Z \\ & \| \\ & P-X \\ R' & \end{matrix}$$

where Z = O or S, R and R' are alkyl, alkoxy, alkylthio or amide groups, and X is a group which can be easily split from the phosphorus. Insecticides of this type include parathion, which is effective against a very large number of insects but is also *very* poisonous to human beings. These compounds, as we saw in Chapter 19, are anticholinesterase poisons. One of their most important properties, however, is that they are readily hydrolyzed to less toxic substances which are not residual poisons.

$$\begin{matrix} C_2H_5O & S \\ & \| \\ & P-O-\bigcirc-NO_2 \\ C_2H_5O & \end{matrix}$$

Parathion

In a study made under the sponsorship of the U.S. Department of Health, Education and Welfare from 1964 through 1967, 10 municipal water supplies were examined for pesticides. These waters were taken from the Mississippi and Missouri rivers. Over 40 per cent of the samples contained dieldrin, over 30 per cent contained endrin, and 20 per cent contained chlordan. The amounts present were barely detectable and in the parts per billion (ppb) concentration range.

While it might occur to us to forbid completely the use of such insecticides, we would soon find that there are many areas of the United States in which agricultural production would drop precipitously were this to be done. An enlarged human population requires that crops be protected against the ever-present threat of destruction by insects. Furthermore, the use of insecticides is essential if we are to control malaria, plague, and other diseases. It must be recognized that a single insect species, the tsetse fly, alone has retarded the development of over 4,000,000 square miles of Africa by means of the diseases which it transmits to men and cattle.

The choice of solutions to our problems with insects is not an easy one. The use of insecticides introduces them into our environment and our water supplies. A refusal to use insecticides means that man must tolerate malaria, plague, sleeping sickness, and consumption of a large part of his food supply by insects. It is obvious that continuing research is needed on new methods and materials for the control of insect populations by both chemical and biological means. From the chemical point of view it is entirely reasonable to believe that chemical agents can be found which will control pests and then in turn be quickly biodegraded in the environment.

INDUSTRIAL WASTES

Industrial wastes can be an especially vexing sort of pollution problem because they often are not removed or are removed very slowly by naturally

449

TABLE 20–5   A Partial List of Industrial Wastes that Have Been Dumped into Natural Waters

Acids
Bases
Salts
Various metal solutions
Oils
Bacteria
Emulsifying agents
Grease
Scrap plant and animal wastes
Dyes
Wastes solvents
Poisons such as cyanides and mercury
Numerous chemicals from washing operations

occurring purification processes and are generally not removed at all by a typical municipal water treatment plant. Table 20–5 lists some of the large variety of industrial pollutants that have been put into our natural waters on a large-scale basis.

The technology necessary to remove industrial wastes from the water before it is returned to natural waters is available now. The limiting factor is the cost. Society will simply have to decide if it is willing to spend a greater fraction of its productivity in order to have an unpolluted environment. The control of this pollution is made very difficult by the lack of national and even international standards for all parties to observe. For example, steel mills have been in the past, and to a lesser extent still are, notorious polluters of water. One company would find it very difficult to compete if it had to clean up all of its wastes while other companies went free. Concerted action, whether voluntary or enforced, is necessary. Very large sums of money have recently been spent by some industrial concerns to control or limit pollution. Dow Chemical Company built a $1.3 million plant to control wastes in a plant producing organic chemicals at Midland, Michigan. However, with the population explosion and the resulting industrial expansion, our society is still losing ground in industrial water pollution. A much greater effort is needed.

An example will now be given of an industrial pollution problem that can be solved in a classical as well as in a recently developed way. Metal processing and finishing operations produce large quantities of ions in solution. Chromium plating, for example, produces a waste solution containing chromium in the form of chromate ($CrO_4^=$) and cyanide (CN^-), and other, less objectional ions. Such wastes formerly were emptied into the natural waters. Chemical treatment has been understood for many years and has been used when efforts were made to clean up these wastes. The cyanide can be oxidized by an alkaline chlorine solution,

$$\text{Cyanide} + \text{Chlorine} \xrightarrow{\textit{Base}} \text{Nitrogen} + \text{Bicarbonate} + \text{Chloride}$$

The chromium-containing ion ($CrO_4^=$ [chromate]) is reduced by sulfur dioxide (SO_2).

$$\text{Chromate} + \text{Sulfur dioxide} \longrightarrow \text{Chromium(III) salts} + \text{Sulfate ions}$$

The products of these reactions either are insoluble or are not objectionable, being common to the natural environment. However, it must be admitted that the waste is not chemically pure, since it is a solution of NaCl containing a high concentration of CO_2. It has been known for some time that an electrolytic reduction would produce chemically pure water. This method, however, has not been used because water is such a poor conductor of electricity. Excessive amounts of time and power would be required to complete the electrolysis. Recently, Resource Control, Inc., of West Haven, Conn., developed an electrolytic cell packed with an insoluble carbonaceous material (Figure 20-6). In this cell, the carbonaceous material provides good electrical conductivity between the electrodes; carbon is a good conductor of electricity. The electrical resistance is not a function of the salt concentration, which can be reduced essentially to zero in a short period of time with minimal power requirements. While the theory is as yet not well understood, there appear to be many microcells on the surface of the carbon. The positive ions (cations) are reduced at each cathode and the negative ions are (anions) oxidized at each anode. Table 20-6 gives before-and-after results for a copper plating waste solution.

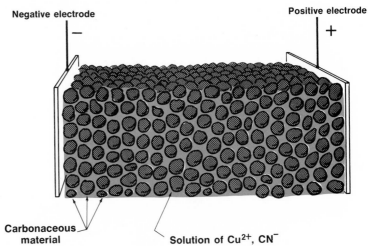

Negative electrode −   **Positive electrode** +

Carbonaceous material   **Solution of Cu^{2+}, CN^-**

FIGURE 20-6   Carbonaceous cell for the electrolytic purification of salt solutions.

Another industrial pollution problem, which does not have a simple economic and chemical solution, is the acid drainage from coal mines. Iron pyrite (FeS_2) is oxidized in the presence of water to give an acidic solution.

$$2FeS_2 + 7O_2 + 2H_2O \longrightarrow 2Fe^{2+} + 4SO_4^{2-} + 4H^+$$

Sulfuric acid

TABLE 20-6   PURIFICATION OF COPPER PLATING WASTE SOLUTION BY CARBONACEOUS ELECTROLYTIC CELL

	CONCENTRATIONS (PPM)	
	Before	*After*
Copper	40	less than 1
Cyanide	20	less than 0.5

Groundwater flushes this acid into surrounding streams, and aquatic life in those streams is destroyed. It was once thought that if the coal mines were sealed, the oxygen would be used up and the reaction would stop. However, bacterial action catalyzes an oxidation process (not as yet well understood) in mines in which the oxygen concentration is nil, and the acid continues to flow. The only apparent solution at this time is to treat the drainage to remove the pollutants.

Heavy metals such as lead, mercury, cadmium, chromium, manganese, copper, zinc, etc., are generally toxic to life forms in greater than trace amounts. Millions of pounds of these metals are "consumed" each year in the United States, and there is no clear understanding of what happens to a considerable portion of that used. Mercury, which has gained widespread attention in recent years, will serve as an example; it should be understood, however, that other metal-pollution problems also exist. The United States government has estimated that 163 million pounds of mercury have been consumed in this country since 1900, with 6 million pounds being the consumption figure in 1969. A breakdown of uses is given in Table 20–7.

TABLE 20-7   MERCURY CONSUMPTION IN THE UNITED STATES TOPS 6 MILLION POUNDS°

	1969 CONSUMPTION (THOUSANDS OF POUNDS)
Electrolytic chlorine	1,572
Electrical apparatus	1,382
Paint	739
Instruments	391
Catalysts	221
Dental preparations	209
Agriculture	204
General laboratory use	126
Pharmaceuticals	52
Pulp and paper making	42
Amalgamation	15
Other	1082
TOTAL	6035

° U.S. Department of Interior, Bureau of Mines.

We do not know where all of the waste mercury goes, but we do know that widespread mercury pollution has been found in most industrialized regions throughout the United States. One chlorine plant is known to have been discharging as much as 200 pounds of mercury a day. Downstream from another plant, silt and clay deposits contained 560 ppm mercury. Four miles downstream the figure was still 50 ppm, and fish in a 35-mile stretch of the river had mercury levels as high as 1.13 ppm. The upper limit in water set by both the United States and Canada for mercury is 0.5 ppm. Water in such polluted regions has much less mercury actually in solution, the figure of 0.03 ppm or less being given. The mercury in solution is often in the form of methyl mercury, $(CH_3)_2Hg$. Evidently the metal either goes into the water in elemental form, or—since it is easily reducible—it is reduced to the element.

452

$$Hg^{++} + \xrightarrow{\text{\textit{Reducing agent}}} Hg + \text{Oxidized form of the reducing agent}$$

Slime deposits containing elemental mercury evidently serve as a source of methyl mercury. It is generally believed that mercury enters the food chain through small organisms that feed on the bottom. These in turn are food for bottom-feeding fish. Game fish in turn eat these fish and accumulate the largest concentration of mercury, the accumulation of poison building up as the food chain progresses.

Stopping mercury pollution appears to be achievable. For example, the tiny particles of mercury in the outflow of chlorine plants can be held in settling tanks. The mercury will be deposited on the bottom, and the brine is concentrated by evaporation and used again.

Another pollution problem, summarized in Table 20–8, concerns pickle liquor, a solution of HCl used to clean steel.

TABLE 20–8   THE FATE OF SPENT PICKLE LIQUOR

Content of Pickle Liquor:	From 0.5 up to 10% acid, 12% iron
Pollution Problem:	Lowers pH of stream and produces solid slime
Scope of Problem:	8 to 15 gallons produced per ton of steel; 50 million tons of steel produced in U.S. per year.
Method of Disposal Used:	(1) Discharge into natural waterway.
	(2) Discharge into deep well.
	(3) Neutralization of acid before release to waterway. Sludge is produced and iron is released.
	(4) Hauling away liquor—a dilution approach.
	(5) Recovery. Cool liquid from 180°F. Iron crystallizes as $FeSO_4 \cdot 7H_2O$. Acid is concentrated and recycled.
	(6) Regeneration: One step beyond step 5. Commercial HCl and Fe_2O_3 recovered.
Immediate Costs:	$6 > 5 > 4 > 3 > 2 > 1$
Long-range Costs:	$1 > 2 > 3 > 4 > 5 > 6$

SILT

Silt, which is finely divided solid material, is picked up by flowing water and is carried along by it until the velocity of the water is reduced as it enters a lake, dam area or the sea. The dispersion of the silt is stabilized by two factors: (1) the kinetic energy of the moving water keeps at least some of the silt moving against gravity, and (2) static charges build up on the silt particles, causing them

453

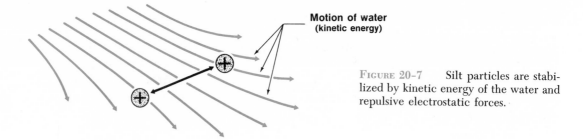

FIGURE 20-7   Silt particles are stabilized by kinetic energy of the water and repulsive electrostatic forces.

to repel each other and remain small (Figure 20–7). In the quiet water of a lake, silt tends to settle out. The salt water of the sea is very effective in removing silt; hence, the great river deltas at the river mouths. The ions in salt water neutralize the electrostatic charge on the particles and then collisions between them result in coagulation. The larger particles settle readily.

Silt is undesirable in municipal water because it is esthetically unappealing and also because the silt particles can be carriers of chemicals or microorganisms, or both, that are harmful to man. Silt is readily removed in holding tanks by sedimentation and by filtration beds. Sedimentation is greatly enhanced if aluminum sulfate and calcium hydroxide (lime) are added to the settling (holding) tank; this results in the formation of aluminum hydroxide, which is a very sticky precipitate. The salt solution tends to neutralize static charges, and the sticky $Al(OH)_3$ collects the silt particles by physical attraction. Most of this settles and filtration removes the rest.

WATER PURIFICATION: MODERN PROCESSES

The commonly used method of municipal water purification has been developed and refined over the past 100 years. The process involves sedimentation, filtration, aeration, and chlorination as is outlined in Figure 20–8.

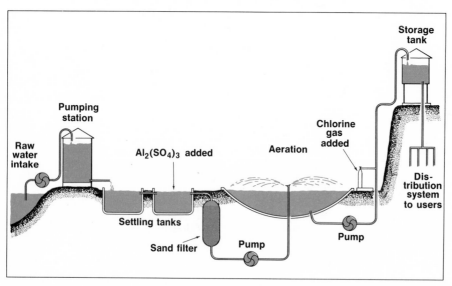

FIGURE 20-8   Schematic drawing of the purification of raw water.

HARD WATER

The presence of Ca^{2+}, Mg^{2+}, Fe^{2+}, or Mn^{2+} will impart "hardness" to waters. Hardness in water is objectionable because (1) it causes precipitates to form in boilers and hot water systems, (2) it causes soaps to form insoluble curds (this reaction is not found with most synthetic detergents), and (3) it can impart a disagreeable taste to the water.

Iron present as Fe^{2+} and manganese present as Mn^{2+} can be removed from water by oxidizing them by air (aeration) to higher oxidation states if the pH of the water is 7 or above (which is generally the case). These reactions produce the insoluble compounds $Fe(OH)_3$ and $MnO_2(H_2O)_x$.

Hardness consisting of calcium or magnesium, present as their bicarbonates, is produced when water containing dissolved carbon dioxide trickles through limestone or dolomite:

$$CaCO_3 + CO_2 + H_2O \longrightarrow Ca^{2+} + 2HCO_3^-$$
Limestone

$$CaCO_3 \cdot MgCO_3 + 2CO_2 + 2H_2O \longrightarrow Ca^{2+} + Mg^{2+} \ 4HCO_3^-$$
Dolomite

Such "hard water" can be softened by removing these compounds. The principal processes for achieving this are (1) the lime-soda process and (2) ion-exchange processes.

The lime-soda process is based on the fact that calcium carbonate $(CaCO_3)$ is much less soluble than calcium bicarbonate $[Ca(HCO_3)_2]$ and that magnesium hydroxide is much less soluble than magnesium bicarbonate. The raw materials added to the water in this process are hydrated lime $[Ca(OH)_2]$ and soda (Na_2CO_3). In the system, several reactions take place, which can be summarized as:

$$HCO_3^- + OH^- \longrightarrow CO_3^{2-} + H_2O$$
$$Ca^{2+} + CO_3^{2-} \longrightarrow CaCO_3 \downarrow$$
$$Mg^{2+} + 2OH^- \longrightarrow Mg(OH)_2 \downarrow$$

If other salts of calcium are present, they are precipitated as calcium carbonate by the sodium carbonate (Na_2CO_3).

$$Ca^{2+} + 2Cl^- + 2Na^+ + CO_3^{2-} \longrightarrow CaCO_3 \downarrow + 2Na^+ + Cl^-$$
Soda

and any salts of magnesium are likewise transformed to magnesium hydroxide:

$$Mg^{2+} + SO_4^{2-} + Ca(OH)_2 + Na_2CO_3 \longrightarrow Mg(OH)_2 \downarrow + CaCO_3 \downarrow + 2Na^+ + SO_4^{2-}$$
Hydrated   Soda
lime

The overall result of the lime soda process is to substitute salts of sodium for those of calcium or magnesium present in the water.

CHLORINATION

Chlorine is introduced as the gaseous free element (Cl_2), and it acts as a powerful oxidizing agent for the purpose of killing bacteria that remain in water

455

after preliminary purification. The great majority of the bacteria in a water supply are removed physically by aluminum hydroxide precipitation. When the filtered water is allowed to stand, the bacterial population often decreases still more since its food supply is inadequate.

When water is heavily contaminated by sewage, the possibility exists that preliminary treatment will not be sufficient to remove *all* of the bacteria capable of causing disease. As cities grew during the late nineteenth century, their water supplies came more and more from waters that had already been used by other cities or from wells which were very close to cesspools or sewers. This meant that the possibility increased for an epidemic caused by a water-borne disease. The principal diseases which are spread in this fashion include cholera, typhoid, paratyphoid, and dysentery.

The fact that cholera is caused by drinking polluted water was proved before the discovery of bacteria. An outbreak of cholera in London in 1854 was examined in great detail by Dr. John Snow. This was a relatively "small" epidemic, in which about 700 died. Dr. Snow made a very careful study of the last few weeks in the lives of about 600 of those killed in the epidemic. Where information was available, he was able to show that in every case, all the victims of the plague had drunk water from a particular well. An examination of the well itself showed a crack in its wall through which sewage from persons sick with

FIGURE 20-9   This apparatus adds chlorine in sufficient amounts to meet health standards (1 ppm residual) for a 60-million gallon per day water treatment plant. (Courtesy of the Robert L. Lawrence, Jr., Filtration Plant.)

cholera had drained before the outbreak of the epidemic. This study showed that polluted water sources are a source of disease.

Subsequent identification of bacteria and proof that they can cause disease led to the chemical treatment of drinking water to kill bacterial populations. The chemical first used for this purpose was chloride of lime [a crude form of calcium hypochlorite, $Ca(OCl)_2$]. This chemical was first used to treat the water supply of Louisville, Kentucky, in 1896. Within a decade and a half, the use of hypochlorites of various sorts spread and became quite common. In 1910, Major C. R. Darnall of the U.S. Army Medical Corps demonstrated that the addition of small amounts of chlorine to partially purified water brought about almost complete elimination of bacteria. Because of the convenience of using elemental chlorine and its ability to remove bacteria even from waters heavily polluted with sewage, chlorine is now widely used in the purification of municipal water supplies. Both bromine and iodine can also be used for this purpose, but these are usually less convenient and more expensive.

Chlorine acts as an oxidizing agent, not only on bacteria and other living microscopic inhabitants of water, but also on dead organic matter. For this reason, chlorine is added to a water supply in sufficient quantity to kill the bacteria, oxidize the remains, oxidize any other organic material the water may contain, and leave a small residual amount of free chlorine. The presence of this small residual amount is a guarantee that the bacterial population has been destroyed. Although there is no direct evidence on this point, it is possible that chlorine is also capable of destroying many, if not all, potentially disease causing viruses that may be present in water drawn from a polluted source.

FLUORIDATION

Fluorine is added to water in the form of solid sodium fluoride or sodium fluorosilicate to introduce the fluoride ion, F^-, or its derivatives, such as SiF_6^{2-} (fluorosilicate), for the purpose of reducing tooth decay.

Widespread fluoridation of water followed the discovery that certain cities in the western United States, where the water supplies contained abnormally high concentrations of natural fluoride, had populations with a less than average incidence of tooth decay. A thorough investigation showed that the ingestion of such fluorides by young children was especially effective in protecting their teeth against cavities. As a consequence, the addition of small amounts of fluoride to drinking water to bring the fluoride concentration up to about 1 part per million was instituted on an experimental basis in some cities. The results bore out the original observation, that small amounts of fluoride in the drinking water could be effective in reducing cavities to about 40 per cent of their normal incidence. Excessive amounts of fluoride, however, can cause damage to teeth.

The mineral content, or the hard part, of bones and teeth consists of two compounds of calcium. Calcium carbonate ($CaCO_3$) is present in bones and teeth in the crystalline form, known to mineralogists as aragonite. The second calcium compound found in teeth is calcium hydroxyphosphate, $[Ca_5(OH)(PO_4)_3]$, or apatite, and it is this crystalline structure that is modified and made more inert in the presence of fluoride ions (F^-). Both aragonite and apatite are susceptible

457

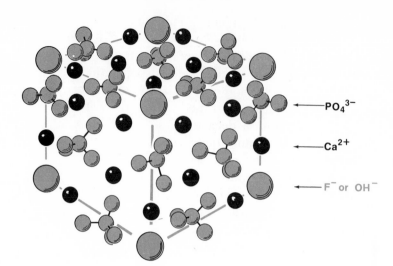

FIGURE 20-10 Structure of apatite and fluoroapatite. The dark circles denote Ca^{2+} ions, the groups of four circles tied together by lines represent the PO_4^{3-} groups, and the largest circles represent OH^- groups in apatite and F atoms in fluoroapatite.

to acid attack from food decay, but apatite is the more so because of its basic characteristics associated with the OH^- ion. If fluoride ions are present in sufficient concentrations during the formative period of the teeth, fluoroapatite $[Ca_5F(PO_4)_3]$ is laid down instead of apatite. The two apatites have the same general structure, so the tooth structure is not physically changed (Figure 20–10).

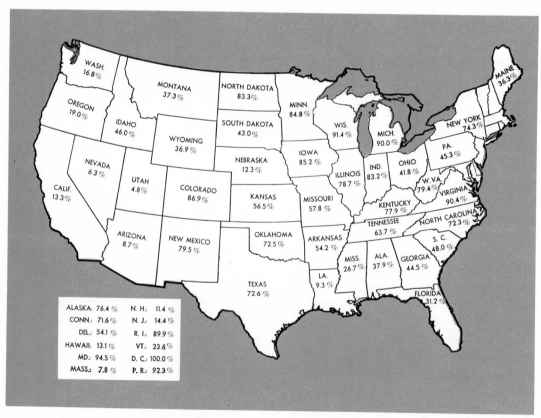

FIGURE 20-11 Fluoridation of U.S. public water supplies as of December 31, 1967. The number within each state refers to the percentage of its population having fluoridated water.

Since fluoroapatite does not have the base properties of apatite and is not readily attacked by mouth acids, the teeth consequently are less likely to develop cavities.

By the end of 1967, nearly 50 per cent of the United States population had water supplies that were fluoridated (Figure 20–11). It should also be noted that fluoride toothpastes and direct applications by dentists were also helpful.

FRESH WATER FROM THE SEA

As men look for new supplies of water, their attention is inevitably drawn to the sea as well as to the vast supplies of brackish water frequently found by drilling in arid lands. Sea water contains 3.5 per cent salts; brackish waters have smaller salt concentrations. It is obvious that processes which can convert such water to potable (drinkable) water are potentially very valuable. Of course, the value of such processes is actually determined by the cost of the pure water produced. Thus far, the cost of purification of salt water is high in comparison with the cost of water supplied by a typical European or American municipal water system. One can expect the cost of fresh water purification to increase if pollution continues to increase, and the cost of purifying salt water to decrease as technology is improved.

The purification of salt water can be divided into two general approaches. Water may be separated from the salt by a change in state, either by evaporation to the gas or by freezing to the solid. A second approach is to cause the ions in the salt water to move, under the influence of electrical charge, chemical attraction, or selective membranes, out of the stream of water flow. If this process continues long enough, pure water results. Bringing about a change of state in water is expensive because of the large amount of energy that must move either in or out of the water. Recall that 540 cal is required for the vaporization of 1 g of water and 80 cal must be removed for the freezing of each gram of water. Moving ions out of the water is expensive because it is relatively slow for large flows of water and involves new techniques that are still under development.

DISTILLATION

The use of the distillation processes to obtain fresh water from sea water has been developed to a considerable state of refinement. Distillation in its simplest sense is illustrated in Figure 1–2. In view of the large amounts of heat required, various methods have been devised to heat the water with "waste heat" prior to entering the still. One method is to cool the steam with raw water as it enters the still. Since the steam has to lose 540 cal per gram in order to liquify, this energy can be used to heat up the raw water to near the boiling point. Solar energy has also been used to preheat the water. Distillation plants are now the most common method in use to refine sea water and can produce water at a cost of less than $1 per 1000 gal.

FREEZING

When cold sea water is sprayed into a vacuum chamber, the evaporation of some of the water cools the remainder and ice crystals form in the brine.

Heat is required to vaporize the cold water, and there is no source for the heat except the cold water and the walls of the chamber. If the chamber is well insulated, the water is cooled to 0°C and some of it even freezes. When the crystals of ice form, they tend to exclude the salt ions. Even though the separation of salt and water is not complete in one step, the ice has less salt than the same weight of liquid solution. The ice crystals can be collected on a filter, washed with a small amount of fresh water, and then melted to obtain "pure" water. The process is repeated until the degree of purity desired is achieved. Plants have been built and successfully operated which produce as much as 250,000 gal of pure water per day by this purification technique.

HYDRATE FORMATION

Propane (C_3H_8), a simple hydrocarbon, offers an interesting approach to the purification of water. In comparison to water, propane molecules have relatively little attraction for each other. Recall that water molecules interact with each other through hydrogen bonding, while weaker Van der Waal's forces are the only attractive forces available between propane molecules. Consequently, theoretical considerations predict the heat changes in the freezing and melting of propane to be considerably less than for water. Experiment shows this to be the case. Experiment also shows that when propane is solidified in the presence of water, a hydrate is formed which is a crystalline solid containing water molecules trapped in the "holes" between the molecules of the solid propane. The salt ions which are attached to a hydrated sphere of water would be, according to theory, too large to fit into the "holes." Experiments confirm this. If the solid hydrate is initially separated from the salt water and then heated, it first melts; the propane then vaporizes, leaving the "pure" water behind. This purification method for salt water has not been widely used because of the higher costs; these costs result from the relatively complex equipment required and the lack of experience in this new technology.

We now turn our attention to methods of purification in which salt ions are removed from the water flow.

ION EXCHANGE

In this process, brackish water or sea water is first passed through a cation exchange resin to replace the cations with H^+ and then through an anion exchange resin to replace the anions with OH^-; these two ions then neutralize each other. Unfortunately, the amount of sea water which can be purified by a given amount of ion exchange resin is quite small, and the resulting water is relatively costly.

Modern ion-exchange resins are high-molecular-weight polymers which contain firmly bonded groups which can exchange one ion for another present in solution. *Cation* exchangers, or positive-ion exchangers, swap their hydrogen ions for the positive ions present in solution. They are usually organic derivatives of sulfuric acid, and their action can be depicted as:

$$R\text{---}SO_3^- \ H^+ + Na^+ + Cl^- \longrightarrow RSO_3^- \ Na^+ + H^+ + Cl^-$$

When the polymer is saturated with cations, it can be regenerated by treatment with strong acid, which reverses the above reaction. Because they generate weakly acidic solutions, *cation* exchange resins themselves do not do a complete job. When they are followed by resins that exchange negative ions (anions), they provide a good route to very pure water. An *anion* exchange resin can replace the anions in solution by the hydroxide ion. These resins are again high molecular weight polymers, but now the polymer contains a nitrogen atom bonded to four other groups. They function as shown:

$$\text{Polymer} \quad -\overset{|}{\underset{|}{N}}-^{+}, OH^{-} + \underbrace{H^{+} + Cl^{-}}_{\substack{\textit{From cation} \\ \textit{exchanges}}} \longrightarrow \text{Polymer} \quad -\overset{|}{\underset{|}{N}}-^{+}, Cl^{-} + H_2O$$

When the anion exchange resin has exchanged all its hydroxide, it can be regenerated by treatment with a strong solution of sodium hydroxide:

$$\text{Polymer} \quad -\overset{|}{\underset{|}{N}}-^{+}, Cl^{-} + Na^{+} + OH^{-} \longrightarrow \text{Polymer} \quad -\overset{|}{\underset{|}{N}}-^{+}, OH^{-} + Na^{+} + Cl^{-}$$

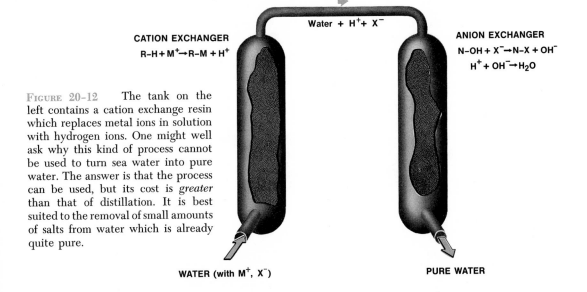

Water + H^{+} + X^{-}

CATION EXCHANGER
$R–H + M^{+} \rightarrow R–M + H^{+}$

ANION EXCHANGER
$N–OH + X^{-} \rightarrow N–X + OH^{-}$
$H^{+} + OH^{-} \rightarrow H_2O$

FIGURE 20-12   The tank on the left contains a cation exchange resin which replaces metal ions in solution with hydrogen ions. One might well ask why this kind of process cannot be used to turn sea water into pure water. The answer is that the process can be used, but its cost is *greater* than that of distillation. It is best suited to the removal of small amounts of salts from water which is already quite pure.

WATER (with M^{+}, X^{-})

PURE WATER

ELECTRODIALYSIS

We have learned that in the electrolysis of salt water, cations migrate toward the negative electrode and anions toward the positive electrode. If an electrolysis cell is divided into three compartments by semipermeable membranes, one permeable to cations and one permeable to anions (Figure 20–13), the process is *electrodialysis*. Dialysis is the passage of selected species in solution through membranes while other species are excluded. Electrodialysis is the special case in which the passage of ions is influenced by an electrical potential gradient.

461

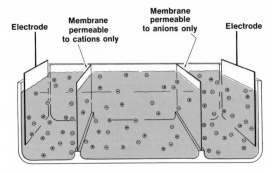

All three compartments filled with brackish water

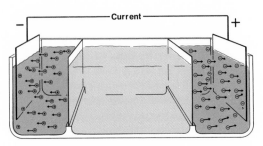

Salts removed from water
in center compartment

FIGURE 20-13 The basic electrodialysis process. Each compartment of the cell contains brackish water. Application of an electrical potential across the cell causes the cations to move into the left compartment and the anions to move into the right compartment. The salt content of the water in the center compartment is thus reduced.

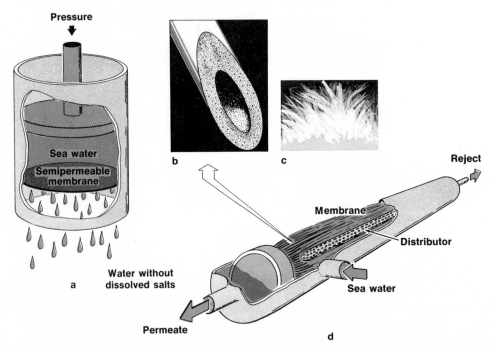

FIGURE 20-14 Reverse osmosis. (a) Mechanical pressure forces water against osmotic pressure to region of pure water. (b) Enlargement of individual membrane. (c) Mass of many membranes. (d) Industrial unit; feed water (salt) that passes through membranes collects at left end (permeate). The more concentrated salt solution flows out to the right as the reject.

In Figure 20–13, note that cations can move to the left out of the center compartment but cannot move through the anion membrane from the right compartment to the center one. In a similar way, anions move out of the center compartment to the right. Thus, the ionic concentration of the water in the central compartment is reduced. If the process is continued long enough, the water in the central compartment is freed from its salt content.

Reverse Osmosis

The elementary science student has learned that osmosis is the process whereby water will move through a semipermeable membrane from a region of relatively pure water into a region containing a concentrated solution. For example, water will move through a cell membrane into the protoplasm, causing the cell to become turgid. The resulting pressure inside the cell is often very large. If enough mechanical pressure is brought to bear on the solution inside the membrane (in other words, if a pressure greater than the osmotic pressure is applied in reverse), the water can actually be made to flow from the concentrated solution inside to the region outside, Figure 20–13.

Reverse osmosis is a very promising process. Costs are down to 25¢ per 1000 gal for a plant producing 30,000,000 gal of pure water per day. This compares to 7.5¢ for the typical municipal treatment and 11¢ for the activated sludge treatments. Membranes used thus far are mostly cellulose acetate and work very well for salt water. Organic materials in the water tend to foul these films when used for municipal sewage. However, much effort is being made in developing new films for a wide variety of applications of reverse osmosis.

CONCLUSION

There is much to be gained by conserving our natural supply of water. Cloud seeding may allow us to distribute water better on this planet: however, we are not very close to this. Proper storage of water in lake systems can facilitate the efficient use of water. It must be admitted candidly, however, that conservation and storage can never solve man's need for water if pollution goes unchecked on a global basis. Generally, to use water is to pollute water. The natural purification processes cannot keep up, and in some cases they are incapable of purifying man-made pollution. The wide variety of pollutants multiplies the difficulties experienced in purifying water. In almost all cases, it is the better part of good sense to purify polluted water just after it is used and return it to nature in a pure form.

QUESTIONS

1. What are some processes that *decrease* the amount of dissolved oxygen in a stream? What are some processes that *increase* the amount of dissolved oxygen in a stream? Which ones are most readily subject to man's control?

2. Explain why each of the following introduces a pollution problem when its wastes are emptied into a stream.

   (a) A slaughterhouse or meat-packing plant.
   (b) A paper mill.
   (c) An electric generating plant burning oil or coal.
   (d) An agricultural area which is intensively cultivated.

3. An old rule of thumb is, "Water purifies itself by running two miles from the source of incoming waste." What processes are active in purifying the water? Is this adage foolproof? Explain.

4. To render a pollutant like DDT harmless, is it necessary to convert the carbon to carbon dioxide, the hydrogen to water, and the chlorine to chloride? Explain.

5. What are some consequences of requiring by law that all water put into our national waters be clinically and chemically pure H_2O? Do you think this is the proper solution to water pollution? If not, what is the proper compromise?

6. What are some ecological consequences of thermal pollution?

7. What pertinent facts would you try to gather if it is your responsibility to vote on a bill to regulate water pollution?

8. Which generally comes first: theoretical solutions to water pollution problems based on an understanding of the reaction of the molecules or technological solutions based on field or laboratory experiences? From the material in this chapter, select an example of each of these two approaches.

9. After thinking about water pollution, at what point do you think pollutants should be removed from used water? Who should be responsible for this removal? Would you distinguish between industrial wastes and household wastes?

10. The major elements in organic compounds are carbon, hydrogen, oxygen, and nitrogen. What are the oxidation products for these elements in the natural decomposition that occurs in nature?

11. Relate molecular structure to biodegradability for detergent molecules.

12. Classify water pollutants into as few major groups as you can. Relate these groups to the topics presented in this chapter.

13. From your study of biochemistry, Chapters 17 and 18, explain why you think there is no question concerning the biodegradability of the proposed detergent builder, sodium citrate.

14. In your judgment, what are the most serious water pollution problems? Be ready to defend your points in class discussion.

15. What is natural osmosis? Explain the significance of the word "reverse" in reverse osmosis.

16. What is a detergent builder? What property should it have to increase the cleansing power of the detergent? What property must it have to avoid environmental problems?

17. From your experience, see if you can add one additional example for each of the classes of pollutants listed in Table 20–1.

18. If electrical energy costs 1¢ per kilocalorie, what will be the cost to distill 4 liters (approximately 1 gal) of water from sea water if no attempt is made to utilize the heat of condensation of the water vapor?

19. What will be the B.O.D. of a stream that contains the equivalent of 0.0001 per cent C as oxidizable organic matter?

20. If the biological half-life of DDT is 8 years, how long will it take to reduce the amount of DDT in a human from 100 mg/kg to 3.1 mg/kg?

SUGGESTIONS FOR FURTHER READING

Behrman, A. S., "Water is Everybody's Business," Doubleday & Co., Garden City, N.Y., 1968.

Encyclopedia of Chemical Technology, Vol. 14, p. 926. John Wiley and Sons, Inc., New York, 1970.

Hills, E. S. (ed.), "Arid Lands: A Geographical Appraisal," Methuen & Co. (New York), 1969.

Howells, G. P., and Kneipe, T. J., "Water Quality in Industrial Areas: Profile of a River," *Environmental Science and Technology*, Vol. 4, p. 26 (1970).

Keller, E., "Fish Kills," *Chemistry*, Vol. 41, No. 9, p. 8 (1968).

Keller, E., "Nuclear-Powered Desalting in the Middle East," *Chemistry*, Vol. 42, No. 2, p. 7 (1969).

Keller, E., "The DDT story," *Chemistry*, Vol. 43, No. 2, p. 8 (1970).

Kirk, R. E., and Othmer, D. F., "Insecticides," *Encyclopedia of Chemical Technology*, Vol. 11, pp. 677–738 (1970).

Leopold, L. B., and Langheim, W. B., "A Primer on Water," U.S. Government Printing Office, Washington, D.C., 1960.

Newman, Frank, "A Water Pollution Study," *Chemistry*, Vol. 42, No. 1, p. 28 (1969).

Overman, M., "Water," Doubleday & Co., Garden City, N.Y., 1968.

Slabaugh, W. H., "Clay Colloids," *Chemistry*, Vol. 43, No. 4, p. 8 (1970).

AIR POLLUTION

You are probably breathing polluted air this very minute. Over 43 million Americans live in 300 cities described by the United States Public Health Service as suffering from "major" air pollution. The PHS also reports that every community of over 50,000 population has an air pollution problem of some sort. If you just now had a breath of clean, fresh air, you are in a rapidly shrinking minority.

Increasing and widespread urbanization, industrialization, and use of the automobile have spread air pollution across our country. It is increasingly becoming a cause of concern to the general public. Isolated incidences of devastating air pollution, however, have been known for centuries in several communities of the world.

Air pollution was so bad in 17th century London that John Evelyn published a treatise called *The Smoake of London* (Figure 12–4). The title page of this book bears a quotation from Lucretius, a Roman poet, translated, "How easily the heavy potency of carbons and odors sneaks into the brain!", indicating air pollution was a problem in the 1st century B.C., too. Shakespeare was perhaps impressed by the pestilence of early 17th century London air when he wrote,

> ". . . this most excellent canopy, the air,
> look you, this excellent o'erhanging
> firmament, this majestical roof fretted
> with golden fire, why, it appears
> no other thing to me but a foul
> and pestilent congregation of vapors."
>
> —*Hamlet,* Act II, Scene II

London experienced its worst air pollution in 1952, and lesser crises in 1956, 1957, and 1962. The mixing of the nearby warm Gulf Stream and the frigid Arctic currents creates frequent fog in the London area. The fog of December 5 to 8, 1952, however, was unusual. It was especially thick and cold. To counteract the cold, Londoners fired their coal heaters hotter and hotter. The smoke spread out near the ground instead of vanishing into the upper atmosphere. The fog grew thicker; the percentages of sulfur dioxide, particulate matter, and other pollutants climbed to 10 times the average readings.

FIGURE 21-1   Air pollution covers
the base of the St. Louis arch. (World
Wide Photos, Inc.)

For five days 12 million people filled their lungs with this poison. A state
of frenzy resulted as deaths from bronchitis jumped to nine times the usual rate
and deaths from pneumonia increased by fourfold. Over 4000 deaths were
attributed to the air pollution. Autopsies on approximately 1000 Londoners who
died in their homes or on the streets showed, almost without exception, that
the victims were already suffering from serious heart or lung ailments. Those
fatally affected were mainly adults aged 45 and older, but mortality of infants
under one year also rose. Many young and apparently healthy persons were made
violently ill.

A government investigation revealed that coal smoke reacting with the
atmosphere generated the deadly pollutants. Sulfur oxides and tarry particles
in contact with the lungs caused severe irritation, a starvation of oxygen, and
finally heart failure. Although the report offered suggestions for preventing
subsequent disasters, another London smog in January, 1956, killed 500 Lon-

467

FIGURE 21-2   Copperhill Basin (Ducktown), Tennessee, as photographed in 1943, shows what can happen to the ecology of an area from poisonous smelter fumes. This plant converted the poisonous fumes to useful acids in 1917, but such damage had already been done that a hundred square miles or more continued to erode. In 1943, the vegetation had started to come back very slowly (notice the cactus in the foreground). (U.S. Forest Service photograph.)

doners, and a third in December, 1956, caused an estimated 400 deaths. In December, 1962, a $3\frac{1}{2}$ day smog accounted for 133 more deaths.

Neither has the United States been spared its share of air pollution disasters. In late October, 1948, Donora, Pennsylvania, had a five day siege of extreme air pollution. Before rain cleansed the air, over 800 domestic animals died and 43 per cent of the population became ill; 5910 people became sick; 20 died, where the normal rate was 2 every 5 days. The spasmodic air pollution crises in New York City and the normal smoggy conditions in Los Angeles attract appreciable attention.

Air pollution damage near Copperhill, Tennessee, is as dramatic in scope as a natural disaster such as an earthquake or a tidal wave (Figure 21-2). Copper ore has been mined and smelted in this area since 1847. In the early years, large quantities of sulfur dioxide, a by-product, were discharged directly into the atmosphere:

$$Cu_2S \quad + O_2 \xrightarrow{\text{Heat}} 2Cu + SO_2$$

Copper ore,   *Air*
containing
copper (I)
sulfide

Before the situation was corrected, the SO_2 killed all of the vegetation for miles around the smelter. Today, the sulfur is reclaimed in the exhaust stacks to make sulfuric acid, but the denuded soil remains a monument to the misuse of the atmosphere.

The widespread interest in air pollution today stems not from the momentary isolated tragedies but from the fairly recent realization that our whole atmosphere is becoming polluted. The long-range effects of these primarily invisible pollutants on materials and the health of plants, animals, and human beings is just beginning to be understood. The lung cancer rate in large metropolitan areas is twice as great as the rate in rural areas, even after full allowance is made for differences in cigarette smoking habits. The serious pulmonary disease, emphysema, shot up *eightfold* during the decade of the sixties. Esthetic, psychological, and weather effects of air pollution are being evaluated. How depressing it is to be enshrouded by smog day after day. How repulsive it is for a new building to become clothed in dirt, filth, and decay. However, only after we comprehend the role and extent of air pollution can we evaluate wisely its relative importance. A logical beginning is to understand some basic facts about our atmosphere.

THE ATMOSPHERE

The atmosphere is a mixture of an estimated 5500 trillion tons of gases, mostly nitrogen (78 per cent) and oxygen (20 per cent) with small quantities of water vapor and carbon dioxide and still smaller quantities of other materials. About 99 per cent of the total mass is below an altitude of 19 miles, and there is sufficient oxygen to sustain life only 4 miles above sea level (Figure 21-3). The region that contains most of the oxygen and supplies most of our weather is the troposphere, a name coined by British meteorologist Sir Napier Shaw (Greek: *tropos*, turning). The troposphere extends to an average altitude of 7 miles.

A representative composition of dry air near sea level is given in Table 21-1. Almost any substance listed in Table 21-1 can be considered a pollutant if it is present in great excess or causes some bad effect. Often near industrial complexes and communities, some of the substances, normally present in very small amounts (less than 1 part per million), increase to abnormal amounts of 50 or more ppm. This radical change alone can qualify substances as pollutants, which already include the nitrogen oxides, sulfur dioxide, carbon monoxide, organic

TABLE 21-1   COMPOSITION OF CLEAN, DRY AIR NEAR SEA LEVEL (PPM)

COMPONENT	CONTENT	COMPONENT	CONTENT
Nitrogen	780,900	Nitrous oxide	0.25
Oxygen	209,400	Carbon monoxide	0.10
Argon	9,300	Xenon	0.08
Carbon dioxide	318	Ozone	0.02
Neon	18	Ammonia	0.01
Helium	5.2	Acetone	0.001
Methane	1.5	Nitrogen dioxide	0.001
Krypton	1.0	Sulfur dioxide	0.0002
Hydrogen	0.5	Lead	0.00000013

469

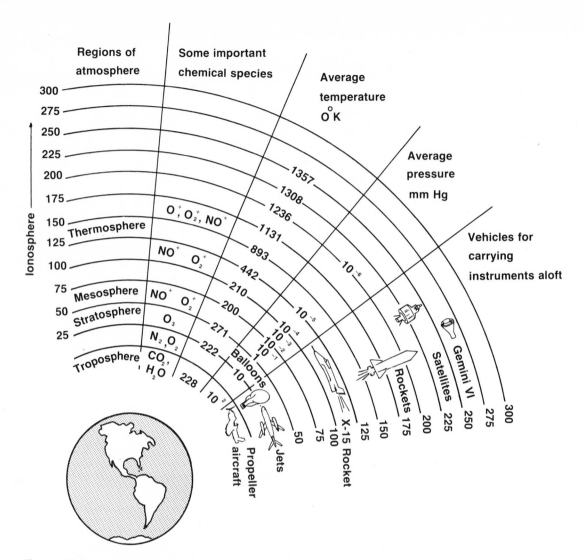

FIGURE 21-3   Some facts about our limited atmosphere. (The troposphere was named by British meteorologist Sir Napier Shaw from the Greek word *tropos*, meaning turning. The stratosphere was discovered by the French meteorologist Leon Philippe Teisserenc de Bort, who believed that this region consisted of an orderly arrangement of layers with no turbulence or mixing. The word *stratosphere* comes from the Latin word *stratum*, meaning layer.)

substances (represented by methane and acetone in Table 21-1), lead, ozone, carbon dioxide, and particulates such as soot, smoke, and dust.

Pollutants litter the troposphere, mixed both vertically and horizontally, often reacting chemically with themselves or with materials on the surface of the earth, and in due time generally return to land or water. The troposphere—the rug in the sky under which we try to hide our wastes—receives more than 140,000 tons of combustion products per day from the United States alone. How pollutants behave chemically while in the atmosphere and what their ultimate fate will be are interesting chemical questions, which we shall deal with later. For now we can say that only rarely do pollutants go beyond the heights of the troposphere.

470

One of the most important parts of the atmosphere is the part we actually breathe. With every breath, the average adult exchanges about 2 liters of air. This quantity contains about 5×10^{22} (50 sextillion) molecules. Most of these molecules leave the lungs just as they entered (Table 21-2). Only those that linger can influence our health.

TABLE 21-2   COMPOSITION OF INHALED AND EXHALED AIR

	INHALED AIR (%)	EXHALED AIR (%)
Oxygen	20.96	15.8
Carbon dioxide	.04	4.0
Nitrogen and other gases	79.00	80.2

Although pollution occurs in relatively small amounts compared to oxygen, nitrogen, and carbon dioxide, about 200 quadrillion pollutant molecules are inhaled per breath on a *clear* day in Los Angeles where the average breath would contain:

Carbon monoxide	175,000,000,000,000,000 molecules
Hydrocarbons	10,000,000,000,000,000
Peroxides	5,000,000,000,000,000
Nitrogen oxides	4,000,000,000,000,000
Lower aldehydes	3,500,000,000,000,000
Ozone	3,000,000,000,000,000
Sulfur dioxide	2,500,000,000,000,000

On a smoggy day, the numbers increase by a factor of five or more. The breath you are now inhaling contains pollutant molecules in comparable amounts, give or take a few quadrillion.

SOURCES OF POLLUTANTS IN THE AIR

Automobiles, industry, and electric power plants are the main sources of air pollutants from man-controlled processes. Volcanic action, forest fires, and

TABLE 21-3   AIR POLLUTANT EMISSIONS IN THE UNITED STATES IN 1965 (millions of tons per year)°

	TOTALS	% OF TOTALS	CARBON MONOXIDE	SULFUR OXIDES	HYDO-CARBONS	NITROGEN OXIDES	PARTICLES
Automobiles	86	60%	66	1	12	6	1
Industry	23	17	2	9	4	2	6
Electric power plants	20	14	1	12	1	3	3
Space heating	8	6	2	3	1	1	1
Refuse disposal	5	3	1	1	1	1	1
TOTALS	142		72	26	19	13	12

°From: Cook, L. M. (ed.): *Cleaning Our Environment: The Chemical Basis For Action* (ACS Committee on Chemistry and Public Affairs), American Chemical Society, 1970.

dust storms are natural sources of air pollutants, but these contribute very little compared to the man-made sources. A summary of the principal sources of emissions in the United States in 1965 is shown in Table 21–3.

POLLUTANT PARTICLE SIZE

The number of molecules lumped together in a solid particle determines the chemical reactivity of the chemicals involved and the effect of gravity on the particle. For these reasons it is important to know the sizes of pollutant particles before attempting to remove them from air.

Pollutant particles may be grouped in three sizes: fundamental (such as single molecules, ions, or atoms), aerosols (a thousand up to about a million atoms, ions, or small molecules per particle), and particulates (suspended particles composed of more than a million atoms, ions, or molecules). Aerosols range in size from 0.001 to 1 micron in diameter (1 micron = 10^{-7} centimeter or 10^{-9} meter); molecules are smaller; particulates are larger. Pollutants dispersed as molecules, ions, or atoms are invisible. Aerosols and particulates serve as carriers and collectors of the chemically active SO_2, nitrogen oxides, ozone, hydrocarbons, and other pollutant molecules.

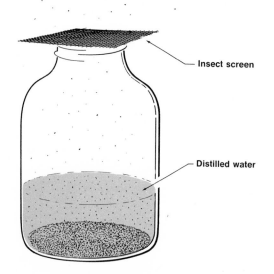

Insect screen

Distilled water

FIGURE 21–4   A dust sample collector for use in the backyard. The open jar containing water is exposed to the air for a known interval. The water is then evaporated and the residue weighed. Fifteen days is sufficient exposure time. Balances sensitive to 0.001 g should be used; the area (cm^3) of the water must be known.

AEROSOLS

Aerosols can be either liquid or solid particles, and are small enough to remain suspended in the atmosphere for long periods of time. They cannot be seen individually by even the most powerful optical microscopes. Smoke, dust, clouds, fog, mist, and sprays are typical aerosols. Aerosols have much greater surface area than a single lump of the material. Consider a cube of coal 1 centimeter on an edge. Such a cube would have a surface area of 6 square centimeters. If this cube is subdivided into a sextillion (1,000,000,000,000,000,-000,000) smaller cubes, each cube would be 1 micron on an edge, about the size of an aerosol particle. The surface area is now 60,000,000 square centimeters—about 1.5 acres. Because of this large surface area, aerosols have enormous capacities to absorb and concentrate gases on their surfaces. Many times aerosols absorb toxic gases, provide the water medium for a reaction to occur, and cause devastating results when breathed. Specific examples and the important role that aerosols play in smog formation will be cited later.

PARTICULATES

Particulates are the sum total of air pollution to most people because these particles are large enough to be seen. They range in size from 1 to 10 microns in diameter. Over 130 million tons of soot, dust, and smoke particulates were deposited into the atmosphere of the United States in 1968 by automobiles and industry. Trash burning, forest fires, and jet aircraft deposited another 30 million tons. Average suspended particulate concentrations in the United States range from 0.00001 gram per cubic meter of air (g per m^3) in remote, rural areas to about six times that value in urban locations. In heavily polluted areas, concentrations up to 0.002 g per m^3, or 200 times the usual values, have been measured.

Particulates may cause physical damage to certain materials. Particles whipped by the wind grind exposed materials by abrasive action. Particles settling in electronic equipment can break down the resistance and foul contacts and switches. Dust settling in paint detracts from its beauty and hastens its

FIGURE 21-5   The General Electric CF6-6D engines that power this McDonnell Douglas DC-10 were designed to eliminate exhaust smoke and reduce by one-half the noise level at takeoff. American Airlines advertises that it spent in a ten year period an amount equal to 43 per cent of its profits on noise and air pollution control systems. (Courtesy of American Airlines.)

473

deterioration by allowing water to reach the surface underneath. Particles may interfere physically with one or more of the clearance mechanisms in the respiratory tract of man and animals (inhibiting the ciliary transport of mucus, for example). People who have asthma or emphysema know that heavy concentrations of particles in the air increase discomfort. In extremely polluted regions, these diseases often lead to death.

Particulates may also injure humans or animals because they are intrinsically toxic. The effects of lead and arsenic were described in Chapter 19. Lead compounds are emitted in automobile exhaust. Arsenic compounds are used as insecticides to dust plants. Particles containing fluorides, commonly emitted from aluminum-producing and fertilizer factories have caused weakening of bones and loss of mobility in animals which have eaten plants covered with the dust. Leaves of bean plants dusted with cement kiln particles (0.00047 g per cm² per day) have been known to wilt significantly.

Particulates may cause damage because they adsorb toxic substances. Sulfur dioxide, nitrogen oxides, hydrocarbons, and carbon monoxide do their greatest damage when concentrated on the surface of particles. Examples will be cited when the chemistry of these pollutants is discussed later in this chapter.

Particulates absorb and scatter light; consequently, they can reduce the amount of light reaching the earth to warm it. The amount of particulate matter in the air has increased since 1964, and this may explain why the mean global temperature has been decreasing over the past few years.

Particles are removed from the atmosphere by gravitational settling and by rain and snow. They can be prevented from entering the atmosphere by treating industrial emissions by one or more of a variety of physical methods

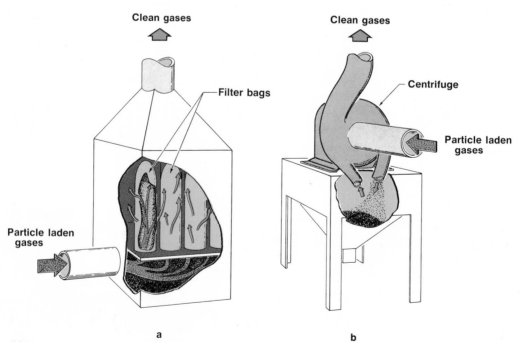

a   b

FIGURE 21-6   Physical methods of removing particulates from emissions.

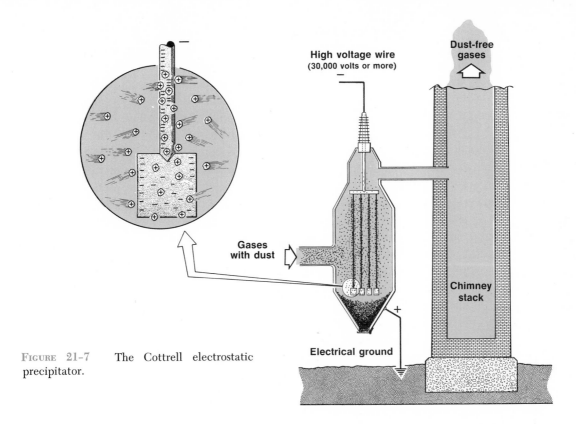

High voltage wire
(30,000 volts or more)

Dust-free
gases

Gases
with dust

Chimney
stack

Electrical ground

FIGURE 21-7   The Cottrell electrostatic
precipitator.

such as filtration, centrifugal separation, spraying, ultrasonic vibration, and
electrostatic precipitation. Some of these methods are illustrated in Figures 21–6
and 21–7.

Glass fiber or silicone treated textile bags filter hot exhaust gases (550°F
maximum); cotton, nylon, acrylics, wool, and felts are used to filter out cold
aerosols.

Centrifugal separators whirl the gases and sling the particles against the
walls where they collect. The same result is obtained when fast-moving exhaust
is made to change direction. The dust particles bump against the wall where
they collect.

Spray treatment is utilized by the petroleum industry, among others. The
smoke from a waste gas burner is collected in the water from several small spray
nozzles set a few inches above the tip of the burner.

Ultrasonic vibration operates on the principle that increased collisions of
the particles and aerosols cause adherence and condensation of the suspended
particles. The ultrahigh frequency vibrations cause the increased collisions.

Electrostatic precipitators can remove aerosols and dust particles smaller
than 1 micron from plant exhaust gases. A Cottrell electrostatic precipitator is
shown in Figure 21–7. The central wire is connected to a source of direct current
and high voltage (about 50,000 volts). As dust or aerosols pass through the strong
electrical field, the particles attract ions which have been formed in the field,
become strongly charged, and are attracted to the walls of the tubes. The
precipitated solid falls to the bottom where it is collected.

475

SMOG—INFAMOUS AIR POLLUTION

The poisonous mixture of smoke, fog, air, and other chemicals was first called *smog* in 1911 by Dr. Harold de Voeux in his report on a London air pollution disaster that caused the deaths of 1150 people. Through the years, smog has been a technological plague in many communities and industrial regions.

Two general kinds of smog have been identified. One is the chemically reducing type that is derived largely from the combustion of coal and oil, which contains sulfur dioxide mixed with soot, fly ash, smoke, and partially oxidized organic compounds. This is the *London type,* which is diminishing in intensity and frequency as less coal is burned and more controls are installed. A second type of smog is the chemically oxidizing type, typical of Los Angeles; it is called *photochemical* smog because light—in this instance sunlight—is important in initiating the photochemical process. This smog is practically free of sulfur dioxide but contains substantial amounts of nitrogen oxides, ozone, ozonated olefins, and organic peroxide compounds, together with hydrocarbons of varying complexities.

What general conditions are necessary to produce smog? Although the chemical ingredients of smogs often vary, depending upon the unique sources of the pollutants, certain geographical and meteorological conditions exist in nearly every instance of smog.

There must be a period of windlessness so that pollutants can collect without being dispersed vertically or horizontally. This lack of movement in the ground air can occur when a layer of warm air rests on top of a layer of cooler air.

FIGURE 21-8   Smog over New York City. A heavy haze hangs over Manhattan Island, viewed from the roof of the RCA Building. The Empire State Building is barely visible in the background. (Wide World Photos, Inc.)

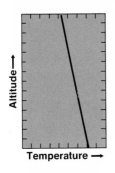
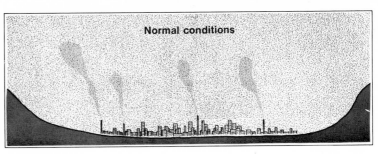
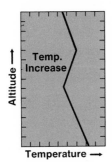
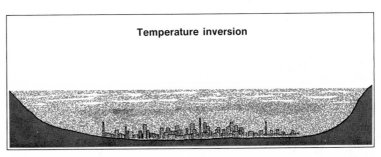

FIGURE 21-9   A diagram of a temperature inversion layer over a city. Warm air over a polluted air mass effectively acts as a lid, holding the polluted air over the city until the atmospheric conditions change. The line on the left of the diagram indicates the relative air temperature.

This sets the conditions for a *thermal inversion,* which is an abnormal temperature arrangement for air masses. If the warmer air is on the bottom nearer the warm earth, which is usual, the warmer, less dense air rises and transports most of the pollutants to the upper atmosphere where they are dispersed. When the warmer air is on top, as in thermal inversion, the cooler, more dense air retains its position nearer the earth and vertical movement is stagnated. If the land is bowl-shaped (surrounded by mountains, cliffs, or the like), horizontal movement of the air mass is also hindered.

When these natural conditions exist, man supplies the killing ingredients by combustion and evaporation in automobiles, electrical power plants, space heating, and industrial plants. The chief pollutants are sulfur dioxide (from burning coal and some oils), and nitrogen oxides, carbon monoxide, and hydrocarbons (chiefly from the automobile). Add to these ingredients the radiation from the sun, and a massive case of smog is in the offing.

A city's atmosphere is an enormous mixing bowl of frenzied chemical reactions. Ferreting out the exact chemical reactions that produce smog has been a tedious job, but in 1951, insight into the formation process was gained when smog was first duplicated in the laboratory. Detailed studies have subsequently revealed that the chemical reactions involved in the smog-making process are photochemical and that aerosols serve as breeders, participants, and products in the formation of the secondary pollutants, which are formed by chemical reaction in the atmosphere. Light provides the energy of activation for the series

477

of photochemical reactions, and ultraviolet radiation from the sun is the energy source for the formation of photochemical smog.

The exact reaction scheme by which primary pollutants are converted into the secondary pollutants found in smog is still not completely understood, but the reactions shown in Figure 21–10 account for the major secondary pollutants in photochemical smog. The process is thought to begin with the absorption of a quantum of light by nitrogen dioxide, which causes its breakdown into nitric oxide and atomic oxygen. The very reactive atomic oxygen reacts with molecular oxygen to form ozone (O_3), which is then consumed by reacting with nitric oxide to form the original reactants—nitrogen dioxide and molecular oxygen. Atomic oxygen, however, also reacts with reactive hydrocarbons—olefins and aromatics—to form chemical radicals (species with unpaired valence electrons). Chemical radicals, in turn, react to form other radicals and secondary pollutants such as aldehydes (formaldehyde, acetaldehyde, acrolein), ketones, and peroxyacyl nitrates (PAN).

$$NO_2 + light \longrightarrow NO + O\cdot$$
$$O\cdot + O_2 + M \longrightarrow O_3 + M$$
$$O_3 + NO \longrightarrow NO_2 + O_2$$
$$O\cdot + Hc \longrightarrow HcO\cdot$$
$$HcO\cdot + O_2 \longrightarrow HcO_3\cdot$$

$$HcO_3\cdot + Hc \longrightarrow RCHO \text{ or } R\overset{\overset{\displaystyle O}{\|}}{C}R$$
$$HcO_3\cdot + NO \longrightarrow HcO_2\cdot + NO_2$$
$$HcO_3\cdot + O_2 \longrightarrow O_3 + HcO_2\cdot$$

$$HcO_3\cdot + NO_2 \longrightarrow R-\overset{\overset{\displaystyle O}{\|}}{C}-O-O-N\overset{\diagup O}{\diagdown O} + \text{other products}$$
$$\text{(PAN)}$$

FIGURE 21–10 Simplified reaction scheme for photochemical smog. Hc is hydrocarbon (olefin or aromatic); M is a third body to absorb the energy released from forming the ozone; among many possibilities, M could be a N_2 molecule, O_2 molecule, or solid particle. A species with a dot, as HcO, is a chemical radical—a very reactive chemical species.

It is evident that further research is needed to clear up several obscure features of the overall scheme. While it is known that a few tenths of a part per million of nitrogen oxides and 1 ppm of reactive hydrocarbons suffice to initiate the process, the sources of all of the hydrocarbons that enter the smog-forming process cannot be accounted for. Furthermore, the exact nature and behavior of the radicals is not well understood at all. The ultimate fate of the substances involved is also not known.

On the other hand, it is known that ozone, aldehydes, ketones, and PAN, at least, are found in the air but are not emitted by any identifiable sources. Measurements made in several United States cities reveal that the amount of airborne aerosols increased tenfold during the 1960's. Despite the more strict controls effected during the last decade, Los Angeles did not show a downward trend in the occurrence and amount of smog, which is present about 300 days a year. A comparison of the levels of pollutants on a clear day and on a smoggy day is presented in Table 21–4.

The type of smog formed in London and around some industrial and power plants is thought to be caused by sulfur dioxide. Laboratory experiments have shown consistently that sulfur dioxide increases aerosol formation, particularly in the presence of irradiated hydrocarbon-nitrogen–oxide-air mixtures. For example, irradiated mixtures of 3 ppm olefin, 1 ppm NO_2, and 0.5 ppm SO_2 at

TABLE 21-4   APPROXIMATE CONCENTRATIONS OF POLLUTANTS IN LOS ANGELES ON A CLEAR DAY (VISIBILITY 7 MILES) AND ON A SMOGGY DAY (VISIBILITY 1 MILE)

POLLUTANT	CONCENTRATION IN CUBIC CENTIMETERS OF VAPOR PER CUBIC METER OF AIR		
	Clear Day	Smoggy Day	Increase
Carbon monoxide	3.5	23.0	×6.5
Hydrocarbons	0.2	1.1	×5.5
Peroxides	0.1	0.5	×5.0
Oxides of nitrogen	0.08	0.4	×5.0
Lower aldehydes	0.07	0.4	×6.0
Ozone	0.06	0.3	×5.0
Sulfur dioxide	0.05	0.3	×6.0

50 per cent relative humidity forms aerosols which have sulfuric acid as a major product. Even with 10 to 20 per cent relative humidity, sulfuric acid is a major product. Without the SO_2, olefins and nitrogen oxides do not cause appreciable aerosols.

How is the sulfuric acid formed? Most likely SO_2 is oxidized to SO_3. The SO_3 has less energy per mole than SO_2 (about 24 kcal per mole; 94.45 kcal are lost in forming one mole of SO_3 and only 70.96 kcal for a mole of SO_2). The oxidation of SO_2 to SO_3 is spontaneous and should go readily if oxygen is present and the energy of activation is available. Air supplies the oxygen and sunlight supplies the energy, but the mechanism is unknown. It has been confirmed experimentally that SO_2 radiated by sunlight forms a sulfuric acid aerosol. The speed of this reaction is increased by the presence of metals in the aerosols, hydrocarbons, and nitrogen oxides. Ferric oxide, Fe_2O_3, is known to catalyze the conversion of SO_2 to SO_3, and there is a considerable amount of rust available. But the step-by-step mechanism of the conversion of SO_2 to SO_3 is still elusive.

Once the SO_3 is formed in the presence of water, it is readily converted to sulfuric acid:

$$SO_3 + H_2O \longrightarrow H_2SO_4$$

The sulfuric acid formed in this kind of smog is very harmful to people suffering from bronchial diseases, asthma, or emphysema. At a concentration of 5 ppm for one hour, SO_2 can cause constriction of bronchial tubes. A level of 10 ppm for one hour can cause severe distress. In the 1962 London smog, readings as high as 1.98 ppm of SO_2 were recorded. The sulfur dioxide and sulfuric acid are thought to be the primary causes of deaths in the London smogs.

We shall now examine some of the principal components of air pollution and their chemistry. Emphasis will be placed on how the compounds are produced and how they can be eliminated from our atmosphere.

SULFUR DIOXIDE

Sulfur dioxide is produced when sulfur or sulfur-containing substances are burned in oxygen:

$$S + O_2 \longrightarrow SO_2(g)$$

Most of the sulfur dioxide in the atmosphere comes from sulfur-containing fuels burned in power plants, from smelting plants treating sulfide ores, and from sulfuric acid plants. Coal contains sulfur in several forms: as elemental sulfur (S), iron pyrites (FeS_2), and as organic compounds, such as mercaptans (compounds containing—SH groups). According to the National Air Pollution Control Association (NAPCA), the burning of fuels by the nation's power plants accounted for 46 per cent (23 million tons) of the SO_2 emitted into the atmosphere in 1966. One study found a concentration of 2200 ppm SO_2 in the stack of a coal-burning power plant. Concentrations as high as 2.9 ppm have been recorded up to one-half mile from power plants. If coal and petroleum containing up to 5 per cent sulfur are burned, a 1000-megawatt electric power plant could emit 600 tons of SO_2 per day. Much of the SO_2 pollution in this country is isolated in seven industrialized states (New York, Pennsylvania, Michigan, Illinois, Indiana, Ohio, and Kentucky), which account for almost 50 per cent of the nation's total SO_2 output.

In spite of the large output of SO_2 in the United States per year, its concentration rarely exceeds a few parts per million. This is because SO_2 has a relatively short atmospheric life. In the presence of oxygen, sunlight, and water vapor, SO_2 is oxidized and converted to sulfuric acid (H_2SO_4):

$$2SO_2 + O_2 \longrightarrow 2SO_3$$

$$SO_3 + H_2O \longrightarrow H_2SO_4$$

The SO_2 and sulfuric acid are readily dissolved in rivers, lakes, and streams and can increase the acidity considerably.

$$H_2O + SO_2 \rightleftharpoons \underset{\substack{\textit{Weak} \\ \textit{acid}}}{H_2SO_3} \rightleftharpoons H^+ + HSO_3^-$$

or
$$H_2SO_4 \longrightarrow H^+ + HSO_4^-$$

Often streams contain ions that help to offset the acidity increase. Sulfite is one such ion.

$$\underset{\textit{Sulfite}}{SO_3^{2-}} + H^+ \longrightarrow \underset{\textit{Bisulfite}}{HSO_3^-}$$

If the pH of streams varies too much, aquatic life suffers. Salmon, for example, cannot survive if the pH is as low as 5.5. The lower limit of tolerance for most organisms is a pH of 4.0. In the late 1950's and early 1960's, certain sections of the Netherlands had precipitation with a pH less than 4.

The corrosion of steel is promoted by sulfur dioxide. In a study involving 100-gram panels of steel exposed at several sites in Chicago in 1964, the amount of weight loss by corrosion was very definitely related to the SO_2 concentration in the air. Particulate matter, high humidity, and temperature are also factors which determine the extent of corrosive action.

480

FIGURE 21-11   Rose leaves in Independence, Missouri, show marginal and inter-veinal necrotic injury from sulfur dioxide. (From publication no. AP-71, National Air Pollution Control Administration, HEW Public Health Service, 1970.)

The effects of SO_2 on vegetation are manifested as bleached spots, suppression of growth, leaf drop, and reduction in yield. Plant damage has been noticed 52 miles downwind from a smelting operation which emitted large quantities of SO_2. The SO_2, and SO_3 formed from it, enter the stomata of the leaves, hydrolyze, and dehydrate the tissue as sulfuric acid; sulfates are then formed, causing yellowing and leaf drop.

There are many ways to control SO_2 pollution of the air. For example, tall stacks (as high as 1000 feet) emit the SO_2 into the higher atmosphere and away from the ground. But this is a compromise of the sort: "one man's solution is another man's pollution."

Chemical means are more efficient than high stacks in removing SO_2. One method involves heating limestone to produce lime (calcium oxide) and reacting this with the SO_2 to produce calcium sulfite, which can be removed from the stack by an electrostatic precipitator.

$$CaCO_3 \xrightarrow{\text{Heat}} CaO + CO_2$$
Limestone   Lime

$$CaO + SO_2 \longrightarrow CaSO_3 \text{ (solid)}$$
Calcium sulfite

Another efficient method involves passing the combustion gases through molten Na_2CO_3 (sodium carbonate), in which solid sodium sulfite is formed.

$$SO_2 + Na_2CO_3 \xrightarrow{800°C} Na_2SO_3 + CO_2$$

481

FIGURE 21-12 (A) When air pollution is minimal, tall stacks help to keep the pollution away from people. This photograph of the stacks of Consolidated Edison in New York City makes it all too clear that tall stacks are no real solution when a temperature inversion exists. Note the United Nations Secretariat Building in the right-hand corner. (Wide World Photos, Inc.) (B) TVA's Bull Run Steam Plant, five miles east of Oak Ridge in east Tennessee, has one of the largest generating units in the world, with a capacity of 950,000 kilowatts. At full operation the plant burns 7584 tons of coal (a large trainload) every day. Nearly 600 million gallons of water a day flow through the plant from Melton Hill Lake to condense spent steam from the turbines. Bull Run has an 800-foot chimney. The plant went into regular operation in 1967.

The use of sulfur-free fuels is one obvious method of eliminating SO_2 emissions. However, most low-sulfur coals are far from major metropolitan areas, and removal of the sulfur from coal can cost up to one dollar per ton; who will pay the cost?

Most of the sulfur in coal is a part of organic matter or iron pyrites (FeS_2) called fool's gold. Pulverized coal, with a consistency of talcum powder, can be cleansed partially of iron pyrites by magnetic separation, electrostatic separation, froth flotation, and dry centrifugation. But none of these methods completely eliminates the sulfur dioxide emissions.

Much of the petroleum used in the Eastern United States, where SO_2 pollution is a problem, is Caribbean residual fuel oil, which has an average sulfur content of 2.6 per cent. While technology has developed refining techniques for reducing the sulfur content to 0.5 per cent, this procedure would increase the cost of the oil by about 35 per cent. The process involves the formation of hydrogen sulfide by bubbling hydrogen through the oil in the presence of metallic catalysts (platinum-palladium, for example). Residual fuel oil produces less than 10 per cent of the total utility power. In 1968, New York's Consolidated Edison switched to low-sulfur oil from Liberia and Nigeria at a cost of $3 million and an increase of $7.5 million in a $63 million fuel bill.

Despite all of these methods, and more under development, the equilibrium concentration of SO_2 in the air is increasing nationwide on an average of about 6 to 7 per cent per year. Some cities—Washington, D.C., for instance—have reduced SO_2 emissions, while in others the emissions are on the increase. The SO_2 concentration depends, of course, upon the amount of sulfur-containing coal burned and the output of SO_2 by chemical industries.

NITROGEN OXIDES

There are eight known oxides of nitrogen, three of which are recognized as important components of the atmosphere. These are: nitrous oxide (N_2O), nitric oxide (NO), and nitrogen dioxide (NO_2).

Most of the nitrogen oxides emitted are in the form of NO, a colorless reactive gas. In a typical combustion process involving air as the oxidant, some of the atmospheric nitrogen reacts with oxygen to produce NO:

$$N_2 + O_2 + \text{Heat} \longrightarrow 2NO$$

Since the formation of nitric oxide is endothermic (21,600 cal per mole of NO), it follows that a higher combustion temperature would produce relatively more NO. This will be an important point to consider later in our discussion of the automobile and its nitrogen oxide emissions, since one way to achieve greater burning efficiency of fuels in automobile engines is to operate at higher temperatures.

In the atmosphere NO reacts rapidly with atmospheric oxygen to produce NO_2:

$$2NO + O_2 \longrightarrow 2NO_2$$
Nitrogen dioxide

Nitrogen dioxide, a brown, choking gas, is a necessary component of photochemical smog (Figure 21–10). Normally its atmospheric concentration is a few parts per billion (ppb) or less.

If NO_2 does not react photochemically, it can react with water vapor in the air to form nitric and nitrous acids:

$$2NO_2 + H_2O \longrightarrow \underset{\substack{Nitric \\ acid}}{HNO_3} + \underset{\substack{Nitrous \\ acid}}{HNO_2}$$

In addition, nitrogen dioxide and oxygen yield nitric acid:

$$4NO_2 + 2H_2O + O_2 \longrightarrow 4HNO_3$$

These acids in turn can react with ammonia or metallic particles in the atmosphere to produce nitrate or nitrite salts. For example,

$$\underset{Ammonia}{NH_3} + HNO_3 \longrightarrow NH_4NO_3$$

The acids or the salts, or both, ultimately form aerosols, which eventually settle from the air or dissolve in raindrops. Nitrogen dioxide, then, is a primary cause of haze in urban or industrial atmospheres because of its participation in the process of aerosol formation. Normally nitrogen dioxide has a lifetime of about three days in the atmosphere.

Tremendous quantities of nitric oxide are formed during electrical storms. A bolt of lightning provides the energy for the endothermic reaction between nitrogen and oxygen:

$$N_2 + O_2 \xrightarrow{Lightning} 2NO \qquad 2NO + O_2 \xrightarrow[\substack{high\ temp. \\ of\ lightning \\ bolt)}]{Fast\ (in} 2NO_2$$

Man's emission of nitrogen oxides to the atmosphere is only a minor part of nature's nitrogen cycle, which has been in operation for millions of years. (Figure 21–14). The effect of nitrogen oxides on this cycle is thought to be insignificant at this time. However, there is some justifiable concern about other pollutants, such as SO_2, that can disrupt the cycle by destroying significant key bacteria.

In laboratory studies, nitrogen dioxide in concentrations of 25 to 250 ppm inhibits plant growth and causes defoliation. The growth of tomato and bean seedlings is inhibited by 0.3 to 0.5 ppm NO_2 applied continuously for about 10 to 20 days.

In a concentration of 3 ppm for 1 hour, nitrogen dioxide causes bronchioconstriction in man, and short exposures at high levels (150 to 220 ppm) produce fibrotic changes in the lungs that produce fatal results.

FIGURE 21-13 Photochemical smog (a brown haze) enveloping the city of Los Angeles. (Los Angeles Times photo.)

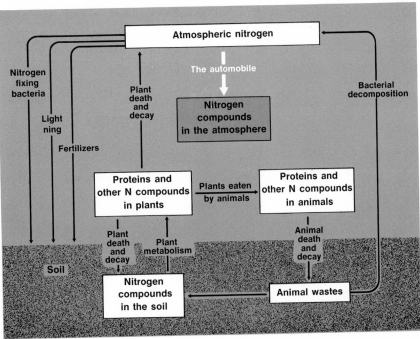

FIGURE 21-14 The nitrogen cycle.

485

CARBON MONOXIDE

Carbon monoxide (CO) is the most abundant and widely distributed air pollutant found in the atmosphere. It appears to be almost exclusively a man-made pollutant because more than any other air pollutant, carbon monoxide is closely related to the total population of an area.

It is produced in combustion processes when carbon or some carbon-containing compound is burned in an insufficient amount of oxygen:

$$2C + \underset{\substack{\text{Limited} \\ \text{supply}}}{O_2} \longrightarrow 2CO$$

In the United States combustion sources of all types dump about 2.1×10^{14} grams of carbon monoxide per year into the atmosphere (about 5×10^{21} grams). In terms of local emission, each person accounts for about 1400 pounds per year in Los Angeles and 600 lbs per year in New York City. For every 1000 gallons of gasoline burned, 2300 pounds of CO are emitted.

The huge emissions of CO are a mystery. There is enough CO emitted to raise the concentration 0.03 ppm yearly, but apparently it is not increasing at all. Although data on background concentrations of CO are still very limited, best estimates of the maximum global level are of the order of 0.1 ppm. The background concentration in cities is higher; in heavy traffic sustained levels of 50 or more ppm are common; and instantaneous concentrations of 150 ppm and higher have been found. National Air Pollution Control Association (NAPCA) data collected from 1964 through 1966 in five major cities show an average of 7.3 ppm background for off-street sites. The variation of CO levels over a 24-hour period is shown in Figure 21–15. However, in spite of increased emission of CO, the global level does not seem to be rising.

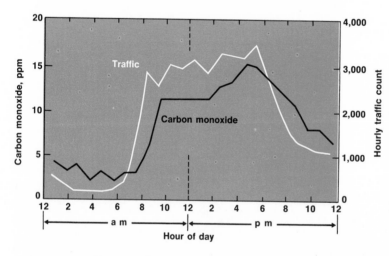

FIGURE 21–15   Hourly average carbon monoxide concentration and traffic count in midtown Manhattan. (From Johnson, K. L., Dworetzky, L. H., and Heller, A. N.: "Carbon Monoxide and Air Pollution from Automobile Emissions in New York City." Science, Vol. *160*, No. 3823, p. 67 [1968].)

It is not known for certain where all the carbon monoxide goes once it gets into the atmosphere. There are a number of possible explanations. One of these is that carbon monoxide is slowly oxidized by molecular oxygen (O_2) in the lower atmosphere:

$$CO + O_2 \longrightarrow CO_2 + \quad O$$

Oxygen
atom

The oxygen atom then reacts with other species. This reaction has a high energy of activation (51,000 cal per mole), however, and only occurs readily at temperatures above 500°C or with suitable catalysts.

At ordinary temperatures and without catalysts, some other method must exist. Since the concentration of carbon dioxide is definitely known to be increasing, the conversion of CO to CO_2 is a likely process, but the method is still a mystery. Other possible mechanisms involve biological processes. For example, the microorganisms *Methanosarcina barkerii* and *Methanobacterium formicum* are known to reduce CO to methane in the presence of water:

$$4CO + 2H_2O \longrightarrow CH_4 + 3CO_2$$

Although our earth is plentifully supplied with soil bacteria, it is not known at what rate and to what extent CO is destroyed by this means.

The oceans could serve as a sink for much of the carbon monoxide produced; however, more than 100 times the amount of carbon monoxide that could dissolve in the oceans is produced in the world each year. It is known that some algae, green plants, and siphonophores (small, free swimming sea creatures) are capable of utilizing CO and thus provide for some removal. Collectively, these are thought to be a relatively small sink for CO, but data are lacking to verify its effect.

It is conceivable that much of the accumulated CO from the various sources in the lower atmosphere may eventually migrate by atmospheric mixing to a potential sink in the upper atmosphere, where it is oxidized to CO_2 in the presence of high intensity, very short wave solar ultraviolet radiation.

The danger of carbon monoxide to man was discussed in detail in Chapter 19. By combining with hemoglobin, CO starves the body of oxygen. This produces listlessness and can lead to death. About 30 ppm for 8 hours is sufficient to cause headache and nausea. Oxygen tanks have been installed at busy intersections of some cities for use of traffic policemen who must stand for hours in the swirl of auto exhaust fumes. Every half hour, the men have to take an "oxygen break."

Carbon monoxide from cigarette smoke can be eliminated only by abstinence. It is reduced in any combustion process by providing an ample supply of air. It can be reduced in the automobile by burning natural gas rather than gasoline, reburning the exhausts, and using catalytic mufflers. These will be discussed in more detail later.

HYDROCARBONS

As we have seen in Chapter 14, hydrocarbons come in all shapes and sizes from methane, CH_4, to octane, C_8H_{18}, and beyond to molecules containing many more carbon atoms. Some have all single bonds, some have double bonds, and a few have triple bonds. Aromatic hydrocarbons have bonds that can be described as intermediate between a single and a double bond. Literally hundreds of these hydrocarbons and their oxygen, sulfur, nitrogen, and halogen derivatives find their way into the atmosphere.

Trees and plants silently breathe turpentine, pine oil, and thousands of other fragrances into the air. These are hydrocarbons or hydrocarbon derivatives. Bacterial decomposition of organic matter emits very large amounts of marsh gas, principally methane. Man contributes his share of 15 per cent (of the total global emissions; a greater quantity in urban areas) through incomplete incineration, leakage of industrial solvents, unburned fuel from the automobile, incomplete combustion of coal and wood, and through petroleum processing, transfer, and use.

In Chapter 19 we saw that polynuclear hydrocarbons like benzo(α)pyrene are capable of causing cancer in mice and, under certain circumstances, in humans.

In the late 1950's, the U.S. Public Health Service, Division of Air Pollution, surveyed 103 urban and 28 nonurban areas of the United States for polynuclear hydrocarbons in the atmosphere. The air in all of the 103 urban areas contained benzo(α)pyrene (BaP). Concentrations ranged from 0.11 to 61 micrograms per 1000 cubic meters of air, with the average concentration being 6.6. In 1967, the estimated annual emissions of BaP in the United States were 422 tons from burning coal, oil, and gas, 20 tons from refuse burning, 19 tons from industries (petroleum catalytic cracking, asphalt road mix, and the like), and 21 tons from motor vehicles. British researchers report that lung cancer in nonsmokers closely parallels the 10 times greater amount of BaP in city air than in rural air; there is 9 times more lung cancer in cities than in rural areas. A resident of a large town may inhale 0.20 g BaP a year. If he is a heavy smoker (two packs a day), add another 0.15 g for a total of 0.35 g. This is about 40,000 times the amount of BaP necessary to produce cancer in mice. Coal smoke contains about 300 ppm BaP. The million tons of coal burned in England in 1958 produced 2.5 million tons of smoke laden with 750 tons of BaP. Many authorities attribute England's high lung cancer rate to this enormous production of BaP.

Other polynuclear aromatics have shown less extensive carcinogenic activity. For example, particulates from the atmosphere around Los Angeles, London, Newcastle, Liverpool, and eight other urban sites were extracted with organic solvents, usually benzene. These extractions produced cancer in mice. In addition to BaP, benz(α)anthracene and 7H-benz(de)anthracene-7-one showed carcinogenic activity.

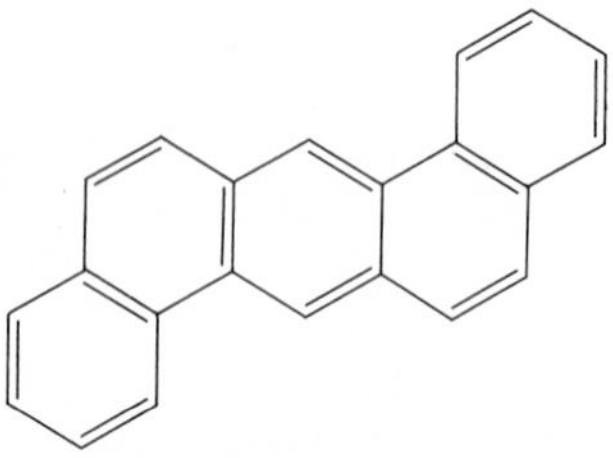

| *1,2,5,6-Dibenzanthracene* | *3,4-Benzpyrene (BaP)* | *Methylcholanthrene* |

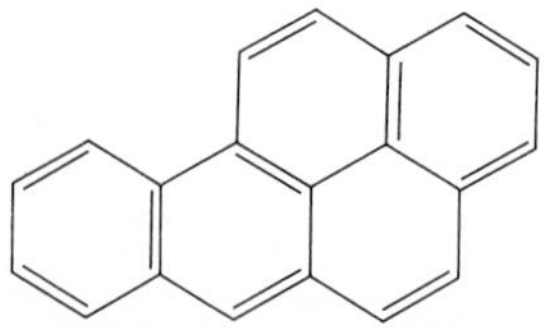

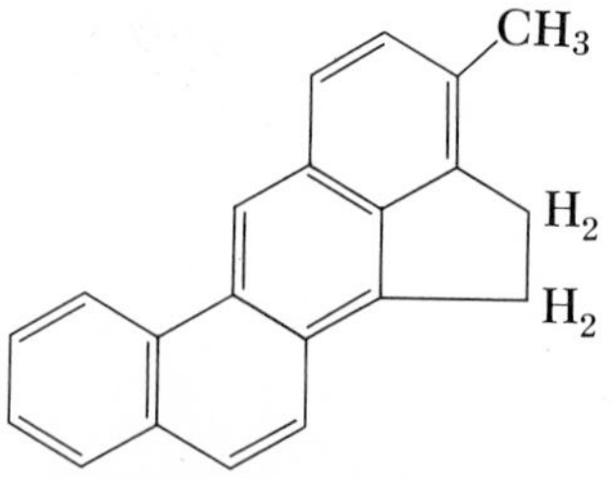

488   FIGURE 21-16   Some carcinogenic polynuclear hydrocarbons found in air.

Although BaP and its polynuclear counterparts have received considerable publicity, other hydrocarbons and hydrocarbon derivatives play equally important roles in air pollution. The chlorinated hydrocarbons, widely used as pesticides, at least partially counteract their beneficial aspects by killing birds and fish, and by generally polluting our streams. Some of these, such as DDT, are found in the atmosphere.

In the section on smog we discussed in detail the role of olefins and aromatic hydrocarbons in photochemical smog formation. In a study made in Los Angeles in 1970, an average of 0.106 ppm (maximum of 0.33 ppm) aromatics was found in that city's atmosphere. (About 38 per cent of the total was toluene and 40 per cent the more reactive dialkyl- and trialkylbenzenes.) These compounds are about as reactive as propylene and higher-molecular weight olefins in causing smog formation. The automobile is responsible for emitting most of these hydrocarbons to the atmosphere. In fact, the automobile emits more than 200 different hydrocarbons and hydrocarbon derivatives. Most of these come from unburned gasoline and from reactions among hydrocarbons, oxygen, and nitrogen during the combustion process. Aromatic compounds constitute an average of 15 to 30 per cent of the components of regular and premium grade gasoline sold in the United States. Several nonleaded gasolines contain as much as 40 per cent aromatics; their exhaust products have about the same percentages. About 85 per cent of these have been shown to be photochemically reactive.

OZONE

Ozone is the pungent smelling gas often noticed around electric motors. A concentration of only 0.02 to 0.05 ppm is required to detect the acrid odor. Sparking and even silent electric discharges convert oxygen into ozone, which is a more reactive form of oxygen:

$$68,400 \text{ cal} + 3O_2 \longrightarrow 2O_3$$
Ozone

Ozone attacks mercury and silver, which are not affected by oxygen at room temperature. Even with all of the electric sparks from lightning, electric motors and such, practically no ozone is emitted into the air. It reacts too quickly to leave its source. A typical reaction might be

$$6Ag + O_3 \longrightarrow 3Ag_2O$$

but

$$Ag + O_2 \longrightarrow \text{No reaction}$$

Ozone is found in the lower troposphere only as a secondary pollutant; that is, it is formed from other substances, as in photochemical smog (Figure 21–10). When sunlight impinges on automobile exhaust fumes, considerable ozone is produced. The stratosphere contains about 10 ppm ozone, and this has the important function of filtering out some of the ultraviolet light.

Pure oxygen can be breathed for weeks by man and animals without apparent injurious effects. Several studies have shown that concentrations of 0.3 to

1.0 ppm ozone, well within the recorded range of photochemical oxidant levels, after 15 minutes to 2 hours causes marked respiratory irritation accompanied by choking, coughing, and severe fatigue. For these reasons, outdoor recreation classes in Los Angeles public schools are cancelled on days when the ozone level reaches 0.35 ppm. Ozone at these levels for 1 hour depresses the body temperature, perhaps by an impairment of the brain center that regulates body temperature or by opening the pores of the skin. These levels (0.2 to 0.5 ppm) cause a considerable decrease in night vision in addition to other effects on vision. Also, track times in long distance running events have been shown to increase as the ozone level increases.

CARBON DIOXIDE

Carbon dioxide is not usually considered a pollutant because it is a normal component of the air and directly involved in the give and take between animal life and plant life (a product of respiration, a reactant in photosynthesis). There is concern, however, because the concentration in the atmosphere is increasing (Figure 21-17). Studies at Scripps Institute of Oceanography have shown that the concentration of CO_2 in the atmosphere increased by 1.3 per cent (or 3.7 ppm) from 1958 to 1962. Estimates of annual emissions by man-made sources predict an 18-fold increase between 1890 and 2000. Electric power plants, internal combustion engines, and the manufacture of cement are the principal technological sources of CO_2 emission, but there are numerous other sources—home heating, trash burning, and bacterial oxidation of soil humus, for example. These emissions have been entering the atmosphere faster than oceans and plants can remove them.

In some ways an increase in CO_2 can be very beneficial. Experiments at Michigan State University showed an increase of 115 per cent in the dry weight of lettuce and tomatoes when the CO_2 was increased from the normal 325 ppm (0.032 per cent by volume) to 800 to 2000 ppm during daylight hours. Fruit

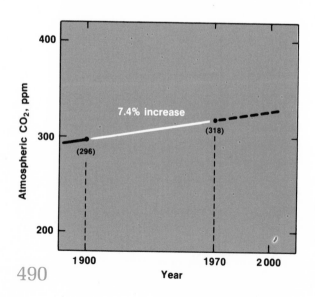

FIGURE 21-17   Atmospheric carbon dioxide. It is estimated that doubling the concentration would increase the average world temperature by 3.8°C.

size increased and quality was enhanced; reproduction improved; tomatoes had higher vitamin C and sugar content; and plants were more resistant to some fungi, virus diseases, and insects. Plants have even responded favorably to 20,000 to 30,000 ppm CO_2 levels.

A substantial increase of CO_2 in the air generally can be detrimental in two ways; it can increase the temperature of the atmosphere and the acidity of the oceans.

Carbon dioxide is not the only substance that can increase the air temperature. Particulates, hydrocarbons, and molecules such as ozone and water that can absorb more heat than oxygen and nitrogen will cause a similar heating effect. At room temperature a mole of CO_2 molecules will absorb 8.96 cal to increase the temperature one degree Celsius; oxygen absorbs 7.05 cal, and nitrogen 6.94 cal. This means that carbon dioxide can absorb more heat energy and will emit more energy per molecule than N_2 or O_2. Carbon dioxide can absorb the long wavelengths of radiation (infrared) emitted from the warm earth, but it allows the shorter wavelength energy from the sun to pass through unabsorbed. The reason for this is that the atoms in the CO_2 molecule have several vibrational movements (Figure 21-18), each of which can absorb a different quantum of infrared energy. By contrast, N_2 and O_2 have only one vibrational mode to absorb energy, and these are not in the infrared region.

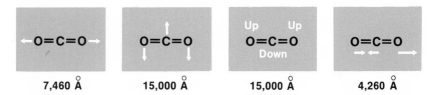

| 7,460 Å | 15,000 Å | 15,000 Å | 4,260 Å |

FIGURE 21-18   Vibrational modes of the carbon dioxide (linear) molecule. Frequencies of infrared radiation correspond to the frequencies of these molecular vibrations; hence the CO_2 molecules absorb this light energy, and then emit this energy to add heat energy to the atmosphere in the process.

Carbon dioxide in the atmosphere acts similarly to the glass (or plastic) on a greenhouse. Both let the shorter wavelengths of light through but absorb the longer wavelengths as they are emitted from the surfaces of objects. The carbon dioxide (and the glass) in turn emit the absorbed heat radiation, some of which returns to earth and some of which energizes other molecules via collisions. The net effect is an increase in temperature of the atmosphere, a phenomenon known as the "greenhouse effect."

In contrast, solid and liquid aerosols scatter incoming sunlight of all wavelengths, thus decreasing the amount of solar energy that reaches the earth. The net effect is a cooling of the atmosphere. Calculations show that a 25 per cent increase in aerosols (turbidity) would counteract a 100 per cent increase in carbon dioxide concentration.

It is generally agreed that the global temperature increased about 0.4°C between 1880 and 1940 and then decreased by nearly 0.2°C by 1967. The CO_2 concentration is believed to have increased by 7.4 per cent during a com-

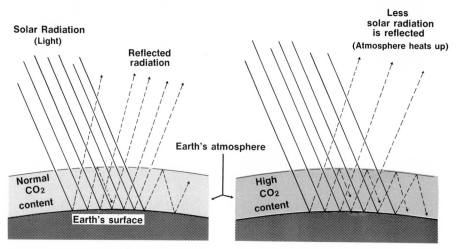

FIGURE 21-19   The greenhouse effect. Owing to a balance of incoming and outgoing energy in the earth's atmosphere, the mean temperature of the earth is 14.4°C (58°F). Carbon dioxide absorption of light and the reemissions as heat energy cause less of the sun's energy to be reflected from the earth's surface.

parable interval of time—from 296 ppm in 1900 to 318 ppm in 1969, a net gain of 7.4 per cent. The cooling during the 1950's possibly was due in part to the particulates emitted by gigantic volcanic eruptions of Mt. Spurr in Alaska in 1953 and the Bezymyannaya (Kamchatka, U.S.S.R.) eruption in 1956. While an increasing CO_2 concentration could explain the increase in temperature and the increasing turbidity could explain the decrease in temperature, other factors may also have contributed significantly. For example, it is assumed in this explanation that solar radiation has been constant over this period of time. Actually, it may not have been; we simply do not know. Carbon dioxide is the only substance whose global background concentration is known to be rising.

What can we expect if present trends continue? Predictions call for a 180 per cent increase in the CO_2 level from 1950 to 2020 and an increase in the global temperature of 9°C. This would melt enough of the earth's ice caps to cause a 4-foot rise in the oceans. The shift of this enormous weight from the land to the oceans would engender earthquakes and volcanic activity. There is no evidence at this time that the ice caps have begun to melt, and the sea level has not changed for about 5000 years.

Three mechanisms account for the removal of CO_2 from the atmosphere. These are photosynthesis, dissolving in surface water (principally the oceans), and weathering of silicate rocks. None of these seems adequate because the CO_2 level is on the rise. While a higher level of CO_2 increases plant activity and growth, it is apparent that the photosynthesis process has not picked up enough to maintain the status quo. It is known that pollutants such as SO_2 hinder photosynthesis and that much of our forest area is being cleared for housing and development. Perhaps these partially explain why photosynthesis cannot absorb enough CO_2. At the rates prevailing in 1962, we would need to add 2.7 billion acres of new, growing forest (more than the area of the United States) to absorb the CO_2 generated by the combustion of fossil fuel.

However, all of the CO_2 emitted by burning does not have to be consumed by the plants. Perhaps as much as half is absorbed by the surface waters. The oceans serve as a huge sink for carbon dioxide, which reacts with water to form carbonic acid. The acid ionizes to raise the acidity and produce bicarbonate ions:

$$H_2O + CO_2 \rightleftharpoons H_2CO_3 \rightleftharpoons H^+ + HCO_3^-$$

Carbonic acid can disrupt the limestone-forming process in the oceans. Limestone is mostly calcium carbonate, $CaCO_3$. The insoluble $CaCO_3$ reacts with carbonic acid to form soluble calcium bicarbonate:

$$\underset{Insoluble}{CaCO_3} + H_2CO_3 \longrightarrow \underset{Soluble}{Ca^{2+} + 2HCO_3^-}$$

The overall effect of the solution of CO_2 in the oceans is to reduce its concentration in the air and, consequently, to reduce the atmospheric temperature; but this raises the acidity, dissolves limestone, or both. The process of removing CO_2 from the atmosphere to the oceans is surprisingly slow. In fact, laboratory studies show that for every six molecules of CO_2 dissolved in the ocean, only one should be in the air above the ocean. Actually the ratio is about 1 to 1. A study made on the Columbia River in the state of Washington on December 14, 1968, revealed that three parts of CO_2 are dissolved in the river while one part is in the atmosphere above the river. More CO_2 would dissolve in the ocean if the surface waters—the upper 200 meters—would mix well with the lower waters. Thorough mixing of the oceans is estimated to take several thousand years. Therefore, a new level of CO_2 could not be adjusted very quickly.

Some CO_2 is dissipated by weathering silicate rocks to carbonates:

$$CaAl_2(SiO_3)_4 + 4CO_2 \longrightarrow CaCO_3 + Al_2(CO_3)_3 + 4SiO_2$$

This process is very slow and could not help to relieve a relatively sudden increase in CO_2 emissions.

Using nuclear power or, alternatively, pumping atmospheric gases into the ocean appear to be ways to diminish the CO_2 level as well as the level of many other pollutants. Nuclear power is considered a good substitute for burning fossil fuels. It has just begun to be applied, however. The feasibility of pumping gases into the ocean has not been explored at this time.

When the CO_2 level increases, it does so at the expense of the oxygen supply. While this is another disadvantage of an increasing CO_2 concentration, the increase is far more detrimental than the equivalent decrease in oxygen. For example, approximately 60 per cent of our oxygen need is being restored by photosynthesis within the United States. The rest comes from bacteria that reduce sulfates in anaerobic environments and from plankton upwellings in the sea. Some upwellings—containing literally tons and tons of diatoms and bacteria (called plankton)—are found off the coasts of California, Peru, Morocco, and Southwest Africa. Others are found in the North Sea and off the Grand Banks of Newfoundland. While it appears that an increase in CO_2 should be removed soon by increased photosynthesis activity, the accompanying decrease in atmospheric temperature might alter the climate and destroy vegetation.

Carbon dioxide is not normally considered detrimental to man's health. Obviously our systems contain it at all times as a product of respiration. Prolonged exposure to about 5000 ppm CO_2 (15 times present levels), however, is considered unsafe primarily because the heavier CO_2 diminishes the O_2 concentration in our lungs.

THE AUTOMOBILE—A SPECIAL CASE

When we think about cleaning up the atmosphere, our first concern should be aimed at the automobile. More than 100 million automobiles jam our roadways, each adding its share of pollutants to the atmosphere. Like it or not, the automobile, as we know it, is *the* major source of air pollution and something must be done about it.

Automobile air pollutants enter the atmosphere in three major ways: from the exhaust, from the crankcase blowby (gases that escape around the piston rings), and by evaporation from the fuel tank and carburetor.

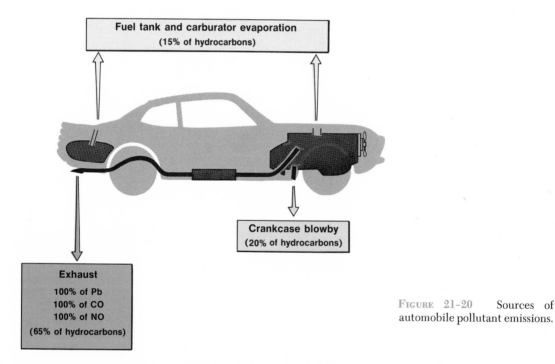

Fuel tank and carburator evaporation
(15% of hydrocarbons)

Crankcase blowby
(20% of hydrocarbons)

Exhaust
100% of Pb
100% of CO
100% of NO
(65% of hydrocarbons)

FIGURE 21-20   Sources of automobile pollutant emissions.

Gasoline is a mixture of hydrocarbons (isomers of octane, for example), tetraethyl lead, ethylene dibromide, ethylene dichloride (in a molar ratio of 2 to 1 in ethyl gasoline) and various "additives" such as TCP-tricresyl phosphate. When this mixture is compressed with air in an automobile cylinder and ignited, many of the hydrocarbon molecules combine with oxygen to form CO, CO_2, and H_2O. Some hydrocarbons react with oxygen, other hydrocarbons, or both,

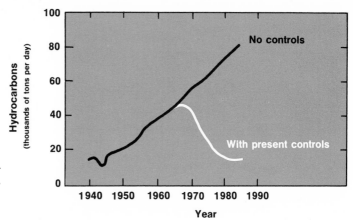

FIGURE 21-21   Automotive hydrocarbon emissions in the United States.

to form a large variety of organic substances. Other hydrocarbons do not react at all. Researchers at the Deepwater, New Jersey, Du Pont plant have found more than 200 different hydrocarbons in automobile *exhaust*.

Nitrogen combines with oxygen to form a variety of nitrogen oxides, principally NO and a little NO_2. The tetraethyl lead, $Pb(C_2H_5)_4$, forms PbClBr, $PbCl_2$, or $PbBr_2$ by reacting with ethylene dibromide and ethylene dichloride. These combustion products, added to the evaporation losses from the carburetor and gas tank, comprise the noxious waste products of the automobile.

WAYS TO CONTROL EMISSION

What can be done to abate these emissions? Before discussing the specific methods, we should be aware that technology over the past two decades has developed many ways to solve this complicated air pollution problem. But every method increases the cost of the automobile, and some improvements increase the costs substantially.

Governments, both state and federal, have recently passed legislation regulating the emissions from the automobile. California has taken the lead in legislating standards for emissions on domestic cars. In fact, the California Senate, in 1969, voted to ban the sale of gasoline- and diesel-powered internal combustion engines starting January 1, 1975. The bill failed to pass the California House of Representatives, but its significance was not overlooked by the automobile manufacturers and the general public. Legislative landmarks for emissions control are summarized in Table 21–5.

Two approaches are presently being developed to meet these standards: mechanical changes in automobiles and changes in fuel. Mechanical changes are made at relatively low cost, but fuel modifications offer advantages over mechanical devices. The most obvious advantage is that fuel changes would apply to all cars, old and new.

Methods for reduction of hydrocarbon and carbon monoxide emissions in the exhaust are being developed along three lines: (1) the adjustment of fuel-air ratio, spark timing, and other variables; (2) the injection of air into the hot exhaust gases; and (3) the use of a catalytic converter. The first two methods have physical limits that will prevent complete conversion of hydrocarbons and

495

TABLE 21-5   GOVERNMENT REGULATIONS ON AUTOMOBILE EMISSIONS°

	EXHAUST	HYDROCARBONS (CRANKCASE BLOWBY)	EVAPORATION	CO	NO_x
Emissions from typical car— no controls	9.7 g/vm	4.3 g/vm	2.5 g/vm	75 g/vm	6.5 g/vm
Emissions allowed—California state government					
1961	—	0.0	—	—	—
1966	3.2	0.0	0.3	33	—
1971	2.2	0.0	0.3	23	4.0
1972	1.5	0.0	0.3	23	3.0
1974	1.5	0.0	0.3	23	1.3
Emissions allowed—Federal government					
1963	—	0.0	—	—	—
1968	3.2	0.0	—	33	—
1970	2.2	0.0	—	23	—
1971	2.2	0.0	0.3	23	—

° g/vm is grams of pollutant per mile the vehicle travels on a test machine; the 0.3 g/vm evaporation standard corresponds to a 6-gram loss in the test cycle.

carbon monoxide to carbon dioxide and water. Ideally, if the stoichiometric amount of oxygen is present, complete conversion to CO_2 is expected. For example, the combustion of a mole of octane requires 12.5 moles of oxygen:

$$C_8H_{18} + 12.5O_2 \longrightarrow 8CO_2 + 9H_2O + heat$$

But the time for reaction in an automobile engine is so rapid that the octane and oxygen molecules cannot react completely. In the chaotic split second of ignition and reaction, many undesirable side reactions occur. These reactions produce the 100 or so hydrocarbons found in an automobile's exhaust which were not originally in the gasoline.

An important factor to consider is that less nitrogen is oxidized to oxides of nitrogen if the temperature within the combustion chamber is reduced to the lowest practical limit. One way to do this is to recycle part of the exhaust gas back into the engine. This reduces the peak combustion temperature and the amount of nitric oxide that forms. Prototypes of such recycling systems have reduced nitrogen oxides emissions by 80 per cent without causing the exhaust to exceed 275 ppm hydrocarbons and 1.5 per cent CO. Coincident with the recirculation of about 15 per cent of the exhaust and the reduction of nitrogen oxides by 88 per cent is a cut in power output by 16 per cent and a decrease in fuel economy by 15 per cent. Changes in timing can increase power and economy, but the amount of nitrogen oxides produced also increases. Fuel injected specifically at the spark plug gap will produce low hydrocarbon, CO, and NO_x emissions, but then particulate emissions become a problem.

Considerable research is being done to develop catalysts for converting CO, hydrocarbons, and NO_x to less harmful products. For example, in a laboratory

496

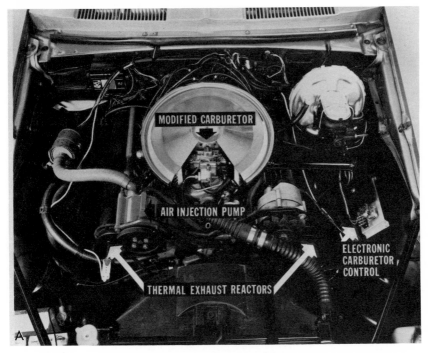

1975 EMISSION SYSTEM

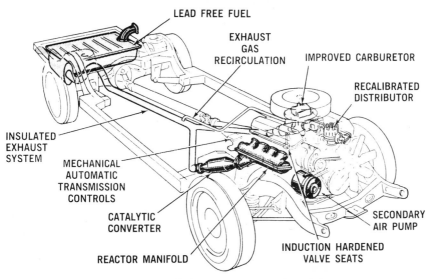

NOTE: THE 1976 (76-1) EMISSION SYSTEM WILL USE
1975 SYSTEM COMPONENTS BUT WILL BE
TOTALLY RECALIBRATED FOR NOx REDUCTIONS.

FIGURE 21-22   Automotive hardware for pollution control.

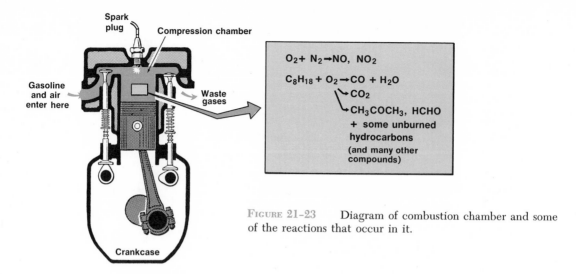

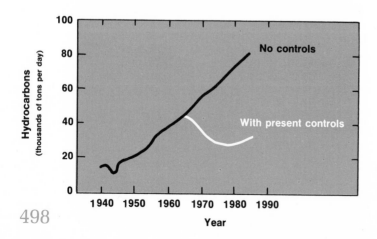

FIGURE 21-23 Diagram of combustion chamber and some of the reactions that occur in it.

mixture of 5 per cent NO, 5 per cent CO, and 90 per cent helium run over a barium-promoted copper chromite catalyst at 200°C, 92 per cent of the NO and CO were converted to N_2 and CO_2, neither of which is considered a pollutant in the ordinary sense. At 350°C, 98 per cent conversion is effected.

$$2CO + 2NO \xrightarrow{\quad\quad} 2CO_2 + N_2$$
$$Ba\text{-}CuCrO_2$$

Using the exhaust from a 1947 Ford V-8 engine, complete reaction to CO_2 and N_2 can be achieved with a copper chromite catalyst if the temperature is between 370° and 460°C and if the reactants are in the stoichiometric amounts—a most important condition. The perfect balance is difficult to achieve because an engine produces the most nitric oxide under conditions that make very little carbon monoxide (high temperatures, much oxygen). Furthermore, if too much oxygen is present in the exhaust, CO reacts with O_2 in preference to NO.

FIGURE 21-24 Automotive carbon monoxide emissions in the United States.

The cost of a catalytic converter such as this would be rather high—$175 to $200—and the unit would require regeneration from time to time because of lead poisoning. Thus, cost and maintenance present two imposing obstacles at this time.

Perhaps the best method of removing lead from the exhaust is to remove it from the gasoline. One company has marketed a lead-free gasoline for several years. Their gasoline contains about 30 per cent aromatics—very reactive, smog-producing compounds. The average amount of lead in gasoline (as tetra-ethyl and tetramethyl lead) is 2.4 grams of metal per gallon. About one-third of this amount becomes airborne from the exhaust as $PbBrCl$, $PbCl_2$, or $PbBr_2$. The tetraethyl lead is put into gasoline to prevent the combustion of low molec-ular weight hydrocarbons while the piston is compressing the gases. It is thought to provide free radicals which in turn react with these hydrocarbons instead of the hydrocarbons reacting with oxygen. Increased capacity for making high octane gasoline and for removing the low molecular weight fractions makes the lead less necessary. In 1967, the four U.S. producers of lead alkyls made 685 million pounds of tetraethyl and tetramethyl lead, valued at about $254 million. At an estimated cost of $4,250 million—roughly 40 per cent of present gross investment in refining equipment in the United States—refining equipment could be updated throughout industry so that lead would no longer be necessary.

New Kinds of Cars?

Changes in the types of automobiles on the streets of the United States are inevitable, and some are taking place now. For example, several thousand automobiles in this country are now burning natural gas instead of gasoline. Natural gas, which is mostly methane, CH_4, burns cleanly to CO_2 and water. A necessary device, positioned over the carburetor, costs about $350. The prob-lem here is that our surplus of natural gas in this country is rapidly diminishing, a potential long range problem.

One of the very few ways to eliminate combustion processes is to use electricity in some way to power our transportation devices. Early attempts to develop electric cars in the United States were made in Boston in 1888, 28 years after the development of the first storage battery in 1860. By 1912, about 6000 electric passenger and 4000 commercial vehicles were being manufactured annually in this country. The electric car lost out in competition with the internal combustion engine during the 1920's. The problems of short range (about 20 miles), low speeds (20 miles per hour maximum), 8 to 12 hours recharge time for the batteries, and relatively high price combined to eliminate the electric car from the race for leader in transportation.

At present, more than 100,000 rider-type, battery powered, materials-handling vehicles are operating in U.S. plants and warehouses where it is vital to prevent air pollution from internal combustion engines. Despite this positive start, attempts to employ electric vehicles for street and highway use generally have been commercial failures. The energy storage capacities of conventional batteries (lead-acid, nickel-iron, and silver-zinc) are too limited or expensive to provide an acceptable energy source for electric passenger cars.

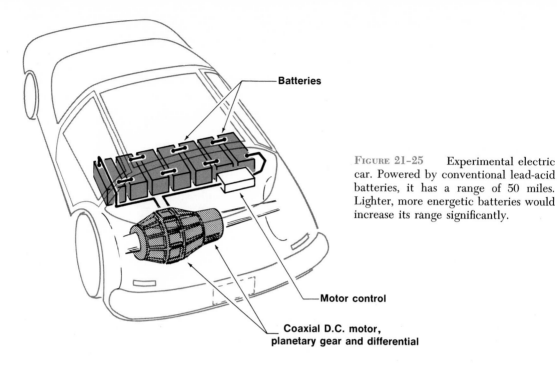

Batteries

Motor control

**Coaxial D.C. motor,
planetary gear and differential**

FIGURE 21-25   Experimental electric car. Powered by conventional lead-acid batteries, it has a range of 50 miles. Lighter, more energetic batteries would increase its range significantly.

Many new systems are in the development stage. For example, zinc-air, lithium-chlorine, sodium-sulfur, sodium-air, lithium-copper fluoride, lithium-nickel halide, and various kinds of fuel cells show some promise of being perhaps 10 or 15 times more powerful than the battery systems now in use.

Perhaps after 1985, the electrically powered car may be most attractive, but sufficiently powerful and durable batteries and fuel cells for this purpose do not now exist.

WHAT DOES THE FUTURE HOLD?

There will undoubtedly be an abatement of air pollution in the future; the sheer pressures of population increase will demand it. But life also will undoubtedly have to be different. Perhaps the first major change will be the disappearance of the automobile from the city, followed by a gradual modification of the power plant of the automobile until it is relatively nonpolluting.

The costs of abating pollution are already high, and they are likely to increase much more. It has been estimated that during the first half of the 1970's, all levels of government will spend about $1.6 billion on programs aimed at cleaning the air. Added to this will be $2.75 billion to upgrade fuels for autos, $0.5 billion for chemical industry to lower its emissions, and $2.64 billion for motorists to control the emission from their own automobiles. All this amounts to $7.49 billion, and most of this cost will be paid for directly by the public.

Most pollution exists because we demand the benefits of a technology which, for the most part, has given little consideration to the long range effects of its products. When industry, automobile manufacturers, or power plant operators add equipment to stop noxious waste products from getting into the air, the costs are added to the already considerable manufacturing expense without adding one cent to the market value of the product being made. The cost to the consumer, however, will go up and will be reflected in the increased price of consumer goods, automobiles, and electrical rates. This is a high price, of course; but it is not too high if we are to have a clean air.

New and more stringent laws can be expected. The Federal Clean Air Act of 1963 was designed to provide financial assistance to state and local agencies for developing, establishing, or improving air pollution control programs. The Air Quality Act of 1967 set up 57 air quality regions within the United States for the purpose of monitoring the amounts of various pollutants in the air, designating an air quality standard for each pollutant, and developing a plan by which the standard will be implemented. New laws will be more specific, regulating the amounts of pollutants that can be emitted by various industries. Special courts will be required to handle pollution cases.

In the final analysis, we all pollute the atmosphere, and much of the pollution is due to the misapplication of chemical techniques. Although we are becoming aware of the problems and although it is within the capabilities of chemical technology to eradicate most forms of air pollution, the process will be very slow. We have time . . . perhaps.

QUESTIONS

1. The formation of photochemical smog involves very reactive free radicals. What structural feature makes free radicals very reactive?

2. Write a balanced chemical equation for the burning of iron pyrite (FeS_2) in coal to sulfur dioxide and Fe_2O_3, iron(III) oxide.

3. What conditions are necessary for thermal inversion?

4. What are the basic chemical differences between New York (or London) smog and Los Angeles smog?

5. What are the major sources of the following compounds?
   (a) carbon monoxide
   (b) sulfur dioxide
   (c) nitrogen oxides
   (d) ozone

6. What is a photochemical reaction? Give an example.

7. If air pollutants rise from earth into the atmosphere, why do they not continue on into space?

8. Why is it so difficult to avoid the formation of either carbon monoxide or nitrogen oxides in the combustion process in an automobile engine?

9. What effects does weather have on local air pollution problems? on regional air pollution problems?

10. Knowing the chemistry of photochemical smog formation, list some ways to prevent its occurrence.

11. What part do aerosols play in the formation of smogs?

12. Report on a recent article on air pollution.

501

13. Describe the antipollution devices on your car.

14. Describe an air pollution problem in your community. How can this problem be solved?

15. Discuss the merits of abatement versus eradication of air pollution.

16. Of the air pollutants—particulates, sulfur dioxide, carbon monoxide, carbon dioxide, ozone, and nitrogen dioxide—

    (a) which is in greatest concentration in the air normally?
    (b) which is normally a secondary pollutant?
    (c) which is emitted almost exclusively by man-controlled sources?
    (d) which is definitely on the rise?
    (e) which is not normally harmful to the human being?
    (f) which can be removed from emissions by centrifugal separators?

17. Define:

    (a) micron
    (b) ppm
    (c) PAN

18. What compounds give "ethyl" gasoline its name?

19. What two products are produced by the perfect combustion of hydrocarbons in gasoline?

20. With a population of 200 million in the United States, how much can we expect air pollution to cost us individually during the first five years of the 1970's?

21. Which is more effective in producing smog, paraffinic (all single bonds) or olefinic (some double bonds) hydrocarbons? Why?

22. Why is natural gas a good substitute for oil and coal as far as air pollution is concerned? What problem is related to its substitution?

23. What is the approximate concentration of air pollutants in the atmosphere during smoggy conditions?

24. Describe an effective way of eliminating carbon monoxide from the exhaust of an automobile.

25. What do you think the future holds for our atmosphere?

SUGGESTIONS FOR FURTHER READING

Beard, R. R., and Wertheim, G. A., "Behavioral Impairment Associated with Small Doses of Carbon Monoxide," *American Journal of Public Health,* 57:2012 (1967).

Cook, L. M. (ed.), "Cleaning Our Environment—A Chemical Basis For Action," A. C. S. Committee on Chemistry and Public Affairs, American Chemical Society, Washington, D.C., 1970.

Hall, H. J., and Bartok, W., "NO_x Control From Stationary Sources," *Environmental Science and Technology,* Vol. 5, No. 4 (1971), p. 320.

Newell, R. E., "The Global Circulation of Atmospheric Pollutants," *Scientific American,* Vol. 224, No. 1 (1971), p. 32.

"Rivers of Dust," *Chemistry,* Vol. 42, No. 11 (1969), p. 24.

Wolf, P. C., "Carbon Monoxide-Measurement and Monitoring in Urban Air," *Environmental Science and Technology,* Vol. 5, No. 3 (1971), p. 212.

CONSUMER CHEMISTRY —FOOD AND MEDICINE

CHAPTER 22

INTRODUCTION

The major portion of a household budget is spent on consumer products. We buy chemical products to feed, cleanse, disinfect, wax, deodorize, paint, remove spots, relieve pain, cure disease, fertilize, preserve, destroy, and do hundreds of other things for us. You already know about the undesirable qualities of certain chemicals as pollutants (Chapter 21) and as toxic substances (Chapter 19). You are also familiar with many of the chemical products that make our lives more pleasant, healthy, and convenient, provided we use them in the proper amounts and for their intended purposes. You are aware that some, if not all, consumer products ultimately become pollutants, although none of us wants to do away with the conveniences we now enjoy.

To reap the full benefit of the products we purchase, we as consumers, and as students of chemistry, should know the types of raw materials in the products, the way in which the products perform their jobs, and the precautions to be taken when using the products. Many formulations are essentially the same (extremely similar in composition) and cost about the same to produce. Competitive products vie for sales through the cleverness of the name, the attractiveness of the container, and the effectiveness of newspaper, radio, television, and store advertisements. These costs are generally added to the price without, of course, increasing the true value of the product. When you have some knowledge of chemistry, the small print on the label becomes important. A comparison of lists of ingredients sometimes uncovers better formulations, and harmful substances. If we know a little chemistry, we can often make better use of the product we buy.

Because there are hundreds of different consumer products, we will discuss the chemistry of only a few types. The discussion is divided into three chapters. In each chapter, emphasis will be placed on the purposes of the specific chemicals in the formulations. Sometimes a chemical is a part of a formulation because it undergoes a desired chemical reaction; sometimes it is simply there to dilute or dissolve the active ingredients. Chemicals are added to improve color, taste, viscosity, texture, hardness, crispness, dryness, and many other properties. As

503

the various formulations are discussed, the purpose of the individual chemicals will be given, and molecular theory will be used to explain the effects of the chemicals.

We should enter this discussion of consumer chemistry with the notion that most consumer products are mixtures and that mixtures of chemicals are not necessarily blends of individual properties of the chemicals. In some formulations, the individual properties of certain chemicals are reinforced and enhanced by other chemicals in the mixture. This is called *synergism*, defined as the co-operative action of discrete agencies such that the total effect is greater than the sum of the effects of each used alone. Citric acid has very little antioxidant effect on foods; butylated hydroxyanisole (BHA) has considerable effect on preventing oxidation of foods. When these two substances are used together in foods, the antioxidative powers of the BHA are increased several fold. The citric acid is said to have synergistic action on BHA. The reasons behind this will be discussed in the section on food chemistry.

A closely related term is *potentiation*. Potentiators do not have a particular effect themselves but exaggerate the effect in other chemicals. The 5′-nucleo-tides, for example, have no taste, but they multiply the flavor of meat or the effectiveness of salt. This phenomenon will also be discussed in the next section.

FOOD

Chemistry of Cooking

Many times the cooking process involves the partial depolymerization of proteins or carbohydrates by means of heat and hydrolysis (Figure 22–1). The polymers which must be degraded if cooking is to be effective are the carbohy-drate cellular wall materials in vegetables and the collagen or connective tissues in meats. Both types of polymers are subject to hydrolysis in hot water or moist heat. In either case, only partial depolymerization is required. A pan of glucose instead of baked bread or a skillet of amino acids in place of broiled steak is not the goal of cooking. However, the partial hydrolysis of starch releases some glucose, which gives the food a sweeter taste. The partial depolymerization into smaller fragments also breaks many of the bonds which must be broken before food can be digested. As a consequence, cooked food is easier to digest—that is, to break down into its monomer monosaccharide or amino acid units. These basic units were discussed in Chapter 17.

FIGURE 22–1   The hydrolysis of starch during the cooking of foods such as potatoes and rice.

A primary purpose of cooking is to destroy harmful bacteria. While it is virtually impossible to be 100 per cent effective, the heat is sufficient to rupture the cell walls of most bacteria at ordinary cooking temperatures.

Sometimes cooking causes a chemical reaction that releases carbon dioxide gas. The carbon dioxide causes breads and pastries to rise. Yeast has been used since ancient times to make bread rise, and remains of bread made with yeast have been found in Egyptian tombs and the ruins of Pompeii. The metabolic processes of the yeast furnish gaseous carbon dioxide, which creates bubbles in the bread and makes it rise. When the bread is baked, the CO_2 expands even more to produce a light airy loaf.

Carbon dioxide in cooking can be generated by other processes. For example, baking soda (which is simply sodium bicarbonate, $NaHCO_3$, a base) can react with acidic ingredients in a batter to produce CO_2 (Figure 22–2):

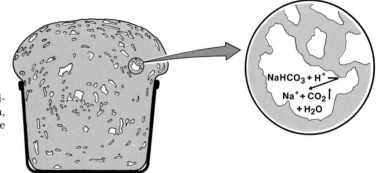

FIGURE 22-2   The carbon dioxide released from baking soda, $NaHCO_3$, expands and causes the bread to rise.

$$NaHCO_3 + H^+ \longrightarrow Na^+ + H_2O + CO_2(gas)$$

Baking powders contain sodium bicarbonate and an added acid salt or a salt which hydrolyzes to produce an acid. Some of the compounds used for this purpose are potassium hydrogen tartrate, $KHC_4H_4O_6$, monocalcium hydrogen phosphate, $Ca(H_2PO_4)_2 \cdot H_2O$, and sodium acid pyrophosphate, $Na_2H_2P_2O_7$. The reactions of these white, powdery salts with sodium bicarbonate are similar, although the compounds all have somewhat different appearances. For example:

$$KHC_4H_4O_6 + NaHCO_3 \longrightarrow KNaC_4H_4O_6 + H_2O + CO_2(gas)$$

Most people look on cooking primarily as a technique for improving the flavor of food. Flavors are usually caused by very small amounts of a large number of volatile constituents. The flavors of meats, for example, are due in significant measure to carbohydrate and protein derivatives of the compounds purine and pyrimidine (Figure 22–3).

Purine   *Pyrimidine*

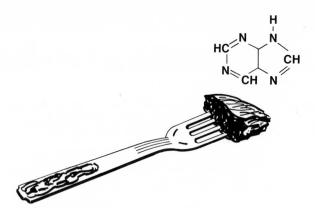

FIGURE 22–3   The flavor of meat is due primarily to the presence of derivatives of the compounds purine and pyrimidine. The structure of purine is shown.

Nucleoproteins contain the basic purine structure (see adenine and guanine structures in Chapter 17), but animals (including man) can synthesize purines from the simplest nitrogen and carbon compounds, such as carbon dioxide, acetic acid, and glycine. The combinations of purine derivatives present in the meat contribute to the flavor. Purines are readily extracted from meat with water, as in the preparation of broth.

The smell of rotting fish is due to the presence of trimethylamine oxide in the flesh of fish:

$$H_3C-\overset{\overset{\displaystyle CH_3}{|}}{\underset{\underset{\displaystyle O^-}{\downarrow}}{N^\pm}}-CH_3$$

As the fish begins to deteriorate, this compound is slowly transformed to the foul-smelling trimethylamine:

$$\underset{\underset{\displaystyle \overset{..}{N}}{}}{H_3C}\overset{\displaystyle CH_3}{\diagdown|\diagup}CH_3$$

In recent years several precooking additives have become popular; the *meat tenderizers* are a good example. These are simply enzymes which catalyze the breaking of peptide bonds in proteins via hydrolysis at room temperature. As a consequence, the same degree of "cooking" can be obtained in a much shorter heating time. Meat tenderizers are usually plant products such as papain, a proteolytic (protein-splitting) enzyme from the unripe fruit of the papaw tree. Papain has considerable effect on connective tissue, mainly collagen and elastin, and shows some action on muscle fiber proteins. On the other hand, microbial protease enzymes (from bacteria, fungi, or both) have considerable action on muscle fibers. A typical formulation for the surface treatment of beef cuts contains 2 per cent commercial papain or 5 per cent fungal protease, 15 per cent dextrose, 2 per cent monosodium glutamate (MSG), and salt. The monosodium glutamate has a synergistic effect on the flavor of the meat. The details of its action will be discussed in the section on food additives.

The few examples discussed here give only a very abbreviated view of the vast amount of chemical knowledge which has been accumulated on the chemistry of food preparation. Further information can be obtained from the literature cited at the end of this chapter.

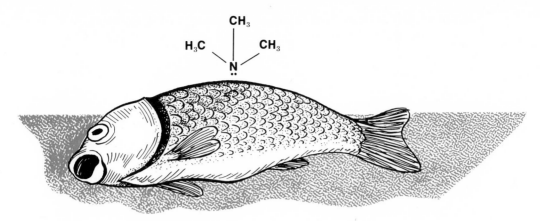

FIGURE 22-4   The bad-smelling odor of decayed fish is due principally to trimethylamine.

PRESERVATION OF FOODS

The major causes of food deterioration are oxygen and microorganisms. Both are aided by water and room temperatures. Any process that prevents the growth of or presence of microorganisms and excludes oxygen or water, or both, is generally an effective preservative process for food. Perhaps the oldest technique is the drying of grains, fruits, fish, and meat. Water is necessary for the growth and metabolism of microorganisms. Without water they dehydrate and die. Water is also important in food oxidation and in this capacity will be discussed in the next section. Dryness thwarts both oxidation and microorganisms.

Salted meat and fruit preserved in a concentrated sugar solution are protected from microorganisms in a similar fashion. The abundance of sodium chloride and sucrose in the immediate environment of the microorganisms forms a hypertonic condition in which water flows by osmosis from the microorganism to its environment (Figure 22-5). The salt and sucrose have the same effect on

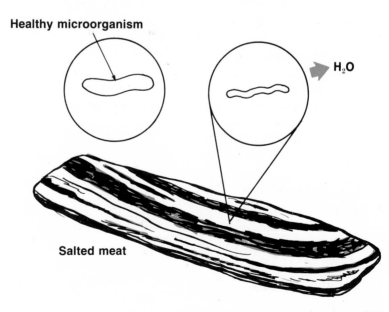

FIGURE 22-5   A salt solution dehydrates microorganisms by osmosis. Without water, microorganisms cannot carry on the chemical reactions required for growth and reproduction.

507

the microorganism as does dryness. All dehydrate the microorganism and thus destroy its ability to grow and reproduce and exude distasteful products into the food.

The canning process was developed by 1810 and involves first the heating of food to kill all bacteria and then sealing it in bottles or cans to prevent the access of other microorganisms and oxygen. This technique is widely used for fish, meat, and vegetables, and some canned meat has been successfully preserved for over a century.

With the increasing growth of urban population and the need for preserving food for longer periods of time, the necessity of developing improved techniques for food preservation has become urgent. As a result of this need, newer techniques have developed such as dehydration, freezing, pasteurization, cold storage, irradiation, and chemical preservation. Before these methods came into general use, a very large percentage of many kinds of foods perished between the harvest and the consumer. Today we can expect food to be fresh for at least several days after it has been processed.

CHEMICAL ADDITIVES

There are many reasons to add chemicals to foodstuffs: additives have control over such diverse factors as color, texture, taste, flavor, moisture content, pH, and nutritive value as well as oxidation and microbial action. They not only preserve the food, and thus increase enormously the total available supply, but make food more appealing and nutritional. We can scarcely find a food item in the market today that has escaped the addition of one or more chemical additives which increase its value in some way. More than 700 million pounds of additives are consumed by Americans each year.

Some people do not agree that chemical additives make food better. Their arguments vacillate between the extremes "nature made the food the way the body should receive it" and "chemical additives cause cancer." The second argument has had some partial basis in a few rare instances. Some notable examples have been *safrole,* which at one time was used to flavor root beer; *butter yellow,* a dye used extensively to color margarine during World War II; and more recently *calcium cyclamate,* which was widely used in dietary soft drinks and foods. Each of these compounds was found to cause cancer in laboratory animals when amounts similar to those in food were used. Fortunately, similar incidents have become even more rare as our knowledge of synergistic effects and toxic levels increases.

Safrole   *Butter yellow*   *Calcium cyclamate*

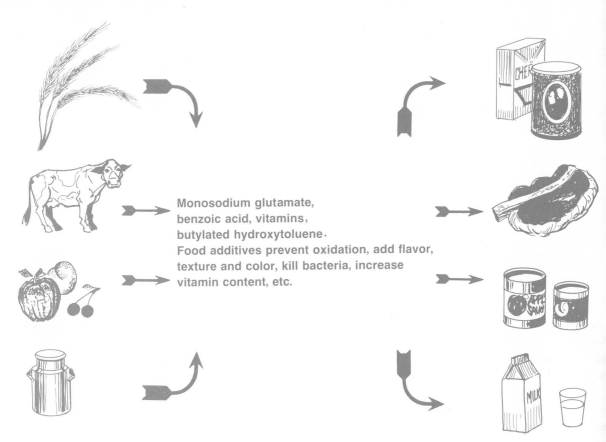

Monosodium glutamate,
benzoic acid, vitamins,
butylated hydroxytoluene.
Food additives prevent oxidation, add flavor,
texture and color, kill bacteria, increase
vitamin content, etc.

FIGURE 22-6   Between the harvested and the consumer-ready food is the addition of a large variety of food additives.

To protect the food consumer, the U.S. Government has enacted many pieces of legislation called Food, Drug, and Cosmetic Acts. The Federal Food and Drug Administration (FDA) administers these laws and, in this sense, is the guardian of our health. For example, the FDA is required by law (Delaney Amendments of 1958) to withdraw sanction on or ban any food additive which "is found to induce cancer when ingested by man or animal." To recognize a food additive as safe, the substance must not be harmful in the amounts used in the food.

The most difficult point to check is the synergistic action of a food additive. An additive may produce no harmful effects by itself. In combination with other substances in the foodstuff, a sleeping giant may be awakened and effects could be amplified much beyond those of the individual chemicals. Synergism and potentiation are the most difficult checks to make because of the scope of the testing. This is being done, however; our increasing understanding of molecular structure and interaction facilitates the predictability of synergism.

At this writing the FDA recognizes over 300 chemical substances "generally recognized as safe" for their intended use. This is known as the GRAS list. A small portion of this list is given in Table 22–1. It must be emphasized that an additive on the GRAS list is safe only if it is used in the amounts and foods specified. The FDA sets permissible levels on all food additives, and these should be followed carefully. To qualify for the GRAS list, food additive must first pass

509

TABLE 22-1   A Partial List of Food Additives Generally Recognized as Safe°

FOOD COLORS
Annatto (yellow)
Carbon black
Carotene (yellow-orange)
Cochineal (red)
Food dyes and colors
Red No. 2
Red No. 3
Blue No. 1
Titanium dioxide (white)

ACIDS, ALKALIES, AND BUFFERS
Acetates: Ca, K, N_2
Acetic acid
Calcium lactate
Citrates: Ca, K, Na
Citric acid
Fumaric acid
Lactic acid
Phosphates, CaH, Ca_3, Na_2, Na_3, NaAl
Potassium acid tartrate
Sorbic acid
Tartaric acid

SURFACE ACTIVE AGENTS
Cholic acid
Glycerides: mono- and diglycerides of fatty acids
Polyoxyethylene (20) sorbitan mono-palmitate
Sorbitan mono-stearate

POLYHYDRIC ALCOHOLS
Glycerol
Sorbitol
Mannitol
Propylene glycol

NONNUTRITIVE SWEETENERS
Saccharin: NH_4^+, Ca, Na

FLAVOR ENHANCERS
Monosodium glutamate (MSG)
5'-nucleotides
Maltol

PRESERVATIVES
Benzoic acid
    Na benzoate
Methylparaben
Oxytetracycline
Propylparaben
Propionic acid
    Ca propionate
    Na propionate
Sorbic acid
    Ca sorbate
    K sorbate
    Na sorbate
Sulfites, Na, K

ANTIOXIDANTS
Ascorbic acid
    Ca ascorbate
    Na ascorbate
Butylated hydroxyanisole (BHA)
Butylated hydroxytoluene (BHT)
Lecithin
Propyl gallate
Sulfur dioxide and sulfites
Nordihydroguaiaretic acid (NDGA)
Trishydroxybutyrophenone (THBP)

SEQUESTRANTS
Citrate esters:
    isopropyl, stearyl
Citric acid
EDTA, Ca and Na salts
Pyrophosphate, Na
Sorbitol
Tartaric acid
    Na tartrate

STABILIZERS AND THICKENERS
Agar-agar
Algins: NH_4^+, Ca, K, Na
Carrageen
Gum acacia
Gum tragacanth
Sodium carboxymethyl cellulose

FLAVORINGS
Acetanisole (slight haylike)
Allyl caproate (pineapple)
Amyl acetate (banana)
Amyl butyrate (pear-like)
Bornyl acetate (piney, camphor)
Carvone (spearmint)
Cinnamaldehyde (cinnamon)
Citral (lemon)
Ethyl cinnamate (spicy)
Ethyl formate (rum)
Ethyl propionate (fruity)
Ethyl vanillin (vanilla)
Eucalyptus oil (bittersweet)
Eugenol (spice, clove)
Geraniol (rose)
Geranyl acetate (geranium)
Ginger oil (ginger)
Linalool (light floral)
Menthol (peppermint)
Methyl anthranilate (grape)
Methyl salicylate (wintergreen)
Orange oil (orange)
Peppermint oil (peppermint) (menthol)
Pimenta leaf oil (allspice) (eugenol, cineole)
Vanillin (vanilla)
Wintergreen oil (wintergreen) (methyl salicylate)

° For precise and authoritative information on levels of use permitted in specific applications, the regulations of the U.S. Food and Drug Administration and the Meat Inspection Division of the U.S. Department of Agriculture should be consulted.

tests on a variety of laboratory animals, as described in Chapter 19. The additive must be harmless by itself, in combination with its foodstuff, and in the amounts used.

Each food additive is included in a consumer foodstuff for a specific, usually chemical, purpose. Table 22–1 is organized according to the function of the additive. Some additives have more than one function; only the major function is given in this table. An understanding of some of the chemistry, among other things, can make the morning's reading of the cereal box more enlightening.

Food colors are generally large organic molecules having several double bonds and aromatic rings. Consistent with the discussion of dyes in Chapter 15, these conjugated structures can absorb only certain wavelengths of light and pass the rest, which gives the substance its characteristic color. β-Carotene is an orange-red substance that occurs in a variety of plants, the carrot in particular, and is commonly used as a food color. It has a conjugated system of delocalized electrons. Other food colors have similar conjugated systems.

β-Carotene

Weak organic acids are added to such foods as cheese, beverages, and dressings to give a mild acidic taste. They often mask undesirable aftertastes. Weak acids and acid salts, such as tartaric acid and potassium acid tartrate, react with bicarbonate to form CO_2 in the baking process.

Some acid additives control the pH of food during various stages of processing as well as of the finished product. In addition to single substances, there are several combinations of substances that will adjust and then maintain a desired pH. These materials are called buffers. An example of one type of buffer is potassium acid tartrate, $KHC_4H_4O_6$, which is used to adjust the pH in candymaking. Most sugar syrups—maple, cane, corn, molasses, and so on—are acidic. Hydrogen ions accelerate the inversion of ordinary sugar (sucrose) to noncrystalline invert sugar (glucose and fructose) (Chapter 17). Since invert sugar inhibits the crystallization of sugar, the candymaker must be sure that the pH of his syrup is such that crystallization will be prevented in the case of creamy fondants and will not be interfered with in the case of hard candies. Potassium acid tartrate achieves a particular pH by means of the following equilibrium:

$$HC_4H_4O_6^- + H_2O \rightleftharpoons H_3O^+ + C_4H_4O_6^{2-}$$

If the syrup is too acidic, the hydrogen ions of the syrup (as H_3O^+) force the equilibrium to the left, and the acidity is reduced. It is unlikely that the syrup would be too basic; but if it is, the equilibrium is shifted to the right to resupply the hydrogen ions used in the neutralization of the basic hydroxide ions, OH^-.

511

$$H_3O^+ + OH^- \longrightarrow 2H_2O$$

To react with either acid or base, and hence to shift the equilibrium in either direction, there must be an ample supply of both $HC_4H_4O_6^-$ and $C_4H_4O_6^{2-}$ ions. In a .003 molar solution of potassium acid tartrate, there are about four $HC_4H_4O_6^-$ ions for every one $C_4H_4O_6^{2-}$, and the pH is between 3 and 4. Under these conditions .0006 mole of acid would use up the 0.0006 mole of $C_4H_4O_6^{2-}$ in a liter of the 0.003 M solution. The FDA limit of 0.25 per cent potassium acid tartrate is usually sufficient to adjust the acidity of most syrups. If adjustment to another pH range is desired, other buffers must be used. The requirement is that the buffer ionize sufficiently to give the desired hydrogen ion concentration.

Adjustment of the pH of a fruit juice is allowed by the FDA. If the pH of the fruit is too high, it is permissible to add acid (called an acidulant). Citric acids and lactic acids are the most common acidulants used since they are believed to impart good flavor; but phosphoric, tartaric, and malic acids are also used. These acids are often added at the end of the cooking time to prevent extensive hydrolysis of the sugar. In the making of jelly they are sometimes mixed through the hot product immediately after pouring. To raise the pH of a fruit which is unusually acidic, buffer salts such as sodium citrate or sodium potassium tartrate are used.

The versatile acidulants also function as preservatives to prevent growth of microorganisms, as synergists and antioxidants to prevent rancidity and browning, as viscosity modifiers in dough, and as melting point modifiers in such food products as cheese spreads and hard candy. The role of acidulants in these areas will be discussed as each topic is treated later.

Anti-caking agents are added to hygroscopic foods—1 per cent or less—to prevent caking in humid weather. Table salt (sodium chloride) is particularly subject to caking unless an anti-caking agent is present. The additive (magnesium silicate, for example) incorporates water into its structure as hydrated water and does not appear wet as sodium chloride does when it absorbs water physically on the surface of its crystals. As a result, the anti-caking agent keeps the surface of sodium chloride crystals dry and prevents crystal surfaces from codissolving, which joins the crystals together.

Antimicrobial preservatives are widely used in a large variety of foods. For example, in the United States sodium benzoate is permitted in nonalcoholic beverages and in some fruit juices, fountain syrups, margarines, pickles, relishes, olives, salads, pie fillings, jams, jellies, and preserves. Propionates are legal in bread, chocolate products, cheese, pie crust, and fillings. Depending on the food, the weight of the additive permitted ranges up to a maximum of 0.1 per cent for sodium benzoate and 0.3 per cent for the propionates.

Food spoilage caused by microorganisms is a result of their mass population, excretion of toxins, and decay. A preservative is effective if it prevents multiplication of the microbes during the shelf-life of the product. Of course, this can be attained by complete sterilization (usually by heat or radiation) or by total inactivation of the microorganisms (freezing). It is seldom achieved by safe chemical agents.

Despite the fact that chemical preservatives of various kinds have been used for several decades, their mode of action remains largely unexplained. Postulated mechanisms could be grouped into three possible categories: (1) interference with the permeability of cell membranes, (2) interference with genetic mechanism, and (3) interference with intracellular enzyme activity.

Interference with the cell membrane and cell wall permeability can have vast effects on the flow of cell nutrients into and waste products out of the cell. For example, propionic acid is known to have an antimicrobial action on certain types of bacteria. It is thought that the propionic acid coats the cell surface with a substance that reduces its permeability.

The coat effectively blocks passage of materials into and out of the cell and the cell ceases to function. On the other hand, oxytetracycline, an antibiotic added to meats, may destroy bacteria by creating a leakage in the cell membrane. Some antibiotics, such as penicillin, are known to prevent cell wall synthesis. Benzoic acid and salicylic acid are known to accumulate on the cell membrane and are thought to inhibit the microorganisms by rendering the cell wall less permeable.

There is no evidence that the generally accepted food preservatives interfere with genetic mechanisms of microbes. However, certain antibiotics such as streptomycin and chloramphenicol are known to exert their effects by chemically combining with the ribosome and inhibiting protein synthesis (Chapter 17). The comparatively simple preservatives with less complex structures are expected to be less reactive with cellular sites of attachment, and even less specific.

There is some evidence that food preservatives interfere with cellular enzymes in microbes. They could interrupt the metabolism of enzymes, and thus destroy them. For example, sorbic acid is known to interfere with cellular dehydrogenases, which normally dehydrogenate fatty acids as the first step in metabolism of molds that grow on cheese. This interrupts their use of food and renders the mold inactive. Several food preservatives are known to inhibit enzyme systems isolated from the cell, but such observations cannot necessarily be extrapolated to cellular conditions. The preservative either may not penetrate the cell wall or may not attain sufficient concentration at the enzyme site. Furthermore, cellular life comprises a host of interdependent enzyme systems, and inhibition or death is the result of interference with only the more sensitive systems.

While spoilage by microbial activity is one of the most important factors to be considered in preserving carbohydrate and protein portions of food products, *atmospheric oxidation* is the chief factor in destroying fats and fatty portions of foods. Chemically, oxygen reacts with the fat to form a hydroperoxide (R—OOH).

Portion of an unsaturated   Hydroperoxide
fat molecule

The mechanism involves the formation of very reactive free radicals (Chapter 14) in a chain reaction process. For example:

$$RH + O_2 \longrightarrow R\cdot + HO_2\cdot$$

(Fat)   *(Free radicals)*

$$R\cdot + O_2 \longrightarrow ROO\cdot$$

$$ROO\cdot + RH \longrightarrow ROOH + R\cdot$$

(Hydroperoxide)

Foods kept wrapped, cold, and dry are relatively free of oxidation. An antioxidant added to the food can also hinder oxidation. Antioxidants most commonly used in edible products contain various combinations of butylated hydroxyanisole (BHA), butylated hydroxytoluene (BHT), or propyl gallate:

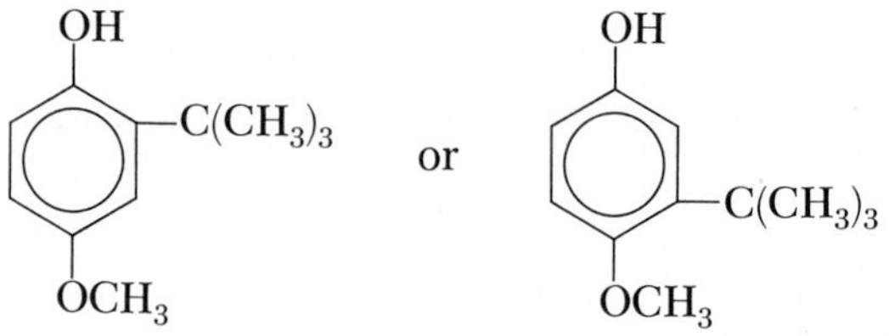

BHA

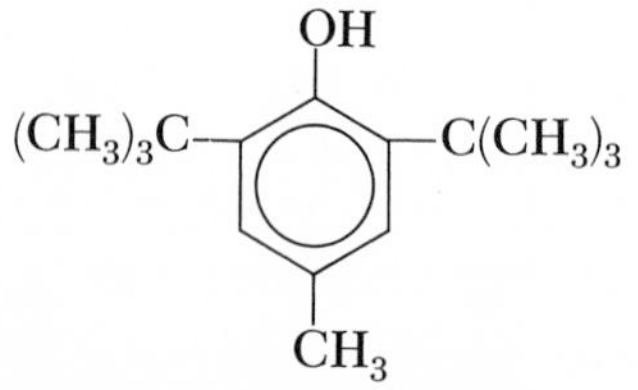

BHT 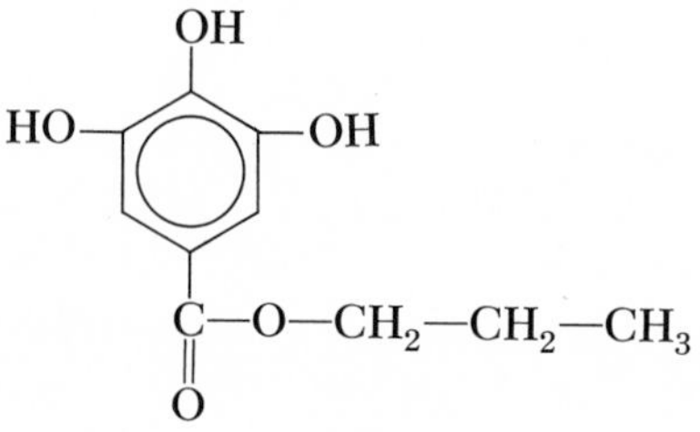   Propyl gallate

The word *butylated* is not widely used in organic chemical nomenclature. It is applied here because the usual names for BHA and BHT include the words *cresol* or *phenol,* and these have a history of being toxic. To avoid consumer rejection of these "safe" compounds—added within the prescribed limits (maximum of 0.02 per cent or less, depending on the food)—the names were made "safe" too.

To prevent the oxidation of fats, the antioxidant can donate its phenolic hydrogen atoms (—OH) to the free radicals and stop the chain reaction. The bulky aromatic radicals formed are relatively stable and unreactive; they add the free electrons to their supply of delocalized electrons:

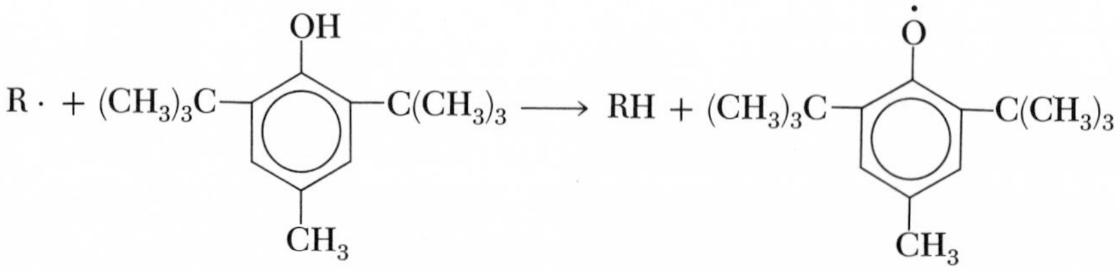

514

If antioxidants are not present, the hydroperoxy group will attack a double bond. This reaction leads to a complex mixture of volatile aldehydes, ketones, and acids, which cause the odor and taste of rancid fat.

As mentioned earlier in this chapter, citric acid is a synergist for antioxidants. It deters oxidation by complexing (or chelating; Chapter 17) traces of metal ions in foods. Metals get into the food from the machinery during the processing. Copper, iron, and nickel, and their ions, catalyze the oxidation of fats. A molecule of citric acid wraps itself around the metal ion, thereby rendering it ineffective as a catalyst. With the antagonist, competitor metal ions tied up, antioxidants such as BHA and BHT can accomplish their task much more effectively.

Citric acid belongs to a class of food additives known as *sequestrants*. (To sequester means "to withdraw from public use.") For the most part sequestrants react with trace metals in foods, tying them up in complexes so the metals will not catalyze the decomposition of food. They are used in shortenings, mayonnaise, lard, soup, salad dressings, margarine, cheese, vegetable oils, pudding mixes, vinegar, confectionery, and other foods. Sodium and calcium salts of EDTA (ethylenediamine tetracetic acid) are sequestrants permitted in beverages, cooked crab meat, salad dressings, margarine, and vinegar from 0.0025 to 0.15 per cent. The bonding pattern of EDTA is shown in Figure 22–7. Note the five-membered rings (five- or six-membered rings are usually required for stable bonding angles), the coordinate-covalent bonds between N and M, and the octahedral arrangement of the donor atoms about the central metal atom (or ion), M.

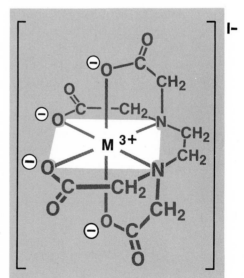

FIGURE 22-7   The structural formula for the metal chelate of ethylene diaminetetraacetic acid (EDTA).

Stabilizers and thickeners improve the texture and blends of foods. The action of carrageenin (a polymer from edible sea weed) is shown in Figure 22–8. Most of this group of food additives are polysaccharides (Chapter 17) having numerous hydroxyl groups as a part of their structure. The hydroxyl groups form hydrogen bonds with water to prevent the segregation of water from the less polar fats in the food, and to provide a more even blend of the water and oils

515

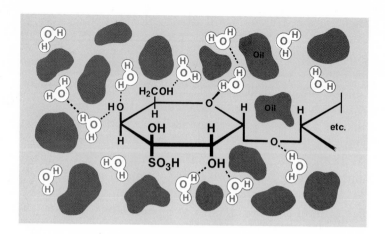

FIGURE 22-8   The action of carrageenin to stabilize an emulsion of water and oil in salad dressing. An active part of carrageenin is a polysaccharide, a portion of which is shown above. The carrageenin hydrogen-bonds to the water, which keeps it dispersed. The oil, not being very cohesive, disperses throughout the structure of the polysaccharide. Gelatin (a protein) undergoes similar action in absorbing and distributing water to prevent ice crystals in ice cream.

throughout the food. Stabilizers and thickeners are particularly effective in icings, frozen desserts, salad dressing, whipped cream, confectionery, and cheeses.

Surface active agents are similar to stabilizers and thickeners in their chemical action. They cause two or more normally incompatible (nonpolar and polar) chemicals to mix. If the chemicals are liquids, the surface active agent is called an emulsifier. If the surface active agent has a sufficient supply of hydroxyl groups, such as cholic acid has, the groups form hydrogen bonds to water. The cholic acid and its associated group of water molecules are distributed throughout dried egg in a manner quite similar to that of carrageenin and water in salad dressing.

Some surface active agents have both hydroxyl groups and a relatively long nonpolar hydrocarbon end. Examples are diglycerides of fatty acids, polyoxyethylene(20) sorbitan monopalmitate, and sorbitan monostearate. The hydroxyl groups on one end of the molecule anchor via hydrogen bonds in the water, and the nonpolar end is held by the nonpolar oils or other substance in the food. This provides tiny islands of water held to oil. These islands are distributed evenly throughout the food.

Polyhydric alcohols are allowed in foods as humectants, sweetness controllers, dietary agents, and softening agents. Their chemical action is based on their multiplicity of hydroxyl groups that hydrogen-bond to water. This holds water in the food, softens it, and keeps it from drying out. Tobacco is also kept moist by the addition of polyhydric alcohols. An added feature of polyhydric alcohols is their sweetness. Two particularly effective alcohols added to sweeten sugarless chewing gum are mannitol and sorbitol.

$$\underset{\text{D-}\textit{Sorbitol}}{\text{HOCH}_2-\overset{\overset{\text{OH}}{|}}{\underset{\underset{\text{H}}{|}}{\text{C}}}-\overset{\overset{\text{H}}{|}}{\underset{\underset{\text{OH}}{|}}{\text{C}}}-\overset{\overset{\text{OH}}{|}}{\underset{\underset{\text{H}}{|}}{\text{C}}}-\overset{\overset{\text{OH}}{|}}{\underset{\underset{\text{H}}{|}}{\text{C}}}-\text{CH}_2\text{OH}}
\qquad
\underset{\text{D-}\textit{Mannitol}}{\text{HOCH}_2-\overset{\overset{\text{H}}{|}}{\underset{\underset{\text{OH}}{|}}{\text{C}}}-\overset{\overset{\text{H}}{|}}{\underset{\underset{\text{OH}}{|}}{\text{C}}}-\overset{\overset{\text{OH}}{|}}{\underset{\underset{\text{H}}{|}}{\text{C}}}-\overset{\overset{\text{OH}}{|}}{\underset{\underset{\text{H}}{|}}{\text{C}}}-\text{CH}_2\text{OH}}$$

Compare the structures of these alcohols with the structure of glucose presented in Chapter 17. It is not surprising that the very similar structures of these polyhydric alcohols and the isomers of glucose produce a similar taste sensation.

Sweetness is characteristic of a wide range of compounds, many of which are completely unrelated to sugars. Lead acetate, $Pb(CH_3COO)_2$, is sweet but poisonous. A number of *artificial sweeteners* are allowed in foods. These are primarily used for special diets such as those of diabetics. Artificial sweeteners have no known metabolic use in the body and therefore are not involved in the insulin balance.

The most common artificial sweetener is saccharin. Its sweet taste was discovered accidentally when the chemist who synthesized it noted that a piece of bread he ate had an obvious sweet taste. He traced the taste back to the saccharin he had prepared. The material was found in the course of studies on the oxidation of orthotoluenesulfonamide, which is the starting material used for its preparation today:

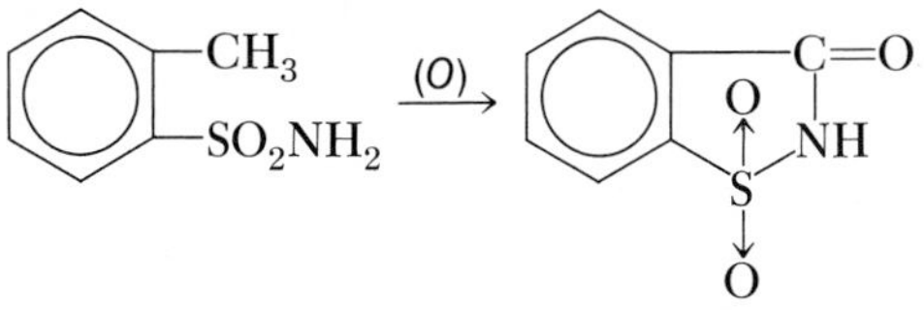

o-Toluenesulfonamide   Saccharin

Saccharin is at least 200 times sweeter than ordinary sugar. Glycine is often used with saccharin to cut the bitter aftertaste. The production of saccharin in the United States exceeds 5,000,000 pounds per year. Its utilization increased after "cyclamates" were taken off the market by the Food and Drug Administration.

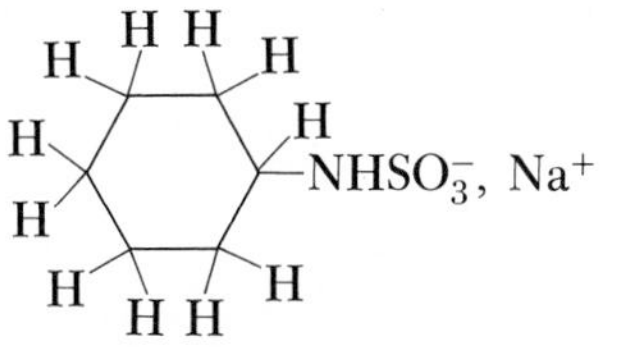

Sodium cyclamate

The sweetness of cyclamate was also discovered accidentally. Cyclamate is about 30 times sweeter than sugar. Because cyclamate does not have the after-taste of saccharin and because it could be used in cooked or baked products, it rapidly replaced saccharin in a wide variety of dietary products. It was removed from the market as a result of suspicions that large doses could be carcinogenic, although no such results have been reported in man.

Because of the great potential usefulness of synthetic sweeteners, other suitable compounds will probably be developed which meet the high FDA requirements for safety.

The sweetness factor is determined by panels of tasters who sample different dilutions of solutions of the sweeteners in water. If a 0.01 M artificial sweetener solution provides the same sweetness sensation as a 1 M solution of sucrose, the artificial sweetener is said to be 100 times sweeter than sucrose.

Most *flavor* additives originally came from plants. The plants were crushed and the compound extracted with various solvents such as ethanol and carbon tetrachloride. Sometimes a single compound was extracted; more often, a mixture

517

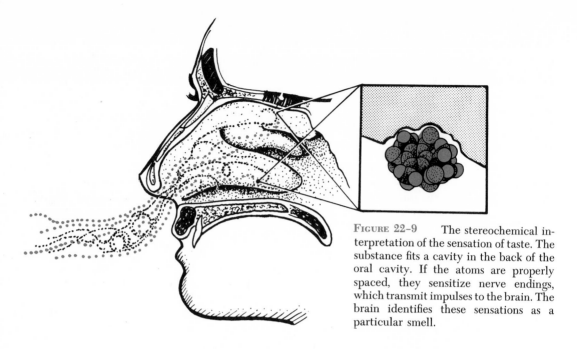

FIGURE 22-9   The stereochemical interpretation of the sensation of taste. The substance fits a cavity in the back of the oral cavity. If the atoms are properly spaced, they sensitize nerve endings, which transmit impulses to the brain. The brain identifies these sensations as a particular smell.

of several compounds occurred in the residue. By repeated efforts, relatively pure oils were obtained. Oils of wintergreen, peppermint, orange, lemon, and ginger, among others, are still obtained this way. These oils, alone or in combination, are then added to foods to obtain the desired flavor. Gradually analyses of the oils and flavor components of plants revealed the active compounds responsible for the flavor. Today, the synthetic preparation of flavors actively competes with natural sources.

Flavor enhancers have little or no taste of their own but amplify the flavors of other substances. They exert synergistic and potentiation effects. Potentiators were first used in meats and fish. Now they are also used to intensify the flavor and cover unwanted flavors in vegetables, bread, cakes, fruits, nuts, and beverages. Three common flavor enhancers are monosodium glutamate (MSG), 5'-nucleotides (similar to inosinic acid; Chapter 17), and maltol.

MSG is the best known and most widely used flavor enhancer. More than 220 million pounds were sold world-wide in 1966. Glutamic acid was isolated in 1866 by the German chemist Ritthausen and converted to a sodium salt, monosodium glutamate, by another chemist. Neither had any interest in flavor. In 1908, a Japanese chemist at the University of Tokyo, Dr. Kikunae Ikeda, discovered the flavor enhancing properties of MSG. Japanese cooks had used the seaweed *Laminaria japonica* for centuries to improve the flavor of soups and certain other foods. Dr. Ikeda discovered that the ingredient in the seaweed that made the difference was MSG and that it had an unusual ability to enhance or intensify the flavor of many high protein foods.

MSG imparts no flavor of its own in the concentrations allowed in foods. There are several theories on how MSG, or any enhancer or potentiator, works. Some chemists say that it acts on certain nerve endings to make the taste buds more sensitive and, therefore, increase the flavor of food. Others claim that it increases salivation and that this leads to increased flavor perception. The detailed chemical and biological action of flavor enhancers is only partially understood. They are very effective and we use them without apparent harm. In some people, however, MSG causes the so-called "Chinese restaurant syndrome," an unpleasant reaction that includes headaches, sweating, and other symptoms, usually occurring after an MSG-rich Chinese meal.

The flavor enhancing property of the 5′-nucleotides was discovered in 1913 by Dr. Shintara Kodama of Tokyo University, an associate of Dr. Ikeda. The 5′-nucleotides were discovered to be the ingredients responsible for the flavor enhancing qualities of bonita tuna. They were approved by the FDA in 1962. While MSG is effective in enhancing the flavor of foods in parts per thousand, the 5′-nucleotides significantly enhance the flavor of foods in concentrations of parts per billion and even less. In liquid foods they create a sense of increased viscosity.

A food additive that has received considerable attention is the *nutrient supplement*. The addition of iodide (as KI) to common table salt to prevent goiter is one example. Many food products now contain added vitamins: vitamin D is added to milk, the B vitamins are added to wheat flour, vitamin C is added to certain beverages, and necessary minerals are added to many cereal products. Since these compounds are needed by the body, they fall into a somewhat different category from the other nonessential additives. The purposes of the various vitamins are given in Table 22–2.

Professor Linus Pauling's announcement, in 1971, that vitamin C appears to prevent the common cold has focused much attention on this vitamin. However, since some vitamins (D, for example) can cause adverse effects when taken in excess, *the unlimited use of vitamins is not advised*. Here, as with other food additives, the amount, the substance itself, and its synergistic effects must all be considered before the additive can be labeled safe to use.

A chemical widely consumed in coffee and tea is the stimulant caffeine, a natural constituent of coffee beans and tea leaves. It is not normally added to foods, although it has been isolated and is used in tablets taken to prevent sleep. Caffeine is a stimulant of the central nervous system and a diuretic (promotes secretion of urine). It has been implicated in a carcinogenic sequence of sarcoma production in the RNA base-pairing operation (Chapter 19).

Caffeine

The flavor of coffee is determined largely by its aroma, and this in turn is due to a very complex mixture of over 100 compounds. The duplication of a particular flavor is usually attained by mixing coffee from different sources rather than trying to match it chemically. Because roasted coffee loses its volatile compounds and, consequently, its flavor, in air, it must be kept sealed to prevent staleness. The caffeine content of tea is considerably higher than that of coffee, but coffee contains a variety of other ingredients, including chlorogenic acid,

519

TABLE 22-2   THE PURPOSES OF VITAMINS°

VITAMIN	BIOCHEMICAL FUNCTION	DEFICIENCY EFFECTS
A	Regeneration of rhodopsin (visual purple)	Excessive light sensitivity; night blindness; increased susceptibility to infection
B_1 (Thiamine)	Nerve activity; carbohydrate metabolism	Beriberi; serious nervous disorders; muscular atrophy; serious circulatory changes
B_2 (Riboflavin)	Coenzyme; affects sight	Sores on the lips; bloodshot and burning eyes; excessive light sensitivity
B_3 (Pantothenic acid)	Growth factor; component coenzyme A	Retarded growth
B_4 (Choline)	Source of transferrable methyl groups	Hepatosis, may allow alcoholic cirrhosis of the liver; dermatosis; anemia
B_5 (Niacin; Nicotinic acid)	Metabolism of ATP	Stunted growth; pellagra
B_6 (Pyridoxine)	Coenzyme; metabolism of fatty acids	Retarded growth; anemia; leukocytosis; insomnia; lesions about eyes, nose, mouth
B_7 (Biotin)	Growth factor; affects scaly and greasy skin; CO_2 fixation; coenzyme	Scaly and greasy skin
B_9 (Folic acid)	Coenzyme; tyrosine metabolism	Anemia
B_{12}	Growth factor; involved in synthesis of DNA	Degeneration of the spinal cord
C	Coenzyme; reducing agent; cholesterol metabolism	Hemorrhages; lesions in the mouth; muscular degeneration; sterility; scurvy
D	Ca and P metabolism	Abnormal development of bones and teeth; rickets
E	Antioxidant; cofactor between cytochromes b and c	Sterility
K	Synthesis of prothrombin, coenzyme Q	Hemorrhage; slow clotting of blood

° These vitamins participate in biochemical changes concerned with the utilization of foodstuffs.

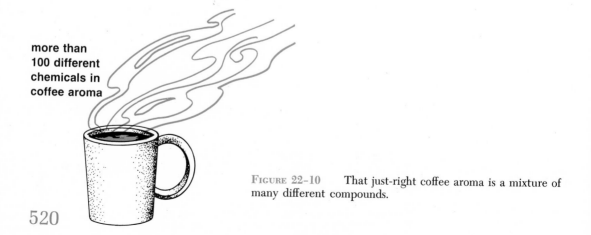

more than 100 different chemicals in coffee aroma

FIGURE 22-10   That just-right coffee aroma is a mixture of many different compounds.

and generally produces more side effects than tea. Roasted chickory, which is often mixed in with coffee, contains no caffeine or other active principle, but it does provide color and flavor.

Tea is made by steeping the dried and fermented tea leaves in hot water for 3 to 5 minutes. This extracts water-soluble compounds, primarily caffeine and products made by the oxidation of phenols (catechins), from the leaves. Because of the way tea is made, it ordinarily does not contain any insoluble matter suspended in it, as coffee does. Tea absorbs flavors and odors readily, and this fact can be used to advantage in the preparation of specially flavored teas. It also means that tea should not be stored near onions, garlic, or soaps since it will also absorb these aromas.

MEDICINE

The average life expectancy in the United States has risen from age 49 in 1900 to 70 in 1970. It is expected to go to beyond the age of 79 by the year 2000. The major contributing factor is the widespread use of a large assortment of new medicinal compounds. Thanks to sulfa drugs, antibiotics, vaccines, and numerous other medicines, quarantine signs warning of scarlet fever have disappeared, polio epidemics no longer exist, and the scourge of smallpox is gone. Tuberculosis, high blood pressure, mental illness, pneumonia, and diabetes are but a few of the many other common afflictions that can now be treated successfully with new drugs.

The contents of the medicine cabinet have changed drastically in the past few years. As the parade of new drugs continues, the indispensable drugs of one decade frequently become obsolete in the next. A survey of physicians shortly before World War I revealed the ten most essential drugs (or drug groups) to be ether, opium and its derivatives, digitalis, diphtheria antitoxin, smallpox vaccine, mercury, alcohol, iodine, quinine, and iron. When another survey was made at the end of World War II, at the top of the list were sulfonamides, aspirin, antibiotics, blood plasma and its substitutes, anesthetics and opium derivatives, digitalis, antitoxins and vaccines, hormones, vitamins, and liver extract. Today there is such a wide array of medicinal chemicals that similar surveys have not been significant statistically, but drugs for (1) reducing fever, (2) relieving pain, and (3) fighting infection still head the list in all areas of medical practice.

Morphine and a few other pain relievers were discussed in the chapter on toxic substances (Chapter 19). Aspirin, antibotics, and a few other medicinals will be discussed in this brief section.

ASPIRIN AND HEADACHES

Americans spend more than $400 million a year on headache remedies. More than 200 kinds of tablets and powders to relieve the problem are on the market. According to the National Health Service, 1 out of every 12 Americans has severe headaches regularly.

The basic ingredient of most headache formulations is aspirin (acetylsalicylic acid). The preparation of this important compound was described in Chapter 15. Headache remedies may also include caffeine, antacids, extra painkillers, antihistamines, vitamins, and tranquilizers.

Acetylsalicylic acid

Headaches are triggered by emotional problems (tension), heredity (migraine), and, less frequently, by eye strain, acute sinus conditions, inflammation of the lining of the brain, infection of a cranial nerve, carbon monoxide poisoning, and poorly positioned teeth.

In tension headaches, muscles are tightened and strained. An overworked head muscle can ache just as much as an overworked arm or leg muscle. The tight muscle may also squeeze arteries and reduce blood flow through the muscle, adding to the pain (Figure 22–11).

Aspirin is a very common and reliable analgesic (painkiller) and antipyretic (fever-reducing) drug. Experimentation has established that aspirin produces its effect on the central nervous system and that salicylic acid (hydrolyzed aspirin)

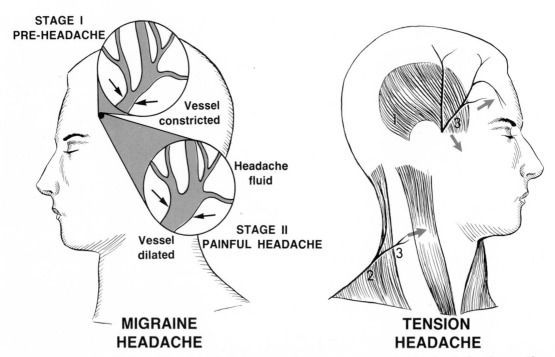

FIGURE 22-11   The physiological action of a headache. In a tension headache the muscles constrict the arteries. The muscles tire and ache, and this causes pain which spreads throughout the head. In a migraine headache the arteries constrict (this is painless) and then expand (this is painful). (In McGraw-Hill Yearbook of Science—Adapted by permission from The New York Times.)

is the active chemical. The hypothalamus gland, attached to the pituitary gland near the center of the brain, is the thermostat of the body. Fever is thought to be produced by a chemical secreted by white blood cells when they engulf bacteria. The chemical enters lymph vessels, migrates to the brain, and resets the thermostat to raise the body temperature. Aspirin in some way adjusts the body temperature back to normal.

The pain of a migraine headache appears to be caused by a headache "fluid" and a constriction of the arteries in the head followed by expansion. Each time the heart pumps, the arteries expand further and more pain from the pulsing is produced. The headache fluid contains two small proteins, bradykinin and neurokinin, which are believed to make nerves sensitive to pain (Figure 22–12). When the substances are extracted from the headache fluid and injected elsewhere in the body, the person will sense pain in the new area, according to Dr. Arnold P. Friedman, physician-in-chief at the Montefiore Headache Unit in New York and a recent chairman of the World Commission for the Study of Headache. The constriction and subsequent expansion of the arteries may be initiated by the release of serotonin, which is known to contract smooth muscle. Normally the serotonin is bound, but its sudden release into the bloodstream could cause the initial contraction of the arteries. An enzyme, monoamine oxidase, metabolizes the serotonin, causing the arteries to relax and expand. When the normal supply of serotonin is depleted, the arteries remain expanded until a new supply is made available. Australian scientists reported in 1967 that the blood content of serotonin does fall sharply at the onset of a migraine attack.

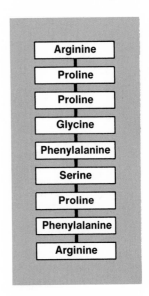

FIGURE 22–12   The amino acid sequence in brady-kinin, a pain-causing protein. The structures of the amino acids are given in Table 17–1.

The migraine headache can be relieved by ergotamine tartrate if it is taken early enough. It acts by constricting the muscles of the blood vessels, preventing painful stretching. This property of ergotamine tartrate has been known for over 40 years. A newer drug, methysergide, is effective in about 70 per cent of patients. It is taken between headaches. Neither drug is always effective; both are sometimes associated with serious side effects when used for long periods.

The hereditary nature of migraine headaches is indicated by a study made by Dr. Adrian M. Ostfield of the University of Illinois. Dr. Ostfield reports that there is a 70 per cent chance that the child will have migraine if both parents have migraine. If one parent has migraine, there is a 45 per cent chance; if neither parent has migraine but there is a history of it in the family, a 25 per cent chance.

ANTISEPTICS AND DISINFECTANTS

An antiseptic is a compound which prevents the growth of microorganisms. It now has the legal meaning "germicide," or a compound which *kills* microorganisms. A disinfectant is a compound which destroys pathogenic (illness-causing) bacteria or microorganisms, but usually not bacterial spores. These materials are generally only suitable for external use, as on the skin on a wound. When taken by mouth they are usually poisonous.

Compounds commonly used for this purpose include those listed in Table 22–3. Some of these, such as the halogens, sodium hypochlorite, hydrogen peroxide, and potassium permanganate, are effective because of their oxidizing properties. This is a general property and allows them to oxidize any kind of cell, including human cells. For this reason they are used mostly as disinfectants in destroying the germs on nonliving objects. Phenol (carbolic acid), which is no longer widely used, is readily absorbed by cells and is a general poison. The quaternary ammonium compounds are surface active agents, and their bactericidal ability seems to be related to their ability to make the bacterial cell material pass through the cell wall.

TABLE 22–3   SOME OF THE MORE COMMON ANTISEPTICS

Iodine	Mercurochrome
Sodium hypochlorite	Metaphen
Potassium permanganate	Merthiolate
Hydrogen peroxide	Pine oil
Iodophors	Soap
Ethanol	Hexylresorcinol
Quaternary ammonium compounds	Hexachlorophene
Chloramine-T	Mercuric chloride
Phenols	

One of the newer developments solves the problem of applying antiseptics to children. You may remember the sting of "iodine" when applied to a scratch or a wound. Old-fashioned iodine is a solution of iodine (I_2) in alcohol with a little potassium iodide (KI) to increase the solubility of the iodine. The alcohol causes most of the pain. Now there are polymers, such as polyvinyl pyrollidone, which complex iodine molecules (Figure 22–13); the products (iodophors) are soluble in water. The resultant solution is a very efficient and painless disinfectant. The iodophors are active ingredients in a popular mouthwash and in restaurant glassware disinfectants.

Because most of these compounds are generally toxic to living matter, it is necessary to utilize only dilute solutions and then only on the skin. While

524

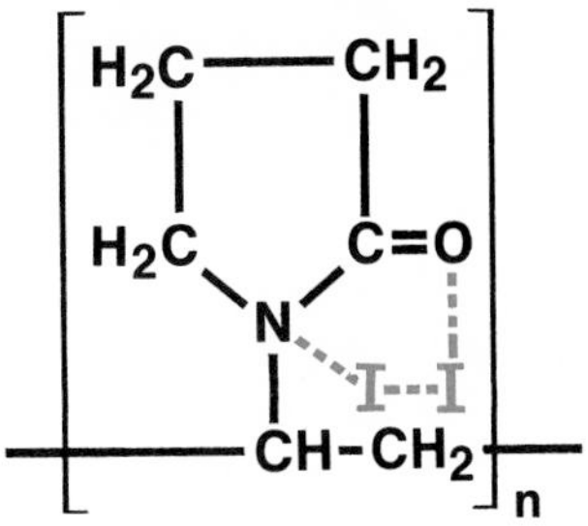

$$\left[\begin{array}{c} H_2C \text{———} CH_2 \\ | \quad\quad | \\ H_2C \quad\quad C=O \\ \diagdown N \diagup \quad\quad \vdots \\ | \quad\quad \text{'I---I} \\ \text{———} CH\text{-}CH_2 \text{———} \end{array}\right]_n$$

FIGURE 22-13   Complex of polyvinyl pyrollidone (PVP) and iodine.

they help to prevent the spread of disease, they are practically useless in its treatment because they act nonspecifically against all cells with which they come in contact. They are to be distinguished from antibiotics, which act more selectively against infecting bacteria than against "organic" cells within the human body.

ANTIBIOTICS

In our own time, the quest for drugs to wipe out disease has virtually been fulfilled by the antibiotics. Since these drugs are so efficient, they were the first of what came to be called "miracle" drugs. Their job generally is to aid the white blood cells by stopping bacteria from multiplying. When a person is sick or is killed by a disease, it means that the bacteria have multiplied faster than the white blood cells could devour them, and that the bacterial toxins increased more rapidly than the antibodies could neutralize them. The action of the white blood cells and antibodies plus an antibiotic is generally enough to repulse an attack of the germs.

The first of the antibiotics was penicillin, and the story of its discovery is

FIGURE 22-14   Sir Alexander Fleming

525

almost as fantastic as its antibacterial action. In 1928, Alexander Fleming, a bacteriologist at the University of London, was working with cultures of *Staphylococcus aureus*, a germ that causes boils and some other types of infections. In order to examine the cultures with a microscope, he had to remove the cover of the culture plate for a while. One day as he started work he noticed that the culture was contaminated by a blue-green mold. Many people would have discarded the contaminated culture, but Fleming's trained eye had noticed something else.

For some distance around the mold growth, the bacteria colonies were being destroyed. Upon further investigation, Fleming found that the broth in which the mold was grown had an inhibitory or lethal effect against many pathogenic organisms. The mold was later identified as *Penicillium notatum* (the spores sprout and branch out in pencil shapes, hence the name).

Pencillin, the name given to the antibacterial substance produced by the mold, apparently had no toxic effect on animal cells, and its activity was selective. Because Fleming's extracts from the mold were crude, clinical results were discouraging, and his brilliant discovery made little impact on the medical world for almost a decade. Eventually, as a result of the interest and further research by Howard Flory and Ernst Chain, the wonder drug, penicillin, was developed. In 1941, penicillin was used for the first time on a human being, a London policeman who was hospitalized with a serious case of blood poisoning contracted from a shaving cut. Since he could not recover by normal means, the doctors decided to try the new drug. The effect was immediately favorable.

Because of wartime needs, a large supply of penicillin was urgently needed. Through cooperation between American and British firms, the supply was provided and thousands of lives and limbs were saved.

Fleming, Flory, and Chain shared a richly deserved Nobel Prize in medicine and physiology in 1945.

The structure of penicillin has now been determined.

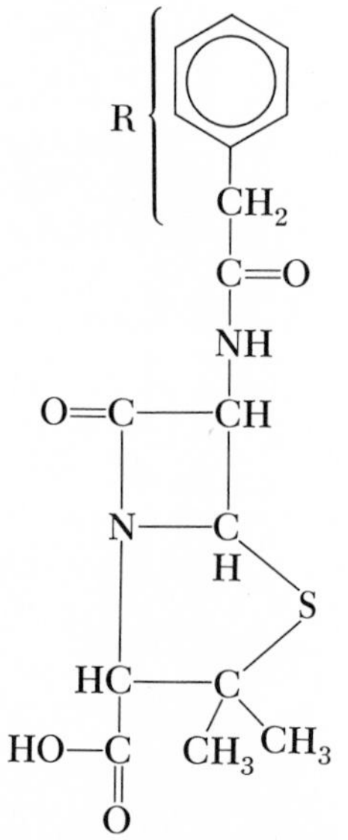

Penicillin G

Many different penicillins exist, differing in the structure of the R group. Penicillin G is the most widely used in medicine.

The development of a series of antibiotics followed soon after the success of penicillin. Most of these were discovered after exhaustive studies of the microorganisms in the soil. In 1943, Dr. Selman Waksman discovered streptomycin; in 1945, aureomycin was discovered by a team led by Dr. Benjamin Duggar; in 1947, Dr. Paul Burkholder discovered chloromycetin. Many others have since been synthesized or discovered. Only relatively few have been found to be both antibacterial and nontoxic. Some of the successful ones are Acramycin, Terramycin, Oramycin, Bacitracin, Tegopen, Garamycin, Syncillin, Polycillin, and Cleocin. The yearly use of all antibiotics is given in Table 15–4 (Chapter 15). Despite the many antibiotics now in use, penicillin, the first to be discovered, is still the most widely used and most generally satisfactory.

The fact that antibiotics specifically destroy bacterial cells and not human cells is the basis of their success. The action of streptomycin was discussed in the section on antimicrobial action of food additives. Its blocking action of the ribosome surface of bacteria is typical of the specificity of the antibiotics. In each case in which the mode of action is known, the antibiotic interferes with some specific metabolic process in specific bacteria.

Several antibiotics, such as penicillin and bacitracin, are known to prevent cell-wall synthesis. The cell wall of some pathogenic bacteria is composed of mucoprotein. Mucoprotein is a combination of proteins and mucopolysaccharides, in which some of the monosaccharide units (usually glucose or galactose) contain a $-NHCOCH_3$ group substituted for a hydroxyl group ($-OH$). Only bacterial cells have mucopeptide walls.

Penicillin interferes with the synthesis of the mucoprotein cell wall of the bacteria by interfering with the formation of cross links between layers of the wall. The cell wall protects and supports the delicate cell components enclosed within it. The cytoplasmic membrane, immediately inside the cell wall, regulates

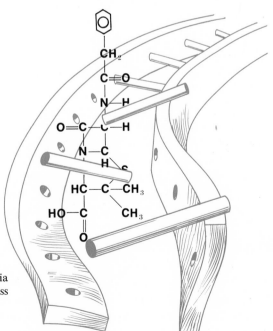

FIGURE 22–15   Penicillin kills bacteria by interfering with the formation of cross links in the cell wall.

527

the flow of nutrients and water in and out of the cell. The layers are reinforced by a series of chemical crosslinks connecting one layer to another. When the crosslinks are not formed properly, the weakened wall, unable to hold its size and shape, expands as water comes into the cell by osmosis. Eventually the cytoplasmic membrane bursts, causing the cell to die.

Of several thousand antibiotics that have been prepared or discovered, only a very few have been found to be relatively nontoxic. Even penicillin is toxic to some people. Streptomycin, which was on the market for more than ten years, has been removed because its toxicity is greater than some equally effective, newer antibiotics. The search continues for antibiotics that are less toxic, yet more effective on more kinds of bacteria. The success of antibiotics in obliterating disease is already renowned. The antibiotics are truly "miracle drugs."

DIET PILLS

There are many kinds of special diets—diets to lower cholesterol, diets to correct some inborn error of metabolism, diets for ulcers, diabetes, obesity, underweight, hormone deficiencies, high blood pressure, kidney disorders, and many others. The most popular diets in our affluent society are those designed for overweight. Only a very small percentage of obese individuals can attribute their overweight to endocrine gland deficiencies. All other overweight people have only their lack of activity and excess food intake to blame for their obesity.

The desired rate of weight loss for most people ranges from about 1 to 2 pounds per week. This rate may vary depending on such factors as water balance, activity, heat loss, and the presence of disease. For most women, diets ranging between 1200 and 1500 calories (a food calorie is actually a kilocalorie, or 1000 ordinary calories) a day will bring about a satisfactory weight loss. For men, the range is between 1500 and 2000 calories a day. An intake of less than 1200 calories is generally not advised. Young children vary greatly from one individual to the next; no general rule can be applied to them, and calorie levels should be prescribed individually by a dietitian or doctor.

A variety of special, rather unusual diets have been reported to be effective in weight reduction. These diets have been known under such names as the grapefruit diet, the water diet, the 10-day diet, and others. The problem with the vast majority of these diets is that they are often made up of only a few foods, so that they are not nutritionally adequate. Five basic types of nutrients are essential. These five types of nutrients are protein, fat, carbohydrate, vitamins, and certain minerals. These must be included in the diet for normal health. If the diet neglects any one category, the body suffers. Therefore, a balanced diet plus water is required, even if the amounts are reduced.

Many individuals have difficulty staying on a diet, and they must be highly motivated in order to maintain a caloric level low enough to bring about a loss of weight. Appetite-depressing drugs, known as anorexigenic drugs, have been used by some physicians to curb the appetite. More than two billion diet pills are distributed per year. The ingredients include amphetamines (which suppress the appetite), digitalis (which affects the heart), various diuretics (which increase the amount of urine), thyroid, and prednisone. Some of the drugs are potentially addictive; others tax the heart or cause potassium loss in the body.

FIGURE 22–16   A limited understanding of the biochemistry of smoking, foods, and medicine produces after-the-fact problems. (Editorial cartoon by Hugh Haynie of the Louisville Courrier-Journal. Copyright, Los Angeles Times Syndicate. Reprinted with permission.)

By themselves anorexigenic drugs will not control obesity. Reliance on such drugs, rather than proper diet, leads to failure in a weight-reduction plan, and there are indications that their unsupervised use may be harmful.

In addition to the appetite-depressing drugs, there are many other products that are advertised to help overweight people lose weight. These products include methyl cellulose (remember from Chapter 17 that the body cannot digest cellulose), vitamins, iron and calcium compounds, benzocaine (a local anesthetic), dextrose, potassium p-aminobenzoate, caffeine, flavorings, and a few other substances. By themselves, however, none of these products will cause a reduction in weight.

ALLERGENS AND ANTIHISTAMINES

Some people are allergic to pollen, mold, food, cosmetics, penicillin, and even cold, heat, and ultraviolet light. In the United States about 5000 people die yearly from bronchial asthma, at least 30 from the stings of bees, wasps, hornets, and other insects, and about 300 people die from ordinary doses of penicillin. The reason: *allergy*. About one in 10 suffers from some form of allergy; more than 16 million Americans suffer from hay fever. This means that many of you already are familiar with the symptoms of headaches, inflamed eyes, congested sinuses, sneezing, and the raw, endlessly running nose that accompanies hay fever.

An allergy is an adverse response to a foreign substance or to a physical condition that produces no obvious ill effects in *most* other organisms, including man. An *allergen* (the substance that initiates the allergic reaction) is, in most cases, a highly complex substance—usually a protein. Some are polysaccharides or complexes of a protein and a polysaccharide. Usually allergens have a molecular weight of 10,000 or more.

What is really the chemical cause of an allergy such as, say, hay fever? The details are at best sketchy now. But the overall process and a few of the details have been worked out reasonably well.

For a patient with pollen allergy, a pollen grain enters his nose and clings to the mucous membrane. The nasal secretions, acting on the pollen grain, release the grain's allergens and other soluble components, which penetrate the outer layer of the mucous membrane. The principal allergen of ragweed pollen, a major allergy-producer, has been isolated and is named ragweed antigen E. It is a protein with a molecular weight of about 38,000; it represents only about 0.5 per cent of the solids in ragweed pollen, but contributes about 90 per cent of the pollen's allergenic activity. A mere 1×10^{-12} gram of antigen E injected into an allergic person is enough to induce a response. The reason why antigen E, of all the ragweed pollen proteins, is so unusually reactive is as yet not understood.

The allergens come in contact with special cells in the nose and breathing passages to which are already attached a particular type of antibody present in unusually high concentration in allergic persons. We have a number of antibodies in our cells and bloodstream to react with and protect us from harmful parasites, bacteria, and the like. Before 1966, three antibodies, designated IgG, IgA, and IgM (Ig for immunoglobulin), were known. These are proteins with molecular weights of about 150,000, 180,000, and 950,000, respectively. In 1966, scientists isolated a previously unknown antibody, IgE. This globulin protein comprises only 0.001 per cent to 0.01 per cent of the globulin proteins in blood serum, and it escaped discovery for a long time by hiding as an impurity in IgA antibodies. The molecular weight of 196,000 for IgE is very close to the molecular weight of 160,000 to 200,000 for IgA. The antibody IgE is formed in the nose, bronchial tubes, and gastrointestinal tract, and binds firmly to specific cells, called mast cells, in these regions. An allergic person has 6 to 14 times as much IgE in his blood serum as a nonallergic person. In nasal secretions of allergics, the concentration of IgE is 100 or more times greater than in the serum of the same person.

Antigen E from ragweed reacts with the IgE antibody attached to the mast cells, forming antigen-antibody complexes. Some think that the ragweed antigen E forms a bridge between two or more IgE molecules attached to a mediator-containing cell. This bridge, in turn, alters the configuration of parts of the attached IgE molecules.

By a series of events that are only crudely understood, the formation of these antigen-antibody complexes leads to the release of so-called "allergy mediators" from special granules in the mast cells. Some believe that the changed configuration of the cell caused by the antigen-antibody complex activates enzymes that ultimately cause the cell to release the allergy mediators. One of these mediators, and the most potent one found so far, is histamine, which is formed by the breakdown of the amino acid, histidine. Although it is widely

$$H_2NCH_2CH_2 \underset{\substack{N \\ H}}{\overset{N}{\bigsqcup}}$$

Histamine

distributed in the body, it is especially concentrated in the 250 to 300 granules of the mast cells. Histamine accounts for many, if not most, of the symptoms of hay fever, bronchial asthma, and other allergies. This compound causes dilation of blood capillaries. In addition, it makes the capillaries more permeable to blood fluids. Thus, these fluids can readily leak out of the capillaries and cause swelling of the tissues. The compound causes contraction and spasm of smooth muscles (a serious difficulty in the bronchial tubes in asthma). It can produce skin swellings (hives) and stimulate the glands that secrete watery nasal fluids, mucus, tears, saliva, and so on.

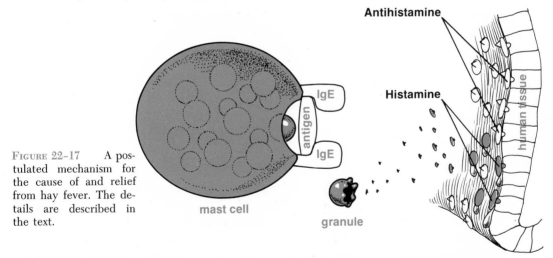

FIGURE 22-17 A postulated mechanism for the cause of and relief from hay fever. The details are described in the text.

The chemical mediators such as histamine must be released from the cell to cause the symptoms of allergy. The release mechanism is an energy-requiring process in which the granules may move to the outer edge of the living cell and, without leaving the cell, discharge their contents of histamine through a temporary gap in the cell membrane. This sends histamine on its way to produce the toxic effects of hay fever.

Although histamine, an allergenic compound, has been isolated and some understanding of the mechanism that leads to allergy is now known, many questions are still unanswered. Since IgE is a class of several proteins, what structural relationships cause an IgE to react, for example, with the allergen in ragweed pollen and another IgE to react with the allergen in lobster? Does

531

heredity play a part in allergy? Are people allergic because their membranes are inherently thinner or weaker than those of nonallergics? If antigen E causes 90 per cent of the ragweed allergic activity, what causes the other 10 per cent? What chemical reactions does histamine undergo in order to produce its symptoms? These are but a few questions, and there are many others. Although hay fever, bronchial asthma, and other allergies may not be conquered, they are better understood and better treated today.

Treatment consists of three procedures: avoidance, desensitization, and drug therapy. The idea of avoidance is simple enough. If you are allergic to strawberries, do not eat strawberries; if you are allergic to pollen, move, say, from Decatur, Illinois (ragweed pollen index: 114), to Seattle, Washington (index: 0.02). Sometimes, however, avoidance is impractical.

Desensitization therapy is costly and inconvenient since 20 or more injections are required to get what usually is a partial cure. One chemical idea of desensitization is to inject a blocking antibody that preferentially reacts with the allergen so that it cannot react with the IgE allergy-sensitizing antibody. This breaks the chain of events leading to the release of histamine or other allergy-producing mediators. Many small injections, spaced in time, are required to build up a sufficient level of the blocking antibody.

Epinephrine (adrenalin), steroids, and antihistamines are effective drugs in treating allergies. The first two are particularly effective in treating bronchial asthma, while the antihistamines, introduced commercially in the United States in 1945, are the most widely used drugs for treating allergies. More than 50 antihistamines are offered commercially in the United States. Many of these contain, as does histamine, an ethylamine group, ($-CH_2CH_2N<$):

Pyribenzamine
(an important antihistamine)

These drugs act competitively by occupying the receptor sites normally occupied by histamine on cells. This, in effect, blocks the action of histamine. A new drug, disodium cromoglycate, acts not by blocking the action of histamine (as do the antihistamines), but by blocking the release of histamine and other mediators from the granules. Efforts are being made to find drugs with less troublesome side effects than those often associated with antihistamines. The most troublesome side reactions are drowsiness, mental confusion, dizziness, headache, nervousness, rapid pulse, nausea, depression, blurred vision, and dryness of the mouth. No proved drug is available now that will relieve the allergy and completely avoid the side effects.

BIRTH CONTROL PILLS

One of the most revolutionary medical developments of the 1960's was the worldwide introduction and use of the Pill. An estimated 8.5 million women in the United States use birth control pills. The basic feature of oral contraceptives for women is their chemical ability to simulate the hormonal processes resulting from pregnancy and, in so doing, prevent ovulation. Ovulation, the production of eggs by the ovary, ceases at the onset of pregnancy because of hormonal changes. This same result can be produced by the administration of a variety of steroids, some of which are effective when taken orally, although the mechanism of their action and their long-term effects are not known in detail.

The active ingredients of the Pill are the hormones progesterone and estrogen, or their derivatives.

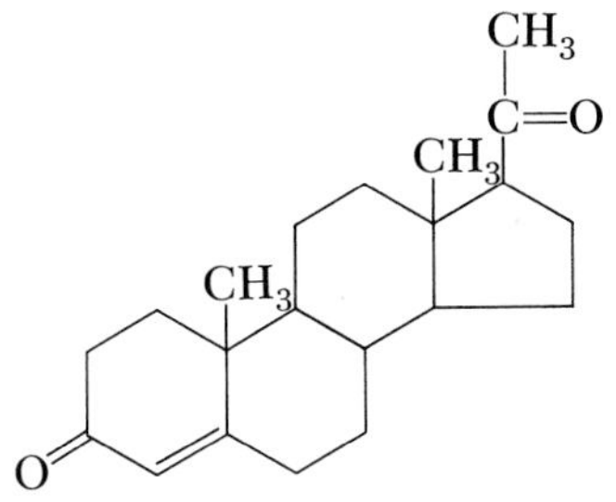

Progesterone

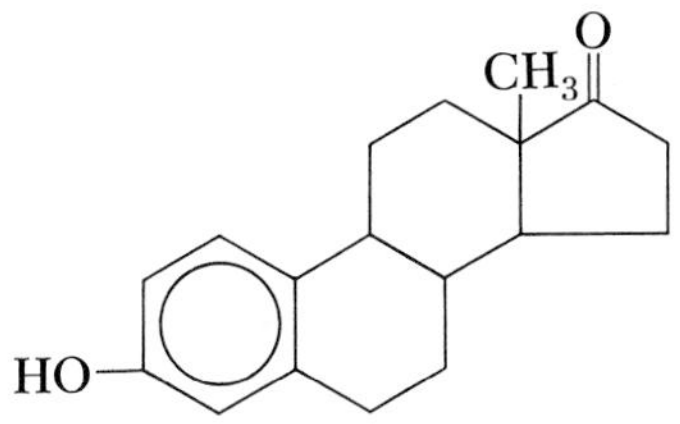

Estrone
(an estrogen)

Envoid, a product of this sort made by G. D. Searle and Co., is a mixture of Norethindrone and Mestranol. Notice the structural similarities between these compounds and progesterone and estrogen:

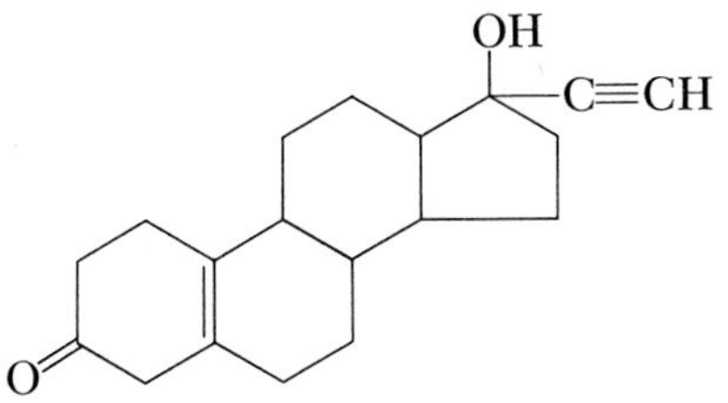

Norethindrone
(major constituent)

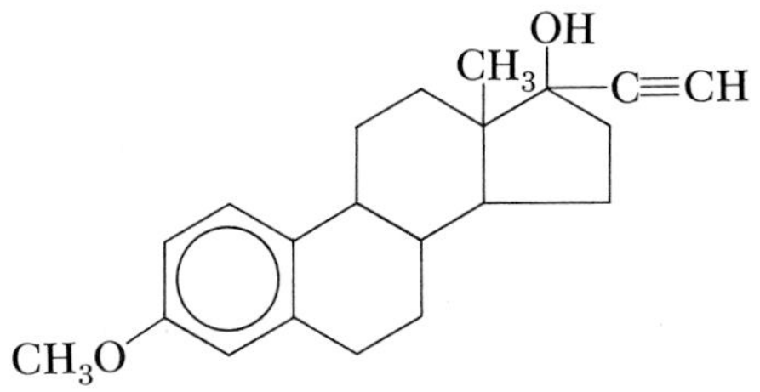

Mestranol
(minor constituent)

Do these pills cause cancer? After the first 10 years of usage, there was no conclusive evidence that oral contraceptives are carcinogenic. Nor was there evidence that they cause diabetes, sterility, eye disorders, mental illness, or any of a number of other diseases to which they have been linked by critics. Since the latency period for cancer is thought to range from 10 to 20 years, it seems probable that if oral contraceptives do cause cancer, this will become evident during the 1970's.

There is, however, one serious disorder that has been linked with the Pill: thromboembolic (blood clotting) disease. Such clots are potentially lethal. If they block a major blood vessel in a limb, amputation may be necessary; if they block a vessel to the lungs or brain, death may result. Studies done in Great Britain and the United States indicate that blood clotting is the cause of about 3 deaths per 100,000 users each year. Statistics indicated that this risk was considerably less than the risk of thromboembolism that would accompany the number of pregnancies averted. However, because there is a danger, all women who use the Pill should have medical checkups at least once every 6 months.

Perhaps the major problem facing our world is the population explosion. 533

As the human population doubles every 40 years and associated problems intensify, wider efforts are being made to study procedures for controlling human fertility. Antifertility drugs for males are under active study, and there is every reason to believe that these will be on the market in the near future. Some of the first drugs of this sort to be studied act by temporarily suppressing the formation of sperm cells in the male.

ANTACIDS

Antacids are compounds used to decrease the amount of hydrochloric acid in the stomach. The normal pH of the stomach ranges from 1.2 to 0.3. Various compounds can accomplish this by one or more of the following processes: neutralization, buffering, absorption, or retention in ion exchange resins. These compounds are widely used in the treatment of gastric hyperacidity (excess stomach acid) and peptic ulcers. Since they treat *symptoms,* and not the underlying causes, they should be used with some caution by laymen.

Some compounds used for this purpose and their modes of action follow:

Magnesium oxide, MgO (a base), is a white insoluble powder that reacts with hydrochloric acid:

$$MgO + 2H^+ \longrightarrow Mg^{2+} + H_2O$$

Milk of magnesia, an aqueous suspension of magnesium hydroxide, $Mg(OH)_2$, which neutralizes hydrochloric acid because it is a base:

$$Mg(OH)_2 + 2H^+ \longrightarrow Mg^{2+} + 2H_2O$$

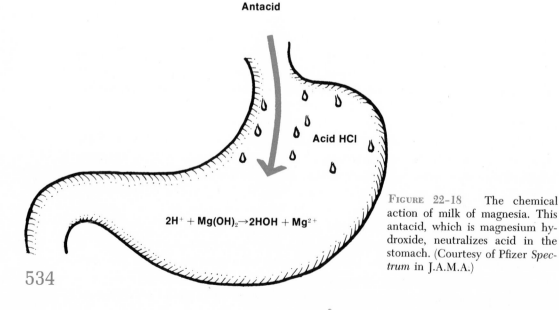

Antacid

Acid HCl

$2H^+ + Mg(OH)_2 \rightarrow 2HOH + Mg^{2+}$

FIGURE 22-18   The chemical action of milk of magnesia. This antacid, which is magnesium hydroxide, neutralizes acid in the stomach. (Courtesy of Pfizer *Spectrum* in J.A.M.A.)

Calcium carbonate, $CaCO_3$, a white insoluble powder which reacts slowly with stomach acid:

$$CaCO_3 + 2H^+ \longrightarrow Ca^{2+} + H_2O + CO_2$$

Sodium bicarbonate, $NaHCO_3$ (baking soda), a soluble white powder, neutralizes hydrochloric acid and produces carbon dioxide gas:

$$Na^+ + HCO_3^- + H^+ \longrightarrow Na^+ + H_2O + CO_2$$

Magnesium trisilicate, $Mg_2Si_3O_8 \cdot nH_2O$, is a fine white powder. Its reaction is

$$Mg_2Si_3O_8 \cdot nH_2O + 4H^+ \longrightarrow 2Mg^{2+} + 3SiO_2 \cdot nH_2O + 2H_2O$$

Aluminum hydroxide gel, $Al(OH)_3$, is a white solid without taste or odor. It reacts like magnesium hydroxide:

$$Al(OH)_3 + 3H^+ \longrightarrow Al^{3+}_{(aq)} + 3H_2O$$

This compound has the advantage that it cannot make the stomach become alkaline.

Dihydroxyaluminum sodium carbonate, $NaAl(OH)_2CO_3$, is a combination of aluminum hydroxide and sodium carbonate (sold as Rolaids). The reaction with acids is:

$$NaAl(OH)_2CO_3 + 4H^+ \longrightarrow Na^+ + Al^{3+}_{aq} + 3H_2O + CO_2$$

This compound will not normally cause the stomach pH to go above 5.

Sodium citrate, $Na_3C_6H_5O_7 \cdot 2H_2O$, is the sodium salt of citric acid, a weak acid. It forms colorless crystals, soluble in water, which react with stomach acid to form sodium chloride and citric acid:

$$Na_3C_6H_5O_7 \cdot +3H^+ \longrightarrow 3Na^+ + H_3C_6H_5O_7 + 2H_2O$$

As can be seen from these examples, antacids are generally insoluble bases or salts with weak anionic bases. The strong bases remove acidity by neutralization:

$$H^+ + OH^- \longrightarrow H_2O$$

while the anionic weak bases reduce the acidity by reacting with the protons:

$$A^- + H^+ \longrightarrow HA$$

The preferred antacids are those which do not reduce the stomach acidity very much. If the reduction of acidity is too great, the stomach secretes an excess of acid. This is called "acid rebound."

535

DRUGS IN COMBINATION

Drugs, like some food additives, can have enhanced effects when placed in certain chemical environments. Sometimes the effects are harmful; sometimes helpful. Take the case of an aging business executive who took an antidepressant and then ate a meal that included aged cheese and wine. The antidepressant is an inhibitor of monoamine oxidase, an enzyme that helps to control blood pressure. Both the aged cheese he ate and the wine he drank contained pressor amines, which raise blood pressure. Without the controlling effect of the enzyme, these amines skyrocketed his blood pressure and caused a stroke. Neither the amines nor the antidepressant alone would likely cause the stroke, but the combination did.

Likewise, people who take digitalis for heart trouble and to cut down on the sodium level should take aspirin only under medical supervision. Aspirin can cause a 50 per cent reduction in salt excretion for three or four hours after it is taken.

Alcohol increases the action of many antihistamines, tranquilizers, and drugs such as reserpine (for lowering blood pressure) and scopolamine (contained in many over-the-counter nerve and sleeping preparations), making such combinations extremely dangerous. Staying away from dangerous alcohol-drug combinations is not as easy as it may seem. Many people fail to realize that a large number of over-the-counter preparations—such as liquid cough syrup and tonics—contain appreciable amounts of alcohol.

Not all drug combinations are bad. Doctors have been highly successful in prolonging the lives of leukemia and other cancer victims with combinations of drugs that individually could not do the job. Resistant kidney disease has also responded to drug combinations in cases in which single drugs were ineffective.

Perhaps the best advice is to take medicine only when you are sick, making sure that the doctor knows what you are taking.

QUESTIONS

1. What weight of MgO is required to neutralize all the acid in a stomach which contains .05 mole of hydrochloric acid?

2. Which of the following food additives should be avoided? Give your reasons.
   Butter yellow   Glycerol
   Propionic acid   Sodium cyclamate

3. Why does it take less time to cook food in a pressure cooker than in an open pot of boiling water?

4. What would happen to your ability to digest protein if you kept the acid in your stomach neutralized all the time? Is acid bad for your stomach?

5. Why is saccharin preferable to chloroform as an artificial sweetener?

6. A label on a brand of breakfast pastries contains the following additives: dextrose, glycerine, citric acid, potassium sorbate, Vitamin C, sodium iron pyrophosphate, and BHA. What is the purpose of each substance?

7. Choose a label from a food item and try to identify the purpose of each additive.

8. Describe some of the chemical changes that occur during the cooking of

   (a) a carbohydrate   (b) a protein   (c) a fat

9. What causes fat in foods to become rancid? How can this be avoided?

10. What causes bread to rise?

11. What are the pros and cons of eating "natural" foods as opposed to foods containing chemical additives?

12. Why were cyclamates taken off the market?

13. Do you think it is wise to use animals in safety tests for drugs and food additives? Should mental patients and prisoners be used for this purpose?

14. Many consumer products are almost identical in chemical composition but are sold at widely different prices under different trade names. Do you think the products should be identified by their chemical names or their trade names? Why?

15. See what you can find out about correlation between taste and smell. Are they the same sensation? Are they independent of each other?

16. Which has more caffeine, tea or coffee?

17. What contribution did Joseph Lister make to science?

18. Bixin (annatto extract from the seeds of the tropical tree *Bixa orellana*) is added to food to give it a yellow color. What part of the structure is primarily responsible for the color? If white light is incident on the substance, why is the substance yellow?

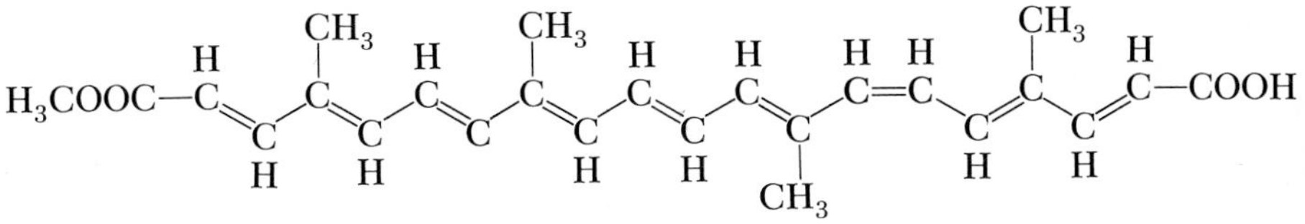

Bixen

19. Write a chemical equation that will show how the tartrate ion $(C_4H_4O_6^{2-})$ can raise the pH when added to an unusually acidic fruit juice.

20. Suggest a way that citric acid sequesters metals.

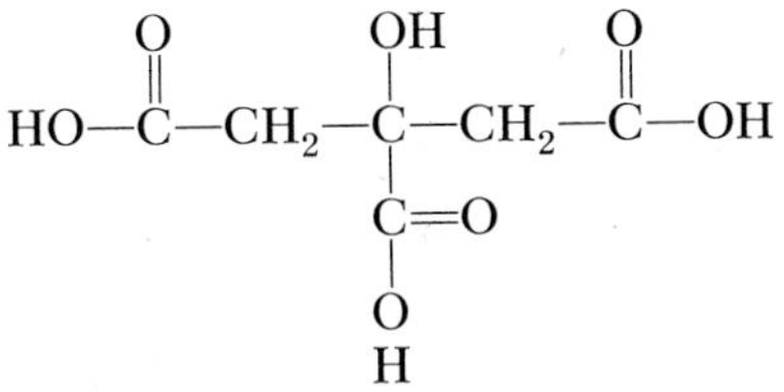

Citric acid

21. Using the structure of a protein given in Chapter 17, show how cooking can affect its structure.

22. Describe the method of testing drugs and medicines for safety.

23. How is the action of antibiotics different from the action of antiseptics in killing bacteria?

24. What is a synergist? Give an example.

25. What part, if any, does hydrogen bonding play in the activities of surface active agents, humectant action of polyhydric alcohols, and stabilizing effect of gelatin in ice cream?

537

SUGGESTIONS FOR FURTHER READING

Amoore, J. E., Johnston, J. W., Jr., and Rubin, M., "The Stereochemical Theory of Odor," *Scientific American*, Vol. 210, No. 2, p. 42 (1964).

"Chinese Restaurant Syndrome," *Chemistry*, Vol. 42, No. 8, p. 4 (1969).

"Flavor of a Potato Chip," *Chemistry*, Vol. 43, No. 7, p. 2 (1970).

Furia, T. E. (ed.), "Handbook of Food Additives," The Chemical Rubber Co., Cleveland, 1968. (The GRAS chemicals are listed and discussed on pages 565 to 751.)

Gates, M., "Analgesic Drugs," *Scientific American*, Vol. 215, No. 6, p. 131 (1969).

Hodge, H. C., and Smith, R. P., "Clinical Toxicology of Commercial Products," Third Edition, Gleason, M. N., Gosselin, R. E., The Williams & Wilkins Co. Baltimore, 1969. (Contains a wealth of information on the toxic aspects of various commercial products.)

"How Penicillin Kills Bacteria," *Chemistry*, Vol. 41, No. 7, p. 44 (1968).

Kirk, R. E., and Othmer, D. F., "Encyclopedia of Chemical Technology," Second Edition, Interscience Publishers, New York, 1963. (Contains detailed information on various aspects of applied chemistry.)

"New Artificial Sweeteners," *Chemistry*, Vol. 43, No. 6, p. 23 (1970).

Pyke, M., "Man and Food," McGraw-Hill Book Co., New York, 1970.

Pirie, N. W., "Orthodox and Unorthodox Methods of Meeting World Food Needs," *Scientific American*, Vol. 216, No. 2, p. 27 (1970).

Sanders, H. J., "The Tasteless Condiment," *Chemistry*, Vol. 40, No. 1, p. 23. (1967).

Schubert, J., "Chelation In Medicine," Scientific American, Vol. 214, No. 5, p. 40 (1968).

Solmssen, U. V., "The Chemist and New Drugs," *Chemistry*, Vol. 40, No. 4, p. 22 (1967).

"The Allergenic Mechanism," *Chemistry*, Vol. 44, No. 5, p. 23 (1971).

CONSUMER CHEMISTRY —BEAUTY AIDS AND CLEANING AGENTS

CHAPTER 23

The millions of dollars spent in the United States each year on cosmetics and cleansers are spent without the consumer knowing very much about the chemical composition of the products. Even less is known about the properties and chemical actions of the various chemicals in the products. Such knowledge could be very important in choosing the best product for a specific application, in avoiding unwanted "side effects," and in paying the lowest price. When the

FIGURE 23-1   Beauty is only skin deep—plus a few layers of cosmetics. The small assortment of cosmetics shown here represents many thousands more that we use to adorn, freshen, and protect ourselves.

choice is possible, it is rather ridiculous to pay more for a product that has the same chemical composition as a cheaper brand, or to choose a product that does not have the chemicals to do the job. It is our purpose in this chapter to describe some aspects of chemistry that will aid you in understanding and buying cosmetic and cleaning products.

COSMETICS

Cosmetics are materials other than soap that are applied to the skin or other surface areas of the human body, to obtain a cleansing action or to make the appearance more attractive. They include an enormous range of products, from hair tonic to polish for toenails. Cosmetics all have sets of properties determined by the chemicals they contain, and these properties are often amazingly sensitive to the molecular structures of the constituent compounds. An obvious case is in the area of perfumes, where even small variations in molecular structure can give compounds with very different odors (see also Chapter 22). Here we will examine only a few of the different types of cosmetic preparations.

Permanent Waving

Human hair is a complex polymeric protein fiber whose shape is determined by hydrogen bonding and the presence of disulfide linkages between adjacent molecular strands (as was discussed in the chapter on biochemical structures, and shown in Figure 23–2.) The basic chemical process in cold permanent waving is the disruption of the disulfide linkages by a chemical reducing agent and the subsequent reforming of at least some of these bonds by an oxidizing agent. While the structure of the hair is disrupted, it is shaped on curlers and then reoxidized. The chemical reactions in simplified form are shown in Figure 23–2.

The most commonly used reducing agent is thioglycolic acid. The common

$$\underset{\displaystyle \overset{|}{SH}}{CH_2}-C\overset{\displaystyle O}{\underset{\displaystyle OH}{\diagup\diagdown}}$$

Thioglycolic acid

oxidizing agents used include hydrogen peroxide, perborates ($NaBO_2 \cdot H_2O_2 \cdot 3H_2O$), percarbonates ($Na_2C_2O_6$), and sodium or potassium bromate ($KBrO_3$). A typical neutralizer solution contains one or more of the oxidizing agents dissolved in water. The presence of water and strong base in the oxidizing solution also helps to break and reform hydrogen bonds between adjacent protein molecules.

An example in which hydrogen peroxide is the oxidizing agent and thioglycolic acid is the reducing agent is the reaction between the two compounds to form a disulfide and water.

$$2HOOCCH_2SH + H_2O_2 \longrightarrow HOOCCH_2S-SCH_2COOH + 2H_2O$$

Reducing agent	*Oxidizing agent*	*Disulfide linkage*

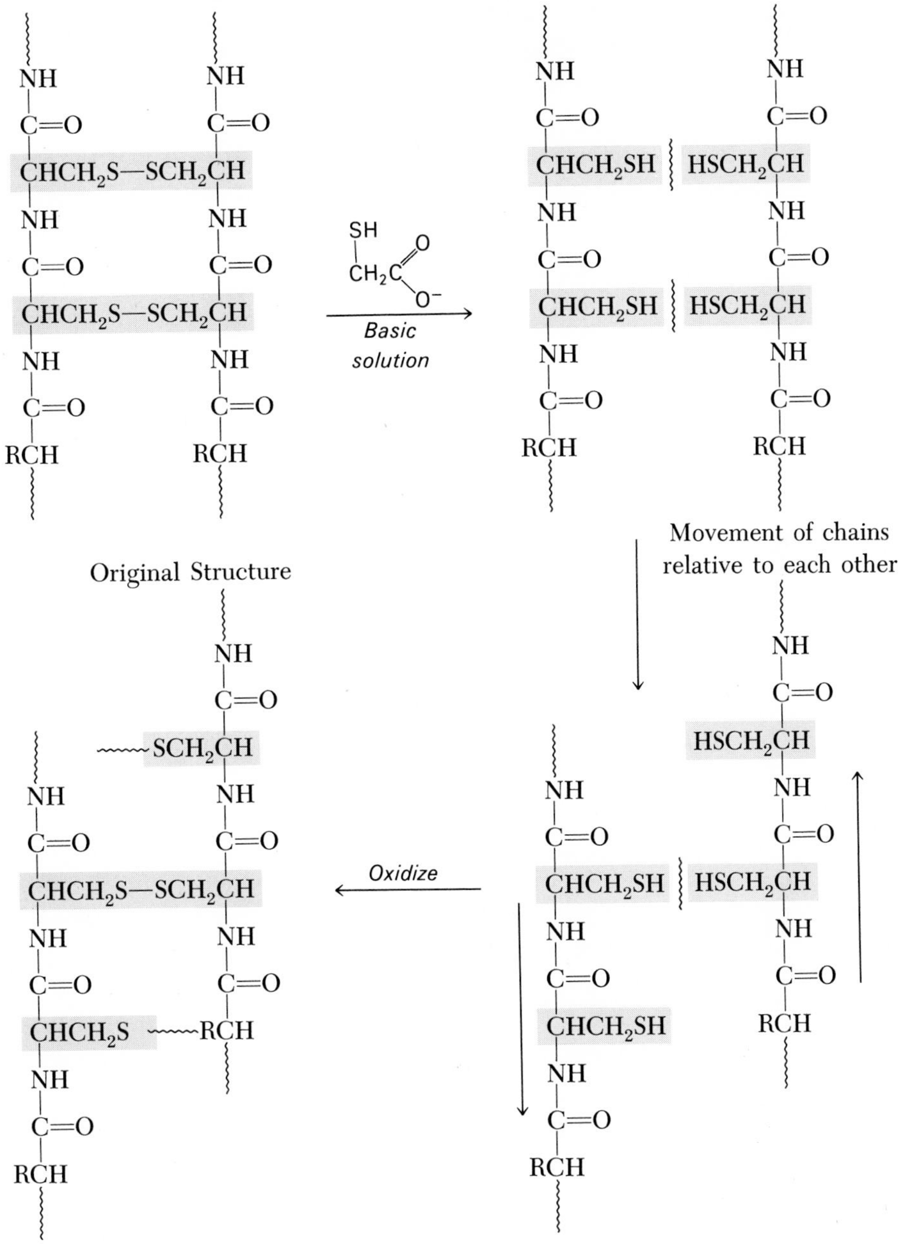

FIGURE 23-2   Structural changes that occur in hair during a permanent wave.

Various additives are present in both the oxidizing and the reducing solutions in order to control pH, odor, and color, and for general ease of application. A typical waving lotion contains 5.7 per cent thioglycolic acid, 2.0 per cent ammonia, and 93.3 per cent water. In some preparations, borax [$Na_2B_4O_5(OH)_4 \cdot 8H_2O$] or sodium lauryl sulfate [$CH_3(CH_2)_{11}OSO_3Na$, an anionic detergent] are added as wetting agents.

All surfaces (including the hair) are covered with layers of either gaseous, liquid, or solid films. The problem of displacing the oil film left on freshly

541

shampooed hair so that the thioglycolic acid (polar) can enter the hair is very important. Substances that can displace these adhering materials are called *wetting agents*. The detergent action of a soap was described and illustrated in Chapter 15.

Borax performs an extra step when used as a wetting agent. The borate ion first hydrolyzes to form a hydroxide ion.

$$B_4O_7^{2-} + H_2O \rightleftharpoons HB_4O_7^- + OH^-$$

The hydroxide ion then reacts with grease to form a soap as described in Chapter 15. The soap may then remove other grease molecules or coat the surface of the hair.

Hair can be straightened by the same solutions. It is simply "neutralized" (or oxidized) while straight (no rolling up).

Depilatories

The purpose of a depilatory is to remove hair chemically. Since skin is sensitive to the same kind of chemical attack as hair (both contain the same types of proteins), such preparations should be used with caution, and even then, some attack on the skin is almost unavoidable. Because of this, the interval between applications of a depilatory should be of the order of a week or so. They should never be used on skin which is infected or which has a rash, and they should not be followed by application of a deodorant with its astringent (contracting) action. If the sweat pores are closed by the deodorant, the caustic chemicals are retained and can do great harm in contact with very sensitive and non-skin protected tissue. If the sweat pores are open, the body fluids will dilute and wash the caustic chemicals to the outside of the body.

The chemicals used as depilatories include sodium sulfide, calcium sulfide, strontium sulfide (water-soluble sulfides), and calcium thioglycolate [$Ca(HSCH_2COO)_2$], the calcium salt of the compound used to break S-S bonds between protein chains in permanent waving. A typical cream depilatory contains calcium thioglycolate (7.5 per cent), calcium carbonate (filler and acid neutralizer, 20 per cent), calcium hydroxide (acid neutralizer, 1.5 per cent), cetyl alcohol [$CH_3(CH_2)_{15}OH$] as skin conditioner, 6 per cent sodium lauryl sulfate (detergent, 0.5 per cent), and water (64.5 per cent).

The water-soluble sulfides are all strong bases in water, as indicated by the hydrolysis of the sulfide ion.

$$S^{2-} + H_2O \longrightarrow HS^- + OH^-$$

For example, a 0.1 M solution of Na_2S has a pH of about 13—a strongly basic solution. The compounds act chemically on the hair to disrupt bonds in the protein chains and cause it to "cure," that is, to hydrolyze to soluble amino acids and small peptides, which may be wiped off with a damp cloth.

The area on which a depilatory has been used should be washed with soap and water, dried, and then treated with small amounts of talcum powder. Talcum is a mixture of finely ground talc [$Mg_3(OH)_2Si_4O_{10}$] and a perfume. Talc has a porous structure and a great capacity for adsorbing liquids. Its adsorbing power

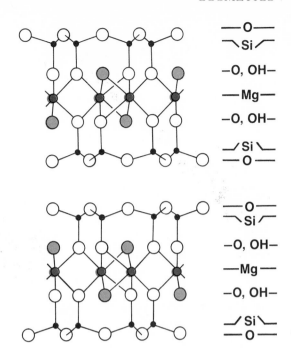

FIGURE 23-3   A partial structure of talc ($Mg_3Si_4O_{10}(OH)_2$). The open circles are atoms of oxygen; the small black circles are silicon; shaded circles are OH groups; brown-filled circles are magnesium. Only a portion of two sheets is shown. (After Evans, R. C.: Introduction to Crystal Chemistry, 2nd ed. New York, Cambridge University Press, 1964.)

results in large part from the tremendous surface area of a finely ground powder. This provides very many opportunities for liquid molecules to "rest" on the surface via physical adsorption. In addition, smaller molecules can be absorbed into the open spaces of the structure of talc (Figure 23-3). Regardless of the mechanism, talc is an effective collector of liquids, such as the serum and mucus secreted after the use of depilatories. Talc is relatively inert chemically and, consequently, it does not undergo chemical reactions that might irritate exposed nerve endings. It forms a protective cover that excludes air and other nerve-irritating chemicals.

Depilatories can cause considerable damage to the skin and should be used very carefully.

Hair Coloring and Bleaching

In recent years hair bleaches and colors have become more popular, largely because of the development of newer formulations which can produce a much more uniform coloration of human hair. The principles utilized are basically the same as those used to dye other fibers, such as textiles. Some of these principles were described in Chapter 15.

The formulations vary from temporary coloring (removable by shampoo), which is usually achieved by means of a water-soluble dye in a suitable solution dyeing the surface of the hair, to semi-permanent dyes, which penetrate the hair fibers to a great extent and often consist of cobalt or chromium complexes of dyes dissolved in an organic solvent. Permanent dyes are generally "oxidation" dyes. These are applied to the hair, which they penetrate, and then are oxidized to give a color which is permanently attached to the hair by chemical bonds or which is much less soluble than the initial molecule. This is the principle of ingrain dyeing described in Chapter 15. These hair dyes generally are derivatives of phenylenediamine.

543

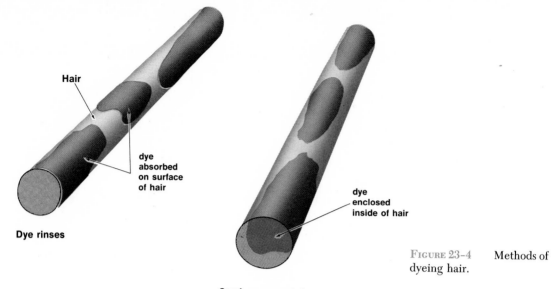

Hair

dye
absorbed
on surface
of hair

Dye rinses

dye
enclosed
inside of hair

Semi-permanent dye

FIGURE 23–4   Methods of
dyeing hair.

This molecule itself dyes hair black. A blond dye can be formulated with
p-aminodiphenylamine-sulfonic acid,

or p-phenylenediamine-sulfonic acid.

One blond formulation contains p-phenylenediamine (0.3 per cent),
p-methylaminophenol (0.5 per cent), p-aminodiphenylamine (0.15 per cent),
o-aminophenol (0.15 per cent), pyrocatechol (0.25 per cent), resorcinol (0.25 per
cent), and inert solvent (98.40 per cent). These compounds are applied in an
aqueous soap or detergent solution containing ammonia to make the solution
basic. The dye material is then oxidized, using hydrogen peroxide to develop
the desired color. The amines are oxidized to nitro compounds.

$$-N\begin{smallmatrix}H\\\\H\end{smallmatrix} + 3H_2O_2 \xrightarrow{oxidation} -N\begin{smallmatrix}O\\\\O\end{smallmatrix} + 4H_2O$$

Amine   *Nitro*

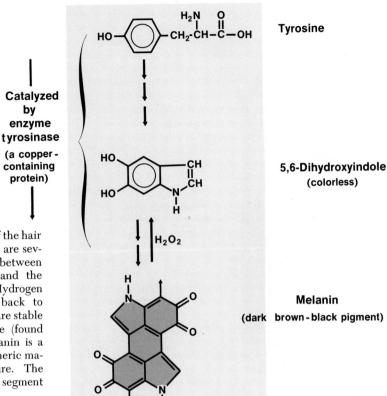

Tyrosine

5,6-Dihydroxyindole
(colorless)

Catalyzed
by
enzyme
tyrosinase

(a copper-
containing
protein)

Melanin
(dark brown-black pigment)

FIGURE 23-5   Bleaching of the hair by hydrogen peroxide. There are several chemical intermediates between the amino acid—tyrosine—and the hair pigment—melanin. Hydrogen peroxide oxidizes melanin back to colorless compounds, which are stable in the absence of tyrosinase (found only in the hair roots). Melanin is a high molecular weight polymeric material of unknown structure. The structure shown here is only a segment of the total structure.

Hair is bleached by hydrogen peroxide, which destroys hair pigments by oxidation. The solutions are made basic with ammonia to enhance the oxidizing power of the peroxide. Parts of the chemical process are given in Figure 23–5. This drastic treatment of hair does more than just change the color. It can destroy sufficient structure to render the hair brittle and coarse.

Hair Sprays

Hair sprays are essentially solutions of some plastic in a very volatile solvent whose purpose, when sprayed on hair, is to furnish a plastic film with sufficient strength to hold the hair in place after the solvent has evaporated. After early experiments with shellac, the introduction of the aerosol can allowed the use of a wider variety of resins and solvents and provided greater control over the application of the product. A typical aerosol can is shown in Figure 23–6.

A very common resin in hair sprays is the addition polymer, polyvinylpyrrolidone (PVP)

Polyvinylpyrrolidone, PVP

545

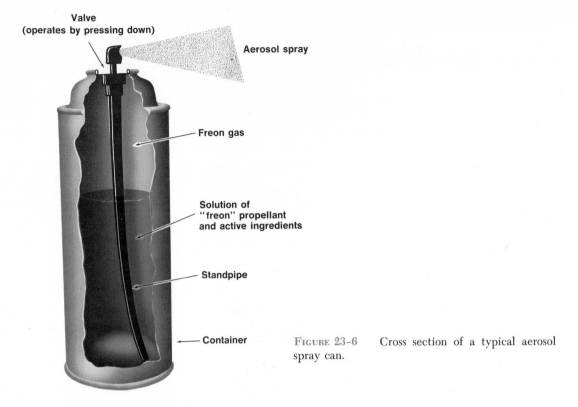

FIGURE 23-6 Cross section of a typical aerosol spray can.

The resin is blended in hair spray formulations with a plasticizer, a water repellant, and a solvent-propellant mixture. The plasticizer makes the plastic more pliable, as described and illustrated in Chapter 16. The solvent-propellant system is a solvent such as anhydrous ethyl alcohol mixed with a liquified propellant, such as

$$
\begin{array}{cc}
\text{F} & \text{Cl} \\
| & | \\
\text{Cl—C—F} & \text{F—C—F} \\
| & | \\
\text{Cl} & \text{F}
\end{array}
$$

Cl—C—F   or   F—C—F

Dichlorodifluoromethane (Freon 12)   *Chlorotrifluoromethane (Freon 13)*

The resin concentration of hair sprays is of the order of 1.5 per cent with a ratio of 30 per cent ethanol to 70 per cent propellant for the liquid phase. The hair spray may also contain sorbitol (a polyhydric alcohol, used as a wetting agent; see *Food Additives*, Chapter 22), lanolin (emollient), silicones (principally methyl silicones for sheen and waterproofing; very resistant to deterioration via ultraviolet radiation; see Chapter 13), vegetable gums (acacia, tragacanth, etc., as adhesive agents), and perfume.

Lanolin is an excellent skin softener and is a component of many cosmetics. It is a complex mixture of esters from hydrated wool fat. The alcohols in the esters have up to 33 carbon atoms, and the fatty acids have up to 36 carbon atoms. A common alcohol in lanolin is cholesterol (Figure 23–8). Cholesterol is found both free and in esters. Cholesterol is the most abundant sterol in animal tissues. (The four-ring structure is the fundamental steroid structure common to

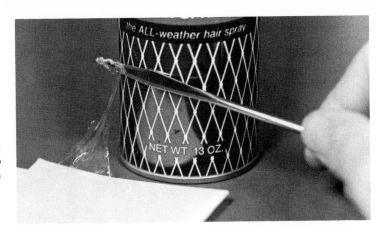

FIGURE 23-7   Film of hair
spray. Hair spray was allowed to
dry on white surface and was then
pulled up to reveal film.

FIGURE 23-8   Cholesterol.

many natural products; the presence of the —OH group makes the structure a
sterol). It is abundant in the plasma membranes of many animal cells. Nomi-
nated as the villain in certain kinds of heart attacks, it is thought to build up
sufficiently in blood vessels to stop the flow of blood. Cholesterol appears to en-
dow fat mixtures with the property of absorbing water. This is one factor that
makes lanolin an excellent emollient (a substance applied to soften or soothe
irritated skin). With its high proportion of free alcohols, particularly cholesterol,
and hydroxy acid esters, lanolin has the structural groups (—OH) to hydrogen
bond water (to keep the skin moist) and to anchor within the skin (the fatty
acid and ester hydrocarbon structures). See Figure 23–9.

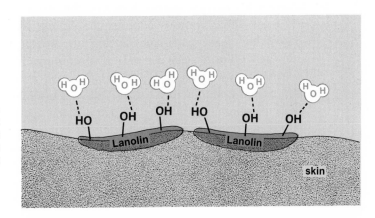

FIGURE 23-9   The hydroxyl
groups of lanolin form hydrogen
bonds with water and keep the skin
moist. The fat parts of the molecule
are "soluble" in the protein and fat
layers of the skin.

547

A typical hair spray formulation contains the following ingredients and amounts.

PVP	(Resin)	4.70 parts by weight
Dimethyl phthalate	(Plasticizer)	0.20
Silicone	(Sheen)	0.10
Ethanol	(Solvent)	25.00
Freon 11	(Propellant)	45.00
Freon 12	(Propellant)	25.00
Perfume	(For effect)	_____
		100.00 parts by weight

Since PVP tends to pick up moisture, other, less hydroscopic polymers are beginning to replace PVP in hair sprays. For example, significantly better moisture control is obtained with a copolymer made from a 60/40 ratio of vinyl pyrrolidone and vinyl acetate.

$$
\begin{array}{c}
\quad\quad\quad O \\
\quad\quad\quad \| \\
H \quad O-C-CH_3 \\
| \quad | \\
C=C \\
| \quad | \\
H \quad H
\end{array}
$$

Vinyl acetate

Deodorants

The 2,000,000 sweat glands on the body surface are primarily used to regulate body temperature via the cooling effect produced by the evaporation of the water they secrete. The evaporation of the water leaves the solid constituents, mostly sodium chloride, plus smaller amounts of other materials, including some organic compounds. The odor results largely from amines and hydrolysis products of fatty oils (fatty acids, acrolein, etc.) emitted from the body and from bacterial growth on the body. Perspiration is both normal and necessary for the proper functioning of the human body.

There are three kinds of deodorants: those which directly "dry up" perspiration or act as astringents, those which have an odor to mask the odor of sweat, and those which remove odorous compounds by combining with them. Among those which have astringent action are hydrated aluminum sulfate, hydrated aluminum chloride, $(AlCl_3 \cdot 6H_2O)$, aluminum chlorohydrate [actually aluminum hydroxy chloride, $Al_2(OH)_5Cl \cdot 2H_2O$ or $Al(OH)_2Cl$ or $Al_6(OH)_{15}Cl_3$], alcohols, tannins and tannic acid. Of these, the aluminum salts are the most widely used. Those compounds which act as deodorizing agents include zinc peroxide, essential oils and perfumes, and a variety of mild antiseptics. Zinc peroxide removes odorous compounds by oxidizing the amines and fatty acid compounds. The essential oils and perfumes absorb or otherwise mask the odors, and the antiseptics are generally oxidizing or reducing agents.

The most widely advertized deodorants contain aluminum salts as the active ingredients. Other materials are added to assist the application or to provide a fragrance. Aluminum salts are astringents in that they can reduce or close up the openings of the sweat glands by affecting the hydrogen bonds that hold protein molecules together or (less often) by precipitating skin proteins. The ones that precipitate skin proteins give skin irritating effects. A typical spray deodorant will have the aluminum salt and minor ingredients dissolved in an alcoholic solution. A typical cream antiperspirant deodorant contains aluminum hydroxy chloride (astringent), 20 per cent; sorbitan monostearate (hydrogen-bonds water and absorbs it), 5 per cent; polyoxyethylene sorbitan monostearate (hydrogen-bonds water and absorbs it), 5 per cent; stearic acid (precipitates amines), 15 per cent; propylene glycol (precipitates fatty acids), 5 per cent; and water (for desired consistency), 50 per cent.

Face Powder

Face powder is used to give the skin a smooth appearance by covering up any oil secretions which would otherwise give it a shiny look. The powder must have some hiding ability, but if it is too opaque it will look too obvious. It usually requires several ingredients to obtain a powder which has the proper appearance, sticking properties, absorbance for oily skin secretions, and spreading ability. A typical formula is

Talc	65%	to which are added small amounts of perfume and coloring matter
Precipitated chalk	10%	
Zinc oxide	20%	
Zinc stearate	5%	

The absorbing properties of talc were discussed earlier. Precipitated chalk is $CaCO_3$, formed by precipitation from aqueous solutions of calcium chloride and sodium carbonate:

$$Ca^{2+} + 2Cl^- + 2Na^+ + CO_3^{2-} \longrightarrow CaCO_3 + 2Na^+ + 2Cl^-$$

Zinc oxide has astringent properties; zinc stearate is a solid soap added for a binder.

Compact powders are formulations of a pigment blend, similar to face powder, with mineral oil (or lanolin) and organic hydroxy compounds as binders. They are pressed into cakes after thorough mixing.

Lipstick

Lipstick consists of a solution or suspension of coloring agents in a mixture of high-molecular-weight hydrocarbons or their derivatives, or both. The material must soften to produce an even application when pressed on the lips, yet the film must not be too easily removed, nor may the coloring matter run. Lipstick is perfumed to give it an odor and pleasant taste. The color usually comes from a dye or lake from the eosin group of dyes. A *lake* is a precipitate of a metal ion (Fe^{3+}, Ni^{2+}, Co^{3+}) with an organic dye. The metal ion enhances the color or changes the color of the dye in a manner similar to that of a mordant (Chapter 15). Two dyes, used in admixture and with their lakes, are dibromofluorescein

549

(yellow-red) and tetrabromofluorescein (purple):

Tetrabromofluorescein
(sodium salt)

The ingredients in a typical formulation include:

Dye	Furnishes color	4–8%
Castor oil, paraffins or fats	Dissolves dye	50%
Lanolin	Emollient	25%
Carnauba wax	Makes stick stiff by	18%
Beeswax	raising the melting point	18%
Perfume	Imparts attractive taste	1.5%

Carnauba wax and beeswax are high-molecular-weight esters. Representative alcohols and acids split out of the esters of carnauba wax by hydrolysis have 32 to 34 carbon atoms. A typical ester in carnauba wax is $CH_3(CH_2)_{30}COO(CH_2)_{33}CH_3$. This wax is extracted from the leaves of the Brazilian palm, where it occurs as the thin external coating of these leaves.

Beeswax comes from the honeycomb of bees. The alcohols and acids hydrolyzed from esters of beeswax contain 26 to 28 carbon atoms. A typical ester is $CH_3(CH_2)_{24}COO(CH_2)_{27}CH_3$. Beeswax melts between 62 and 65°C; carnauba wax melts between 83 and 88°C.

In the manufacture of lipstick, the dye is first dispersed in the castor oil, and then the waxes, lanolin, and perfume are added, while the whole is heated and stirred to obtain a homogeneous mixture. The molten mass is then cast into suitable forms and subsequently inserted into holders, passed momentarily through a flame to obtain a smooth surface, and then packaged.

Eye Makeup

There are several types of eye makeup: eyebrow pencils, mascara for eyelashes, and shading, among others. Their use varies with current fashion, though the variations can usually be traced back anywhere from 100 to 4000 years. Eyeshadow, which currently is popular, was also very popular in ancient Egypt.

Eyebrow pencils are very much like lipstick, but they contain a different coloring matter. The coloring matter is a pigment such as lampblack; the other ingredients include fats, oils, petrolatum, and lanolin, blended to give the desired melting point, which may be raised by the addition of beeswax or paraffin. Petrolatum is a semisolid mixture of hydrocarbons (saturated, $C_{16}H_{34}$ to $C_{32}H_{66}$; and unsaturated, $C_{16}H_{32}$, etc; melting point, 34 to 54°C). Brown pencils are made by using iron oxide pigments in place of lampblack.

Mascara is used to darken eyelashes and give them a longer appearance. The same colors as in eyebrow pencils are used, as well as other mineral coloring matters such as chromic oxide (dark green) and ultramarine (blue pigment of variable composition; probably a double silicate of sodium and aluminum silicate with some sodium sulfide). The coloring matter is suspended in a foundation that is a mixture of a soap, oils, fats, and waxes. The mascara may be water-soluble or water-resistant, depending upon the composition of the foundation. A typical formula consists of about 40 per cent wax (beeswax, carnauba wax, and paraffin, adjusted for hardness), 50 per cent soap (such as triethanolamine, an emulsifier and humectant), 5 per cent lanolin and 5 per cent coloring matter.

In the discussion of depilatories, the danger of the sulfide ion was emphasized. Through hydrolysis, the sulfide ion can produce a hydroxide ion, the destructive species in lye. Everyone knows what lye can do to tender, mucous tissue. If very much mascara containing sodium sulfide gets into the eye, it can cause blindness.

Take the case of *Lash-Lure,* a product sold many years ago to dye eyebrows and eyelashes. It was very toxic and the *Journal of the American Medical Association* reported at least 17 cases of *Lash-Lure* causing blindness. The active ingredient of *Lash-Lure* was a paraphenylenediamine derivative (see *Hair Coloring and Bleaching*). The blindness (or death in at least one case) was a disaster, but in addition, the severe pain was almost unbearable. Ulcers developed around the eyes: the cornea peeled off. The eyes drained constantly.

Paraphenylenediamine derivatives are still used in hair dyes. Extreme caution should be used when these dyes are applied. They should be kept away from the eyes and out of scrapes and cuts.

Eye shadow is a formulation of a coloring matter suspended in an oily-fatty-waxy base. A formula which has been used for this purpose is 60 per cent petroleum jelly, 6 per cent lanolin, 10 per cent fats and waxes (approximately equal amounts of cocoa butter, beeswax, and spermaceti), and the balance zinc oxide (white) plus tinting or coloring dyes. Cocoa butter is composed of glycerides of stearic, palmitic, and lauric acids. It is obtained from natural products by compression of cacao seed. The melting point is 30 to 35°C. Spermaceti is chiefly cetyl palmitate, $CH_3(CH_2)_{14}COO(CH_2)_{15}CH_3$. It is taken from the solid fat from the head of the sperm whale. Its melting point is 45°C.

In all these formulations, modern synthetic waxes, soaps, and fats of fixed and reproducible properties are being used successfully to obtain more standardized products. For example, the Carbowaxes, which are polyethylene glycols, are used in the formulation of protective hand creams, astringent cream, hair conditioners, and shaving creams. The molecular weights range from 1000 to 6000. They have low toxicity and are soluble in water. Since they are water-soluble, they are easy to remove after using. Their preparation begins by adding water to ethylene glycol in the presence of sodium hydroxide. The process continues until the ethylene oxide is exhausted or the NaOH is neutralized.

$$H_2O + CH_2 \!-\! CH_2 \xrightarrow{NaOH} HOCH_2CH_2OH$$

$$HOCH_2CH_2OH + CH_2 \!-\! CH_2 \xrightarrow{NaOH} HO(CH_2CH_2O)_2H$$

$$HO(CH_2CHO)_2H + CH_2 \!-\! CH_2 \xrightarrow{NaOH} HO(CH_2CH_2O)_3H$$

$$HO(CH_2CH_2O)_3H + NCH_2 \!-\! CH_2 \xrightarrow{NaOH} HO(CH_2CH_2O)_{n+3}H$$

551

Perfume

A perfume is a material containing one or more volatile constituents which can produce a desired aroma. The sense of smell is quite complex and the nose is able to distinguish a truly amazing number of different molecules. The stereo-chemical theory of smell was described and illustrated in Figure 22–10. The chemistry of perfumes is quite complex, since it involves up to 5000 different natural or synthetic materials. A typical perfume will have at least three compo-nents of somewhat different volatility and molecular weights. (Recall that lower-molecular-weight compounds are generally more volatile.) The first, called the *top note*, is the most volatile and is the obvious odor when the perfume is first applied. The second, called the *middle note*, is less volatile and is generally a flower extract (violet, lilac, etc.). The last, or *end note*, is least volatile, and is usually a resin.

Most perfumes contain many components, and chemically they are often complex mixtures. As the analysis of natural perfume materials has progressed, the use of pure synthetic organic compounds to duplicate specific odors has become very common. An example is the isolation of civetone, a cyclic ketone $[CO(CH_2)_7CH—CH(CH_2)_7]$ from civet, a secretion of the civet cat of Ethiopia. It is highly valued for perfumes. The disgustingly obnoxious odor of civet becomes pleasant in extreme dilutions. The secretion, composed of free NH_3, resin, fat, and volatile oil, is located in a double pocket pouch under the skin of the animal's abdomen, with an opening near the tail. The animal uses its musk glands for scenting tree trunks, the ground, and similar places as a means of communication so that members of the species will be able to find each other at night in the forest. The musk is also used as a means of self-defense when the civet is attacked by dangerous carnivorous animals. The foul-smelling, burning secretion is dis-charged into the enemy's face; this momentarily stuns the attacker, giving the civet time to escape.

Civetone is now available in a synthetic form. It is prepared by forming 8-hexadecene-1,16-dicarboxylic acid into a ring. The thorium ion (Th^{4+}) cata-lyzes the closure.

$$
\begin{array}{ccc}
\underset{\parallel}{\overset{\displaystyle HC—(CH_2)_7COOH}{}} & & \underset{\parallel}{\overset{\displaystyle HC—(CH_2)_7}{}} \\
HC—(CH_2)_7COOH & \xrightarrow[\Delta]{Th^{4+}} & HC—(CH_2)_7
\end{array} \diagdown C{=}O
$$

Other compounds used in perfumes include high molecular weight alcohols and esters. An example is geraniol (b.p. 230°C), a principal component of Turkish geranium oil.

$$
\begin{array}{l}
CH_3 \\
\diagdown C{=}CH—CH_2—CH_2—\underset{\underset{HC—CH_2OH}{\parallel}}{C}—CH_3 \\
CH_3
\end{array}
$$

Esters of this alcohol are used to make synthetic rose aromas for perfumes. For example, the ester formed by reaction between geraniol and formic acid has a rose type odor.

$$\underset{\substack{\text{Formic}\\\text{acid}}}{\text{H}-\overset{\displaystyle\text{O}}{\overset{\|}{\text{C}}}-\text{OH}} + \underset{\text{Geraniol}}{\text{HOCH}_2-\overset{\displaystyle\text{H}}{\overset{|}{\text{C}}}=\text{R}} \longrightarrow \underset{\text{Geranyl formate}}{\text{H}-\overset{\displaystyle\text{O}}{\overset{\|}{\text{C}}}-\text{O}-\text{CH}_2\text{CH}=\overset{\displaystyle\text{CH}_3}{\overset{|}{\text{C}}}(\text{CH}_2)_2\text{CH}=\text{C}(\text{CH}_3)_2}$$

Typical perfumes are 10 to 25 per cent perfume essence and 75 to 90 per cent alcohol. Perfumes are added to most cosmetics to give the product a desirable odor; they also mask the natural odor of other constituents. They are often mildly bactericidal and antiseptic. Some of the synthetic compounds used as fragrances in perfumes are given in Table 23–1.

Nail Polish

Nail polish is essentially a nail lacquer or varnish. It can be made of nitrocellulose, a plasticizer, a resin, and solvent. The nitrocellulose can be replaced by another polymer molecule, which possesses similar qualities. The evaporation of the solvent leaves a film of nitrocellulose, plasticizer, resin, and dye. The nitrocellulose furnishes the shiny film; the plasticizer is added to make the film less brittle; and the resin is added to make the film adhere to the nail better and prevent flaking. Perfumes are added to cover the odor of the other constituents. A typical formulation is

Nitrocellulose	15%
Acetone (solvent)	45%
Amyl acetate	30%
Butyl stearate (plasticizer)	5%
Ester gum (resin)	5%

Perfumes and colors are added as needed

Ester gum is a combination of esters—mainly glyceryl, methyl and ethyl esters of rosin. Rosin is the resin remaining after distilling turpentine from the exudate of various species of pines. It is 80 to 90 per cent abietic acid.

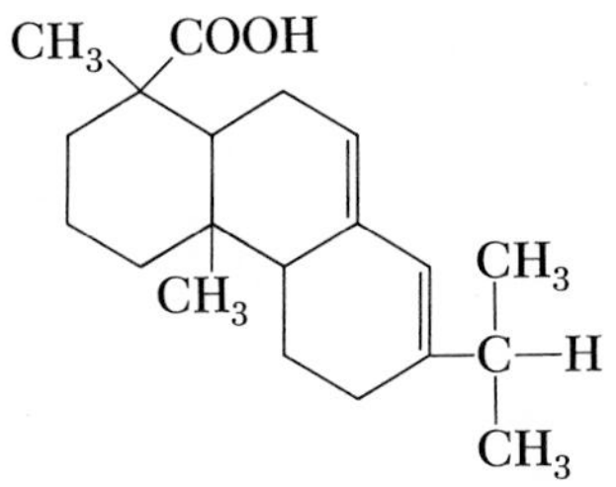

FIGURE 23-10   Abietic acid.

Abietic acid

Rosin is slightly toxic to mucous membranes and slightly irritating to the skin. Its sticky nature is well known. The ester gum is prepared by heating rosin and

TABLE 23-1 COMPOSITION OF A TYPICAL PERFUME AND COLOGNE

	FRENCH LILAC-TYPE PERFUME	
Material	*Source or Formula*	*Percent by Weight*
Jasmine, artificial	Mixture a (see below)	6.70
Pelargonic aldehyde (noranal)	$CH_3(CH_2)_7CHO$	1.02
Terpineol	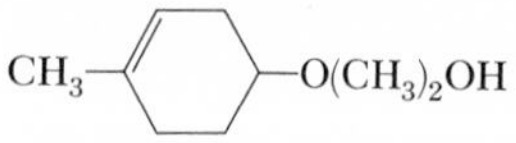 CH_3—⬡—$O(CH_3)_2OH$	3.35
Musk ketone	Musk deer found principally in mountains of Northern India and Central Asia: $CH_3COC_6(C_4H_9)(CH_3)_2(NO_2)_2$	3.35
Hydroxycitronellal	$(CH_3)_2COH(CH_2)_3CH(CH_3)CH_2CHO$	67.00
Geraniol	$CH_3C(CH_3)CH(CH_2)_2C(CH_3)CHCH_2OH$	1.75
Phenyl acetaldehyde, 50%	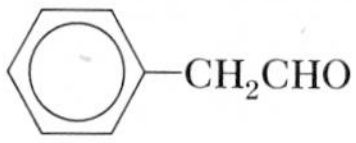 ⬡—CH_2CHO	1.75
β-Phenyl ethyl alcohol	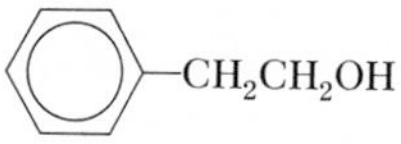 ⬡—CH_2CH_2OH	10.05
Anisic aldehyde	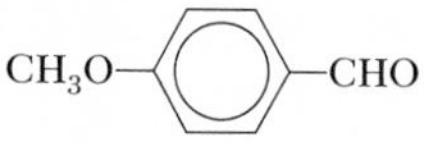 CH_3O—⬡—CHO	0.66
Rose, artificial	Mixture b (see below)	1.02
Labdanum	Cistus bushes of Cyprus Mixture c (see below)	3.35
		100.00

Mixture a

Benzyl acetate	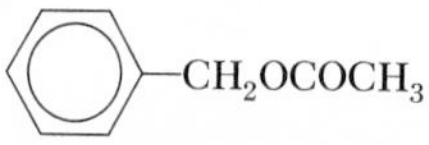 ⬡—CH_2OCOCH_3	65.0
Linalyl acetate	CH_2=$CHC(OCOCH_3)(CH_3)CH_2CH_2CH$=$C(CH_3)_2$	7.5
Linalool	CH_2=$CHCOH(CH_3)CH_2CH_2CHC(CH_3)_2$	15.5
Benzyl alcohol	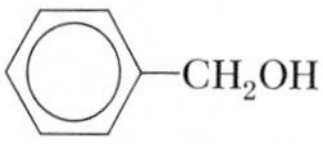 ⬡—CH_2OH	6.0
Indole	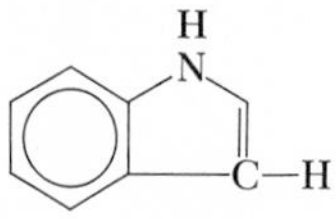 ⬡—(N-H ring)—C—H	2.5
Jasmine	A ketone 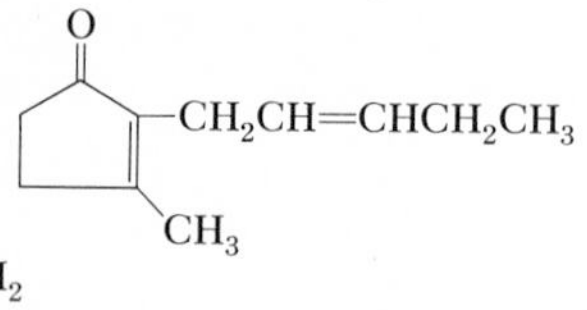 (cyclopentenone)—CH_2CH=$CHCH_2CH_3$, CH_3	3.0
Methyl anthranilate	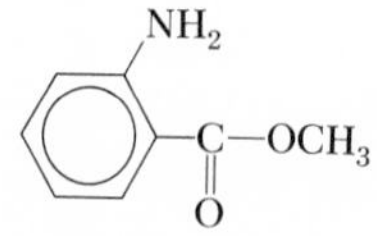 ⬡(NH_2)—C(—OCH_3)(O)	0.5
		100.00

Mixture b

A mixture of geraniol, linalool, higher homologues of phenyl ethyl alcohol, citronellol, and esters of phenyl ethyl alcohol and phenyl acetic acid.

Mixture c

A mixture of two ketones (acetophenone and 1,5,5-trimethyl-6-cyclohexanone) and other, still unidentified compounds.

TABLE 23-1   COMPOSITION OF A TYPICAL PERFUME AND COLOGNE (*Continued*)

EAU DE COLOGNE

Material	Source or Formula	Amount
Ethyl alcohol	C_2H_5OH	3 liters
Oil of bergamot	Peel of fruit of *Citrus aurantium*. Principal odorous constituent (36–45%) is linalyl acetate (above), but also contains other substances, such as 6% linalool, limonene, dipertene, and bergaptene	7 grams
Oil of lemon	Peel of lemons {90% limonene + terpene + phellandrene + pinene / 6% citral + citronellol + geranyl acetate + sesquiterpenes}	17 grams
Oil of neroli	Flowers of bitter orange. *Citrus aurantium;* subspecies, *sinensus* {90% limonene + citral + decylaldehyde + methylanthranilate + linalool + terpineol}	20 grams
Oil of rosemary	Herbs of the rosemary plant {10% borneol + 2.5% esters (bornyl actate) + camphor + eucalyptol + pinene + camphene}	7 grams

the alcohol under pressure until the esterification occurs. The gums are soluble in nonpolar solvents.

Nail polish removers are simply solvents which dissolve the film left by the nail polish. They consist largely of acetone or ethyl acetate, or both, to which small amounts of butyl stearate and diethylene glycol monomethyl ether

$$CH_2-CH_2OCH_2-CH_2$$
$$OH \qquad\qquad OCH_3$$

have been added. However, some formulations contain combinations of amyl acetate, butyl acetate, ethyl acetate, benzene, olive oil, lanolin, and alcohol. Both nail polish and nail polish removers are very flammable, and care should be taken never to use them in the presence of open flames or lighted cigarettes.

Cuticle softeners are primarily wetting agents and alkalies used to soften skin around the fingernail so it can be shaped as desired. The use of alkali to soften and swell protein is well known. A typical cuticle softener contains potassium hydroxide (3 per cent), glycerol (12 per cent), and water (85 per cent). It may also contain sodium carbonate (an alkali), triethanol amine (a detergent, used as wetting agent), and trisodium phosphate (an alkali).

Creams

Creams are generally emulsions of either an oil-in-water type or a water-in-oil type. An emulsion is simply a colloidal suspension of one liquid in another. The oil-in-water emulsion has tiny droplets of an oily or waxy nature dispersed throughout a water solution (homogenized milk is an example). The water-in-oil emulsion has tiny droplets of a water solution dispersed throughout an oil (natural petroleum and melted butter are examples). An oil-in-water emulsion can be washed off the hands with tap water, while a water-in-oil one gives the hands a greasy, water-repellant surface.

Cold cream originally was a suspension of rose water in a mixture of almond

555

oil and beeswax. Subsequently, other ingredients were experimented with in order to get a more stable emulsion. An example of a modified composition is the following cold cream: almond oil, 35 per cent; beeswax, 12 per cent; lanolin, 15 per cent; spermaceti (from whale oil), 8 per cent; and strong rose water, 30 per cent. Other oils can be substituted for some or all of the almond oil. The lanolin stabilizes the emulsion.

Vanishing cream is a suspension of stearic acid in water, to which a stabilizer is added to make the suspension more stable. The stabilizer may be a soap, such as potassium stearate. These creams do not actually vanish, they merely spread as a smooth, thin covering over the skin.

Creams of various sorts may be used as the base for other cosmetic preparations, in which case further ingredients are added, to give additional properties to the creams. As an example, hydrated aluminum chloride can be added to prepare a cream deodorant.

Suntan Lotions

The variety of suntan products ranges from lotions, which selectively filter out the higher energy ultraviolet rays of the sun, to preparations which essentially dye the skin a tan color. The lotions which filter out the ultraviolet rays are more accurately described as sunscreens, and their ingredients are often mixed with other materials, to give a lotion which both screens and tans. A common ingredient in preparations used to *prevent* sunburn is p-aminobenzoic acid.

$$H_2N-\langle\bigcirc\rangle-COOH$$

p-Aminobenzoic acid

Like most aromatic compounds, it absorbs strongly in the ultraviolet region of the spectrum (Figure 23–11). The p-aminobenzoic acid is emulsified with a mixture of alcohols, an oil, and water by a high-molecular-weight fatty acid ester. A leading suntan lotion contains monoglyceryl p-aminobenzoate (3 per cent), mineral oil (25 per cent), sorbitan monostearate and polyoxyethylene sorbitan monostearate (10 per cent; used as emulsifiers), and water (62 per cent). Perfume is added to give the material a pleasant odor.

Newer suntan lotions contain 2-ethoxyethyl-p-methoxycinnamate as the absorber of ultraviolet light. It is said to absorb the peak tanning and burning

$$CH_3O-\langle\bigcirc\rangle-CH=CHCOOCH_2CH_2OC_2H_5$$

wavelength, which is at 308 nanometers. These lotions, like all the others, do not prevent sunburn completely. They merely prolong the possible length of exposure to the sun before severe sunburn occurs.

Tanning is a process in which the skin is stimulated to increase its production of the pigment, melanin (see *Hair Coloring and Bleaching* in this chapter). At the same time, the skin thickens and becomes more resistant to deep burning.

Preparations for the relief of the pain of sunburn are solutions of local anesthetics (such as benzocaine), plus other ingredients for color, odor, and

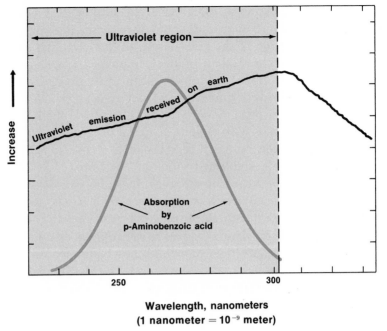

FIGURE 23-11 Absorption spectrum of p-aminobenzoic acid. Maximum absorption occurs at 265 nanometers, although it absorbs at other wavelengths as shown. The maximum of the deep-burning ultraviolet radiation received on earth is at about 308 nanometers. One nanometer is 10^{-9} meter.

softening of burned tissue. As you might expect, lanolin is especially suited for this purpose and it is used often. By softening the tissue, the emollient provides a ready access of the anesthetic and of blood plasma.

SOAPS, SYNTHETIC DETERGENTS, AND HOUSEHOLD CLEANSERS

Dirt has been defined as matter in the wrong place. Tomato catsup is esteemed as a palatable and nutritious food, but on your shirt it is dirt and you try to remove it. There is a large number of cleansing, or surface-active, agents capable of removing the dirt without harm to the shirt. Indeed, radio and television advertising might lead us to believe that the soaps and detergents we have today are unique and vastly superior to the products of a year ago. Soap, for example, has been made for a long time by a method very similar to that described in Chapter 15. Lye and lard, or fat, are heated to make soap and glycerol. This time-tested recipe was used at least as long ago as the second century of the Christian era. Galen, the great Greek physician, mentions that soap was made from fat, ash lye, and lime. Moreover, Galen stated that soap not only served as a medicament but also removed dirt from the body and clothes.

What *is* new is the greater purity of soap, the improvement in its cleaning action by numerous additives, and the advent of the relatively new synthetic detergents. The soap-making industry was revolutionized by two events. The first was the discovery of the process of making soda ash (Na_2CO_3; its water-softening and alkali properties will be discussed later) from ordinary salt (NaCl) by Nicolas LeBlanc in 1791, and the second was the epoch-making work of the celebrated French chemist Chevreul, whose researches into the chemical constitution of the natural fats extended from 1813 to 1823. As a result of these two

557

advances, the soap-makers of the 19th century were provided with ample quantities of sodium carbonate at a reasonable price and were armed with the knowledge of the true nature of fats and fatty acids. Accordingly, they made rapid progress; their products improved steadily and grew in diversity, until little by little the soap industry attained its present gigantic proportions.

The first true synthetic detergent was made by a Belgian chemist named Reychler, who, as early as 1913, reported the laboratory preparation of several compounds with definite soaplike properties. Although he reported his work in a scientific journal, it received little attention. No great need was seen to improve on a well-established product. The impetus for the development of detergents came when the Allied Forces blockaded Germany during World War I. Germany, cut off from access to imported fats and oils, sought to relieve the situation by developing soap substitutes not requiring the use of fats. The first patent for a synthetic detergent was filed on October 23, 1917, by Dr. Fritz Gunther of the company then known as the Badische Anilin und Soda Fabrik. The raw materials were derived from coal tar, with which Germany was well supplied, and from other nonfat sources. The product was finally marketed in 1925 under the trade name "Nekal." It is still being sold today and is the prototype of the largest class of synthetic detergents, the aryl-alkyl sulfonates. The major component of Nekal is sodium alkyl naphthalene sulfonate.

$$\text{(naphthalene)}-(CH_2)_x-\overset{\displaystyle O}{\underset{\displaystyle O}{S}}-O^-,\ Na^+$$

The values of x are generally 12 to 14.

Since the preparation and detergent nature of soap were described earlier, in Chapter 15, we shall limit our discussion in this section to some of the additives and other properties of cleansing agents that should be of interest to the consumer.

SOAPS

Soaps, you will recall, are salts of long-chain fatty acids. The nature (length of the chain and number of double bonds) of the fatty acid chain is largely determined by the glyceride used, which ultimately determines the unique properties of the soaps. The principal glycerides used for soap-making are derived from both plants and animals:

(1) Tallow or animal fat from beef or mutton is primarily an ester of stearic acid [$CH_3(CH_2)_{16}COOH$]. It is usually mixed with coconut oil in making soap to prevent the product from being too hard.

(2) Coconut oil is a low melting solid. It is primarily an ester of lauric acid [$CH_3(CH_2)_{10}COOH$]. A soap made from coconut oil alone is very soluble in water and will lather even in sea water.

(3) Palm oil contains a very high concentration of free fatty acids, about 45 to 50 per cent of which is oleic acid [$CH_3(CH_2)_7CHCH(CH_2)_7COOH$]. It is an important constituent in toilet soaps.

(4) Olive oil is used in making Castile soap. It has a larger percentage (70 to 85 per cent) of esters of oleic acid than palm oil.

(5) Bone grease is an animal fat of somewhat lower melting point than tallow, and it comes from a variety of sources. It is a relatively cheap source of fat. The esters of oleic acid (41 to 51 per cent) are most prominent.

(6) Cottonseed oil is also a cheap source of glycerides for making soap. Its esters are mostly of linoleic acid $[CH_3(CH_2)_4(CHCHCH_2)_2(CH_2)_6COOH]$.

The length and degree of saturation of the fatty acid chain influence the solubility and hardness of the soap. A saturated, long-chain fatty acid makes a harder, more insoluble soap.

In large soap-making factories, after most of the glycerol is separated for sale, salt (NaCl) is added to facilitate the separation of the soap from the water present (Figure 23–12). The hot soap can then be poured into molds, cooled, and cut into bars, which contain about 30 per cent water.

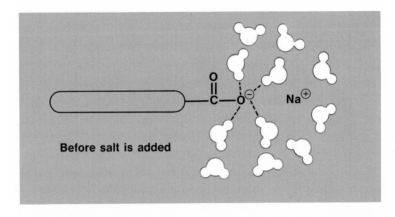

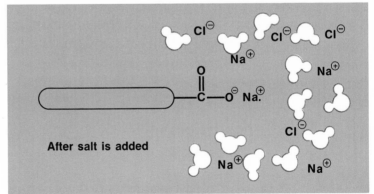

FIGURE 23–12 Salting out (shown diagrammatically). Without salt, there is sufficient water of hydration to keep the soap dissolved. When the salt is added, the equilibrium between the soap anions and the positive ions is shifted toward more neutral soap molecules, which decreases the hydration and renders the soap less soluble.

Fillers or Builders

A number of materials are added to soap powders, especially if they are to be used for laundry purposes. These materials are often quite basic, and their addition gives the soap a greater detergent action. Commonly added materials include sodium carbonate, sodium phosphates, sodium polyphosphates, and sodium silicate. Rosin neutralized with sodium hydroxide is also commonly added to laundry soaps in large amounts. Such soaps are not to be recommended for use on the human skin. Phosphates, carbonates, and silicates hydrolyze to give OH^- ions, which react with grease to make soaps.

559

$$PO_4^{3-} + H_2O \longrightarrow HPO_4^{2-} + OH^-$$

$$CO_3^{2-} + H_2O \longrightarrow HCO_3^- + OH^-$$

$$SiO_3^{2-} + H_2O \longrightarrow HSiO_3^- + OH^-$$

The rosin is mostly abietic acid (Figure 23–10). The neutralized acid has the nonpolar (hydrocarbon) part and polar end required for a soap.

The effect of phosphate as a pollutant is discussed in Chapter 20. The other ions are not considered serious pollutants.

TYPES OF SOAPS

The variety of soaps available result from relatively minor variations in the soap-making process or in the ingredients added.

Toilet Soaps

These generally have little or no filler and a minimal amount of free base, if any. Often much of the glycerol released in the saponification process is left in the soap. Perfumes, dyes, and medicinal agents may be added prior to casting the soap into a solid form. Floating soaps have air beaten into them as they solidify. A hard soap is obtained if there is a high percentage of a sodium salt of a relatively long-chain fatty acid, such as stearic acid, present. A soft or liquid soap is obtainable by saponification with potassium hydroxide, with the liquidity increasing as the chain length of the fatty acid decreases. Fatty acids with chains as short as C_{12} or shorter are not used because the resultant soaps irritate the skin. They are more volatile and create an odor problem. Fatty acids with chains longer than stearic acid tend to give very insoluble soaps.

Shampoos

These are often mixtures of several ingredients designed to satisfy a number of requirements. In addition to soaps, condensation products from diethanolamine and lauric acid are often used. These are essentially a type of detergent obtained by the reaction

$$\underset{\text{Diethanolamine}}{HN(CH_2CH_2OH)_2} + \underset{\text{Lauric acid}}{CH_3(CH_2)_{10}COOH} \xrightarrow{\Delta} \underset{\text{An amide (detergent)}}{CH_3(CH_2)_{10}\overset{\overset{\text{O}}{\|}}{C}-\overset{\overset{\text{H}}{|}}{N}(CH_2CH_2OH)_2}$$

Shampoos also contain compounds to prevent the calcium or magnesium in hard water from forming a precipitate; EDTA is often used for this purpose. (For a discussion of its chelating action, see Chapter 22.) Lanolin and mineral oil are often added, and these give the product a cloudy appearance, in addition to keeping the scalp from drying out and scaling.

Talcum Powders

Face powders and other talcum powders are generally composed of a relatively high percentage of talc, a naturally occurring hydrated magnesium silicate. The most common other ingredient is zinc stearate. While zinc stearate is a "soap," it is quite different from the common soaps in that it is water repellant and is generally used as a fine, soft, bulky powder. Other materials, such as dyes, perfumes, and boric acid (mildly germicidal), are added to obtain the desired properties in the final product.

Other Soaps

Zinc stearate is only one of a large number of metallic soaps manufactured and used for special purposes. These are soaps of metal ions which are not used in regular consumer soaps. For example, lead stearate, a very poisonous soap, is used as a high pressure lubricant and as a catalyst to accelerate the drying of varnishes. Cadmium soaps are used in waterproofing and copper soaps in the manufacture of fungicides. In general, the soaps made with metallic ions with a charge of $+2$ or greater are very insoluble in water and are more like a grease (for which purpose some of them find considerable application).

SYNTHETIC DETERGENTS

Synthetic detergents ("syndets") are derived from organic molecules which have been designed to have the same cleansing action, but not the same reaction, as soaps with the cations found in hard water, such as Ca^{+2}, Mg^{+2} and Fe^{+3}. As a consequence, synthetic detergents are equally effective in hard and soft water, while a soap used in hard water will give a precipitate of the Ca^{+2}, Mg^{+2} or Fe^{+3} salt of the long-chain fatty acid. Since such precipitates have no cleansing action and tend to stick to laundry, their presence is very undesirable.

There is an enormous number of different synthetic detergents on the market. Their molecular structure consists of a long oil-soluble (hydrophobic) group and a water-soluble (hydrophilic) group. The hydrophilic groups include the sulfate ($-OSO_3-$), sulfonate ($-SO_3-$), hydroxyl ($-OH$), ammonium ($-NH_3$), and phosphate [$-OPO(OH)_2$] groups.

The great majority of synthetic detergents were originally of the sodium alkyl sulfate type. The preparation of sodium lauryl sulfate, the essential ingredient of some early common household synthetic detergents, illustrates the chemical processes involved. This amplifies the discussion given in Chapter 15. The principal starting material is a suitable vegetable oil, such as cottonseed oil or coconut oil. The first step is to treat the dry oil, dissolved in toluene, with sodium dispersed in toluene:

$$\text{Coconut oil} + \underset{\text{Dispersion}}{\text{Na}} \longrightarrow \text{Glycerol} + \text{Alcohols} + H_2$$

(Water free in toluene) [Mainly lauryl alcohol, $CH_3(CH_2)_{11}OH$]

The second step involves putting the more polar hydrogen sulfate group on the end of the hydrocarbon chain. This is accomplished by treating the lauryl alcohol with sulfuric acid.

$$CH_3(CH_2)_{11}OH + H_2SO_4 \longrightarrow CH_3(CH_2)_{11}OSO_3H + H_2O$$

Lauryl alcohol   Sulfuric acid   Lauryl hydrogen sulfate

The final step involves neutralizing the acidic lauryl hydrogen sulfate with sodium hydroxide:

$$CH_3(CH_2)_{11}OSO_3H + NaOH \longrightarrow \underset{\textit{Sodium lauryl sulfate}}{CH_3(CH_2)_{11}OSO_3Na} + H_2O$$

Another group of synthetic detergents (also called "surfactants," from *surface active agents*) is the alkylbenzene sulfonates. These are prepared by putting large alkyl groups on a benzene ring and then sulfonating the benzene ring with sulfuric acid or a related reagent. Before use, they are transformed into their sodium salts. The reactions involved are:

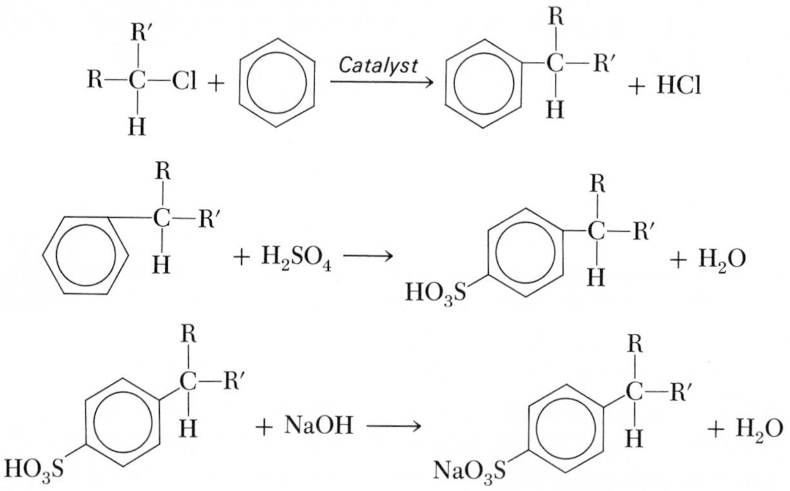

Sodium alkylbenzenesulfonate

These molecules, like all others in this class, consist of a long hydrophobic chain and a highly polar group which interacts strongly with water.

In addition to the anionic (negatively charged) synthetic detergents already described, in which the polar group is an anion, there are also detergents in which the polar group at the end of the hydrocarbon chain is positive or neutral.

Cationic (positively charged) detergents are almost all quaternary ammonium halides:

$$\begin{array}{c} R_2 \\ | \\ R_1{-}N{-}R_3{}^{+} \\ | \\ R_4 \end{array} \quad X^-$$

where one of the R groups is a long hydrocarbon chain and another frequently includes an —OH group. In these the water-soluble portion is positively charged; so they are sometimes called invert soaps (in soaps the water soluble portion is negatively charged). They are prepared by treating the appropriate amine with an alkyl chloride:

$$\begin{array}{c} R_2 \\ | \\ R_1{-}N{-}R_3 \end{array} + R_4Cl \longrightarrow \begin{array}{c} R_2 \\ | \\ R{-}N{-}R_3{}^{+} \\ | \\ R_4 \end{array} \quad Cl^-$$

They frequently exhibit pronounced bactericidal qualities. Cationic detergents are incompatible with anionic detergents. When they are brought together, a high molecular weight insoluble salt precipitates out, and this has none of the desired detergent properties of either starting material:

$$R_1 \overset{\overset{\displaystyle R_2}{|}}{\underset{\underset{\displaystyle R_4}{|}}{N}} -R_3^+ \quad Cl^- + Na^+ \quad O_3SR_5^- \longrightarrow R_1 \overset{\overset{\displaystyle R_2}{|}}{\underset{\underset{\displaystyle R_4}{|}}{N}} -R_3^+, \ O_3SR_5^- + Na^+ + Cl^-$$

Precipitate

Nonionic detergents bear a polar, but not an ionic, grouping attached to a large organic grouping of low polarity. A typical example is a material prepared by the reaction of an acid with ethylene oxide:

$$RCOOH + (x + 2)CH_2 \overset{\displaystyle O}{\underset{}{=\!\!\!\triangle\!\!\!=}}CH_2 \longrightarrow RC \overset{\displaystyle O}{\overset{\|}{}}-O-(CH_2)_2O(CH_2-O)_x-CH_2CH_2OH$$

In a typical nonionic detergent, $R = C_{12}H_{23}$ and $x = 2$. The large number of weakly polar C—O—C bonds has an effect similar to that of a single ionic group, and this end of the molecule provides the water solubility.

The nonionic detergents have several advantages over ionic detergents. Since they contain no ionic groups, they cannot form salts with calcium and magnesium ions and are, consequently, unaffected by hard water. For the same reason, they do not react with acids and may be used even in strong acid solutions.

In general, the nonionics foam less than ionic surface active agents, a property which is desirable where nonfoaming detergents are required, as in dishwashing. Nonionics do suffer from one drawback. They cannot be dried to solid powders. They are heavy liquids with melting points below room temperature. Consequently, nonionic detergents must be used in the form of water solutions which are far less convenient to handle than powdery materials.

Enzyme detergents are simply detergents to which enzymes capable of degrading the molecules of typical stains have been added. For example, lipases hydrolyze relatively insoluble fats to mixtures of glycerol and fatty acids. These have been used in laundering cotton for some time. The glycerol is water-soluble, and the fatty acids are converted to water-soluble soaps by fillers in the detergents. Some stains, such as blood, are proteins. Proteases are included in detergents and presoaks for decomposing protein stains into smaller fractions that either are water-soluble or can be mechanically washed away because their attachment to the fabric has been broken. A mechanism whereby enzymes can facilitate the hydrolysis of fats and proteins is described in Chapter 18.

Enzyme detergents and presoaks are being examined very carefully for possible health hazards. Several cases of dermatitis (irritation and breaking out of the skin) on the hands have been tentatively associated with the use of enzyme detergents. At this time, there is no conclusive evidence that enzyme detergents present a hazard to the consumer which would justify label warnings or other action under the Federal Hazardous Substances Act.

Biodegradability

After a detergent molecule has been used for cleansing, it is washed out with rinse water and enters a sewage system or river. If the normal microorganisms present in such systems can transform the detergent molecule to CO_2 and H_2O in a reasonable time, the detergent is said to be *biodegradable*. Soaps are typical biodegradable molecules. Many of the newer synthetic detergents are not readily biodegradable. We find that straight chain hydrocarbon derivatives are easily biodegraded, but when we build up substituents on the chain, the biodegradability is decreased. Many such detergent molecules which contain aromatic hydrocarbon rings are very resistant to bacterial destruction. The ionic group on the detergent may also influence its biodegradability. The interest in pollution has prompted the removal of several detergents from the market which are not readily biodegradable.

Other Cleansers

A very large number of special cleaners or cleansing agents is available. Simple abrasive cleansers contain a large percentage of an abrasive such as silica (SiO_2) or pumice (65 to 75 per cent SiO_2, 10 to 20 per cent Al_2O_3), a variable amount of soap, and generally some polyphosphates. They may also contain some synthetic detergent and a bleaching agent. All-purpose solid cleansers may contain one or more of a variety of salts which react with water to produce a basic solution: trisodium phosphate, sodium carbonate, sodium bicarbonate, sodium pyrophosphate, or sodium tripolyphosphate, plus a detergent and perhaps pine oil to give an attractive odor. Metal cleansers may contain strong acid or strong base to dissolve impurities. Many cleaning liquids contain organic solvents such as perchloroethylene, 1,1,1-trichloroethane, and the like. The vapors of these can be quite toxic when the cleaners are used in an unventilated space.

BLEACHING AGENTS AND OPTICAL BRIGHTENERS

Bleaching agents are compounds which are used to remove color from textiles. Most commercial bleaches are oxidizing agents such as sodium hypochlorite. Optical brighteners are quite different since they act by converting a portion of the invisible ultraviolet light, which impinges on them, into visible blue or blue-green light, which is emitted.

In earlier times textiles were bleached by exposure to sunlight and air. In 1786, the French chemist Berthollet introduced bleaching with chlorine, and subsequently this process was carried out with sodium hypochlorite, an oxidizing agent prepared by passing chlorine into aqueous sodium hydroxide:

$$2Na^+ + 2OH^- + Cl_2 \longrightarrow \underline{2Na^+ + OCl^-} + Cl^- + H_2O$$

Sodium hypochlorite

Shortly after this, hydrogen peroxide was introduced as a textile bleach. Subsequently, a number of other compounds which contain oxidizing agents based on chlorine were developed and introduced. Regular laundry bleaching solutions still consist of sodium hypochlorite and water, but solid compounds which release hypochlorite on hydrolysis have been developed for use in detergents. Two of these which are used are trichloroisocyanuric acid and potassium dichloroisocyanurate:

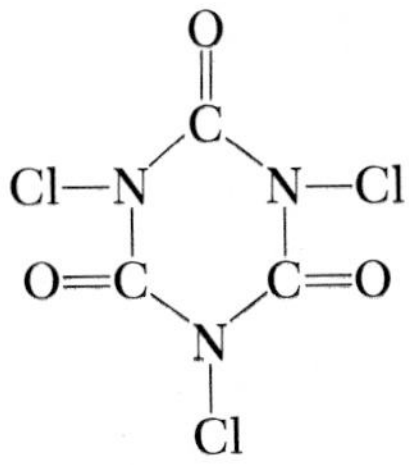

Trichloroisocyanuric acid 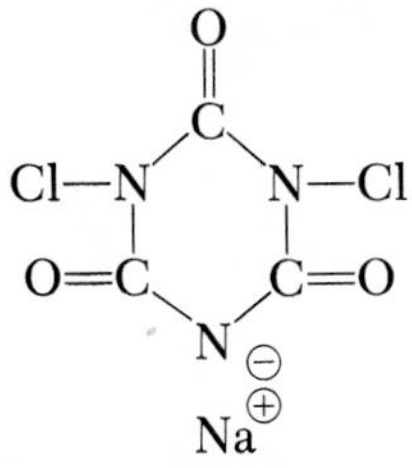   *Potassium dichloroisocyanurate*

The reaction by which hypochlorite is generated by these compounds is:

$$N-Cl + HOH \longrightarrow N-H + OCl^- + H^+$$

One way to decolorize materials is to remove or immobilize those electrons in the material which are activated by visible light. The hypochlorite ion is capable of removing electrons from many colored materials. In the process hypochlorite is converted to chloride and hydroxide.

$$ClO^- + H_2O + 2e^- \longrightarrow Cl^- + 2OH^-$$

Optical brighteners are compounds which transform incident ultraviolet light into emitted visible light. This phenomenon is a type of fluorescence. When optical brighteners are incorporated into textiles or paper, they make the material appear brighter and whiter. An example of such a brightener has this structure:

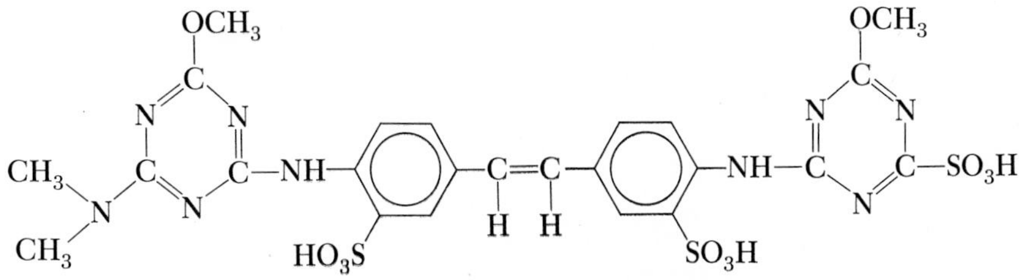

and its absorption and emission spectra are presented in outline form in Figure 23–14.

ORGANIC SOLVENT CLEANSERS

Most cleansing problems involve dissolving fats, greases, or organic stains, and these can often be solved by the appropriate choice of organic solvents. Dry cleaning involves the use of such solvents as chlorinated and fluorinated hydrocarbons and a petroleum distillate of hydrocarbons (boiling point 177 to 210°C) related to gasoline called "Stoddard solvent." Carbon tetrachloride, CCl_4, was very popular for drycleaning at one time, but it is quite toxic to the liver and kidneys and has now been replaced by perchloroethylene, $Cl_2C=CCl_2$, and

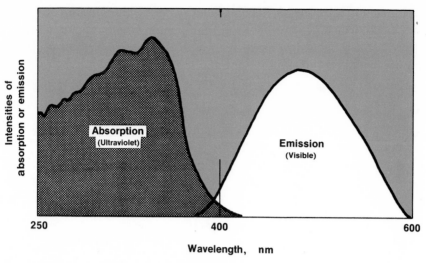

fluorinated hydrocarbons. One mixture that boils at 47°C is very compatible with speciality fabrics and dyes and is often used in coin-operated drycleaning units.

In a typical drycleaning operation, the cloth is tumbled with the solvent, which dissolves out fat and grease that binds the dirt to the fabric. Subsequent steps remove the solvent, which is recovered and reused. Because drycleaning solvents do not interact much with wool, cotton, or other polar fibers, the fibers *do not* absorb the solvent and swell during the process as they do during laundering in water. Oily stains can also be removed by turpentine in some cases, though this has been replaced in commercial operations by much more effective solvents, such as TCE (1,1,1-trichloroethane, Cl_3CCH_3), "Stoddard solvent," and the fluorinated hydrocarbons.

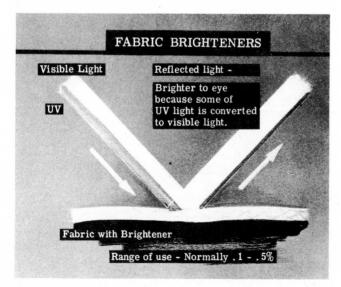

FIGURE 23-14   An optical brightener converts ultraviolet energy to visible light; hence, more light can be detected by the eye.

Stain Removers

To a large extent stain removal procedures are based on solubility patterns or chemical reactions. Many stains, such as those due to chocolate or other fatty foods, can be removed by treatment with the typical drycleaning solvents.

Stain removers for the more resistant stains are almost always based upon a chemical reaction between the stain and the essential ingredients of the stain remover. A typical example is an iodine stain remover, which is simply a concentrated solution of sodium thiosulfate. The reaction here is

$$I_2 + 2Na_2S_2O_3 \longrightarrow \underbrace{2\ NaI + Na_2S_4O_6}$$

Soluble in water
(colorless)

Iron stains are removed by treatment with oxalic acid, which forms a soluble complex with the iron:

$$Fe_2O_3 + 6\ H_2C_2O_4 \longrightarrow \underbrace{3H_2O + 2Fe(C_2O_4)_3{}^{3-} + 6H^+}$$

Oxalic acid
Soluble in water

Mildew stains can be removed by hydrogen peroxide or laundry bleach (sodium hypochlorite), which oxidizes the fungus responsible for the mildew. Blood stains on cotton can be removed by hypochlorite solution. Bleach should

TABLE 23-2   SOME COMMON STAINS AND STAIN REMOVERS[*]

STAIN	STAIN REMOVER
Coffee	Sodium hypochlorite
Lipstick	Isopropyl alcohol, isoamyl acetate, Cellosolve $(CH_2OHCH_2OCH_2CH_3)$, chloroform
Rust and ink	Oxalic acid, methyl alcohol, water
Airplane cement	50/50 amyl acetate and toluene or acetone
Asphalt	Benzol (benzene) or carbon disulfide
Blood	Cold water, hydrogen peroxide
Berry, fruit	Hydrogen peroxide
Grass	50/50 amyl acetate and benzol or sodium hypochlorite or alcohol
Nail polish	Acetone
Mustard	Sodium hypochlorite or alcohol
Antiperspirants	Ammonium hydroxide
Perspiration	Ammonium hydroxide, hydrogen peroxide
Scorch	Hydrogen peroxide
Soft drinks	Sodium hypochlorite
Tobacco	Sodium hypochlorite

[*] Before any of these stain removers are used on clothing, the possibility of damage should be checked on a portion of the cloth that ordinarily is hidden.

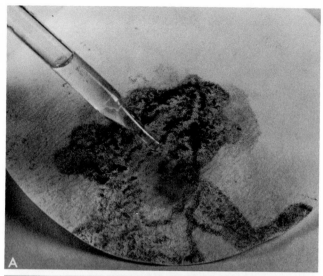

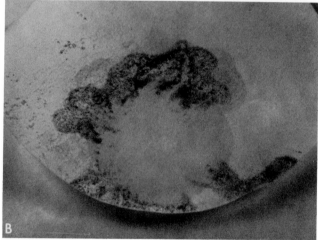

FIGURE 23-15   Oxalic acid can easily remove iron rust stains as these photos illustrate. (a) Concentrated oxalic acid is dropped on a rust-stained piece of paper. (b) In less than a minute the stain is gone where the oxalic acid solution was applied. CAUTION: Oxalic acid is a poison and must be handled with care.

not be used on wool because it reacts chemically with the nitrogen atoms present in the peptide chains. The chemicals used to remove a few common stains are listed in Table 23–2.

QUESTIONS

1. You read in the newspaper about a new compound which will break disulfide bonds in proteins. What potential use might it have?

2. (a) What is the purpose of an emulsifier? (b) In which of the following cosmetics is an emulsifier important: suntan lotion, fingernail polish remover, face powder, hair spray, cold cream? (c) What structural qualities qualify sorbitan monostearate as an emulsifier?

3. Which of the following properties would be appropriate for a hair spray propellant?

    (a) High or low boiling point?
    (b) Soluble or insoluble in the active ingredients?
    (c) Capable of chemical reaction with the active ingredients?
    (d) Odor?
    (e) Toxicity?

4. What is the monomer unit in polyvinylpyrrolidone?

5. What is the purpose of the following:

   (a) Freon in hair sprays?
   (b) Polyvinylpyrrolidone in hair sprays?
   (c) Aluminum chloride in deodorants?
   (d) Paraffin in mascara?
   (e) p-Aminobenzoic acid in suntan lotion?
   (f) Trisodium phosphate in detergents?

6. What specific substance is broken down during the bleaching of hair?

7. A detergent label indicates that the material contains 20 per cent $Na_2CO_3 \cdot 10H_2O$. What is the percentage of water in the detergent from this source?

8. Why are detergents better cleansing agents than soaps in regions where the water supply contains calcium or magnesium salts?

9. How much sodium hydroxide is needed to saponify 10 grams of glyceryl tristearate? The reaction here is

10. What quantity of oxalic acid is needed to remove a stain of Fe_2O_3 which weighs 0.01 gram? The reaction is

$$Fe_2O_3 + 6H_2C_2O_4 \longrightarrow 2Fe(C_2O_4)_3^{3-} + 3H_2O + 6H^+$$

11. Suggest ways of removing the following from clothing: (a) motor oil, (b) iodine stain, (c) lard, (d) copper sulfate.

12. Explain why vinegar is able to remove some stains which are soluble in weak acids.

13. If you prepared a shoe wax, which of the following chemicals would you use? Specify a reason for your choice. Lanolin, an ester, spermaceti, carbon black, water, para-aminobenzoic acid, polyvinylpyrrolidone, silicones.

14. Use the structures of the constituents of lanolin to justify its ability to emulsify face creams.

15. Why is a soap made from coconut oil more soluble in water than a soap made from palm oil?

16. Search recent issues of *Science, Chemical and Engineering News,* and other scientific journals for the present status of enzyme detergents as regards their safety and extent of use.

17. (a) Write an equation for the chemical reaction between the calcium ion (Ca^{2+}) and soap. (b) Write an equation for the chemical reaction between the calcium ion (Ca^{2+}) and the water softener, Na_2CO_3. (c) A washtub contains 40 liters of water. The water has run through a limestone region and picked up 0.1 mole calcium ion per liter. What is the maximum amount (in grams) of soap powder [$NaOCO(CH_2)_{16}CH_3$, molecular weight: 306] required to precipitate the calcium ions?

569

18. Hydrogen-bonding is a very handy theoretical tool. Name three applications of hydrogen bonding in consumer products.

19. Explain how optical brighteners work.

20. If you were going to formulate a suntan lotion, what particular spectral property would you look for in choosing the active compound?

21. A perfume contains paracresyl acetate [$CH_3C_6H_4OCOCH_3$, molecular weight: 150], musk ketone (see Table 23–1, molecular weight 294), and oil of carnation. Based on the data given and the discussion in this chapter, identify the top note, middle note, and end note (fixative).

22. A typical hair spray for women *or* men has a formula:

resin (like PVP)	3.0%
plasticizer	0.2%
ethyl alcohol	26.0%
propellant (Freon 11/12)	70.8%
perfume (to suit application)	negligible
	100.0%

Costs on these ingredients vary with the number of cans of hair spray a manufacturer produces. Assume the resin costs $.90 per lb., the plasticizer, $1.10 per lb., the ethyl alcohol, $0.15 per lb., and the Freon 11/12 mixture $0.45 per lb. Compute the costs of the ingredients of a 16 oz. (wt.) can of hair spray.

SUGGESTIONS FOR FURTHER READING

Bennett, H., "The Chemical Formulary," Chemical Publishing Co., New York, 1933–1965. (Twelve volumes of formulas for making soaps, cosmetics, perfumes, and so forth.)

Bennett, H., "Chemical Specialities," Chemical Publishing Company; New York, 1969. (A detailed practical guide on how to set up a chemical specialities business to manufacture and sell cosmetics, herbicides, and so forth.)

"Enzymes in Detergents," *Chemistry*, Vol 43, No. 2, p. 25 (1970).

Gleason, M. N., Gosselin, R. E., Hodge, H. C., and Smith, R. P., "Clinical Toxicology of Commercial Products" Third Edition, The Williams & Wilkins Co. Baltimore, 1969. (Contains a wealth of information on the toxic aspects of various commercial products as well as their composition.)

Harry, R. G., "Modern Cosmetology," Chemical Publishing Co., New York, 1947. (A source of formulas and manufacturing techniques for almost all standard types of cosmetics.)

Kirk, R. E., and Othmer, D. F., "Encyclopedia of Chemical Technology," Second Edition, Interscience Publishers, New York, 1963. (This contains a wealth of detailed information on various aspects of applied chemistry.)

"On Getting a Suntan," *Chemistry*, Vol. 42, No. 3, p. 5 (1969).

Sax, N. Irving, "Dangerous Properties of Industrial Materials," Third Edition, Reinhold Book Co., New York, 1968.

"The Smell of Detergents," *Chemistry*, Vol 40, No. 5, p. 10 (1967).

CONSUMER CHEMISTRY
—POTPOURRI

== **CHAPTER 24**

As you study the chemistry of automotive products, paints, and photography in this chapter, you will see numerous examples of the chemical similarities among these three important categories of consumer products and also between these products and food, medicine, cosmetics, and cleansers. While each area is unique in the intended use of its chemicals and, to some degree, unique in its assortment of chemicals, the whole of consumer-product chemistry is blended together by the underlying thread of molecular interpretations of the actions of the various agents in the products. Antioxidants, wetting agents, emulsifying agents, dyes, perfumes, oxidizing agents, reducing agents, and chelating agents, to name only a few, are found in many consumer products. While the choice of the antioxidant in potato chips and motor oils varies, the mechanism of action of the two antioxidants will quite likely be very much alike. Each product usually claims its uniqueness by the *degree* to which its components accomplish their intended purpose and less often by the molecular mechanisms of its chemical reactants.

AUTOMOTIVE PRODUCTS

A significant amount of chemistry is involved in the production and operation of the automobile. Before an automobile can exist, metals must be won from their ores; plastics must be synthesized and fabricated; paints have to be formulated; sulfuric acid for the battery must be made; rubber must be synthesized, vulcanized, and formed into shape; and glass has to be mixed, fired, molded, and cut. Many chemical reactions are required to power the automobile. The two most familiar are the combustion of the fuel and the electrochemistry of the battery. In its wake the automobile leaves its chemical exhaust to undergo a variety of chemical reactions, such as smog formation (Chapter 21). In this section, we will discuss some of the chemicals that consumers buy for their automobiles.

571

Gasoline

The amount of gasoline burned in the United States is astronomical. During the summer months, on the average, more than 12 million gallons are consumed per hour. This large amount will probably increase in the years to come as it has over the past several years.

The ideal gasoline would be one that burns completely to carbon dioxide and water at the right time and leaves no deposits in the engine. It would have sufficient vapor pressure (from lower boiling fractions) to ignite when the spark plug sparks, but not enough vapor pressure to cause a vapor lock in the fuel lines or fuel pump and stop the flow of gasoline. Moreover, it would have low gum and sulfur content. Gum formation is the result of oxidation and polymerization of unsaturated fuel components, particularly hydrocarbons with two or more double bonds per molecule. Most motor gasolines contain less than 3 mg of gum per 100 ml. Thiols, or mercaptans (—SH groups), present in crude oil sometimes find their way into the finished gasoline to provide the sulfur.

Distillation alone does not produce a gasoline with all of the desired properties. Modern gasolines contain a blend of additives that supposedly bolster its qualities. Antiknock agents such as tetraethyl lead and tetramethyl lead are added—up to 4.23 g Pb per gallon of gasoline. These additives prevent the uneven burning of gasoline caused by the ignition of low molecular weight hydrocarbons. The abrupt energy release from precombustion reactions and detonation of the low molecular weight fractions can cause high-frequency pressure fluctuations throughout the combustion chamber, which register on the ear as the sharp metallic noise called "knock." Tetraethyl and tetramethyl lead combine with the smaller molecular fragments and prevent their reaction before the advancing flame front from the spark plug reaches them. Chemical scavengers (for example, ethylene dibromide and ethylene dichloride) are added to gasoline to remove the small amount of lead that might remain in the engine after combustion of the fuel.

A typical antiknock mixture for gasolines consists of about 62 per cent tetraethyl lead, 18 per cent ethylene dibromide, 18 per cent ethylene dichloride, and 2 per cent other ingredients, such as dye, petroleum solvent, and stability improver. Gasolines with appreciable amounts of aromatics (20 to 30 per cent) are less prone to produce knocking. It is important to reduce the knock in automobiles because prolonged knock overheats valves, spark plugs, and pistons and can substantially shorten their lives.

Miscellaneous Gasoline Additives

Other chemicals are added to gasoline to prevent the ignition of new fuel by glowing particles from a previous ignition. These additives are called *deposit modifiers*. Phosphorous compounds, such as tricresyl phosphate and, more recently, boron compounds, have been used for this purpose. These alter the composition of the deposits and make them less likely to glow. The phosphorous compounds also prevent spark plug deposits from becoming so electrically conductive that the charge leaks away instead of firing the plug. Boron compounds raise the octane number of the gasoline by preventing sulfur compounds from rendering tetraethyl lead ineffective.

Antioxidants such as phenylenediamine, aminophenols, dibutyl-p-cresol, and orthoalkylated phenols, are added to prevent the formation of peroxides that lead to knock and gum formation. About 2 or 3 pounds of these additives are added to every 1000 barrels of gasoline. The mechanism of antioxidation of automotive oils is very similar to the mechanism described for BHA and BHT in food additives (Chapter 22). Because copper ions catalyze gum formation, metal ion scavengers such as ethylenediamine are added to chelate the trace amounts of copper ions and render them ineffective. The copper gets into the gasoline from the copper tubing used for fuel lines and from brass (a Cu and Zn alloy) parts of the engine.

To inhibit water from corroding and rusting storage tanks, pipelines, tankers, and fuel systems of engines, *antirust agents* are added. Four compounds used to prevent corrosion are trimethyl phosphate, sodium and calcium sulfonates, and N, N′-di-sec-butyl-p-phenylenediamine. These agents coat metal surfaces with a very thin protective film that keeps water from contacting the surfaces, as shown in Figure 24–2. This also helps prevent gummy deposits in the carburetor and combats carburetor icing during cold weather.

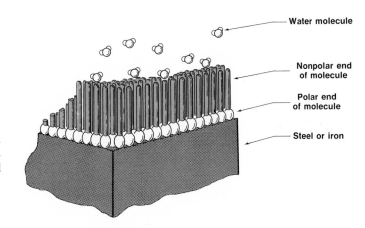

Water molecule

Nonpolar end of molecule

Polar end of molecule

Steel or iron

FIGURE 24-1   The action of surface active agents, such as rust inhibitors, mild antiwear agents, and some de-icing agents.

Anti-icing agents coat metal surfaces, as do antirust agents, and thus prevent ice particles from accumulating on surfaces, and/or depress the freezing point. The freezing point depressants, which include alcohols and glycols, act in the same manner as the antifreeze in the engine's cooling system. The small ice particles pass harmlessly through the carburetor and into the engine where the heat converts them to water vapor that eventually exits with the exhaust gases.

Detergents, which include alkylammonium dialkyl phosphates, are added to prevent the accumulation of high-boiling components on the walls of the carburetor. These deposits interfere with the air flow into the carburetor and cause rough idling, frequent stalls, poor performance, and increased fuel consumption. The effectiveness of these detergents stems from their surface active properties, as shown in Figure 24–1. The film of detergent provides a thin, nonpolar coating on the metal surfaces which prevents high molecular weight, nonpolar gums from forming thick deposits on the surfaces.

Upper cylinder lubricants are sometimes blended into gasoline. These are usually light mineral oils or low viscosity naphthenic distillates (such as cyclopentane) that dissolve away deposits in the intake system from the carburetor, cylinders, top piston rings, and valves. (Nonpolar deposits dissolve in nonpolar solvents.)

At one time *dyes* were added to distinguish antiknock "ethyl" gasolines from nonantiknock gasolines. Always present in very small amounts, they are now used to distinguish grades and brands.

A common misconception about gasolines is the relationship between *energy content* and *octane number* (Chapter 15). All grades of gasoline, whether regular or high test, have almost identical energy contents, about 29,000 kilocalories per gallon if burned completely. Power losses are caused by incomplete burning and mechanical knocking due to preignition. The difference is not their energy content but their octane rating and completeness of burning, which are determined by the amount of low molecular weight hydrocarbons as well as the kind and amount of additives present. Regular gasoline is about 91 to 94 octane and high test is about 97 to 101.

LUBRICANTS AND GREASES

Lubricants have been used to separate moving surfaces, and thus minimize friction and wear, for a long time. Even before 1400 B.C., animal tallow was used to lubricate chariot wheels. Petroleum lubricating oils and greases came into wide-scale use after the famous Drake well was drilled at Titusville, Pennsylvania, in 1859. The use of additives in lubricants has progressed rapidly since about 1930; synthetic lubricants have developed largely since World War II.

Lubricating oils from petroleum consist essentially of complex mixtures of hydrocarbon molecules. These generally range from low viscosity oils, having molecular weights as low as 250, to very viscous lubricants with molecular weights as high as about 1000. The viscosity of an oil can often determine its use. For example, if the oil is too viscous (too resistant to flow), it offers too much resistance to the metal parts moving against each other. On the other hand, if the oil is not viscous enough, it will be squeezed out from between the metal surfaces, and consequently offer insufficient lubricating power. For these and other reasons, motor oil is often a mixture of oils with varying viscosities. The common 10W-30 oil, for instance, combines the low-temperature viscosity of the Society of Automotive Engineers (SAE) 10W classification for easy low-temperature starting with SAE 30 high-temperature viscosity for better load capacity in bearings at the normal engine running temperature.

The SAE scale for rating motor oils is based on the viscosity of the oils. The viscosity criteria for the SAE scale are given in Table 24–1.

During distillation of petroleum crude oils, the lubricating crude oil fractions boil off after the lower boiling gasoline, kerosene, and fuel oils are removed. Most of the aromatic compounds are then removed from the lubrication oil fraction by solvent extraction to prevent the formation of sludge during high-temperature operation. Paraffin wax is then removed by low-temperature filtration to give the oil better flow characteristics. The final refining step is contact

TABLE 24-1   VISCOSITY DATA AT 0°F AND 210°F FOR THE
SAE METHOD OF RATING MOTOR OILS

| | VISCOSITY | | | |
| | 0°F | | 210°F | |
MOTOR OIL	min. (SUS)	max. (SUS)	min. (SUS)	max. (SUS)
5W		4,000	39	
10W	6,000	12,000	39	
20W	12,000	48,000	39	
20			45	58
30			58	70
40			70	85
50			85	110

° SUS is the Saybolt Universal Second, which is the time in seconds required for 60 ml of oil to empty out of the cup in a Saybolt viscometer through a carefully specified capillary opening.

with an activated clay such as Fuller's earth, which absorbs many of the colored particles and provides an oil with a light color. Lubricating oils with the desired properties are then made by blending one or more refined stocks with the proper additives.

After the motor oil has been separated and refined, its usefulness is improved by the addition of substances such as antiwear agents, oxidation inhibitors, rust inhibitors, detergents, viscosity improvers, and foam inhibitors.

When two metal surfaces contact under heavy load and high temperature, as in the differentials of most cars, the friction produces intense heat which renders organic lubricant films ineffective. Under extreme conditions the metals can weld together. To combat this, *extreme pressure lubricants* were developed. These contain organic compounds as additives which react at the high contact temperatures to form high-melting inorganic lubricant films, such as lead sulfide and iron sulfide, on the metal surfaces; the presence of these films inhibits massive welding and breakdown. These additives generally consist of sulfur, chlorine, phosphorous, and lead compounds which act either by providing layers that are hard and difficult to wear away or by serving as fluxing agents to contaminate the metal surface and prevent welding.

Under conditions of less severe friction, mildly polar organic acids, such as the alkyl succinic type, and organic amines are often added as *antiwear agents*. These compounds provide an adherent, adsorbed film over metallic surfaces and reduce shearing of the metal. In somewhat more severe conditions about 1 per cent tricresyl phosphate (TCP) or zinc dialkyldithiophosphate is widely used.

Rust inhibitors, such as the mild antiwear agents and antirust agents in gasoline, are preferentially adsorbed as a film on iron and steel surfaces to protect them from attack by moisture (Figure 24–1). If only a little water is present in a large amount of oil, mildly polar organic compounds such as alkyl succinic acids and organic amines are often used. Where severe conditions are anticipated, more strongly adherent organic phosphates (TCP, for example), polyhydric alcohols, and sodium and calcium sulfonates are used.

Oil oxidation is thought to involve a chain reaction mechanism with hydroperoxide formation (—OOH) as the initiating process which eventually leads to the formation of organic acids and other products. *Oxidation inhibitors*

575

appear to interrupt the chain reaction by tying up the hydroperoxide. This action delays the formation of sludge, varnish, and acids for extended operating periods and minimizes corrosion problems with the zinc, cadmium, and copper bearing alloys, which are corroded by organic acids in oxidized oils. Zinc, barium, and calcium thiophosphates are frequently used to prevent oxidation of the oils.

Detergents are widely used in a 2 to 20 per cent concentration in motor oils to prevent or remove deposits of oil-insoluble sludge, varnish, carbon, and lead compounds. They are adsorbed on the insoluble particles, keeping them suspended in the oil so as to minimize deposits on rings, valves, and cylinder walls. The action is similar to the action of soap (or detergent) in removing grease from clothes or hands, as described in Chapter 15. A basic difference exists, however, because the polar end of the oil detergent generally is attached to the particle, and its hydrocarbon end extends into the medium (oil). In soapy water, the hydrocarbon end of the soap is in the oil or grease particle and the polar end is in the medium (water). Barium and calcium sulfonates and phenoxides are used extensively as detergents in automotive motor oils.

The viscosity of motor oils can be adjusted with additives. Polymethacrylate polymer is added in small amounts (1 per cent or less) to prevent wax, which condenses out at low temperatures, from forming a network of crystals that would immobilize the oil. The additive appears to adsorb on crystal faces, which prevents the interlocking crystal growth. These additives are ineffective in normally high viscosity oils. The viscosity of oils can be increased by adding linear polymers in the molecular weight range of about 5000 to 20,000. The three types most commonly used are polyisobutylenes, polymethacrylates, and polyalkylstyrenes (Chapter 16). The entanglement of the long chains prevents easy flow of the oil. With use, the chains are broken into smaller fragments and the oil assumes its base viscosity.

Severe churning and mixing of oil with air may cause foam and an oil overflow from the engine; failure of the machine may eventually ensue. Methyl silicone polymers (Chapter 13) in concentrations of only a few parts per million are effective for defoaming oil. Since the silicone additive is not completely soluble in the oil, it functions by forming minute droplets of low surface tension which aid in breaking up foam bubbles to release the trapped air.

An increasingly large number of synthetic lubricating oils continue to be developed which offer properties not easily obtained from natural petroleum oils. For example, methyl silicone oils have low volatility, stability at high temperatures, and the smallest change in viscosity with temperature of any fluids known. As a result they find wide application as hydraulic fluids (brake fluids, for example). Silicone oils and greases consist of a silicon-oxygen polymer chain to which are attached either methyl or phenyl hydrocarbon groups (Chapter 13).

$$\left(\begin{array}{c} R \\ | \\ -Si-O- \\ | \\ R \end{array} \right)_n$$

Greases are essentially lubricating oils thickened with a gelling agent such as fatty acid soaps of lithium, calcium, sodium, aluminum, or barium (Chapter

23). The fatty acids are usually oleic, palmitic, stearic, or other carboxylic acids derived from tallow, hydrogenated fish oil, castor oil, or, less often, wool grease and rosin. The soaps form a network of fibers that entrap the oil molecules within the interlacing fiber structure. Carbon black, silica gel, and clay are also used to thicken petroleum greases. Chemical additives similar to those used in lubricating oils and gasolines are added to greases to improve oxidation resistance, rust protection, and extreme pressure properties. Synthetic greases are being developed that deteriorate so slowly that longer intervals between grease jobs are now possible. Silicone greases have a useful life of up to 1000 hours at 450°F (232°C). Unfortunately, silicone greases provide relatively poor lubrication for gears and other sliding devices. Diester greases such as di(2-ethylhexyl) sebacate have found extensive use among synthetic greases. Lithium soaps dissolve well in the diester oil and form a grease with equal or better lubrication characteristics and a considerably longer use life than petroleum greases. Blends of silicone oil and diester oil provide greases with good low-resistance lubricating power even at low temperatures (−100°F). Greases are graded 0, 1, 2, and so forth, on the basis of their softness, grade 0 being the softest. Grade 2 greases have the most common and widest applications in automobiles.

SOLID LUBRICANTS

Perhaps the most tenacious and wear resistant solid lubricant is molybdenum disulfide, MoS_2. Like graphite, a very common solid lubricant, MoS_2 has a layered structure that enables one layer to float over another and provide lubrication (Figure 24-2). Strong bonds exist within each S—Mo—S layer, while weak S—S bonds between the layers allow easy sliding of one layer over another. MoS_2 is extensively used in greases for automotive chassis lubrication. When used alone, MoS_2 suffers like any other solid lubricant from having poor wear resistance and being unable to heal any breaks in its surface coating.

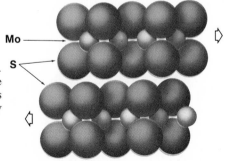

FIGURE 24-2   The structure of molybdenum disulfide, MoS_2. The small spheres represent molybdenum atoms. Note how the sulfides are adjacent to sulfides in every other layer. This allows one layer to slide over another, accounting for the slippery nature of MoS_2.

ANTIFREEZE

An antifreeze is a substance which is added to a liquid, usually water, to lower its freezing point. Although a variety of substances has been used as antifreezes in the past, nearly all of the current market is supplied by ethylene glycol and methyl alcohol.

$$
\begin{array}{cc}
\text{OH\ OH} & \text{OH} \\
\text{H—C——C—H} & \text{H—C—H} \\
\text{H\ \ H} & \text{H}
\end{array}
$$

Ethylene glycol   *Methyl alcohol*

More than 95 per cent of the antifreeze on the market is "permanent" antifreeze, having ethylene glycol as the major constituent. The largest use of antifreeze is in protection of water-cooled automobile and truck engines. Water has been selected as the coolant for these engines because of its universal availability, low cost, and good heat transfer properties; but it has two serious disadvantages. It has a relatively high freezing point and, under normal operating conditions, it is corrosive. Modern antifreeze mixtures effectively counteract these problems.

The temperature in the United States seldom, if ever, falls below $-40\,°F$ $(-40\,°C)$. Both ethylene glycol and methyl alcohol can prevent water from freezing at these temperatures, as shown in the graph of Figure 24–3. Methyl alcohol more effectively lowers the freezing point, but because of its volatility and combustibility it is seldom used.

For ethylene glycol, methyl alcohol, and other nonelectrolytes, the depression of the freezing point in dilute solutions is proportional to the concentration of the solute and nearly independent of the nature of the solute. That is,

$$\Delta T_f = K_f m$$

where ΔT_f is the freezing point depression, K_f is the proportionality constant peculiar to each solvent ($1.86\,°C$ for water), and m is the number of moles of solute per 1000 grams of water (molal concentration). In concentrated solutions this relationship can be used to obtain a rough estimate of the freezing point.

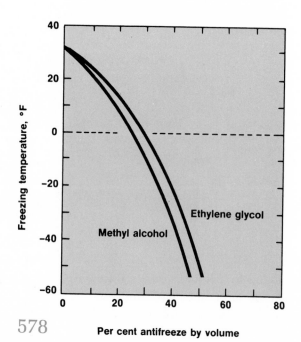

FIGURE 24–3   Freezing point depression of water by ethylene glycol and methyl alcohol.

If a gallon of antifreeze is added to four gallons of water, the actual freezing point of the mixture is 16°F.

Antifreeze protection charts always show the temperatures at which the first ice crystals form. Below this temperature the antifreeze solution turns to slush. If the slush is unable to circulate through the radiator, overheating, boiling and engine damage can result. For this reason it is best to add sufficient antifreeze to prevent the formation of the first ice crystals even at the lowest anticipated temperature (Figure 24–4).

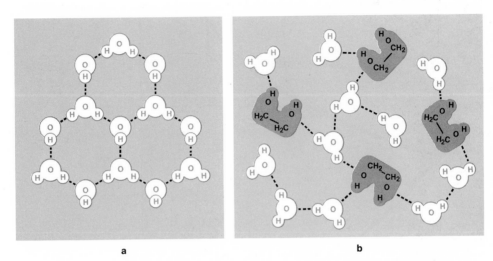

a   b

FIGURE 24-4   How foreign molecules prevent water from forming its normal crystal structure. For example, by forming hydrogen bonds to water molecules, ethylene glycol prevents water molecules from assuming their places in the ice structure.

A service station attendant measures the effectiveness of the antifreeze in your car by reading the position at which a hydrometer floats in a portion of a radiator solution. He is really measuring the solution's density, which varies with the amount of antifreeze present. An assortment of densities, concentrations, and freezing point lowerings are given in Table 24–2.

TABLE 24-2   DENSITY AND FREEZING POINT DEPRESSION
FOR VARIOUS SOLUTIONS OF ETHYLENE GLYCOL IN WATER

PERCENTAGE OF ETHYLENE GLYCOL (BY WEIGHT)	DENSITY (GRAMS/ML) (20°C)	FREEZING POINT OF SOLUTION
2.0	1.0025	−.60°C
4.0	1.0049	−1.24
6.0	1.0074	−1.92
8.0	1.0100	−2.64
10.0	1.0125	−3.41

In 1960, 13 to 17 pounds-per-square-inch pressure caps were installed on automobile radiators, which allowed a 35 to 50°F increase in maximum coolant temperatures. Thermostats operate now between 185 and 210°F, whereas in the past the range was 140 to 160°F. Ethylene glycol raises the boiling point of water as well as lowers its freezing point, as shown in Figure 24-5. Ethylene glycol reduces the vapor pressure of water, requiring a higher temperature for the solution to boil. Therefore, it is good policy to keep the antifreeze in the radiator year round. If the car has an air conditioner, the antifreeze serves a dual role in the summer time, for it prevents freezing in the heater core and prevents boiling in the radiator.

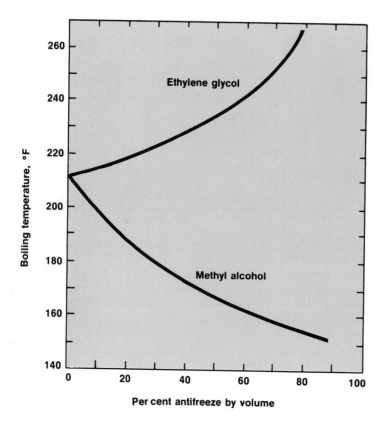

FIGURE 24-5   Boiling points of aqueous antifreeze solutions.

Commercial antifreeze contains various additives to prevent corrosion, leaks, damage to rubber, and foaming. A typical permanent antifreeze contains more than 95 per cent ethylene glycol, several reducing agents to prevent corrosion, a substance to stop small leaks in the cooling system, and an antifoaming agent.

The prevention of corrosion is the second most important job of antifreeze. Metals in the cooling system which are subject to corrosion are copper, steel, cast iron, aluminum, solder (Pb and Sn), and brass (Cu and Zn). The presence of oxygen, along with high temperatures, pressures, and flow rates, increases the possibility of general corrosion. Although corrosion can "eat" through the walls of the cooling system, the most general trouble is overheating caused by flakes of metal clogging the radiator. A large assortment of substances is used to inhibit corrosion, including amines (triethanolamine), mercaptans (sodium mercapto-benzothiazole), borates, nitrites, phosphates, molybdates, arsenites, and carbon-ates. Some of these inhibit corrosion by acting as reducing agents (e.g., nitrites), some as ion scavengers (e.g., phosphates), and some as surface active agents (detergents; e.g., triethanolamine). Most antifreezes contain two or more inhibitors for the different metals. All inhibitors are depleted with use.

Radiator sealants have been on the market for years. Modern sealants in antifreeze include asbestos fiber (in Union Carbide's Prestone) and polystyrene spheres (in Du Pont's Zerex). When a radiator springs a leak, the coolant penetrates the crack because the pressure inside the cooling system is greater than atmospheric pressure. Asbestos fibers are often too big to squeeze through and get caught in the crack, plugging the hole in the radiator. On the other hand, the larger polystyrene particles initially plug up most of the leak while the smaller ones build up behind them. The spheres fuse together under the pressure and temperature conditions that exist and form a solid plug in the crack (Figure 24–6). This is effective in stopping up holes or cracks up to 0.5 millimeter width, which, according to Du Pont, covers about 90 per cent of all radiator leaks.

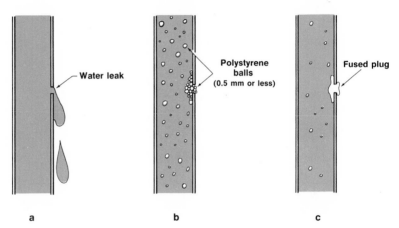

a   b   c

FIGURE 24–6 How polystyrene works to plug radiator leaks. Details are given in the text.

Foaming is caused by one or more of several factors: air leaks in hoses, water pump, or radiator; exhaust gas leaking into the cooling system; failure to drain out cooling system cleansers; or extended use of antifreeze. Foaming can be corrected by tightening the system or by use of antifoam additives, such as silicones, polyglycols, mineral oils, high molecular weight alcohols, organic phosphates, alkyl lactates, castor oil soaps, and calcium acetate. These substances reduce the surface tension so bubbles cannot hold together.

De-icing fluids are a type of antifreeze, but they also will melt ice and frost. They are chiefly employed in removing ice and frost from parked aircraft and car windows. The glycol-alcohol type used on cars came on the market in 1959. The better formulations contain the following ingredients: ethylene or propylene glycol (for protection against refogging); one or more of the lower alcohols, such as 2-propanol, denatured alcohol, etc. (for low viscosity and good spray pattern); water (to minimize inside fogging by diminishing the evaporative-cooling effect

581

of the formulation); nitrite or other inhibitor (to prevent corrosion of the container); and wetting agent, such as ethanolamine (to break surface tension so fluid will attack the ice readily).

A simple alcohol-water mixture works satisfactorily for quick frost removal and to defog the inside of glass, but it evaporates quickly and refogging occurs unless glycol is present. Before a de-icer formulation is applied, the snow should be removed and the ice should be scored to permit faster penetration.

MISCELLANEOUS AUTOMOTIVE PRODUCTS

Consumers spend millions of dollars each year on a large variety of specialized automotive products. The formulations of some typical examples of these products are given in Table 24–3.

TABLE 24–3   TYPICAL COMPOSITION OF SOME AUTOMOTIVE PRODUCTS

ENGINE AND MOTOR CLEANERS

Ethylene dichloride	63%
p-Cresol	25%
Oleic acid	7.2%
Potassium (or sodium) hydroxide	1.4%
Water	3.0%

RADIATOR CLEANERS

(1) Oxalic acid	40%
Boric acid	60%
(2) Sodium carbonate	85%
Potassium dichromate	15%
May contain also:	
sulfamic acid	
1-butanol (n-butyl alcohol)	
2-propanol (isopropyl alcohol)	

RADIATOR STOP LEAK

Dextrin	5–10%
Cellulose gum	0.5%
Asbestos	5%
Sodium carbonate	0.8%
Isopropyl alcohol	10–15%
Water	to 100%

SHOCK ABSORBER FLUIDS

(1) Petroleum ether	97%
Kerosene	3%
(2) Mineral oils	90–100%
Fatty oils	0–5%
May contain also:	
viscosity improvers	0–5%
(polymethacrylate esters)	
dyes	0–150 ppm

AUTOMATIC TRANSMISSION FLUIDS

Mineral oils	75–100%
Oxidation inhibitors and detergents	0–20%
Viscosity improvers	0–5%
Polymethacrylate esters	
Polyisobutylenes	
Anti-wear agents	0–2%
Organic borates	
Antifoam agents	0–15%
Polysiloxanes	
Sealant	to 100%
Triaryl phosphate	
Dyes	0–200 ppm

BRAKE FLUIDS

Lubricant	20–25%
Castor oil	
Butyl or glyceryl ether of polyoxyethylene propylene glycol	
Polypropylene glycol	
Solvent	80–85%
Methyl, ethyl and butyl ethers of ethylene glycol and related glycols	
May contain also:	
Inhibitors	
Amine soaps	
Potassium soaps	
Borax	
Antioxidants	
Hydroquinone	
Dyes	

PAINTS, VARNISHES, AND SURFACE FINISHES

About $3 billion is spent on paints in the United States each year. This sum is about equally divided between industrial coatings, such as the paint on automobiles and package cans, and trade sale products, such as off-the-shelf house paints. Paints and industrial coatings usually constitute less than 1 per cent of the total cost of the object or installation and serve two obvious purposes: protection and appearance.

BASIC COMPONENTS OF PAINTS

Paints are generally considered to be composed of three basic components: *pigment, nonvolatile vehicle* (often referred to as the binder), and a *volatile vehicle.*

The *pigment* supplies the desired color and provides a shield against high energy ultraviolet radiation. In a given paint the pigment will often absorb the ultraviolet radiation and release it as thermal energy. As a result, the film of paint lasts many years. Without the pigment, the ultraviolet energy attacks the organic film, breaks bonds within the film, and the paint cracks and peels within a year. Pigments also are employed as rust inhibitors on metals and as antifoulants on ship bottoms. Paint pigments may be either organic or inorganic. Carbon black, C, white titanium dioxide, TiO_2, yellow lead chromate, $PbCrO_4$, and red or brown iron oxide (actually a mixture of oxides) are examples of inorganic pigments. The lead pigments in old fashioned "lead" paints are the cause of lead poisoning of ghetto infants who eat paint chips. Many organic pigments are intensely colored because of chromophoric, or color-absorbing, groups within the molecule. An example is Pigment Red 49, a barium salt:

Pigment Red 49

The *nonvolatile vehicle* is a drying oil, a resin, or some combination of the two. A drying oil is an oil which, on exposure to the atmosphere, becomes a solid. Linseed oil is perhaps the best-known drying oil. However, others are well known and widely used, such as soybean, sunflower, hemp, and tung oils. A typical linseed oil would yield, on hydrolysis, the following fatty acids:

4–7%	palmitic acid (16 carbon atoms), a saturated acid
2–5%	stearic acid (18 carbon atoms), a saturated acid
9–38%	oleic acid (18 carbons, 1 double bond)
3–43%	linoleic acid (18 carbons, 2 double bonds)
25–58%	linolenic acid (18 carbons, 3 double bonds)

Two reactions are probably involved in the drying of the oil: (1) The metal 583 ion or drier in the formulation forms salts of the fatty acids (solid soaps):

$$Ca^{2+} + \text{Stearic acid} \longrightarrow \text{Calcium stearate (a salt)}$$

(2) Air oxidation of the double bonds results in reactive intermediates which react further, building up a complex web of organic molecular structure. The result is a tough solid film which tends to be impermeable to both water and air. While the dried film from a drying oil can be considered a resinous material, a resin is a more general type of material—an amorphous solid or semi-solid mixture of organic compounds. The word *amorphous* means that there is no tendency to crystallize into a definite and rigid structure and implies that the melting point is indefinite. Resins may be animal, vegetable, or synthetic in origin and may be composed of a wide variety of chemical structures. The bulk of a dried paint film is the nonvolatile vehicle. It has three purposes: (1) to provide adhesion to the surface, (2) to provide a moisture or a gaseous barrier when needed, and (3) to provide a medium in which the pigment is held.

The *volatile vehicle* is sometimes merely a solvent which lowers the viscosity of the liquid paint. A viscous liquid such as the nonvolatile resinous vehicle is difficult to spread. A mixture of hydrocarbon solvents is used as the vehicle to thin oil paints. Sometimes ketones and esters are used along with hydrocarbons in particular oil-base formulations. The vehicle in water-based paints is, of course, water. In the largest group of water paints, the nonvolatile vehicle is emulsified into the water medium. Thus, the water makes the paint easy to spread in a unique way: the low viscosity of water allows the minute droplets of resin to be spread evenly with almost the ease of spreading water itself. Figure 24–7 presents a simplified view of such an emulsion. The volatile vehicle may make up to 50 per cent of the volume of the paint, but it completely vaporizes into the atmosphere when the paint is dried or cured.

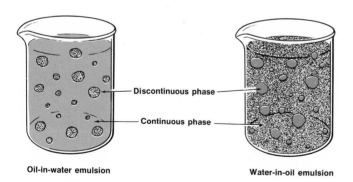

Discontinuous phase

Continuous phase

Oil-in-water emulsion

Water-in-oil emulsion

FIGURE 24–7   Two kinds of emulsions. An emulsion is composed of two immiscible liquids, one dispersed as tiny droplets in the other. An emulsifying agent is required to stabilize an emulsion.

OIL-BASED PAINTS

Perhaps the simplest oil-based surface coating is a *varnish*. Varnish is a mixture of a drying oil, such as linseed oil, rosin, and a thinner. Rosin, the natural resin from pine trees, has as its chief fatty acid component abietic acid, $C_{20}H_{30}O_2$. Its structure is given in Chapter 23. On drying, the rosin mixed with the solidified drying oils gives a hard, relatively transparent and shiny surface.

An *enamel* is a pigmented varnish, and, consequently, it has a hard smooth surface with a high luster. Of course, enamels can be much more complicated than this when formulated for special purposes.

The first white, oil-based paint which found wide usage was simply white lead, $Pb(OH)_2 \cdot 2PbCO_3$, dispersed in linseed oil. Because of the low opacity of this pigment, three coats were required to get a smooth white surface. An additional problem with this paint is its severe sensitivity to fumes. For example, sulfide fumes in concentrations as low as a few parts per million give enough sulfide to form black lead sulfide in the paint.

$$Pb(OH)_2 + H_2S \longrightarrow \underset{\substack{\text{Black} \\ \text{lead sulfide}}}{PbS} + 2H_2O$$

The paint is slow to dry and has a surface texture that tends to hold dirt. It eventually fails by developing cracks in its surface. Lead paints are toxic if eaten and are illegal in many places (Chapter 19).

The pigment of a modern oil-based paint contains titanium dioxide (TiO_2), "titanium calcium" (30 to 50 per cent TiO_2 on the surface of calcium sulfate particles), zinc oxide (ZnO), and various extender pigments. Titanium dioxide has greater opacity and whiteness and is less toxic than white lead. The different crystalline forms of TiO_2 catalyze the decomposition of the binder (such as linseed oil) at different rates. The deterioration of the binder forms a white powder, which is easily rubbed off. This process is called chalking. The rate of chalking of the dried paint can be controlled with different crystalline forms of TiO_2. One form, rutile, chalks slowly while another, anatase, chalks more rapidly. Controlled chalking is desirable because it gives a constantly renewed surface over a period of years.

Zinc oxide is a hardener in oil-based paints; zinc and other divalent ions (e.g., Ca^{+2}) form insoluble soaps with the fatty acids from the linseed oil, making the dried surface more durable. Extender pigments do not contribute much to opacity but are added to control the hardness of the paint film and to reduce the cost of the paint. If the relative amount of the total pigment is too low, the film is likely to be too soft. Talc (magnesium silicate; Figure 23–2), limestone (calcium carbonate, $CaCO_3$) and silica (silicon dioxide, SiO_2) are examples of extender pigments. (A typical outdoor white oil-based paint could contain the following ingredients:

	Percent by weight
Titanium dioxide (TiO_2)	29.0%
Zinc oxide (ZnO)	2.2%
Calcium carbonate ($CaCO_3$)	12.0%
Silica (SiO_2)	1.3%
Linseed oil	19.6%
Drier	1.7%
Mineral spirits	34.2%

Raw linseed oil has the desired low viscosity for the nonvolatile vehicle, but it does a poor job of wetting the pigments. Boiled linseed oil has a higher viscosity, but wets the pigments well, so a combination is usually used. The thinner is often mineral spirits (naphtha), a hydrocarbon mixture that boils between 93 and 150°C. Modern paints are not discolored by hydrogen sulfide gas. Since both ZnS and ZnO are white, sulfide or oxide formation would not discolor the paint. Titanium dioxide does not react with atmospheric H_2S.

Colored paints are prepared by using relatively small amounts of pigments in the place of, or in addition to, the white pigments.

Interior and exterior oil paints are very similar. Titanium dioxide is added in greater amounts to give whiter whites. Natural resins or greater concentrations of pigments are added to give a quicker drying film. Many additives are used in both exterior and interior oil-based paints to hasten the drying of the paint, provide better dispersion of the pigment, resist mildew and fungi formation, and prevent flooding, settling, skinning, yellowing, and destruction by ultraviolet light.

FIGURE 24-8   Photographers photographing a painter painting—in this picture the John Quincy Adams birthplace in Quincy, Massachusetts. The Sears Great American Home Series of advertisements illustrates the importance of surface coatings as preservatives. (Courtesy of Sears, Roebuck and Co., Chicago, Illinois.)

The *metallic driers* are Co, Fe, Zn, Mn, and Ca salts of carboxylic acids, which catalyze the oxidation of the oils in the paint. Metal atoms and ions seem to catalyze the breakdown of the intermediate hydroperoxides, thus accelerating the rate of the drying action.

Pigment-dispersing agents promote the suspension of the fine particles of pigment in the oils and thinner. Since some pigments are hydrophilic and others hydrophobic, no one dispersing agent will work for all pigments. Certain inorganic materials such as phosphates (e.g., $Na_4P_2O_7$) and silicates (e.g., Na_2SiO_4) can act as dispersants for pigments. They are adsorbed on the surface of the pigment, producing an electrostatic charge on the particles which causes them to repel each other.

586

Mildew and fungi can feed on the organic materials in paints. To prevent this, *mildewcides* and *fungicides* are added to paints. Among these are phenyl mercuric salts, copper 8-hydroxyquinolate, and tetrachlorophenol. Zinc oxide is also effective. For mildew resistance, 0.35 to 0.50 per cent mercuric phenyl acetate, $Hg(OCOCH_2C_6H_5)_2$, is required per total weight of the paint. Mercury compounds sometimes show discoloration in atmospheres rich in hydrogen sulfide. The mercuric ion reacts with the sulfide to form black mercuric sulfide.

$$Hg^{2+} + H_2S \longrightarrow HgS + 2H^+$$
$$\text{\textit{black}}$$

Flooding in paints results when the color components separate on drying and concentrate in streaks or patches in the surface of the film. *Antiflooding agents* are generally wetting agents, such as aluminum stearate or stearic acid. The polar end of the wetting agent bonds to the pigment particle, and the nonpolar hydrocarbon chain entangles itself compatibly in the oil vehicle.

Ultraviolet radiation can cause paint to crack, fade, and depolymerize. Protection against ultraviolet rays can be afforded by (1) pigmentation with suitable colors, which absorb the rays without being broken down (e.g., zinc oxide, titanium dioxide, carbon blacks, and iron oxides), (2) pigmentation with substances which reflect the rays (e.g., aluminum flakes), or (3) incorporation of small amounts of ultraviolet absorbers in the first coat (e.g., derivatives of benzotriazoles or benzophenone). Good absorbers have little color and absorb without being broken down themselves.

The yellowing of paint is observed often, particularly with interior paints. It has been confirmed experimentally that paints tend to yellow more in the dark and in the presence of moisture. Darkness favors the condensation of atmospheric moisture on surfaces; light bleaches and tends to evaporate the moisture. The role (or roles) of moisture in the atmospheric yellowing of paints is not clearly understood. Water may hydrolyze linseed oil into acids that have stronger colors. It may catalyze the oxidation of the oils to other compounds that are highly colored. Linolenic glycerides are now considered responsible for the greatest discoloration. The three double bonds per linolenic acid molecule are preferred sites for hydroperoxide formation, which can lead to ketol formation. A ketol is a combination of an alcohol group (R—OH) and a ketone group (RCR), such as an alpha ketol:

$$
\underset{\text{(R—OH)}}{} \qquad \overset{\displaystyle O}{\underset{\text{(R\overset{\|}{C}R)}}{}}
$$

$$
R\overset{\displaystyle OH}{\underset{\displaystyle O}{\overset{|}{C}H\overset{\|}{C}R}}
$$

Some ketols are yellow. Yellowing is prevented by the inclusion of antioxidants.

WATER-BASED PAINTS

The newer water-based paints are an outgrowth of the synthetic rubber program of World War II. Synthetic rubbers, copolymers containing a 25 to 75

587

ratio of styrene to butadiene (Chapter 16), were developed from large stock piles of these two monomers. It was found that if the ratio of styrene to butadiene was reversed to values such as 85:15 down to 60:40, a latexlike material similar to natural rubber latex resulted. This "latex" was easily emulsified in water, and latex, or modern water-based, paints were born. Subsequently, many other resinous materials have been similarly employed in a wide variety of water-based emulsion paints.

Table 24–4 provides some information concerning other, older water-based paints; in use before World War II.

TABLE 24–4   WATER-BASED PAINTS (PRIOR TO WORLD WAR II)

NAME	COMMENTS
Whitewash	Calcium hydroxide, $(Ca(OH)_2$, dispersed in water. Only cheapness recommends it.
Casein paint	Purchased as a powder, a protein mixture. When mixed with water and lime (CaO), insoluble calcium caseinate forms, which is the coating material. Pigments can be added.
Cement paints	Used on masonry. A powder that contains Portland cement along with alkali-resistant pigments.
Linseed oil, emulsion paints	Emulsifiers are used to disperse linseed oil in water. Metallic driers are required to harden the film. Used infrequently.

Advantages and disadvantages of emulsion paints are given in Table 24–5.

TABLE 24–5   ADVANTAGES AND DISADVANTAGES OF EMULSION PAINTS

ADVANTAGES	DISADVANTAGES
(1) Low-cost, continuous phase-water	(1) Tends to corrode some metals
(2) Not a fire hazard	(2) Tends to swell and raise the grain of wood and composite wall materials
(3) No odor	(3) Sensitive to temperature extremes. Storage under freezing conditions should be avoided
(4) Less volatile	
(5) No air pollution	
(6) Low viscosity	
(7) Easy clean-up with soap and water	

The formulation of emulsion paints consists basically of combining pigment with latex, the latex being the emulsion of the resin in water. Ideally, the pigment will enter into the dispersed resinous phase and become homogenously dispersed in it. Table 24–6 gives additional information concerning some of the additives that are also blended into the mixture.

The polymers that have found greatest use in emulsion paints are given in Table 24–7.

The styrene-butadiene resin is the least expensive material used, but it has a relatively long curing period, relatively poor adhesion, and a tendency to yellow with age. Polyvinylacetate is only a little more expensive and is an improvement over the styrene-butadiene resin. It quickly captured 50 per cent of the latex market for interior paints. A type with rapidly growing popularity,

TABLE 24-6   ADDITIVES USED IN EMULSION PAINTS

(1) Dispersing agents for pigments	Example: tetrasodium pyrophosphate ($Na_4P_2O_7$). Same principle of like-charged particles repelling as in oil-based paints.
(2) Protective colloids and thickeners	A thicker paint is slower to settle and drips and runs less. A protective colloid tends to stabilize the organic-water interface in the emulsion. Examples: sodium polyacrylates, carboxymethylcellulose, clays, gums. (Same mechanism as soap-dispersing oil in water; Chapter 15.)
(3) Defoamers	Foaming presents a serious problem if not corrected. The surface tension of the water must be decreased. Chemicals used: tri-n-butylphosphate, n-octyl alcohol and other higher alcohols.
(4) Coalescing agents	As the water evaporates and the paint dries, a coalescing agent is needed to stick the pigment particles together. As the resin film forms, the coalescing agent evaporates. Coalescing agents must volatilize very slowly. Examples: hexylene glycol and ethylene glycol.
(5) Freeze-Thaw-Additives	Freezing will destroy the emulsion. Antifreezes such as ethylene glycol are used.
(6) pH Controllers	The effectiveness of the ionic or molecular form of the emulsifier depends on the acid or alkaline conditions (pH). The wrong pH will break down the emulsion. Most paints tend to be too acidic. Ammonia, NH_3, is added to neutralize the acid.

TABLE 24-7   POLYMERS USED IN EMULSION PAINTS

POLYMER	GENERAL FORMULA
Polystyrene	$\left[CH_2-CH\right]_n$ (with benzene ring)
Poly(styrene-butadiene)copolymers (x and y may vary)	$\left(\left[CH_2-CH\right]_x\left[CH_2-CH=CH-CH_2\right]_y\right)_n$ (with benzene ring)
Polyvinylacetate (also copolymers)	$\left[CH_2-CH \quad O-\underset{\underset{O}{\parallel}}{C}-CH_3\right]_n$
Polymeric esters of acrylic acid ($R=CH_3$ or C_2H_5)	$\left[CH_2-CH \quad O=C-OR\right]_n$

though about one-third more expensive, are the acrylic resins and the "acrylic latex" paints. These are more washable and much more resistant to light damage. They are especially useful as exterior paints.

The fluoropolymers, similar to Teflon, are especially promising as surface coatings because of their great stability. Fluorine atoms are substituted for hydrogen atoms in the organic structure. Metals covered with polyvinylfluoride carry up to a 20-year guarantee against failure from exposure.

A typical formulation for an exterior latex paint is given in Table 24–8.

TABLE 24-8   COMPOSITION OF A TYPICAL EXTERIOR LATEX PAINT°

FORMULATION A	Pounds	Gallons	FORMULATION B	Pounds	Gallons
Dispersing agent	15	2.0	Acrylic latex† (46% nonvolatile)	605	63.9
Defoamer	2	2.2	Preservative	9	1.0
Water	50	6.0	Defoamer	2	0.2
Titanium dioxide	250	7.2	Water	8	1.0
Extender pigments	117	5.0	Ammonia‡	1	0.1
Hydroxyethylcellulose (a 2% aqueous solution)	50	6.0			
Ethylene glycol	25	2.6			

° Formulation A is mixed and ground on a high-speed stone mill, and then Formulation B is added.

† A blend of acrylic esters (e.g., methyl methacrylate–ethyl acrylate–methacrylate acid terpolymer) emulsified in water to make a suspension containing 46% resin particles.

‡ More or less, as determined by test.

THE CHEMISTRY OF PHOTOGRAPHY

BLACK AND WHITE PHOTOGRAPHY

Many chemical substances are known to be photosensitive, that is, they change in light; some of these, the halide (chloride, bromide, or iodide) salts of silver, have the useful ability of storing an image. The various processes known as *photography* are based upon this property.

Photography as we know it today is the invention of no single individual, but rather the result of the efforts of many chemists, beginning as early as 1727, when J. H. Schulze noticed that a mixture of silver nitrate and chalk darkened on exposure to light. These early images were not permanent, however, for there was no known way to rid the exposed material of unexposed silver compound.

In the early 1830's, a Frenchman, Louis Daguerre, discovered by accident that mercury vapor was capable of developing an image from a silver-plated copper sheet that had been sensitized by iodine vapor. The *daguerreotype* image was rendered permanent by washing the plate with hot concentrated salt solution. Daguerreotype portraits were very popular in the mid-nineteenth century.

In 1839, an Englishman, Fox Talbot, developed a way to "fix," or make permanent, an image by adding a solvent for the nonexposed silver salt. This process was the forerunner of our present photography processes. The Talbot process involved reducing light-sensitive silver iodide with gallic acid (the developer) to produce a free silver image (Figure 24–9). This image turned

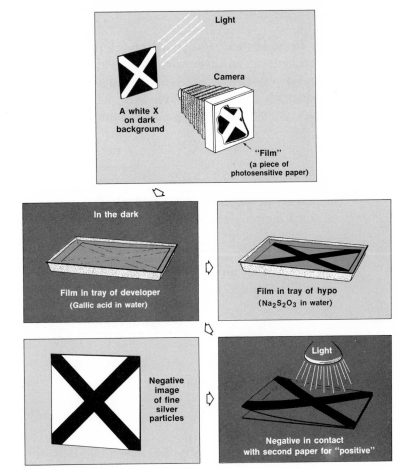

FIGURE 24-9 Talbot's process.

out to be an exact reverse of the original, which led one of Talbot's contemporaries, Sir J. W. F. Herschel, to name the image a "negative." With semi-transparent paper, Talbot's negative was then contact printed onto another piece of photographic paper to yield a "positive," or direct copy of the original image.

By 1870, various workers had found that silver salts suspended in collodion yielded good results and were quite practical, and by 1880, photographic plates with silver salts suspended in gelatin were commercially available; these were much like the plates used in modern portrait photography. Today roll films are common; these are prepared by coating a gelatin emulsion of silver salts on a plastic film which can be easily handled (Figure 24–10).

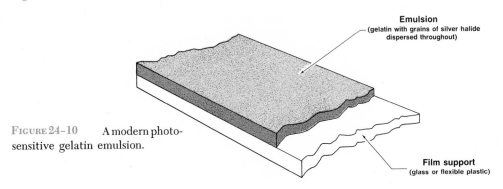

FIGURE 24-10 A modern photosensitive gelatin emulsion.

591

Photochemistry of Silver Salts

To understand the chemistry of photography, we must first look at the photochemistry of silver salts. A typical photographic gelatin emulsion is made up of tiny crystallites which are called *grains*, of some very slightly soluble silver halide, such as silver bromide, $AgBr$. These are formed when solutions of silver nitrate, $AgNO_3$, and potassium bromide, KBr, are mixed.

$$Ag^+ + NO_3^- + K^+ + Br^- \longrightarrow AgBr + K^+ + NO_3^-$$

The solid $AgBr$ is then "ripened" to produce grains of the right size. After being suspended in gelatin and washed, the gelatin emulsion is melted and applied as a coating on glass plates or plastic film.

When light of an appropriate wavelength (blue or light of shorter wavelength) strikes one of these grains, a series of reactions begins which leaves in the grain a small amount of free silver. Initially a free bromine atom is produced when the bromide ion absorbs the photon of light:

$$Ag^+Br^- \xrightarrow{\text{Light absorption}} Ag^+ + Br^0 + e^-$$

The silver ion-electron pair now may do one of two things: (1) it may recombine to form a free silver atom:

$$Ag^+ + e^- \longrightarrow Ag^0$$

(but this is unlikely in a typical silver halide grain) or (2) the silver atom may combine with a neighboring silver ion to form an aggregate of silver atoms when finally neutralized by the available electrons in the grain:

$$Ag^0 + Ag^+ \longrightarrow Ag_2^+ + e^- \longrightarrow Ag_2^0$$
$$Ag_2^+ + Ag^0 \longrightarrow Ag_3^+ + e^- \longrightarrow Ag_3^0$$
$$Ag_3^+ + Ag^0 \longrightarrow Ag_4^+ + e^- \longrightarrow Ag_4^0$$

It is the presence of this free silver in the exposed silver bromide grains that provides the latent image which will be brought out later by the developer. It appears that a minimum of four silver atoms (Ag_4^0) are needed to initiate development (this process will be discussed later).

The free bromine atom, Br^0, tends to destroy the latent image by reactions which reduce either the number of free electrons

$$Br^0 + e^- + Ag^+ \longrightarrow Ag^+Br^-$$

or the size of the free silver particle

$$Br^0 + Ag_3^0 \longrightarrow Ag_2^0 + Ag^+Br^-$$

Hastening this process is the fact that the bromine atom, being neutral, effec-

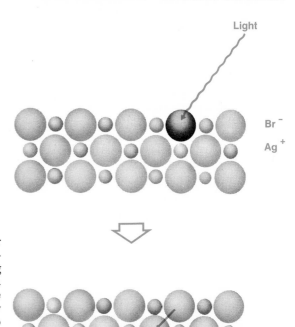

Light

Br⁻

Ag⁺

FIGURE 24-11   How light affects the silver bromide crystal. In the upper part of the illustration a photon strikes a bromide ion, producing a bromine atom. The bromine atom appears as a "positive hole" in the lattice of negative bromide ions. In the lower portion, the "hole" migrates by the exchange of electrons from a bromide ion to the "hole."

tively acts as a "positive hole" in the Ag^+Br^- lattice. Other neighboring bromide ions can and do give up electrons to this "hole," producing an effective migration of the image-destroying Br^0 throughout the grain (Figure 24–11).

In order to overcome the problem of "migrating bromine atoms" within the exposed grain, chemical sensitizers are added. S. E. Sheppard discovered, in 1925, that these compounds were present in gelatin emulsions as impurities, principally sulfur compounds. Now, compounds such as silver sulfide (Ag_2S) are added to a photographic emulsion. It appears that the wandering "positive holes" (bromine atoms) react with the Ag_2S molecules. The net reaction is:

$$2Br^0 + Ag_2S \longrightarrow 2Ag^+Br^- + S^0$$
$$\textit{Free sulfur}$$

This mechanism renders a silver halide grain more sensitive than it would be ordinarily since the latent image-destroying bromine atoms are used up in another process.

Upon development, the silver halide grains which contain the tiny nuclei of free silver, Ag_n^0, are completely reduced by the developer.

Our study of silver halide photochemistry would not be complete without mentioning "grain" from the photographer's viewpoint. As the grain size in the emulsion increases, the effective light sensitivity of the film increases (up to a point). The reason for this is that the same number of silver atoms are needed to initiate reduction of the entire grain by the developer despite the grain size. If the grain gets too large, the chances of four silver atoms forming a nucleus are not as great, so that very large-grained emulsion would actually be less sensitive.

593

Spectral Sensitivity

Probably the most important ingredients in a black and white photographic emulsion, other than the silver halide salts themselves, are the spectral sensitizing dyes. Silver halides are most sensitive to blue light or higher energy electromagnetic radiation such as UV light (Figure 24–12). A film manufactured with only silver halides as the photosensitive agents will be only blue-sensitive and will not "see" reds, yellows, greens, and so on, as ordinary colors.

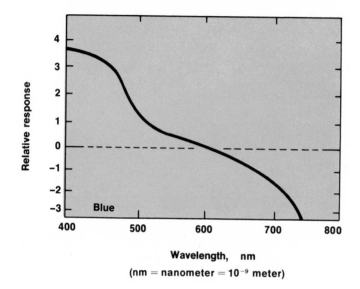

FIGURE 24–12   Spectral sensitivity of a typical AgBr emulsion.

In 1873, while trying to eliminate light scattering problems in photographing the solar spectrum, W. H. Vogel, a German chemist, added a yellow dye to his emulsion. To his surprise he discovered that he could now record images in the green region of the visible spectrum. Later, in 1904, another German, B. Homolka, discovered a dye, pinacyanol, which when added to a silver halide emulsion rendered it sensitive to the entire visible spectrum (Figure 24–13). Films of this type are called *panchromatic* or "pan" films.

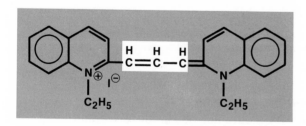

FIGURE 24–13   Pinacynol, a cyanine-cyan dye. The conjugated group (white block) is the chromophore (Chapter 15). Other cyanine dyes have more CH groups, absorb longer wavelengths of light, and shift the film sensitivity toward the red.

The mechanism by which a dye molecule can impart spectral sensitivity to silver halide grains seems to involve initially the absorption of a photon of light by the dye molecule. Next, the excited molecule ejects an electron into the silver halide grain where a free silver atom is formed. The electron-deficient dye molecule then oxidizes the bromide ion, producing a bromine atom or positive hole:

594

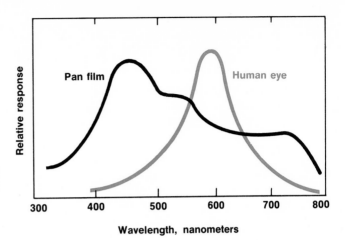

FIGURE 24-14   Spectral sensitivity of a panchromatic film compared to that of the human eye.

$$\text{Dye} + h\nu \longrightarrow \text{Dye}^+ + e^-$$

$$\text{Ag}^+ + e^- \longrightarrow \text{Ag}^0 \longrightarrow \text{(silver nuclei)}$$

$$\text{Dye}^+ + \text{Br}^- \longrightarrow \text{Dye} + \text{Br}^0 \text{ (positive hole)}$$

Thus, the process is effectively the same as a photon striking the bromide in the grain itself.

Amplification of the Latent Image—Development

Silver halides are not the most photosensitive materials known. Why, then, are they such good image producers? The answer to this question lies in the fact that the impact of a single photon on a silver halide grain, producing a nucleus of at least four silver atoms, is amplified as much as a billion times by the action of a proper reducing agent (developer).

When an exposed film is placed in developer, the grains containing silver atom nuclei are reduced faster than those grains which do not. The more nuclei present in a given grain, the faster the reaction. The reduction reaction is

$$\text{Ag}^+ + e^- \longrightarrow \text{Ag}^0$$

Factors such as temperature, concentration of the developer, pH, and the total number of nuclei in the grains determine the extent of development and the intensity of free silver (blackness) deposited in the film emulsion.

Not only must the developer be capable of reducing silver ions to free silver, but it must be selective enough not to reduce the unexposed grains, a process known as "fogging." Table 24–9 lists some substances that are used as developing agents.

Most developers used for black and white photography are composed of hydroquinone and metol or hydroquinone-phenidone. A typical developer consists of a developing agent (or two), a preservative to prevent air oxidation, and an alkaline buffer to prevent the actual reduction reaction from being retarded (Table 24–10). Other components might be added but are not absolutely necessary.

595

TABLE 24-9   SOME COMPOUNDS USED AS PHOTOGRAPHIC DEVELOPERS

NAME	FORMULA
Gallic acid	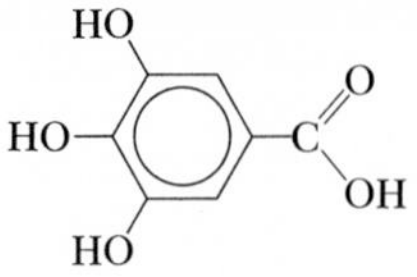
o-Aminophenol	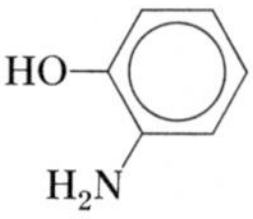
Hydroquinone	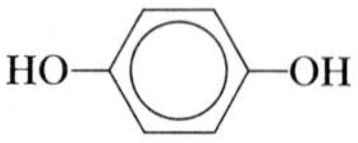
p-Methylamino-phenol (metol)	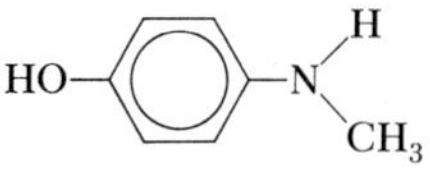
1-Phenyl-3-pyrazolidone (phenidone)	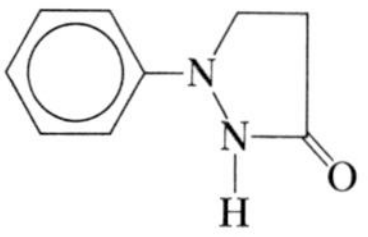

TABLE 24-10   FORMULA FOR A TYPICAL DEVELOPER FOR BLACK AND WHITE FILMS

750 ml water at 50°C; dissolve in this water:

Metol	2.0 g
Hydroquinone	5.0
Sodium sulfite	100.0
Borax, $Na_2B_4O_7 \cdot 10H_2O$	2.0

When hydroquinone acts as a developer, quinone is formed. Two protons are also produced for every two silver atoms:

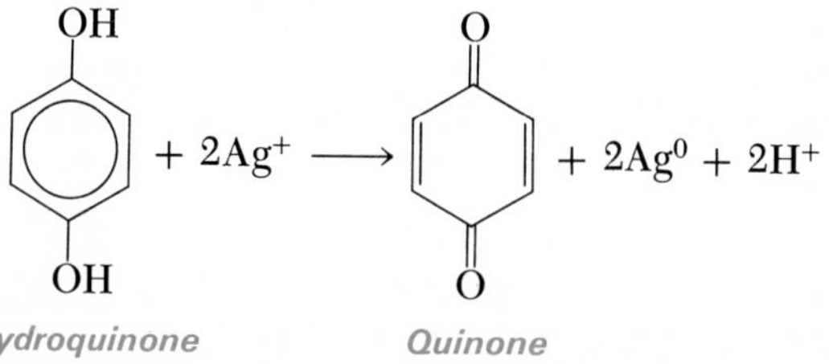

Hydroquinone   *Quinone*

Owing to the reversibility of this reaction, a buildup of either protons or quinone would impede the development process. The sodium sulfite added will react with quinone:

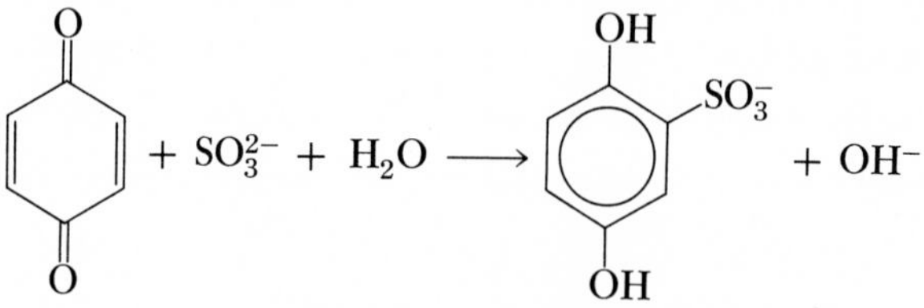

596

This, by the way, is the way sulfite acts as a preservative in the developer since air oxidation also produces quinone, and its buildup in solution prior to use would render the developer virtually useless.

When dissolved in water, a buffer such as borax yields the following products:

$$Na_2B_4O_7 \cdot 10H_2O \longrightarrow \underset{\substack{\text{Boric} \\ \text{acid}}}{2H_3BO_3} + 2Na^+ + \underset{\substack{\text{Borate} \\ \text{anion}}}{2B(OH)_4^-} + 3H_2O$$

The presence of both the boric acid and the borate anion makes the solution a buffer solution (that is, it does not change pH easily). In the developer solution, protons produced in the reduction reaction are used up by the borate anion:

$$H^+ + B(OH)_4^- \longrightarrow B(OH)_3 + H_2O$$

A stop bath in photography stops the development process by decreasing the pH and usually contains a weak acid such as acetic acid. If development proceeds too long, sufficient fogging occurs to render a negative useless. In addition, the action of a developer is very much dependent upon the temperature. Since the rates of these development reactions increase with increasing temperature, development for 8 minutes at 20°C will produce *less* free silver on a negative than the same development time at 25°C. In the darkroom, the photographer usually controls the temperature of the development bath very carefully.

Fixing

One of the principal problems in the early development of photography was the permanence of the image. If development takes place with free silver being deposited where the light intensity was greatest and nothing further is done to the negative, the undeveloped silver halide will be exposed the instant it is taken into the light. After that, almost any reducing agent will completely fog the negative. In order to overcome this problem, a suitable solvent to remove the unreduced silver halides (AgX) had to be found. The most commonly used fixing agent in black and white photography is the thiosulfate ion ($S_2O_3{}^{2-}$) (solutions of sodium thiosulfate, known as "hypo," were first used by Talbot), which forms a stable complex with the silver ion in aqueous solution:

$$\underset{\substack{\text{Insoluble} \\ \text{salt}}}{AgBr(s)} + 2S_2O_3^{2-} \longrightarrow \underset{\substack{\text{Water-soluble} \\ \text{complex}}}{Ag(S_2O_3)_2^{3-}} + Br^-$$

After fixing is complete, sufficient washing to remove the last traces of thiosulfate from the emulsion is necessary. Otherwise, the thiosulfate ion will decompose to produce free sulfur which will then react with the free silver to produce darkly colored silver sulfide and mar the quality of the image.

$$S_2O_3^{2-} + 2H^+ \longrightarrow S^0 + SO_2 + H_2O$$

$$S + 2Ag \longrightarrow Ag_2S \text{ (brown)}$$

597

When certain types of black and white films are processed, as much as 60 to 80 per cent of the original silver in the emulsion is removed by fixing. This silver can be recovered before the fixing bath is discarded, but the recovery only pays for itself when one has available large quantities of solution. For example, a photofinisher processing 100,000 rolls of black and white film a year would realize about $1240 of silver, with silver selling at $2.00 a troy ounce.

Printing

Photographic paper for black and white photography is essentially like film and is developed and fixed in similar ways. The principal difference in photographic paper is the presence of silver chloride, which is a less photosensitive material requiring a longer exposure time.

Special precautions must be taken during the washing to eliminate the thiosulfate ions from the paper or it will decompose and stain the print with silver sulfide (see preceding section). A *hypo eliminator,* containing hydrogen peroxide, generally is used to oxidize the thiosulfate to the stable sulfate ion:

$$S_2O_3^{2-} + 4H_2O_2 \longrightarrow 2SO_4^{2-} + 2H^+ + 3H_2O$$

Hydrogen peroxide   Sulfate ion

Instant Black and White Pictures

In 1947, Dr. Edwin H. Land invented a process that produced a finished picture in one minute. Since its introduction, the Polaroid process has been popular for just this unique feature.

After exposure, a Polaroid film is brought into contact with a piece of receiver paper; at the same time a pod of developer-hypo is broken and spread over the film (Figure 24–15). As the developer reduces the exposed silver halide grains in the film emulsion, the hypo complexes the unexposed silver ions, which then diffuse across the boundary onto the receiver paper. There, in contact with minute grains of silver already in the paper, the developer reduces the silver in the hypo complex to free silver and a *positive* image. This positive image forms since black areas in the original image photographed exposed no silver grains in the emulsion, and it was this silver, as Ag^+ ions, which was then carried onto the receiver paper and reduced (Figure 24–16).

COLOR PHOTOGRAPHY

The chemistry of color photography, which dates back to 1861, is a good deal more complicated than that of black and white photography. James Clerk Maxwell, the famous English physicist, was the first to photograph an object in color. He used three exposures through three primary color filters and then resynthesized the color of the object by projection of the image through the same filters.

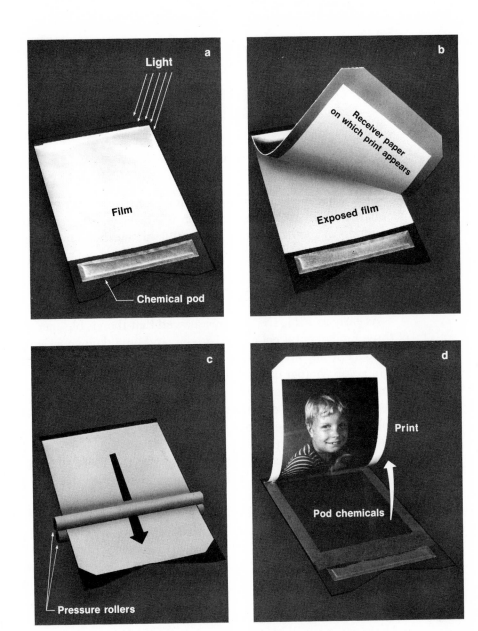

FIGURE 24–15 Schematic diagram of the Polaroid process for black and white photographs.

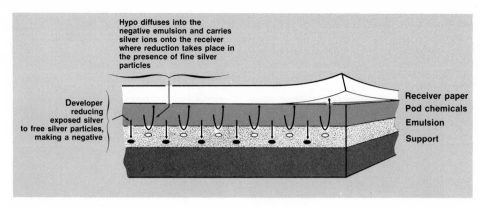

Hypo diffuses into the
negative emulsion and carries
silver ions onto the receiver
where reduction takes place in
the presence of fine silver
particles

Receiver paper
Pod chemicals
Emulsion
Support

Developer
reducing
exposed silver
to free silver particles,
making a negative

FIGURE 24-16   The chemistry of the Polaroid process.

This experiment actually predated panchromatic film, but owing to a peculiarity of his particular emulsion it was essentially panchromatic anyway. The important point was that the results fit the then emerging theory of color vision and led to other more significant results.

Additive and Subtractive Primary Colors

As early as 1611, De Dominis showed that the visible spectrum is composed of three fundamental colors: red, green, and blue (known as additive primaries). This concept has since proved useful in the development of color vision theory and in color photography. After 1861, the idea slowly evolved that in order to reproduce color images, a film with three different layers, each sensitive to one of the three primary colors, would have to be developed. After the discovery of color-sensitizing dyes and panchromatic black and white film, several different techniques for color photography were developed, but not until 1935, when Kodachrome was placed on the market, did the process reach the consumer. The Kodachrome process produced transparencies (or slides) which are viewed as transmitted light.

If additive primary colors are used to form an image by superposition, such as we might expect in a color transparency, problems with light transmittance arise (see color plate VI). In order to overcome these problems and obtain a color slide (or color negative, for that matter) another system of primary colors was developed (see color plate VII) known as *subtractive primary* colors. These colors are produced by dyes that *absorb* the additive primary colors. Thus a dye that absorbs red light transmits or reflects the remainder of the spectrum and appears greenish blue (*cyan*). Absorption of blue light renders a dye *yellow*, and absorption of green light makes the dye appear bluish red (*magenta*).

Mixtures of these subtractive primary dyes in a photographic emulsion, formed in the appropriate proportions in the development process, produce an image of the desired color. For example, a mixture of magenta and cyan dyes would appear *blue*, since the magenta dye absorbs green light and the cyan dye absorbs red light, leaving only blue.

The use of subtractive primaries in color photography was suggested in 1869, but it was much later before the chemistry was worked out in detail to yield

600

good results. The problem is to get the right amount of the correct subtractive primary dye in the right place to reproduce the correct true-to-life color. White is produced by the absence of all three subtractive primaries, and black is produced by their presence (Plate VII).

Color Film

Generally, a color film consists of a support and three color-sensitive emulsion layers. The blue-sensitive layer is usually on the top since silver halides are inherently blue-sensitive. Next, a yellow colored filter layer is added. This layer absorbs blue light and serves to protect the lower emulsion layers from blue light. A green-sensitive layer is added and followed by a red-sensitive layer and the support (Figure 24–17). These layers are rendered color-sensitive by dyes similar to those in the cyanine class, which render black and white film panchromatic. It should be realized, however, that the color-sensitizing dyes are *not* generally involved in producing the final primary colors responsible for the color of the image. It is the final processing of the color film that yields the color image.

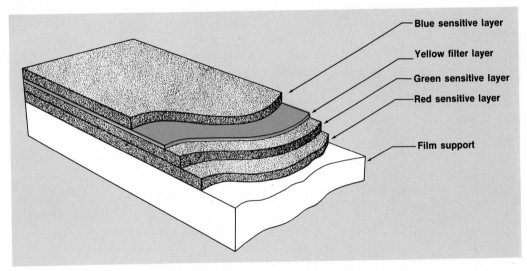

FIGURE 24-17   A typical arrangement of color-sensitive emulsion layers in a color film.

Color Development

Most color films are developed with the aid of a dye-forming color process first introduced by a German chemist, R. Fischer, in 1912. The basis for this process is the oxidation of the developer to a dye-forming substance, which is then reacted with a molecule called a *coupler* to form the dye.

In some color films such as Kodachrome II, the coupler is dissolved in the developer, and the two react together in the presence of the silver halide grain. In other color films, notably, Kodacolor, Ektachrome, and Anscochrome, the couplers are distributed evenly in the appropriate emulsion layers in which the desired dye is to be formed.

601

Color developers are generally substituted amines, such as N,N-diethyl-p-phenylenediamine, like the one below:

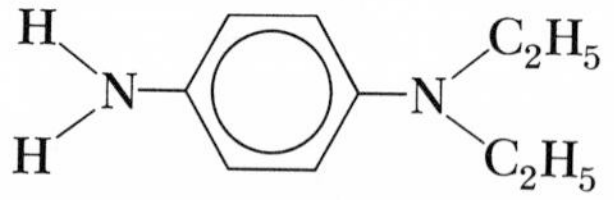

N,N-Diethyl-p-phenylenediamine
(*a color developer*)

Minor changes in the structure of the developer are designed by the photographic chemist to vary development speed and solubility. Perhaps the most bothersome aspect of these compounds is that they are *allergenic,* causing a skin irritation in most people similar to that caused by poison ivy. Amateur color developing is relatively rare, but is possible with specially designed developing agents which have low toxicity.

To form a cyan dye, a phenol compound such as α-naphthol acts as a coupler:

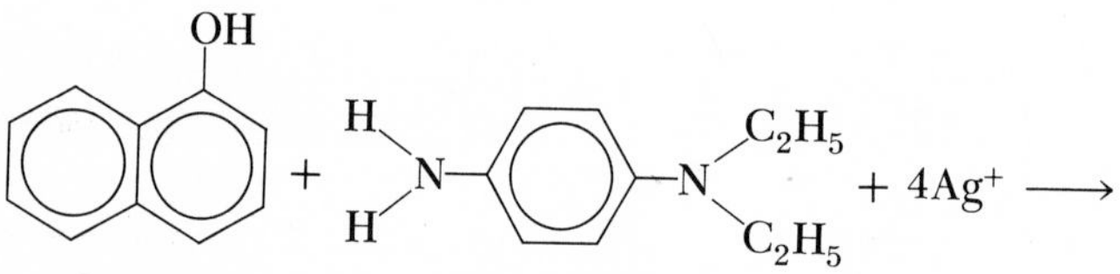

Coupler
(*α-naphthol*)   *Developer*

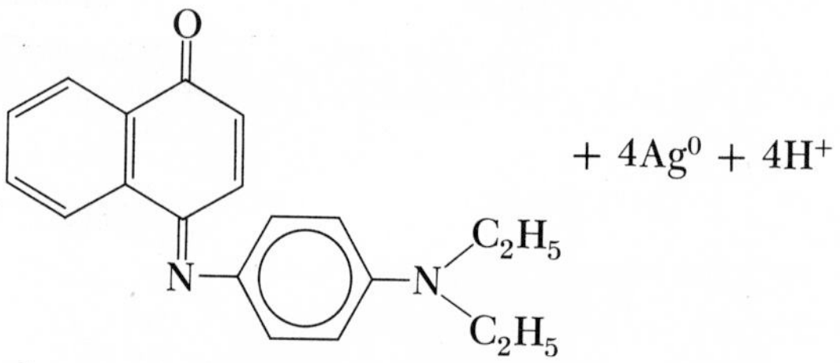

$+ 4Ag^0 + 4H^+$

A cyan dye (absorbs at 630 nm)

Thus, in the development of an exposed silver halide grain in the red-sensitive emulsion layer, a small amount of cyan dye is produced. The free silver must be bleached out prior to finishing.

The Kodachrome Process

An interesting example of a widely used color photography system is the Kodachrome process of Eastman Kodak Co. The Kodachrome process is a *reversal* process; this means colors are reproduced in terms of their correct values and not their negative or complementary colors. The first developer in the Kodachrome process is a black and white developer. By careful temperature control, development of the exposed silver halide is made essentially complete.

The remaining unexposed silver halide in the three color-sensitive emulsions is a positive record of the original exposure. For example, blue light striking

602

the film would, upon black and white development, leave free silver in the blue-sensitive layer (Figure 24–18). Since no other color-sensitive layers were exposed by the original image, they contain no information. Now, selective reexposure and color development will produce free silver throughout the emulsion layers, along with the colored dyes, *except* where the blue light originally struck the film. No dye could form there since the silver was reduced with a black and white developer earlier.

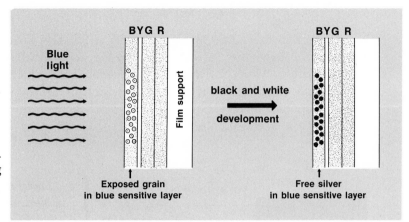

FIGURE 24–18 Simplified color image-forming process.

Next, all the silver in the three emulsion layers, as well as the yellow-colored protective layer, is bleached out of the emulsion with an oxidant such as ferricyanide ion, $Fe(CN)_6^{3-}$:

$$Ag^0 + Fe(CN)_6^{3-} \longrightarrow Ag^+ + Fe(CN)_6^{4-}$$

Once oxidized, the silver is treated with hypo and washed from the emulsion. The resulting emulsion is colored, but transparent. Considering the result of the blue light on the film, we see that the transmitted light will appear blue (Figure 24–19).

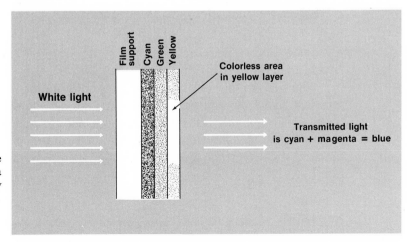

FIGURE 24–19 White light passes through a three-layer transparency produced by the light.

Color Pictures in a Minute

As everyone knows the Polaroid Camera can produce an almost "instant" color picture. The chemical processes involved in making this type of color image (Polacolor) are similar to those mentioned earlier in this chapter, but there must be a delicate balance of properties of the developers, couplers, and dyes in order to obtain a good picture. Figure 24–20 shows a schematic of the Polacolor system. Prior to development, the receiver and photosensitive layers are separated. Light reaching the film strikes the blue-sensitive emulsion first, as in other color films. After exposure, development is initiated by pulling the film past pressure rollers, which also break the alkali-containing pod. The Polacolor dyes forming the negative image are "precipitated" in the photosensitive layer, which becomes a negative, while the dye-developer molecules which have not encountered exposed silver halide diffuse in the alkali up onto the receiver. There the dye-developer molecules react with the mordant to form the proper colors for a positive image (see color plate VI).

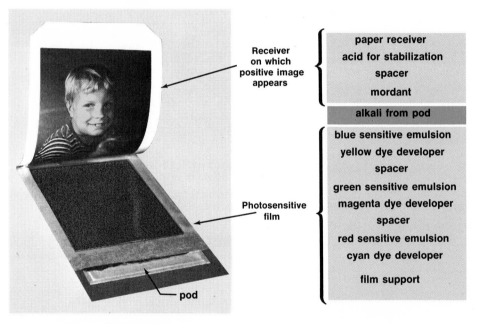

FIGURE 24-20   The Polacolor system.

The spacers serve to slow down the diffusion of the alkali into the photo-sensitive layer, allowing enough time for the developers to work. Even so, one minute is a relatively fast reaction time. Once the alkali has served its purpose of moving the dye-developer molecules onto the receiver, it is neutralized by the acid stabilizer. The finished Polacolor picture requires no further stabilization.

QUESTIONS

1. What are the essential differences between high and low octane gasolines with respect to energy content, additives, and completeness of burning?

2. Write a balanced equation for the complete combustion of 2,2,4-trimethylpentane.

3. What is meant by high detergency (HD) gasoline? Specifically, what do the detergents do for a gasoline engine?

4. What is the purpose of each of these substances in gasoline?

   (a) ethylenediamine   (c) ethylene glycol
   (b) TCP   (d) dyes

5. What is the meaning of 10W-30 oil?

6. How do the antiwear agents in motor oils work?

7. What is the difference in the composition of motor oils and vegetable oils, such as corn oil or palm oil?

8. What is the basic difference in the composition of motor oils and automotive greases?

9. Suppose 186 grams of ethylene glycol (molecular weight: 62) is added to 1000 grams of water. What is the freezing point of the mixture?

10. Referring to the structure of ethylene glycol, explain why this compound is so soluble in water.

11. Why would it be undesirable to have a methanol-type antifreeze in your radiator during the summer?

12. Explain how a de-icer melts ice.

13. Why is the methanol-type antifreeze called temporary and the ethylene glycol-type called permanent antifreeze?

14. Write a chemical equation for the rusting of iron in a radiator.

15. List three typical additives in antifreeze formulations and describe the action of each.

16. Suggest a reason why certain metals in paint pigments would retard biological growth on the paint surface.

17. What is the purpose of each of the following in paints? TiO_2, $CaCO_3$, linseed oil, polyvinyl acetate, $MnCl_2$, mineral spirits, $Na_4P_2O_7$, phenyl mercuric acetate, ethylene glycol, carboxymethylcellulose, ZnO.

18. What is meant by the "drying" of paints? Is the process any different for latex paints than for oil-based paints? Explain.

19. Do modern "latex" paints contain rubber latex (Chapter 14)? Explain.

20. What is the chemical cause of the following phenomena in paints:

   (a) chalking?   (c) yellowing?
   (b) gloss of enamel paints?   (d) flooding?

21. What agents in the environment destroy paint?

22. Use the structure of PVA and a polyacrylate to show why they soften when soaked in water.

23. Look up the chemistry of the Daguerreotype process. How is it similar to modern photography? How is it different?

24. A small amount of developer is sometimes added to the chemicals making up a photographic emulsion (i.e., the silver halide salts, gelatin, and so on). Why is this done?

605

25. What is the purpose of sodium sulfite in a developer solution?

26. What would be the effect of low pH on a typical developer?

27. Explain the term *fixing*. What is it chemically and why is it important in photography?

28. What are the subtractive primary colors?

29. Explain this statement: "There is no silver in that color slide!"

30. Explain how a red dot would be photographed with a color film such as Kodachrome II.

SUGGESTIONS FOR FURTHER READING

Eaton, G. T., "Photographic Chemistry," Morgan and Morgan, Publishers, New York, 1965. (A good, elementary monograph.)

Geller, I., "The World's First Oil Mine," *Chemistry*, Vol. 41, No. 8, p. 10 (1968).

Hillson, P., "Photography, A Study in Versatility," Doubleday and Co., Inc., Garden City, N.Y., 1969.

Keller, E., "Photography, Part I: Images in Silver," *Chemistry*, Vol. 43, No. 9, p. 6 (1970).

Keller, E., "Photography, Part II: Images in Color," *Chemistry*, Vol. 43, No. 11, p. 8 (1970).

Kirk, R. E., and D. F., Othmer, "Encyclopedia of Chemical Technology," Interscience Publishers, New York, 1967.

Newhall, B., "Latent Image—The Discovery of Photography," Doubleday and Co. Inc., Garden City, N.Y., 1967.

Schaar, B. E., "Chance Favors the Prepared Mind. Part Five—Photography," *Chemistry*, Vol. 39, No. 5, p. 34 (1966).

Weissberger, A., "A Chemist's View of Color Photography," American Scientist, 58(6):648 (1970).

NUCLEAR ENERGY

The 19th century Daltonian atom pictured each element with its own characteristic, indestructible atom. Becquerel, with his discovery of natural radioactivity, and later students of this phenomenon showed that some of the atoms not only are destructible, but are actually unstable and decompose spontaneously. Such studies revealed that the decomposition of massive nuclei of heavier elements resulted in less massive nuclei of lighter elements. Reactions involving changes in nuclear structure are called *nuclear reactions*. Nuclear reactions are fundamentally different from chemical reactions since the nucleus is undergoing change; as learned earlier, chemical changes involve only changes in the outer electronic structure of the atom.

It has been pointed out that the making or breaking of chemical bonds involves the release or absorption of energy. We have, therefore, come to expect energy changes to be associated with chemical changes. The energy change associated with a nuclear change may be dramatically larger, atom for atom, than the largest possible energy change for a chemical reaction. For example, when one mole (6.02×10^{23} molecules or 16 grams) of methane from natural gas is burned, about 200 kilocalories of heat are liberated:

$$CH_4 + 2O_2 \longrightarrow CO_2 + 2H_2O + 200 \text{ kilocalories (kcal)}$$

In contrast, a lithium nucleus can be made to react with a hydrogen nucleus to form two helium nuclei in a nuclear reaction. The energy released per mole of lithium in this reaction is 23,000,000 kilocalories.

$$_3^7\text{Li} + _1^1\text{H} \longrightarrow 2\,_2^4\text{He} + 23,000,000 \text{ kcal}$$

One of the basic needs of man is energy. The world population explosion raises serious questions concerning the present energy supply, particularly so in certain areas of the world. In addition to the problem of an insufficient food supply, fossil fuels (coal and petroleum) and hydroelectric energy appear inadequate for modern man's projected energy needs for warmth and mechanical work. This is especially true if the living standards of the Western World are to be enjoyed by the entire human race. The amazing amounts of energies released in certain nuclear reactions have thus far been only slightly tapped in man's quest for cheap, controlled energy. The destructive power of the nuclear bomb illustrates the magnitude of the energies involved, but this, of course, is in a virtually uncontrolled form. The nuclear power plants in ships and electrical power stations are what may be just the beginning in man's control of this energy mammoth. In this chapter we shall study how nuclear reactions are detected and studied, and some of the major classes of nuclear reactions and how they are initiated, giving emphasis to their energy potentialities.

NUCLEAR PARTICLES AND REACTIONS

As in the case of atomic and molecular theory, we must depend on circumstantial evidence to establish the identity of the particles involved in nuclear reactions. The study is somewhat more difficult with nuclear reactions because many of the reactions of interest lead to products which can exist for only a very short period of time. For such nuclear reactions it is quite impossible to collect molar amounts of the reaction products, to study the products, and thereby deduce characteristic properties of the submicroscopic particles involved. It is evident then that successful methods for the study of nuclear reactions must have two characteristics: (a) rapid observations, and (b) the ability to record these observations for later study.

Three methods for the detection of particles of nuclear reactions were briefly discussed in Chapter 6. Alpha, beta, and gamma rays from the spontaneous nuclear decays in uranium ore can be detected by the darkening produced on photographic film or the light produced on a phosphor (scintillation) screen. The scintillation screen is so sensitive that a single impact of a high energy atomic particle can be seen as a flash of light by observing the screen with a magnifying lens. A third method which can "see" only charged particles (not gamma rays) is the collector plate in a mass spectrometer (Chapter 6). As the charged particles hit the plate, a current is induced, the size of which is a measure of the number of charged particles. Like the photographic exposure, a large number of particles are needed for detection in a mass spectrometer.

Instruments that have been designed specifically for the detection of high-energy particles or photons from nuclear transitions include the Geiger counter and the scintillation counter. The Geiger counter is composed of a Geiger tube (Figure 25–1) and electrical equipment to amplify the current signal from the tube so it can be heard in the form of a click, seen as a flash of a light bulb, or stored for later study in a recording. The Geiger tube is made of a metal case with a window of thin mica (a mineral having suitable strength and transparency). Running through the center of the tube and insulated from the metal case is a wire charged at $+1500$ volts relative to the metal case at zero potential. The tube is filled with argon, an inert gas, at 0.05 atmosphere pressure. Under these conditions the charge is so great that the tube is on the verge of discharge. A slight increase in voltage would cause the argon atoms to ionize

$$Ar \longrightarrow Ar^+ + e^-$$

and the tube would discharge; the Ar^+ ions would then rush to the negative

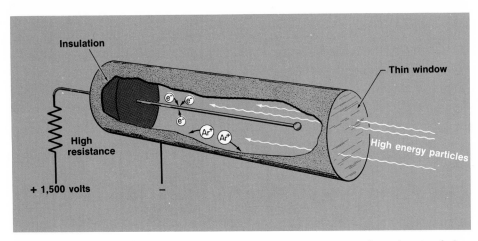

FIGURE 25-1   Schematic drawing of a Geiger tube. The voltage is adjusted to just below the discharge potential. Under these conditions high energy particles which penetrate the thin window cause the argon atoms to ionize. The result is a cascade effect, in which newly formed charged particles, accelerated by the electric field, produce more ions, and a massive and sudden discharge occurs. The high resistance in series with the high voltage electrodes prevent a large current for more than an instant, thus making the high voltage device safe to handle.

electrode (tube) and the free electrons to the positive electrode (wire). If the voltage on the tube is adjusted to just below the discharge potential, a high energy particle entering through the mica window will cause one or more argon atoms to ionize. The resulting charged particles cause other argon atoms to ionize in a cascade effect. Consequently there is a sudden and massive discharge, all produced by the single high energy particle. The number of discharges per second can be measured by the current output of the tube. It is even possible that each pulse (discharge) can be counted with counting circuits. As a result, the Geiger counter not only can detect products of nuclear decay, but can actually measure their number as well. The Geiger counter is rugged and dependable, but it suffers from one disadvantage: in order to be detected, the particle must have sufficient energy to penetrate the mica window.

A number of interesting nuclear changes do not produce particles with sufficient energy for Geiger counter detection, so a more sensitive radiation detector is needed to study these cases. The scintillation counter is sensitive to such "soft-nuclear" emissions. The outside of a glass window in a scintillation tube is covered with a phosphor coating. The minute flash of light produced by a relatively low energy nuclear emission passes through the glass to a very sensitive photoelectric detector. The resulting electric current is amplified and either displayed or recorded.

Scientists were not able to characterize the electron, alpha particle, beta particle, and positive ray particles until it was learned how to observe their behavior in electric and magnetic fields (Chapter 6). With the charge-to-mass ratios determined by Thomson and the charge per particle determined by Millikan, investigators were quickly able to study those particles which could be produced (electrical discharge) or observed (natural radioactivity) in relatively large and steady streams. At this point, it was reasonable to assume that other nuclear particles escaped detection because they were produced in insufficient

609

quantities to be detected. Subsequently, C. T. R. Wilson invented the *cloud chamber*, which could actually "see" a *single*, high energy, nuclear particle in flight, the collision of such a particle with another nuclear particle, and the path of the products of such a reaction. A single nuclear event could be observed! Furthermore, when the cloud chamber is placed in a magnetic field, the charged particles will follow a curved path, and the individual particles, as a result, can be characterized.

The structure of a simple cloud chamber is illustrated in Figure 25–2. Pressure is exerted on a closed system containing a nuclear particle source, such as an alpha emitter, air, and a layer of water. The invisible alpha particles are constantly being emitted from the alpha source. If pressure is exerted on this system, the concentration of water vapor is increased in the air space around

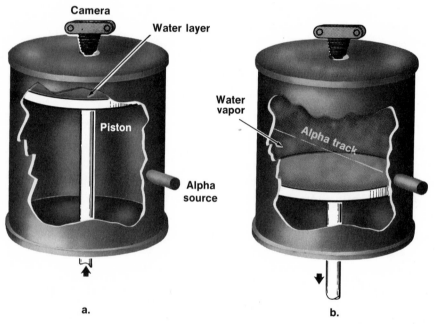

FIGURE 25–2   A simple Wilson cloud chamber. Figure 25–5 is a photograph of alpha cloud tracks.

the alpha emitter. Now, if the pressure is suddenly reduced, the temperature of the air drops, and the air will contain more water vapor than it can normally hold. Consequently, there will be a strong tendency for the water vapor to condense (precipitation). In such a supersaturated system the water molecules readily condense on charged particles. Now, the alpha particle, because of its high energy, ionizes air particles in its path. As a result, there will be a visible path of condensed water (a cloud track) tracing the alpha particle pathway. Any charged particles with sufficient energy to ionize the molecules of the air can thus be observed. It is a relatively simple matter to photograph such cloud tracks and record nuclear events, such as a collision between an alpha particle and a nitrogen molecule, for later study. The cloud track is somewhat analogous to the vapor trails of a high-flying jet airplane; even when the plane is too high to be seen itself, the condensed water from the exhaust clearly marks its pathway.

An even more sophisticated device for photographing the path of high energy particles is the *bubble chamber*. Using basically the same principle as the cloud chamber, the nuclear event occurs in a bath of liquid hydrogen. The particle path is observable because bubbles of hydrogen gas form along the track.

TRANSMUTATION IN NATURE

Armed with the ability to detect and characterize high energy particles in flight from nuclear reactions, the nuclear scientist began collecting information concerning the nuclear reactions that occur in substances that are naturally radioactive. There are many such reactions since all of the elements above bismuth in atomic number and a few below have one or more naturally occurring radioactive isotopes. This means, of course, that each of these isotopes is an emitter of alpha, beta, and/or gamma rays, and with each emission a nuclear reaction spontaneously occurs.

The isotope of uranium with atomic mass 238 is an alpha emitter. The atomic number of uranium is 92, which means that uranium-238 has 92 protons and 146 (i.e., $238 - 92$) neutrons in the nucleus. When the uranium-238 nucleus gives off an alpha particle, made up of 2 protons and 2 neutrons, it necessarily loses 4 units of atomic mass and 2 units of atomic charge. The resulting nucleus would then have a mass of 234 and a nuclear charge of 90. Now, atoms containing 90 protons in the nucleus are atoms of thorium, not uranium. This spontaneous nuclear reaction then has changed an atom of one element into an atom of another element, and is an example of the *transmutation* of elements. The transmutation of cheap elements into gold was the dream of ancient alchemists and was discarded only with the acceptance of Dalton's indestructible atoms. Interestingly enough, transmutations were going on all the while. The decomposition of the U-238 nucleus is stated briefly by the nuclear equation:

$$^{238}_{92}U \longrightarrow {}^{4}_{2}He + {}^{234}_{90}Th$$

In this equation, the *mass number* of the particle is given by the *superscript*, and the *nuclear charge* (atomic number) is given by the *subscript*. If the characterized alpha emission was not proof enough that this reaction occurs, additional evidence is supplied by the fact that Th-234 is always found with U-238 in natural ore deposits and almost always in just that concentration predicted by the speed of the reactions involved.

Thorium-234 is also radioactive. However, this nucleus is a beta emitter. This poses an interesting question: How can a nucleus containing protons and neutrons emit an electron? It has been established that an electron and a proton can combine outside the nucleus to form a neutron. Therefore, the reverse process is proposed to occur in the nucleus. A neutron decomposes, giving up an electron and changing itself into a proton.

$$^{1}_{0}n \longrightarrow {}^{1}_{1}p + {}^{0}_{-1}e$$

Since the mass of the electron is essentially zero compared to that of the proton and neutron, the nucleus would maintain essentially the same mass but it would

611

now carry one more positive charge (a proton instead of one of the neutrons). This nucleus is no longer thorium since thorium can have only 90 protons in the nucleus; it is now a nucleus of element 91, protactinium (Pa). The reaction is:

$$^{234}_{90}\text{Th} \longrightarrow ^{234}_{91}\text{Pa} + ^{0}_{-1}\text{e}$$

Gamma radiation may or may not be given off simultaneously with alpha or beta rays, depending on the particular nuclear reaction involved. Since gamma rays involve no charge and essentially no mass, it is evident that the emission of a gamma photon cannot alone account for a transmutation event.

The uranium decay is extremely slow compared to the thorium decay. However, each of these, and every other radioactive decay reaction, is found to have a characteristic *half-life*. The half-life represents the period of time required for half of the radioactive material originally present to undergo transformation. The half-life is independent of the amount of radioactive material present and is determined only by the type of radioactive nucleus present in the sample. For example, in the reaction above, half of the thorium will be left in 24 days. In another 24 days one-half of the half ($\frac{1}{4}$) will be left. This process continues indefinitely with one-half of the thorium-234 that exists decaying each 24 days. The time taken for one-half of the amount of a radioactive isotope to decay is called the half-life of the isotope (Figure 25–3). Some half-lives are extremely long and others are extremely short. The half-life for the U-238 alpha decay is 4.5 billion years. As one would expect, relatively large amounts of U-238 can be found in nature while only trace amounts of Th-234 occur.

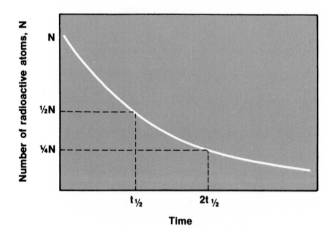

FIGURE 25–3   Half-life. The rate of decay for a radioactive atom depends in a very special way on the number of those atoms present. The rate is such that in a given period of time—the half-life for the species—one-half of the original number of atoms will be gone regardless of the number present at the start. In this graph the number of atoms remaining is plotted with time. $t\frac{1}{2}$ is the half-life, and at the end of one $t\frac{1}{2}$ period the original number, N, is reduced to $\frac{1}{2}$N. After two of these periods, the number is reduced to one-half of $\frac{1}{2}$N, or $\frac{1}{4}$N.

The radioactive decay of thorium-234 into protactinium-234 is the second step in a *series* of nuclear decays that starts with U-238 and after 14 decays ends up as a stable (nonradioactive) isotope of lead, Pb-206. This decay series is called the *uranium series* (Figure 25–4). Two other natural decay series exist which are similar to the uranium series except that they start out with a different parent isotope and proceed downward through a different set of radioactive nuclei. The *thorium series* begins with Th-232 (a different isotope from the two thorium

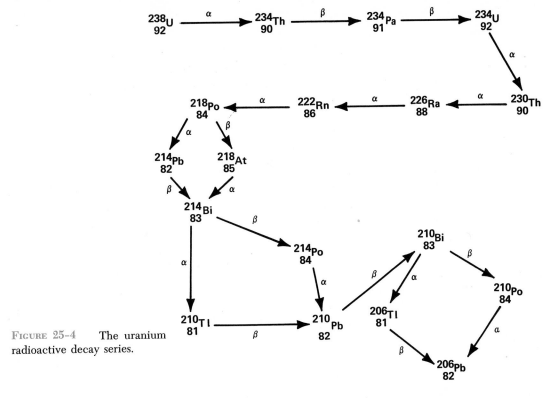

<figure data-ref="fig-25-4">FIGURE 25-4   The uranium
radioactive decay series.</figure>

isotopes that occur in the uranium series) and ends with stable lead-208. A third
series, called the actinium series, begins with uranium-235 and ends with lead-
207. Most of the naturally occurring radioactive isotopes are members of one
of these three decay series.

If the assumption is made that a uranium ore originally contained uranium
compounds and no other members of the uranium decay series, one can measure

TABLE 25-1   HALF-LIVES OF THE NATURALLY OCCURRING
RADIOACTIVE ELEMENTS IN THE URANIUM-238 SERIES

ISOTOPE	TYPE OF DISINTEGRATION	HALF-LIFE
U-238	α	4.5 billion years
Th-234	β	24.1 days
Pa-234	β	1.18 minutes
U-234	α	250,000 years
Th-230	α	80,000 years
Ra-226	α	1620 years
Rn-222	α	3.82 days
Po-218	α,β	3.05 minutes
Pb-214	β	26.8 minutes
Bi-214	α,β	19.7 minutes
Tl-210	β	1.32 minutes
Pb-210	β	22 years
Bi-210	β	5 days
Po-210	α	138 days
Pb-206	stable	

the present concentrations of the various isotopes in the rock and thereby determine the age of the rock. This is possible since the half-life for each decay reaction is known. For example, 1 gram of uranium-238 in its half-life of 4.5 billion years would leave 0.5000 g of U-238 and in the process produce 0.4326 g of lead-206. The amount of lead-206 can be calculated by using the known half-life for each step in the decay series. Now if one finds in a uranium ore the ratio of 0.5000 g uranium to 0.4326 g lead, it would follow that the rock is 4.5 billion years old. Ages determined in this fashion for various rocks taken from different parts of the world are approximately 3 billion years. Some meteorites have been determined to be 4.5 billion years old. In general, radioactive dating techniques determine the age of the planets in the solar system to be approximately 4.5 billion years.

ARTIFICIAL NUCLEAR REACTIONS

After it was realized that the nuclei of some of the heavier isotopes were unstable, scientists wondered if nuclear reactions could be initiated with other nuclei that were apparently stable. In order to explain the stability of any nucleus, one must postulate the existence of short-range, attractive, nuclear forces between *positive* particles (protons) and/or *neutral* particles (neutrons). Such forces must be stronger than the electrostatic forces that would tend to make the positive particles fly apart. It seemed reasonable to believe that two nuclei might possibly react to form new nuclear species if they could be brought so close together that the short-range nuclear forces could be operative. In order to achieve this, the two nuclei would have to approach each other with sufficient kinetic energy to overcome the electrostatic repulsion. In such a case, one could postulate an unstable *compound nucleus* that would emit particles, energy, or both, in seeking a stable structure.

In 1919, Rutherford was successful in producing the first artificial nuclear change explained by man. He placed nitrogen gas in a cloud chamber and directed helium nuclei (alpha particles) into the box. The penetrating power of the alpha particles is determined by their kinetic energy, which in turn is determined by the reaction producing them. Since, for a given alpha source, the alpha particles all have about the same energy, a photograph of such alpha tracks in a cloud chamber would show tracks of essentially the same length (Figure 25–5). Under the conditions of this experiment, Rutherford found some tracks that were much longer than the typical alpha track. Furthermore, these longer tracks did not appear to start at the origin of the alpha tracks but seemed to begin at the termination of an alpha track. When these tracks were studied in a magnetic field, their curvature indicated a particle with a charge-to-mass ratio identical to the value already established for the positive-ray particle, the proton. It was concluded then that the tracks were produced by high energy protons which, because of their smaller size and charge, are more penetrating than an alpha particle for a given amount of energy. All of the results of the experiment could be explained if one assumed the nuclear reaction to be:

$$^{14}_{7}\text{N} + ^{4}_{2}\text{He} \longrightarrow ^{18}_{9}\text{F} \longrightarrow ^{17}_{8}\text{O} + ^{1}_{1}\text{H}$$

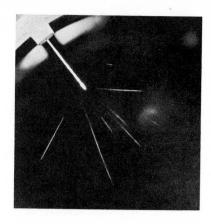

FIGURE 25-5   Alpha tracks photographed in a cloud chamber. Although the alpha particles do have different energies, owing to collisions with other particles in escaping the source, the apparent difference in path lengths is exaggerated because they are not all moving in the same plane. Under the same conditions proton tracks from the reaction, $^{14}_{7}N + {}^{4}_{2}He \longrightarrow {}^{1}_{1}H + {}^{17}_{8}O$, would tend to be considerably longer. (From Fundamentals of College Chemistry, by J. H. Wood et al., First Edition, Harper and Row, 1963.)

Although there is no physical evidence for the existence of the F-18 compound nucleus, it seems reasonable to believe that it has at least a momentary existence before breaking down into the more stable oxygen and hydrogen nuclei. Since both the oxygen-17 and the hydrogen nuclei are stable, the products show no further tendency to undergo nuclear change.

Following Rutherford's original transmutation experiment, there was considerable interest in subjecting isotopes to high energy particles to discover new nuclear reactions. As you might guess, numerous reactions were found. For example, beryllium can be converted to carbon when subjected to an alpha ray bombardment:

$$^{9}_{4}Be + {}^{4}_{2}He \longrightarrow {}^{13}_{6}C \longrightarrow {}^{12}_{6}C + {}^{1}_{0}n$$

Although the carbon-12 produced in this reaction is stable, the neutron is given off with sufficient energy to provoke additional nuclear reactions in nuclei with which it collides. As stated in Chapter 6, it was this nuclear reaction that was used by Chadwick in 1932 to finally prove the existence of the previously postulated neutron.

Not all nuclear reactions produce stable isotopes as in the cases cited above. If magnesium-25 is bombarded with an alpha source, a radioactive isotope of aluminum, Al-28, is produced which does not exist in nature.

$$^{25}_{12}Mg + {}^{4}_{2}He \longrightarrow {}^{29}_{14}Si \longrightarrow {}^{28}_{13}Al^{\circ} + {}^{1}_{1}H$$

The symbol $^{\circ}$ is often used to denote a radioactive isotope in nuclear equations. Radioactive isotopes, such as Al-28, have characteristic half-lives just as do the naturally occurring ones. The half-life of Al-28 is relatively short, only 2.3 minutes. The Al-28 nucleus gives off a beta particle and becomes a stable isotope of silicon.

$$^{28}_{13}Al^{\circ} \longrightarrow {}^{28}_{14}Si + {}^{0}_{-1}e$$

Nitrogen can be bombarded with neutrons from the Chadwick reaction to produce radioactive carbon-14.

$$^{14}_{7}N + {}^{1}_{0}n \longrightarrow {}^{15}_{7}C \longrightarrow {}^{14}_{6}C^{\circ} + {}^{1}_{1}H$$

615

This is a very interesting "artificial" radioactive isotope for three reasons. First, it is not actually artificial since it occurs naturally in the earth's atmosphere as a result of neutrons produced by cosmic rays. Second, this radioactive isotope is used in very small amounts in the form of carbon dioxide by green plants along with nonradioactive CO_2. Third, the half-life of C-14 is 5570 years. Since a plant (or an animal that eats the plant) stops ingesting C-14 at death, and since the C-14 loses its radioactivity at a known rate, the activity present in the remains of a historical object containing carbon from the air can be used to date that object in history. Crude cloth and other objects found in Egyptian pyramids, for example, have been dated in this way. However, it should be noted that the age of objects dating back more than one half-life (5570 years) are subject to a significantly increasing degree of error in the age estimate.

An interesting question arises as to why the alpha particles were primarily scattered by the gold foil in Rutherford's gold-foil experiment (Chapter 6), and yet the same alpha source can produce a nuclear change with nitrogen-14 nuclei. The answer lies in the fact that the charge on the gold nucleus is $+79$, whereas on the nitrogen nucleus the charge is $+7$. Most of the naturally occurring alpha particles from radioactive decay simply do not have enough energy to penetrate to a heavy, positively charged nucleus such as that of gold. Therefore, if artificial nuclear reactions are to be studied for the heavier elements, it is necessary to find ways of increasing the kinetic energy of the subatomic projectile particles.

SOURCES OF HIGH ENERGY PARTICLES

Monuments to man's recent technological skill are the intricate devices that have been developed to increase the kinetic energy of charged particles. Linear accelerators, cyclotrons, betatrons, synchro-cyclotrons, and synchrotrons are capable of accelerating electrons, protons, deuterons ($_1^2H$), alpha particles, and similarly charged particles to tremendous speeds. In the case of electron accelerators, the speed of the stream of electrons has been brought very close to the speed of light (186,000 miles per sec). Even with this costly hardware, there is still no way to increase directly the kinetic energy of uncharged particles like neutrons and the neutral atoms. The reasons for this will be evident after it is understood how these devices work.

Although the electronic gear is quite elaborate and different for each apparatus, the basic principles of operation have been developed in earlier chapters and are comparatively simple, namely, opposite charges attract and the path of a charged particle is curved as it passes through a magnetic field. When these effects are combined with the fact that electrical fields do not penetrate to the inside of a charged metal container (the energy is absorbed by the metal), you have the fundamental facts necessary to explain how particle accelerators work.

Sufficient kinetic energy for a particle to penetrate the electrical fields of an atom and to enter a large nucleus cannot be gained by accelerating the particle in one step between two electrical poles. This would require an impossibly large potential difference of millions of volts. However, if the acceleration is done in several thousand steps, with readily obtainable potential differences of 2000 to 10,000 volts being applied to each step, sufficient energy can be imparted to the particle. The linear accelerator and the cyclotron illustrate the two primary ways the stepwise acceleration is accomplished.

The operation of the linear accelerator is less complicated in that it does not require a magnetic field. The principle of operation is outlined in Figure 25-6. A source of electrons or protons is provided by ionization (as in a canal ray tube) at one end of the line of hollow metal cylinders and the target material is at the other end. The entire device is enclosed in a vacuum so that the accelerated particles may move in a straight path without the possibility of collisions with molecules in the air. Adjacent cylinders are charged oppositely so that when the charged particles traverse the space between the two cylinders, the particles are repelled by the previous tube and attracted by the upcoming one. When a given particle enters a cylinder, it is necessary to change suddenly the sign of the charge on that cylinder so that the particle will be repelled as it moves from the tube. Simultaneously, the signs of the charge on all the other cylinders are also changed. The result is that the particle is successively attracted to and repelled by each cylinder. At each gap, then, the particles increase their speed. The cylinders are increasingly longer so the increasingly greater velocities of the particles will not disrupt the timing of the changing of the charge on alternate cylinders. When the particles reach the target sample, if they have received sufficient energy to penetrate the fields, they can enter the target nuclei. Since the nuclei of atoms are very small compared to the cross-sectional area of an atom, only relatively few of the high energy particles enter target nuclei.

The largest linear accelerator in the world went into operation in 1967 at Stanford, California. It is two miles long and accelerates electrons to energies

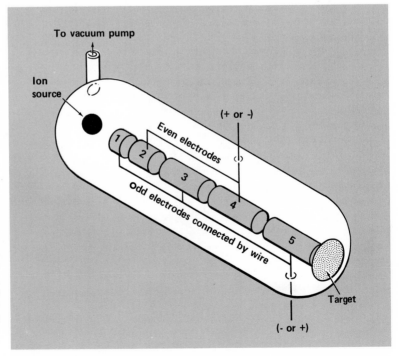

FIGURE 25-6   Linear accelerator diagram. Charged particles are produced at the ion source and are attracted toward an oppositely charged electrode, cylinder 1. When passing through cylinder 1, the charge on the cylinder is changed to the same sign as that on the particle being accelerated. At the same instant, the sign on cylinder 2 becomes opposite to the charge on the particle. Thus the particle is accelerated as it crosses the gap between cylinders. Note that the tubes become successively longer because of increased speeds of the particle.

of from *20 to 40 billion electron volts (bev)*. An electron volt is a unit of energy; it is the amount of energy gained by an electron when it is accelerated by an electric field, the electric potential across the field being one volt.

The cyclotron was developed in 1931, by Ernest O. Lawrence (element 103, a synthetic element, is named in his honor) and M. S. Livingston. In addition to the electronic gear and the huge magnets, the instrument consists of two hollow D-shaped metal containers enclosed in a vacuum as shown in Figure 25–7. Charged particles are formed near the center of the gap between the D's and begin their acceleration to high energy by being attracted into one of the D's. While the particles are in a D, the influence of the magnetic field causes their path to curve. By the time the particles have completed the semicircle, the signs on the D's have changed and the repulsion of the previous D and the attraction by the upcoming one accelerate the particles across the gap between the D's. Each time the particles traverse the gap, they are accelerated to a greater speed.

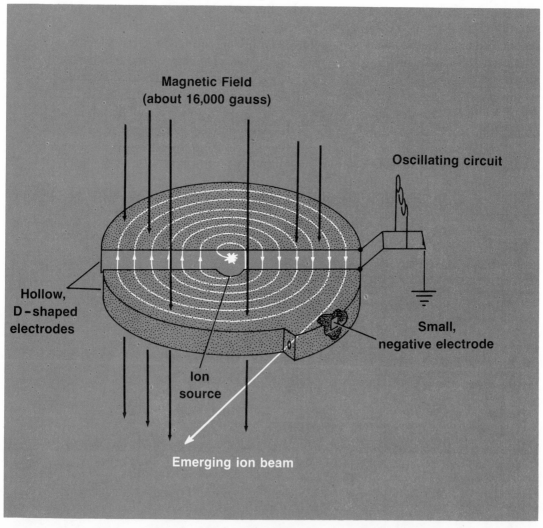

FIGURE 25-7   Schematic diagram illustrating the operation of the cyclotron. (Adapted from College Chemistry, Third Edition, by Linus Pauling. W. H. Freeman and Company. Copyright © 1964.)

The increased speed causes the particles to move in a wider arc on each revolution. After many accelerations and revolutions, the arc is sufficiently large so the charged deflector plate can repel the particles through a window and onto the target sample. In some cyclotrons, the particles go too fast to be directed outward by the deflector plate and the target sample is placed inside one of the D's.

The earlier models of the cyclotron had D's with diameters of about 2 feet and accelerated protons to energies of about 0.5 million electron volts (mev). Modern cyclotrons have diameters up to about 10 feet and accelerate protons to energies of 250 mev. The voltage across the gap between D's may have any value from 10,000 to 200,000 volts.

The cyclotron is a source of relatively low energy particles. More sophisticated accelerators capable of much higher energies include the betatron, synchro-cyclotron, and bevatron. These instruments are based on the same basic principles as the linear accelerator and the cyclotron, but are capable of producing particles with energies up to billions of electron volts. The United States expects to complete the world's largest particle accelerator in Weston, Illinois, during the mid 1970's. This installation, under the direction of the Atomic Energy Commission, is expected to cost $395 million and will employ 2000 scientists and technicians. The annual budget for this operation is to be approximately $60 million. This accelerator will achieve energies per particle up to 200 billion electron volts (bev).

TABLE 25-2   SUBATOMIC PARTICLES PRODUCED IN ARTIFICIAL NUCLEAR REACTIONS

PARTICLES PRODUCED FROM COLLISIONS INVOLVING PROJECTILE PARTICLES HAVING ENERGIES BELOW 100 MEV INCLUDE:

	Mass in multiples of the mass of an electron	Charge
Electron	1	−1
Positron	1	+1
Proton	1837	+1
Neutron	1838	0
Neutrino	0	0

PARTICLES PRODUCED FROM COLLISIONS INVOLVING PROJECTILE PARTICLES HAVING ENERGIES ABOVE 100 MEV INCLUDE:

Mesons	Mass in multiples of the mass of an electron	Charge
μ mu (muon)	206	+1 or −1
$\pi^{\pm}$ pi	273	+1 or −1
π° pi	264	0
κ kappa (kaon)	967	0 or +1 or −1
Baryons (or hyperons)		
Λ lambda	2183	0
Σ sigma	2330	0 or +1 or −1
Ξ cascade	2580	−1 or 0

When a beam of such high energy particles (electrons, protons, deuterons, alpha particles, or other ions) bombards nuclei, a wide variety of reactions take place. A massive nucleus, having captured a high energy particle, will usually emit one or more subatomic particles before reaching a stable state. If the bombarding particles have energies less than about 100 mev, the emitted particles can be described in terms of relatively simple subatomic particles such as protons, neutrons, electrons, positrons, neutrinos, and antineutrinos. However, when the bombarding particles have energies of thousands of mev or greater, new and strange particles are created; some of these are briefly described in Table 25–2.

TRANSURANIUM ELEMENTS

The heaviest known element prior to 1940 was the element uranium. The invention of the cyclotron and other devices to obtain high energy particles made it possible to react these particles with heavy nuclei to obtain even more massive nuclei. Thus, *transuranium* elements with atomic numbers greater than 92 were prepared.

In 1940, McMillan and Abelson prepared element 93, the synthetic element neptunium (Np). The experiment involved directing a stream of high energy deuterons ($^2_1$H) onto a target of uranium-238. A deuteron is the nucleus of an isotope of hydrogen containing one neutron as well as one proton. The initial reaction involved the conversion of uranium-238 to uranium-239.

$$^{238}_{92}\text{U} + ^2_1\text{H} \longrightarrow ^{239}_{92}\text{U} + ^1_1\text{H}$$

Uranium-239 has a half-life of 23.5 minutes, converting spontaneously to the new element, neptunium, by the emission of beta particles.

$$^{239}_{92}\text{U} \longrightarrow ^{239}_{93}\text{Np} + ^{\ 0}_{-1}\text{e}$$

Neptunium itself, however, is unstable, with a half-life of 2.33 days; it converts into a second new element, plutonium.

$$^{239}_{93}\text{Np} \longrightarrow ^{239}_{94}\text{Pu} + ^{\ 0}_{-1}\text{e}$$

Plutonium-239, like its parent atom, is radioactive, but its half-life is 24,100 years. Because of the relative values of the half-lives, very little neptunium could be accumulated, but the plutonium could be obtained in larger quantities. The plutonium-239 is important as fissionable material since atomic bombs (see Fission Reactions, the next section) can be made with it as well as with naturally occurring uranium-235. The names of these two elements were taken from the mythological names, Neptune and Pluto.

Although Neptune and Pluto are the last of the known planets in the solar system, their namesakes are not the last in the list of elements. The rush of transuranium experiments that followed produced additional elements: americium (Am), curium (Cm), berkelium (Bk), californium (Cf), einsteinium

TABLE 25-3   NUCLEAR REACTIONS USED TO PRODUCE
TRANSURANIUM ELEMENTS

ELEMENT	ATOMIC NUMBER	REACTION
Neptunium, Np	93	$^{238}_{92}U + ^{1}_{0}n \longrightarrow ^{239}_{93}Np + ^{0}_{-1}e$
Plutonium, Pu	94	$^{238}_{92}U + ^{2}_{1}H \longrightarrow ^{238}_{93}Np + 2^{1}_{0}n$
		$^{238}_{93}Np \longrightarrow ^{238}_{94}Pu + ^{0}_{-1}e$
Americium, Am	95	$^{239}_{94}Pu + ^{1}_{0}n \longrightarrow ^{240}_{95}Am + ^{0}_{-1}e$
Curium, Cm	96	$^{239}_{94}Pu + ^{4}_{2}He \longrightarrow ^{242}_{96}Cm + ^{1}_{0}n$
Berkelium, Bk	97	$^{241}_{95}Am + ^{4}_{2}He \longrightarrow ^{243}_{97}Bk + 2^{1}_{0}n$
Californium, Cf	98	$^{242}_{96}Cm + ^{4}_{2}He \longrightarrow ^{245}_{98}Cf + ^{1}_{0}n$
Einsteinium, Es	99	$^{238}_{92}U + 15^{1}_{0}n \longrightarrow ^{253}_{99}Es + 7^{0}_{-1}e$
Fermium, Fm	100	$^{238}_{92}U + 17^{1}_{0}n \longrightarrow ^{255}_{100}Fm + 8^{0}_{-1}e$
Mendelevium, Mv	101	$^{253}_{99}Es + ^{4}_{2}He \longrightarrow ^{256}_{101}Mv + ^{1}_{0}n$
Nobelium, No	102	$^{246}_{96}Cm + ^{12}_{6}C \longrightarrow ^{254}_{102}No + 4^{1}_{0}n$
Lawrencium, Lr	103	$^{252}_{98}Cf + ^{10}_{5}B \longrightarrow ^{257}_{103}Lr + 5^{1}_{0}n$
Kurchatovium, Ku	104	$^{242}_{94}Pu + ^{22}_{10}Ne \longrightarrow ^{260}_{104}Ku + 4^{1}_{0}n$
Hahnium, Ha	105	$^{249}_{98}Cf + ^{15}_{7}N \longrightarrow ^{260}_{105}Ha + 4n$

(Es), fermium (Fm), mendelevium (Md), nobelium (No), lawrencium (Lr), kurchatovium (Ku), and hahnium (Ha), the new elements being named after countries, states, cities, and men. Reactions employed in the production of the transuranium elements are given in Table 25–3. As accelerators with greater and greater energy capabilities are produced, even more nuclear reactions should be available for study.

FISSION REACTIONS

Perhaps no manifestation of matter has captured the awe of people to quite the extent that atomic fission has in recent years. Fission, or the splitting of atoms, is the basis of the atomic bomb. Since 1938 when Otto Hahn, Fritz Strassmann, Lise Meitner, and Otto Frisch discovered that $^{235}_{92}U$ is fissionable, the power of atomic fission has given the world new hope for a new energy source but, on the other hand, has caused fear and respect for the terrific power for devastation contained in this new giant.

Fission can occur when a thermal neutron (with a kinetic energy about the same as gaseous molecules at ordinary temperatures) enters certain heavy nuclei (^{235}U, ^{233}U, ^{239}Pu). The splitting of the heavy nucleus produces two smaller nuclei, two or more neutrons (an average of 2.5 for each ^{235}U atom), and energy. Typical nuclear fission reactions are written as follows:

$$^{235}_{92}U + ^{1}_{0}n \longrightarrow ^{141}_{56}Ba + ^{92}_{36}Kr + 3^{1}_{0}n + energy$$

$$^{235}_{92}U + ^{1}_{0}n \longrightarrow ^{103}_{42}Mo + ^{131}_{50}Sn + 2^{1}_{0}n + energy$$

Note that the same nucleus may split in more than one way. The fission products   621

$^{141}_{56}$Ba, $^{92}_{36}$Kr, and so forth, emit beta particles and gamma rays until stable isotopes are reached.

The emitted neutrons can cause fission of other heavy atoms. For example, the three emitted neutrons of the first reaction above could produce fission in three more U atoms; the nine emitted neutrons could produce nine more fissions; the 27 neutrons from these fissions could produce 81 neutrons; the 81, 243; the 243, 729; and so on. This process is called a chain reaction (Figure 25–8) and occurs at a maximum rate when the sample of uranium is large enough for each neutron to be captured by a nucleus before passing out of the sample. Sufficient sample to sustain a chain reaction is called the *critical mass*.

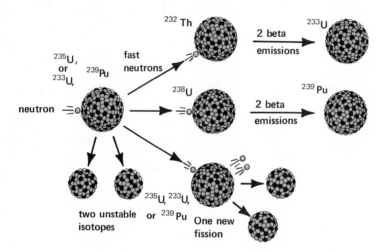

FIGURE 25–8   A chain reaction. A thermal neutron collides with a fissionable nucleus and the resulting reaction produces three additional neutrons. These neutrons can either convert non-fissionable nuclei, such as ^{232}Th, to fissionable ones or cause additional fission reactions. If enough fissionable nuclei are present, a chain reaction will be sustained.

In the atomic bomb the critical mass is kept separated into several smaller subcritical masses until detonation, at which time the masses are driven together. It is then that the tremendous energy is liberated and heats the gases in the vicinity up to as high as 10,000,000 degrees. The sudden expansion of gases literally explodes everything nearby and scatters the radioactive fission fragments over a wide area. In addition to the tremendous movement of gases which is, to a lesser degree, associated with conventional bombs, there is the vaporizing heat and the radioactive products that make the atomic bomb so devastating.

There is no danger of an atomic explosion in the mineral deposits on earth for two reasons. First, uranium is not found pure in nature; it is found only in compounds which in turn are mixed with other compounds. Second, less than 1 per cent of the uranium found in nature is ^{235}U which is fissionable. The other 99 per cent is ^{238}U which is not fissionable by thermal neutrons. For these two reasons, ^{235}U is not found naturally in concentrations greater than its critical mass.

MASS DEFECT

What is the source of the tremendous energy of the fission process? It ultimately comes from mass being converted into energy according to Einstein's famous equation, $E = mc^2$, where E is energy in ergs, m is the mass in grams, and c is the speed of light in cm per sec. If separate neutrons, electrons and protons are combined to form an atom, there is a loss of mass called the *mass defect*. For example, the mass of a proton and a separate electron is 1.007825 awu; the mass of a neutron is 1.008665 awu. The calculated mass of one $_2^4$He atom from these masses is

$$2 \times 1.007825 = 2.015650 \text{ awu, mass of 2 protons and 2 electrons}$$
$$2 \times 1.008665 = \underline{2.017330} \text{ awu, mass of 2 neutrons}$$
$$4.032980 \text{ awu, calculated mass of } _2^4\text{He atom}$$

Since the measured mass of a $_2^4$He atom is 4.002604, the mass defect is

$$4.032980 \text{ awu}$$
$$\underline{4.002604}$$
$$0.030376 \text{ awu, mass defect}$$

Since the atom is more stable than the separated neutrons, protons and electrons, the atom must correspond to a lower energy state. The 0.030376 awu lost per atom is released in the form of energy which provides for the lower energy state of the atom. The energy equivalent of the mass defect is called the *binding energy*. The binding energy is analogous to the earlier concept of bond energy in that both are the energy necessary to separate the package (nucleus or molecule) into its parts.

Atoms with atomic numbers between 30 and 63 have a greater mass defect per nuclear particle than very light elements or very heavy ones. A summary, in

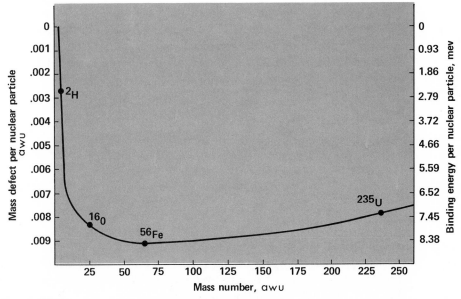

FIGURE 25-9   Mass defect for different nuclear masses. The most stable nuclei center around ^{56}Fe, which has the largest mass defect per nuclear particle.

the form of a graph, for all of the mass numbers is shown in Figure 25-9. This means that nuclei having lost the greatest amount of energy per nuclear particle are also found in this region. Consequently, elements of intermediate atomic number require the *most* energy per nuclear particle to separate a nucleus into its separate protons and neutrons. Conversely, elements with intermediate atomic numbers lose the most energy per nuclear particle in the hypothetical formation of the element from separate neutrons, protons, and electrons and are consequently the most stable atoms. This is similar in principle but not in quantities of energy to the formation of a chemical bond between two atoms; the more energy given off, the more stable the bond.

Because of the relative stabilities, it is in the intermediate range of atomic numbers that most of the products of nuclear fission are found. Therefore, when fission occurs and smaller, more stable nuclei result, these nuclei will contain less mass per nuclear particle. Therefore, some mass must be changed into energy. Note that, theoretically, no particles are lost but the same particles have less mass. This energy obtained from mass gives the fission process its tremendous energy. It takes only about 2.2 lbs of ^{235}U or ^{239}Pu undergoing fission to be equivalent to the energy released by 20,000 tons (20 kilotons) of ordinary explosives like TNT. The energy content in matter is further dramatized when it is realized that the atomic fragments from the 2.2 lbs of fuel would weigh *almost* 2.2 lbs! The fission bombs dropped on Japan during World War II were approximately this size.

FUSION REACTIONS

When very light nuclei such as H, He, Li, and so forth, are combined or fused to form an element of intermediate atomic number, energy must be given off consistent with the greater stability of the elements in this intermediate atomic number range (Figures 25-9 and 25-10). This energy comes from a decrease in mass and is the source of the energy released by the sun and by hydrogen bombs. Typical examples of fusion reactions are:

$$4^1_1H \longrightarrow {}^4_2He + 2\,{}^0_{+1}e + energy$$

$${}^2_1H + {}^2_1H \longrightarrow {}^3_2He + {}^1_1H + energy$$

$$3\,{}^4_2He \longrightarrow {}^{12}_6C + energy$$

$${}^{12}_6C + {}^4_2He \longrightarrow {}^{16}_8O + energy$$

$${}^3_1H + {}^2_1H \longrightarrow {}^4_2He + {}^1_0n + energy$$

The first reaction (${}^0_{+1}e$ is a positron) just mentioned is thought to be the overall reaction which is taking place on the sun and which is responsible for the tremendous amount of energy released. This emission of energy results from the conversion of five million tons of the sun into energy each second.

Fusion reactions take place rapidly when the temperature is of the order of 100 million degrees or more. At these high temperatures atoms do not exist as such; instead there is a plasma of nuclei and of electrons. In this plasma nuclei merge or combine. In order to achieve the high temperatures required for the fusion reaction of the hydrogen bomb, a fission bomb (atomic bomb) is initiated first. One type of hydrogen bomb depends on the production of tritium ($^3_1$H) in the bomb. In this type lithium deuteride (^{6}Li^{2}H, a solid substance) is placed around an ordinary ^{235}U or ^{239}Pu fission bomb. The fission is set off in the usual way. Lithium-6 absorbs some of the neutrons produced and splits into tritium and ^{4}He.

$$^6_3\text{Li}^2_1\text{H} + {}^1_0\text{n} \longrightarrow {}^3_1\text{H} + {}^4_2\text{He} + {}^2_1\text{H}$$

The temperature reached by the fission of ^{235}U or ^{239}Pu is sufficiently high to bring about the fusion of tritium and deuterium:

$$^3_1\text{H} + {}^2_1\text{H} \longrightarrow {}^4_2\text{He} + {}^1_0\text{n} + 17.6 \text{ mev}$$

A 20-megaton bomb usually contains about 300 lbs of lithium deuteride as well as a considerable amount of plutonium and uranium.

CONTROLLED ENERGY

The fission of a uranium-235 nucleus by a slow-speed neutron to produce smaller nuclei, three neutrons, and large amounts of energy suggested to Fermi, and others, not only the possibility of an uncontrolled nuclear explosion but also that if the number of neutrons could be controlled, the reaction could proceed at a moderate rate. If a neutron control could be found, the concentration of neutrons could be maintained at a level sufficient to keep the fission process going but not high enough to allow an uncontrolled explosion. It would then be possible to drain the heat away from such a reactor on a continuing basis to do useful work. In 1942, Fermi, working at the University of Chicago, was successful in building an atomic reactor, called an *atomic pile*.

An atomic reactor has three essential components. First, the *charge material* (fuel) must be made of or contain significant concentrations of a fissionable isotope such as U-235, Pu-239, or U-233. Ordinary uranium, which is mostly the nonfissionable U-238, can be used since it has a small concentration of the U-235 isotope. Second, a *moderator* is required to slow the speed of the neutrons produced in the fusion reactions. Graphite, water and other substances have been used successfully for this purpose. Third, a substance that will absorb neutrons, such as cadmium or boron steel, is present in order to have a fine control over the neutron concentration. A schematic diagram of the now famous X-10 graphite pile in Oak Ridge, Tennessee, is illustrated in Figure 25–10. Here the fuel is packed in openings in the graphite pile. Unlike the atomic bomb, the reaction cannot lead to an atomic explosion if more than the critical mass (the minimum mass needed for a self-sustaining reaction) is assembled. The material of which the pile is made absorbs neutrons. The control rods made of cadmium or boron steel offer a final and complete control to stop the fission reactions. Even if the critical mass of U-235 came together without pressure to hold it together for a brief moment, it would simply expand owing to the heat. The concentration

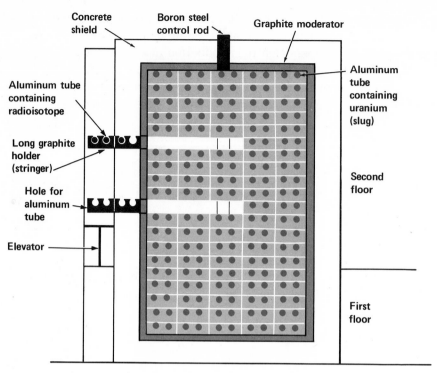

Concrete shield · Boron steel control rod · Graphite moderator · Aluminum tube containing uranium (slug) · Aluminum tube containing radioisotope · Long graphite holder (stringer) · Hole for aluminum tube · Elevator · Second floor · First floor

FIGURE 25-10   Schematic diagram of X-10 reactor at Oak Ridge, Tennessee.

of mass would then be less than critical, although it would be dangerously hot and radioactive. With the control rods properly placed, the reaction can be maintained at various safe levels of activity. Radiation shielding, while not necessary to the reactor operation, is necessary for human safety and tends to make reactors bulky installations.

Once the heat is produced in a nuclear reactor and safety measures are employed to protect against radiation, conventional technology allows this energy to be used to generate electricity, power ships, or operate any device that operates via heat energy. A system for the nuclear production of electricity is illustrated in Figure 25–11.

Numerous nuclear power stations for the generation of electricity are now in operation around the world. However, this source of energy had trouble competing in cost with hydroelectric power, where such power is plentiful, during the two decades after World War II. Such costs have prohibited the utopia of cheap energy on a world-wide basis for such purposes as desalting sea water and the irrigation of deserts. Recent developments indicate that design size is a critical factor and that, with large enough installations, nuclear electricity may become the world's cheapest electricity.

The real dream of large supplies of cheap energy for all men cannot be realized with the fission-type reaction; there is simply not enough fuel of this type in the world supply. The hope for such energy lies in the yet-to-be-harnessed fusion reaction since there is no practical limit to the availability of light nuclei for such reactions. As pointed out earlier, the fusion reactions require tempera-

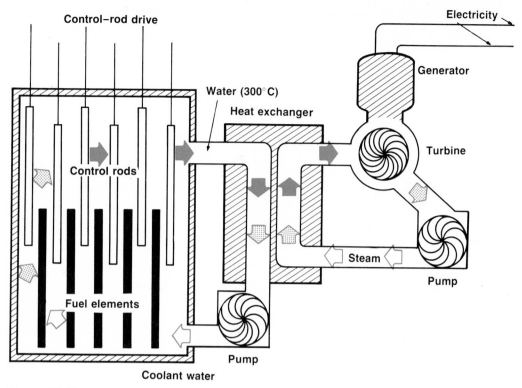

FIGURE 25-11   Electricity produced by atomic energy.

tures and pressures obtainable on earth only in the midst of an atomic explosion. Even if these conditions could be obtained, there is no known material that could be used to contain fusion reactions. One approach to the harnessing of the energy from fusion reactions is to find ways to maintain the reaction under less stringent conditions. These efforts have not, as yet, been successful. Magnetic bottles have been devised, which are made of no matter at all, to contain a plasma, a mixture of atoms under fusion conditions. The magnetic bottle is an enclosure bounded by a magnetic field and holding its contents by magnetic forces. Major breakthroughs in the control of earth-bound fusion are still in the future. However, the possibility is real that the ingenuity of man will tame a nearly unlimited source of energy.

QUESTIONS

1. Which theoretically gives greater energy per mole?

   (a) the burning of gasoline
   (b) the fission of ^{235}U
   (c) the fusion of ^{56}Fe

2. Describe the operation of a Geiger counter.

3. Why is the cloud chamber (or the bubble chamber) more useful than other types of nuclear particle detectors?

4. Define or illustrate: deuteron, transmutation, radioactive decay series, mass defect, megaton.

5. In general, how have the synthetic transuranium elements been produced?

6. What does the symbol $^{11}_5B$ mean?

7. (a) Describe the discovery of the proton.
   (b) Describe the discovery of the neutron.

8. Complete or supply the following nuclear reactions:
   (a) $^1_1H + ^{35}_{17}Cl \longrightarrow ^4_2He + ?$
   (b) beta emission of $^{60}_{27}Co$
   (c) alpha emission of $^{228}_{90}Th$
   (d) $^1_0n + ^{60}_{28}Ni \longrightarrow ? + ^1_1H$
   (e) $^2_1H + ^1_1H \longrightarrow ? + ^0_0\gamma$
   (f) $^{238}_{92}U + ^{12}_6C \longrightarrow ? + 4^1_0n$

9. (a) The half-life of ^{218}Po is 3 minutes. How much of a 2-gram sample of this isotope remains after 15 minutes?
   (b) Suppose you wanted to buy some of this isotope and it requires 5 hours for it to reach you. How much should you order if you want to use 0.001 gram?

10. What produces the tremendous energy of a fission reaction?

11. Why is it difficult to fuse two $^{52}_{24}Cr$ nuclei? Explain.

12. What is meant by a chain reaction?

13. What is meant by radiocarbon dating?

14. How can ages of rocks be determined? What assumptions are made in using this method?

15. Use data given in the chapter to calculate the mass defect of the isotope ^9_4Be, which has an atomic mass of 9.01219 awu.

16. What major problem is associated with harnessing the energy from a fusion reaction?

17. We speak of "seeing" a nuclear event with a cloud or bubble chamber. Explain why we are more truthful in using the quotation marks about the word *seeing*.

18. Look up the origin of the word *mutation* and explain why the word *transmutation* was an apt word choice to describe the changing of one element into another.

19. If radium atoms ($^{226}_{88}Ra$) lose one alpha particle per atom, what element is formed? What is its atomic weight? What is its atomic number?

20. Suppose you were given $1000.00 and told that you could spend one-half of it the first year, one-half of the balance the second year, and so on. If you spent the maximum allowed, at the end of what year would you have $31.25 of the original $1000.00 left? How is this problem analogous to radioactive dating techniques? Can you think of ways in which it is not?

21. What is the difference between a thermal neutron and a high energy neutron resulting from cosmic radiation?

SUGGESTIONS FOR FURTHER READING

Chappin, G. R., "Nuclear Fission," *Chemistry*, Vol. 40, No. 7, p. 25 (1967).
Clark, H. M., "The Origin of Nuclear Science," *Chemistry*, Vol. 40, No. 7, pp. 8–11 (1967).
"Dendrochronology," *Chemistry*, Vol. 43, No. 7, p. 26 (1970).
"Detecting Forgeries in Paintings," *Chemistry*, Vol. 41, No. 5, p. 5 (1968).

"Element 105," *Chemistry*, Vol. 43, No. 6, p. 20 (1970).

Flerov, G. N., and Zvara, I., "Synthesis of Transuranium Elements," *Science*, Journal, Vol. 4, No. 7, p. 63, July (1968).

"Fossil-Track Dating," *Chemistry*, Vol. 43, No. 7, p. 27 (1970).

Gordus, A. A., "Neutron Activation Analysis of Almost Any Old Thing," *Chemistry*, Vol. 41, No. 5, pp. 8–15 (1968).

Hudis, J., "Nuclear Reactions," *Chemistry*, Vol. 40, No. 7, pp. 20–24 (1967).

"Ionium Dating," *Chemistry*, Vol. 43, No. 7, p. 22 (1970).

Johnsen, R. H., "Radiation Chemistry," *Chemistry*, Vol. 40, No. 7, pp. 31–36 (1967).

Libby, W. F., Radiocarbon Dating, University of Chicago Press, Chicago, 1955.

"Nuclear Power and By-Product Plutonium," *Chemistry*, Vol. 40, No. 3, pp. 7–8 (1967).

"Otto Hahn," *Chemistry*, Vol. 39, No. 12, p. 5 (1966).

Overman, R. T., Basic Concepts of Nuclear Chemistry, Reinhold Publishing Co., New York, 1963. (paperback)

"Radiocarbon Dating," *Chemistry*, Vol. 43, No. 7, p. 24 (1970).

Seaborg, G. T., Man-Made Transuranium Elements, Prentice-Hall, Inc., New York, 1963. (paperback)

Seaborg, G. T., "Some Recollections of Early Nuclear Age Chemistry," *Journal of Chemical Education*, Vol. 45, p. 278 (1968).

"University of California's Berkley Campus: 1941–42," *Chemistry*, Vol. 40, No. 3, pp. 25–26 (1967).

APPENDIX A

The Metric System

I. **UNITS OF LENGTH.** The standard unit of length is the *meter*. It was originally meant to be one ten-millionth of the distance along a meridian from the north pole to the equator. However, the lack of precise geographical information necessitated a better definition. For a number of years the meter was defined as the distance between two etched lines on a platinum-iridium bar kept at 0°C (32°F) in the International Bureau of Weights and Measures, at Sèvres, France. The inability to measure this distance as accurately as desired prompted a recent redefinition of the meter as being a length equal to 1,650,763.73 times the wavelength of the orange-red spectrographic line of $^{86}_{36}Kr$.

The meter (39.37 inches) is a convenient unit with which to measure the height of a basketball goal (3.05 meters), but it is unwieldy for measuring the parts of a watch or the distance between continents. For this reason, prefixes are defined in such a way that, when placed before the meter, define distances convenient for man's purposes. Some of the prefixes with their meanings are:

> micro—1/1,000,000 or 0.000001
> milli—1/1000 or 0.001
> centi—1/100 or 0.01
> deci—1/10 or 0.1
> deka—10
> hecto—100
> kilo—1000

The corresponding units of length with their abbreviations are:

> micrometer (μm)—0.000001 meter
> millimeter (mm)—0.001 meter
> centimeter (cm)—0.01 meter
> decimeter (dm)—0.1 meter
> meter (m)
> dekameter (dkm)—10 meters
> hectometer (hm)—100 meters
> kilometer (km)—1000 meters

Since the prefixes are defined in terms of the decimal system the conversion from one metric length to another involves only shifting the decimal point. Mental calculations are quickly accomplished.

How many centimeters are in a meter? Think: Since a centimeter is the one-hundredth part of a meter, there would be 100 centimeters in a meter.

Conversion of measurements from one system to the other is a common problem. A systematic approach to this type problem is presented in Appendix B. Some commonly used English-metric equivalents (conversion factors) are given in Table A-1.

TABLE A-1   English-Metric Equivalents* (Conversion Factors)

Length:	1 inch (in)	=	2.54 centimeter (cm)
	1 yard (yd)	=	0.914 meter (m)
	1 mile (mi)	=	1.609 kilometers (km)
			(1,609 meters)
Volume:	1 ounce (oz)	=	29.57 milliliters (ml)
	1 quart (qt)	=	0.946 liter (l)
	1.06 quart (qt)	=	1 liter (l)
	1 gallon (gal)	=	3.78 liters
Mass (weight)°°:	1 ounce (oz)	=	28.35 grams (g)
	1 pound (lb)	=	453.4 grams (g)
	1 ton (tn)	=	907.2 kilograms (kg)

° Commonly used English units are used.
°° Mass is a measure of the amount of matter, whereas weight is a measure of the attraction of the earth for an object at the earth's surface. The mass of a sample of matter is constant, but its weight varies with position and velocity. For example, the space traveler, having lost no mass, becomes weightless in earth orbit. Although mass and weight are basically different in meaning, they are often used interchangeably in the environment of the earth's surface.

II. UNITS OF MASS.

The primary unit of mass is the kilogram (1000 grams). This unit is the mass of a platinum-iridium alloy sample deposited at the International Bureau of Weights and Measures. One pound contains a mass of 453.4 grams (a five-cent nickel coin contains about 5 grams).

Conveniently enough, the same prefixes defined in the discussion of length are used in units of mass, as well as in other units of measure.

III. UNITS OF VOLUME.

The volume capacity used most frequently in chemistry is the liter. The liter is defined as 1 cubic decimeter (1 dm^3). Since a decimeter is equal to 10 centimeters (cm), the cubic decimeter is equal to $(10\ cm)^3$ or 1000 cubic centimeters (cc). One cc, then, is equal to one milliliter (the thousandth part of a liter). The ml (or cc) is a common unit that is often used in the measurement of medicinal and laboratory quantities.

Factor-Label Approach to Metric-English Conversion Problems

To convert a measurement from either the English or metric system to the other, the following method is straightforward and does not require the decision of whether to divide or multiply to obtain the proper answer. This method makes the decision for you, a decision which is sometimes difficult when dealing with new units with which there is little experience.

Example Problem.   How many liters are in 6 quarts?

Factor-Label Method of Solution:

1. Write down the unit to which you are converting. The question mark indicates the number of liters to be determined.

1. ? liters =

2. On the right hand side of the equal sign, write down the quantity given. Write both the number and the name or label.

2. ? liters = 6 quarts

3. Now look at the two units. Recall a conversion between these two units. These conversions must be learned or looked up.

3. 1 liter = 1.06 quarts,
   1 liter per 1.06 quarts
   or $\dfrac{1 \text{ liter}}{1.06 \text{ quarts}}$

4. Write the conversion factor on the right hand side so that the unwanted units will cancel; that is, a unit in the numerator will cancel the same unit in the denominator, and only the unit you want will remain. Do not, of course, cancel the numbers, just the units.

4. ? liters = 6 q̶u̶a̶r̶t̶s̶ $\times \dfrac{1 \text{ liter}}{1.06 \text{ q̶u̶a̶r̶t̶s̶}}$

5. Do the indicated multiplication and division. The line, of course, means divided by. Check the units on both sides of the equation. The units should be the same and in the same position (numerator or denominator).

5. ? liters $= \dfrac{6}{1.06}$ liters

$= 5.69$ liters

Now, suppose you needed to convert 3 pints to ml, and you do not know a conversion factor which will make the conversion in one step.

Proceed as before:

$$? \text{ ml} = 3 \text{ pints}$$

Note this is a volume conversion from the English system to the metric system as before. Recall the volume conversion factor which you know between these systems.

$$1 \text{ liter} = 1.06 \text{ quarts}$$

Convert in the system of the given quantity until you reach the unit in the conversion factor for this system. Then write down the conversion factor between systems so like units in separate factors will cancel.

$$? \text{ ml} = 3 \text{ pints} \times \frac{1 \text{ quart}}{2 \text{ pints}} \times \frac{1 \text{ liter}}{1.06 \text{ quarts}}$$

Cancel units and recall a conversion factor which converts the units you have left to the unit you want. Do the indicated multiplication and division.

$$? \text{ ml} = \frac{3 \times 1 \times 1 \text{ liter}}{2 \times 1.06} \times \frac{1000 \text{ ml}}{1 \text{ liter}}$$

$$= \frac{3 \times 1 \times 1 \times 1000}{2 \times 1.06 \times 1} \text{ ml}$$

$$= 1420 \text{ ml}$$

In brief, the method is very simple:

1. Write the units you want, an equals sign, and the quantity you have given.
2. Write conversion factors so the unwanted factors will cancel.
3. Keep in one system until you come to units in a familar intersystem conversion factor.
4. Do the indicated arithmetic and check the units.
5. Examine the size of the answer for reasonableness.

Note: This approach to problem solving has wide applicability to many other types of problems. It is especially useful in problems pertaining to weight relationships in chemical reactions. See Appendix D.

Centigrade or Celsius
Temperature Scale

The system of measuring temperature which is used in scientific work is based on the centigrade or Celsius temperature scale. This temperature scale was defined by Anders Celsius, a Swedish astronomer, in 1742. The Celsius scale is based upon the expansion of a column of mercury which occurs when it is transferred from a cold standard temperature to a hot standard temperature. The cold standard temperature is the temperature of melting ice and is defined as 0°C. The hot standard temperature is the temperature at which water boils under standard conditions of pressure; it is defined as 100°C. The expansion of the mercury is assumed to be linear over this range, and the distance through which the mercury column expands between 0°C and 100°C is divided into 100 equal parts, each corresponding to one degree.

On the scale commonly used in the United States, the Fahrenheit scale (Daniel Fahrenheit), the freezing mark is 32° and the boiling mark is 212°. Obviously, the ice is no hotter when a Fahrenheit thermometer is stuck into the system than when a Celsius thermometer is used. The marks of the tubes are simply different names for the same thing.

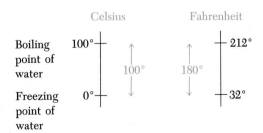

Because both systems are used in our environment, it is necessary at times to convert from one system to another.

Note that 100 Celsius degrees = 180 Fahrenheit degrees
or 1 C degree = 1.8 or $\frac{9}{5}$ F degrees.

But if we wish to convert temperature from one scale to another, we must remember that 0°C is the same as 32°F. Therefore, 32 must be added to the calculated number of degrees Fahrenheit in order to revert back to the start of the counting in the Fahrenheit system.

Therefore,

$$°F = \frac{\frac{9}{5}°F}{1°C} (°C) + 32,$$

which can be rearranged to

$$°C = \frac{5}{9} (°F - 32)$$

when the factor, $\frac{9}{5}°F/1°C$ is written as $\frac{9}{5}$.

Example Problem. Convert 50°F to the corresponding Celsius temperature:

$$\begin{aligned} °C &= \frac{5}{9} (°F - 32) \\ &= \frac{5}{9} (50 - 32) \\ &= \frac{5}{9} (18) \\ &= 10°C \end{aligned}$$

Stoichiometric Calculations

The bases for stoichiometric calculations were presented in Chapter 4. Problems of a more complex nature and a systematic approach to their solution are presented in the following examples. Finally, a list of exercise problems are given for further study.

Example 1
Balanced Equations Express Particle-Number Ratios

In the reaction of hydrogen with oxygen to form water, how many molecules of hydrogen are required to combine with 19 oxygen molecules?

Solution: A chemical equation can only be written for a reaction if the reactants and products are identified and the respective formulas determined. In this problem, the formulas are known and the equation is:

$$H_2 + O_2 \longrightarrow H_2O.$$

It is evident that one molecule of oxygen contains enough oxygen for two water molecules and the equation, as written, does not account for what happens to the second oxygen atom. As it is, the equation is in conflict with the conservation of atoms in chemical changes. This conflict is easily corrected by balancing the equation:

$$2\,H_2 + O_2 \longrightarrow 2\,H_2O$$

Now, all atoms are accounted for in the equation and it is obvious that two hydrogen molecules are required for each oxygen molecule. In other words, two hydrogen molecules are equivalent to one oxygen molecule in their usage. This can be expressed:

$$2 \text{ hydrogen molecules } \left\{ \begin{array}{c} \text{are} \\ \text{equivalent} \\ \text{to} \end{array} \right\} 1 \text{ oxygen molecule;}$$

or,

$$2\,H_2 \text{ molecules} \sim O_2 \text{ molecule;}$$

or,

$$\frac{2 \text{ H}_2 \text{ molecules}}{\text{O}_2 \text{ molecule}},$$

which can be read as two hydrogen molecules per one oxygen molecule.

Using now the factor-label approach developed in Appendix B, the solution is readily achieved.

? H$_2$ molecules = 19 O$_2$ molecules

? H$_2$ molecules = 19 O$_2$ ~~molecules~~ $\times \dfrac{2 \text{ H}_2 \text{ molecules}}{\text{O}_2 \text{ ~~molecule~~}}$ = 38 H$_2$ molecules

Note 1: The reader is likely to say at this point that the method is cumbersome and that he can quickly see the answer to be 38 H$_2$ molecules without "the method." However, problems to follow are made much easier if a systematic method of approach is used.

Example 2
Laboratory Mole Ratios Identical With Particle-Number Ratios

How many moles of hydrogen molecules must be burned in oxygen (the reaction of example 1) to produce 15 moles of water molecules (about a glassful)?

Solution: The balanced equation,

$$2 \text{ H}_2 + \text{O}_2 \longrightarrow 2 \text{ H}_2\text{O},$$

tells us that two molecules of hydrogen produce two molecules of water;

or,

2 molecules hydrogen $\sim$ 2 molecules water,

and,

1 molecule hydrogen $\sim$ 1 molecule water.

It is obvious then that the number of water molecules produced will be equal to the number of hydrogen molecules consumed regardless of the actual number involved. Therefore,

6.02×10^{23} molecules of hydrogen $\sim 6.02 \times 10^{23}$ molecules of water.

Since 6.02×10^{23} is a number called the mole, it follows that one mole of hydrogen molecules will produce one mole of water molecules. The general conclusion, then, is: the ratio of particles in the balanced equation is the same as the ratio of moles in the laboratory. The solution to the problem logically follows:

$$\left. \begin{array}{l} \text{? moles of hydrogen} \\ \text{molecules} \end{array} \right\} = \left\{ \begin{array}{l} 15 \text{ ~~moles~~} \\ \text{~~water molecules~~} \end{array} \right\} \times \frac{1 \text{ mole hydrogen molecules}}{1 \text{ ~~mole water molecules~~}}$$

$$= \quad 15 \text{ moles hydrogen molecules}$$

Note 2: Again the solution to the problem looks simple enough without resorting to the factor-label method. However in examples 3 and 4, the numbers become such that a quick mental solution is not readily achieved by most students.

Example 3

Mole Weights Yield Weight Relationships

How many grams of oxygen are necessary to react with hydrogen to produce 270 grams of water?

Solution: From the balanced equation,

$$2 \, H_2 + O_2 \longrightarrow 2 \, H_2O,$$

the mole ratio between oxygen and water is immediately evident and is one mole of oxygen molecules per two moles of water molecules, or

$$\frac{1 \text{ mole oxygen molecules}}{2 \text{ moles water molecules}}.$$

This mole ratio can be changed into a weight ratio since the mole weight can be easily calculated from the atomic weights involved. One molecule of oxygen (O_2) weighs 32 awu (16 awu + 16 awu). Therefore a mole of oxygen molecules weighs 32 grams. Similarly, two moles of water weigh 36 grams [2(16 + 2 × 1)]. Therefore the weight ratio is:

$$\frac{1 \text{ mole oxygen molecules} \times \dfrac{32 \text{ grams oxygen}}{\text{mole oxygen molecules}}}{2 \text{ moles water molecules} \times \dfrac{18 \text{ grams water}}{\text{mole water molecules}}}$$

or,

$$\frac{32 \text{ grams oxygen}}{36 \text{ grams water}}.$$

This weight relationship is exactly the conversion factor needed to answer the original question:

$$? \text{ grams oxygen} = 270 \text{ grams water} \times \frac{32 \text{ grams oxygen}}{36 \text{ grams water}}$$

$$= 240 \text{ grams oxygen}$$

Note 3: It should be observed that a weight relationship could be established between any two of the three pure substances involved in the reaction, regardless of whether they are reactants or products.

Example 4

Gas Volumes Can Be Calculated by Knowing the Molar Volume

What volume of hydrogen measured at 0°C and 1 atmosphere of pressure is required to react with 240 grams of oxygen?

Solution: The mole ratio from the balanced equation,

$$2 \, H_2 + O_2 \longrightarrow 2 \, H_2O$$

is:

$$\frac{2 \text{ moles hydrogen molecules}}{1 \text{ mole oxygen molecules}}.$$

When we remember that one mole of gas molecules (any gas) occupies 22.4 liters at "standard conditions" (0°C and 1 atmosphere), we can see that the mole ratio can be changed to a volume-gram ratio in the following way:

$$\frac{2 \text{ moles hydrogen molecules} \times \dfrac{22.4 \text{ liters hydrogen}}{\text{mole of hydrogen molecules}}}{1 \text{ mole oxygen molecules} \times \dfrac{32 \text{ grams oxygen}}{1 \text{ mole oxygen molecules}}}$$

or,

$$\frac{2\ (22.4) \text{ liters hydrogen}}{32 \text{ grams oxygen}}.$$

The solution to the problem is:

$$? \text{ liters hydrogen} = 240 \text{ grams oxygen} \frac{2\ (22.4) \text{ liters hydrogen}}{32 \text{ grams oxygen}}$$

$$= 336 \text{ liters hydrogen}$$

The following example problems are similar to the ones given above and a brief solution is given to each. After you have mastered the solutions to these problems, you should be able to work through the exercise list that follows.

Example 5.   How many molecules of water are produced in the decomposition of 8 molecules of table sugar? The equation is:

$$C_{12}H_{22}O_{11} \longrightarrow C + H_2O$$

Solution: Balance the equation,

$$C_{12}H_{22}O_{11} \longrightarrow 12\,C + 11\,H_2O$$

$$? \text{ molecules of water} = 8 \text{ molecules sugar} \times \frac{11 \text{ molecules water}}{1 \text{ molecule sugar}}$$

$$= 88 \text{ molecules water}$$

Example 6.   How many grams of mercuric oxide are necessary to produce 50 grams of oxygen by the reaction:

$$2\,HgO \longrightarrow 2\,Hg + O_2$$

Solution:

(a) weight of two moles of HgO = 2(201 + 16) = 2(217) = 434 g

(b) weight of 1 mole of O_2 = 2(16) = 32 g

(c) ?g HgO = 50 g oxygen $\times \dfrac{434 \text{ g mercuric oxide}}{32 \text{ g oxygen}}$ = 678 g mercuric oxide

Example 7.   How many pounds of mercuric oxide are necessary to produce 50 pounds of oxygen by the reaction:

$$2 \, HgO \longrightarrow 2 \, Hg + O_2$$

Solution: Note that the problem is the same as Example 6 except for the
units of chemicals. Also note that the conversion factor of Exercise 6,

$$\frac{434 \text{ g mercuric oxide}}{32 \text{ g oxygen}}$$

can be converted to any other units desired,

$$\frac{434 \text{ g mercuric oxide } \frac{\text{pound}}{454 \text{ g}}}{32 \text{ g oxygen } \frac{\text{pound}}{454 \text{ g}}} = \frac{434 \text{ pounds mercuric oxide}}{32 \text{ pounds oxygen}}$$

It is evident that the ratio, $\frac{434}{32}$, expresses the ratio between weights of mercuric
oxide and oxygen in this reaction regardless of the units employed.

$$? \text{ pounds mercuric oxide} = 50 \text{ pounds oxygen} \times \frac{434 \text{ pounds mercuric oxide}}{32 \text{ pounds oxygen}}$$

$$= 678 \text{ pounds of mercuric oxide}$$

Example 8.   Carbon dioxide can be produced by the action of hydrochloric acid on calcium carbonate, the principal chemical in limestone. The
equation is:

$$2 \, HCl + CaCO_3 \longrightarrow CaCl_2 + H_2O + CO_2$$

What volume of carbon dioxide at 0°C and one atmosphere pressure can be
produced from 75 grams of calcium carbonate, $CaCO_3$?
    Solution:

(a) One mole of $CaCO_3$ produces one mole of carbon dioxide.

(b) Therefore 100 g of calcium carbonate (molecular weight = 100) pro-

    duces 22.4 liters of carbon dioxide.

(c) ? liters CO_2 = $75 \text{ g CaCO}_3 \times \frac{22.4 \text{ liters CO}_2}{100 \text{ g CaCO}_3}$ = 16.8 liters of
    carbon dioxide

Example 9.   In the problem of Example 8, what volume of gaseous HCl
at 0°C and 1 atmosphere would be required to prepare the acid solution?

$$? \text{ liters HCl} = 75 \text{ g CaCO}_3 \times \frac{44.8 \text{ liters HCl}}{100 \text{ g CaCO}_3} = 33.6 \text{ liters HCl}$$

or

$$? \text{ liters HCl} = 16.8 \text{ }\cancel{\text{liters CO}_2} \times \frac{44.8 \text{ liters HCl}}{22.4 \text{ }\cancel{\text{liters CO}_2}} = 33.6 \text{ liters HCl}$$

EXERCISES

1. What weight of oxygen is necessary to burn 28 g of methane, CH_4? The equation is:

$$CH_4 + 2\,O_2 \longrightarrow CO_2 + 2\,H_2O$$

2. What volume would the oxygen in Problem 1 occupy at 0°C and 1 atmosphere pressure?

3. Potassium chlorate, $KClO_3$, releases oxygen when heated according to the equation:

$$2\,KClO_3 \longrightarrow 2\,KCl + 3\,O_2.$$

What weight of potassium chlorate is necessary to produce 1.43 grams of oxygen? What volume would this weight of oxygen occupy at 0°C and 1 atmosphere?

4. Fe_3O_4 is a magnetic oxide of iron. What weight of this oxide can be produced from 150 grams of iron?

5. Steam reacts with hot carbon to produce a fuel called water gas; it is a mixture of carbon monoxide and hydrogen. The equation is:

$$H_2O + C \longrightarrow CO + H_2$$

What weight of carbon is necessary to produce 10 grams of hydrogen by this reaction?

6. Iron oxide, Fe_2O_3, can be reduced to metallic iron by heating it with carbon.

$$2\,Fe_2O_3 + 3\,C \longrightarrow 4\,Fe + 3\,CO_2$$

How many tons of carbon would be necessary to reduce 5 tons of the iron oxide in this reaction?

7. How many grams of hydrogen are necessary to reduce 1 pound (454 grams) of lead oxide (PbO) by the reaction:

$$PbO + H_2 \longrightarrow Pb + H_2O$$

8. How many liters of hydrogen at 0°C and one atmosphere pressure can be produced from one pound of water?

$$2\,H_2O \longrightarrow 2\,H_2 + O_2$$

9. Hydrogen can also be produced by the reaction of iron with steam.

$$4\,H_2O + 3\,Fe \longrightarrow 4\,H_2 + Fe_3O_4$$

What weight of iron would be needed to produce one-half pound of hydrogen?

10. Tin ore, containing SnO_2, can be reduced to tin by heating with carbon.

$$SnO_2 + C \longrightarrow Sn + CO_2$$

How many tons of tin can be produced from 100 tons of SnO_2?

INDEX

645

649

659

PERIODIC CHART OF THE ELEMENTS

IA	IIA	IIIB	IVB	VB	VIB	VIIB	VIII			IB	IIB	IIIA	IVA	VA	VIA	VIIA	O
1 H 1.00797 ±0.0001																	2 He 4.0026 ±0.00005
3 Li 6.939 ±0.0005	4 Be 9.0122 ±0.00005											5 B 10.811 ±0.003	6 C 12.01115 ±0.00005	7 N 14.0067 ±0.0005	8 O 15.9994 ±0.0001	9 F 18.9984 ±0.00005	10 Ne 20.183 ±0.0005
11 Na 22.9898 ±0.00005	12 Mg 24.312 ±0.0005											13 Al 26.9815 ±0.0005	14 Si 28.086 ±0.001	15 P 30.9738 ±0.00005	16 S 32.064 ±0.003	17 Cl 35.453 ±0.001	18 Ar 39.948 ±0.0005
19 K 39.102 ±0.0005	20 Ca 40.08 ±0.005	21 Sc 44.956 ±0.0005	22 Ti 47.90 ±0.005	23 V 50.942 ±0.0005	24 Cr 51.996 ±0.001	25 Mn 54.9380 ±0.00005	26 Fe 55.847 ±0.003	27 Co 58.9332 ±0.00005	28 Ni 58.71 ±0.005	29 Cu 63.54 ±0.005	30 Zn 65.37 ±0.005	31 Ga 69.72 ±0.005	32 Ge 72.59 ±0.005	33 As 74.9216 ±0.00005	34 Se 78.96 ±0.005	35 Br 79.909 ±0.002	36 Kr 83.80 ±0.005
37 Rb 85.47 ±0.005	38 Sr 87.62 ±0.005	39 Y 88.905 ±0.0005	40 Zr 91.22 ±0.005	41 Nb 92.906 ±0.0005	42 Mo 95.94 ±0.005	43 Tc (99)	44 Ru 101.07 ±0.005	45 Rh 102.905 ±0.0005	46 Pd 106.4 ±0.05	47 Ag 107.870 ±0.003	48 Cd 112.40 ±0.005	49 In 114.82 ±0.005	50 Sn 118.69 ±0.005	51 Sb 121.75 ±0.005	52 Te 127.60 ±0.005	53 I 126.9044 ±0.0005	54 Xe 131.30 ±0.005
55 Cs 132.905 ±0.0005	56 Ba 137.34 ±0.005	57 La 138.91 ±0.005	72 Hf 178.49 ±0.005	73 Ta 180.948 ±0.0005	74 W 183.85 ±0.005	75 Re 186.2 ±0.05	76 Os 190.2 ±0.05	77 Ir 192.2 ±0.05	78 Pt 195.09 ±0.005	79 Au 196.967 ±0.0005	80 Hg 200.59 ±0.005	81 Tl 204.37 ±0.005	82 Pb 207.19 ±0.005	83 Bi 208.980 ±0.0005	84 Po (210)	85 At (210)	86 Rn (222)
87 Fr (223)	88 Ra (226)	89 †Ac (227)	104 Ku (257)	105 Ha (260)													

Lanthanum Series

58 Ce 140.12 ±0.005	59 Pr 140.907 ±0.0005	60 Nd 144.24 ±0.005	61 Pm (147)	62 Sm 150.35 ±0.005	63 Eu 151.96 ±0.005	64 Gd 157.25 ±0.005	65 Tb 158.924 ±0.0005	66 Dy 162.50 ±0.005	67 Ho 164.930 ±0.0005	68 Er 167.26 ±0.005	69 Tm 168.934 ±0.0005	70 Yb 173.04 ±0.005	71 Lu 174.97 ±0.005

Actinium Series

90 Th 232.038 ±0.0005	91 Pa (231)	92 U 238.03 ±0.005	93 Np (237)	94 Pu (242)	95 Am (243)	96 Cm (247)	97 Bk (247)	98 Cf (249)	99 Es (254)	100 Fm (253)	101 Md (256)	102 No (253)	103 Lw (257)

Atomic Weights are based on C^{12} —12.0000 and Conform to the 1961 Values